COLLECTED SCIENTIFIC PAPERS
OF
SIR WILLIAM BATE HARDY

COLLECTED
SCIENTIFIC PAPERS
OF
SIR WILLIAM BATE HARDY

FELLOW OF THE ROYAL SOCIETY
FELLOW OF GONVILLE AND CAIUS COLLEGE, CAMBRIDGE

Published under the
auspices of the Colloid Committee of the
Faraday Society

CAMBRIDGE
AT THE UNIVERSITY PRESS
1936

CAMBRIDGE
UNIVERSITY PRESS

32 Avenue of the Americas, New York NY 10013-2473, USA

Cambridge University Press is part of the University of Cambridge.

It furthers the University's mission by disseminating knowledge in the pursuit of
education, learning and research at the highest international levels of excellence.

www.cambridge.org
Information on this title: www.cambridge.org/9781107475083

© Cambridge University Press 1936

First published 1936
First paperback edition 2014

A catalogue record for this publication is available from the British Library

ISBN 978-1-107-47508-3 Paperback

PREFACE

THESE, the collected works of the late Sir William Hardy, are published under the auspices of the Colloid Committee of the Faraday Society. This is not the time or the place to enter in detail into the many and various reasons which prompted the Committee to take this action; it may however be appropriate to trace the steps which led to the foundation of that Committee.

A number of excellent accounts have been published of the scientific life of Sir William Hardy, but no better epitome of his aims has been given us than that expressed by Sir Hugh Anderson, the late master of Hardy's College, Gonville and Caius: "Hardy once observed a cell divide under a microscope, and wondered why." The wonder excited in Hardy by the observation of cell division set him on the path of these investigations on physico-chemical properties of the proteins and the behaviour of matter in the boundary state. Many believed that, if he had been spared, he would at long last have returned to the problem of why and how a cell divides. This is the key to the plan of his scientific life work, though it represents only one side of his many activities, a side that in the numerous appreciations that have appeared has been little touched on. (Those who are stimulated by this collection of Hardy's work to enquire into his wide and varied interests are directed to the notices in *The Cambridge Review*, *The Times* and *The Proceedings of The Royal Society* of 1934.)

In about 1920 he found that his explorations were leading him into contact with others who were not primarily biologists, but who were occupied with problems similar in many ways to those with which he was becoming increasingly occupied. There existed at that time in this country no scientific body like the Kolloid Gesellschaft in Germany whose primary function it was to assist and stimulate enquiry into such subjects; private discussion however with a number of people led Hardy to conclude that such an organisation would be warmly supported. From such discussions the idea emerged that it might be possible under the ægis of the Faraday Society to hold a series of meetings devoted to consideration of these very problems, so interesting alike to the biologist, and to the physicist and chemist. It was decided eventually to bring into existence a colloid committee of the Faraday Society whose function should be to arrange them. Whilst at all such meetings the problems of the colloid state in its widest sense should receive consideration, every alternate meeting should be devoted to some phase of the work which might assist eventually in solving Hardy's problem—"Why does a cell divide?"

I have to thank on behalf of the Colloid Committee of the Faraday Society all those who have contributed towards the cost of this publication, especially the Dominion Governments of Australia, South Africa and New Zealand; the Master and Fellows of Gonville and Caius College; the Royal Society; the Chemical, Biochemical, Physiological and Faraday Societies. I am indebted to Professor A. V. Hill, F.R.S. for permission to use the excellent photograph in his possession. And lastly I wish to thank the Cambridge University Press, both for their grant in aid of this work and for the care and trouble they have exercised in making this volume a worthy memorial of one about whom a friend has written

χαῖρε· θανὼν ἔτ᾽ ἔχεις τέχνης κλέος ἐν δ᾽ ἑτάροισι
λῆμα φιλάνθρωπον μήνεσιν οὐ φθινύθει.

ERIC K. RIDEAL

Laboratory of Colloid Science, Cambridge
1935

CONTENTS

PAGE

LIST OF PLATES

ON SOME POINTS IN THE HISTOLOGY AND DEVELOPMENT OF *MYRIOTHELA PHRYGIA*

[*Quart. Journ. Micr. Sci.* (1891), XXXII, 505]

While working at the Marine Biological Laboratory at Plymouth during the summer of 1888 my attention was directed to the remarkable hydroid *Myriothela phrygia*. I was acquainted with Prof. Allmann's memoir, and it appeared to me that there were several points which demanded further study—notably the development of the gonophore and the structure of the endoderm, especially in relation to the physiological processes taking place in the enteric cavity. My work was continued, so far as circumstances would permit, during the ensuing winter; but on the appearance of Korotneff's account of the growth of the gonophore, published in the *Archives de Zoologie Expérimentale* for that year (1888), I discontinued it, in the hope that at some future time I might find the opportunity of examining with something like completeness the physiology of *Myriothela*. That opportunity unfortunately has not arisen, and I am induced to publish my results as they stand, believing that though incomplete they contain certain points of interest.

The first complete account of the anatomy and development of *Myriothela* is that of Prof. Allmann, which was published in the *Philosophical Transactions* in 1875. This was followed by a long monograph illustrated with abundant figures, and published at Moscow in 1880, unfortunately in Russian, by Korotneff, who was the first to study *Myriothela* with the aid of properly prepared sections.

In 1881 Korotneff published a further paper dealing with the same subject. A copy of this I have unfortunately not been able to obtain. The further literature of the subject will be referred to as occasion demands in the following pages.

The work was carried out partly in the Marine Biological Laboratory at Plymouth, and I am grateful for the many kindnesses experienced there. But the bulk of the work was done in the Morphological Laboratory of Cambridge University, and I would here thank Mr Sedgwick for placing the resources of his department at my service.

The general structure of *Myriothela* will be best learned by reference to Allmann's monograph, and to that I must refer my readers, merely stating here that it is a solitary attached hydranth. The proximal part of the body is usually bent at right angles to the rest, is covered with a thick perisarc, and gives origin to short processes by which it is attached to the under side of large stones. The perisarc of the foot is represented over the rest of the body by a delicate cuticle. Following on the foot is the middle region of the body, whence spring numerous blastostyles. Each of these bears gonophores, male and female, on its proximal portion, while short capitate tentacles spring from the distal extremity. The blastostyles are without a mouth.

The distal or oral region of the body of the hydranth is the longest, and is studded with very numerous, small, capitate tentacles to within a short distance from the mouth.

The Early Stages in the Formation of the Gonophore and the Origin of the Sexual Elements

The points noticed under this heading may be regarded as an addendum to Korotneff's account of the development of the gonophore alluded to above. His brief and somewhat diagrammatic account is illustrated by figures which unfortunately show little attempt at histological detail.

The structure of the ectoderm of the blastostyle. In each blastostyle two regions may be distinguished; a distal and shorter region bearing several small capitate tentacles, and a proximal region, on the whole free from tentacles, on which the gonophores are developed. The latter embraces two-thirds to four-fifths of the whole length of the blastostyle. The ectoderm of the distal portion resembles that of the body generally, with the exception that the muscle-fibres, which form its deepest layer, are not so strongly developed. Its structure is shown in Pl. I, fig. 1. Like the ectoderm of the animal generally, it is covered superficially by a well-marked cuticle, which, when stripped off and examined with a high power, is seen to be divided into irregular areas, doubtless corresponding to the subjacent cell layer. This latter is composed of long columnar cells, tapering at their lower end. Each cell is about $25\,\mu$ long, though their dimensions possibly vary according as the whole blastostyle is in a condition of extension or contraction. Lying here and there between the pointed bases of these cells are scattered ganglion cells, which are connected with a rich plexus of nerve-fibrils situated in the deepest part of the ectoderm and immediately over the muscle-fibres. The latter run longitudinally, and immediately overlie the supporting lamella. Fibrils from the basal nerve plexus pass between the columnar cells towards the surface, and, in tangential sections, are seen to form a superficial plexus between the columnar cells.

The same type of nucleus is found throughout the whole ectoderm. It is characterised by the presence of a distinct nucleolus linked by scattered fibrils with irregular patches of chromatin on their course to a deeply staining envelope.

The cell substance of the columnar cells is granular and turbid, the granulation mostly being fine. Picro-carmine or hæmatoxylin stains it only slightly. The ganglion cells, on the other hand, stain well with picro-carmine.

The ectoderm of the proximal or gonophore-bearing region is vastly different from that just described. In the first place it is much thicker and more complex, being composed of more varied elements. The ectoderm of the distal region is about 30–$35\,\mu$ thick, of the body 40–$50\,\mu$, while that of the gonophore-bearing region varies from 50 to $70\,\mu$ in thickness. The only other region in which the ectoderm at all approaches it in thickness is in the foot. To this fact we will return later.

The second striking feature of the proximal ectoderm is that its characters are not constant. It is most complex and thickest in specimens killed in spring and early summer, while in autumn it is not only much thinner (30–$35\,\mu$), but also presents the appearance of being exhausted. A comparison of fig. 2 with fig. 4 will render this abundantly evident. The following description applies to specimens killed in March, April and May.

Starting from the outside we have first a well-developed cuticle, which overlies cells resembling the columnar cells of the distal region, but, for the most part, shorter and broader. They are composed of the same ill-staining granular protoplasm, and the border between cell and cell is often so indistinct that we might almost call this with Allmann a nucleated protoplasmic layer. The proximal region is abundantly armed with nematocysts, which, as was pointed out by Allmann, are of two kinds. Here and there these may be seen wedged between the columnar cells. The next following layer is a fairly distinct one of large rounded cells engaged in the manufacture of nematocysts. The protoplasm of these cells, when stained with osmic acid, appears granular under moderate magnification. With high powers (Zeiss $\frac{1}{18}$th ob.) this is seen to be due to the presence of numberless minute vacuoles. Each cell has embedded in its substance a hyaline mass of uniform texture (fig. 2), which at first is deeply placed in the cell, but as growth proceeds it is pushed to one side (fig. 5 *b*), and eventually forms one of the curious rhabdite-like nematocysts which stain so deeply with hæmatoxylin and osmic acid. The first formed amorphous hyaline masses, however, stain only lightly with osmic acid, and, whatever the chemical change may be which alters so profoundly their behaviour towards that reagent, it does not occur until a relatively late period in their development. I am not at all certain that the hyaline masses found in these cells are, in every case, early stages in the development of nematocysts. For many reasons I am inclined to think that they may sometimes be of the nature of reserve nutritive material. My researches on this point are, however, incomplete, and I will merely content myself here with noting the fact that the very earliest change, in that part of the ectoderm where a gonophore is about to be developed, is amongst other things a local accumulation of the large cells bearing hyaline masses, sometimes one, sometimes two (fig. 8), while at the same time there is a total disappearance of nematocysts from that area.

The deepest layer of the ectoderm of the gonophore-bearing region is the most remarkable. It consists for the most part of small rounded cells (fig. 6), either scattered about fairly evenly, or, and more frequently, gathered into smaller or larger clusters. If one of these clusters be examined with a high power, the cells composing it are seen to present considerable differences in size, varying from 8 to 12μ in diameter. In most of the clusters, and more especially if the blastostyle under examination be one which has scarcely commenced to produce gonophores, few or many of the cells will be found with two small nuclei, and the cluster will betray other evidence of the active proliferation of its constituents (fig. 2).

In those cells with a single nucleus, that nucleus is uniformly of about 4μ in diameter, and, like the nuclei generally, is characterised by the possession of an exceedingly distinct nucleolus. The smallest of these cells are generally isolated (fig. 6), and each consists of a nucleus surrounded by a delicate pellicle of exceedingly finely granular protoplasm. The whole cell may be only 5μ in diameter. They form the distinctive feature of the ectoderm of the proximal region of the blastostyles, and, by their number, give to it its great thickness. What part they play we shall see later.

The remaining constituents of the proximal ectoderm as yet unnoticed are its nervous and muscular elements. These I propose to mention very briefly, for they lie to a certain extent outside the limits of the present paper. One of the most striking features of osmic acid preparations, whether sections or teased, are tufts of branching filaments with curious deeply staining matter disposed on them in irregular patches and granules (fig. 3). In teased preparations these filaments are seen to largely end in a thick plexus between the columnar cells (fig. 3), but some of them also end in the cells which are developing or have developed a nematocyst (fig. 5). Not infrequently a filament may be seen having on its course a mass of granular protoplasm containing a nucleus (fig. 2). Traced downwards these filaments appear to be connected with a deeply placed nerve network,[1]

---

[1] Fig. 5 is an accurate drawing made with the aid of a camera lucida of a portion of this nerve complex, which by good fortune was isolated in a teased osmic acid preparation. It exhibits a remarkable fact in the arrangement of this primitive nervous system, namely, that some of the ganglion cells are enclosed in a fine reticulum, formed by the breaking up of filaments derived from the general nerve network. Since this drawing was made the remarkable researches of Golgi, Ramon y Cayal, Kölliker, and others have demonstrated the existence of similar structures in the central nervous system of the higher animals.

which in turn is in close relation to the muscle-fibres which are placed immediately upon the supporting lamella.

If the ectoderm of the gonophore-bearing region of a specimen killed in the autumn be examined, it will be found to consist of externally a cuticle with columnar cells underlying it, and then an indefinite number of gang-lion cells and cells concerned in the manufacture of nematocysts (fig. 4). In some sections scanty patches of small cells similar to those described above may be seen, but they are now the exception and not the rule. On the one hand, over considerable regions the ectoderm may be even more decrepit than that described above, for even the columnar cell layer may be im-perfect, while the basal cells are represented largely by irregular spaces. On the other hand, the muscular layer and supporting lamella are pro-portionately better developed. Measured from the external surface of the cuticle to the external surface of the supporting lamella the ectoderm at this period is from 30 to 35μ in thickness.

We thus see that as the season advances and the reproductive period comes to a close the ectoderm of the gonophore-bearing region becomes more and more exhausted of those small cells which are such a characteris-tic feature of it in the spring, and I think there can be little doubt that their disappearance is connected with the active formation of gonophores during the summer months.

So far I have spoken of them merely as the small cells of the proximal ectoderm. It is necessary to see whether these cells are all alike, or whether they may be divided into one or more sets differing in their function and connections. In tracing the process of formation of the nematocysts I have said that they first appear as a rounded hyaline mass embedded in the protoplasm of a cell which then lies in the deeper part of the ectoderm.

The smallest cells containing these masses may be only 10μ in diameter; but though they do not differ markedly from the bulk of the small cells in point of size, they do differ in one very important particular, namely, in the fact that they are always connected by a delicate process with the nerve network (fig. 5).

On the other hand, we have in addition to these small cells which are in connection with the nerve network, and so often contain some trace of a developing nematocyst, other still smaller rounded cells, which appear, even with the highest powers, to be entirely free; and these are the elements which occur so characteristically in little groups, the cells of which so frequently betray signs of active proliferation. These histological facts, together with the absence of these free cells in other parts of the body and their peculiar relation to the gonophores, entitle us, I think, to regard them as preformed sexual elements.

The Earliest Stages in the Formation of the Gonophore and its Relation to the Process of Budding in Myriothela

In early spring, and before sexual reproduction has taken place to any marked extent, specimens of *Myriothela* may be found which bear buds in various stages (Pl. II, fig. 13). These appear to be always developed just at the junction of stolon and body. Once only have I met with a bud formed elsewhere, namely, in the lower tentacular region. This had, however, more the appearance of a permanent growth than of a bud to be cast off.

The process of budding, so far as I have followed it, is a rather remarkable one. The first stage is a modification of the character of the ectoderm, which in the stolon and lower part of the body is composed of very long columnar cells, resembling the columnar cells of the blastostylar ectoderm in all particulars save in their inordinate length. Lying between the bases of these columnar cells are interstitial cells, characterised by the fact that they stain more deeply with picro-carmine. These cells appear to be partly nervous and partly concerned in the formation of nematocysts, which, curiously enough, are produced in limited number even under the thick and dense perisarc of the upper part of the foot. Where a bud is about to be formed the ectoderm cells lose their defined characters, proliferate, and a bulging mass of amorphous tissue results. At the same time the thick supporting lamella becomes absorbed, and the endoderm cells likewise proliferate and take on an amorphous character. The result is a kind of blastema in which the limits of ectoderm and endoderm are undistinguishable. This grows while at the same time its elements lose their distinctness and become highly charged with spherical masses of stored nutriment, resembling in many particulars the nutritive spheres of the general endoderm. As it grows it pushes the perisarc before it, and ultimately forms a rounded egg-like mass attached to the parent body by a short thick pedicle (fig. 13). From this the young *Myriothela* is developed (fig. 13). All connection with the body of the parent is lost at a very early period, almost before the bud has re-formed its ectoderm and endoderm and enteric cavity. It remains attached to the perisarc, however, by a sucker-like arrangement at the aboral pole until it is fully formed.

As will be seen, the formation of a gonophore is, in its earliest stages, essentially similar to this method of budding. In other words, the gonophore is a true bud which, like the other buds, is derived from a blastema formed by a fusion of ectodermal and endodermal elements. The difference, however, lies in the fact that in the case of the gonophore bud, after it is a well-formed structure, a group of the primitive germ cells make their way into it.

The first stage in the growth of a gonophore is shown in fig. 8. The ectoderm of the gonophore-bearing region becomes thickened over a small surface, the increase in thickness being due largely to an accumulation of the primitive germ cells, but partly to an increase in the cells carrying hyaline masses. At the same time nematocysts disappear in that region of the ectoderm, though they may occur in their usual profusion in close proximity. In the next stage the basement membrane is absorbed or ruptured (figs. 7 and 9), I cannot determine which, and a tongue of endoderm cells pushes its way into the ectoderm and through the deepest layer. Thus the cluster of primitive germ cells come to lie not on its apex, but, generally, asymmetrically disposed on one side.

The removal of the supporting lamella is, I am inclined to think, mainly a process of solution, since scattered rounded fragments may sometimes be seen; and further because the muscular elements are absorbed for some little space about the point where the rupture takes place (figs. 7–10).

The tongue of endoderm-cells rapidly becomes a tubular outgrowth, and the cells at its apex lose their nutritive spheres and become small, dense, on the whole ill-staining cells.

At this stage it is perhaps impossible to distinguish the limit between ectoderm and endoderm, while at the same time a fusion of cell-substance has taken place (fig. 10), so that we have a stage closely resembling the blastema which gives rise to the bud as described above.

Fig. 10 represents this stage, and is an accurate drawing, made with the aid of a camera lucida, of a preparation from a specimen killed with corrosive sublimate and stained with picro-carmine; the whole specimen being remarkable for the good preservation and clear definition of its histological elements. The section figured passes rather obliquely through the young gonophore, but the next in the series shows that the primitive germ cells have now travelled in under the superficial columnar cells of the ectoderm to form a cap for the fused ectoderm and endoderm.

The next stage to be noticed is in many respects remarkable. By a fresh formation of supporting lamella the whole bud with its contained germ cells becomes separated from the maternal tissue, while at the same time a fold of supporting lamella becomes formed which separates the ectodermal elements with the primitive germ cells from what is usually known as the endoderm lamella (fig. 11). The endoderm lamella, therefore, from this time onwards, is separated from the endoderm of the parent by a well-defined and permanent supporting lamella; and it is noticeable that up to this stage and until they degenerate the cells of the endoderm lamella are not of the ordinary endoderm type, but *resemble in every detail the ectodermal cells of the gonophore.*[1]

---

[1] The columnar cells of the maternal ectoderm remain undisturbed and unaltered by these various changes. I regard them as belonging to the maternal tissues, and not to the gonophore bud. They are, therefore, not included in the above statement.

In the further history of the gonophore I am only concerned with one point, namely, the fate of the primitive germ cells. I entirely agree with Allmann that in their earliest history it is impossible to distinguish between male and female gonophores. Position does not help one, for, as has been recognised before, both male and female elements are produced on the same blastostyle, and to a certain extent indiscriminately, the same transverse section frequently passing through both male and female gonophores. Very soon, however, the male gonophores become distinguished by the rapid proliferation of their generative elements.

In the female gonophore at some period, often relatively late, two or three of the generative cells become larger and more prominent than the others. The period at which this happens does not appear to be fixed; and whatever factor it may be, whether something inherent or accidental, that determines which of these struggling cells shall obtain the mastery and eat up its fellows, it sometimes does not come into play until the gonophore has become a well-formed structure. But it is quite late in the history of the gonophore, when that structure is large and already swollen with yolk, before these two, three or four cells, which, so to speak, have succeeded in attaining to the final heat, decide who is the winner.

In the facts which have been set down above with regard to the structure of the ectoderm of the blastostyles, and the formation of the gonophores, two main points appear to me to be of special significance. These are (1) that the gonophore appears to be a curiously modified bud, and (2) that the generative elements pre-exist as free cells having lodgment in the tissues of the adult, and only travel into the abortive bud, which is their place of final development.

On Certain Points in the Structure and Functions of the Endoderm of *Myriothela*

In any animal three main problems concerning the manipulation of its food-stuffs present themselves. These are (1) the disintegration and solution of its food; (2) the absorption of the dissolved or liberated and unchanged material; and (3) the distribution of the products of digestion. We might also add to these the storage of prepared food-stuff. Many and diverse reasons justify the statement that a cell is hampered or, better, limited in the range of its activity by being loaded with indifferent reserve nutriment; and it therefore becomes almost the duty of a special tissue to store material, either as a provision for some extraordinary metabolic effort, or as a consequence of an infrequent and uncertain food-supply. In the case of *Myriothela* we shall see that this last problem—the storage of reserved nutriment—occupies a large part of the endoderm.

The disintegration of the food in all animals not possessed of a masticatory apparatus is a process of solution differing only in degree from the final solution of the smaller particles. Yet I think we are justified in speaking of the whole act as a process of disintegration and solution, because of the very general tendency of animals to divide the process into those two stages, and to differentiate the alimentary tract into a special region where disintegration of the food by solution takes place, *e.g.* the stomach of Vertebrates, and another tract where solution is completed, and where solution and absorption go on hand in hand.

In an animal of any considerable size the distribution of the products of digestion becomes a problem of the utmost importance, not only physiologically, but also from its far-reaching influence on the morphological characters of animals. It has been solved in three main ways:

1. By the utilisation of the enteric space itself, which functions in part as a digestive cavity, and in part as a common space in close communication with all parts of the organism, and containing not only the results of the solution of the food, but also material discharged from the lining cells of one region, and destined for the nutrition of other parts of the organism. It is as an example of this class that I wish to consider *Myriothela*.

2. By the development of a system of spaces or a common space round the gut, into which the results of digestion can be discharged, and from which the tissues can directly derive their nutriment. Such a space would be the hæmocœl of morphologists. The physiological significance of the cœlom is still, I think, very much under judgment.

3. By the aid of a closed vascular system. The true relations of hæmocœl and cœlome to one another and the relations of both to the vascular system of Annelids, which appears to be initially respiratory; and to the vascular system of Arthopods, which is in its first inception a mechanism for the circulation of the fluids of the circum-enteric space, are questions which do not concern us here, but the interrelation of cases 1 and 2 demands brief notice.

It is necessary at the outset to distinguish clearly between the digestive functions of the enteric space, and the part it may play in the distribution of the nutritive material. The researches of Miss Greenwood on the digestive process in *Hydra* justify the conclusion that in that animal the enteric space is used mainly, if not solely, for digestion. The endoderm cells forming its walls absorb and largely store the products of digestion. There our exact knowledge ends, but the uniform character of the endoderm throughout the entire animal, the fact that its cells everywhere, even in the tentacles, absorb and store nutriment, renders it probable that in the physiology of *Hydra* the discharge of elaborated nutritive material into the enteric space to form a common nutritive fluid akin to the blood of higher animals plays little part. But in the higher Cœlenterates, in the colonial forms, in Medusæ, and in Ctenophora especially, we have no reason to doubt that such a fluid does exist, and that it forms the metabolic link between the different regions of those animals in such a way that the demands of one part may be met by the discharge of stored nutritive material from other regions. Such a fluid, which Allmann has called the "somatic fluid", would not only contain the immediate results of digestion, but also elaborated material from the store of reserve nutriment possessed by certain cells discharged in response to any special demand in some particular region. In other words, it is not merely a fluid for the distribution of the immediate products of digestion; it is more than that, and we shall see reason to think that it is a true metabolic link between one part of the body and another strictly comparable to the blood of higher forms.

In Allmann's account of digestion in Hydroids he thus describes the "somatic fluid": "Its basis is a transparent colourless liquid, and in this solid bodies of various kinds are suspended. These consist partly of disintegrated elements of the food; partly of solid coloured matter which has been secreted by the walls of the somatic cavity; partly of cells, some of which have undoubtedly been detached from those walls, though it is possible that others may have been primarily developed in the fluid; and partly of minute irregular corpuscles, which are possibly some of the effete elements of the tissues."

Leaving the Cœlenterates and turning to the Turbellaria, we have a striking instance of how the enteric cavity, the gut, may serve as an organ for the distribution of nutriment. The case of the Turbellaria also enables us to contrast cases 1 and 2, for in the Rhabdocœls we have animals with a simple gut surrounded by a tissue, the mesenchyme, with numerous cleft-like spaces, which may be so far developed as to form a fairly well-marked space round the gut, as in *Mesostoma tetragonum*, and to a less extent in *Microstomum lineare*. That such a space or system of spaces facilitates the distribution of material derived from the gut hardly needs stating, whether we consider the distribution to be brought about by diffusion or by the agitation of the contained fluid as a result of the muscular movements of the animal.

In the Triclads and Polyclads, on the other hand, the same problem has been solved in a rather different fashion. The development of spaces round the gut is not the most obvious fact, but the gut itself ramifies into a network of tubes which penetrate to every part of the body.

The structure of the endoderm. Leaving these general considerations, we will pass at once to a consideration of the structure of the endoderm in *Myriothela*.

The mouth is bounded by a thin membranous lip, or hypostome, which is exceedingly muscular and sensitive, and is probably of considerable use in seizing the prey. It is figured by Allmann and Hincks in a condition of extension. In fig. 20 it is shown as it appears when the animal is retracted.

In the lip the supporting lamella is either reduced to the thinnest film or is entirely absent. It is always absent from about half the breadth of the lip from the free edge. The ectoderm of the external surface is rich in nerve cells, and overlies a well-developed layer of radial muscles. At the free edge there is no loss of histological continuity, the ectoderm being continued on to its under surface for a certain distance, and having the same structure. Passing downwards towards the attachment of the lip, there appears at the base of this ectodermal epithelium a more and more defined layer of highly vacuolate cells, which are the first commencement of the endoderm. In other words, the ectoderm for a short distance overgrows the endoderm, thus forming a distinct zone of mixed character.

At the point where the lip merges into the tentacle-bearing region the ectoderm, as a distinct structure, has finally disappeared, and we have the arrangement shown in fig. 14. Three kinds of cells may be distinguished: (1) a superficial layer of elongated cells, staining well and uniformly with picro-carmine. Each possesses a nucleus and nucleolus in its basal portion, which is tapering and wedged in between the subjacent cells. In favourable preparations these cells appear covered with short, fine cilia. These are, however, not always visible, mainly because the animal in dying usually retracts this region to such an extent that the epithelium is thrown into deep folds, and the free surfaces of the cells apposed. This ciliated zone forms only a narrow band. Between the bases of the ciliated cells are (2) numerous rounded and deeply staining cells, which are more numerous the nearer the lip. Delicate processes may be sometimes seen passing from them to the surface, and they are probably sense cells. Below these, and forming by far the greater portion of the epithelium, are strikingly characteristic palisade-like cells. Each has a very scanty and ill-staining protoplasm, which surrounds a large irregular vacuole occupying the bulk of the cell. Two nuclei, each with a small nucleolus, are usually present, and may be either close together about the middle of the cell, or one at either end. There is not the slightest evidence that the presence of these twin nuclei indicates cell division.

At the lower edge of the ciliated zone conical cells appear between and rapidly replace the ciliated cells, while at the same time the deep-staining sense cells disappear. Each of these conical cells resembles in its general appearance a goblet cell of an ordinary mucous membrane. We can, therefore, conveniently style the next region *the goblet-cell zone*. This embraces a considerable portion of the tentacle-bearing region. It is, however, impossible to reduce the relative dimensions of these various zones to numerical exactness because of the extreme extensibility of this portion of the animal.

Each goblet-cell is, as its name implies, flask-shaped, and consists of an expanded part which stains lightly with picro-carmine (fig. 16) or osmic acid (fig. 15). The contents of this part are turbid from the presence of ill-defined, granular masses. The expanded portion of the flask is continued downwards into a tail, which contains a small nucleus embedded in deeply staining granular protoplasm. The numerous and coarse granules of the basal portion stain deeply with osmic acid, and less so with picro-carmine. These cells are undoubtedly of the nature of gland cells, and the well-formed basal granules may be regarded as the first stage in the elaboration of the secretion product which occupies the expanded portion of the cell. In the upper part of the goblet-cell zone the endoderm is composed of goblet cells lying wedged between the apices of palisade-like cells exactly resembling those occurring in the ciliated zone, except in the fact that they now abut on the free surface. When this portion is retracted the deep folds present the appearance in sagittal sections of long tubular glands, which, from the character of the abundant goblet cells, closely simulate the crypts of the large intestine.

In the middle and lower part of this region the endoderm is thrown up into low conical villi. These structures are characteristic of the whole of the endoderm with the exception of the part already described, that is in the neighbourhood of the mouth, and in the foot. They vary very much in length in the different regions of the animal, but are usually longest in the lower portion of the tentacle-bearing region, where they are long filiform structures, sometimes branched, and may measure 0·3–0·5 mm. from base to apex. Generally speaking, the villi are not muscular, but in the goblet-cell region they appear to have a distinct muscular axis.

Structure of a Villus from the Goblet-cell Zone

A section through the axis of one of these is shown in fig. 16. On the sides of each villus, and between their bases, we have the same arrangement of palisade cells and goblet cells as that described above. At the apex, however, are a group of cells presenting many new features. These I propose to call the *apical cells*, and they form not only the apex of the villus, but are also continued downwards as its muscular axis.

Each apical cell has the following general structure. The protoplasm is abundant, and stains deeply, thus offering a marked contrast to the other cells of the villus. It is also turbid and opaque, but differs very much in this and in its behaviour with stains in different parts of the cell. In the muscular stem, however, the protoplasm is always of a uniform texture, and behaves towards osmic acid and other stains in a manner closely resembling the muscular elements of the ectoderm. The protoplasm of the expanded end of the cell encloses the following structures. One, rarely two, nuclei, which vary so much in position as to suggest an extreme mobility of the contents of the cell. A varying amount of pigment, dark brown in colour, and disposed in scattered grains near and on the free surface of the cell, but usually in little heaps of grains in the deeper parts. One or more large vacuoles. And, lastly, turbid masses of substance, some of which are certainly the remains of material which has been ingested by the cell, and all of which are in more or less obvious relation to the vacuoles.

These factors sum up the constituents of the apical cell as usually seen; but the extent and character of the vacuoles and the turbid masses of enclosed matter vary very much at different periods and in adjacent cells. To this, however, we will return later.

In the varying position of the nuclei we have some indication of the mobility of the protoplasm of the apical cells; and this mobility finds further expression in the fact that from the free surface of the cells pseudo-podial extensions are pushed out, especially from those cells which are fairly free from enclosed masses (cf. Allmann's *Memoir*, Pl. LVI, fig. 2).

In the middle tentacular region the endoderm assumes a different character. The goblet cells disappear, and the palisade cells gradually pass into a shorter and broader type of cell, mostly with only one nucleus. These cells I will call vacuolate cells, adopting the term applied by previous observers to similar cells occurring in *Hydra*. These cells usually contain numerous round hyaline corpuscles, which vary in their characters but stain always with osmic acid and many aniline dyes (such as methyl blue and green), but typically take no coloration with hæmatoxylin and little or none with carmine stains. These are, without doubt, identical with the sphere-like masses of reserve nutriment described by Miss Greenwood as occurring in the vacuolate cells of *Hydra* under the name of "nutritive spheres". The vacuolate cells of *Myriothela* when free from nutritive spheres may be seen to possess a large vacuole surrounded by scanty protoplasm. In this condition they recall the palisade cells of the oral region, and the inter-mediate types are so numerous that I am disposed to regard these cells as fundamentally the same. The most constant differences are that the palisade cells have almost always twin nuclei, and only rarely contain nutritive spheres.

Wedged here and there between the vacuolate cells are other and smaller dark-staining cells (fig. 21), which occupy the same position but are not disposed with the same regularity as the goblet cells. These cells are as variable in size and appearance as the vacuolate cells, and they correspond to the "gland cells" of Nussbaum and Jickeli. Miss Greenwood has shown reasons for considering that cells similar to these occurring in the endoderm of *Hydra* are concerned in the formation of the digestive enzymes, and I find that her conclusions are warranted by the different appearances presented by these cells under varying conditions in *Myriothela*. To this point we will return later.

Rarely a gland cell may be seen apparently bearing a deliberate pseudo-podium or flagellum, and having the appearance shown in fig. 18 *b*. Nussbaum similarly describes cilia on the gland cells of *Hydra*.

These gland-cells are very widely dispersed throughout the endoderm. They occur, perhaps, in greatest abundance on the sides of the villi; sometimes, however, one or two may occur at the apex of a villus. Rarely, at the apex of a villus, a group of two, three, or four small cubical darkly staining cells is found (fig. 19). Whether these are or are not stages in the development or multiplication of gland cells I was unable to determine. That they are, however, the antecedents of the free corpuscles which at certain periods occur in the somatic fluid I see no reason to doubt.

The gland cells of *Myriothela* have a rather wider distribution than those of *Hydra*, where they are restricted to the body. In *Myriothela* they occur in their greatest abundance in the endoderm of the lower half of the tentacle-bearing region. But they are also numerous in the middle region of the body whence the blastostyles spring, and may even occur in limited numbers in the endoderm of those structures near their points of attachment.

In the middle and lower regions of the body the endoderm is to a certain extent different from that already described. The villi, as a rule, become less muscular, while at the same time their apical cells change their characters and become more and more akin to the vacuolate cells. The endoderm of the body-wall from which the villi spring, and which in the lower tentacle-bearing region is composed of cells in no wise distinguishable from those lining the villi, changes its character in the blastostyle-bearing region. There it is composed of long columnar cells, each with a single nucleus, and each composed of dense well-staining protoplasm free from vacuoles.

The endoderm maintains these characters in the foot; that is to say, the villi, which here are of the nature of broad flange-like folds, are covered by vacuolate cells with rarely a gland cell, while the endoderm of the body-wall is composed of the deeply staining columnar cells. At the extreme end of the animal, however, the supporting lamella ceases to exist, and to a certain extent the limits between ectoderm and endoderm become obscured, and a kind of growing point to the creeping stolon is the result.

The endoderm of the blastostyles will receive special mention later. To make this general account of the whole endoderm complete, however, I will merely state here that the villi in the blastostyles are low conical structures almost exclusively composed of vacuolate cells. A few gland cells lie scattered in the proximal third of each blastostyle.

Taking the most general view of the structure of the epithelium lining the enteric space of *Myriothela* as described in the preceding pages, we see that it may be divided into different regions. These are (1) An oral region characterised by the presence of sense cells and cilia in its upper part, and of numerous glandular cells, the goblet cells, in its lower part. (2) A middle zone comprising the middle region of the entire animal, and characterised by the presence of numerous gland cells. (3) The blastostyles and the foot region, where the endoderm is almost exclusively composed of vacuolate cells, usually loaded to the full with stored nutritive material in the form of nutritive spheres.

Of the function of the goblet cells I can say little. From their position I had supposed that their stored material was discharged when the food was first received, and that they were concerned in the elaboration of a digestive ferment or ferments. But I have found them apparently unaltered in animals which have just taken in their prey (a crustacean). The glairy, sticky appearance of the contents of the expanded portion of the goblet, however, suggests the idea that they form a strongly adhesive surface to what may be called the prehensile portion of the endoderm. The gland cells, on the other hand, present no especial difficulties. Fig. 18 *a* represents one shrunken and discharged as seen in an animal at the close of a digestive act, that is with merely the detritus of a meal in its enteric cavity. Fig. 18 shows one taken from a fasting animal. It is fully loaded with granules. Fig. 18 *b* represents an intermediate condition. The granules are large and coarse, and appear to be formed in the deeper portions of the cell. Their discharge is characteristic, and may be witnessed in preparations from an animal which has just ingested its prey. The granules are extruded, apparently unchanged, into the somatic fluid, there to be dissolved. That is to say, they do not break down to form the digestive enzymes until they are free from the cell in which they were formed (fig. 21).

The process of digestion. For a long time I was unable to determine the natural food of *Myriothela*, while at the same time all endeavours to induce it to ingest pieces of raw meat or fragments of Molluscs and Crustacea completely failed. I therefore, in the meantime, turned my attention to carmine and sodium sulphindigotate, and with a fine pipette, inserted through the mouth, injected a drop of sea water containing the one in suspension or the other in solution. Later, however, I was more fortunate, and succeeded in obtaining specimens with food in the enteric cavity. In one case the prey was a Crustacean of some considerable size, so that it produced a very obvious bulging of the animal. I did not witness the capture and ingestion of the prey, but the specimen was killed before the digestive fluid had produced any change in the tissues. Fig. 21 is taken from this specimen. It also contained the remains of a previous meal in the lower part of the enteric space. Other specimens furnished other stages, and in this way, as a result of an examination of a large number of animals, I was enabled to obtained a series representing, to a certain extent, the various stages of digestion.

Myriothela is carnivorous, and captures small Crustacea. In one case the meal consisted of a half-digested egg, either derived from the gonophores of the individual in question, or from those of its neighbours.[1]

Digestion is carried on at first in the lower portion of the tentacle-bearing region—that is, in that region where the gland cells are most abundant, and results in a disintegration of the prey, brought about by the agency of the digestive fluid.

The gland cells which first discharge their contents are those in the immediate neighbourhood of the meal, but even there all the cells are not affected at once, those which are most loaded with granules probably being discharged first. Some of the gland cells in the proximal region of the blastostyles and in the foot may be found undischarged until nearly the close of a digestive act. The disintegration of the prey is accompanied by a great amount of solution, so that after digestion has been in process for some time the enteric cavity contains a number of fragments of the prey floating in a fluid rich in proteids, which form a deeply staining granular precipitate after treatment with corrosive sublimate. The disintegrated fragments of the meal find their way into the foot and blastostyles, the somatic fluid probably being circulated by the active movements of the animal, which include the extension and retraction of the blastostyles, and possibly by the cilia, which appear to be borne here and there by the endoderm cells. The process of digestion is therefore largely extracellular; but this is not all, intracellular digestion takes place to a marked, but undoubtedly a subordinate extent, and is mainly, perhaps entirely, confined to the amœboid and mobile apical cells of the villi in the tentacular region.

If these cells be examined in an animal killed towards the close of digestion they will often be found to contain the more or less imperfect capsules

[1] The huge yolk-laden eggs, when set free from the gonophores, are taken by tentacle-like bodies, the "claspers", which hold them while development proceeds and until the actinula larva is fully formed. In March and April, however, the claspers are not always present, and the eggs formed then when ripe are shed and sink to the bottom, where they become attached. I think that further research will show that these early eggs have a more direct development than those formed later, and pass at once to the form of the adult without the intervention of the free-swimming actinula stage. It was one of these free eggs which apparently had found lodgment in the enteric cavity.

of thread cells or irregular fragments of hyaline material, which recall the
cuticle of a Crustacean, and probably are fragments of that structure.

In fig. 17 such a nematocyst is shown embedded in a turbid mass of
darkly staining material, which occupies a vacuole in the cell protoplasm.
At *n* is another nematocyst, now no longer lying in a vacuole, but embedded
in the cell protoplasm. The digestion of the food-mass with which this
nematocyst was associated having been completed, the vacuole has filled
up.

Other interesting points may be mentioned which point to this intracellular digestion. In some cases an enclosed mass may be found embedding a more or less perfect nucleus. By a careful examination of different
cells we may see such nuclei in various stages of disintegration, until they
are finally resolved into chromatin granules which are scattered through a
pseudopodial-like lobe of protoplasm, which appears to be thrust into a
vacuole as the digestion of its contents becomes complete.

Carmine grains injected into the enteric cavity are eagerly ingested by
the apical cells, both in the body of the animal and in the proximal region of
the blastostyles.

Towards the close of digestion a few free cells appear in the somatic fluid.
Each is rounded and composed of dark-staining protoplasm embedding a
nucleus with contained nucleolus.

Before turning to the further fate of the food, that is to say before
considering the process of absorption, it will be well to describe more
particularly the contents and characters of the vacuolate cells.

These when unloaded with nutritive spheres present the appearance
shown in fig. 18, where the cell is seen to consist of a thin pellicle of protoplasm surrounding a large central vacuole and embedding a nucleus. The
protoplasm stains only slightly. The outline of the cell is exceedingly sharply
defined by some staining material which has almost the appearance of a
cuticle. The process of loading with nutritive spheres is remarkable, and
essentially similar to that described by Miss Greenwood as occurring in
Hydra. The protoplasm at one point develops a small vacuole which increases in size, and bulges into the large vacuole. In this the nutritive
sphere is formed from the turbid semi-fluid material which first fills it. This
process continues until the whole of the cell becomes occupied by small
vacuoles, each containing a nutritive sphere. The size of the cell, therefore,
does not necessarily vary according to the amount of reserve nutriment it
contains. This, however, only holds good for the vacuolate cells of the body.
In the blastostyles they are slightly different. In the first place the large
central vacuole is not developed, with the consequence that when the cell
discharges its nutritive spheres it frequently shrinks to the condition of a
cell with dense, non-vacuolated protoplasm, which stains deeply with

picro-carmine. In fig. 22, at 1 is a cell without nutritive spheres, and at 2 one partly filled with vacuoles containing nutritive spheres.

But the discharge of the nutritive spheres does not always leave the cells smaller and solid. The vacuoles may persist (fig. 22, 4), and the result is a remarkable cell with bubbly protoplasm. Such cells form a very striking feature of osmic acid preparations. The fluctuations in the size of the individual cells in the endoderm of the blastostyles naturally leads to a corresponding fluctuation in the total bulk of that tissue. In specimens taken in May or June it sometimes almost fills the cavity of the blastostyles.

We therefore have in the blastostylar endoderm, in addition to the scattered gland cells, the following: (1) Small dense cells with nucleus and nucleolus which stain deeply. (2) Cells whose protoplasm is completely occupied by vacuoles, each of which, some of which, or none of which contain nutritive spheres. And there are numerous intermediate stages between (1) and (2), with either a small or a large portion of the cell-substance occupied by vacuoles. The protoplasm of cells (1) and (2), or of the intermediate stages, appears in osmic acid preparations to be remarkably dense and almost glassy, and the vacuolation is limited to the large vacuoles which embed the nutritive spheres. But other vacuolate cells occur, forming a third class, whose protoplasm is not of this character, but is so occupied by vacuoles of all sizes as to give the whole cell a very characteristic appearance (fig. 22). The vacuoles always differ in size in different parts of the cell, being larger near the free surface where they may contain young nutritive spheres. These cells, forming the third type of vacuolate cell to be seen in sections through the blastostyle, may be regarded as cells which are actively forming nutritive spheres.

Before considering the processes going on in the vacuolate cells generally, it will be well to turn to the nutritive spheres themselves. The endoderm generally contains three kinds of formed bodies—that is to say, bodies which are not of the nature of material merely ingested by the cells, but rather are new formations resulting from the activity of those cells.

Of these the most common are small spherical bodies, generally about $3\,\mu$ in diameter. These crowd the endoderm cells in great numbers, but are always most numerous in the foot and blastostyles. They stain uniformly, but not so deeply as to become opaque, with osmic acid, and when perfect appear singularly hyaline and structureless even with the highest powers; they do not stain at all with picro-carmine or hæmatoxylin, but take a deep tinge with aniline blue and methyl green.

They are of a complex character chemically, with probably a proteid basis. That they are not wholly proteid is, I think, shown by the fact that the coloration obtained with the xanthroproteic reaction, and with acidulated ferrocyanide of potassium and ferric chloride, is never intense though it is sufficiently distinct. Neither iodine nor iodine and sulphuric acid give distinctive results. They are also not of a fatty nature, for exposure for many hours to hot turpentine fails to change them materially. These are not only the most abundant, but the most permanent nutritive spheres.

The second class of bodies is markedly distinct from those just mentioned, and they occur sometimes in considerable abundance. They are perhaps most noticeable at the close of a digestive act. They measure as a rule $12\,\mu$ in diameter, and are spherical bodies, each embedded in a vacuole of a vacuolate cell, which may or may not contain at the same time the small nutritive spheres. Like the latter they are sometimes homogeneous bodies showing no internal structure, and then they stain very intensely with picro-carmine (fig. 27). But they may be also found composed half of intensely staining homogeneous material, and half of more turbid material which stains scarcely at all (fig. 28). In yet other cases the intensely staining material is reduced to a small spherical nodule or patch placed excentrically on the surface of the sphere. There can be no doubt that these are various stages in the formation or destruction of the same bodies, but I do not think that the evidence at my disposal justifies me in deciding which stage is which. As a general rule, however, the smallest of these bodies, perhaps but little larger than the first-mentioned type of nutritive sphere, are homogeneous and deeply staining, while without exception the largest forms are those with the deeply staining material reduced to an excentrically placed bleb. These bodies resemble in some respects the yolk spherules of the ripe egg, and they are also found in some abundance in the young buds. I can throw no light either on their special significance or on their relation to the small nutritive spheres. I might finally add that they are rarely widely distributed throughout the endoderm, but usually occur in localised patches as though they were related to some localised and special metabolic process. They are perhaps never completely absent.

Though the term nutritive sphere can, I think, be applied with justice to both the preceding bodies, its application to the third class of endodermal products would be misleading, since they are merely a specialised product of the endoderm of the tentacles.

When abundant, each cell of the endoderm of a tentacle may contain one of these bodies, and then they doubtless give rise to that opacity which was noted by Allmann. They form, however, a very variable element, for while one tentacle may be fully charged with them its neighbour may contain few or none. Each of these bodies is, when fully formed, $10\,\mu$ in diameter, and consists of a sphere which stains intensely with picro-carmine (fig. 26), and is embraced by a cup-shaped capsule with expanded edge.

That these bodies play any great part or are at all concerned in the general nutrition is extremely improbable. Under certain circumstances they are discharged from the endoderm cells of the tentacles and find their way into the enteric space, and are ingested and digested by the apical cells of the adjacent villi.

In what I have to say concerning the formation and fate of the nutritive spheres in the following pages, I shall refer exclusively under that title to the small nutritive spheres which are so abundant and numerous. The method of formation of these bodies has been already described, and it was seen that they develop in vacuoles formed in the protoplasm of the vacuolate cells; and I see no reason to doubt, but rather every reason for agreeing with Miss Greenwood's view, that these bodies are formed from material absorbed in the fluid form from the results of digestion in the enteric cavity. The fate of sodium sulphindigotate when introduced into the enteric cavity is interesting in this connection, for though a considerable portion of that pigment is rapidly decolorised, some may be found after a short period in the vacuolate cells associated with the nutritive spheres in the various stages of their formation in such a way that these bodies are tinged with blue in irregular patches. We may conclude, therefore, that the substances resulting from the solution of the tissue of the prey are mainly absorbed by the vacuolate cells, while at the same time ingestion of disintegrated fragments, possibly of a resistent nature, takes place to a limited extent, and is carried on by the apical cells of the villi in the middle region of the body. And the distribution of the products of digestion to the more remote portions of the endoderm is effected by the agency of the general somatic fluid.

But, as I said before, there are reasons for believing that the somatic fluid is more than a vehicle for the distribution of the immediate results of digestion. Let us now turn to a consideration of those reasons.

In the blastostyles the extensive accumulation of nutritive spheres is in obvious relation to the active and important processes carried on in those structures. But the nutritive spheres are stored in equal abundance in the endoderm of the foot, where, during the summer and autumn, there appears to be no great call for this enormous reserve of nutritive material. On the other hand, throughout the whole tentacular region and usually in the endoderm of the tentacles themselves no abundant reserve of food-stuff is present. It is, as I pointed out above, highly doubtful whether the peculiar bodies sometimes found in the endoderm of the tentacles are of the nature of simple reserve nutriment. On the other hand, it is equally certain that the actively motile and sensitive tissues of the tentacular and oral regions are not supplied directly and solely from the immediate products of digestion. We are, therefore, compelled to conclude that the nutritive material stored in one region of the body may be conveyed as occasion demands to regions quite remote. In other words, we must suppose that the endoderm is not merely a collection of cells each fighting for a share of the nutritive material to be ultimately used by itself or by the ectoderm with which it is anatomically in immediate relation, but rather that the metabolic activities of the units of the endoderm throughout its whole length are linked together into one consistent and interdependent whole.

What is the nature of that link, and how is the interchange of material which it implies between widely separated regions brought about? That it is effected entirely by the laborious passage of material from cell to cell throughout, perhaps, a considerable length of the animal is, I think, an impossible suggestion. Yet this process undoubtedly takes place to a certain extent, and I think we may see in the palisade cells of the oral region, from the contents of whose enormous vacuoles amorphous masses are precipitated by the action of corrosive sublimate, a mechanism for facilitating such a process, and whereby nutritive material may find its way even to the lip, a region which during the early stages of the digestion of prey of any considerable size must be more or less cut off from the general somatic fluid. This same method of distribution of nutritive material must also obtain between the endoderm and the ectoderm, the interchange being facilitated by the numerous pores which may be seen to penetrate the supporting lamella when horizontal sections of that structure are examined with the highest powers. But the stored nutritive material of the foot can only be rendered available for the body generally through the agency of the somatic fluid, and to a certain extent histological facts support this conclusion.

If we attempt to follow the further fate of the nutritive spheres we find that two things may happen to them. They may either undergo a gradual disintegration, their substance becoming at the same time studded with pigment grains which, after the nutritive sphere as such has ceased to exist, remain as a little heap of dark granules, bound together by a scanty amount of unstaining substance (fig. 25), while, as the disintegration of the sphere approaches completion, the vacuole in which it lay becomes obliterated; or they may be rapidly and entirely dissolved and discharged from the cells, the vacuoles meanwhile persisting and leaving the striking honeycombed appearance described above (fig. 22). Whatever may happen to the constituents of the sphere in the first case, which agrees with the fate of the nutritive spheres of *Hydra* as described by Kleinenberg and Greenwood, we can only conclude that in the second case they have been dissolved and discharged into the enteric cavity. It is even possible that we may divide the endoderm cells, other than the gland cells, into two sets: (1) Those which are concerned in the elimination of waste matter from the nutritive spheres or from the somatic fluid directly. These are the apical cells of the villi and the vacuolate cells in their immediate neighbourhood. (2) Those which discharge their stored material, leaving, so far as can be detected, no residue. These lie towards and between the bases of the villi in the blasto-styles and middle regions of the body and foot.

This conclusion, which may be accepted as a provisional hypothesis until the processes taking place in the endoderm shall have been worked out more fully, is based upon two facts: namely, that the pigment, as was noted by Allmann, is located only in the cells near the free ends of the villi, and that the bubbly cells in all their various conditions of incomplete or complete discharge always lie between or towards the bases of the villi.

Another point of evidence in favour of the view that the somatic fluid conveys stored nutriment from one part of the body to another is derived from a study of the histology of the spadix of the gonophores.

Structure of the spadix of a gonophore. This, in the completely formed female gonophore, is composed of a considerable number of tongue-shaped villi, which have their apices turned towards the axis of the spadix, and project a considerable distance downwards towards the centre of the blastostyle. The cells between their bases, and therefore forming the apex of the spadix, are long, narrow, and columnar in character; and their protoplasm is filled with numerous small vacuoles, the contents of which become precipitated by preserving reagents, thus conferring a character-istically turbid character on the entire cell. That end of the cell which abuts on the basement membrane separating the endoderm of the spadix from the developing generative elements is excavated to form a vacuole. If we examine the cells which form the villi we find that they have fundamentally the same structure, except that the basal vacuole is now so much elongated that we may almost speak of that portion of the cell as being canalised (fig. 23). In other words, the whole spadix is a specialised structure for the absorption of nutriment from the somatic fluid of the blastostyle, which nutriment is doubtless largely derived from the stored material of the vacuolate cells, through the help of the somatic fluid.

The absorbed nutriment, probably after it has undergone important changes at the hands of the cells of the spadix, is discharged into the vacuoles at their base which abut on the supporting lamella, whence it passes to supply the remarkably abundant fluid present in the entocodon of the female gonophore.

These different points—(1) the general distribution of the nutritive spheres, (2) the method of the discharge of those bodies, and (3) the fact that the gonophore possesses an organ, the spadix, the histological cha-racters of which lead us to suppose that it is designed to absorb nutriment from the somatic fluid (which, especially in the autumn, when *Myriothela* appears to exist largely at the expense of its stored material, and probably throughout the year, must be largely recruited from the reserve material of the vacuolate endoderm cells)—although when taken singly they are of slight value, yet when considered together and as mutually supporting one another they justify the statement that the metabolic activities of the different parts of the endoderm are brought into relation with one another through the agency of the somatic fluid.

If no such link existed, and if, therefore, the animal were unable to direct its entire resources towards the accomplishment of any metabolic act, we should expect to find evidence of the fact in the more marked exhaustion of the endoderm during starvation in the immediate neighbourhood of a gonophore as compared with the other parts of the blastostyle or of the body generally. But such evidence appears to be wanting.

EXPLANATION OF PLATES I AND II

Fig. 1. Section through the ectoderm of the distal portion of a blastostyle. [Animal killed in May.]

Fig. 2. Section of the ectoderm of the gonophore-bearing region. [Animal killed with osmic acid in May.] $\frac{1}{15}$th ob.

Fig. 3. Teased preparation from the same specimen. The primitive germ cells have dropped out. $\frac{1}{15}$th ob.

Fig. 4. Section through the generative region of an exhausted animal killed in autumn.

Figs. 5, 5*a*, 5*b*. Ectoderm elements isolated by teasing. Osmic acid. In 5 and 5*a* are represented parts of the nerve network.

Fig. 6. Isolated primitive germ cells. Osmic acid. $\frac{1}{15}$th ob.

Fig. 7. Section showing the process of absorption of the supporting lamella.

Fig. 8. First stage in the formation of a gonophore. Ectoderm thickened and containing a cluster of primitive germ cells in its lowest part.

Fig. 9. The second stage in the formation of a gonophore.

Fig. 10. The third or blastema stage.

Fig. 11. A completely formed young gonophore.

Fig. 12. Piece of the cuticle which covers the ectoderm stripped off. $\frac{1}{15}$th ob.

Fig. 13. A *Myriothela* bearing buds in various stages.

Fig. 14. Section through ciliated cell zone of the endoderm.

Fig. 15. Goblet cell. Osmic vapour.

Fig. 16. Villus of goblet-cell zone.

Fig. 17. Two apical cells.

Fig. 18. Loaded gland cell and empty vacuolate cell.

Fig. 18*a*. Discharged gland cell.

Fig. 18*b*. Intermediate condition of gland cells.

Fig. 19. Group of small cells from apex of villus.

Fig. 20. Oral end of *Myriothela* with lip retracted.

Fig. 21. Villus from middle region of the body, with gland cells just commencing to discharge.

Fig. 22. Group of cells from the endoderm of a blastostyle. Osmic acid.

Fig. 23. Villus from the spadix of a gonophore, showing canal-like vacuoles in the lower portion of the cells. At a and a' these are cut across.

Fig. 24. Two cells from the endoderm of a blastostyle. Osmic acid.

Fig. 25. Nutritive spheres containing pigment granules. Osmic acid. $\frac{1}{18}$th ob.

Fig. 26. Body from the endoderm of a tentacle.

Fig. 27. Endoderm cell containing the large type of nutritive sphere.

Fig. 28. Large nutritive sphere.

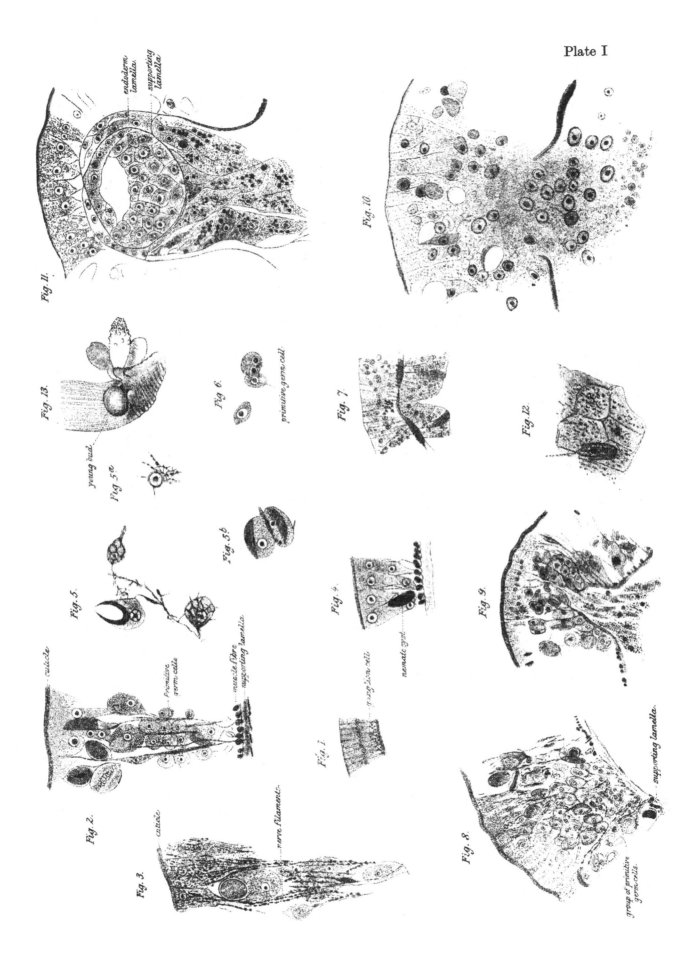

Plate I

endoderm lamella.

supporting lamella

Fig. 11.

Fig. 10.

Fig. 13.

young bud.

Fig. 6.

primitive germ cell.

Fig. 7.

Fig. 12.

Fig. 5a

Fig. 5.b.

Fig. 5.

cuticle

Primitive germ cells

muscle fibre

supporting lamella.

Fig. 4.

ganglion cell

nematocyst

Fig. 9.

Fig. 1.

Fig. 2.

cuticle

nerve filaments.

Fig. 3.

supporting lamella.

Fig. 8.

group of primitive germ cells.

Plate II

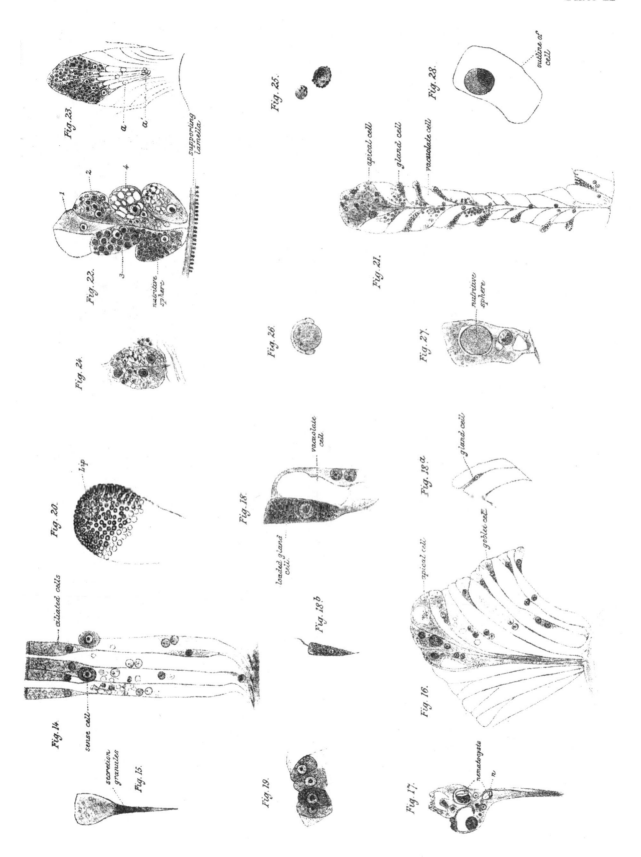

Fig. 23.

Fig. 25.

Fig. 28.

outline of cell

a

a

supporting lamella

apical cell

gland cell

vacuolate cell

Fig. 22.

1

2

4

3

nutritive sphere

Fig. 21.

Fig. 24.

Fig. 26.

Fig. 27.

nutritive sphere

Fig. 20.

Lip

Fig. 18.

vacuolate cell

loaded gland cell.

Fig. 18 a.

gland cell

Fig. 14.

ciliated cells

sense cells

apical cell

goblet cell

Fig. 18 b.

Fig. 16.

secretion granules

Fig. 15.

Fig. 19.

Fig. 17.

nematocysts

n

2

ON THE REACTION OF CERTAIN CELL-GRANULES
WITH METHYLENE BLUE

[Proc. Camb. Phil. Soc. (1891), VII, 256]

In 1878, Prof. Ehrlich pointed out the fact that the granule-containing cells of the body, whether found free in the body fluids or elsewhere, could be distinguished from one another by the character of the reaction of their granules with different aniline dyes.[1] He distinguished five classes of granules characterised by staining with acid, basic, or neutral dyes, or indifferently with acid or basic dyes (amphophil). The present communication deals with a further subdivision of the basophil granules into two groups, characterised by very distinct colour reactions with the basic dye methylene blue, the one class of granules staining a deep blue, the other a bright rose.

The distinction of tint depends, for some reason not at present at all obvious, on illumination with yellow artificial light. Under these circumstances the colour contrast is one of extraordinary brilliancy. With daylight, or with gaslight after it has been filtered through neutral-tint glass, the rose colour either appears a blue like that shown by the blue-staining granules, or is dulled to a blue-violet tint. The explanation of this change may be found in the fact that the yellow gaslight is relatively richer in red rays (or poorer in blue rays) than is daylight, or the phenomenon may be of a more complex nature. Be this as it may, the abrupt transition from bright rose to bright blue produced by directing the mirror from the gas flame to the window is a striking feature of this colour reaction.

The discrimination of basophil granules into rose-staining or blue-staining varieties may depend upon the dichroic nature of methylene blue. Thin films produced by running a minute quantity of a strong solution under a coverslip appear rose with artificial light, while more dilute solutions appear blue.

Methylene blue is a salt having the composition of a chloride, the base being a pigment of the aniline series, and it has already been noticed that alkalis produce a rose or reddish modification possibly by decomposing this salt and liberating the pigment-base. This suggests that the rose tint may not be solely a physical phenomenon but may depend upon a chemical

---

[1] The various papers dealing with this subject have been republished by Dr Ehrlich in pamphlet form under the title *Farbenanalytische Untersuchungen zur Histologie und Klinik des Blutes,* Berlin, 1891.

action of the granule substance on the dye, or a chemical change produced by the osmosis of the pigment through the cell protoplasm to the granule.

That the reaction does not depend simply upon the thickness of the film of methylene blue is shown by the fact that the rose tint may be developed by granules of sizes varying from mere points to spherules $2-4\mu$ in diameter. Nor does it depend upon the solidity or fluidity of the staining substance; for the contents of large vacuoles in the ectoderm cells of *Daphnia* frequently stain an intense rose, the rest of the cell appearing blue. Lastly the rose tint may be produced neither in the cell nor in the animal, but (in the case of *Daphnia*) by the action on the dye of a substance poured by the ectoderm cells into the surrounding water.

The hypothesis that the reaction is really of a chemical nature is favoured (1) by the general fact that the imbibition of dyes by the fresh unfixed cells is determined by the chemical nature of those dyes—whether the pigment be basic, or acid; and (2) by the peculiar method of imbibition of dyes by fresh cells. If still living cells, such as basophil blood corpuscles, are treated with either an organic fluid or normal salt solution, in which a small quantity of methylene blue is dissolved, it is noticed that imbibition of the dye is coincident with the onset of death. So long as the cell remains fully alive it resists infiltration by the pigment, and the granules remain uncoloured. This condition may, especially with eosinophil cells, last for hours. With the first onset of death the dye makes its way through the protoplasm and the granules become coloured. Later, when rigor mortis has become thoroughly established, the nucleus and cell-body absorb the dye, and appear blue. In other words the first imbibition of the dye occurs at a period when the complex cell protoplasm is commencing to disintegrate, and when therefore profound chemical changes are taking place.

In order to determine whether granules are of the rose- or blue-staining varieties it is necessary to apply the stain in some relatively innocuous fluid to the living cells; and subsequent treatment with fixing reagents entirely obliterates the reaction. This is because all fixing agents with which I have experimented have some action on methylene blue. Thus corrosive sublimate produces a rose-coloured modification, and converts blue staining into violet or rose. Ammonium picrate produces a violet tint except in the case of very intensely blue granules. Osmic acid converts the rose into a blue tint.

Rose-staining basophil granules have been found by me in free basophil cells of *Astacus*, and of Vertebrates, in the ectoderm of *Daphnia*, and of the Ammocœte larva of Lampreys, and in the alveolar cells of salivary glands.

The last two instances are of a specially suggestive nature, as affording instances of cells containing at the same time blue- and rose-staining granules. In the cells lining the alveoli of the submaxillary gland of a rabbit I have seen, after treatment with dilute methylene blue in normal salt solution, a zone of rose-coloured granules surrounding the lumen and extending about half-way towards the basement membrane, while outside this there was a zone of blue-staining granules. This suggests that the rose-staining condition is a final stage in the elaboration of the constituents o the granules of these cells.

A still more instructive example is found in the ectoderm cells of the Ammocœte larva which I examined at the request of Dr Gaskell. These cells are each overlaid by a thick cuticle perforated by coarse canaliculi which lead from the cell protoplasm to the external surface of the animal. Miss Alcock has shown that these cells, under appropriate stimuli, discharge on to the general surface a viscid substance which has the power of rapidly digesting fibrin in an acid medium. If these cells are treated with methylene blue we find (1) that the extruded secretion gives the rose reaction, (2) that the pores in the cuticle may appear as rose-coloured rods, owing to their being filled with the secretion, and (3) that the cells themselves are occupied by rose-coloured granules which lie in the half of the cell next to the cuticular border, and by blue-coloured granules which occupy their deeper portions.

In the ectoderm of *Daphnia* rose-coloured *granules* are scanty, while, under certain circumstances to be detailed elsewhere, the cells may include a number of large vacuoles, the contents of which give a brilliant rose reaction. In connection with the presence of these vacuoles we find that *Daphnia* possesses the power of extruding on to its surface, through cuticular pores, a substance which swells up to form a jelly in water, and stains brilliant rose. This particular case will receive more detailed description on some future occasion. For the present I will only say that the secretion is used by the animal to prevent parasitic vegetable or animal growths obtaining a foothold on the shell.

I have never yet found a blood or lymph cell with both blue- and rose-staining granules. It may be regarded as probable that blue-staining granules are absent from wandering cells. The cells with rose-staining granules have a remarkable distribution. In *Astacus*, as I have noted elsewhere,[1] they occur normally lodged in the spaces of a peculiar tissue which forms an adventitia to some of the arteries. They are only discharged into the blood as a result of special stimuli. In Vertebrates they occur to a marked extent in the peculiar adventitia of the blood vessels of the spleen.

It is noticeable that I have so far failed to find rose-staining granules in endoderm cells, though I have examined the lining cells of the alimentary canal and of its glands in very diverse animal types.

The cells of the excretory organ (end-sac of *Daphnia*) contain granules which have a remarkable affinity for methylene blue and stain a deep opaque blue.

[1] *Journ. Physiol.* (1892).

3

THE BLOOD CORPUSCLES OF THE CRUSTACEA, TOGETHER WITH A SUGGESTION AS TO THE ORIGIN OF THE CRUSTACEAN FIBRIN-FERMENT

[Journ. Physiol. (1892), XIII, 165]

PART I

In this paper it is proposed to deal exclusively with the histology of the blood of two forms of the Crustacea representing the extremes of development in that group. These forms are *Astacus* and *Daphnia*. For many reasons it is advisable to deal first with the higher of these two forms.

ON THE CORPUSCLES OF THE BLOOD OF *ASTACUS*

Methods of Observation

Specimens of the blood of *Astacus* may be readily obtained by pushing the point of a fine pipette through the soft skin which joins the dorsal portion of the cephalo-thoracic shield to the tergum of the first abdominal segment. This soft skin forms the roof of the posterior portion of the pericardial sinus and by perforating it a sample of pericardial blood is obtained.

The method of operation is as follows. A glass pipette is drawn out to a fine point and fitted with a strong india-rubber bag, by the aid of which fluid may be sucked up into the tube. The abdomen of a Crayfish is now firmly pushed downwards in order to stretch the thin tough skin between thorax and abdomen, and the point of the pipette is inserted from the side and horizontally in order to avoid injuring the heart and superior abdominal artery. A drop or two of blood sufficient for examination is sucked up, the pipette withdrawn, and the abdomen quickly bent upwards so as to mechanically close the tiny aperture. In a short time a blood clot forms round the wound and the animal may be replaced in the tank with absolutely no further loss of blood.

The blood should, for reasons which will appear later, remain in the pipette for as brief a time as possible. The blood so obtained may be examined in the fresh condition by placing it on a slide and covering with a coverslip in the ordinary way, or by suspending a drop on the underside of a coverslip over a cell. Permanent preparations may be made by inverting a drop of blood over a 2 per cent. solution of osmic acid, care being taken that the air is saturated with the vapour. After 15 min. the blood can be stained and mounted in glycerin or balsam.

It is advisable to use very dilute stains in order that the action may be as selective as possible. Otherwise the solidified plasma may become too deeply coloured. A reagent of the very greatest value in the study of the corpuscles is iodine. Its use in special cases will be mentioned in the subsequent pages.[1]

The number of the corpuscles in a cubic millimetre of blood was measured by the aid of a Gower's Hæmatocytometer. The corpuscles were counted both in undiluted blood in the fresh condition, and after fixation by osmic vapour.

The *relative* number in different samples of blood was found as follows: Shallow cells were made by fastening down rings of equal size cut from a sheet of tinfoil. A glass pipette was drawn out into a fine tube on which a mark was made so that the amount of blood contained in the pipette when filled up to this mark just sufficed to fill one of these cells. In this way each preparation contained the same quantity of blood which was fixed by osmic vapour and stained. The corpuscles were then counted in 50 "fields" of the microscope with Oc. 4, Ob. D Zeiss, and an average taken.

Specimens of blood obtained from the blood spaces in the neighbourhood of the antero-ventral and postero-ventral arteries were also examined.

General Features

If a drop of blood taken from an active well-fed *Astacus* is examined with a Zeiss Ob. D or E it is found to contain a large number of actively amœboid corpuscles characterised by the possession of a great number of extremely large, highly refractive granules, or rather spherules (Pl. III, fig. 1). If the preparation of the specimen has occupied a few seconds, there will be seen, in addition to the spherule-bearing cells, a number of large, distinct, rounded nuclei which float free in the plasma. These latter elements belong to blood cells markedly distinct from the spherule-bearing corpuscles, and characterised by such an extreme sensitiveness to certain stimuli that contact with a foreign body, such as glass, causes an explosive disruption of their protoplasm. These cells, which I propose to call "explosive corpuscles", can be fixed by osmic vapour or iodine, and the study of such preparations (fig. 3) makes very evident the fact that, as was pointed out by Heitzmann (9)[2] and Frommann (10), there are two distinct kinds of cells present in the blood of *Astacus*, both kinds being amœboid and both resembling the white rather than the red corpuscles of mammalian blood.

Though the large refractive globules which are such a remarkable feature

---

[1] The solution used contained 0·5 per cent. of iodine, with sufficient potassium iodide to ensure solution. An equal volume of the iodine solution should be added to the blood.

[2] See Bibliography, p. 48.

of the non-explosive cells are more correctly designated by the term "spherule" than by the word "granule", yet, having in view the wide application of the latter word by Prof. Ehrlich, and the fact that these bodies are identical with his *a*- or eosinophile granules, it will be better to adhere to this word, and to call the cells the "eosinophile cells". In this way we may avoid the cumbrous, though morphologically the more accurate designation, "spherule-bearing".

The number of corpuscles in one cubic millimetre. This was estimated by counting the corpuscles (that is, free nuclei and intact corpuscles) in 500 squares of a Gower's Hæmatocytometer. This was done in the case of ten animals which were active and appeared to be normal. To test the accuracy of the method which involved the assumption that each explosive corpuscle left one and only one nucleus, the following experiment was made. Two drops of blood were taken from the same animal, one was placed on the hæmatocytometer, and of the other an osmic vapour preparation was made. The relative number of eosinophile and explosive corpuscles was then carefully estimated in each case. In the blood of the hæmatocytometer it was found that there were 4·6 free nuclei of explosive corpuscles to each eosinophile corpuscle, while in the osmic preparation the ratio of the (preserved) explosive corpuscles to the eosinophile corpuscles was 4·7 to 1. That is to say, the error in this particular case was a little less than 2 per cent.

The mean of the different countings made with the blood of normal animals gave an average of 286 corpuscles in the cubic millimetre. There were however wide departures from this mean, the numbers falling as low as 250 and rising as high as 400 in the cubic millimetre.

These very wide variations are possibly explained by reference to peculiar variations in the condition of the blood which are more readily recognised in *Daphnia*. Not infrequently, apparently healthy specimens may be found in which the blood corpuscles are peculiarly adhesive, so that a considerable number, or even the bulk of the corpuscles, adhere to the walls of the blood spaces and thus are withdrawn from active circulation.

The relative number of the two kinds of corpuscles. The explosive corpuscles are always more numerous than the eosinophile corpuscles in the normal animal, but, like the total number of corpuscles, the ratio of granular to explosive cells varies within wide limits.

The normal ratio is one granular to three explosive cells, the variations lying between 1 to 1·22, and 1 to 5.

The causes of this variation are, at present, extremely obscure and complex. The relative number of the two kinds of corpuscles is not directly related to the total number of cells present in the blood. Animals with above 250 corpuscles in the cubic millimetre, or those with a small total number, may furnish the extreme ratios.

The relative abundance of the corpuscles differs slightly in blood drawn from different parts of the same animal. The granular cells are always slightly more numerous in blood obtained by perforating the sternum in the thoracic region as compared with blood from the pericardial sinus.

What may remain to be said on this point is best deferred to some future time when the function of the corpuscles is discussed.

We will now proceed to a more detailed consideration of the two kinds of cells found in the blood of a normal animal.

The Explosive Corpuscles

These cells can only be adequately studied after treatment with osmic vapour or iodine solution. If a drop of blood be examined under the microscope within a few seconds from the time of its removal from the animal, it will be seen to contain a large number of pale oval corpuscles, containing a few granules of various sizes but all small. To see the corpuscles in the fresh state in this condition requires rapid manipulation.

If attention be fixed on one of these corpuscles it will be seen to undergo remarkable changes.

Extremely fine pseudopodia are shot out and blebs of cell-substance travel rapidly along these, expand into a vesicle or bubble, and burst. Sometimes only short blunt processes are formed, which swell into little bladders, remain an instant, and then burst. In other cases again the surface of the cell, without forming processes, develops vesicles which expand until they burst. In any case the process is the same and results in what may be described as an explosive solution of the cell-substance.

While the cell-substance is undergoing dissolution the nucleus also suffers remarkable changes. When the cell is first seen the nucleus is entirely invisible; but as the disintegration proceeds it rapidly comes into view. The optical effect may best be likened to the rapid development of the image on a photographic plate. With almost the first stage in the dissolution of the cell protoplasm the nucleus becomes visible as a faintly defined oval body. It rapidly acquires distinctness, the outline becomes firmer, and highly refractive, yellowish-green masses appear, some of which join one another to form the periphery or "capsule" of the nucleus, while the others are distributed as rounded spherules or "nucleoli" in its interior (fig. 2). At the same time traces of a network with large meshes may appear and rapidly vanish. This change, which appears to be of the nature of a rigor or clotting, marking the death of the nucleus, converts the substance of the nucleus from an invisible matter, whose refractive index differs only slightly from its surroundings, into one of remarkable visibility, and is accompanied by an alteration in shape, the nucleus changing from an ellipsoid to a sphere, and expanding slightly (figs. 8 and 9).

The change in the physical condition of the nucleoplasm may be beautifully demonstrated by allowing a drop of freshly drawn blood to fall on to about an equal volume of a very dilute solution of methylene blue in normal salt solution. The pigment does not appear to affect the rate of dissolution of the cells but renders slightly more permanent the nuclear network. The nucleus as it comes into view presents the appearance of an oval bladder, containing a coarse network the strands of which are of uniform diameter. After a few seconds the network breaks up and some of its substance runs into the spherules noticed above. As these spherules are forming they imbibe the stain with great rapidity, so that by the time the rigor has thoroughly set in both they and the nuclear capsule are coloured an intense blue. This rigor mortis of the nucleus so far resembles the rigor of muscle in that in time, it may be many hours, it passes away. Some of the material is dissolved, the refractive appearance vanishes, and there is left only a shell with faintly marked outlines, which may persist unchanged for days or weeks.

The nucleus of the explosive corpuscle when rigor has set in thus resembles the type of nucleus so commonly seen in preparations of the tissues of the Crustacea, in its well-marked capsule and numerous and well-marked nucleoli.

The various stages in the disintegration or explosion of the explosive corpuscle may be fixed for prolonged examination by taking advantage of the preservative action of iodine. If a drop of the iodine solution be added to a drop of fresh blood the corpuscles are absolutely fixed, showing the shape and for a time the granulations of life. If, however, a drop of iodine solution and a drop of blood are placed side by side on a slide and a coverslip lowered on to them so as to avoid blending them, we have, at the contact between the two drops under the coverslip, a narrow zone where they have mixed. In this zone the living and unchanged corpuscle has been fixed and preserved. Beyond this region and towards the blood the iodine has taken a longer and longer time to penetrate, and we thus get the corpuscles fixed at longer and longer intervals as we penetrate into the blood region. It was in this way that the stages shown in figs. 7, 8 and 9 were obtained.

From what has been said above it is obvious that the direct study of the living explosive corpuscle is an impossibility; we can only proceed by inference from preserved specimens, and it is advisable to consider the action of the two preserving agents used—iodine and osmic vapour.

Action of osmic vapour. When a drop of blood is suspended over osmic acid it rapidly sets to a firm jelly, the process resembling and probably being identical with normal clotting. Exposure for fifteen minutes to half-an-hour to the reagent so completely and permanently fixes the corpuscles that the prolonged action of water, alcohol, or dilute glycerine fails to sensibly alter them. After fixation they may be stained with dilute hæmatoxylin, picro-carmine, methyl violet, methylene blue, etc.

On studying the stained or unstained osmic preparation with the microscope we see that the morphological differences between the two types of cells are rather increased than diminished by the reagent. The explosive cell is not only the smaller but its cell-substance is affected by the reagent in a way markedly different from that of the eosinophile cell. The cell-substance is clear, faintly visible and is marked by very fine granulations, or appears quite hyaline. In each cell is an oval nucleus containing an ill-defined and irregular network. The nuclear network is never perfectly preserved, the bulk of its substance has broken down to form nucleoli (figs. 3 and 11). The most striking feature of these cells however is the great irregularity and diversity of form they exhibit. Most commonly they are elongated spindles, or the spindle may be bent to form a crescent, or may be thrown into undulations, or may give off at its ends or sides short or long processes. In any case the long axis of the oval nucleus coincides with the long axis of the cell.

We may safely conclude from these appearances that the explosive cells are, during life, extremely mobile.

In the fresh and still fluid blood the explosive cells appear as flattened ellipsoids, which are biconvex when seen edgeways, the central bulging being produced by the nucleus.

Action of iodine. This reagent furnishes us with a remarkable and suggestive series of facts. Iodine, when about 0·25 per cent. is present, greatly delays and profoundly modifies the post-mortem changes in the explosive cells. The formation of explosive bullæ and the general solution of the protoplasm no longer occur, but, if we watch the cell after treatment with the reagent, we see a gradual disappearance of certain of its elements, namely, the large number of fine discrete granules. With 0·5–1 per cent. of iodine the cells resemble the specimen shown at fig. 7. The nucleus, as in the fresh cell when first seen, is invisible, being obscured by the large number of fine granules which are imbedded in the cell protoplasm and confer on it a characteristically solid, dense appearance. If only 0·25 per cent. of iodine be present these fine granules are at first present, but in from 10 to 15 min. they are dissolved, the cell protoplasm assuming the clear hyaline character so noticeable in preparations fixed by osmic vapour (fig. 15). The impression conveyed by the brief glimpse one is able to have of the cell in the fresh state, that the protoplasm contains fine granules in considerable abundance, is confirmed by the appearances presented after treatment with iodine.

The size of the explosive corpuscles. Owing to their diversity of shape it is difficult to obtain exact measurements. Only rarely are even approximately rounded cells found. Still by measuring the two main axes and taking an average, values were obtained which were found to hold good for a considerable number of specimens. These numbers are: 25–$30\,\mu$ for the long axis and 10–$11\,\mu$ for the short axis. The nucleus was found to measure 13–$14\,\mu$ on the long axis and 8–$10\,\mu$ along the short axis.

Histological Characters of the Eosinophile Corpuscles

These remarkable cells differ from the explosive corpuscles in the fact that they are readily studied in the fresh and living condition. They persist unchanged and may even retain their powers of amœboid movements for a long time, sometimes for hours, in a drop of blood placed under a coverslip. They therefore persist for a long time after the blood has clotted. In one case a corpuscle was found to be actively amœboid two hours after the blood had been placed on the slide. These elements present greater structural differences than are found in the explosive cells.

There is a marked distinction into ectosarc and endosarc, the former being clear and optically structureless, and frequently thrust out into pseudopodia of variable form. The endosarc on the other hand is more or less completely occupied by large granules or spherules. Sometimes it is partly occupied by granules and partly by vacuoles.

These cells present great diversity of shape, and their irregularity is not merely due to the formation of pseudopodia but affects the general outline of the cell and involves the endosarc. Sometimes, and generally when the granules are present in very great abundance, the endosarc is spherical and the ectosarc forms a thin mobile envelope with pseudopodia of diverse forms, either long and thin or short and blunt, and thrust out in various directions. At other times the ectosarc does not form pseudopodia but merely streams round the endosarc. In some cases these cells assume a remarkable external resemblance to a Heliozoon such as *Actinophrys*, filiform and immobile pseudopodia radiating from the body of the cell. The resemblance is much heightened when the endosarc is relatively free from granules. The numerous vacuoles which may be present under those circumstances recalls with singular exactness the "bubbly" or vesicular nature of the Heliozoon cell body (fig. 5). The pseudopodial movements of the ectosarc, though distinct, are generally slow, exceedingly so in the case of large cells much loaded with spherules; on the other hand smaller cells less loaded may be exceedingly active. It is impossible without occupying undue space to indicate the variations of form presented by these cells, one or two things however call for special mention.

The division between ectosarc and endosarc is not always clear and exact;

usually in the smaller less loaded cells it is difficult to make out. Very shortly after the blood has been placed on the side the active movements cease; the pseudopodia still remain protruded but they are apparently fixed in position by the solidification of the plasma which occurs when the blood clots. Faint and apparently ineffectual movements may still be seen. In the course of time however the corpuscle dies and a series of events then take place which recall those marking the dissolution of the explosive corpuscle. The first change heralding death is the loss of definiteness in the outline of the endosarc and the extrusion of some or all of its granules into the ectosarc and sometimes thence to the exterior. The granules then rapidly dissolve, leaving absolutely no visible trace.

Whilst these events are happening, and when the solution of the granules is advanced, the cell passes into rigor mortis. The previously almost invisible outline of the ectosarc becomes very clear and visible, and the nucleus comes into view and acquires that intense distinctness which we noticed as such a remarkable feature of the rigor mortis, or clotting, of the nucleus of the explosive cells.

The *granules* are remarkable bodies, extraordinarily refractive and very large. The refractive nature is so striking that Haeckel[1] was led by it to regard them as fat. This supposition is untenable since (1) exposure for 16 hours to osmic vapour failed even to stain them, and (2) they are readily soluble in the serum. When not pressed against one another they appear as spherules of $1-3\mu$ in diameter; Haeckel described them as being $2-3\mu$ in diameter. They are therefore much larger than the granules of the wandering cells of vertebrates. A distinctive microchemical reaction of these bodies was found by following the lines of Prof. Ehrlich's classical researches on cell granules.

Reaction with eosin. If a grain or two of eosin be stirred up with a drop of fresh blood it readily dissolves. At first the eosinophile corpuscles are absolutely resistant to the stain, but later, as their vitality decreases, the granules rapidly absorb it, the rest of the cell remaining for some time completely uncoloured. This reaction, as Ehrlich pointed out, appears to depend upon the affinity of the substance of the granule for acid rather than for basic dyes, that is to say, for those dyes in which the colour is associated with the acid and not with the basic portion of the molecule.

Reaction with sodium sulphindigotate. This pigment serves as another instance of an acid pigment but of different chemical character. If it be dissolved in normal salt solution and added to blood, the granules are stained an intense blue (fig. 10). The imbibition of the dye, when the living cells are exposed to its influence, takes place only at a critical point in the degradation of the molecular structure of the cell-substance. If the cell is possessed of great vitality it may be some hours before this point is reached.

3-2

Reaction with basic dyes. Exposure for hours to a basic dye such as methylene blue fails to stain the granules, and so long as they remain there is no trace of coloration in the cell. Immediately on the onset of rigor mortis the cell and especially the nucleus are coloured blue. The granules dissolve without being affected by the stain.

Reaction with iodine. The granules are preserved and stain readily. The coloration is deep brown and cannot, I think, be taken as evidence of a carbohydrate nature.

Reaction with osmic vapour. This reaction is especially valuable since it furnishes us with a striking differential characteristic between the *cell-substance* of these cells on the one hand, and that of the explosive corpuscles and of certain basophile cells to be presently described on the other hand. In an osmic vapour preparation, whether stained or unstained, the cell-substance of the eosinophile cells is found to have reduced the osmic acid and to have assumed an opaque glassy appearance; the eosinophile granules are no longer visible, their place being apparently taken by vacuoles. The cell-substance of the explosive corpuscles on the other hand does not reduce the osmic acid but remains clear and uncoloured.

Eosin after fixation with osmic vapour affords us an interesting and valuable histochemical test for the presence of the eosinophile cells of *Astacus*. It is possible that the spaces in the dark refractive protoplasm which, after treatment with osmic vapour, alone represent the previously very distinct granules are really spaces formed by the solution of the granular substance. At any rate, whether this is so or not, treatment of the cells in this condition with strong eosin fails to render any granular substance visible. Instead of that we find that the whole cell-substance has become strongly eosinophile, so that if a drop of blood fixed by osmic vapour be treated with a solution of eosin in glycerin the eosinophile cells become deep opaque red, while the explosive cells even after many hours or days remain uncoloured and, from the general coloration of the plasma, become obscured and appear to vanish.

Reaction of osmic vapour after *impregnation with sodium sulphindigotate.* By taking advantage of the fact that the granules are less soluble after impregnation with the dye, osmic vapour preparations may be made in which the granules are preserved as deep blue bodies. I have only tried this reaction once and it then succeeded perfectly, and it may furnish another valuable test.

Methylene blue after osmic vapour stains both kinds of cells.

After treatment with osmic vapour the ectosarc of the eosinophile cells cannot be distinguished as a separate structure (fig. 3) even though the preparation be deeply stained. The cells may however by fixation with a saturated solution of corrosive sublimate, be preserved with pseudopodia still extruded and the ectosarc clearly shown as a hyaline layer surrounding the more prominent endosarc.

My reasons for setting forth the histochemical characters of the eosinophile cells at such length are two. (1) The tests given above and those to be presently mentioned in connection with the basophile cells render evident the fact that these different elements, explosive, eosinophile, and basophile cells, differ not merely in the kind of granules they produce but also in the very nature of their cell-substance. (2) That in the long run the success of any investigations undertaken with the object of determining what physiological meaning we must attach to the presence in the bodies of animals of very diverse position in the animal kingdom of the same different types of wandering cells, must depend upon the accuracy and completeness of the tests applied to determine the presence or absence of these different kinds of cells in any particular fluid or tissue.

Very frequently the cell is not completely charged with spherules. When this is the case the spherules generally occupy a more or less definite region of the endosarc, the rest of the substance of which is excavated by vacuoles which so closely correspond to the fully formed spherule as to suggest strongly that they are intimately related to the formation of those bodies.

We have a parallel case of the formation of spheres of elaborated material in vacuoles in the endoderm cells of the Cœlenterata. Miss Greenwood(7) has carefully followed the method of their formation in *Hydra*, and I have detected similar bodies enclosed in the vacuoles of the endoderm cells of another Hydroid, *Myriothela*(8).

Sometimes when at the death of cells the spherules are dissolved while still within the endosarc they leave behind a very obvious vacuole. When, as mostly happens, they are actively extruded, the definite contraction of the endosarc which causes the extrusion obliterates the vacuole.

It follows therefore that these spherules are not permeated by and do not form part of the cell-substance, but that they form rather a distinct material lodged in spaces in that cell-substance.·

It is, I think, worth while emphasising the fact that the extrusion of the spherules may be due to (1) an active contraction of the cell-substance, or (2) to a process of solution, depending upon changes in the molecular constitution of that cell-substance which allows the surrounding fluid to gain ready access to the spherules. These facts have an obvious general bearing on the extrusion of stored granules from gland cells.

If the spherule has been impregnated with eosin before solution commences we see as it gradually diminishes in size an increase in the intensity of the red coloration owing to an increase in the density of the pigment. This proceeds until an almost opaque red spot alone remains, which in turn finally and completely disappears. The increasing density of the pigment however appears to affect the rate of solution which becomes increasingly slower as the red tint intensifies; the unstained granule dissolves much more quickly than the stained granule. In neither case does any previous swelling due to imbibition of water take place; the granule merely diminishes in size until it vanishes. We may thus conclude that the compound formed by the granule with the eosin is less soluble than the pure granule. Sodium sulphindigotate behaves in a way precisely similar to eosin.

Size of the eosinophile cells. They vary much in this respect but may be said to be from 18 to 27 μ in diameter. They have a higher *specific gravity* than that of the explosive cells; they sink much more rapidly.

The eosinophile cells have, from their permanence in fresh specimens of blood, been readily recognised by previous observers. Haeckel figured them in 1857, though he failed to detect the explosive cells. He found the dimensions of the nucleus to be 8–12 μ broad by 10–24 μ long. Heitzmann (1873) and Frommann (1884) detected the existence of more than one kind of corpuscle in Crustacean blood, and the two types of cells as seen after treatment with osmic acid were clearly figured by Prof. Löwit (see his figs. 1 and 2)[11]. The last-mentioned observer also noticed the disruption of the cells which occurs when the blood is shed and to this process he gave the name "Plasmoschise". Neither from his paper nor from his figures however does one gain a clear conception of the distinctive parts played by the two forms of cells. Geddes[3,4] briefly mentions two kinds of corpuscles in *Pagurus* which are respectively coarsely and finely granular. "The former are much elongated when freshly drawn, but rapidly become oat- or egg-shaped, and throw out blunt pseudopodia from any part of their surface." The finely granular corpuscles send out filamentous pseudopodia and alone possess the power of uniting with one another to form plasmodia. There can be little doubt that the coarsely granular cells of *Pagurus* correspond with the eosinophile cells of *Astacus*, and it is at any rate possible that the finely granular cells of the former also correspond with the more evanescent explosive cells of the latter. Geddes figures the two types of cells as about the same size, and Lebert and Robin[2] in 1846 found that the blood corpuscles of *Pagurus streblonyx* measured 10–12·5 μ in diameter with a nucleus 7·5–10 μ in diameter.

That the presence of corpuscles of divergent types in the same animal is a widespread phenomenon in the animal kingdom is shown by the fact that no less than four types of cells are found in the perivisceral fluid of echinoderms. In the perivisceral fluid of the sea urchin Geddes found (1) a flagellated cell which is detached from the living epithelium, (2) a cell charged with brown pigment granules and probably connected with respiration; in addition to these however were (3) finely granular amœboid cells, and (4) less numerous cells full of coarse granules and forming short blunt pseudopodia. Dr Griesbach(12) also detected the existence of coarse and finely granular cells in the blood of molluscs.

Other Cells found infrequently and in small numbers in the Blood of Astacus. Basophile Cells

The granular cells, even in the blood of a normal animal, are not always of the same type. Very rarely in the normal animal, but always in blood drawn from animals which have been poisoned by certain substances, cells are found which, although they are charged with granules closely resembling those possessed by the eosinophile cells, yet may by appropriate reagents be shown to profoundly and entirely differ from those cells.

These cells we will call at once with Prof. Ehrlich *basophile* cells, that is, cells whose spherules betray a profound affinity for basic pigments such as methylene or methyl blue.

Reaction with methylene blue. The basophile cell may be differentiated from the eosinophile cells by irrigating the blood with a dilute solution of methylene blue in normal saline. While the spherules of the eosinophile cells absolutely refuse to stain and the whole cell resists, as long as it remains alive, the entrance of the pigment, the spherules of the basophile granules stain deeply in the course of a few minutes without the cell generally undergoing any obvious changes.

Reaction with osmic vapour. A second and more valuable differential reagent is found in osmic vapour. As has been noted previously the eosinophile granules absolutely refuse to stain in the vapour, while the cell protoplasm becomes markedly distinct, so as to produce the appearance shown in fig. 3.

If fig. 3 is contrasted with the appearance of a basophile cell shown at fig. 12, as it appears after treatment with osmic vapour, the profound difference of the two types of cells, a difference which exists in spite of their close similarity in the fresh condition, is at once obvious. The granules of the basophile cell after treatment with osmic vapour stand out sharply distinct from the still clear hyaline cell-substance.

The basophile cells are only rarely found in the blood of normal animals, that is to say, they are either present in very small numbers so that one overlooks them, or they are really and entirely absent. It is most probable that the latter hypothesis is correct, for their presence has always been found to be associated with an abnormal condition of the blood, such as a general paucity of corpuscles.

The basophile cells are not only characterised by the reactions of their spherules and cell-substance with basic pigment and osmic vapour; they also present certain definite and peculiar structural features. They are of irregular shape, being either square, triangular, or rounded, and in size are larger than the eosinophile cells. The basophile cell shown at fig. 12, and detected in blood drawn from the pericardial sinus, is squarish in outline. Each side of the square measured $38\,\mu$, and the nucleus was $12\,\mu$ in diameter. I have never detected any trace of amœboid movement in a basophile cell, though it is probable that there are sluggish movements, which from being very slow and slight and occurring at several points of the surface at the same time, escape observation. There does not appear to be any formation of distinct pseudopodia; and associated with the absence of pseudopodial movement we find no trace of an external granule-free motile layer of cell-substance. There is no indication of any division into ectosarc and endosarc, the granules extending right to the surface of the cell.

Another most distinctive feature of these cells is the common presence of one or more large vacuoles. There usually is also in addition to the granules an irregular "amorphous" mass of refractive substance.

The *granules* of the basophile cells differ very markedly from those of the eosinophile cells. In a fresh cell they generally resemble the eosinophile granules in their distinctness, but even under natural conditions, they are colourless and slightly less refractive, while the fresh eosinophile granule has a peculiar yellowish green tinge. The basophile granules also present greater diversity in size, they may be mere points or they may measure over $2\,\mu$ in diameter. The same cell frequently contains granules of all sizes. They are much less soluble than the eosinophile granules.

Reaction with methylene blue. The granules possess the peculiar property of staining a bright rose tint with methylene blue. This colour reaction which is excessively distinctive is only clearly seen when yellow artificial light is used as an illuminant. If the fresh cells be treated with quite dilute methylene blue and viewed with a yellow gaslight we have the basophile granules appearing bright rose, and nuclei and stained particles, or cell-substance generally, a brilliant blue. With natural light the whole reaction vanishes, the spherules appearing blue like the rest.

The *amorphous masses* are apparently derived from the granules by their fusion. This is well shown in fig. 13, where the outlines of the granules are still imperfectly retained. This fusion of the granules is part of the normal process going on in the cell, the process however is not a simple blending of the granules but their substance at the same time largely loses its basophile reaction, staining only very slowly and with difficulty. When the staining does take place however the tint is pronounced and of a rose colour with yellow light.

On the Origin of the Basophile Cells

In *Daphnia*, as will be pointed out later on, basophile cells are commonly present in the blood-stream. In *Astacus*, on the contrary, they have as it were been withdrawn from the general circulation and only appear there under special conditions and as a result of special stimuli. In this animal under normal conditions they appear as an important constituent of a special and peculiar tissue which forms a thick external sheath or adventitia to some of the arteries, notably the antero-ventral artery. Haeckel has described the main histological features of this tissue under the name of the "Zellgewebe". The tissue may be regarded as consisting of (1) a supporting framework, and (2) a number of enclosed cells. The supporting framework when seen in optical section has the appearance of a network with nodal thickenings containing nuclei. On further examination the network appearance is found to be due to the optical or actual section passing through a number of spherical spaces or chambers, the walls of which are to a certain extent incomplete, thus placing them in communication with one another. The framework therefore is not disposed as anastomosing filaments forming a network but rather as a series of lamellæ, which bound incomplete chambers.[1] These are united one to the other by a transparent cementing substance. On the wall of each chamber is a thickening which encloses a nucleus. Leaving the connective-tissue framework, which is of no immediate interest, we pass to the second constituent of the tissue, the enclosed cells.

The cells are large and irregular in shape. They vary in size, the one shown at fig. 14 was found to have the extreme dimensions of $65\,\mu$ in length and $37\,\mu$ in breadth. In their general character and reactions these cells accurately resemble the basophile cells occasionally found free in the blood. Frequently however they contain large spheres, which by their reaction with osmic acid and from the fact that they are soluble in such reagents as turpentine we may conclude to be fat.

If this tissue, immediately after its removal from the animal, be slightly teased in a very dilute solution of methylene blue and then examined in the course of half-an-hour to an hour, with strong yellow light the cell granules will appear as shown in fig. 14, that is, they show a marked rose colour, at the same time the nuclei of the cells and the nuclei of the connective tissue appear blue, like the nuclei of the basophile cells when stained. In other words we see that the spherules of this tissue, like the spherules of such basophile cells as may be found free in the general circulation, give this peculiar "rose reaction" with methylene blue and yellow light.

[1] A chamber may measure $\frac{1}{10}$th of a millimetre in diameter.

On the origin of Crustacean Fibrin-ferment from a Substance
stored as Granules in the Explosive Cells

I purpose to state here very briefly certain facts which tend to show that the explosive cells are specially concerned in the production of that substance, the liberation of which causes the blood to clot.

Halliburton(5), Fredericque, Löwit and others have discussed the causes of the clotting of Crustacean blood, and the first-mentioned brought forward good reasons for believing that there is present in the plasma a proteid to which he gave the name of "crustacean fibrinogen" and which is the mother-substance of the fibrin of the clot. He further came to the conclusion that the substance ("fibrin-ferment") which caused the formation of the fibrin is yielded by the corpuscles of the blood. Prof. Löwit studied the changes which the blood undergoes on being shed with the aid of cold, and in this way he was enabled to show that "diese Abtrennung von Zellprotoplasma von den weissen Blutkörperchen"—to which he applied the term Plasmoschise—is, at any rate in point of time, a precursor of the clotting of the plasma.

The following considerations lead to the conclusion that this substance, fibrin-ferment, is a special product of the explosive cells:

(1) All writers who have dealt with the blood of the higher Crustacea have mentioned the extreme rapidity with which clotting follows the shedding of the blood. In describing the changes which the blood corpuscles undergo it was noticed that the explosive cells disintegrate with very great rapidity when the blood is shed. In other words there is a marked correspondence in time between the solution of these corpuscles and the solidification of the plasma.

(2) On the other hand the only other cellular elements normally present, the eosinophile cells, remain unchanged and alive for a considerable time after the blood has clotted. We are therefore compelled to refer the origin of the fibrin-ferment to the explosive cells.

In what relation does the fibrin-ferment stand to the explosive cells? Is it a special product of their activity? Though the answer to these questions cannot be given with the definiteness and certainty that is desirable, yet I think that the following facts entitle us to suggest, as a provisional hypothesis, that the substance which determines the clotting of the plasma is a special product of the activity of these cells and is a result of the solution of the cell granules before noticed. I have mentioned above that if iodine be present in not too great quantity (0·25 per cent.) it delays the post-mortem changes in the explosive corpuscles so that one is enabled to watch the gradual solution of the granules. Now iodine in about this quantity does not prevent the clotting of the blood, it merely delays it, and we find that the time of solution of the granules coincides with the time of clotting of the blood. It may be noticed in passing that 0·25 per cent. iodine prevents indefinitely the explosion of the corpuscles, the shell which remains after solution of the granules (fig. 5) persisting unchanged. Also the iodine permanently fixes the granules and cell-substance of the eosinophile cells.

Further evidence is obtained by comparing the action of osmic vapour. This reagent does not appear to interfere with the normal clotting of the plasma. A drop of blood suspended over osmic vapour passes into the jelly condition in much the same time and way as a drop simply exposed to the air. On the other hand blood which is slow in clotting, or which refuses to clot, is equally slow in clotting when exposed to the vapour, and exposure to osmic vapour fails to produce solidification of a drop of blood which would not otherwise clot. If this be a normal clotting, from whence is the fibrin-ferment derived since the explosive cells are preserved by the reagent? The answer to this question is found in the fact that the protoplasm of the explosive cells after treatment with osmic vapour is singularly clear and free from granules. The reagent, while it permanently fixes the general substance of the cell, permits or aids the solution of these granules as we have seen reason to believe that it does in the case of the eosinophile granules.

Causes determining the explosion of the explosive cells and the clotting of the plasma. A few points of interest may be noticed under this heading. The explosion of the corpuscles when blood is shed appears to be mainly due to either (1) contact with foreign solid bodies, or (2) by the presence in the plasma of the products of the disintegration of other explosive corpuscles. This may sometimes be beautifully seen by placing a large drop of blood on a coverslip and inverting over a cell. The corpuscles in contact with the glass explode practically at once, while as one focuses further and further into the hanging drop one sees a larger and larger number of intact cells, which in their turn explode.

On the Changes produced in the Corpuscles by the injection of Solid Particles and on the production thereby of a Special Phagocyte Modification

I do not propose to deal in this paper with the function of the various types of cells, nor with the modifications in their absolute or relative abundance which may be produced by the introduction into the blood of various poisons. My experiments in that direction are still incomplete. The reason therefore for bringing forward this isolated case is that any account of the histology of the blood would be defective which did not mention the phagocyte modification of the explosive cells which occurs as a result of the presence of an abundance of foreign solid particles in the plasma. I used for the purpose of the experiments indian ink, and it was injected suspended in sterilised (boiled) normal salt solution.[1] Different amounts of salt solution were injected in different experiments, the amount varying from 3 minims to $\frac{1}{3}$ of a c.c., the solution in each case being very highly charged with suspended indian ink.

I may remark in passing that the general effect is to produce a considerable drop in the total number of corpuscles free in the blood. The number is smallest a few hours after the injection, then they commence to increase in quantity, so that in about twenty-four hours they are more numerous than at the commencement of the experiment.

If the blood be examined in the fresh condition about an hour after the injection the plasma will be found to be still highly charged with ink particles. Here and there in the preparation large cells are to be seen (fig. 16) of an irregular shape, with numerous irregular and exceedingly fine pseudopodia, and with striking vacuoles of varying size in their substance, which is of a very clear transparent nature. These cells are highly charged with ink particles which are embedded in patches in the cell-substance, in the neighbourhood of a vacuole. I have seen a group of particles which lay at the edge of a vacuole suddenly discharged into it, whereupon they at once exhibited active Brownian movements. One also sees explosive cells containing ink particles, and this the more readily because the presence of the included foreign matter appears to delay their disintegration. On the other hand one notices the striking fact that in no case can ink particles be detected in the interior of eosinophile cells.

What is the origin of the above-mentioned large hyaline cells? The study of osmic vapour preparations leads to the conclusion that they are modified explosive cells, formed in some cases at any rate by the fusion of two or more to form a plasmodium. In the osmic vapour preparation we notice that the great number of the explosive cells are charged with ink particles (fig. 18). Where this is not the case we see a striking incipient vacuolation (fig. 17). On the other hand the eosinophile cells show no included particles.

We thus have the remarkable fact that power of ingesting solid particles of the nature of indian ink appears to reside only in one of the two kinds of cells normally found in the blood of *Astacus*. On the other hand, and the fact has a certain bearing on the function of the basophile cells, the cells of the basophile "Zellgewebe" take in the particles of indian ink to a limited extent. Every cell in the tissue may, and probably does, contain ink particles but never in great abundance. In the place of dense clumps of aggregated granules we find scattered, isolated ink particles which are grouped round the basophile granules.

[1] Control experiments were carried out to determine the effect of normal salt solution alone when injected in varying amounts. They do not affect what is stated above and therefore will be passed by for the present.

The Blood Corpuscles of *Daphnia*

The blood cells of *Daphnia* may be easily observed while still in the unchanged blood stream owing to the generally transparent nature of the animal. They are best seen in the blood space in the shell, where they may be noticed as granular amœboid bodies hurrying past in the blood stream. It is at once obvious, and prolonged examination confirms the impression, that the blood cells of *Daphnia* are all of one kind. Corresponding to the generally archaic character of the animal we find that its blood cells show no trace of differentiation into markedly different types such as are found in the greatly more advanced form *Astacus*. On the contrary they all appear as more or less amœboid cells, about $7-8\,\mu$ in diameter, and more or less charged with distinct spherules, about $0\cdot5\,\mu$ in diameter.

In the relatively sluggish stream which is found in the blood spaces in the shell the corpuscles may be easily and carefully watched.

The first property of the cells which strikes one is their *adhesiveness*. A cell as it streams along may be seen to touch the walls of the space, when it usually sticks for a moment. The adhesion is usually most persistent at one particular spot on the surface of the cell. The rest is readily detached and streams away in the current, remaining anchored as it were by a hyaline process which may attain an extraordinary length. After a while either this is ruptured or the adhesion parts, and the cell passes on in the stream.

The adhesiveness of the cells is a curiously variable property. At one time the slightest contact suffices to fix them, at another it is exceptional to see a stationary corpuscle. Their adhesiveness may become so marked that of the very numerous corpuscles present only a few may remain in circulation, and this condition is not necessarily connected with the existence of any specific disease or obviously pathological condition of the blood. On the other hand, and in this respect the blood cells of *Daphnia* resemble the colourless corpuscle of Vertebrates, any local irritation, such as a slight injury to the shell or the presence of an irritant, is sufficient to immediately fix any corpuscles in the neighbourhood. Similarly, weak induction shocks, while not affecting the heart beat, may so far modify the condition of the corpuscles in the direction of increased adhesiveness that none remain free in the circulation.

The blood cells not only show variations in their adhesiveness but also in the extent to which they exhibit amœboid movements. When the cells are free in the circulation and when therefore their adhesiveness is slight they are usually rounded, and the increase in adhesiveness is, up to a certain point, associated with an increase in their irregularity of form and activity of movement.

The most striking fact concerning the blood corpuscles of *Daphnia* is that they are extremely generalised and primitive in their characters. There is no sharp distinction into two kinds of blood cells such as exists in *Astacus*, and further, the blood cells of *Daphnia* not only represent the blood cells of *Astacus* but they also perform functions which in the higher animal are relegated to cells which are no longer wandering but are fixed in a particular tissue, namely, the basophile tissue. It appears to me to be a point of remarkable morphological interest that the archaic features of the anatomy of *Daphnia* should extend to its blood cells. It is also an important gain to be able to regard the blood cells of the more complex type as a specialised tissue, or tissue sharply defined morphologically, and probably also possessed of equally sharply defined physiological characteristics. In other words the wandering cells of the body are not merely an amorphous residue of the primitive mesoblast, that is to say not merely mesoblast cells which have had no share in the morphological changes of the body, but a portion of that mesoblast which though remaining free in the body spaces has taken part in the general increase in complexity of form and specialisation of function so far as to become differentiated into parts which play very distinct rôles in the economy of the animal. This point can only be briefly mentioned here, it will receive I hope more complete discussion in connection with what I may have to say at some future time in connection with the functions of the eosinophile and basophile cells of *Astacus*. It will be sufficient here to notice certain of the primitive features of the blood cells of *Daphnia*.

Most important is the absence of any division of the cells into different types. Each cell of *Daphnia* includes within itself the characteristics of the explosive and eosinophile blood corpuscles and of the basophile cells of the cell tissue of *Astacus*. They also all of them may, under appropriate conditions, ingest solid masses.

The explosive nature of the cells. The blood corpuscles of *Daphnia* may, as was pointed out by Metschnikoff[6], disintegrate with explosive rapidity under certain conditions, namely, under the influence of certain poisons formed by the *Torula* found by him in the blood of *Daphnia*. "Auch konnte ich mehrmals sehen, dass die Blutkörperchen in der Nachbarschaft mit zahlreichen Pilzzellen sich allmählich auflösten und vollständig verschwanden, was darauf hinweist, dass die Conidien irgend eine für Blutkörperchen schädliche Flüssigkeit absondern." That is to say, just as in the shed drop of *Astacus* blood suspended on the under surface of a cover-slip the explosive cells remote from contact with the glass appear to be exploded by the presence in the plasma of the products of the explosion of those cells in contact with the glass, so in *Daphnia* the blood cells may be exploded by the presence in the plasma of the powerful irritant poison produced by the *Torula*.

Similarly, one sees a very rapid disintegration of the blood corpuscles of

Daphnia if the animal be teased on the slide, the disintegration being I think enormously hastened by the liberation of the contents of the gut.

The ingestion of solid particles by the blood cells of *Daphnia* is a remarkable phenomenon and clearly illustrates the statement that the cells are all alike. Metschnikoff has shown that the corpuscles ingest and digest the spores of the *Torula*, and further, they may form plasmodia, recalling the special phagocytes produced by fusion and modification of the explosive cells of *Astacus*: but a striking light is thrown on the general part played by these corpuscles in the economy of the animal when we find that they may be loaded with ingested particles derived from the alimentary canal. These substances are fat and carmine, and to clearly understand the part played by the corpuscles it will be advisable to summarise briefly the fate of fat in the animal.

The digestion of fat by Daphnia. *Daphnia* may be fed readily with almost any substances, since it appears to possess little power of selecting its food. Particles of carmine, indian ink, indigo-carmine, alizarine, methylene blue, or globules of fat are readily taken into the alimentary canal. It thrives remarkably well on fresh yolk of egg, which constitutes a diet rich in fat and fat-forming constituents. The fresh yolk is mixed with sufficient aquarium water to produce a creamy emulsion, this is then poured with a fine pipette over the bottom of a dish containing water in which are some *Daphnias*. The emulsion spreads, forming a kind of mud and mixes scarcely at all with the supernatant water in which duckweed or some green plant should be kept to supply oxygen. The *Daphnias* for the most part swim about in the clear water but frequently settle for brief periods into the emulsion. The conditions closely resemble the normal condition of *Daphnia* which feed on the vegetable mud at the bottom of the aquarium.

In the course of a few hours the gut walls are found to be studded with fat globules which are embedded in the columnar cells; in about ten to twelve hours *every blood corpuscle in the body is found to contain a fat globule or globules*.

Daphnia possesses a special fat-holding tissue which is composed essentially of rounded cells anchored by fine processes. Under the circumstances just mentioned these become charged with fat particles. I further convinced myself that this tissue may, when a great quantity of fat is absorbed, be recruited from the blood cells which fix themselves as now stationary fat-holding cells. In fig. 24 a recently fixed corpuscle is shown.

The presence of fat globules in the blood corpuscles of *Daphnia* cannot be regarded as a pathological phenomenon. In well-fed individuals actively engaged in egg building it is ordinarily seen. The huge summer eggs are highly charged with fat globules, and their production entails a large fat-forming activity on the animal.

Inert substances, such as *carmine*, if mixed with a food substance, such as fresh yolk of egg or milk, also pass through the gut wall and appear to a relatively limited extent in the corpuscles, eventually passing into the cells of the end-sac of the shell gland.

The *granules* of the blood corpuscles of *Daphnia* are largely of a basophile nature and give the rose-colour reaction which is so typical of the basophile granules of *Astacus*. The blood corpuscles of *Daphnia* thus in every respect include the characters and functions of the basophile cells and tissue of the higher form, and in the occasional wandering of the fixed basophile cells of *Astacus* into the general circulation we see a partial return to a more archaic condition.

The histological characters of the basophile cells of *Astacus*, and the close connection of the basophile blood corpuscles of *Daphnia* with events taking place in the alimentary canal, lead to the conclusion that the basophile cells of the Crustacea are intimately related to the normal metabolism of these animals.

The blood corpuscles of *Daphnia* reproduce by direct fission. This fact has been previously noted by other workers. The blood corpuscles of *Astacus* also reproduce in a similar way. Fig. 6 shows an explosive corpuscle undergoing division.

LIST OF PAPERS REFERRED TO IN THE PRECEDING PAPER

(1) HAECKEL. "Ueber die Gewebe des Flusskrebses." *Müllers Archiv* (1857), XVIII and XIX, 469.

(2) LEBERT and ROBIN. "Kurze Notiz über allgemeine vergleichende Anatomie niederer Thiere." *Müllers Archiv* (1846), p. 120.

(3) P. GEDDES. "On the Coalescence of Amœboid Cells into Plasmodia." *Proc. Roy. Soc.* (1880), p. 252.

(4) P. GEDDES. "Obs. sur Le Fluide Pèrivisceral des Oursines." *Arch. d. Zool. Exp.* (1880), VIII, 183.

(5) HALLIBURTON. "On the blood of Decapod Crustacea." *Journ. Physiol.* (1885), VI, 300.

(6) METSCHNIKOFF. "Ueber eine Sprosspilz Krankheit der Daphnien." T. IX and X. *Virchows Archiv* (1884), XCVI, 177.

(7) M. GREENWOOD. "On Digestion in Hydra." *Journ. Physiol.* IX, 317.

(8) HARDY. "On the Histology and Development of *Myriothela*." *Quart. Journ. Micr. Sci.* (1891), p. 505.

(9) HEITZMANN. "Unters. ueber das Protoplasma." *Sitz. d. k. Akad. Wiss. in Wien* (1873), Bd. LXVII.

(10) FROMMANN. "Unters, über Struktur, Lebenserscheinungen, und Reaktionen thierischer u. pflanzlichen Zellen." *Jen. Zeits.* (1884), Bd. XVII.

(11) LÖWIT. "Ueberd. Beziehung d. weissen Blutkörperchen zur Blutgerinnung." *Zieglers Beiträge* (1889), Bd. V.

(12) GRIESBACH. "'Beitrage zur Kenntniss des Blutes." *Arch. f. Mikr. Anat.* Bd. XXXVII.

Other papers dealing with the histology of the blood of the Crustacea are:

HEWSON. *Sydenham Societies' Publications* (1846), p. 233.

T. WHARTON JONES. In *Phil. Trans.* (1846), p. 90.

CARUS. *Von den aüssern Lebensdingungen der weiss- und kalt-blutigen Thiere*, Leipzig, 1824, p. 85.

WAGNER. *Nachträge zur vergl. Physiol. des Bluts*, 1838, p. 40.

LEYDIG. *Naturgeschichte der Daphnien*, 1860.

CLAUS. "Zur Kenntniss d. Organisation w. d. feineren Bau d. Daphnien." *Zeits. f. Wiss. Zool.* (1876), p. 388.

WEISSMANN. "Naturgeschichte d. Daphnien." *Zeits. f. Wiss. Zool.* (1880), XXXIII, 188.

Also Gruithuisen, Lievin and others on *Daphnia* are referred to by Claus.

EXPLANATION OF PLATE III

Fig. 1. Living eosinophile cell fully loaded with granules. Oc. 4, Ob. D.[1] Cam. luc.

Fig. 2. Nucleus surrounded by the remains of the cell-substance of an explosive cell. Fresh preparation. Oc. 4, Ob. E. Cam. luc.

Fig. 3. Group of eosinophile and explosive cells after fixation with osmic vapour. The eosinophile cells show the characteristic vacuolate appearance (as seen in optical section), which results from the disappearance of the granules. Oc. 4, Ob. D. Cam. luc.

Fig. 4. Eosinophile cell from which, as a result of its removal from the body, the greater number of the granules have been dissolved, leaving vacuoles. Fresh preparation.

Fig. 5. Eosinophile cell showing vacuolate endosac. The blood of the animal from which this cell was taken contained numerous eosinophile cells, which, like the example figured, had discharged their granules while still within the circulation and as the result of some special stimulus. Fresh preparation. Oc. 4, Ob. D. Cam. luc.

Fig. 6. Explosive cell undergoing division.

Figs. 7, 8 and 9. Stages in the explosion of an explosive cell studied by the aid of iodine. Oc. 4, Ob. D. Cam. luc.

Fig. 10. Eosinophile cell showing a stage in the coloration of the granules with indigo-carmine. Cam. luc.

Fig. 11. Explosive cell. Osmic vapour. Oc. 4, Ob. E. Cam. luc.

Fig. 12. Basophile cell from blood. Osmic vapour and hæmatoxylin. Oc. 4, Ob. D. Cam. luc.

Fig. 13. Basophile cell from blood to which a slight quantity of methylene blue has been added. Illuminated with ordinary gaslight. Cam. luc.

Fig. 14. Basophile cell lying in a chamber of the basophile tissue. Fresh tissue treated with dilute methylene blue in normal salt solution. Ordinary gaslight. Oc. 4, Ob. D. Cam. luc.

Fig. 15. Explosive cells treated with dilute iodine solution. The one to the left is the cell as first seen. The one to the right is the same cell as it appears about 15 min. after fixation. Oc. 4, Ob. D. Cam. luc.

---

[1] Oculars and Objectives referred to are those made by Zeiss.

Fig. 16. "Phagocyte" from blood. Indian ink particles in heaps. Fresh preparation. Oc. 4, Ob. E. Cam. luc.

Fig. 17. Vacuolate explosive cell from blood one hour after injection of indian ink. Osmic vapour. Oc. 4, Ob. D. Cam. luc.

Fig. 18. Explosive cell from blood one hour after injection of indian ink. It contains little clusters of particles. Osmic vapour.

Fig. 19. Blood corpuscle of *Daphnia* as it appears in the living animal. Oc. 4, Ob. $\frac{1}{18}$th, Zeiss. Cam. luc. Owing to the very flat field of the objective used, only a few of the granules present are seen.

Fig. 20. Blood corpuscle of *Daphnia* containing a small heap of carmine grains. From living animal. Oc. 4, Ob. D. Cam. luc.

Fig. 21. Blood corpuscle of *Daphnia* with granules stained by methylene blue. Ordinary gaslight. Oc. 4, Ob. E.

Fig. 22. Blood corpuscle of *Daphnia* with small fat drops. From living animal.

Fig. 23. Blood corpuscle of *Daphnia* with large fat drops. Oc. 4, Ob. D.

Fig. 24. Blood corpuscle of *Daphnia* containing fat globules immediately after its attachment to abdominal fat tissue. Oc. 4, Ob. D. Cam. luc.

Fig. 25. Basophile cell from blood of *Astacus*. Unstained.

Plate III

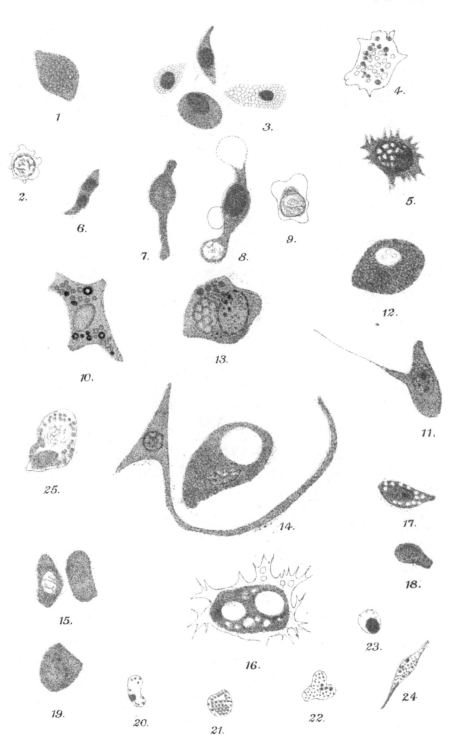

4

THE PROTECTIVE FUNCTIONS OF THE SKIN OF CERTAIN ANIMALS

[Journ. Physiol. (1892), XIII, 309]

At an early period in the course of a somewhat prolonged study of certain members of the Crustacea, the remarkable absence of parasitic vegetable or animal growths on the carapace of these animals appealed to me as a notable fact. I was the more struck with this as there appeared to be no mechanism obviously able to produce this cleanliness. Every animal which inhabits water or has an habitually moist external surface is confronted with the same problem which has so long puzzled the shipbuilding world. Larvae and spores are continually settling on to the surface where many of them would develop into growths which would impede the movements, or injure the tissues of the animal unless some method of removing or destroying them were adopted. The shipmaster meets the difficulty in one of two ways. He either coats the ship's bottom with a highly poisonous paint, or with some substance which is slowly soluble in water, such as copper or zinc. In the second case a fresh surface is continually being formed and the germs have no sooner settled than they lose their hold and are washed away.

In many animals the way in which the surface is kept clean is at any rate partly obvious. Commensal forms feeding on matter adherent to the skin, as for instance the ciliated protozoon *Trichodina* which is found on the surface of *Hydra*, contribute to this result. Similarly the periodical shedding of an external cuticle is a radical means of ridding the animal of growths which may have obtained a hold in spite of other preventive devices, and the continuous desquamation which is characteristic of the stratified epithelium of the higher vertebrates is a more subtle device of the same nature. The appendages are more or less useful in different animals for removing at any rate the larger forms of adherent bodies.

It has been shown that in the Mammalia, the moist mucous surfaces are kept clean in a way which is strictly comparable to the case of *Hydra*. Multitudes of phagocytic wander cells are poured on to the surface to ingest foreign particles whether organic or inorganic (7), (8). Lastly, when all these devices have failed, absolute attrition of the body surface against surrounding objects such as rocks or sand may be resorted to. The writer had many opportunities of witnessing this in the summer of 1883 when the salmon fishery in the River Dee was almost destroyed by the infection of the fish with the fungus *Saprolegnia ferox*. The rocky pools frequently presented the spectacle of a fish persistently rubbing the infected patch against some angle of rock.

But the conspicuous cleanliness of the carapace of Crustacea of very diverse habits is not accounted for by any of the above-mentioned devices. In most forms, *e.g. Daphnia* and the crabs, the appendages can only effectively sweep a limited portion of the surface, while the casting of the shell occurs at far too great intervals when we remember that the bottom of a ship, even when moored in a strong tideway, may in hot summer weather become coated in a period to be reckoned by days. Some special mechanism must be present else they would have to learn to revel in a host of ectoparasites as does, for instance, *Cyclops*. *Daphnia*, as a conspicuously clean Crustacean, seemed to me to be the most suitable for study and to that I mainly directed my attention.

The Structure of the Skin of Daphnia

The skin of *Daphnia* is composed of an external thin hyaline cuticle or, as I shall henceforth term it, carapace which is seen by the aid of the microscope to be divided up into a series of areas by fine lines forming a pattern on its surface. Immediately below this is the cellular ectoderm. There does not appear to be anything of the nature of a mesodermal cutis such as is found in the higher Crustacea where a fibrous and cellular connective tissue forms a distinct zone of fine textured tissue under the ectoderm.

The ectoderm of *Daphnia* has already been studied by Leydig[1] and Claus[2] apparently without the aid of sections and merely in the living, unstained animal. It is therefore not to be wondered at that there should have been a certain amount of uncertainty even with regard to such an elementary fact as the relation of the disposition of the ectoderm cells to the pattern on the carapace. Claus however, in opposition to the views of Leydig, arrived at the just conclusion that each area of the pattern corresponds to a single subjacent ectoderm cell. I have followed the methods of these earlier observers in so far as I have largely examined fresh, living animals, but I have called in to my aid the remarkable power possessed by certain of the anilin dyes of colouring fresh and even living tissues without modifying their morphological characteristics.

The ectoderm of *Daphnia* may be most successfully studied after treatment with methylene blue. The animal is placed on a microscope slide in a drop of normal salt solution (0·6 per cent.) lightly coloured by the addition of a small quantity of the pigment. The preparation is then covered by a coverglass which is supported by a few slips of moistened paper so that it only just touches the animal. If the coverglass is now *very* lightly pressed with a mounted needle the brittle cuticle of the upper half of the shell is ruptured, and a crack a tenth of a millimetre or so long is made through which the pigment slowly makes its way, and a most beautiful differential staining occurs, the various histological elements coming into view one by one as the stain penetrates further and further from its point of entrance.

The first structures to become coloured are the basophil granules of the blood corpuscles. These I have described in an earlier paper (3). Then the cells of the ectoderm take up the colour one by one and stand out with singular distinctness. The cells of the ectoderm are, for the most part, thin and plate-like in character, and each corresponds in size and shape to a superjacent area of the pattern on the carapace. Like those areas therefore they differ in size in different regions, being smallest in the head region where they form irregular polygons, and most regular and largest on the external surface of the shells. Here the cells are elongated with parallel edges and disposed in fairly parallel rows. In addition to these flat, plate-like cells there are also thicker cells which bear processes extending into the setæ-like tactile hairs which form a fringe to the free edges of the shells, and occur also on the appendages.

The flat ectoderm cells are either rhombohedral or polygonal in shape. Though they differ in size in different regions, yet their histological characters are everywhere the same. Each cell forms a thin plate closely adherent to the carapace. A large rounded nucleus with one marked nucleolus occupies a central position. When treated with methylene blue in the manner indicated above this nucleus appears as a clear unstained space in the blue cell-substance, and situated somewhat eccentrically in this is the spherical nucleolus which takes an intense blue colour. The cell-substance shows some remarkable histological characters. It is coloured a transparent but distinct blue by the dye and immersed in this blue matrix are a number of round vacuoles, which are apparently occupied by fluid matter. The contents of these vacuoles give the peculiar rose reaction with methylene blue to which I have drawn attention elsewhere (4). That is to say when the preparation is illuminated with ordinary (yellow) gaslight the vacuoles appear as rose-coloured spaces in the blue cell protoplasm. The nucleolus also appears blue while the rest of the nucleus remains uncoloured. The only other structures in the body of *Daphnia* which give the rose reaction are the basophil granules of the blood corpuscles. The alimentary canal, muscles, nervous system and excretory organs stain blue. The cells lining the end-sacs of the last-mentioned structures are characterised by containing basophil granules of an irregular shape which take an intense and opaque blue coloration. Each ectoderm cell generally appears to be separated from the adjacent edges of the neighbouring cells by a slight interval due probably to a faint contraction of the cell-substance consequent on the stimulus of the methylene blue.

The vacuoles in the cell-substance may be the receptacle of solid particles which have by some process been ingested by these cells. I have succeeded in infecting a patch of the ectoderm in this way with anthrax bacilli though, owing to the difficulty experienced in introducing these bodies into *Daphnia*, only a few vacuoles were occupied by one or rarely more rods. Whether few or many foreign bodies enter the cells does not affect the value of the experiment which conclusively points to a certain phagocytic power exercised by the ectoderm cells. In order to introduce anthrax into the body of *Daphnia* I proceeded as follows. A small quantity of an active growth was taken up on the point of a very finely pointed needle and, with the aid of a Zeiss dissecting microscope, a very small puncture was made in the carapace just posterior to the upper portion of the shell gland. In successful cases a little of the growth gets carried into the body. No vital organ need be injured by the operation, and in control experiments, animals similarly treated were kept alive, isolated in a small quantity of water with a little duckweed floating on it, for many days. I may mention here, though the point hardly concerns us at present, that in summer time, when a relatively high temperature prevails, anthrax is fatal both to *Daphnia* and to *Astacus*.

Not infrequently one finds *Daphnias* whose ectoderm presents dark, irregular patches where the cells are infested by a parasite, probably of a vegetable nature.

In transverse sections through the animal the plate-like ectoderm cells are seen to form a layer closely adherent to the carapace. They are now found to have a very appreciable thickness, namely about $4–6\,\mu$, and form an apparently continuous sheath of a very finely granular nature. The division between cell and cell is quite invisible in all my sections. From the study of such preparations we therefore obtain additional evidence that the slight interval between the cells which is seen so clearly in fresh preparations stained with methylene blue is due to a faint contraction of the cell-substance.

The peculiar rose reaction is not confined to the contents of the vacuoles found in the ectoderm cells and to the basophil granules of the blood corpuscles of *Daphnia*. It is also seen to be a prominent characteristic of some substance of the nature of a slime which this animal has the power of casting on to its surface. The existence of this slime may be demonstrated by placing a *Daphnia* under a coverslip supported in the usual way in water or normal salt solution to which a trace of methylene blue has been added. The substance has a strong affinity for the dye giving a brilliant rose reaction and we may, under the microscope, watch the formation of a film some-

times of considerable thickness over the entire surface. It appears to be
formed in larger quantities than normal as a result of some stimulus
furnished either by the pressure of the coverslip, the pigment itself, or the
normal salt solution when that fluid is used. In contact with water the
substance first swells up and then slowly dissolves. It may I think be
regarded as a true secretion, or product of the activity of the ectoderm cells,
and this view of its nature is supported by the fact that we may see as a
result of its formation a distinct diminution in the intensity of the rose
reaction of the ectoderm cells. A parallel case in which the source of the
substance thrown on to the surface of the animal could be more directly
traced was found in the case of the Ammocœtes stage of *Petromyzon Planeri*.
Miss Alcock, working under the direction of Dr Gaskell, discovered that the
slime which, under appropriate stimuli, is formed in very large quantities
on the surface of the body of these animals contains an active proteolytic
ferment capable of rapidly digesting fibrin in the presence of a small
quantity of free hydrochloric acid (5). I examined the skin and found that it
is covered by cells the free surface of each of which bears a cuticular plate
perforated by a number of relatively large canaliculi. These canaliculi
therefore pass from the exterior through the cuticle which covers the
general surface of the animal to the cell-substance of the subjacent cells.
The cell-substance itself is charged with basophil granules of which those
occupying the upper half of the cell give a marked rose reaction with
methylene blue. The mode of discharge of these secretory granules could
also be observed, many of the canaliculi being filled with staining matter so
that they presented the appearance of rose-coloured rods.[1]

In other words, we have in the case of the Ammocœtes as a peculiar
function of its ectoderm the discharge on to the general surface of the body
of a complex substance which is a secretion of the cells of the ectoderm. I
see little reason to doubt that the slime found on the surface of the carapace
of *Daphnia* is generally similar to that produced by the Ammocœtes, and
we may probably regard the vacuoles of the ectoderm of the former as the
analogues of the basophil granules of the latter. In both cases the process
is essentially the discharge of a basophil substance, but to the question
whether this substance is similar to the material of the rose-staining granules
of the blood corpuscles I can return no satisfactory answer. The substance
is discharged in the case of the Ammocœtes through cuticular pores, and
similar pores probably perforate the carapace of *Daphnia*. This is borne
out by the fact that in sections through the decalcified carapace of *Astacus*
such pores are readily seen perforating the lamellae.

The real existence of a slime on the surface of *Daphnia* is impressed on

[1] The cell-substance of these cells only imbibes the dye after a prolonged exposure.
The granules stain rapidly.

one's mind by placing the animal in water containing suspended carmine particles, these not only adhere in spite of currents but also are seen not to be attached to the carapace, but to be removed a little way from it. In this way we can outline a zone of unstaining and perfectly transparent jelly round the animal as definitely as by staining that jelly with methylene blue.

I completely failed to demonstrate the existence of a slime on the shells of *Cyclops* which are notoriously infested by ectoparasitic growths. If the basophil slime is a true secretion, formed by the ectoderm cells, and designed in some way to aid the animal in its conflict with the germs of parasitic animal or vegetable forms, then we might expect that the power of producing the substance would be increased up to a certain stage if the conflict became more acute, and, further, that the victory of the parasites would be associated with the destruction of this rose-staining substance and of the power of producing it. I was enabled to satisfy myself as to the correctness of these predictions.

Daphnias were infected with anthrax either by direct inoculation as described above or by feeding them on yolk of egg, with which some of the growth had been mixed.[1] I also kept individuals in water to which had been added some of the isolated poison produced by the anthrax, and kindly given to me by my friend Mr Hankin. In this last case the poison would attack the animal from the surface and also, from the peculiar feeding habits of *Daphnia*, through the alimentary canal. The effect on the animals was (1) That during about the first ten hours the power of producing the rose-staining slime was enormously increased. This took place to such an extent that, while in the normal animal the excessive stimulation due to placing it on the slide for examination exhausted the entire store, in the pathological individuals on the contrary a second (and sometimes a third and fourth) copious secretion took place when the first was washed away by a stream of water. (2) The blood corpuscles discharged their rose-staining granules at the same time becoming more adhesive, and then, but at a later period, the corpuscles disintegrated. The discharge of the rose-staining substance into the plasma sometimes took place to such an extent that that fluid itself took a rose tint, a phenomenon never observed in a normal animal. (3) Lastly, and this proves that the rose-staining substance is not a product of the microbes themselves, just before the animal died the whole rose reaction both inside and outside the body was destroyed.

We are now in a position to point out the general similarity in the behaviour of both ectoderm cells and blood cells towards microbes. Metschnikoff has shown that all the blood cells of *Daphnia* are phagocytes (7), and

---

[1] I am unfortunately unable to give the details of experiments as fully as I should wish, owing to the fact that the complete record of them was lost by shipwreck.

the same observer has also furnished us with an instance of phagocytic action on the part of the ectoderm cells of certain polyps (9). Similarly the inclusion of anthrax rods in the ectoderm cells of *Daphnia* show that these too have a limited phagocytic activity. Also both blood cells and ectoderm cells, as a feature of the conflict of the animal with the microbic poisons, discharge a rose-staining substance previously stored in their cell-substance.

The power of forming a slime on the surface of the body is widely diffused in the animal kingdom, and if we review what is known as to the manner of its formation the similarity pointed out above between ectoderm cells and blood cells obtain additional interest. In many animals, such as Platy-helminthes and Vertebrates, it appears like that of *Daphnia* to be entirely of ectodermal origin, being formed by uni- or multi-cellular ectodermal glands, in others however mesodermal elements in the form of wander cells laden with granules contribute to it.

Durham has shown that in Echinoderms the slime is formed by the disintegration of wander cells which have migrated through the thin walls of the dermal branchiae on to the surface (6), and I would emphasize the fact that these cells are loaded with granules. Durham styles them "spheru-liferous", and quotes the description of Hamann "Eiförmige, stark lichtbrechenden Körnchen erfüllten Zellen". It is most unfortunate that no attention was paid to the histochemical character of these granules. Wander cells may also furnish some as yet unknown constituent of the slime formed on various mucous membranes of the higher Mammalia. Metschnikoff (7) has shown that the cells which wander into the mucus covering the tonsils are phagocytes, but this does not exclude them from possibly furnishing some substance which is dissolved in the slime and is inimical to the microbes. I have examined the cells found in the nasal discharge during catarrh with the object of determining their histological reactions. The discharge was exceedingly copious, very viscid, and colourless. In spite of its transparency it contained countless multitudes of cells which were almost entirely of one type, namely, small cells with readily staining nuclei and recalling the microphages of the tonsils. The cell-substance contained ill-defined granules *which gave a marked rose reaction with methylene blue.* Preparations stained with the object of demonstrating the presence of microbes showed various cocci and short bacilli in insignificant numbers. With such a copious discharge so continually being removed, this is of course what might have been expected.

What has been said in the foregoing pages allows us to offer two sugges-tions as to the way in which the surface slime of animals protects them:

(1) It may have a mechanical action. I think we may take it for granted *that the presence of a film of soluble slime on the surface of an animal immersed*

in water would, like the copper sheathing of ships, mechanically prevent the occurrence of parasitic growths by continually forming a fresh surface.

(2) *The slime may have a specific poisonous power directed mainly perhaps against the more minute and subtle forms of vegetable parasites.* We have seen that exposure to the poisons produced by anthrax causes an increased elimination of a rose-staining substance both on to the external surface and into the blood plasma, and Hankin has shown that the blood serum of animals contains a substance which has a bactericidal action and to which the name "alexine" has been applied. This alexine he believes has its origin in the blood corpuscles.

By this brilliant suggestion he has given fresh force to the conception of Ranvier, who long ago styled the granule-bearing wander cells unicellular glands. May we not also suggest that the rose-reacting substance of *Daphnia*, whether derived from the ectoderm, or from the blood cells, has a specific bactericidal action and contains an alexine? It is always produced in increased quantities when the animal is infected with pathogenic growths, and it cannot be regarded as merely a product of the microbes in or on the body of the animal since the first change which occurs when the microbes finally have the upper hand in the conflict is the entire destruction of the rose-staining matter. Lastly, may not the increased production of the rose-reacting substance upon an appropriate stimulus be compared with the increased production of ferments by the glands of the alimentary canal which occurs when these glands are stimulated by the presence in the blood of the absorbed products of digestion?

Such a conception as this finds its support in the wonderful phagocyte theory of Metschnikoff. Miss Greenwood has shown that an Amoeba digests its food in vacuoles by a ferment-like action, similarly the killing and digestion of a microbe in the vacuole of a phagocyte depends on the capacity of the cell-substance of the latter for producing substances deleterious to the microbe. Having this thought in my mind I sought long and anxiously for *Daphnias* whose blood was infected with parasites.[1] Only once however did I succeed in seeing a foreign body ingested by a blood cell.

I fortunately saw the cell come to rest on the wall of the blood space in the shell. It was kept under observation and figures drawn at intervals. They show a gradual diminution in the number of rose-staining granules until all trace of them had disappeared. The cell remained actively amœboid for as long as I was able to watch it, the protoplasm streaming round the enclosed body. This point however can only be settled by the study afresh of individuals infected with the classical "Sprosspilzkrankheit". These I

---

[1] An anthrax bacillus would not be readily detected in a living cell moving in the blood stream.

have not been able to obtain. Still there are two observations which lend additional support to the suggestion that the rose-staining granule is related to or forms the antecedent of the ferment by the aid of which the phagocyte digests its prey. They are (1) the fact that the ferment-containing slime of the Ammocœtes is derived from zymogen granules which give the characteristic rose reaction, and (2) that as I have mentioned elsewhere the secretory granules nearest the lumen of the alveoli of salivary glands similarly colour rose with methylene blue.

In *Daphnia* as was shown by Metschnikoff[9] the blood cells are all phagocytes. In the higher type *Astacus* there are three kinds of wander cells, *two only of which appear to be phagocytes*, the third cell contains abundant granules which, like those of the cells which wander out to form the surface slime of Echinoderms, contain "stark lichtbrechenden körnchen". In fact, in studying the blood corpuscles of the Crustacea we find every stage in their specialisation. The primitive wander cell is found in the larvæ of *Daphnia* while still within the brood pouch, where it appears as a small amœboid mass, free from granules and about half the size of the corpuscle of the adult. But though there are no granules the *entire cell* gives a faint but distinct rose reaction. Next, in the adult, the cell may be said to have attained to the status of a gland which stores the products of its activity as spherules in its cell-substance. The granules now stain rose, the cell-substance remains unstained or becomes blue only after prolonged exposure to the dye. Lastly, in *Astacus* we have different types of wander cells each of the gland type but markedly distinct in form and in their histochemical reactions.

LIST OF PAPERS REFERRED TO IN THE PRECEDING PAGES

(1) Fr. Leydig. *Naturgeschichte der Daphniden*, Tübingen, 1860.

(2) C. Claus. "Zur Kenntniss der Organisation u. des feinern Baues der Daphniden u. verwandter Cladoceren." *Zeits. f. wiss. Zool.* (1876), xxvii, 362.

(3) Hardy. "The Blood Corpuscles of the Crustacea." *Journ. Physiol.* (1892), xiii, 165.

4) Hardy. "On the Reaction of Certain Cells with Methylene Blue." *Proc. Camb. Phil. Soc.* (1892).

(5) R. Alcock. "The Digestive Processes of Ammocœtes." *Proc. Camb. Phil. Soc.* (1892).

(6) Durham. "On Wandering Cells in Echinoderms." *Quart. Journ. Micr. Sci.* (1891), vol. xxxiii.

(7) Metschnikoff. *Ann. de l'Institut Pasteur* (1887), No. 7.

(8) Ruffer. "On the Phagocytes of the Alimentary Canal." *Quart. Journ. Micr. Sci.* (1890), xxx, 481.

(9) Metschnikoff. *Arb. aus d. Zool. Inst. in Wien u. Triest.* (1884), Bd. v.

5

ON THE STRUCTURE AND FUNCTIONS OF THE ALIMENTARY CANAL OF *DAPHNIA*

WITH W. McDOUGALL

[*Proc. Camb. Phil. Soc.* (1893), VIII, 41]

So far as we know none of the workers who have investigated the gut of the Crustacea have described the existence in the case of the lowest members of that group of a differentiation of regions corresponding to the processes of digestion, of absorption, and of elaboration of the fæces. As is well known, the gut of the lower Crustacea differs in a very striking way from that organ in the higher forms. Throughout the entire group the existence of the three main divisions, stomodæum, mesenteron and proctodæum, is very clearly seen, but whereas in the former the mesenteron constitutes the greater part of the gut, in the latter it is limited to a very short region into which the bile-ducts open, and which forms only a very small fraction of the total length of the alimentary canal.

The long mesenteron of the lower forms exists as a simple tube offering no obvious distinction into regions or diversity of structure save the two so-called liver diverticula which spring from its anterior end and extend forwards over the brain.

The œsophagus or stomodæum is a short tube running upwards and forwards from the mouth to open on the ventral surface of the mesenteric tube near its anterior end and a little posterior to the points of origin of the liver diverticula which spring from the lateral aspect.

The proctodæum is rather longer than the stomodæum but differs in this and other respects in the different divisions of the Entomostraca. In the Phyllopoda it occurs as a short simple tube.

Though the mesenteron appears as a simple tube without any obvious anatomical expression of differentiation of function, yet a study of the processes of deglutition and digestion, and of the character and arrangement of the cells lining its wall makes evident the fact that in *Daphnia* this apparently simple tube is divided into three regions, an anterior region devoted to the absorption of the products of digestion, a middle region wherein digestion occurs, and a posterior region in which the fæces are formed.

The fact that digestion occurs in a region of the gut *posterior* to that which is occupied in the absorption of the products is so far as we know without a parallel in the animal kingdom, except perhaps among the simplest Cœlenterates.

For purposes of observation *Daphnia* can be readily fed by pouring beaten yolk of an egg, milk or carmine etc. over the bottom of the dish in which it is living, and the phenomena accompanying the taking in and digestion of food may be easily followed in the living *Daphnia* owing to the transparent nature of the animal. The various events will be described in the order in which they occur.

Deglutition is a rapid act. The food particles, *e.g.* carmine, or yolk globules, are carried over the mouth in the current of water which is constantly maintained by the movements of the foliaceous appendages and many of them adhere to the sticky surfaces of the mouth appendages. These adherent particles are formed into a bolus by the movements of the appendages. We have not succeeded however in determining how this is done.

When the bolus is complete it is rapidly carried into the mesenteron by a peristaltic movement of the œsophagus, which in its resting condition is a closed tube flattened dorso-ventrally. The peristaltic movement spoken of above consists of a wave of dilatation which starts at the mouth, and which is followed by a wave of constriction. The bolus is sucked into the œsophagus by the first wave and then thrown with a certain degree of force into the mesenteron.

When the peristaltic wave reaches the mesenteron it is continued backwards over that structure and carries the bolus before it as far as the middle third.

The act of deglutition therefore is brought about by a peristaltic wave which starts at the mouth as a result of the stimulation of the sensory surfaces there and runs backwards carrying the food along the gut to the middle third or digestive region of the mesenteron.

Digestion. The food when swallowed consists of particles, *e.g.* precipitated proteid, fat globules, carmine grains, or unicellular plants, which are glued together by some sticky substance. As digestion proceeds the insoluble particles are slowly liberated and some are carried into the anterior region of the midgut and even into the liver diverticula. There the nutritive particles such as the fat globules of milk or yolk (but *not* the carmine grains) are ingested by the columnar cells lining that region. If the food contain any soluble colouring matter, such for instance as chlorophyll, that also accumulates in the anterior region. The non-nutritive particles do not accumulate in the anterior region but are from time to time driven backwards. Thus, in the case of an animal which has been fed on green algæ we find the anterior third of the mesenteron including the liver diverticula occupied with a bright-green fluid in which are suspended a few solid particles; the middle region occupied by a dark-green mass of still undigested food, while the posterior region contains a mass of brown particles which will form the fæces.

The movements of the gut which bring about this distribution of the contents are twofold. First there is a constantly occurring peristalsis which consists of waves occurring at regular intervals. These start at the junction of mesenteron and proctodæum and run *forwards*, ultimately dying away in the liver diverticula. They are best developed, that is they lead to the most considerable constriction, in the posterior region, and they are apparently conditioned by the presence of food particles or of their soluble products in the lumen of the gut. They, however, are sometimes seen in starving animals.

The second movement consists of a quick contraction of the walls of the liver diverticula which occurs at irregular intervals during the progress of digestion and which is apparently due to the stimulation of the walls of the liver diverticula by innutritious food particles. It is difficult to be certain on this point, but so far as our observations go they agree with this view of the causation of the movement.

It is clear that the first movement is the agent which propels the soluble products of digestion and the freed discrete food particles to the anterior region of the mesenteron. The contractions of the liver diverticula on the other hand serve to drive the innutritious residue of these particles back to the middle and posterior regions.

The effect of these two movements may be made clear by taking the case of an animal fed on a mixture of yolk of egg and carmine. The yolk particles, consisting mainly of proteid precipitate and fat globules, and the carmine grains form at first an agglutinated mass in the middle region of the gut. As solution proceeds the fat globules and carmine grains are liberated and are driven in relatively small numbers into the anterior region, where the former are taken up by the lining epithelium so that the walls become in time thickly studded with fat drops. The carmine particles are not taken up but are at intervals driven back by contractions of the liver diverticula to the middle region.

Absorption. By feeding with food rich in fat globules we can determine the region in which the absorption of solid particles occurs, and we find that fat is most readily taken up by the cells of the anterior region, including the liver diverticula. If the amount of fat given be not great, globules appear practically only in this region. If however a large quantity is ingested, in 24–40 hours the cells of the middle region also become considerably loaded but always to a much less extent than those of the anterior region. The globules of fat in the cells differ in size in the two regions, thus in one animal the globules in the cells of the liver diverticula measured $3–5\,\mu$ in diameter, in the anterior region of the mesenteron they measured 4–7 μ; while in the middle region the globules were relatively scanty and too small to measure.

If the fatty diet be pushed to great excess, fine globules may occur in the cells three quarters of the way back along the mesenteron.

It is however quite clear that the absorption of fat is practically confined to the anterior third of the mesenteron.

Our reasons for supposing that absorption of the dissolved products of digestion is especially the function of the anterior region cannot be regarded as conclusive. The very small quantity of fluid that is found at any time in the middle and posterior regions of the midgut of a well-fed animal points to this conclusion. In the chlorophyll of the algae which form so large a portion of the diet of *Daphnias*, we have a substance whose fate we can to a certain extent trace, and we find that as digestion proceeds the food-mass in the middle region loses its green tint, while the fluid contents of the anterior region become coloured a vivid green. Further there is evidence that this dissolved chlorophyll is absorbed, for the striated border of the epithelium becomes coloured an intense green and the cells charge themselves with yellow pigment masses.

Defæcation is a sudden act: the fæces, which are formed in the posterior region of the mesenteron, are quickly expelled from the intestine passing rapidly through the proctodæum.

Defæcation is carried out by a wave of contraction which appears to start at some undeterminable point in the mesenteron, and a wave of dilatation and contraction passing backwards over the proctodæum are of such a nature that they first aspirate the contents of the posterior region of the mesenteron, and then expel them through the anus.

There is undoubtedly a sphincter muscle at the junction of mid- and hind-gut in *Daphnia*.

The proctodæum also exhibits a rhythmical movement which consists of peristaltic waves starting from the anus and travelling forwards to the junction with the midgut. The interval between successive waves varies (*e.g.* 3–9 sec. in one animal). They have no connection with the forward running peristaltic waves of the mesenteron, for the latter occur at quite regular intervals without any reference to the time of occurrence of the proctodeal wave. Thus in the case given above, the proctodeal waves followed one another at irregular intervals varying from 3 to 9 sec., while the peristalsis of the mesenteron occurred with perfect regularity every $1\frac{1}{5}$ sec.

The rhythmic movements of the proctodæum appear to be independent of the central nervous system, for they are maintained long after all signs of life have vanished and may be shown by the proctodæum when isolated as completely as possible in normal salt solution. They undoubtedly lead to the entrance and exit of water and, in the absence of any definite knowledge on the subject, we may perhaps regard them as respiratory in character.

It is clear that the stomodeal and proctodeal portions of the intestine of *Daphnia*, so far as the manipulation of the food-stuffs is concerned, take part only in the processes of deglutition and defæcation, and the ingesta and egesta are not lodged in them but merely hurried through. They thus differ profoundly in function from the similar structures in the higher forms of the Crustacea.

Minute structure of the gut of Daphnia.

The walls of the gut consist of a single-layered epithelium resting on a muscular membrane.

The mesenteron including the liver diverticula is lined by columnar cells disposed quite regularly and with a well-developed hyaline border with vertical striation. When the gut is in a state of average normal distension the cells are short columnar, being only a little deeper than broad. When the gut is much constricted the epithelium is thrown into longitudinal rounded ridges separated by furrows. The cells of the ridges are then very tall and thin, those of the furrows short and broad. The gut may be so distended by food or perhaps artificially during manipulation that the cells are broader than deep.

The depth of the hyaline border varies in different regions of the gut but bears a fairly constant proportion to the depth of the cells in each region. There can generally be seen embedded in the substance of the border more darkly staining rods. They are most commonly wholly embedded in the border, reaching just to its free surface; and are generally separated by spaces about equal to or slightly greater than their own thickness. The distance which separates them from one another varies with the degree of distension of the gut and the consequent breadth of the cells. When the cells are very broad the border with the rods sometimes appears discontinuous, there being patches of striated border on each cell, separated by intervals.

Sometimes no rods or structure of any kind can be made out in the border; this may be due to complete retraction of the rods; but possibly to faulty preparation.

The rods may be sometimes seen to be shorter than the depth of the border and so not quite to reach to its surface.

In other cases the rods may be seen to project freely beyond the border to at least half their length. In such cases the projection is not due to retraction of the border but to projection of the rods, for the border retains its normal depth while the rods are longer proportionally to the height of the cell; they may be more than half the height of the cell when thus projected.

When the rods project through increase in length, they do not seem to be narrower, but rather thicker; if this be so, there must be a passage into them of substance from the cell-body.

The rods are not always parallel and not always perpendicular, sometimes they lie at an angle to the surface of the cells.

The rods again may appear to stand freely on the surface of the cells without any embedding hyaline substance.

Sometimes the border contains a number of small clear non-staining patches oval or irregular in shape, generally close to the base of the border; these are commonest opposite the intervals between cells.

The borders of adjacent cells are generally in contact and closely adherent. In cases where the cells are at all displaced the borders may be seen stretched out and still adherent. The border is therefore fairly tough and of course extensible as it changes in shape with the alterations in form of the cell bodies.

The mobility of the rods and of the hyaline substance of the border is thus very obvious in *Daphnia*, and Heidenhain has attributed a similar mobility to the hyaline border of the cells lining the small intestine of Vertebrates,[1] and parallel changes have been noticed at times in the ciliated cells of the gut of *Lumbricus*.[2]

Actual measurements show that the depth of this border forms a greater proportion of the total depth of the cell in the diverticula and anterior portions than in the rest of the gut. Thus in one case the measurements were:

	Depth of cell body	Depth of border
Diverticula	5μ	$2\cdot5 \mu$
Anterior region of intestine	10μ	6–7μ
Middle to posterior region	$7\cdot5 \mu$	3μ

Thus we may say generally that the border is best developed in that region in which the absorption of fat was found to take place, and this corresponds to observations made in other animals. Thus in Vertebrates the hyaline border is best developed on the cells of the villi of the small intestine, and in *Lumbricus*[3] it, or rather its homologue, is only found on those cells which ingest fat.

Just posterior to the junction of the stomodæum with the mesenteron, there is a short neck-like region in which the lumen is smaller and the cell border lower in proportion to the depth of the cells than in regions before or behind.

If we turn to the histological characters of the cell-substance of the cells lining the mesenteron we obtain further evidence of a differentiation of the

[1] Heidenhain, *Pflügers Archiv*, Bd. XLIII, Suppl.
[2] M. Greenwood, *Journ. Physiol.* XIII, 239.
[3] *Ibid.*

epithelium into regions corresponding to those which actual observation of the processes of digestion shows to exist.

The cells of the middle region of the mesenteron are specially character-ised by the presence under certain conditions of granules which are pre-served by osmic vapour and to a less extent by corrosive sublimate. Since these granules accumulate in great numbers in starving animals, and since it can be shown that they make their way into the lumen of the gut there to swell up and dissolve, we are justified in regarding them as secretory granules, and the cells which bear them as gland cells, engaged in the elaboration of a digestive ferment or ferments.

In a series of sections of a gut preserved by most reagents (osmic acid and corrosive sublimate excepted) there are usually seen, apparently between the cells, clear spaces reaching from basement membrane to striated border but usually not into the border; they vary in shape, in a gut taken from an animal during digestion, from a narrow slit, straight or curved, according to the shape of the cell, to broadly oval spaces, which seem to compress the adjacent cells to a dice-box shape.

Occasionally the striated border is discontinuous opposite these spaces; being perforated by a small canal which in oblique sections often resembles a small vacuole.

In cells preserved with osmic acid vapour, no such gaps between the cells occur, or if so, very rarely; but there are seen within the cells granules which are not seen in the specimens where the gaps occur.

These granules appear to stain rather differently with osmic vapour in different cases. Sometimes only lightly, at other times more darkly.

In cells preserved with a saturated solution of corrosive sublimate the granules are generally only partly preserved. In some cases each granule seems to be swollen up and to have become clear and non-staining, but not to have fused with the neighbouring granules, or only to a slight extent; the cell-substance then appears as a stained protoplasmic network of very fine meshes, enclosing the clear swollen granules.

The change in places goes further; the granules and the network are no longer seen but in their place appear clear unstained patches chiefly at the bases of the cells.

Absolute alcohol sometimes preserves the granules in part. They then stain deeply with hæmatoxylin. It may then often be seen that in a cell some of the granules are swollen up and coalesced into a clear non-staining mass, while the rest remain and are very clearly seen.

Wherever granules are well preserved they may be seen to be arranged in vertical rows stretching from the basement membrane to the cell border.

In the middle region of a fasting gut preserved by osmic vapour the cells are often very much distended laterally with granules. The rods of the hyaline border then stand much further apart than usual and often do not reach the surface of the border.

In any one section the granules of different cells are rather differently preserved. Some of the cells appear lighter than the rest, and these are generally broader; possibly in them swelling of the granules has begun, while in the others the granules are better preserved.

In sections through fasting animals which have been preserved in absolute alcohol the epithelium shows large clear (*i.e.* unstained) spaces alternating with narrow threads or columns of staining substance. The spherical nucleus may sometimes be seen attached as it were to one side of the staining column and projecting into the clear space. In many cases the clear space contains scattered granules which have not swollen up.

The discharge of the granules into the lumen of the gut may not unfrequently be seen in osmic vapour preparations especially of fasting animals. They appear in places to be streaming from the cells through the border and to form a homogeneous mass in the lumen which sometimes embeds still unaltered granules.

These granule-bearing or gland cells occur throughout the length of the gut, but they are most numerous in the middle or digestive region. They are fewest in the short neck-like region, already alluded to, which is just posterior to the junction of œsophagus and mesenteron.

It is interesting to note that quite at the posterior end of the mesenteron gland cells occur which contain remarkably compact groups of granules which stain deeply with osmic vapour. These like the other granules are best seen in starving animals.

We may correlate the existence of these cells in this particular position with the formation of the fæces which takes place in this region. The fæces are composed of the innutritious detritus of the food-stuffs glued together by some substance which makes its appearance at a very late period in digestion and in the posterior end of the mesenteron. The effect of the production of this substance is to cause the fæces to change from particles scattered through the lumen to a compact mass which occupies the centre of the gut and is not in contact with the walls.

This concludes the main part of the paper, and in it we have endeavoured to show that the apparently undifferentiated mesenteron of *Daphnia* is really divided into regions defined by the processes which take place in them and by the character of the cells forming their walls.

In order to complete the account of the histology of the gut we will briefly describe the structure of the stomodæum, proctodæum, and muscular basement membrane of the mesenteron.

Structure of stomodæum. The stomodæum forms a muscular œsophagus

5-2

leading upwards and forwards into the gut. It consists of a muscular tube lined by a simple low epithelium covered internally by a cuticle.

The muscle fibres are striated; they are arranged in a single layer of annular fibres and a pair of longitudinal muscles.

The proctodœum is the terminal vertical or recurved portion of the gut. It is narrow, flattened laterally and lined by a low cubical epithelium and cuticle. Its basement membrane is especially well developed.

Structure of basement membrane of the gut. This is the actively contractile organ which brings about peristalsis. Its structure is peculiar and it forms a membrane which is continuous over the whole mesenteron. This membrane is thin, hyaline, structureless, very tough, and very elastic, as is shown in teasing, and highly refractive. In it are embedded protoplasmic strands which are presumably the contractile elements. These are narrow flattened bands arranged in two series, a longitudinal and a circular. The longitudinal bands are continuous for at least half the length of the gut, probably the whole; whether the circular bands are annular or spiral we have not determined.

The circular bands are rather broader and more closely set than the longitudinal, being separated by intervals two to three times as broad as themselves. Each band is enclosed by a splitting of the hyaline membrane which sends a sheath both internal and external to the band. In transverse section the latter therefore appears as a dark oval patch lying embedded in the hyaline substance, and resembling a nucleus of the latter.

The bands show indistinct longitudinal fibrillation in places; and also a faint indication of irregular cross striation; these markings however are probably due to folding.

We have not succeeded in finding any traces of nuclei in the basement membrane. Where the longitudinal and circular bands cross one another, they appear to run straight on without being specially connected.

6

ON THE CHARACTERS AND BEHAVIOUR OF THE WANDERING (MIGRATING) CELLS OF THE FROG, ESPECIALLY IN RELATION TO MICRO-ORGANISMS

WITH A. A. KANTHACK

[*Proc. Roy. Soc.* (1892), LII, 267]

ABSTRACT

The paper deals with the results of an investigation of the structure and functions of the wandering (migrating)[1] cells of the frog. Certain preliminary observations on Mammals and Crustacea are also included.

The results may be summarised as follows:

The histology of the wandering cells of the frog is almost identical with that of the wandering cells of *Astacus*. The different cells are very clearly marked off from one another when seen alive or when in preparations. Excluding red blood corpuscles and platelets, which stand on a different footing from all the rest, the following forms are found:

Normal

I. Cells normally free in the blood and in the lymph.

(a) Eosinophile cells; nucleus horse-shoe shaped or lobed; do not ingest particles; are motile unicellular glands.

(b) Hyaline cells, free from specific granulation; nucleus round with central nucleolus. Phagocytic, *i.e.* they possess the power of ingesting and digesting discrete particles.

II. Cells very few in number and small in normal lymph. Normally present in the lacunar spaces of areolar tissue.

(c) Basophile cells, spherical, with scanty protoplasm when small; angular, rounded or flattened when large; cell substance charged with tiny basophile granules, which give a vivid rose colour with methylene blue. Large oval or round vesicular nucleus, sometimes containing irregular chromatin mass and filaments.

Abnormal

III. Large amœboid cells; vacuolate, frequently with ingesta in the vacuoles, multinuclear, very active and phagocytic.

Giant cells formed by fusion of hyaline cells, similar to the large phagocytic cell of *Astacus*.

IV. Small bodies, either round and quiescent or amœboid.

Nucleated cells budded off from the eosinophile or hyaline cells.

Non-nucleated bodies produced by breaking up of red corpuscles.

The hyaline cell is less resistant than the eosinophile cell. Rough manipulation causes a rapid bursting up of the cell, thus recalling the hyaline explosive corpuscles of *Astacus*.

---

[1] This appellation is used in preference to such terms as "leucocyte" or "white corpuscle", since it is more inclusive.

We have studied the functions of these cells in relation to their anti-bacillary action (1) by taking samples of lymph from a frog at varying intervals after the injection of bacilli, etc.; (2) by inoculating hanging drops suspended in moist chambers and kept at different temperatures, the chambers being sufficiently large to afford plenty of oxygen. By the second method we have been able to observe the conflict between cells and bacilli for continuous periods of eight to nine hours. The *same cells and bacilli* have been watched for the whole period.

In the same manner we have also examined the effect of the injection of finely divided coagulated proteid (boiled white of egg solution), indian ink, vermilion, egg albumen, and anthrax spores. At first we used curarised frogs to obtain lymph, and this led to the discovery that curare produces a profound alteration in the wandering cells.

The phenomena of leucocytosis have also been examined, and we find the following:

1. Corresponding with the three different kinds of wandering cells found in the blood and lymph, three kinds of leucocytosis may be distinguished, each characterised by the relatively greater increase in number of one particular kind of cell. This may be illustrated by citing the effect of the injection of finely divided coagulated proteid, which produces a great increase in the number of the hyaline (phagocytic) cells without a correspondingly large increase in the numbers of the other wandering cell forms. Eosinophile leucocytosis, that is, increase in the numbers of the eosinophile cells, occurs with wonderful rapidity after injection of anthrax bacilli or other micro-organisms, and it is then followed by a leucocytosis of the hyaline cells.

2. The leucocytosis, or increase in the number of the cells, is largely due to the proliferation of the cells themselves. Thus eosinophile leucocytosis, followed by hyaline leucocytosis, occurs out of the body in a hanging drop of lymph. Also we have witnessed the division of the cells in a hanging drop. The phenomena classed under the head of chemiotaxis are undoubtedly to be partly explained by the very rapid power of proliferation by fission of the wandering cells.

The behaviour of the cells towards micro-organisms differs according to the nature of the latter. In this abstract we will confine ourselves to the conflict with *Bacillus anthracis*.

The frog at ordinary temperatures is absolutely immune against anthrax. When lymph is treated with anthrax bacilli the following phenomena are seen, and may be grouped as successive stages:

Stage I. The eosinophile cells are strongly attracted to the anthrax. They apply themselves to the chains of bacilli. When contact is absolutely or nearly effected their cell-substance shows the following phenomena:

1. It is profoundly stimulated, and exhibits quick streaming movements. Ordinarily the eosinophile cell is very sluggish.

2. The eosinophile spherules are discharged: those nearest the bacillus fading and dissolving first.

3. If the eosinophile cells are present in sufficient numbers to match the anthrax, in other words, if they are unharmed by the bacilli, they bud off daughter cells, which are at first free from granules. These creep a short way from the point of conflict, and in a short time spherules appear at one end. Later, these daughter cells seek the same or another focus of conflict. Several eosinophile cells will, towards the close of Stage I, and when their numbers have increased, be massed round one chain, and they ultimately fuse, though the endosarc, with its granules, remains distinct. In this way an eosinophile plasmodium is formed, though the fusion is confined to the more mobile peripheral cell-substance. Whether the eosinophile cells or the bacilli win the fight depends largely on their relative numbers. The bacillus is only injured near the eosinophile cell; there the contents become rapidly curdled and irregular in appearance, and may be completely dissolved (it should be noted that Leber has shown that pus dissolves copper, and even platinum, and Kanthack has shown that the pus cell is the eosinophile cell). If the bacillary chains are in great number, then there may not be eosinophile cells enough to attack them all, although the eosinophile cell will extend itself to most attenuated lengths in order to be able to attack as great a length of chain as possible. Even where the chain is not directly attacked, the near presence of eosinophile cells profoundly arrests its development.

If the cells win they early recharge themselves with spherules; *but these are no longer eosinophile—they are amphophile;* that is, they stain with both eosine and methylene blue, and rather more readily with the latter.

During the later portion of Stage I the eosinophile cells are aggregating and fusing round the chains of bacilli.

This fusion, and the later and more complete fusion of the hyaline cells, is a kind of conjugation, the cells ultimately separating.

During State I the hyaline cells, *the phagocytes,* remain quiescent, and *are not attracted towards the bacilli,* though *they may take up indifferent matter such as indian ink.* In the neighbourhood of a healthy bacillus they appear to be paralysed.

Stage II. Hyaline cells have now increased in numbers, and come to the eosinophile cell masses surrounding a bacillus and *fuse with them.* The eosinophile cells probably lie extended along a chain; the hyaline cells work with one object, namely, to draw the long-drawn-out mass into a ball. To this end a hyaline cell will attach itself by a broad attachment, and then, by means of long filiform pseudopodia stretched towards more distant parts, it will bend the chain up into a close U, rolling the eosinophile cells round itself, and fusing superficially with them. The superficial fusion of eosinophile cells with the hyaline cell produces violent streaming movements. Other hyaline cells come and fuse with the now lobate spherical and opaque mass. The impact of each successive cell acts as a stimulus, causing streaming and pseudopodial movements, which fade away, to be re-awakened by the arrival of a fresh cell.

We have now a lobed mass, curiously opaque, and—to take one particular instance—formed by the fusion of seven eosinophile cells and four hyaline cells. Three eosinophile cells originally attacked the chain. (It will be noted that we retain the term eosinophile cells, though the second formed spherules are at first amphophile.) This fusion may persist for one to two hours.

Stage III. The cells of the mass commence to regain their individuality and slowly separate. The separation is in two very distinct stages, and when the individual cells are again to be seen the mass is found to consist of a central giant hyaline plasmodium, formed by the very complete fusion of the four hyaline cells, and enclosed by a crust of eosinophile cells. The first stage in the dissolution of the mass is the separation and wandering away of the eosinophile cells, fully charged with the second set of spherules, which have now become truly eosinophile. A very curious appearance is presented as they shred themselves off the central hyaline mass. This plasmodium or giant cell is now seen to be an amœboid body, with several food vacuoles containing ingesta in the form of the remnants of the chain of bacilli. It pushes out on one side long filiform pseudopodia, which resemble those of the Heliozoa in their sluggish, streaming movements, while from the other side project short round pseudopodia.

The hanging drop contains, at this stage, multitudes of these phagocytic plasmodia, with free eosinophile cells and free hyaline and rose-staining cells.

Stage IV. This is the second stage of the disintegration of the cell masses. The food vacuoles of the plasmodium close up, and the whole structure becomes lobed, taking on the appearance of a heap of hyaline cells, which subsequently separate into the original four cells.

While these stages are in progress the rose-colouring cells are increasing in size and number. They are at first small and spherical, with not very abundant cell-substance. Later they become large, angular, and sometimes vacuolate, and their cell-substance becomes completely filled with baso-phile, rose-staining granules.

The activities of the rose-staining cells are, we believe, directed towards the removal of foreign noxious substance in solution in the plasma. We find that if the bacterial poisons accumulate beyond a certain point they para-lyse the eosinophile cells, and destroy the hyaline cells. This is prevented, in part at any rate, by the action of the rose-staining cells. We correlate the increase in the granulation of these cells, or, in other words, the increase in the amount of rose-staining substance, with the removal of the bacterial products.

The conflict thus consists of, *first*, the maiming of the bacilli by the eosinophile cells; *secondly*, the removal of the remains of the bacilli by means of the ingestive and digestive activity of the hyaline cells; and, *thirdly*, the removal of dissolved foreign substances by the rose-staining cells. We do not propose to deal at present with the further processes of repair.

Action of Urari

It induces extensive leucocytosis.

Stage I. After three hours lymph drawn is found to contain hyaline and *amphophile* cells, the latter in great abundance.

By treating a hanging drop with urari and methylene blue, we were able to watch the granules of the eosinophile cells slowly undergo a slight decrease in size and stain with the methylene blue. The granules of the normal cell never stain with methylene blue.

Stage II, 12 hours. Repair in progress; numerous large cells present charged with ingesta.

Stage III. The normal eosinophile cell reappears. Frogs completely recover from urari in a day or two.

Action of Heat

Frogs are rendered susceptible to anthrax by being warmed. We therefore inoculated hanging drops and watched them on the warm stage.

We found that the first attack of the eosinophile cells was commenced before the temperature had risen, but never carried out, the cells becoming completely paralysed, and showing no movement for five hours. Therefore there was no phagocytosis, *for this can only follow the eosinophile attack.*

Morphology and Comparative Physiology of these Wandering Cell Elements

We are now able to point to three animal forms, the frog and lamprey, types of a complex and highly developed group, and *Astacus*, a complex member of a group containing animals of widely divergent complexity. In all these different forms of wandering cells occur. These we may class as:

Granular eosinophile.	Found free in the body fluids.
Non-granular hyaline.	„ „
Rose-reacting cell, granular.	Wandering cell which is found in the body fluids, but which also inhabits the spaces of connective tissues, though it is not by any means identical with the connective-tissue cell.

Of these diverse forms we see the archetype in the granular, protective digestive, absorptive, and constructive (for it contributes to form the fat tissue and scar tissue) blood cell of the primitive animal *Daphnia*, and the granulation of this primitive cell is amphophile and rose staining, as is also the granulation of the ectoderm of *Daphnia*.

The physiological differentiation we can trace when we see that the eosinophile cell has accentuated the glandular and protective character of the primitive cell; while in its attack by direct contact brought about by pseudopodial activity we see the remnant of the direct pseudopodial and ingestive attack of the primitive cell.

The hyaline cell, or permanently free phagocyte, represents the specialisation of the direct pseudopodial ingestive activity of the primitive cell.

While, lastly, the absorptive powers of the primitive cell are represented by the rose-staining cell of the more differentiated animal forms.

7

ON THE CHANGES IN THE NUMBER AND CHARACTER OF THE WANDERING CELLS OF THE FROG INDUCED BY THE PRESENCE OF URARI OR OF *BACILLUS ANTHRACIS*

With L. B. KENG

[Journ. Physiol. (1893), xv, 361]

The profound alterations in the number and activities of the wandering cells of the frog which are produced by the presence of urari in the body of that animal are of interest in two ways. In the first place they throw some light on what may be called the resolution of that pathological condition of the blood or lymph known as leucocytosis, and in the second place they suggest certain considerations affecting the vexed question whether the different kinds of cells comprised under the term wandering cells are or are not completely distinct from one another, in origin and life-history. We may say at once that the phenomena observed by us do not furnish a final answer to this question.

There can be no question of the histological and functional distinctness of the different kinds of cells, as found in the blood or lymph under normal conditions. The three kinds of wandering cells found in the frog, eosinophile cell, the hyaline cell and the basophile cell, are, as has been pointed out elsewhere, not only sharply marked off from one another by their respective structural characteristics but also in the way they behave when foreign substances, especially micro-organisms, are present in the plasma, and this is true also of the four kinds of wandering cells found in the Mammalia. But although this renders their complete independence probable, yet such *à priori* considerations can have no finality in view of the little exact knowledge we possess at present as to the metaplasic powers of protoplasm. For instance it is still a moot question whether wandering cells can arise by the proliferation of fixed cells, and whether they may or may not become the fixed formative cells of scar tissue.

The view that the varieties of wandering cells are derived from a common parent cell, in other words, that they are really stages in the life-history of the same cell, has been advanced by Biondi[1] and H. F. Müller.[2] Their conclusions however are based upon the study of the cells of the blood

---

[1] Biondi, *Arch. per le scienze mediche*, XIII (ref. *Centr. f. Allg. Path.* (1890), p. 172).

[2] H. F. Müller, *Sitz. d. k. Akad. Wiss. in Wien*, III. Abth. Bd. xcVIII.

in pathological conditions when, as our researches on the histological phenomena of leucocytosis in the frog show, the different forms of cells may depart to a very marked extent from their normal state and produce forms which are seemingly but not really intermediate between the normal forms.

Leucocytosis is essentially a pathological condition of the blood or lymph, the obvious phenomenon of which is an increase in the total number of corpuscles, but in the case of the leucocytosis produced by the introduction of micro-organisms, or their products, or of curari, rapid dissolution of the wandering cells goes hand in hand with rapid production of these bodies, and the former phenomenon is as characteristic of the state into which the body fluids are thrown as the latter.

During these changes, and it is this fact which we wish to insist upon here, cells of one kind may mimic the characters of those of another kind both in nuclear structure and in the nature of the specific granules which they contain; and this may occur to such an extent that an observer who examined the cells at any one period only and was ignorant of the changes which had preceded and which would follow that particular stage, would almost certainly be led to conclude that he was dealing with forms transitional between the clearly marked types characteristic of the normal animal, whether it be frog or mammal. In order to show how this may occur it will be well to refer in this place to one particular case though in so doing we necessarily anticipate what will be described in detail later on.

The dissolution of cells spoken of above which marks the leucocytosis produced by the presence of microbic poisons or of urari is effected in two ways. The cells may simply break up in the plasma, a large part of the cell body apparently being dissolved while the rest, notably the remains of the nucleus, persists for some time as amorphous masses floating in the plasma. This process is characteristic of the earlier stages. At a later period, in what we shall speak of as the second stage, the dissolution is no longer of this passive kind but certain of the cells are actively destroyed by the hyaline cells. The eosinophile cells in particular are destroyed by being bodily ingested by the hyaline cells, and this fate overtakes them while they are apparently quite normal so far as their nuclei and granules are concerned. In this way there come to be introduced into the body of a great number of the hyaline cells large masses of eosinophile granules which, as digestion of the prey proceeds, are scattered as isolated granules through the cell substance of the hyaline cells (Pl. IV, figs. 7 and 8). Thus at this stage cells are found which possess the same nuclear characters and have the same type of cell body and of movement as the hyaline cells but which also show an eosinophile granulation, and in film preparations such cells appear, so far as their structural features are concerned, intermediate in character between the normal eosinophile and the normal hyaline cell. But this is not all, for an observer watching these cells while still alive would find that they were phagocytic and, supposing him to be ignorant of their mode of origin, he would be to a certain extent justified in concluding that under certain conditions the eosinophile cells of the frog ingested foreign bodies.

There is also another result which follows from our work. Metschnikoff in his treatise on inflammation,[1] and again in a recent paper in which he criticises certain recent contributions to the study of the functions of the wandering cells,[2] lays considerable stress on the possession of phagocytic powers by polynuclear leucocytes. We may say at once that in the frog at least all polynuclear cells are not phagocytic. When the eosinophile cells of that animal are increased in number many are found with more than one nucleus, further as a stage in their dissolution these bodies may become markedly polynuclear owing to fragmentation of the nucleus. Under no conditions and in none of these phases do they ever ingest foreign bodies.

The introduction of urari into the body of a frog is followed, not only by lesion of the neuro-muscular mechanism but also by profound changes in the entire vascular system, which may be traced probably to the direct action of the poison on the blood vessels and their contents and to the effect on the constitution of the plasma resulting from the general disturbance of the metabolism of the tissues. The lymph also is affected, the most obvious change being a large increase of fluid in the lymph spaces, caused in part by increased formation of lymph and in part also by an absorption of water by the skin which may occur to such an extent in the case of urarised frogs placed in contact with water as to render them bloated to an extreme degree.

Frogs recover from urari in a few hours to as many days, the period varying with the dose. If the urari be sterilised before injection the animals withstand its action much better and will completely recover from doses sufficient to maintain paralysis for as long as thirteen days.

The poison does not appear to paralyse the wandering cells except perhaps for a very short period, on the contrary their movements are on the whole increased.

The increased production of lymph is accompanied by the migration of a large number of the white blood corpuscles and, when large doses of poison are given, a diapedesis of red corpuscles into the lymph.

---

[1] Metschnikoff, *Leçons sur la Path. comparée de l'Inflammation*, Paris, 1892, pp. 131 *et seq.*

[2] *Ibid.* "La Theorie des Alexocytes," revue critique, *Ann. Pasteur* (Jan. 1893), p. 50.

We may just recall the fact that three kinds of wandering cells are found in the frog; the eosinophile cell, the basophile cell (with fine granules which colour a bright rose tint with methylene blue), and the hyaline cell or phagocyte. The two first-mentioned cells are not phagocytic. In the normal animal all three kinds of cells are mononuclear, the nucleus of the eosinophile cell however being lobed or crescentic.

In the larger number of our experiments a filtered and sterilised $\frac{1}{2}$ per cent. solution of urari was used. In these cases the animals for the most part remained free from intrusive micro-organisms. When the solution was not sterilised previous to injection the lymph became filled with a short bacillus which does not form chains and which is sometimes found in summer infecting frogs. The presence of these bacilli was often followed by the death of the animal. A remarkable difference in the histology of the lymph was found in the two cases, for when the frog remained free from infection with micro-organisms no increase in the basophile (rose-staining) cells took place, on the contrary they were diminished in number, but when bacilli made their appearance these cells increased in numbers.

The histological changes in the lymph occur in two stages, which however are not clearly marked off from one another in point of time. The first event following the introduction of the poison is an increase in the number of the eosinophile cells, or to adopt the recognised term, an eosinophile leucocytosis. While this is in progress the hyaline cells also increase in number. This constitutes the first stage. The second stage is of the nature of a resolution of the disease and is characterised by certain ingestive phenomena of remarkable interest and by a continuous fall in the number of corpuscles, especially of the eosinophile variety. The first stage occupies roughly one-third of the whole period.

Stage I. Changes in the eosinophile cells. The first effect of the poison is to modify the granulation of the eosinophile cell so that it stains with basic dyes. In other words, an amphophile granulation is produced. The granules at the same time increase slightly in size and become spherical and less refractive (fig. 1). The granules of the normal cell are usually slightly spindle-shaped and very lustrous. The cells at the seat of inoculation (always the dorsal lymph sac in our experiments) show this change more completely than those obtained elsewhere but, if the dose be sufficiently large, amphophile cells will be found in the lymph from other parts of the body. In many of these amphophile cells, especially in those which retain this modified granulation the longest, the nucleus loses its lobed character and swells up to form a structureless sphere.

Very early, either as a result of the re-establishment of the normal granulation, or of immigration, cells containing the normal eosinophile granules reappear. At the same time the total number of eosinophile cells is increased. The eosinophile cells however now show important modifications in structure. They continuously increase in size until a cell is produced which has a diameter one-third or more greater than the normal. At the same time cells quite normal so far as their granulation is concerned become common which in preparations appear to contain one, two, and less frequently more than two spherical nuclei. Yet another change in the eosinophile cells has to be recorded, and that is their disintegration. Throughout the whole of the first stage they are found either with obscured granulation or with no granulation. In these disintegrating cells the nucleus is usually in fragments which may be angular, rounded, or curved, and the cell substance usually colours diffusely with eosine. It should however be clearly understood that eosinophile cells in all other respects normal are very frequently found containing two nuclei each of which is a perfectly formed and well-staining structure.

It is important to state afresh the fact that in no case do the eosinophile cells manifest phagocytosis, neither when of the normal size with the typical lobed nucleus, nor when enlarged and multinuclear, nor when they contain only fragmented nuclei; and this is true even when the hyaline cells are laden with ingesta.

The increase in the number of the eosinophile cells present in the lymph may be due to immigration from other parts of the body, or to rapid proliferation of the cells already existing in the lymph spaces.

Ehrlich[1] showed that in the Mammalia the red marrow of bone is rich in eosinophile cells which are considerably larger than those found free in the blood. He further pointed out, and later workers have confirmed his conclusions in this respect, that an increase of the number of free eosinophile cells is usually correlated with an inflammatory condition of the red marrow. This observation led to the conclusion that the free eosinophile cells arise from the proliferation of fixed eosinophile cells which are stationed in the red marrow, and H. F. Müller found the latter undergoing mitotic division.[2] On the other hand there can be little doubt that free eosinophile cells undergo division. Flemming,[3] Wertheim[4] and Muir[5] have found mitoses in blood containing an abnormally large number of eosinophile cells, and the division of a living eosinophile cell has been witnessed in a drop of frog's lymph.[6]

Therefore, in the absence of any evidence to the contrary, we may suppose that the increase of eosinophile cells in the lymph of the frog is in

[1] Ehrlich, *Zeits. f. Klin. Med.* (1880), I, 553.
[2] Müller, *Archiv f. exper. Path. u. Pharmak.* XXIX, 221.
[3] Flemming, *Arch. f. Mikr. Anat.* XX, 57.
[4] Wertheim, *Zeits. f. Heilkunde* (1891), p. 281.
[5] Muir, *Journ. Path.* II, 123.
[6] Kanthack and Hardy, *Proc. Roy. Soc.* LII, 267.

part due to increased production of those elements in the bone marrow, the cells so arising travelling by way of the blood or perhaps by some more direct route.

Muir, who examined the blood in leucocythæmia, found eosinophile cells present which were larger than those occurring in the blood of healthy subjects, and he regards these as being the cells newly arrived from the bone marrow.

The large eosinophile cells which appear during eosinophile leucocytosis in the frog do not however all come from the bone marrow, though of course some may have this origin. They appear to arise mainly by a direct increase in size of the free eosinophile cells already in the lymph.

In order to determine this fact we used a method originally devised by Kanthack.[1] The skin was stripped off the legs of some pithed frogs, all possible precautions being taken to prevent infection of the internal surface. In some of the bags thus obtained a small quantity of a sterile solution of urari was introduced, and into some of the urarised bags pieces of a split femur freshly removed from a frog were placed. In seven hours it was found (1) that marked eosinophile leucocytosis had occurred in those bags treated with urari, while the cells were in comparison still few in number in the control bags which had not been treated with urari; (2) that the large eosinophile cell was present in the urarised bags which did and in those which did not contain bone fragments.

The bags were kept for 54 hours, always in an atmosphere saturated with aqueous vapour, and were then cut open and the contents carefully examined. The contained fluid was found to contain bacilli in such enormous numbers as to make it milky and opaque. It was examined both by means of alcoholic methylene blue and by film preparations. Eosinophile cells were found in all stages and hyaline cells, and the latter showed active phagocytosis, or, to speak more guardedly, bacilli were found in the bodies of these cells either because they were ingested or because they found there a suitable pabulum, but even film preparations failed to furnish a suggestion of phagocyte activity on the part of the former, and this too in spite of the incredible number of bacilli present.

This experiment clearly shows that marked increase in the number of cells may arise as a result of more active proliferation of the free cells of the lymph.

The *hyaline cells* suffer changes during the first stage as striking as those manifested by the eosinophile cells.

In the first place many of them are found with two nuclei, each nucleus preserving the characteristic features of the nucleus of the hyaline cells

---

[1] Kanthack, *Brit. Med. Journ.* (November, 1892).

(figs. 4 *b* and 5). In this case as in the case of the binuclear eosino-phile cells we have no certain facts which point to an explanation of the phenomenon.

Very early in the first stage the hyaline cells become more or less laden with ingesta of various kinds. (1) The ingesta is present in the cells in the form of spheres, each of which is enclosed in a digestive vacuole. In any one cell the spheres are of different sizes and often present varying stages in the digestive act, and they are composed of material which stains deeply with a solution of methylene blue, so that a fully loaded cell in such stained pre-parations may bear some resemblance to an eosinophile cell in the ampho-phile condition, especially to those in which the lobed nucleus has swollen out to form a sphere. The similarity however is only superficial, for the balls of ingested material are of different sizes and each lies in a definite digestive vacuole hollowed out of the cell, whereas the amphophile granules are quite or nearly of the same size and are evenly distributed throughout the cell-substance. The nuclei too present very striking differences; that of the amphophile cell may be of the typical lobed form, or it may be swollen and spherical. In the second case we are probably dealing with a dying cell, for the nucleus shows no trace of structure but stains uniformly with methylene blue and presents a curious glassy appearance (fig. 1). In the cells laden with ingesta the nucleus is of the form so characteristic of the hyaline cells, that is it shows a well-defined nuclear capsule and an equally well-defined nucleolus. We regard these spheres of ingestive material as being formed from minute particles of detritus resulting from the disintegration of the dead cells, and representing perhaps chiefly the remains of their nuclei. Our reasons for this are that precisely similar spheres are formed when protozoa are fed with minutely divided coagulated proteid. Our knowledge of this fact is due to the researches of M. Greenwood on digestion in protozoa. She finds that when *Carchesium* is placed in water holding even a small quantity of coagulated proteid in suspension in such a fine state of division that it cannot be removed by rapid centrifugalising, the animals load themselves with spheres which they form by the assimilation and aggregation of the minute proteid particles into digestive vacuoles. Having this fact in mind we tried the effect of the injection of a small quantity of sterilised finely divided coagulated proteid suspended in normal salt solution into the dorsal lymph sac of a frog. In an hour the hyaline cells were found to contain ingestion spheres accurately resembling those described above.

(2) The second form of ingesta consists of coloured masses or particles. This has three sources.

(*a*) Quite early hyaline cells are found at the seat of the injection of the poison which are laden with finely divided brown pigment. There can be

H   6

little reason to doubt that this represents the absorbed colouring matter of the solution of urari.

(b) At the close of the first stage and throughout the second stage of the phenomena following the poisoning hyaline cells are found more or less completely loaded with black pigment. Sometimes the cells load themselves so heavily that they become opaque, but even in this condition they retain the power of amœboid movement and may be seen to emit one or more long filiform pseudopodia.

The pigment grains which load the cells are picked up by them from the plasma, the latter sometimes containing large numbers of grains in suspension.

The origin of this pigment is not at all obvious, but so far as the cells of the lymph are concerned it seems to be certain that they obtain it ready formed from the plasma in which it occurs as the result of an exaggeration of some process which normally occurs in the body of the animal. We are led to this conclusion by the fact that cells laden with pigment were found in abundance in female frogs but were only rarely met with in males; and during the period occupied by our researches the female frogs were laden with eggs which, as is well known, are produced in great numbers and are deeply pigmented. Therefore, during the period of egg formation, active pigment production must form a part of the processes normally occurring in the female. The production of pigment after urari, and the same thing occurs after inoculation with anthrax, must be very great, and may occur to such an extent that the plasma in hanging drops of lymph appears under the microscope to be studded with minute particles.

(c) The third class of pigment-laden cell contains yellow, yellow-brown, or brown pigment. This is undoubtedly derived from the hæmoglobin of the red corpuscles which sometimes under the influence of the poison are found in considerable numbers in the lymph. Every stage may be found between cells containing yellow masses situated in digestive vacuoles to cells studded with particles of brown pigment. The way in which this pigment is produced may be followed in a drop of lymph removed from a frog one or two days after the injection of urari.

As will be shown in describing the phenomena which occur in the second stage of the poisoning, the powers of the hyaline cells appear to be augmented, so that whereas in the earlier stage they ingested only small particles, in the later stage they engulf entire cells, and this increased capacity is apparently conditioned by a striking increase in the size and activity of the hyaline cells. In the earlier stages the ingestion of a red corpuscle is carried out in a piecemeal fashion, that is to say, the hyaline cell by its movements breaks up the stroma of the red corpuscle, which has become as it were macerated and lost all its elasticity, into irregular fragments each of which is separately engulfed. Following the fate of any one of these fragments we find it at first occupies a digestive vacuole the fluid of which appears to rapidly dissolve the stroma so that the hæmoglobin is set free and the entire vacuole becomes tinged with yellow. The pigment solution so produced is then concentrated, the vacuole at the same time growing smaller, until a brown pigment particle alone remains embedded in the cell-substance.

In the later stages entire red cells are ingested by the enlarged hyaline cells. Each at first occupies a digestive vacuole and, as in the first case, the stroma appears to be dissolved first and the vacuole is then left containing a solution of hæmoglobin. Strands of cell-substance then grow across the vacuole dividing it up into smaller spaces, in each of which the hæmoglobin is dehydrated and a number of pigment particles are thus produced.

The production of a solid particle from a solution of a pigment in a vacuole is a process not confined to hæmoglobin but is one which occurs also when the very different body methylene blue is introduced into a frog, and this is the more striking when we remember that methylene blue has no proteid matter in its molecule. The production of the vacuoles and the events occurring in them when formed may, in the case of the ingestion of fragments of red corpuscle, be regarded as due to the stimulus of contact of the cell with a solid mass, and to the nutritive nature of that mass, composed as it is largely of proteid. None of these conditions are fulfilled by methylene blue, yet an amount of this dye insufficient to produce obvious coloration of the plasma leads to the occurrence of phenomena in the cells closely resembling those seen when fragments of red corpuscles are ingested. Thus if methylene blue dissolved in normal salt solution be injected into a frog (e.g. into the peritoneal cavity), the lymph in all parts of the body will be found to contain cells more or less laden with what appears to be solid blue spheres. Figs. 10 and 11 represent such cells as they appear when alive. If at this period a second injection be made, then four hours afterwards the cells will present the appearance shown in fig. 12. In other words, the vacuoles have been reconstituted round the already absorbed pigment and the fluid they contain is very obviously tinged with the dye.

In the case of the cells represented at fig. 12 both injections were made into the peritoneal cavity and the lymph examined was obtained from the foot.

It is obvious from what has been said above that at the close of the first stage the cells of the lymph appear at first sight to be of very diverse character, and it is reasonable to suppose that an observer who had not studied the changes from the commencement would be puzzled to effect a satisfactory histological analysis. The cells which may be present in such lymph are the following:

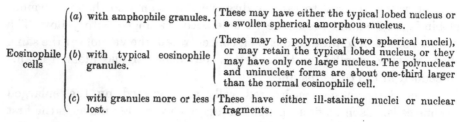

Eosinophile cells

(a) with amphophile granules. { These may have either the typical lobed nucleus or a swollen spherical amorphous nucleus.

(b) with typical eosinophile granules. { These may be polynuclear (two spherical nuclei), or may retain the typical lobed nucleus, or they may have only one large nucleus. The polynuclear and uninuclear forms are about one-third larger than the normal eosinophile cell.

(c) with granules more or less lost. { These have either ill-staining nuclei or nuclear fragments.

In none of these different conditions are the eosinophile cells phagocytes.

Hyaline cells which may possess one or two nuclei, each of the typical form.

(a) Laden with ingestion masses staining with the basic dye, methylene blue.

(b) Laden with the debris of red blood corpuscles which stain with eosin, and therefore in film preparations such cells may appear to contain irregular "eosinophile" masses.

(c) Laden with finely divided yellow pigment.

(d) Laden with black pigment grains.

(e) Laden with brown pigment grains.

(f) Dead hyaline cells with split nucleus.

The Second Stage

As has been already pointed out the hyaline cells continuously increase in size during the first stage. In the second stage they are large cells with abundant finely punctuated cell-substance and a single nucleus of the typical form. The forms with two nuclei continuously become more and more infrequent. The second stage is characterised by the increased ingestive activity of these cells, they no longer engulf only particles but take up entire and apparently intact eosinophile cells and red corpuscles. During this period *the large hyaline cells engulf entire eosinophile cells of all kinds, so that one finds in their digestive vacuoles eosinophile cells with completely intact and normal granules and nuclei. Further, the appearance of freshly enclosed eosinophile cells does not warrant the assertion that they had died before being engulfed by the hyaline cell.* It is obvious from this that during the second stage the number of eosinophile cells in the lymph falls, and their destruction may be pushed so far as to make it difficult to find a single free eosinophile cell. At the same time of course a fall in the total number of corpuscles occurs. What red corpuscles remain are also ingested. The ingestion of the eosinophile cells is carried out with considerable vigour, so that one finds not infrequently a hyaline cell which contains an eosinophile cell still intact and obviously only lately ingested occupying one vacuole, another eosinophile cell in a more advanced stage of digestion in a second vacuole, and remains of other cells, notably the granules, occupying vacuoles scattered through the rest of the cell-substance. One such stained with methylene blue is shown in fig. 7, and a cell with an intact eosinophile cell in a vacuole is shown at fig. 8 as seen in a film preparation.

The digestion of an eosinophile cell by a hyaline cell is a process of great interest. The cell is first of all received into a large digestive vacuole. While there its cell-substance is dissolved, and its nucleus loses its outline and staining properties. The eosinophile granules however appear to resist the digestive action for a long time and persist in the vacuole after the solution of the substance of the cell. The large vacuole is then broken up by strands of cell-substance which grow into it and divide it into a number of small spaces in each of which a single granule is enclosed. Ultimately each isolated granule swells up and becomes so far altered that it stains slightly with methylene blue. Thus at a certain stage in the process we have a hyaline cell containing scattered eosinophile granules derived from an ingested eosinophile cell. Such a cell might ingest other bodies, and in film preparations we might be deceived by finding a cell apparently eosinophile and yet betraying phagocytic powers.

We may summarise the main phenomena so far described as follows:

(1) The introduction of urari leads to an increase in the number and in the size of the cells in the lymph followed by a decrease, so that the number of cells just after recovery from the poison has taken place is below the number in the normal animal.

(2) The increase in the number of cells is due both to immigration and to proliferation.

(3) Throughout the whole course of the disease there is marked destruction as well as marked production of cells. This destruction is at first of the nature of a passive death and dissolution of the cells in the plasma. Later, cells with apparently normal granulation and nuclei are bodily ingested by the enlarged hyaline cells.

(4) The hyaline cells manifest phagocytosis at an early period, and during the period when the resolution of the disease is being effected they ingest entire cells, their activities being then mainly directed to the destruction of the eosinophile cells.

The later stages of the leucocytosis following the introduction of anthrax into the frog. The earlier phenomena manifested by the wandering cells of the frog when anthrax is present have been described elsewhere.[1] In the communication referred to they were followed up to the point where the bacilli are destroyed and the lymph is left with a large number of eosinophile, hyaline and basophile (rose reacting) cells. Both the eosinophile and hyaline cells at this period are actively amœboid. The further phenomena resemble those described above as constituting the resolution of the leucocytosis produced by the presence of urari, that is to say active

[1] Kanthack and Hardy, *loc. cit.*

ingestion and destruction of the eosinophile cells by the hyaline cells occurs. An instance of this, obtained from the lymph of a frog 24 hours after inoculation with anthrax, is shown at fig. 7.

In all that has been said so far it will be noticed that, in spite of the profound departures from the normal condition of the eosinophile and hyaline cells, no real loss of identity was observed; nothing was seen which suggests the formation of an eosinophile from a hyaline cell or vice versa. Not only can this negative evidence be adduced but also the devouring of the eosinophile cells by the hyaline cells until only a very few examples of the former remain points to a divergence of function of so marked a character as to render it difficult to believe in the essential identity of the two types. One cannot conceive on the hypothesis that the eosinophile and hyaline cells represent merely phases in the development of the same cell, why the one should devour the other instead of the one merely developing into the other, and the time has gone by for the suggestion that the eosinophile granulation marks the degeneration of the wandering cells.

EXPLANATION OF PLATE IV

Fig. 1. Amphophile condition of the granules of an eosinophile cell. Nucleus swollen and showing no structure. Third day after injection of urari, methylene blue. Oc. 4, ob. $\frac{1}{12}$th, cam. luc.

Fig. 2. Binuclear eosinophile cell, stained with methylene blue. Third day after urari. Oc. 4, ob. $\frac{1}{12}$th, cam. luc.

Fig. 3. (a) Dead and disintegrating eosinophile cell. (b) Dead and disintegrating hyaline cell, methylene blue. Third day after urari. Oc. 4, ob. $\frac{1}{12}$th, cam. luc.

Fig. 4. Binuclear hyaline cells, stained with methyl green acidulated with acetic acid. Urari 20 hours. Oc. 4, ob. E, Zeiss, cam. luc.

Fig. 5. Hyaline cell laden with amorphous ingestion, methylene blue. (a) 20 hours after urari. Oc. 4, ob. $\frac{1}{12}$th, cam. luc. (b) 24 hours after injection of anthrax. Oc. 4, ob. F, Zeiss, cam. luc.

Fig. 6. Hyaline cell with vacuolated protoplasm. Active ingestion going on, methylene blue, 24 hours after anthrax. Oc. 4, ob. $\frac{1}{12}$th, cam. luc.

Fig. 7. Hyaline cell which has ingested two eosinophile cells and whose protoplasm is studded with eosinophile granules. Methylene blue 24 hours after injection of anthrax. Oc. 4, ob. $\frac{1}{12}$th, cam. luc.

Fig. 8. Hyaline cell which has ingested an eosinophile cell. Film preparation stained with eosin and Lœffler's methylene blue, after urari. Oc. 4, ob. $\frac{1}{20}$th, cam. luc.

Fig. 9. Hyaline cell with the remains of another cell occupying a vacuole. Methylene blue. Oc. 4, ob. $\frac{1}{12}$th, cam. luc.

Fig. 10. Cell from peritoneal cavity of a frog 16 hours after injection into that cavity of methylene blue dissolved in normal salt solution. Drawn when alive.

Fig. 11. Cell from the lymph of the foot of the same animal.

Fig. 12. Cells from the lymph of the foot of the same animal four hours after a second injection of methylene blue into the peritoneal cavity. Living cell drawn.

Plate IV

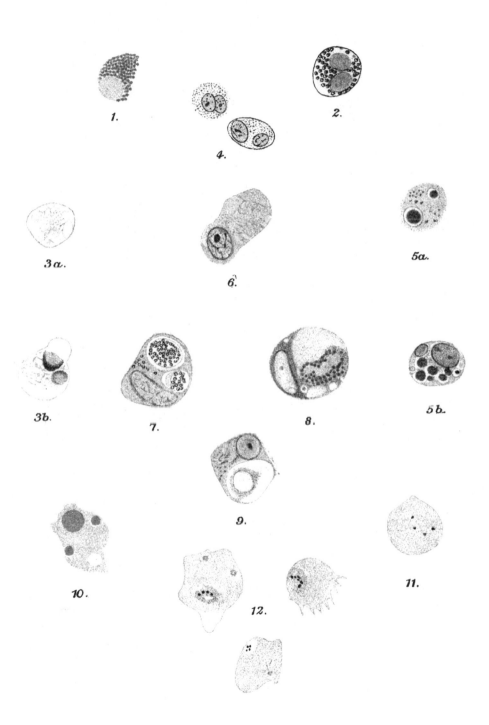

8

THE MORPHOLOGY AND DISTRIBUTION OF THE WANDERING CELLS OF MAMMALIA[1]

With A. A. KANTHACK

[*Journ. Physiol.* (1894), XVII, 81]

PART I. INTRODUCTORY. CLASSIFICATION OF CELLS

A foundation for our knowledge of the morphology of the wandering cells, or sporadic mesoblast, was made by Wharton Jones in 1846.[2] In the first of three memoirs published in the *Philosophical Transactions* for that year, he described the cells of the blood and lymph in cartilaginous and osseous fishes, in amphibia, in birds and in mammals. The second memoir included the cells of the "lymph" of worms, insects, crustaceans, and molluscs. In the third memoir the author compared the cells found in Vertebrates with those found in Invertebrates. Throughout these memoirs the white corpuscles are regarded as the young forms of the red corpuscles. Putting this aside, a great advance was made in the clear recognition of the fact that the white cells of blood and lymph are not all of one kind. Two main classes of cells were described: (i) the "granule cells", and (ii) the "nucleated cells". The granule cells were again divided into a coarsely granular variety, and a finely granular variety. It is interesting, in connection with the purport of the second part of our paper, to note that the author describes himself as having borrowed the term "granule cell" from Prof. Vogel, who first applied it to the form of cell "developed in inflammatory exudations". Streaming movements of the substance of both granule and nucleated cells were noticed, though the author does not appear to have recognised in these the manifestation of independent locomotor activity.

In 1863 Rindfleisch[3] confirmed the existence of granular (Körnchenzellen) and non-granular forms of colourless corpuscles in the frog.

In 1865 Max Schultze[4] carried the knowledge of the wandering cells to the

---

[1] The research has been carried out on man and on rodents (rat, mouse, rabbit and guinea-pig). Since, however, excursions to other mammals (dog and ox) have revealed similarity of structure and function, and not differences, we feel that we may extend our results to the higher mammals. Our thanks are due to Prof. M. Foster, Sec. R.S., who has aided us in all departments of this work, and to Dr Ruhemann who has advised us in all that relates to the chemistry of the dyes employed. Part of the expense of this research has been paid by the Scientific Grants Committee of the Brit. Med. Assoc.   [2] Wharton Jones, *Phil. Trans.* 1846.

[3] Rindfleisch, *Experimentalstudien über die Histologie des Blutes*, Leipzig, 1863.

[4] Max Schultze, *Arch. f. Mikr. Anat.* (1865), I, 1.

furthest point attainable with the methods then in use: namely, the study of fresh living cells. By observing fresh blood on the warm stage, this author was able to identify and describe the following forms:

(i) A small round cell not larger than a red corpuscle, composed of a round nucleus enclosed by a thin film of clear protoplasm which did not exhibit amœboid movement. The diameter of the smallest of these cells was found to be 0·005 mm.

(ii) Cells like these but larger, the increased size being due to the possession of more abundant protoplasm which showed amœboid movements.

(iii) Cells whose protoplasm contains fine granules and embeds one, two, or more nuclei. Diameter 0·009–0·012 mm. These cells are amœboid and from their abundance are to be regarded as the typical colourless blood cell.

(iv) Cells whose protoplasm contains coarse granules and shows amœboid movements.

After Max Schultze no further advance was made or indeed was possible in the histological analysis of the sporadic mesoblast until Ehrlich,[1] in 1878, furnished a rational basis for the use of staining reagents by his far-reaching discovery that the elective affinity of certain constituents of tissues for particular stains could be referred to two factors, the chemical nature of the staining substance employed and, a point too often neglected by workers who have followed his methods, the nature of the medium in which the stain is dissolved. Ehrlich drew particular attention to the granules, the possession of which characterises various forms of wandering cells. These he divided into five classes differing either in their special affinity for basic, acid, or neutral dyes, or in size. The α or eosinophile granulation colours only with acid dyes, the β granulation colours with both acid and basic dyes (amphophile), the γ granulation colours only with basic dyes and the individual granules are large, the δ granulation colours only with basic dyes; but the individual granules are small and the ϵ granulation colours only in neutral dyes. The nomenclature of the granules was extended to the cells bearing them. Thus the various forms of white cells found by Ehrlich in blood were: (i) A small cell free from granules, to which the name *lymphocyte* was given, from the fact that it appears to be developed in lymphoid tissue. This is the small non-amœboid form of Max Schultze. (ii) A cell characterised by possessing fine granules and one or several nuclei. This is by far the most numerous form of white blood corpuscle in Mammalia, and was found by Ehrlich to be neutrophile in man, and amphophile in rabbits and guinea-pigs. (iii) The eosinophile cell, or coarsely granular cell of Wharton Jones and Max Schultze. It occurs only in small

[1] Ehrlich, *Verhandlungen d. physiol. Gesellschaft zu Berlin* (1878–79), No. 8. And *Farbenanalytische Untersuchungen*, Berlin, 1891.

numbers in the blood of Mammalia, but is abundant in the blood of lower Vertebrates. (iv) A basophile cell with fine basophile granules (δ granulation).

The mononuclear amœboid cells of Max Schultze (ii) are apparently grouped with the neutrophile cells by Ehrlich. In addition to these forms Ehrlich describes a basophile cell with coarse granules (γ granulation), occurring mainly in connective tissues and also in the blood of frogs but not in the blood of mammals. These he calls "Mastzellen". Unfortunately neither Ehrlich nor his pupils appear to have attempted the formulation of a connected account of all these forms of wandering cells, in other words, the morphology of the tissues as a whole, the only exception being the attempt to classify the cells as lymphocytes and myelocytes according to whether they have their origin in lymphatic glands or in bone marrow.

From what we have said so far it will be seen that the group of finely granular blood corpuscles described by Max Schultze includes the amphophile and neutrophile, and the finely granular basophile cells of Ehrlich. Since Ehrlich's work no contribution to our knowledge of the morphology of the wandering cells has been made except on points of detail. Mention must, however, be made of the groups of cells recognised by Metschnikoff[1] in his treatise on inflammation (1892). There the term "leucocyte", originally applied by the French school of physiologists, is used to designate wandering cells, and the following varieties are recognised: (i) lymphocytes, (ii) mononuclear leucocytes with abundant protoplasm and a round nucleus, (iii) polynuclear leucocytes or "leucocytes neutrophiles", (iv) eosinophile leucocytes.

Ehrlich's grouping of the granules discussed. Our researches have led us to the conclusion that Ehrlich's grouping of the cells is imperfect in that it has reference solely to the micro-chemical reactions of their specific granules, and none to the relation one to another of the cells which contain these bodies. We have also been led to conclude that granules having a purely neutrophile reaction have not been proved to exist, and that the term "neutrophile cell" is, at any rate when applied to the finely granular and most abundant colourless corpuscle of human blood, a misnomer. For these fine granules have an oxyphile reaction (*i.e.* they stain with acid dyes) quite similar to, but very much less intense than, that shown by the coarse α or eosinophile granules.

The neutrophile granule. The reaction of granules with acid dyes or, as we should perhaps say, the union of the granule substance with the acid pigment, is, as Ehrlich pointed out,[2] dependent for its occurrence or non-occurrence on the nature of the solvent of the dye. He found that the staining power of the dye was greater in an aqueous solution, and less in a

---

[1] Metschnikoff, *L'Inflammation*, Paris, 1892.

[2] *Verhandlungen d. physiol. Gesellschaft zu Berlin* (1878–79), No. 20.

glycerine solution; and on these lines Schwarze[1] classified the acid dyes as weak or strong, the weak dyes staining only when dissolved in water, the strong dyes staining in both aqueous and glycerine solutions though, of course, very much more strongly in aqueous solution. Orange G, for instance, stains only in aqueous solution, eosine stains with overwhelming power in aqueous solution, but has a markedly selective action when dissolved in glycerine, colouring those substances only which are strongly oxyphile. We have carried this distinction a stage further by using solutions of eosine in strong alcohol, for these have an even weaker staining action than glycerine solution.

In this way we can distinguish three grades of granules differing in the intensity of their affinity for acid dyes: (i) those which stain with eosine only in aqueous solution or in alcoholic solutions of a percentage below 60; (ii) those which stain with eosine in both aqueous and glycerine solution but are not coloured by a solution of the dye in strong alcohol (90 to 95 per cent.); (iii) those which stain with aqueous, glycerine and strong alcoholic solutions.[2]

The neutrophile and amphophile granules of Ehrlich constitute the first grade, they are granules which have a minimal attraction for acid dyes or, briefly, a minimal oxyphile reaction. The neutral mixture of Ehrlich is, chemically speaking, not a neutral dye, but an exceedingly powerful though withal an exceedingly differential acid dye, colouring very intensely oxyphile granules of all grades. So far as we can gather, there are no chemical facts which warrant the assumption that a mixture of the two rosaniline derivatives acid fuchsine and methyl green with the acid azo-dye Orange G will form a neutral dye. In point of fact, the only dyes to which, from their chemical constitution, the term "neutral" can be rightly applied are those coloured salts which are formed by the union of colourless acid and a colourless base, and in which, therefore, the whole salt is a chromophore and not simply the acid or the basic radicle. The fuchsines and methyl green are instances of such truly neutral dyes, and, in neutral solutions, they do not colour the granules in question.

Seeing that the fine granule of that form of white corpuscle which is most abundant in the blood has an affinity for acid dyes which, though it be small as compared with the very great affinity displayed by the large eosinophile or α granule, is quite clear and distinctive, and that it has no

[1] "Ueber eosinophile Zellen"; Inaugural Dissertation, 1880. Reprinted in *Farben-analytische Untersuchungen* of Ehrlich, Berlin, 1891.

[2] It is not our purpose to give any complete account of the factors which modify these micro-chemical tests. We may, however, note that the density of the solution of the stain affects the reaction. In order to eliminate this factor we used in all cases a saturated solution.

Eosine throughout this paper refers solely to Grübler's "wasserlösliches Eosin".

affinity to and is not coloured by basic dyes or by truly neutral dyes, we propose to discard the doubtful and possibly misleading term "neutrophile", and in its place to call these granules the fine oxyphile granules, and the cells bearing them the finely granular oxyphile cells, as opposed to the coarse oxyphile granule or eosinophile granule and the coarsely granular oxyphile cells.[1]

The other term which was applied to the finely granular oxyphile cells by Metschnikoff and others, namely, "polynuclear leucocyte", is not satisfactory, seeing that, as Flemming, we believe, first pointed out, the different nuclear masses which are dispersed through the body of the cell and appear at first sight to be distinct nuclei are, in point of fact, joined by threads or bars of nuclear substance so that the cell is really mononuclear with a very much branched nucleus.

The recognition of the fact that the fine granule of the common blood corpuscle of Mammalia is oxyphile gives to Ehrlich's classification of the specific granules of wandering cells greater simplicity and homogeneity, since it enables us to arrange them in two main groups, each group being again sub-divided into two. These are:

I. Oxyphile granules
  (a) Coarse oxyphile granules (eosinophile of most writers).
  (b) Fine oxyphile granules.

II. Basophile granules
  (a) Coarse (Ehrlich's γ granulation).
  (b) Fine (Ehrlich's δ granulation).

And similarly the wandering cells of Mammalia fall into three groups:

I. Oxyphile cells
  (a) Coarsely granular oxyphile cells.
  (b) Finely granular oxyphile cells.

II. Basophile cells
  (a) Coarsely granular basophile cell.
  (b) Finely granular basophile cell.

III. Non-granular cells or, as we have elsewhere termed them, hyaline cells.

To these would be added a fourth group:

IV. Immature cells, or lymphocytes.

METHODS EMPLOYED

The reactions of substances to basic and acid dyes are modified by the manner in which the tissues are prepared for staining. Heat, as Ehrlich pointed out, intensifies the oxyphile reaction of the granules,[2] fixation with corrosive sublimate, osmic acid, and Flemming's fluid also modifies the

[1] This classification has also been adopted by Sherrington, *Proc. Roy. Soc.* LV, 186, etc.

[2] *Farbenanalytische Untersuchungen*, p. 14.

histochemical reactions. It is therefore necessary to adopt some standard method of preparation for obtaining what may be then called the standard oxyphile or basophile reactions.

Standard method. The films, whether of fixed tissue, of blood or of lymph, were allowed to dry at the temperature of the room. The stain most commonly used was eosine in saturated aqueous and glycerine solutions and in alcoholic solutions in which the alcohol formed 50, 75, 80, 85, 90 and 95 per cent. Orange *G* was also used in saturated aqueous, and in saturated alcoholic solutions. The Ehrlich-Biondi mixture as supplied by Grübler and the "neutral" mixture of Ehrlich which is composed of acid fuchsine, methyl green and Orange *G* were also used. The films stained with the aqueous solution of eosine were exposed to it for the shortest possible time and were then washed for 5 min. in 95 per cent. spirit, dried between filter papers and stained with Lœffler's solution of methylene blue. Films stained in glycerine eosine were dipped in the solution and then washed in 95 per cent. alcohol until all the glycerine was removed. They were then immersed for a moment in fresh alcohol, dried between filter papers and stained with Lœffler's solution. To stain in the alcoholic eosines the films were immersed in the solution for 15 sec., washed in two changes of 95 per cent. spirit (total time 1 min.), dried between filter paper and stained in Lœffler's solution. The Lœffler's solution was in all cases washed off with water after which the films were finally dried and mounted in balsam.

It will be noticed that the excess of eosine was always washed off with 95 per cent. spirit, and in order to secure trustworthy results, this point must be attended to with scrupulous care. If, for instance, a film which has been stained in alcoholic or glycerine eosine is washed in water, it is exposed in the process to an aqueous solution of the dye which, though dilute, has a staining power considerably above that of the alcoholic solution, and often above that of the glycerine solution.

Orange *G* in saturated aqueous and saturated alcoholic solutions; hæmatoxyline[1] in neutral saturated aqueous solution, and sodium sulphindigotate in saturated aqueous solution were used as weak acid dyes.

The weak acid dyes (Orange *G*, hæmatoxyline, and sodium sulphindigotate) were washed out with water. The mixtures (Ehrlich-Biondi and the "neutral" mixture) were washed out with water. Sometimes however in order to secure more differential acid staining the Ehrlich-Biondi mixture was washed out with 95 per cent. spirit. This point will be specially noticed whenever reference is made to this stain.

It is obvious that if eosine in aqueous solution has a very high staining power, and in strong alcohol only a very weak staining power, the staining power of a saturated solution of eosine in a mixture of water and alcohol

---

[1] Schwarze, *loc. cit.* and *Farbenanalytische Untersuchungen*, p. 90.

will become less as the amount of alcohol present becomes greater. In other words, that the staining power varies inversely as the amount of alcohol present. Therefore, by using solutions of eosine in spirit of different strengths very delicate and exact discrimination may be made between the intensities of the oxyphile reaction of various substances.

In order to preserve very unstable cells, or to fix cells in the position they occupy when attacking micro-organisms, very rapid fixation by heat was resorted to. Some of the fluid, e.g. peritoneal fluid, was brought on to the surface of a carefully cleaned coverglass which was then at once immersed in the flame of a Bunsen's burner. It is not easy to stain successfully films fixed in this way, because the coagulated plasma colours intensely with methylene blue. We overcame this difficulty by volatilising the excess of methylene blue with the heat of a small spirit flame.

Since film preparations cannot be relied upon to show the true size or shape of cells, measurements and drawings were made from fluid preparations made by quickly adding to fresh blood or other fluid three times its volume of a dilute solution of methylene blue in 40 per cent. alcohol to which a trace of caustic potash and of osmic acid had been added. This solution is especially valuable for the cells of blood; indeed without its aid it would be difficult to study the finely granular basophile cell. In order to make a successful preparation, it is above all things essential that the staining fluid be added to the amount of at least three times the volume of the blood before clotting takes place. It will then be found that the blood is rendered laky and the refractive index of the fluid is so nearly that of the stroma of the red corpuscles that these bodies become quite invisible, so that the wandering cells may be counted or examined with the same ease as in lymph. In preparations of this kind differences in the refractive indices of granules form a striking feature.

In order to determine the relative abundance of the various forms of cells it is very necessary not to use a pipette; the fluid must be allowed to run or drop on to the coverslip or slide. The percentage of the basophile cells is especially affected by using a pipette, since these are very adhesive and remain attached to the sides of the pipette.

DESCRIPTION OF THE CELLS AND OF THEIR DISTRIBUTION IN THE BODY

In this section the cells are dealt with in the order in which they occur in the scheme given on p. 91.

The Oxyphile Cells.

1. *The coarsely granular oxyphile cell, or eosinophile cell* (Pl. V, figs. 1, 2, 6; 7, 8, 10, 12 (*a*), and 14), varies in size in different animals, not only absolutely, but relatively to the dimensions of the other classes of cells. In man it is

larger than either the hyaline cell, the finely granular oxyphile cell, or the finely granular basophile cell. In the rat, rabbit and guinea-pig, on the other hand, it is smaller than the largest hyaline cells, but larger than the finely granular oxyphile and basophile cells.[1]

The *nucleus* is typically an elongated body bent to form a horse-shoe. In the rat the arms of the horse-shoe are carried so far round that in film preparations the ends often overlap, giving to the nucleus the appearance of a circle with a large hole in the centre (figs. 7 and 10a). Sometimes the nucleus is lobed, but we are inclined to regard this appearance as being largely due to the stresses to which the nucleus is subjected when the cell is dying. In the living cell at rest, when it is spherical, the shape of the nucleus, so far as it can be determined by the disposition of the cell-granules, is a simple horse-shoe or crescent. A distinct nuclear network is present.

Cell-granules. The cell-granules are relatively large, spherical, or slightly ovoid bodies, and are sharply marked off from the cell-substance by their very high refractive index, which is so great that in fluid preparations the granules have a brilliant greenish lustre (figs. 1, 8 and 12a). The cell-substance in which they are embedded has the appearance of a clear transparent structureless jelly. The intensity of the oxyphile reaction of these granules differs in different animals but is always high. Thus it is very high in the case of the granules of man; these staining with eosine dissolved in 95 per cent. alcohol; it is lowest in the granules of the rat which do not stain with eosine in strong alcohol, but do colour with eosine in glycerine or in alcohol below 85 per cent. The granules also stain with weak acid dyes, such as Orange *G*, hæmatoxyline and sodium sulphindigotate. Ehrlich-Biondi's mixture (washed out with 95 per cent. spirit) colours these bodies brown-purple, and the "neutral" mixture (washed out with water) stains them a very intense red purple. Corrosive sublimate increases the oxyphile reaction as does also heat when applied to the dried film.[2]

2. *The finely granular oxyphile cell* (figs. 1, 2, 3 and 6b, 11, 12c and 15b) is always smaller than the coarsely granular oxyphile cell. On the other hand, it is in man and the guinea-pig rather smaller than the hyaline cell, and in the rat and rabbit very considerably smaller. The *nucleus* is an exceedingly irregular structure branching throughout the cell (figs. 2 and 6b). The branches swell out here and there into masses of an irregular shape, the intervening or connecting portions being of the nature of slender bars or threads. A fine and close nuclear network can be detected.

The granules are very small spherical bodies which crowd the otherwise

[1] The dimensions of the various cells are given in the table on p. 100.
[2] Cf. Sherrington, *op. cit.* pp. 189 and 190, where a description is given of this cell in the blood of the cat.

clear and optically structureless cell-substance. They have a refractive index only slightly above that of the medium in which they are embedded so that they are scarcely visible when unstained (figs. 1b and 12c). Their oxyphile reaction is much feebler than that shown by the large granules of the coarsely granular oxyphile cells. Thus they do not stain with the weak acid dye Orange G, nor do they stain with eosine, except in an aqueous solution or in an alcoholic solution in which the alcohol is less than 60 per cent. On the other hand, they stain with the fuchsine and orange of the "neutral" mixture and with Ehrlich-Biondi's mixture. The dye, however, can be washed out by strong alcohol. Long exposure to hæmatoxyline stains them and they colour though very feebly with sodium sulphindigotate. In this case they do not retain the colour in presence of water unless it contain a trace of the dye in solution.

The finely granular cells of the rabbit need special mention. The granules of these cells constitute the β granulation of Ehrlich to which he gave the name of "amphophile" granulation. According to the nomenclature of Ehrlich this name can mean only that the granules to which it is applied stain both with acid and with basic dyes. Taken in this sense we cannot see why the fine oxyphile granules of the rabbit were termed amphophile, since so far as we know they fail to stain with any basic dyes. On the contrary, we find them to possess an oxyphile reaction much above that of the similar granules of the other mammals investigated by us, and we are thus led to regard them as especially interesting. Thus the fine oxyphile granules of man have an oxyphile reaction which is slightly less intense than that shown by the red corpuscles, but the fine oxyphile granules of the rabbit possess a reaction considerably above this level, and often verging closely upon that shown by the coarse oxyphile granules of the rat.[1] The red corpuscles in all cases fail to stain with eosine in glycerine solution or in alcoholic solution when the alcohol is above 80 per cent., or with aqueous hæmatoxyline or sodium sulphindigotate; but the fine oxyphile granules of the rabbit stain deeply with eosine in glycerine solution, and in alcoholic solution when the alcohol is not above 90 per cent. They also colour with aqueous solutions of sodium sulphindigotate and hæmatoxyline though very slowly in the last case. The fine oxyphile granules of man, rat, mouse and guinea-pig do not stain with eosine in glycerine solution, nor in alcoholic solution when the percentage of alcohol is above 75.

Corresponding with the higher oxyphile reaction we find that the fine oxyphile granules of the rabbit have a higher refractive index than have the similar granules in the other animals.

The oxyphile reaction and refractive index of the fine oxyphile granules of the rabbit may be very readily raised above the normal by the presence

[1] Cf. Sherrington, op. cit. pp. 186–9.

in the blood of minimal amounts of microbic poison, in other words, by what are commonly known as immunising doses. The rabbit is peculiar in the readiness with which this change may be induced. No confusion of the two types of oxyphile cells, however, follows this change, and we only allude to the fact in passing because it might, and possibly has, led to a certain amount of confusion, especially when we bear in mind that so many apparently healthy rabbits have cocci and bacilli present in the cells and fluid of the peritoneal cavity.

Distribution of the oxyphile cells. In the animals examined the coarsely granular oxyphile cell is abundant in the fluid occupying the pleural, pericardial and peritoneal cavities—in other words, in the cœlomic fluid— in the interstices of areolar connective tissue throughout the body, and in peripheral lymph channels. There are also considerable numbers of these cells in bone marrow and in parts of the spleen and lymphatic glands.

The cœlomic fluid of mammals is very richly supplied with wandering cells, indeed it presents from this cause a cloudy turbid appearance. Of the cells present from 30 to 50 per cent. are coarsely granular oxyphile cells (figs. 7, 8, and 12 *a* and 14). In connective tissue again, the total number of these cells must be very great, for the thinnest film spread out on a slide will frequently show from 5 to 10 in a single field of a Zeiss D, Oc. 4. Thus this cell has an extraordinarily wide distribution outside the vascular system. In sharp contrast to this it occurs in very small numbers only in the blood, forming not more than 2–4 per cent. of the white corpuscles there. Wherever found the cells agree in all their histological features—in size and shape, in the features of the nucleus, and in the histochemical reaction of their granules. We must however except certain of the individuals occurring in bone marrow and in the spleen. Further, under appropriate stimuli the cells of any particular locality may proliferate rapidly. Therefore, we cannot at present point to any part of the body as the headquarters, or special place of origin of these coarsely granular oxyphile cells, and in the present state of our knowledge we can assign no special significance to the accumulation of these cells in the connective tissue of lymphatic glands of the spleen, and of bone marrow.

The relation of the coarsely granular oxyphile cells to the lymphatic system, how far for instance the extra-vascular cells, under normal conditions, find their way by its aid into the blood, is very obscure. For the present we will content ourselves with pointing out that they appear to be always more abundant in the lymph-capillaries and vessels on the far side of lymphatic glands than in the thoracic duct.

The finely granular oxyphile cell has a very limited and precise distribution, for under normal conditions it is entirely absent from extra-vascular spaces and occurs only in the blood, where it is by far the most numerous corpuscle forming from 20 to 70 per cent. of the total number of white corpuscles. The fluctuation in this percentage is probably due in the main to the great periodic variations in the number of lymphocytes present in the blood. Thus the effect of a meal is to cause a considerable increase in the number of lymphocytes in blood and, therefore, a fall in the share of the total white corpuscles due to finely granular cells. If this disturbing factor be eliminated and the percentage of the finely granular oxyphile cells be taken of the adult white corpuscles only, then this is found to be always very high; in man 75–90 per cent., and lowest in the rabbit 50–70 per cent.

Thus the group of oxyphile cells falls into two sub-groups defined by marked histological peculiarities, and by the fact that the cells of the one sub-group inhabit almost exclusively the extra-vascular body spaces, notably those derived from or in communication with the cœlomic spaces; while cells of the other sub-group inhabit solely the vascular channels.

The Basophile Cells.

The coarsely granular basophile cells (figs. 4, 7, 9, 10, 13 and 17) have been described by Ehrlich[1] under the name of "Mastzellen" and their existence in the cœlomic spaces and in the connective tissues was recognised by him and by Ranvier.[2] The cells of the connective tissues and those of the cœlomic fluid differ slightly in size and shape, and we will therefore deal with them separately.

The coarsely granular basophile cell found in the cœlomic fluid is, so far as we know, under conditions not endowed with the power of amœboid movement. It is a large cell having the form of a very much flattened sphere which might not inaptly be compared with a millstone whose edges had been very much rounded off (figs. 7 *b* and 9). The *nucleus* is rounded, occupies a central position and seems to be singularly devoid of chromatin, staining with difficulty. The cell-substance, clear and optically structureless, is charged with a very great number of exceedingly large spherules, the basophile "granules".

The coarsely granular basophile cell of the connective tissue (figs. 10 *b*, 13 *a*, *c*, *d* and 17) differs from the other forms of wandering cells in that it appears to be, at any rate in the normal animal, not only non-amœboid like its cœlomic brother but also stationary. It is therefore, except perhaps under certain conditions, not strictly a wandering cell. It is of a rounded or slightly polygonal shape and usually is more flattened, and so rather larger in its extreme dimensions, than the cell of the cœlomic fluid. In its other histological features, such as the nature of its granules, nucleus and cell-substance, it exactly resembles the cœlomic cells.

[1] Ehrlich, *loc. cit.* [2] Ranvier, *Comptes Rendus*, CX, 768.

The *granules* in both cases are the same, and in the rat and mouse are large spherules of an average diameter of rather more than $1\,\mu$. Like all forms of basophile granules their refractive index differs only slightly from that of the cell-substance, and they are therefore almost undistinguishable when unstained. Their large size enables one to study them in detail, and they are found to present certain features of interest. When microbes or microbic poison or "irritants" are present, these cells are frequently found with their granules so changed that they no longer stain, or stain imperfectly. In place of all the granules in the cell staining, all or some of the granules are now refringent spherules of the same size as the normal granules, but they either completely refuse to stain, or they stain in patches. The phenomenon very strongly suggests that these granules are not entirely composed of the basophile substance, but rather are composed of an unstaining groundwork with which the staining basophile material is associated but from which it may be removed.

The unstable or *explosive nature* of the coarsely granular basophile cells in certain animals is one of their most remarkable characters. In the rat and mouse perfect preparations of these cells may be very easily made, but in the guinea-pig and rabbit they can be preserved only with the most rapid fixation by heat or corrosive sublimate or absolute alcohol. In these animals the mere exposure of the cœlomic fluid to the air, or to contact with a coverslip for a few seconds, is sufficient to cause their complete disappearance. Cells characterised by great instability have been described elsewhere in *Astacus*[1] as the "explosive" cell, of that animal, and the basophile cells of the guinea-pig and rabbit might, with equal justice, be designated the explosive cells of those animals. Under the influence of certain chemical stimuli, as will be seen later, the basophile cells of the rat also become explosive. The instability of the coarsely granular cells of the guinea-pig and rabbit is so marked and constant a feature that Ranvier was unable to detect their presence in the cœlomic fluid of these animals, and came to the conclusion that they were absent.[2] Examples from the guinea-pig are shown in fig. 13, (*a*) being from the wall and (*b*) from the fluid of the peritoneal cavity, while (*d*) and (*c*) represent the appearance produced by the bursting of the cells in a film of connective tissue.

The finely granular basophile cell (figs. 1 *e* and *f*, 6 *b*, 12 *c*, 15 *e*), unlike the last, is a small cell, being usually the smallest of all the wandering cells. It is spherical in shape and possesses a characteristically trilobed nucleus. The cell-substance is clear and optically structureless, and usually contains an immense number of minute granules which often appear as mere points, and are characterised by staining a very opaque blue or purple colour with

[1] Hardy, *Journ. Physiol.* XIII, nos. 1 and 2.
[2] Ranvier, *Comptes Rendus*, CXII, 923 footnote.

methylene blue. Another very characteristic feature is the fact that the cell-substance colours a purple or pink tinge with the methylene blue solution.

These cells may be readily studied in man and the rabbit, in preparations of blood made with the aid of the methylene blue solution. In the rat, however, these cells appear to be very unstable, and the methylene blue solution fails to preserve them. They may, however, be found in films of blood fixed with the greatest rapidity with absolute alcohol. The blood is allowed to drip from a cut vessel on to a carefully cleaned coverslip—the excess blood is shaken off the coverslip which is then plunged into absolute alcohol. Staining may be best effected by immersing the film in 20 to 30 per cent. spirit to which a trace only of methylene blue has been added. In preparations of the blood of the guinea-pig made with the methylene blue solution, cells are found like the one shown in fig. 12 c. The granules are large and stain with the methylene blue, they are therefore, at least under certain conditions, basophile. It is this cell which appears in Table II as the finely granular basophile cell of the guinea-pig. We are not, however, at all certain that it is really identical with the very characteristic and readily detected cell of man and the rabbit. Our knowledge of the finely granular basophile cell, however, is so deficient that it is perhaps wisest for the present to class it as we have done.

Distribution of the basophile cells. The distribution of the coarsely granular basophile cells in the body resembles that of the coarsely granular oxyphile cells in so far that they occur in the cœlomic fluid and in the interstices of the connective tissue. But, unlike the latter, they are not merely rare but completely absent from the blood.[1] Moreover while they form only 10 per cent. (in the rat and mouse) of the cells of the cœlomic fluid, they are exceedingly numerous in connective tissue spaces, where they form sometimes an almost complete sheath for the lymph capillaries.

The finely granular basophile cell. These we have found solely in the blood, and even there they are present only in small numbers (1 to 5 per cent.). Their presence and the extent to which they are charged with granules appears to depend very closely on events occurring in the alimentary canal. Thus, in man six to eight hours after a meal, these cells can be detected in the blood only with difficulty, owing largely to the fact that they contain but few granules. About two hours after a meal, however, they can be found with ease and are then seen to form an appreciable percentage of the white corpuscles. Similarly, they form a very conspicuous feature being relatively speaking abundant, and very heavily laden with granules, in the

[1] Blood obtained by plunging a fine pipette into the heart rarely fails to contain one or two removed from the pericardial space. Blood from a vessel is always entirely free from them.

TABLE I

Showing percentage and size of the various forms of the wandering cells of the blood

	Oxyphile cells		Basophile cells		Hyaline cells	Lymphocytes
	Coarsely granular	Finely granular	Coarsely granular	Finely granular		
Man	10 to 11 μ	8 to 9 μ	Absent	7 μ	8·5 to 10 μ	6 μ
Rat	2 % 10 μ	45 % 7 to 8 μ	Absent	Absent	2 % 8 to 10 μ	50 % 6 μ
Rabbit	1 to 2 % 10 to 11·5 μ	20 to 30 % 8 to 9 μ	Absent	2 to 5 % 8 μ	2 to 6 % 8 to 11 μ	70 to 80 % 6 μ
Guinea-pig	2 to 3 % 10 μ	62 % 8 μ	Absent	0·7 % 8 μ	11 % 9·5 to 10 μ	24 % 6 μ

TABLE II

Showing percentage and size of the various forms of the wandering cells in the peritoneal fluid

	Oxyphile cells		Basophile cells		Hyaline cells	Lymphocytes
	Coarsely granular	Finely granular	Coarsely granular	Finely granular		
Man						
Rat	25 to 40 % 10 μ	Absent	5 to 10 % 18 μ	Absent	13 μ	65 to 80 % { 6·5 μ
Rabbit	10 μ	Absent		Absent	12 to 14 μ	6 μ
Guinea-pig	30 to 50 % 10·5 μ	Absent		Absent	13 μ	50 to 65 % 6 μ

Basophile cells in connective tissue of the rat 23 μ.

blood of a rabbit which has been kept unfed for ten hours and then given a full meal (cf. Table I). To say that these cells are found in the body only in very small numbers, being confined to the blood and scanty even there, is probably only equivalent to saying that we are at present very ignorant as to their history, distribution and significance. However, since we find this cell in the blood but do not find it either in the cœlomic fluid or in the interstitial spaces of the tissues (except perhaps in those of the mucous coat of the alimentary canal), we must until further facts are forthcoming regard it as the basophile cell of the blood.

Thus our present knowledge of the basophile cells leads us to class them in two great groups—a sub-group with coarse granules occurring only in the extra-vascular spaces, and a sub-group with fine granules occurring only within the vascular system, and these sub-groups correspond closely in structural characteristics and distribution with the sub-groups of the oxyphile cells. At the same time we regard the grouping together of the finely granular basophile cell which is so scarce and the finely granular oxyphile cell which is so abundant in blood as blood cells to be based as much upon ignorance as upon knowledge.

The Hyaline Cell. Figs. 1, 2, 3, 6, 7 *c*, 8 and 12 *b*, and 15 *c*.

This is usually of about the same size as the coarsely granular oxyphile cell and rather larger than the finely granular cell. When at rest it is a spherical cell but, owing to its pseudopodial activity, it mostly appears as a somewhat tenuous body of very irregular shape. The nucleus is usually spherical, it is sometimes kidney-shaped probably as a result of the mode of killing the cell. A very fine nuclear network with large meshes spreads through the nucleus and, in most preparations, a nucleolus is present. The cell-substance is always free from discrete granules, it is not, however, very transparent but presents rather the appearance of ground glass.

Distribution in the body fluids. The hyaline cell occurs both in blood and in the extra-vascular spaces and, unlike the granular cells, we are unable to point to any histological differences except perhaps in point of size between the forms found in the blood and those found elsewhere. At the same time, this cell is rare in the blood (2 per cent.), while it is abundant in the cœlomic fluid where it forms about 50 to 70 per cent. of the cells, so that its distribution accords closely with that of the coarsely granular oxyphile cell.

We may here point out certain homologues of the hyaline with the other cells. The cell-substance of the hyaline cells differs, at first sight, from that of the granule bearing cells, in taking a faint diffuse coloration with basic dyes. The highest powers of the microscope, however, frequently resolve this cloudy and apparently continuous colour into a number of stained points (figs. 8 and 12 *b*); and the cell-substance of the dead hyaline cell is then seen to be composed of an optically structureless substance remarkably indifferent and resistant to all stains, embedding a staining substance dispersed throughout it in the form of a cloud of minute particles. We have had occasion again and again to define the cell-substance of the granule bearing cells as optically structureless, and it also resembles the basis of the cell-substance of the hyaline cells in its resistance to stains. Therefore, viewed from the standpoint of the morphology of the cell, the defined and specialised granules of the oxyphile and basophile cells are homologous with the amorphous "points" of staining matter in the hyaline cell. The structural differences are of degree and not of kind.

But the question, What are the homologues of the granules in the less specialised cells? can be carried further towards solution, if we interpret the structure of the cell-substance of a dead hyaline cell by what can be seen in living specimens; though, unfortunately, facts of the order of those which we are now considering can only be intelligibly expressed in terms of some hypothesis, the simpler the better.

A striking physical peculiarity of the cell-substance of a living hyaline cell is its opacity. The whole thickness of a resting cell is small, and when spread out in amœboid movement, it forms extremely thin films; yet, in spite of this, it hides or blurs to a marked extent the outlines of any object over which it may happen to lie. So great an opacity in so thin a colourless membrane can only be produced in two ways, (i) by the formation of refracting surfaces within the membrane, owing to the presence of substances having different refractive indices, and (ii) by the disposition of these surfaces in such a way that they are concave or convex, or, if plane, inclined at some angle other than a right angle to the rays of light. A simple structure of this order would be composed of two substances having different refractive indices, the one forming a continuous basis in which the other is embedded in the form of spheres or rounded masses. If these spheres or spheroids were sufficiently small and sufficiently near together, a very thin membrane having considerable translucency but slight transparency would be the result.

Following out this hypothesis we may suppose the cell-substance of the hyaline cell to be composed of a colourless basis, embedding multitudes of minute spaces or vacuoles filled with a substance, possibly more fluid, of a different refractive power, and we may suppose that the action, for instance of the alcoholic solution of methylene blue, is to precipitate the contents of these tiny spaces to a minute mass of stained matter. Further, following out this hypothesis, the specific granule may be regarded as the specialised homologue of the contents of one or of a multitude of these tiny vacuoles.

IMMATURE FORMS OF WANDERING CELLS PRESENT IN THE BODY FLUIDS

Three forms of immature cells are readily recognised in the various body fluids of a normal animal. These are as follows:

(i) A small round cell characterised by possessing a deeply staining horse-shoe shaped nucleus embedded in scanty cell-substance. The cell-substance is charged with minute granules which give the same reactions as the coarse oxyphile granules (fig. 7 a'), that is to say, using the ordinary terminology, they are undoubtedly eosinophile granules. The diameter of the cell in the rat is $6 \cdot 5\mu$. We have found it only in the cœlomic fluid, and from this fact and from its histological characters we may unhesitatingly regard it as the immature form of the coarsely granular oxyphile cell.

(ii) A round cell, somewhat larger than the last (diameter 8 to 9μ). It possesses a spherical nucleus and the cell-substance is charged with small basophile granules (figs. 7 b', 8 c). It is readily distinguished from the finely granular basophile cell of the blood by its nucleus, which is a simple sphere instead of being trilobed, and by its granules, which are larger, each being a distinct though small spherule. We have found this cell only in the cœlomic fluid and we may without hesitation regard it as the immature form of the coarsely granular basophile cell.

(iii) A small cell which, like the first described, is $6 \cdot 5\mu$ in diameter (rat) and possesses a deeply staining nucleus and scanty cell-substance (figs. 1 d, 5, 7 c', 15 d). The nucleus, however, is spherical and the cell-substance is free from granules. These cells have been called *lymphocytes*, from the fact that they are produced in lymphatic glands. They occur in both blood and cœlomic fluid, and in the lymph of the thoracic duct. Their number in blood varies in different animals and at different times, and is especially dependent on the digestive activity of the alimentary canal, a meal causing a considerable rise. The blood of the rabbit is characterised by containing them in especial abundance, after a full meal they may in this animal form 70 to 80 per cent. of the wandering cells present.[1] Similarly, in man, about two hours after a full meal they form as many as 30 per cent. of the cells. Thus active digestion produces a condition of the blood which may be called lymphocytosis. On the other hand, starvation decreases the number of these bodies relatively to the other cells. The blood of a rabbit deprived of food for 12 hours was found to contain under 20 per cent. of these cells. We have reason to think that lymphocytes are produced in lymphoid tissue no matter whether it be found in lymphatic glands, in the spleen, or elsewhere; and from their histological characters we might infer that they are the young forms of the hyaline cells; and we may advance this view with some

---

[1] Okintschitz, *Arch. f. exp. Path. u. Pharm.* (1893), vol. XXXI.

approach to certainty, for we find in cœlomic fluid and in blood, as Max Schultze showed, not only lymphocytes, but also all stages of increase in size and relative abundance of cell-substance until they merge into undoubted hyaline cells. Beyond this, however, we cannot go, and the further question whether any of the lymphocytes are also the young form of other wandering cells is at present so far from solution that we scarcely have facts at our disposal sufficient to warrant us in discussing the probable answer. Ouskoff,[1] as a result of the study of the histology of blood, arrived at the conclusion that these cells developed into the finely granular oxyphile cells. As the original paper is in Russian we are ignorant of the facts on which he bases this conclusion. Certain observations, however, appear at first sight to discredit this view. Briefly they amount to this, if by irritants we produce a local increase in the number of the leucocytes in, for instance, a lymphatic gland, the result always is the development of a very large number of actively amœboid and phagocytic hyaline cells. The finely granular oxyphile cells are invariably absent, unless the lesion is excessive and involves the blood vessels. Muir[2] also points out that in a form of the disease leucocythæmia which is characterised by a large increase in the number of lymphocytes and hyaline cells in the blood there is not only no increase in the number of the "polynuclear leucocytes" (finely granular oxyphile cells) but very frequently an absolute diminution.

COMPARATIVE MORPHOLOGY

Under this heading but little can be said that can have any pretence to finality, but certain broad conclusions may be pointed out. The great feature of the structure and arrangement of the sporadic mesoblast of the Mammalia is the comparatively sharp division of the structure into two portions, one of which mainly inhabits the blood spaces, while the other is gathered into the cœlomic spaces, and distributed through the body in the interstitial spaces of the connective tissues. Each of these great divisions comprises three classes of cells—an oxyphile cell, a basophile cell, and a hyaline or granule free cell. This feature, namely, the differentiation of the tissue into two distinct and complete portions is, if the frog may be regarded as typical, absent in the similar tissue of the lowest vertebrates, for in the frog we find that the tissue comprises only three forms of cells, an oxyphile cell, a basophile cell, and a hyaline cell, which are found indifferently inhabiting the vascular and extra-vascular spaces.

The more special question of the origin and morphological relation of the several classes of cells which make up the sporadic mesoblast of the

---

[1] *Le Sang comme tissu*, St Pétersbourg, 1890 (in Russian), quoted by Metschnikoff in *L'Inflammation*.
[2] *Journ. Path.* (1892), I, 123.

Mammalia is still obscure. Here again, however, certain though partial conclusions may be drawn. We may assume that the three cells of the extra-vascular, or, as it might conveniently be called, the cœlomic portion, ought to be considered as being entirely distinct from one another in life-history and function, until convincing evidence to the contrary is forthcoming. It is difficult to see how the very peculiar and very specialised non-amœboid basophile cell can be derived from either of the other much smaller forms. And the real and permanent distinctness of the coarsely granular oxyphile cells and the hyaline cells is also rendered probable on three grounds: firstly, the presence in the cœlomic fluid of the young form of each; secondly, the difficulty of conceiving how the hyaline cell can develop into the oxyphile cell since the former is larger and has different functions from the latter; and, thirdly, the difficulty of conceiving the converse change, namely, the passage from a complex glandular cell endowed with the power of storing up granules definite in shape and in their histochemical characters to a comparatively unspecialised cell which has none of these attributes.[1] Yet, although such evidence points to the independence of each other of the several cœlomic cells, we are ignorant of the relations of these cells to those of the blood. We cannot, for instance, determine whether the oxyphile cell of the cœlom may or may not change into that of the blood or vice versa, and we are also largely ignorant of the special seat of origin, if they have any, of these several classes of cells, and whether even if they have entirely different life-histories they may develop from a common type of young cell.

The facts of structure and distribution of the cells composing the sporadic mesoblast of the higher Mammalia may be expressed in schematic form as follows:

	I. Oxyphile	II. Basophile	III. Hyaline
Division I Hæmal cells; character-ised by being relatively small cells and having fine specific granules	Nucleus branched; specific granules small, with rela-tively feeble oxyphile re-action. (The coarsely gra-nular cells also occur to a limited extent in the hæmal system)	Nucleus lobed; specific granules very small	Nucleus round; no specific gra-nules
Division II Cells of cœlomic and inter-stitial spaces. Charac-teristic, large size of cells and granules	Nucleus crescentic; specific granules large with intense oxyphile reaction	Nucleus round; granules very large	Nucleus round; no specific gra-nules

[1] Cf. Sherrington, *op. cit.* p. 203.

PART II. ON CERTAIN ACTIVITIES OF THE CŒLOMIC CELLS

In an earlier paper on the phenomena displayed by the wandering cells of the frog when in the presence of microbes and microbic poisons,[1] we stated at the outset that our researches demonstrated a disparity of function between the different classes of wandering cells comparable to the disparity in form. In what follows an extreme disparity of function will be shown to exist between the varieties of the cœlomic cells of Mammalia, and, as in the case of frog, we would again point to these marked differences of function as affording the strongest reasons for believing that the different forms of wandering cells have each a distinct and separate physiological significance.

LEUCOCYTOSIS AND CHEMIOTAXIS

By the term leucocytosis is meant an increase in the number of wandering cells present in any defined area, or in the body as a whole. We may therefore speak of local leucocytosis or of general leucocytosis. Leucocytosis, whether local or general, may be referred to two possible causes—to an excessive proliferation of the wandering cells already present in the area affected, or to an excessive immigration, usually from the blood. It is obvious from what we said in Part I concerning the distribution of the wandering cells, that a cell accumulation produced by local proliferation must consist solely of cœlomic wandering cells, whereas, if the lesion involves the blood capillaries the boundaries between the vascular and extra-vascular cells will be obliterated and a cell accumulation of mixed character will result. On the other hand a leucocytosis which was limited to the blood would affect solely the hæmal system of wandering cells.

The real distinctness of the two forms of leucocytosis, hæmal leucocytosis and cœlomic leucocytosis as they might for convenience be called, may be demonstrated either by producing very localised cell accumulations in connective tissue, in chambers or fine tubes placed under the skin, or in a cœlomic cavity; or by determining what may be called the order of arrival of the various cells by observing from time to time the character of the cells in any leucocytic focus.

If for instance Ziegler's chambers or capillary tubes which have been filled with bacilli, or their products, or some "irritant" such as nitrate of silver and turpentine are placed under the skin or in the peritoneal cavity and are allowed to remain there for periods up to 24 hours, they will be found to contain a multitude of cells solely of the cœlomic type. If however the irritant is situated in such a position that it appeals to the blood vessels

---

[1] Kanthack and Hardy, *Proc. Roy. Soc.* LII, 267 and *Phil. Trans.* 1894.

of a vascular membrane rather than to the cells of the connective tissue spaces, then the cells will be those of the hæmal system. Even in these cases however we usually find in the earliest stages a preponderance of the coarsely granular oxyphile cells. In cutaneous blistering for instance the irritant appeals immediately to the blood vessels of the dermis; and in over thirty blisters produced on ourselves or on others with the liquor epipasticus we found that the cells present were mainly of the blood type, though the coarsely granular oxyphile cell was always more abundant relatively to the others than in blood.

Experiments of this nature bring to light differences between the different forms of wandering cells as regards their rate of accumulation, or, as it is generally expressed, chemiotaxis. Thus the first cell attracted by micro-organism or irritants is always the coarsely granular oxyphile cell. Indeed in the majority of the tube or chamber experiments although many thousands of cells had crowded in they were almost all of this type.

If we ask ourselves the question where do these cells come from we are able to answer positively, in the case of chambers placed under the skin, that they have their immediate origin in the surrounding connective tissue. Fig. 19 shows the striking appearance presented by a film of connective tissue taken from the immediate neighbourhood of a Ziegler's chamber filled with a dilute broth culture of the cholera vibrio after 7½ hours' sojourn under the ventral skin of a guinea-pig. The chamber was found to contain multitudes of the coarsely granular oxyphile cells, and a comparison of fig. 19 with fig. 20 will show how strikingly abundant were these cells in this leucocytic focus as compared with the normal connective tissue.

The experiments which we have performed bearing on the subject-matter of this section may be divided into four sets. These are: (1) the introduction of chambers or tubes containing irritants or cultures of micro-organisms into the peritoneal cavity; (2) the introduction of similarly prepared chambers under the skin; (3) the introduction of copper into the anterior chamber of the eye; and (4) the production of skin blisters by an irritant (liquor epipasticus). The operations necessary in the first three cases were always performed under anæsthetics, and with antiseptic precautions.

1. *Tubes or chambers in the peritoneal cavity.* Fine capillary tubes were used and were filled with very dilute nitrate of silver, or turpentine, or with the diluted filtrate from a broth culture of pyocyanine which had been passed through a Chamberland's filter. The tubes were placed in the peritoneal cavity of rabbits and in 6 to 18 hours the animals were killed and the contents of the tubes converted into films. In all cases the cells were almost without exception of the coarsely granular oxyphile type and were present in great numbers.

The chambers used were those known as Ziegler's chambers. They were made in the usual way by cementing together two coverslips separated by a circular strip of tinfoil. Two small openings situated at opposite points were made and through these, by means of capillary attraction, the chambers were filled with broth cultures of bacilli which were not more than one to two days old and had been considerably diluted with fresh sterile broth.

The chambers were inserted through a slit about $\frac{3}{4}$ in. long made in the linea alba; the wound was then sewn up and dressings applied. They were allowed to remain in the peritoneal cavity for 2, $2\frac{1}{2}$ or 7 hours. The animal was then killed and the chambers were at once removed and the cells on the external surface and those in the interior were examined at once and without the addition of any reagent. In this way the cells were seen to perform amœboid movements, and to be attacking the bacilli (fig. 21). The chambers were then split open and the cells adherent to the external surface and those in the interior were examined with the aid of the methylene blue solution, and as film preparations.

The statements made as to the relative abundance of the different forms of corpuscles are founded upon countings made with fresh unopened chambers and with stained preparations.

Cultures of *Bacillus ramosus*, *B. anthracis*, and of the comma bacillus were used and the experiments were performed on rabbits and guinea-pigs.

The perforation of the abdominal wall led in all cases to a certain amount of bleeding into the peritoneal cavity, and therefore to the presence in the peritoneal fluid of red corpuscles and of a variable number of the finely granular oxyphile cells, in addition to the normal cells of the peritoneal fluid, namely, the coarsely granular oxyphile cell, the coarsely granular basophile cell, the hyaline cell, and lymphocyte.

After their sojourn in the animal the chambers were in all cases found to contain red corpuscles and in some cases also a few flat epithelial cells with large round flattened nuclei were present. Doubtless these had been detached from the peritoneal membrane by the friction of the edges of the chambers. There were present, in addition to these accidentally introduced bodies, intrusive wandering cells. These, unlike the red corpuscles and epithelial scales which were distributed irregularly over the interior of the chambers, were massed at the openings and spread thence fanwise through the interior; both as isolated cells and as masses aggregated round chains of bacilli.

The cells adherent to the external surface of the chambers included the normal cells of the peritoneal cavity together with often a striking number of the finely granular oxyphile cells.

So far we have set down the phenomena common to all the experiments; differences however appear according to the nature of the bacillus employed when we turn to the number and character of the wandering cells present in the interior of the chambers.

As in the case of the tube experiments the different forms of wandering cells do not invade the chambers indifferently in numbers corresponding to their relative abundance in the external fluid, and if we consider simply the order of arrival of the different forms we find, again as in the tube experiments, that in all cases the coarsely granular oxyphile cells are the first to make their way in and the first to attack the bacilli.

If however we consider the rate of arrival of the different forms, then a striking difference is found between chambers filled with cultures of the virulent forms *B. anthracis*, and the comma bacillus, as compared with those occupied by culture of the non-pathogenic *B. ramosus*, for in the former case the wandering cells other than the coarsely granular oxyphile cells do not succeed in invading the chamber to any marked extent even in 7 hours, the longest period over which our experiments extended, in the latter case the hyaline cells are found to have invaded the chamber in enormous numbers in as short a period as $2\frac{1}{2}$ hours.

A further difference between virulent and non-virulent cultures appears when we consider the total number of cells present in the chamber. For instance, comparing the experiments in which the chambers were allowed to remain in the peritoneal cavity for two hours, we find that in the case of *B. ramosus* the number of cells present can only be expressed by the word "countless"; whereas in the case of the virulent forms the total number though very large might have been counted with the aid of extreme patience.

The examination of the contents of the chambers occupied by cultures of virulent and non-virulent bacilli convinced us of the very unsatisfactory nature of the evidence which has been adduced to prove in migratory cells what is known as "negative chemiotaxis".

In our experiments we compare the influence on the cells of cultures of a non-pathogenic bacillus[1] with that of cultures of bacilli which are rapidly fatal to the animals employed, and yet we find that the phenomena observed differ only in degree, for in both cases the cells are attracted to and invade the area occupied by the bacillus, and in both cases the order in which the different cells make their way in is the same. In both cases therefore positive chemiotaxis is seen.

Further, though, as we have pointed out, a very marked difference is found in the rate of accumulation of the cells in the chambers, yet it is by no means certain that this can be referred to an equally marked difference in the rate of arrival or rate of attraction of the cells. An examination,

---

[1] The injection of the whole of a large and thick broth culture of *B. ramosus* into the peritoneal cavity of a rabbit does not appear to cause the animal any serious inconvenience.

especially in the fresh condition of chambers which have been filled with cultures of virulent bacilli, shows that very rapid destruction and disintegration of cells is in process in the interior; all conditions from cells with imperfect granulation to those which have become reduced to a thin bladder-like vesicle surrounding the swollen and distorted remains of the nucleus are to be seen in especial abundance in the neighbourhood of heaps of bacilli. Until we know something of the rate of destruction of the cells, a certain amount of caution must be exercised in referring the paucity of cells present to the action of negative chemiotaxis.

2. *Ziegler's chambers placed under the skin.* In order to insert the chambers a small slit was made in the loose skin on the ventral surface and then the skin was separated from the subjacent muscle on either side of the slit with a blunt instrument. In this way two spaces were hollowed out of the loose connective tissue and a chamber was placed in each. A slight amount of bleeding into the spaces always occurred. Dilute broth cultures of *B. ramosus* and the comma bacillus were used to fill the chambers, which were then allowed to remain under the skin for from $7\frac{1}{2}$ to $9\frac{1}{2}$ hours.

It is unnecessary to describe in detail the contents of the chambers when removed, since what we have said of similar chambers placed in the peritoneal cavity holds for these experiments. We will therefore briefly state that after 7 to 9 hours' sojourn the chambers were found to contain in the case of the comma bacillus considerable numbers, in the case of *B. ramosus* enormous numbers of the coarsely granular oxyphile cells. On the external surface of the chambers and in the serous exudation which distended the cavities in which they lay were free coarsely granular oxyphile cells and hyaline cells together with a number of cells from the blood—but the total number of cells free in the fluid was not great. The most striking change was found in the connective tissue which formed the walls of the space in which the chambers were placed, for this was packed with an immense number of the coarsely granular oxyphile cells together with a smaller number of the coarsely granular basophile cells. A teased-out film of areolar connective tissue (fig. 20) taken from a normal animal is seen to consist, so far as bulk is concerned, mainly of the fibrous matrix. Fixed in spaces in this matrix are the connective tissue corpuscles and a limited number of the coarsely granular basophile cells, and a few coarsely granular oxyphile cells are also always present. In a similarly prepared film taken from the wall of the space which has lodged a Ziegler's chamber for 7 to 9 hours, the matrix and its connective corpuscles are seen as before, and no evidence of change in either element is observable in these preparations. But these now form only a small proportion of the bulk of the tissue, for adherent to the lamellæ or bands of connective tissue are masses of wandering cells in such countless numbers as to render the thinnest film that can be teased out almost opaque when stained. It would be difficult to exaggerate the physiological significance or the extent of the change produced. In 7 to 9 hours after the chambers have been inserted the coarsely granular oxyphile cells and the coarsely granular basophile cells have accumulated to such an extraordinary extent that there are thousands where there was previously but a single cell, and the walls of the cavity are now no longer mainly inert, dead matrix, but are formed by a continuous mass of packed living cells which constitute a barrier between the micro-organisms and their poisons and the rest of the body possessed of the profound chemical potentialities of protoplasm, which we must therefore look upon as being endowed with powers of destruction and powers of construction—of the destruction of the poisons of the bacillus, and of the construction of bactericidal substances.

The first worker to suggest the connection of the oxyphile granulation with the conflict with microbes was Hankin. He succeeded in isolating *in vitro* a bactericidal substance from the lymphatic glands of cats and dogs,[1] and this discovery led him to the view that the resistance of animals to the growth of microbes is due to the production in their bodies of substances possessing bactericidal properties to which he gave the name of defensive proteids.[2] To these bodies Buchner, who was working on the same lines, gave the general name of "alexines".[3] In seeking the origin of alexines in the body Hankin was led to the conclusion that they are formed by the wandering cells. He found support of this view (1) by showing in conjunction with Kanthack that outside the animal body during fever a rise in the bactericidal power of the blood occurs *pari passu* with the increase in the number of leucocytes present,[4] and (2) by showing that this increased bactericidal power is apparently correlated with a discharge of oxyphile granules, coarse or fine, into the plasma.[5] In this way Hankin associates the bactericidal powers of the body in the first place with the wandering cells in general, and in the second place with the fine and coarse oxyphile granulation in particular. Our observations on the accumulation of coarsely granular oxyphile cells at a centre of infection, and on the discharge of the oxyphile granules during the conflict of particular cells with the microbes, may be said to give further support to Hankin's view.

Nothing is more remarkable than the fixed habit of the cells. As we have already noticed the serous exudation contains comparatively speaking only

---

[1] "A Bacteria-killing Globulin", *Proc. Roy. Soc.* (1890), vol. XLVIII.

[2] "The Conflict between the Organism and the Microbe", *Brit. Med. Journ.* July, 1890.

[3] *Münchener med. Woch.* (1891), No. 25.

[4] Hankin and Kanthack, "On the Fever produced by the Injection of sterilized Vibrio Metschnikovi cultures into Rabbits", *Proc. Camb. Phil. Soc.* vol. VII, pt. VI.

[5] *Centralblatt f. Bakt. u. Par.* (1892), XII, 22 and 23, and also (1893), XIV, 25.

a very few free corpuscles. The application of a coverslip to the connective tissue fails to dislodge the cells, and even the spreading out of a portion of the fresh tissue on a coverslip to form a film frees only a few. Yet an examination of such teased films with high powers convinces one that the cells are not in the lamellæ of connective tissue but are closely attached in layers two, three or several cells deep to their surfaces.

There is yet another inference which may be drawn from the experiments with Ziegler's chambers. In all cases, both when the chambers were placed in the peritoneal cavity and when they were placed in cavities in the subcutaneous connective tissue, the surrounding fluid contained in addition to the cells of the cœlomic portion of the sporadic mesoblast a very appreciable number of finely granular oxyphile cells derived from the blood which had found their way there together with red corpuscles and other elements owing to slight bleeding. In our experiments we always found that the finely granular oxyphile cells either were not found within the chamber or were present in much smaller relative numbers than in the fluid outside.

3. *The introduction of copper into the anterior chamber of the eye of a rabbit.* A small strip of copper foil carefully sterilised was used for this experiment. The operation was carried out on a deeply anæsthetised animal, the cornea being moreover rendered anæsthetic with sterile cocaine, and with all possible antiseptic precautions. The wound healed rapidly and completely. The aqueous humour however became charged with cells and the copper was dissolved. After 2 or 3 days the animal was killed and films were prepared from the aqueous humour. These showed a large number of cells consisting practically of the coarsely granular type (about 95 per cent.).

4. *Blisters.* These were produced on the forearm on ourselves and on others. In the fully formed blister the fluid contained both forms of oxyphile cells, and hyaline cells. In all cases the majority of the cells present were of the finely granular oxyphile type, that is to say the leucocytosis is chiefly hæmal in character.

The proportion of coarsely granular oxyphile cells varied between 6 to 45 per cent., some individuals producing blisters with a large number of these cells, while in others they were always scanty.

In certain cases the blister fluid was examined from time to time, and a variation in the proportion of the cells at different periods was detected. Only a few cases were followed in this way but they agreed in showing the presence of a larger relative number of coarsely granular oxyphile cells in the early stages.

ON THE EFFECT ON THE CELLS OF THE INTRODUCTION
OF CULTURES OF BACILLI INTO THE PERITONEAL
CAVITY AND INTO BLISTERS

In these experiments the bacilli used were either *B. anthracis*, *B. pyo-cyaneus*, or the comma bacillus, and the animals employed were rats or guinea-pigs. The amount injected was usually small.

After a certain interval the animals were killed, the peritoneal cavity carefully opened and the fluid from various regions examined as fluid preparations or as films stained in various ways. The cells were always counted with a $\frac{1}{12}$th oil immersion and Oc. 4. The cells were entered in different columns according to their nature, the extent to which they were charged with granules and whether they were or were not attacking or ingesting bacilli. No possible means suggested itself to us of measuring the total amount of fluid in the peritoneal cavity, and we therefore judged it useless to measure the total number of cells per cubic millimetre. Thus the percentages obtained convey only a limited information.

The first and instantaneous effect of the introduction of cultures of bacilli into the peritoneal cavity is the disintegration of a considerable number of the cells. This effect is of constant occurrence but varies largely in extent. It may probably be referred largely not to the immediate influence of the bacilli, but to the action of the substances injected with them, for the more carefully the bacilli were freed before injection from these products the less was the destruction produced. Therefore the variations in the extent of the destruction in different cases was not found to depend at any rate entirely upon the species of the bacillus injected, but upon the conditions of growth of the bacillus, the nature of the medium used as the vehicle for their injection, and upon the dose.

As these variations are not germane to the main issues of this paper, no attempt will be made to analyse more fully their causes; it is sufficient to mention the fact that an immediate destruction of cells follows the injection of bacilli and their products, and that this destruction involves mainly the basophile and oxyphile cells, so that, if it is extensive free basophile granules and fragments of oxyphile cells may be detected in considerable numbers in the plasma. This disintegration of cells must profoundly alter the chemical constitution of the plasma and therefore may play an important part in the struggle with the bacilli.

In an incredibly short time after the introduction of the bacilli they are attacked by the coarsely granular oxyphile cells. The attack consists in the application of the cells to the bacilli, and it entails the using up of the substances stored as the oxyphile granules.

The rapidity with which the coarsely granular oxyphile cells attack the bacilli may be gathered from the fact that in specimens prepared 5 min. after the introduction of anthrax bacilli into a rat no less than 40 per cent. of these cells were already applied to bacilli (Tables III and IV).

TABLE III

Showing the percentage of each kind of cell attacking bacilli

	Coarsely granular oxyphile cells	Hyaline cells
Rat		
Anthrax 5 min.	39	1
,, 10 ,,	60	27
,, 15 ,,	40	85
,, 30 ,,	44	96
Pyocyaneus 15 ,,	63	65
,, 2 hours	70	85
,, 2½ ,,	27	62
Guinea-pig		
Cholera 30 min.	93	100

TABLE IV

Showing the percentage of each form of cell found to be attacking bacilli or ingesting indian ink where bacilli and ink were injected together

Rat	Attacking bacilli		Cells containing indian ink	
	Coarsely granular oxyphile cells	Hyaline cells	Coarsely granular oxyphile cells	Hyaline cells
Pyocyaneus 15 min.	63	65	0	65
,, 2 hours	70	85	0	80 to 90

At a slightly later period the hyaline cells begin to ingest the bacilli; thus 10 min. after the injection of anthrax bacilli into a rat 60 per cent. of the coarsely granular oxyphile cells were attacking the bacilli and 27 per cent. of the hyaline cells had commenced the ingestive act, while in 15 min. 85 per cent. of the hyaline cells were ingesting, and in 30 min. 96 per cent.

In an earlier communication on the behaviour of the cells of the frog towards micro-organisms we pointed out the difference between the attack of the oxyphile cell, which we described as the application of the cell to the bacillus and the excretion of the granule substance, and the true ingestive or phagocyte activity of the hyaline cells. Similar differences occur between the mode of action of the coarsely granular oxyphile cells and the hyaline cells of mammals.

We have been able to observe the details of the process by watching living cells at work either in hanging drops of the fluid taken from blisters which had been inoculated with a small quantity of a fresh broth culture of *B. anthracis* or *B. ramosus* and were observed on a warm stage, or by examining on a warm stage samples of the fluid of blisters into which a culture of *B. ramosus* had been injected (figs. 22, 23 human, 21 guinea-pig, 27 rat). In this way the coarsely granular oxyphile cells were seen to apply themselves to a chain of bacilli and to extend along it. The granules then travelled, usually in groups, to those portions of the cell in immediate contact with the bacilli and there they were seen to diminish in size to mere points and ultimately to disappear. The rate of loss of granular substance is very rapid as compared with the rate of loss in the gland cells of salivary glands or of the pancreas during the act of secretion, for an appreciable diminution may be observed in a quarter of an hour.

In two experiments an attempt to obtain a numerical expression of the loss of oxyphile granules was made. In the first case a fully formed blister was tapped and films made from the fluid. These films were labelled normal blister fluid. Then a few drops of the sediment from a fresh (20 hours old) broth culture of *B. ramosus* was injected into the blister. 20 min. afterwards film preparations were made of the contents of the blister, and also hanging drops which were kept for 2 hours on the warm stage. In the normal blister fluid the coarsely granular oxyphile cells were found to form 5 per cent., after 20 min. this proportion was not materially changed, but after 2 hours the percentage had dropped to 3·9. In the second experiment anthrax bacilli were used and control drops were kept on the warm stage and at the temperature of the room. A number of hanging drops were made and divided in four sets, two sets were inoculated with anthrax bacilli and two were kept as controls. A set inoculated with anthrax bacilli and a control set were kept on warm stages for one hour, the rest were kept at the temperature of the room for 2½ hours. It was found (1) that the control drops on the warm stage contained at the close of the hour 13 per cent. of coarsely granular oxyphile cells, while those inoculated with anthrax contained only 7·6–8·5 per cent., (2) that the controls kept at the temperature of the room (15° C.) contained at the end of the 2½ hours 12 per cent. of coarsely granular oxyphile cells, while those inoculated with anthrax contained 6 per cent. In both experiments coarsely granular cells were seen under the microscope to attack chains of bacilli and to suffer diminution of granulation.[1]

The hyaline cells unlike the coarsely granular oxyphile cells manifest a true phagocytosis, *i.e.* the intussusception of discrete particles into their substance, and their solution in digestive vacuoles (fig. 24 human).

The difference between the activities of these two kinds of cells finds in a

[1] Sherrington (*op. cit.* pp. 201–203) discusses the diminution or loss of granulation at length and has not been able to convince himself that any change in size or number of the oxyphile granules occurs in the coarsely granular hæmic cells of cats in which he set up an acute local inflammation. His method however differs so much from our own that it is impossible to compare the two sets of experiments.

certain sense numerical expression in the tabular statement of the results obtained by injecting into the peritoneal cavity cultures of bacilli to which indian ink had been added. Table IV shows that the oxyphile cells neither ingest nor do they attach to themselves the particles of the ink but do swiftly attack the bacilli, while on the other hand the hyaline cells rapidly ingest the ink particles and the bacilli. All observed cases of what appeared as mere contact between hyaline cells and bacilli are included as cases of ingestion, and the table therefore takes no account of the fact that the hyaline cells were more heavily laden with ink particles than with bacilli.

The destruction and removal of the bacilli therefore is due to the different activities of the two kinds of cells, the coarsely granular oxyphile cell and the hyaline cell.

The attack of the oxyphile cells on the bacilli must result very frequently in the destruction of the cell. This occurs sometimes to such a marked extent that in film preparation from $\frac{1}{2}$ to 2 hours' experiments a large number of the bacilli are seen to have attached to themselves heaps of still intact granules or amorphous masses of oxyphile substance, while the body of the cell has either broken away or disintegrated.

In a former paper[1] we described the formation of plasmodial masses of wandering cells in the course of the conflict with the micro-organisms. Such bodies are readily seen in hanging drops of blister fluid, in Ziegler's chambers, or in preparations of peritoneal fluid containing bacilli. Fig. 27 shows a striking case, a chain of anthrax being attacked by a coarsely granular oxyphile cell at one place, while at the same time it is being ingested by a phagocyte at another point. Fig. 25 shows an interesting stage, the oxyphile cells being massed in the centre of a number of hyaline cells.

A point which, although it does not bear on the subject of this portion of the paper, namely the activities of the cœlomic wandering cells, must be mentioned in order to complete the picture of the events occurring in the peritoneal cavity, is the extent to which the bacilli and their products present "induce" immigration of cells from the vascular system. The experiments with anthrax were all performed on the rat and did not endure above 30 min. In none of these was any immigration of hæmal cells detected. In the experiments with pyocyaneus on the rat, however, a large number of finely granular oxyphile cells and of red corpuscles were present in 2 and $2\frac{1}{2}$ hours, and a very few even could be detected in 15 min. The contrast is interesting when we remember that the rat is extremely resistant to anthrax and susceptible to infection by the bacillus pyocyaneus. The number of experiments however is much too small and they are not sufficiently contrasted to allow us to draw any conclusions from them.

[1] Kanthack and Hardy, *loc. cit.*

TABLE V

Cultures injected into peritoneal cavity. The table shows the percentage of the different forms of cells found in the peritoneal fluid at varying intervals after the introduction of the bacilli. The percentage in the normal animal is given for comparison.

	Oxyphile		Hyaline	Lymphocytes	Basophile
	Coarsely granular (Granules all gone. Nucleus characteristic)	Finely granular			
Rat					
Normal	25 to 40	—	{ 65 to 80 (combined)		5 to 10
Anthrax. 5 min. Broth culture centrifugalised and 0·2 c.c. of sediment injected	4·5	0	{ 89 (combined)		6
Anthrax. 15 min. Same dose	7·2	0	{ 83 (combined)		8
Anthrax. 10 min. Agar culture; bacilli washed off in normal salt solution. 0·2 c.c. injected	19·4	0	70	(no record)	10
Anthrax. 30 min. Agar culture washed off in broth. 0·2 c.c. injected	3·1	0	79	15	2·4
Anthrax. 24 hours. 1 c.c. of broth culture	10	45	45	29	0
Pyocyaneus. 15 min. Potato culture; bacilli washed off with distilled water. Indian ink added and 0·2 c.c. injected	23	0·9	{ 41 (combined)		—
Pyocyaneus. 2 hours. ditto.	6	70	{ 19 (combined)		3·7
Pyocyaneus. 2·5 hours. The same but broth used in place of water	1	79	{ 14 (combined)		1·9
Pyocyaneus. 45 min. Whole potato culture injected in 1·25 c.c. distilled water	Cells all destroyed except a few basophiles				
Guinea-pig					
Normal	30 to 50	—	{ 50 to 65 (combined)		Doubtful
*Cholera. 5 min. One-sixth agar culture in 0·2 c.c. distilled water	70	4	22	6	(slight bleeding into peritoneal cavity)
Cholera. 30 min. ditto.	13	1	58	28	—

* The guinea-pigs used had been previously immunised to cholera by Haffkine's method. This may account for the very large number of coarsely granular oxyphile cells present.

GENERAL RESULTS OF PART II

1. Two kinds of leucocytosis can be recognised. In the one the wandering cells are entirely or mainly of the cœlomic type, in the other they are entirely or mainly of the hæmal type.

2. In all cases investigated by us the first cells to accumulate at a leucocytic focus were oxyphile cells.

3. Of the two forms of oxyphile cells the coarsely granular form accumulates more quickly than the finely granular form.

4. When the conflict with the bacilli is watched in hanging drops of blister fluid, or in Ziegler's chambers, the coarsely granular oxyphile cells are seen to attack the bacilli and to suffer thereby a diminution of granulation.

5. The attack is very rapidly carried out, and is quickly followed by phagocytosis. The latter process, which is carried out by the hyaline cells, commences at a much earlier period than is usually supposed and is at its maximum in about 25 min. after the introduction of the bacilli.

6. Bacilli or their products in the cases examined by us were always found to attract the wandering cells even when the animal employed was not immune to the particular bacillus employed. At the same time there was destruction of cells at the focus of conflict. This destruction was very much greater in the case of pathogenic than in that of non-pathogenic bacilli. We have been led to think that a great deal of what is known as negative chemiotaxis is based upon those cases where the rate of destruction nearly or quite equals the rate of arrival of cells at a leucocytic focus.

The main interest of the facts set forth in the second part of this paper lies we believe in the marked difference which they show to exist between the activities displayed by the coarsely granular oxyphile cell and the hyaline cell. It may be that these cells change into one another.[1] But for our own part the facts at present known lead us to regard these cells as morphological units as distinct as are the striped and unstriped cells of muscle tissue.

[1] Cf. Sherrington, *op. cit.* p. 203.

EXPLANATION OF PLATE V

All the figures are to the same scale and drawn with Oc. 4, ob. $\frac{1}{12}$th hom. imm. with a camera lucida.

Figs. 1, 2, 3, 4 and 5 human.

Fig. 1. Cells of blood of healthy boy. Fluid preparation made with the methylene blue solution. (a) coarsely granular oxyphile cell, (b) finely granular oxyphile cell, (c) hyaline cell, (d) lymphocyte, (e) and (f) finely granular basophile cell.

Fig. 2. Blood from same boy. Film preparation exposed to a cold saturated solution of eosine in 50 per cent. spirit for about 5 sec., then to Lœffler's methylene blue. (a) coarsely granular oxyphile cell, (b) finely granular oxyphile cell, (c) hyaline cell.

Fig. 3. Blood from healthy adult. Film preparation stained with the "neutral" mixture described on p. 89. (a) coarsely granular oxyphile cell, (b) finely granular oxyphile cell, (c) hyaline cell.

Fig. 4. Coarse granular basophile cell, connective tissue.

Fig. 5. Lymphocyte to show nuclear network. Flemming's fluid. Hæmatoxyline.

Fig. 6. Rat. Cells from blood. Film preparation stained with glycerine eosine and Lœffler's methylene blue. (a) coarsely granular oxyphile cells, (b) finely granular oxyphile cell (note the complete absence of staining in the granules, cp. with fig. 11), (c) hyaline cell.

Fig. 7. Rat. Cells of peritoneal fluid. Film preparation stained with glycerine eosine and Lœffler's methylene blue. (a) adult and (a') young coarsely granular oxyphile cell, (b) adult and (b') young coarsely granular basophile cell, (c) adult and (c') young hyaline cell.

Fig. 8. Rat. Cells of peritoneal fluid. Fluid preparation made with the methylene blue solution. (a) coarsely granular oxyphile cell, (b) hyaline cell, (c) young form of coarsely granular basophile cell. The adult form of the coarsely granular basophile cell is omitted and

Fig. 9. The coarsely granular basophile cell of the mouse is given in its place. From a fluid preparation made with the methylene blue solution.

Fig. 10. Rat. Cells from a film of subcutaneous connective tissue, stained with glycerine eosine and Lœffler's methylene blue. (a) coarsely granular oxyphile cell, (b) coarsely granular basophile cell.

Fig. 11. Rat. Finely granular oxyphile cell from film of blood stained with the "neutral" mixture as described on p. 89.

Fig. 12. Guinea-pig. (a) and (b) coarsely granular oxyphile cell, and hyaline cell from a fluid preparation of peritoneal fluid made with the methylene blue solution; (c) finely granular oxyphile cell from fluid preparation of blood made with the methylene blue solution. To show the differences between the oxyphile cell of the peritoneal fluid and the finely granular oxyphile cell of the blood; (d) coarsely granular oxyphile cell from a film of peritoneal fluid stained with an aqueous solution of sodium sulphindigotate. A very much flattened cell was chosen for drawing in order to show clearly the individual granules and the clear unstained space which represents the nucleus.

Fig. 13. Guinea-pig. Coarsely granular basophile cell in various stages of disintegration. Film preparations stained with glycerine eosine and Lœffler's methylene blue: (a) from wall of peritoneal cavity, fixed with absolute alcohol; (b) from film of peritoneal fluid, fixed by heat; (c) and (d) from film of subcutaneous tissue, fixed by drying at temperature of room.

Fig. 14. Rabbit. Coarsely granular oxyphile cell from film of peritoneal fluid stained with glycerine eosine and Lœffler's methylene blue.

Fig. 15. Rabbit. Cells of blood shortly after a heavy meal. Fluid preparation made with the methylene blue solution. (a) coarsely granular oxyphile cell, (b) finely granular oxyphile cell, (c) hyaline cell, (d) lymphocyte, (e) finely granular basophile cell.

Fig. 16. Rabbit. Blood; fluid preparation made with the methylene blue solution. About 10 hours after a meal: (a) finely granular basophile cell, (b) lymphocyte.

Fig. 17. Rabbit. Coarsely granular basophile cell from subcutaneous connective tissue stained with the methylene blue solution.

Fig. 18. Rabbit. Finely granular basophile cell, inflamed area stained with the methylene blue solution.

Fig. 19. Film of subcutaneous connective tissue taken from the neighbourhood of a Ziegler's chamber filled with diluted broth culture of the comma bacillus 7 hours after introduction into the chamber. Guinea-pig. Eosine in 90 per cent. alcohol, Lœffler's methylene blue, Oc. 4, ob. D, cam. luc. A part of the preparation where the coarsely granular oxyphiles were relatively few in number was chosen for the drawing in order to show the connective tissue elements.

Fig. 20. Film of subcutaneous connective tissue from normal guinea-pig for comparison with fig. 19.

Fig. 21. Living cells attacking a chain of B. ramosus. Ziegler's chamber 2½ hours in peritoneal cavity of guinea-pig. The chamber was removed unopened to warm stage and examined with Oc. 4, ob. D.

Fig. 22. Hanging drop of blister fluid from forearm inoculated with B. ramosus. (a) coarsely granular oxyphile cell at rest, hanging drop at temperature of room. Warm water was then allowed to run into warm stage, cell became active and applied itself to bacilli—(b) and (c). Human.

Fig. 23. Blister fluid inoculated with B. ramosus. (a) chain of bacilli with three cells applied to it, Oc. 2, ob. A, (b) part of the same chain stained with the methylene blue solution, Oc. 4, ob. $\frac{1}{12}$th, cam. luc. Human.

Fig. 24. Hanging drop of blister fluid. Phagocyte with ingested fragments of B. anthracis, Oc. 4, ob. D, cam. luc. Human.

Fig. 25. Peritoneal fluid, rat. 15 min. after injection of B. pyocyaneus. Plasmodial mass of coarsely granular oxyphile cells with surrounding hyaline cells. Oc. 4, ob. $\frac{1}{12}$th, cam. luc.

Fig. 26. Same experiment as last. Single cell attacking bacilli. Oc. 4, ob. $\frac{1}{12}$th, cam. luc.

Fig. 27. Peritoneal fluid, rat. 10 min. after injection of B. anthracis. Fluid preparation stained with the methylene blue solution. Part of much distorted anthrax chain attacked by oxyphile cell and also being ingested by phagocyte.

Fig. 28. Peritoneal fluid, rat. 15 min. after injection of anthrax. Coarsely granular oxyphile cell which has discharged its granules. Fixed by heat. Oc. 10, ob. $\frac{1}{12}$, apochr. Powell and Lealand.

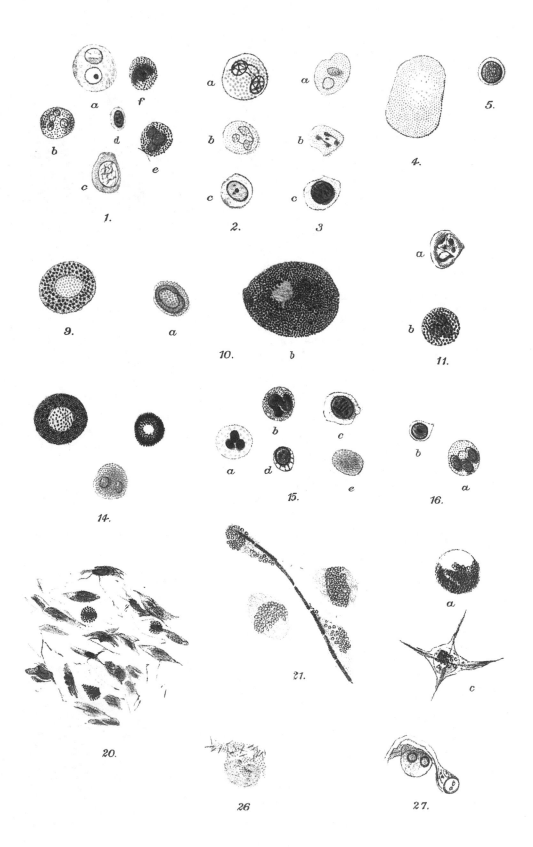

Plate V

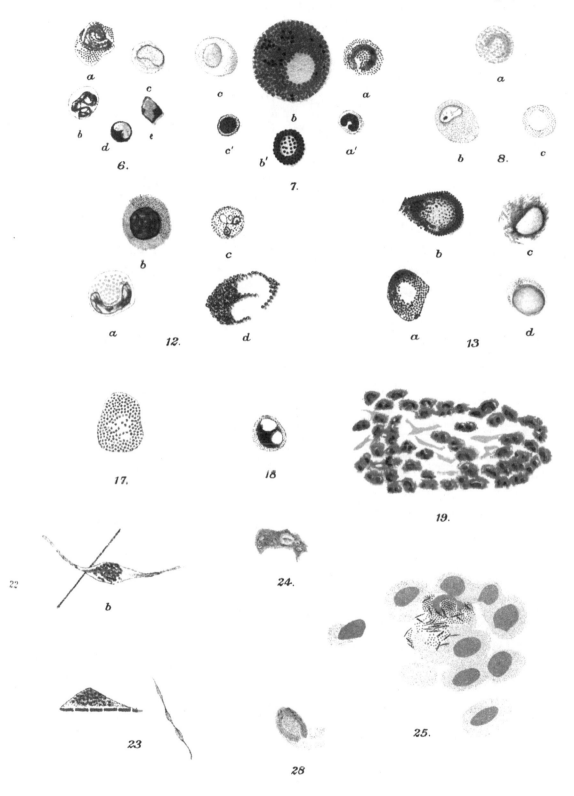

ON SOME HISTOLOGICAL FEATURES AND PHYSIOLOGICAL PROPERTIES OF THE POST-ŒSOPHAGEAL NERVE CORD OF THE CRUSTACEA

[*Phil. Trans.* (1894), B, CLXXXV, 83]

About two years ago, at the suggestion of Dr W. H. Gaskell, F.R.S., I undertook the examination of the minute anatomy of the nervous system of the Crustacea. Certain of the results of the investigation are incorporated in the following paper.

The nervous systems of *Branchippus*, of *Astacus*, and of the Zooea larva have been examined, but the following pages are limited, almost exclusively, to the two first mentioned.

I have to thank Dr Gaskell and Mr Langley for the many suggestions with which they have aided the work, and Prof. Foster, Secretary, Royal Society, for the kindly interest he has shown in my investigations from their commencement, an interest which is extended to all those who have the good fortune to assist him in his professional duties. I am indebted to Mr Bateson, of St John's College, for some carefully preserved specimens of *Branchippus* and *Artemia*.

PART I. THE POST-ŒSOPHAGEAL NERVE CORD OF *BRANCHIPPUS*

The central nervous system of *Branchippus* is of special interest as an example of the conditions found in the most primitive Crustacea. It consists of a well-developed supra-œsophageal ganglion, or brain, connected with a small dorsal median simple eye, and a pair of laterally placed compound eyes. This ganglion also innervates the first pair of antennæ. From the brain a pair of longitudinal ganglionated nerve cords pass backwards. These are connected with one another in each segment by a pair of transverse commissures, except in the most posterior part of the animal. The whole nervous system shows many primitive features, it is connected with the ectoderm in many places, shows little trace of fusion in the mid-ventral line, and has nerve cells very diffusely scattered about it. Generally speaking, both the brain and the nerve cords arising from it consist of a central fibrous core, the fibres of which form an exceedingly fine plexus. This is the "Punktsubstanz" of Leydig,[1] and the nerve cells lie wholly on its surface.

[1] Leydig, *Handbuch d. vergleich. Anatomie*, vol. I, Tübingen, 1864.

Posterior to the œsophagus there are sixteen well-marked ganglia. These are respectively one mandibular, two maxillary, eleven corresponding to the eleven foliaceous swimming appendages, and lastly, two closely approximated ganglia, situated posterior to the appendage-bearing region, and related to the genital organs. Posterior to these again, in the abdominal region, are two very slightly developed ganglia. The œsophagus is surrounded by a ring of nervous tissue which forms a single remarkable ganglion connected with the sensory surfaces of the lips and from it a pair of nerves pass to the second pair of antennæ. There are thus posterior to the brain nineteen ganglia, as follows:

1. The circumoral ganglion connected with the sensory surfaces round the mouth and the second pair of antennæ.
2. The mandibular ganglion.
3. The ganglion of the first pair of maxillæ.
4. The ganglion of the second pair of maxillæ.
5 to 15. The ganglia of the eleven pairs of swimming appendages.
16, 17. The genital ganglia.
18, 19. The abdominal ganglia.

The ganglia are separated from one another by portions of the cords absolutely unconnected by transverse commissural fibres.

In the *inter-ganglionic regions* each cord is rather rounded in transverse section (fig. 1),[1] and free from the ectoderm, but as the ganglia are approached a continually closer connection with the ectoderm obtains. In its free or inter-ganglionic part a delicate sheath with small deeply staining nuclei invests the cord. Passing from the middle region towards the ganglion above or below, nerve cells become more numerous, always immediately underlying the investing sheath. One striking fact thus becomes evident in this primitive nervous system, namely, that there is no sharp line of demarcation between ganglionic and inter-ganglionic, or strictly conducting regions. Further, the fibrous core in this latter region is seen in longitudinal sections to consist of longitudinally arranged fibres which stain well with carmine stains. Connecting these, however, in many cases are more delicate filaments which run obliquely, so that the whole, instead of being merely a bundle of parallel fibres, may be more exactly described as a plexus of which the longitudinal strands have been relatively exaggerated.

The shallow groove on the ventral surface of the animal and between the nerve cords deepens considerably just posterior to the mouth, and is there lined by elongated sense cells, and abundantly supplied with nerve filaments, thus forming a sensory pit between the mandibles. In the thoracic region the groove is very shallow in the inter-ganglionic regions,

---

[1] For figures see Plates VI–IX.

but deepens between the ganglionic enlargements. In this groove is a line of sense cells on either side of the mid-ventral line. They are especially prominent near the ganglia. We may, therefore, say that there exists on the ventral surface a pair of almost continuous sense lines, situated on the ridge bounding the ventral groove, and perfectly segmented in correspondence with the ganglionic segmentation of the cords. They meet just posterior to the mouth in a deep sense pit or groove (fig. 2), which again, in its turn, may be regarded as a backward continuation or thickening of the sense ring round the mouth. The arrangement of these ventral sense cells will be seen to have an important bearing on the disposition of the sensory elements of the cord. In the inter-ganglionic, as well as in the ganglionic regions, filaments pass from these ventral sense cells to the cells of the cord.

STRUCTURE OF A TYPICAL GANGLION

For purposes of description I will choose the ganglion related to the third pair of swimming appendages. It consists of a double thickening of each nerve cord, so that on either side are anterior and posterior swellings, each connected with its fellow of the opposite side by a transverse commissure. Seen in transverse section (figs. 4, 5, and 6), each of these thickenings is oval, the long axis being horizontal, and from the outer border of each a distinct nerve passes. Each ganglionic swelling, besides being intimately connected with the subjacent ectoderm, abuts dorsally against the gut.[1] Between the anterior and posterior swellings on each side, and lying partly wedged in between the lateral extension of the oval nerve cords and the ectoderm, are a pair of large vacuolated cells with large granular deeply staining nuclei (fig. 6). These are the peculiar segmental excretory glands described and figured by Prof. Claus.[2] The anterior and posterior portions of the ganglion present profound differences in the arrangement of their elements, and these differences are constant throughout the entire thoracic region.

THE ANTERIOR HALF OF THE GANGLION (figs. 4 and 5)

At its anterior end the ganglion commences by an increase in the number of the cells on the internal edge of the cord; these rapidly expand into a well-marked group of multipolar cells. The delicate membrane previously described as enveloping the cord becomes discontinuous over these cells, allowing them to extend themselves both underneath the cord, and also beyond the cord laterally, always in most intimate connection with the

[1] Similarly, the pre-œsophageal ganglion or brain lies beneath and in close contact with a pair of anteriorly directed diverticula of the mesenteron.

[2] Claus, "Untersuch. über d. Organisation u. Entwickel. v. *Branchippus* u. *Artemia*", *Arb. aus Wien*, VI, 267.

ectoderm. There is thus formed a *plate of cells* which extends under the entire ganglion, dying out at either end only when the inter-ganglionic cord loses its close and immediate connection with the ectoderm. From the most external of these cells delicate processes pass, which branch and end in various ectoderm cells on the ventral surface external to the ganglia, and furnished with sense hairs, and we have here every intermediate condition between ectoderm cells which are almost continuous with nerve cells, to ectoderm cells connected with the ganglion cells by comparatively long processes (fig. 4). From their connections with the ectoderm this ventral group of nerve cells would appear to be sensory in character, and to form a sensory system which does not give origin to a definite sensory nerve. They are connected with the imperfectly segmented series of sense cells which occur on the ventral surface in close connection with the nerve cords.

On the inner dorsal angle of each of the anterior ganglionic swellings is a cell group (figs. 4 and 5, int. dorsal cells), the cells of which give off processes to the anterior commissure, and are also connected by a dorsal series of arcuate fibres with a third group of cells on the outer dorsal edge of the anterior ganglionic swelling (*ibid.* ext. dorsal cells). These external dorsal cells, though abutting on the external cells of the ventral sense plate, yet differ from them in size and in the fact that from them a bundle of fibres passes through the ganglionic plexus to form the anterior part of the anterior commissure (fig. 4). The posterior part of the anterior commissure is composed of fibres which traverse the ganglionic plexus and pass directly out as part of the outgoing nerve bundle (fig. 5). These decussating fibres thus divide the plexus of each anterior swelling into dorsal and ventral portions, of which the latter is finer and denser in character.

We can therefore distinguish in each half of the anterior portion of the ganglion three cell groups: (1) the ventral group, which extends under the entire ganglion; (2) the external dorsal group, from which decussating fibres pass, forming the anterior portion of the anterior commissure; (3) the internal dorsal group. The external and internal dorsal groups are connected by a bundle of dorsal arcuate fibres.

The *anterior pair of nerves* plunge immediately into the lateral mass of muscles, and, turning dorsalwards and outwards, pass into the upper part of the appendage, giving off motor fibres to the various muscles as they pass them. These end immediately in the neighbourhood of one of the nuclei of the curious granular protoplasmic sheath which, more or less completely, envelopes the transversely striated contractile portion of the muscle fibre (fig. 7). The anterior pair of nerves thus are mainly, if not entirely, motor. Careful searching with a $\frac{1}{15}$th objective failed to discover a case of undoubted connection with the ectoderm. Fibres can be traced into each anterior nerve from both the external and internal dorsal cell groups of its own side. These arise either directly from the cells, or as lateral branches from processes of the cells which are lost in the plexus. A bundle of fibres also passes directly from the posterior portion of the anterior commissure directly across the plexus into each of the anterior nerves.

We have already seen that the fibres of the anterior commissure not only contribute to the formation of the anterior nerves, but also pass to the two dorsal cell groups. It is thus possible that the directly decussating fibres of the anterior nerves are in part connected with the dorsal cell groups of the opposite side.

Some of the fibres of each anterior nerve may be traced directly into the more dorsal portion of the ganglionic plexus. These probably are connected with nerve cells by lateral branches, but the connection cannot be seen in sections. It is probable, from the fact that a large number of the fibres of the anterior commissure end in the plexus, that a crossed connection exists between each anterior nerve and the plexus of the opposite anterior ganglionic enlargement.

The following, then, are the central connections of each anterior nerve:

1. With both external and median dorsal cell groups of each side, but mainly with those of the same side.

2. With the dorsal plexus of each side.

Finally it should be remembered that the dorsal cell groups of the same side are connected by the dorsal arcuate fibres.

These dorsal cells and the dorsal coarser fibrous reticulum I regard as the motor portion of the anterior half of the ganglion.

It may be well here to emphasise the fact that the elements of the entire ganglion are placed in the most intimate connection by means of the complex central plexus. No conception of this primitive nervous system will be an adequate one which does not realise the fact that it is throughout a cell and fibre plexus condensed along certain lines. Even the well-marked bundle of fibres passing from the commissure direct to the anterior nerve is, in its course through the ganglionic plexus, intimately related to, and obviously a part of, that structure, being connected with it by filaments.

The *posterior half of the ganglion* (fig. 6) is very different in its arrangement from the anterior part just described. It consists, like the former, of a swelling on each nerve cord, which is oval in transverse section, and of a pair of nerves, one on each side, arising from their lateral aspects. The posterior ganglionic enlargements, like the anterior swellings, are connected by a transverse commissure. But while the anterior commissure lies in close contact with and is intimately related to the ventral ectoderm (fig. 4), the posterior commissure arches freely above it.

The cells of the posterior ganglionic swelling on each side form three groups:

1. A median group placed at the junction of the commissure with the cord (fig. 6, median group). Each cell gives off a strongly marked process to the commissure, while, on the external side, it contributes processes to the ganglionic plexus. The commissure stretches between these cells, and appears to communicate only with them, neither contributing directly to the formation of the posterior nerves, nor sending fibres directly into the plexus, as is the case with the anterior commissure.

2. An external lateral group of cells which give off marked cell processes into the plexus, while on their outer side they give origin to part of the outgoing nerve.

3. The ventral cells before mentioned which are continuous with the ventral cells found beneath the anterior portion of the ganglion and have the same connections.

The *posterior nerves* pass outwards, but take a course much more ventral than that of the anterior nerves. They pass below the muscles and immediately over the large excretory cells. Each is composed of two bundles (fig. 6) which are respectively dorsal and ventral. The ventral and larger ramus curves downwards and is distributed to the ectoderm of the ventral lobe of the appendage; the dorsal ramus passes outwards and upwards to be distributed to the more peripheral ectoderm. The fibres of the ventral ramus arise from the lateral cell group. The dorsal ramus arises directly from the ganglionic plexus. Within the plexus the processes from the median (commissural) cell group passing outwards, and those from the nerve direct, and from the external cell group radiating inwards, almost meet (fig. 6).

Those fibres of the nerve which pass to the more peripheral ectoderm, that is, the fibres forming the dorsal ramus of the posterior nerves, do not go direct to their destination, but enter, at any rate in some cases, multi-polar cells, the processes from which radiate to the ectoderm cells. Such a cell is shown in fig. 8. This relation is extremely interesting and suggestive when taken in connection with the known relations of sporadic nerve cells in Vertebrates, where such cells are regarded as being distributing or multiplying centres for centrifugal impulses. In this case, however, the contrary occurs, for these cells in *Branchippus* act as *condensing centres for centripetal impulses*. Similar "condensing" cells form a distinct layer in the optic ganglion of *Branchippus*, one process passing from each cell to the brain while several enter the cell from the eyes. Yet another consideration is suggested by these condensing sensory cells, which are mostly suspended in the body cavity by their processes. The dorsal ramus of the posterior nerve, which is characterised by the possession of such cells, passes to parts more removed from the central nervous system than does the ventral ramus. At the same time, its fibres, instead of being given off from cells, arise directly from the ganglionic plexus. Thus we may regard the ganglion cells from which the fibres of the ventral ramus take origin, as having travelled inwards from positions more remote from the nerve cords, and this process has led to their becoming an integral part of the central nervous system. Such a view has this advantage, that it accords with the many features of the primitive system of *Artemia*, which point to a derivation from a much more diffuse cell and fibre plexus by a process of condensation.

Each of the posterior nerves is therefore composed of two parts (1) a dorsal bundle, which springs directly from the central ganglionic plexus, and has ganglion cells on the course of its fibres; and (2) a ventral bundle, which arises from the lateral group of nerve cells.

There is yet another group of nerve cells which, like the ventral sensory cells, cannot be said to belong specially to either the anterior or posterior divisions of the ganglion. In describing the anterior part of the ganglion I mentioned a group of cells, the internal dorsal cells, situated on the median dorsal surface and connected with the anterior commissure on the one hand, and with the external dorsal cell group on the other, and contributing directly to the formation of the motor nerves. With this internal dorsal cell group a column of cells is connected (fig. 6), at any rate in point of position. They extend backwards over the region between the anterior and posterior ganglionic thickenings, and over the latter, to finally die away posterior to the ganglion. From this group of cells scattered fibres pass which are lost amid the body muscles. The latter are situated internal to the muscles of the appendages, and form masses closely applied to the sides of the gut (fig. 6). I often suspected the connection of some of these fibres with the gut itself, and am inclined to believe that a few, at any rate, form the nerve supply of the mesenteron. However that may be, they mainly constitute the motor supply of the body as opposed to the appendage muscles, and form a motor system comparable in its antero-posterior extension to the sensory system connected with the ventral lines of sense cells. I stated above that the ventral sensory system owed its antero-posterior extension to the fact that it is related to longitudinal lines of sense cells. Similarly, this dorsal diffuse motor system appears to me to owe its antero-posterior extension to the fact that it is connected with a group of muscles, the flexors and extensors of the body, which extend from end to end in almost unbroken series, and are not clearly divided into disconnected myomeres related to the successive body segments.

Summary of Part I

1. The central nervous system of *Branchippus*, taken as a whole, consists of two cords of nervous tissue running the length of the body, and connected anterior to the mouth by the brain, in the region of the mouth by the circumoral ganglion, and posterior to the mouth by the transverse commissures which connect the various ganglionic enlargements.

2. Each nerve cord is comprised within a delicate nucleated sheath, and consists of a mass of fine nerve fibres invested more or less completely by nerve cells. The investment of nerve cells is most complete in the brain, and more or less wanting in the inter-ganglionic regions of the cords.

3. The ganglionic regions differ from the inter-ganglionic regions chiefly in the development of a very fine plexus on the ventral aspect of the cords. The dorsal portion of the cords in each ganglion is mainly a direct continuation of the inter-ganglionic fibres, and is, therefore, largely conducting in character. The ganglionic regions are also characterised by the large number of nerve cells.

4. The inter-ganglionic and ganglionic regions are not sharply distinct from one another. Nerve cells extend from the latter to a considerable distance along the former, and a plexiform arrangement of the fibres is not wholly wanting in the inter-ganglionic regions.

5. The nerve cells have the following connections:

(*a*) They send their processes wholly into the plexus. These cells occur mainly in the brain and pre-œsophageal cords. They give off one or several processes into the central plexus. In fig. 3 we have a section through a cord in the pre-œsophageal region, and it will be seen that the cells have a rounded external surface, and give off numerous processes into the central plexus. Where the cell layer thickens, however, those most removed from the plexus fuse, as it were, all their processes into one, and thereby become unipolar cells with one thick process which ultimately breaks up in the plexus.

(*b*) They give off an axis-cylinder filament to a peripheral nerve on the one side, while on the other side they are connected with the central plexus, either by branching filaments, or by a single process which breaks up into that structure. Such cells appear to lie on the course of afferent fibres. Cells belonging to this group are, in many cases, pyramidal in shape, and give off from the base of the pyramid several processes to the skin, while from the apex of the pyramid only one process passes to the central plexus. Such cells act as "condensing centres" for centripetal impulses.

(*c*) They give off a peripheral axis-cylinder process to end in a muscle, while from the principal process, or from the cell itself, other finer processes arise which end in the central plexus. Such cells are disposed dorsally in the central nervous system, and are connected with the outflow of efferent fibres.

(*d*) Pyramidal cells giving off on the one side a single fibre to the posterior commissure, while from the other side filaments pass into the central plexus.

6. The distribution of the elements in a typical ganglion is as follows:

(*a*) On the ventral surface of the ganglion is a group of sensory ganglion cells, the fibres from which are connected with a line of sense cells on the ridges bounding the mid-ventral groove, and which form an almost continuous linear series exaggerated beneath each ganglion. These two lines of sense cells are a backward extension of the folded sense ring which encloses the mouth, and they meet just posterior to the mouth, where the otherwise shallow ventral groove narrows and deepens to form a sense pit.

(*b*) On the internal dorsal angle of the cords is a group of cells, from which are derived the motor fibres to the body muscles. These, like cell group (*a*), extend over both the anterior and posterior portions of the ganglion, and for some distance along the inter-ganglionic cords.

(*c*) On the internal and external dorsal angles of the *anterior* half of the ganglion are two cell groups which are connected with the fibres of the anterior pair of nerves passing to the appendage muscles.

(*d*) On the external angle of the *posterior* half of the ganglion is a group of cells which give origin to the fibres of the short ventral ramus of the posterior nerves. Similar cells lie scattered on the course of the fibres of the longer dorsal ramus. These cells are connected with afferent fibres.

(*e*) On the internal angle of the *posterior* half of the ganglion is a group of cells from which the fibres of the posterior commissure arise.

The anterior and posterior halves of the ganglion present the following striking structural differences:

i. The anterior commissure is closely connected with the skin. The posterior commissure arches freely above the skin.

ii. The anterior commissure does not lie stretched between two cell groups. Its fibres mainly plunge directly into the "Punkt" substance, and many of them continue as a well-defined bundle across that structure to pass out in the anterior nerves. The posterior commissure lies stretched between two cell groups (group *e*).

iii. A large number of the fibres of the anterior commissure continue as a well-defined bundle across the central plexus, to form part of the anterior nerves. Or, in other words, a considerable number of the fibres of the anterior nerves directly decussate. The fibres of the posterior nerve do not decussate.

The identification which has been made above of certain elements as the afferent and efferent mechanisms of the nervous system of *Branchippus* finds further support if we turn to the structure of the circumoral ganglion, or ganglion of the stomodæum and second pair of antennæ. A complete description of this ganglion would lengthen this communication unduly, and, moreover, would be mainly of morphological interest. The facts which lead to the identification of the motor and sensory elements may, however, be briefly set down.

The circumoral ganglion consists of a double ring of nervous matter enclosing the œsophagus and uniting the nerve cords. The double ring is displaced so that it lies at a considerable angle with the horizontal plane of the body, this displacement being due to the very great hypertrophy of the upper lip and the continuation of the œsophagus as a horizontal and posteriorly directed tube which ends posteriorly in the mouth. The double ring is especially developed posterior to the œsophagus, where it lies stretched as a double commissure between a ganglionic enlargement on each nerve cord. In this, the commissural portion of the ring, the outer part runs across from nerve cord to nerve cord, as a bundle of regularly arranged parallel fibres, resembling the commissures of the typical ganglion. The fibres of the commissural portion of the inner ring also run for the most part regularly, but still a plexiform arrangement is very obvious. This inner ring is very closely connected with the sensory surfaces of the mouth, and it also shades off into a plexus of filaments connected with the epithelium lining the œsophagus. In point of fact, it may be very accurately described as a local hypertrophy or condensation of this œsophageal sensory nerve plexus. The outer ring, on the other hand, is connected with the innervation of the well-developed muscles of this region, and may similarly be described as a local development of a nerve plexus connected with the musculature of the œsophagus. Having thus established the fact that the inner ring is sensory and the outer motor, can we proceed a step further and identify the two rings with the two parts, anterior and posterior, of the typical ganglion? The connections of the commissural portions, that is, those thicker portions which lie on the posterior surface of the œsophagus stretched between the ganglionic enlargements on each nerve cord, enable us to do this with certainty. The anterior and posterior commissures of the typical ganglion present, as we have seen, certain well-defined structural characters. The anterior commissure, unlike the posterior commissure, does not lie stretched between two cell groups. On the contrary, its fibres plunge directly into the ganglionic plexus. The commissural portion of the external ring of the circumoral ganglion presents this feature. On the other hand, the posterior commissure of the typical ganglion does lie stretched between two cell groups, and a similar connection characterises the commissural portion of the inner ring of the circumoral ganglion. Thus the outer motor ring of the circumoral ganglion presents the same structural features as the anterior portion of the typical ganglion, and a similar structural agreement is found between the inner sensory ring and the posterior portion.

If we turn to the nerves arising from the circumoral ganglion, further confirmatory evidence is obtained. The circumoral ganglion gives origin to the nerves of the second pair of appendages, the second antennæ. Two nerves pass to each second antenna, and they arise from the cords a short distance anterior to the ganglion. Claus,[1] in his account of the anatomy of *Branchippus stagnalis*, describes these nerves, and traces the anterior nerve on each side to the muscles of the second antennæ, and the posterior nerve to the sense cells of those organs, and I am able to confirm him in this respect. Tracing these two nerves centrally we find that the motor pair are connected with the motor, or external, commissure, in a way precisely similar to the direct connection which we have seen to exist between the anterior commissure and anterior nerves of the typical ganglion. Similarly there is evidence that the fibres of the posterior or sensory nerves to the second antennæ are connected with those portions of the ganglionic enlargements of each cord which correspond in structure to the posterior swellings of the typical ganglion, and which are placed in connection by the inner sensory commissure.

PART II. THE POST-ŒSOPHAGEAL NERVE CORD OF *ASTACUS FLUVIATILIS*

The abdominal region of the central nervous system of *Astacus* contains six ganglia, related respectively to the six metameres into which the abdomen is divided. The first five of these ganglia differ from one another only in certain small details, and a description of one will serve as a description of the whole.

The *second abdominal ganglion* forms a marked enlargement on the cord, the bulging being most prominent ventrally. There arise from it two pairs of nerves which I propose to distinguish respectively as the anterior and posterior nerve pairs. From the external dorsal aspect of the cord just posterior to the ganglion, a nerve arises on each side, thus forming a third pair to be distinguished as the posterior dorsal nerves.

Distribution and character of the nerves. The *anterior pair* pass directly outwards beneath the great flexor muscles of the abdomen (fig. 9). Their point of origin is not only more ventral than that of the other nerves, but they also pursue a more ventral course. They run outwards almost at right angles to the longitudinal axis of the body, in the groove formed by that thickened portion of the sternal carapace which extends transversely under the ganglion, and between the bases of the second pair of abdominal

[1] C. Claus, *Zur Kenntniss d. Baues u. d. Entwickel. von* Branchippus stagnalis u. Apus cancriformis, Göttingen, 1873.

appendages. Immediately anterior to these nerves is a vertical slip of muscle passing down from the great flexors of the abdomen to be inserted into the thickened carapace (fig. 9a). Very shortly after leaving the ganglion each nerve divides into two branches, which, however, continue in the same sheath for some distance. They separate from each other just before the appendage is reached (fig. 9). The posterior and smaller then turns downwards into the appendage, the anterior and slightly larger divides into three main branches on the anterior border of the base of the appendage. These again divide into smaller branches which are distributed to the appendage muscles, pleuron and dorsal skin. The posterior branch is distributed to the comparatively small muscles which lie within the terminal part of the appendage, but by far the greater number of its fibres pass to the skin which, in the males, is extremely sensitive, the first and second pairs of abdominal appendages being modified to form copulatory organs. On their course the anterior nerves give off a few fine branches to the sternal skin.

Fibres. The anterior nerves are composed of two markedly different classes of fibres (figs. 11 and 13):

1. Large bold fibres with nucleated sheaths, and
2. Fine filaments.

The large fibres may be again divided into (*a*) large fibres with thin sheaths, (*b*) small fibres with thick sheaths. In a section through the nerve (fig. 11), it will be seen that these different classes of fibres are segregated into distinct bundles. Each of the large fibres is bounded by a distinct nucleated sheath, which, as was pointed out by Krieger,[1] encloses a substance of such extreme fluidity, that under the influence of preserving agents it shrinks into small clots here and there in the course of the fibre. There is no trace of a sheath resembling the medullary sheath of Vertebrate nerve fibres, either in preparations treated with osmic acid, or in those stained by Weigert's method.

In tracing the anterior nerves into the ganglion a curious point is noticed, namely, that the transverse sectional area of the whole nerve on each side is much reduced, thus forming a narrow neck just before the nerve plunges into the ganglion. If the fresh nerve is examined under the microscope, or if individual fibres are carefully followed in a series of sections through this part, the sudden increase in the bulk of the nerve, which takes place a fraction of a millimetre from the ganglion, is found, as has already been shown by Haeckel,[2] to be due to the fact that the large fibres there divide into large and medium-sized fibres (fig. 12). In the

---

[1] Krieger, "Centralnervensyst. d. Flusskrebses", *Zeits. f. wiss. Zool.* vol. XXXIII.

[2] Haeckel, "Ueber d. Gewebe des Flusskrebses", *Müllers Archiv* (1857), Taf. XVIII, figs. 2 and 8. Cf. also Krieger, *loc. cit.* fig. 9n[3].

example figured it will be seen that while the smaller branch is conspicuously smaller than the nerve fibre before the branching, the large branch is of much the same size.

Both branches of each anterior nerve contain fibres of each of the two classes, as will be seen on reference to fig. 11.

The *posterior ventral nerves* arise directly from the ganglion posterior to, and slightly above, the origin of the anterior nerves. Instead, however, of passing outwards at right angles with the cord, and at a level anterior to the appendages, they trend outwards and backwards (figs. 9 and 10), running on the ventral surface of the great flexor muscles to the region immediately posterior to the attachment of the appendages. Their position in this part of their course is not so ventral as that of the anterior nerves. They lie at a level between the slips of the flexor muscles which pass to the sternum in each segment and the main mass of those organs. Each nerve, shortly after leaving the ganglion, divides into two branches which run in the same sheath for some distance. The anterior and smaller branch separates from the main trunk of the nerve where the latter curves abruptly round the belly of the flexor muscles to pass dorsalwards. It then turns upwards and outwards, to run along the infolding of the carapace, which forms the joint between abdominal segments 2 and 3, and is lost in the extensor muscles (figs. 10, 10*a*, 10*b*). The main trunk of the nerve passes sharply backwards along the outer surface of the belly of the flexor muscles and appears on the dorsal surface of those muscles about the middle of the *third* abdominal segment. It gives off the following branches:

(*a*) To a thin sheet of the external extensor muscles which spreads out into the pleuron of the third segment, and to the dorsal skin in that region.

(*b*) Branches which pass above and end in the external extensor muscles, but also give off fibres to the dorsal skin.

(*c*) Branches which pass beneath the external extensors to end in the coiled internal extensors.

The extensor muscles consist of four distinct bands situated on each side of the mid-dorsal line. The two median bands, or internal extensors, as I have called them, are coiled on themselves so as to form roughly a spiral of muscle; the external extensors, on the contrary, are flat band-like muscles with slips passing off to be inserted into the tergum in each body segment (fig. 29).

The posterior ventral nerves of the second abdominal ganglion, therefore, are distributed to the extensor system of muscles in the posterior two-thirds of the third abdominal segment, and the anterior third of the fourth segment. Prof. Claus[1] describes a similar distribution of the nerves to the body muscles in the abdominal region of *Nebalia*.

[1] Claus, "Ueber d. Organismus der Nebaliden", *Arbeit. Wien* (1889), vol. VIII.

The different branches when they reach the extensor muscles divide continuously, forming horizontal fan-like tufts of fibres, thus forming a plexus which spreads through a considerable length of the muscles.

The posterior ventral nerves are composed of two sizes of fibres like the anterior pair, but the small fibres form only a small part of each nerve (fig. 14).

The *dorsal nerves* are of extraordinary interest. They spring from the dorso-lateral portions of the cord just posterior to the ganglion (fig. 16), and turning backwards are distributed to the great flexor muscles (fig. 26).

Each of the dorsal nerves near its origin, and before any branching has taken place, is composed of only a few, generally ten, very large tubular fibres, about 30–60 μ in diameter.

In order to adequately realise the size of these fibres, it should be remembered that they have only a thin sheath. The dimensions given above should therefore be compared with those of the axis cylinders of vertebrate medullated fibres rather than with the whole fibre. Each fibre is distributed to a relatively enormous mass of muscle by a process of continuous branching. No small fibres occur in the main nerve, but there is usually a small bundle composed wholly of a few of the small thick sheathed class of tubular fibres (fig. 15) which leaves the main trunk immediately after its exit from the cord and curves abruptly ventralwards. I have not followed it further.

The fibres when they enter the cord form ascending and descending columns, seven of the ten fibres passing upwards to the second abdominal ganglion, and three passing downwards to the third ganglion. In other words, the dorsal nerves of the second ganglion, though belonging mainly to that ganglion, yet derive part of their elements from the ganglion next below.

Like the posterior ventral nerves the dorsal nerves branch very extensively and irregularly, and pass to a part of the flexor muscles which lies mainly in the third abdominal segment.

We thus see that the body muscles, both flexors and extensors, have been displaced posteriorly, the relation of the second abdominal ganglion to the second pair of abdominal appendages affording us a fixed point for reference. And this displacement has taken place to such an extent that the main branch of the posterior ventral nerves of the second abdominal ganglion courses on the dorsal surface at the junction of the *third and fourth segments* on its way to the extensor muscles (figs. 10 *a* and 10 *b*).

The same posterior displacement of the muscles is shown when we examine the nerves of the remaining abdominal ganglia. In connection with the general question of metamerism it is interesting to notice that the very definite blood supply of these muscles does not appear to have suffered a similar displacement. The dorsal aorta gives off at about the middle of each segment two pairs of arteries, one on each side to the extensor muscles, and one passing ventrally on each side of the flexor muscles.[1]

Motor and sensory fibres. Before turning to the ganglion itself it will be advisable to discuss the question of the identification of motor and sensory fibres in the nerves.

Evidence derived from the distribution of the different classes of fibres. The chief tactile organs possessed by *Astacus* are long hair-like filaments, each composed of a special expansion of the general cuticle jointed to the carapace, and containing a filamentous process of an ectoderm cell in its axis. These lie mainly on the appendages, but also occur on the edges of the pleura, the telson and the posterior edge of the tergum in each segment. The general surface elsewhere is covered by a thick and, for the most part, hard lamellated horny cuticle, which in sections is seen to be traversed by comparatively infrequent delicate pores in which fine processes of certain elongated ectoderm cells lie. These are especially well seen in sections through the skin near the anus (fig. 17). The tergal surface, where the carapace is especially thick, is relatively insensitive, while the swimmerets, on the contrary, with their numerous tactile hairs, are especially sensitive. In correspondence with this we find that the nerves supplying the swimmerets contain a very large number of fine fibres, while the nerves to the tergal surface contain relatively fewer; and that the branch of the anterior nerve which passes *into* the appendage and to the small muscles therein contained contains relatively fewer large fibres than does the larger branch which innervates the larger muscles of the appendage (fig. 11). Further, the nerves passing to the telson, which contains no muscular elements, and to the sensitive region round the anus, are entirely composed of fine fibres. On the other hand, the nerves passing solely to the flexor muscles, that is, the posterior dorsal pair in each segment, contain only the large tubular type of fibres.

Evidence derived from direct experiment. Seeing that the anterior nerves of the abdominal ganglia contain a considerably larger number of the fine nerves than do the posterior ventral nerves, I determined to try the relative effects of stimulation.

The abdominal cord was laid bare from the ventral surface, the operation requiring great care to avoid injuring the very delicate and superficial anterior nerves. The anterior and posterior nerves were then stimulated with an interrupted, or tetanising current, and the following points noticed:

1. The anterior nerve of the left side was stimulated at a point about midway between the ganglion and the appendage. The effects were double,

---

[1] Compare also Milne-Edwards, *Histoire Nat. d. Crustacés*, Paris, 1834, vol. i, Plate 7.

(*a*) flexure of the limb, and (*b*) enormous reflex disturbance of the animal generally, every appendage being moved. In other words, the main effect of stimulating this nerve was to produce very great sensory disturbances.

2. The posterior ventral nerve on the same side was stimulated with the same strength of current, and the only effect was a faint sensory disturbance and feeble movement of the second abdominal appendage of the same side.

The difference between the central effects produced by stimulating these nerves is very obvious when the animal is exhausted. Then, when stimulation of the anterior nerve on one side will produce reflex movements of all the appendages, stimulation of the posterior nerves leads to no such result.

The movement of the appendage which resulted from stimulation of the posterior nerve I regarded as being due to reflex action, and to settle this point the following experiment was made:

3. The posterior ventral nerve was cut and its peripheral end stimulated. No movement of the appendage resulted, but, on applying the electrodes to the central end, results occurred similar to those described under (2).

4. With respect to the posterior ventral nerves, Marshall[1] states that on stimulating the peripheral end of one of them (in the lobster) no effect was produced. Hence he concluded that "the anterior nerve would seem to be mixed, but the posterior nerve purely sensory". This result was due to the fact that, in his experiments, the animal was fixed in such a position that it was impossible to observe any contraction of the extensor muscles. If, however, the animal be placed in such a position that the extensor muscles can be observed while the posterior nerve is being stimulated near the ganglion, and therefore on the ventral aspect of the body, a marked tetanus of the extensor muscles is found to follow stimulation.

5. The posterior dorsal nerves in the case of *Astacus* are so delicate and deep lying, and their unbranched free portion is so extremely short, that it is impossible to directly stimulate them without escape of the current into the muscles. The experiment was therefore carried out in the following manner: The anterior and posterior ventral nerves were cut on both sides and in each segment of the abdomen. Thus only the posterior dorsal nerves were left intact. The nerve cord was then cut in the thoracic region and lifted on to the electrodes. On opening the circuit the abdomen was sharply flexed.

The facts both of dissection and stimulation thus lead to the conclusion that the anterior nerves are of mixed function, and contain both afferent and efferent fibres. But they mainly supply sensory surfaces. The posterior ventral nerves are also of mixed funtion, but they supply a relatively larger

[1] Marshall, "Some Investigations on the Physiology of the Nervous System of the Lobster", *Studies from Owens College, Manchester*, 1886.

mass of muscles, and a much more insensitive region of skin. In correspondence with this difference between these two pairs of nerves, we find that the anterior nerves are relatively poor in the large tubular fibres, but contain a very great number of the fine fibres. The posterior ventral nerves, on the other hand, are relatively much richer in large tubular fibres.

The posterior dorsal pair differ from the anterior and posterior ventral nerves in being solely efferent in function, and we find that they contain only the large tubular fibres.

We may thus, I think, conclude that the large tubular nerve fibres of *Astacus* are efferent, while the fine nerve fibres are afferent in function; and these two classes of fibres are not only sharply marked off from one another in point of size, but also the gulf between them is unbridged by intermediate forms within the limits of the somatic system.

Summary. Three pairs of nerves arise from each of the first five abdominal ganglia:

1. An anterior pair, which arise directly from the ganglion, and contain a large number of the fine, or afferent fibres, and comparatively few of the large, or efferent fibres. These supply the appendages with motor and sensory fibres, and also the skin of the sternum and pleura.

2. The posterior ventral nerves containing relatively more large fibres. These supply the dorsally placed extensor muscles and the dorsal skin in the third segment, or segment next following.

3. Posterior dorsal nerves which are purely motor and innervate the flexor muscles.

THE GENERAL RELATIONS AND STRUCTURE OF THE SECOND ABDOMINAL GANGLION

Only a very general mention of the main features of the ganglion itself need be made, since Krieger has so fully dealt with them in his paper on the "Centralnervensystem des Flusskrebses" (*Zeits. f. Wiss. Zool.* XXXIII, 527). The ganglia appear as "small knot-like swellings on an apparently single longitudinal commissure". Each ganglion is surrounded by a tough lamellated membrane, which is separated from the nerve substance by a loose reticulum, especially abundant on the ventral face of the ganglion. The spaces in this connective tissue reticulum are large and are filled with blood. We thus have two distinct sheaths, a point not sufficiently insisted on by Krieger (fig. 18), of which the inner one is practically a system of blood spaces bridged by connective tissue, and is only slightly developed in the inter-ganglionic regions (fig. 23).

Owing to the continuity of the investing membrane, the nerve cord at first sight appears to be single, but the nervous elements in reality form two perfectly distinct cords which are united only in each ganglion by a thick transverse bridge. In the inter-ganglionic regions the two halves are separated by a median vertical lamella of connective tissue. As they enter a ganglion either above or below they diverge from one another, and the inner sheath at the same time becomes thicker. There is thus formed at each end of the ganglion a vertical cleft-like space between the lateral masses of nervous tissue, and filled by the inner sheath of the cord; and the anterior cleft, or *anterior fissure*, is separated from the *posterior fissure* by the transverse bridge of nervous tissue which was spoken of above.

The outer sheath. This sheath is composed of one or more lamellæ, each of which is built up of fine fibrils which appear to be bound into bundles (fig. 19). These may either run parallel to one another, or interlace. The lamellæ appear to be entirely free from cellular elements in their substance, but on the surface of the sheath flattened plate-like cells occur (figs. 19 and 20), with large, flattened, deeply staining nuclei.

Quite on the inner face of this sheath these cells are more numerous and their plate-like extensions often overlap one another (fig. 19). Supporting filaments are seen to arise from the outer sheath which are continuous with the inner sheath (fig. 20).

The inner sheath appears in sections as a coarse reticulum, largely cellular in nature. It closely resembles in structure the tissue surrounding the sternal artery, and lying beneath and about the nerve cord. The inner sheath is essentially a blood-containing investment, and into its spaces the arteries supplying the ganglion open. In the inter-ganglionic regions of the cord this sheath is very thin. The branched cells of the inner sheath are continuous, on the one hand, with the flattened cells on the inner face of the outer sheath; and, on the other, with the variously modified cells which form the intimate supporting structure of the nervous elements.

The *supporting tissue of the nervous structures* appears to be mostly, if not entirely, cellular in character, and is best seen by teasing out the fibres of the inter-ganglionic commissures where they diverge into bundles on entering the ganglion. If these have been preserved with Flemming's fluid mixed with an equal volume of 0·5 per cent. solution of osmic acid, and afterwards stained with hæmatoxylin, the individual bundles will be seen to be covered by an imperfect sheath of flattened cells with large, oval, deeply staining nuclei (fig. 21). Cells of this type represent the imperfect attempts at endothelium building found so widely in *Astacus*. They occur, only of much larger individual size, lining certain of the large body spaces, and similar cells have been already noticed in connection with the outer sheath of the ganglion. Within the substance of the ganglion, where the nerve fibres branch and become much finer, similar supporting cells occur. Each consists of a small cell-body, from which arise short plate-like processes which branch into the wildest tangle of exceedingly delicate filaments (fig. 22), thus recalling the neuroglia cells of the Mammalia. There is thus formed a supporting tissue composed, like the neuroglia of Vertebrates, of the delicate processes of much modified cells. Both plates and filaments are of an optically structureless and non-staining character, very different from the finely granular and staining cell-substance of the nerve cells. Their nuclei also stain more deeply than the nerve cell nuclei, and are not so coarsely granular.

Blood supply. The ganglion is supplied with blood by four arteries which arise by a short common stem, which springs from the posterior sternal artery, immediately under the middle of the ganglion. After penetrating the external sheath, two of the arteries curve round the sides of the ganglion between the roots of the anterior and posterior nerves, and, running in the inner sheath, finally end in the dorsal portion of an enlargement of that sheath on the lateral face of the nervous tissue, and above and between the two nerve roots. These may be called the *lateral arteries*. The other two arteries run respectively anteriorly and posteriorly for a short distance, and then make their way upwards through the anterior and posterior fissures to the dorsal surface, where they open into a dorsal median thickening of the inner sheath. They may be called the *anterior and posterior arteries.*

Thus, in the intact ganglion, there must be a stream of blood flowing through the interstices of the inner sheath from the dorsal to the ventral side, where it drains into a large ventral sinus situated in the inner sheath, and incompletely divided into four longitudinal sinuses by septa. The blood finally leaves the ganglion by apertures placed in the mid-ventral line at the posterior end of the ganglion, and leading from the ventral sinus of the ganglion into the ventral abdominal sinus. The arrangements for the nutrition of the large nerve cells which occupy the ventral, and, to a less extent, the lateral surfaces of the nervous tissue of the ganglion, are most interesting. These cells are of the unipolar, pear-shaped type, and are quite removed from the dense nerve substance of the ganglion. Each is covered by a delicate cellular sheath, and this alone separates the cell-substance from the blood; for they may be said to hang in bundles suspended by their processes, and steadied by the reticulum of the inner sheath, in blood spaces of that tissue. Between them, in the mid-ventral line, and below them, in the posterior region of the ganglion, are the venous sinuses mentioned above.

The mass of the ganglion is composed of a fibrous reticulum, coarse in some places, fine in others, and the fibres in the finer reticulum appear to touch one another, so that blood spaces are conspicuous by their absence. In the case of nerves and of commissures the same fact strikes one, whether they are viewed in the fresh condition, or examined by means of sections—

the sheathings of the nerve fibres are contiguous with one another, and, at first sight, no provision appears to have been made for their nutrition. This may, I think, explain the prevalence of "tubular fibres" in the central nervous system and peripheral nerves of *Astacus*. Each of these fibres in section appears as a tubular nucleated sheath, and little more. The contents of the tube, or what corresponds to the axis cylinder of the nerve fibres of Vertebrates, have shrunk into small clots gathered here and there at long intervals on the course of the fibre. In other words, the contents of such a tubular fibre are exceedingly fluid. If they are examined in the fresh state bubbles of air may often be seen, which may be made to move about in the almost fluid contents, as though one were dealing with a fine tube filled with fluid. If these fibres are isolated and watched, their contents will, as was pointed out by Krieger,[1] be seen to undergo a change comparable with rigor mortis. Clotting takes place, the clot appearing in the form of granules, which outline delicate fibrils, which I regard with Krieger as the true axis cylinder fibrils. These, I take it, are suspended in life in an extremely fluid substance, protoplasm or plasma, by the aid of which the transportation of nutritive material or the removal of waste matters can be managed through considerable lengths of these tubular fibres. It is therefore possible that each efferent fibre is the morphological equivalent of a considerable number of afferent fibres, each one of the latter being, without doubt, a single axis cylinder.

Be this as it may, it is, at any rate, abundantly clear that the disposition of the nerve cells on the surface of the dense nervous tissue of the ganglion, and their relation to the blood streams, lends no support to the idea that they are nourished, even in part, by their processes. The distinction of the processes of nerve cells into "nutritive processes" and true nerve or axis-cylinder processes has been advocated by Nansen[2] for the Crustacea, and by Golgi for Vertebrates. In the latter case the distinction is based upon histological facts which appear to me to be adequately explained by the effect of the shrinkage of the tissue in occluding the lymph lacunæ which, we must suppose, surround the nerve cells of the central nervous system and their processes during life.

ARRANGEMENT OF THE NERVOUS ELEMENTS OF THE GANGLION

These consist of cells, fibres, and fibrillar plexus. The relation of the cells and fibrous elements is the same as that which is found in the nerve cord of *Branchippus* where the former lie wholly on the surface of the latter. In the cord of *Astacus*, however, cleft-like spaces filled with the inner or blood-

---

[1] *Loc. cit.*   [2] Nansen, *Bergens Mus. Aarsberet*, 1886.

containing sheath penetrate the nerve substance, and the cells to a certain extent occupy these (fig. 34c). The nervous tissue of the ganglion is composed of fibres and fibrillar plexus and nerve cells. The fibres are of the same tubular character as those already noticed in the nerves. They run in the ganglion for the most part in longitudinal bundles, which, penetrating the ganglionic plexus, divide it up into regions. Many of these fibres, especially in certain of the bundles, run straight through the ganglion without effecting connection with its elements. The other tubular fibres in the ganglion are either commissural between the lateral halves, or derived from the large unipolar cells which occur in the ventral cell-plate mentioned above.

In passing up the cord towards the brain one finds, as in Vertebrates, that the number of fibres passing straight through the ganglia to regions below continuously increases. Such fibres are limited to the more dorsal portion of the cord, and in the upper portion of the thoracic cord they may be readily separated by simple dissection from the ventral bundles of fibres which pass to and through the ganglionic substance. This is well seen in fig. 24, which is a drawing of the ganglion of the chelæ, as seen from the ventral surface.

The ganglionic plexus is separated into regions distinguished both by the size of the fibres forming the plexus and the complexity of their arrangement. We may distinguish two grades:

(a) A coarser plexus, the elements of which, though much smaller than the tubular fibres of the inter-ganglionic cord or nerves, yet show the marked tubular appearance (fig. 36).

(b) A fine plexus, comparable in the extreme tenuity of its elements to that of the nerve cord of *Branchippus* (fig. 32). The fine plexus and the coarse plexus are, respectively, ventral and dorsal, and related, the former to the fine nerve fibres, and the latter to the large nerve fibres. The fine plexus may be said generally to form an irregular plate on the under surface of the fibrous portion of the ganglion.

The large tubular nerve fibres pass at once to the coarse plexus, and we may regard the latter structure as being formed by their branching. I do not, however, regard the coarse plexus as constituting the final ending of the large nerve fibres, but merely as the place where those fibres subdivide before passing to various regions of the more ventral fine plexus. The fine nerve fibres, on the contrary, run in bundles directly to the fine plexus (figs. 32 and 34). The fine plexus, as will be seen later, is not a homogeneous structure, but presents differences in density in its different parts.

The Internal Connections of the Anterior Pair of Nerves

Each of the anterior nerves, as it enters the ganglion, divides into a dorsal and a ventral root, each of which contains both large and small fibres (fig. 35). In each root the large and small fibres form quite distinct bundles, and in the section through the trunk of this nerve, figured in fig. 13, it will be seen that a short distance from the ganglion the fine and large fibres are still arranged in distinct bundles.

The *dorsal root* passes to a distinct mass of plexus which lies in the outer, or more lateral, part of its own half of the ganglion in the anterior region, at the level of the entry of the anterior nerve, and immediately above the ventral root (fig. 33, *d*, *c*). This mass of plexus is very definitely divided into a small round and ventral mass of fine plexus, and a dorsal mass of coarse plexus, which extends to the lateral surface of the ganglion above the entering nerve. (Fig. 33, *d*, *c*. The dotted line bifurcates, and ends respectively in the coarse and in the fine plexus.) The fine fibres of the dorsal root pass to the ventral fine plexus, the large fibres to the dorsal coarse plexus.

I regard this root as being of mixed function, motor and sensory, and its centre as being composed of two parts: (1) a dorsal coarser portion, formed by the first branching of the large efferent fibres; and (2) a ventral, very much finer portion, to which the fine afferent fibres pass directly.

Connection with cells. By far the greater number of the fibres, both large and small, pass directly to break up in the fibrous reticulum. Where the large fibres arch round the ventral ball of "punkt" substance on their way to their own coarser reticulum, I found several distinct T-shaped junctions. One limb of the T could be traced to the most external of the ventral unipolar cells in the region of the entering nerve. These form a distinct group of cells of fairly uniform size, and averaging from 30 to 40 μ in the shortest axis.

In addition to these, there are cells situated between the bundles of entering nerve fibres, where they diverge. These are much smaller. Each gives off one process which proceeds to strike the entering bundles at right angles. I think I may say positively that nerve fibres do not end in these cells, though their processes are connected by T-shaped junctions with certain of those fibres—probably the fine ones (fig. 35, *a*, *a'*).

In longitudinal vertical sections it is seen that these cells lie chiefly in two masses anterior and posterior to the entering nerve, and that their fine processes pass in two main bundles to join the fine fibres before they break up into the "punkt" substance (fig. 35, *a*, *a'*). There is a third very distinct group of cells related to the dorsal root, and lying on the dorsal side of the entering nerve, and on the lateral surface of the coarse nerve network (fig. 36). Each cell receives an axis-cylinder process, and gives off from its internal face processes into the plexus of the ganglion. The small tubular fibres, with thick neurilemma, pass to these cells. This cell group is continuous with the more dorsal cells of the lateral extension of the ventral plate of cells between the point of exit of the anterior and posterior ventral nerve, and, from them, fine tubular processes pass, which curve round as dorsal arcuate fibres (fig. 36).

The *ventral root* also consists of large tubular and fine fibres. It passes in beneath the centre for the dorsal root, and the fine fibres break up in a large mass of fine plexus in the most ventral part of the ganglion near the middle line (fig. 33, *v.c.*). The large fibres pass up anterior to the fine fibres to break up in a mass of coarse plexus dorsal and anterior to the above-mentioned fine plexus (the region it will occupy is indicated by the upper limb of the dotted line in the figure). This root, therefore, also contains motor and sensory elements connected respectively with a coarse and fine plexus.

Cells. The large fibres appear to be connected by short branches given off at right angles, with a group of the ventral unipolar cells immediately underlying them, just after their entrance into the ganglion (fig. 35, *b*). The fine fibres are connected with the group of small nerve cells lying between the divergent roots (fig. 35, *a, a'*).

To sum up the internal connections of the anterior pair of nerves, we see that each nerve divides into two parts on entering the ganglion. Each part, or root, is composed of large tubular and fine fibres which pass respectively to a coarse and fine ganglionic plexus. There are thus two distinct centres related to the two roots. These centres are connected with one another by zones of plexus, and by unbranched fibres in such a way that the coarse plexus of the external centre (that of the dorsal root) is connected with the coarse plexus of the more median centres, and, similarly, the fine plexus of the external centre is connected with the fine plexus of the median centre. Further, the centres of the two sides of the ganglion are placed in communication with each other by two commissures, one of which consists of the finest fibrous elements, and stretches between the median masses of fine plexus. The other consists of coarser elements and connects the median coarse plexus of the one side with that of the other. The completely dual nature of the internal connections of the anterior pair of nerves, that is, the nerves to the appendages, I take to be of especial significance. In the thoracic region, where the nerves to the appendages are large and predominate, the dual nature of their internal connections is even more strikingly shown. In the thoracic region, also, the two roots of the appendage nerves, instead of being fused in the neighbourhood of the ganglion, and only diverging within its sheath, form two long roots which remain quite distinct as far as the second joint, *i.e.* a considerable distance into each appendage (fig. 25).

These two roots differ markedly in size, and, as a result of experiments carried out on the lobster, Marshall (*loc. cit.*) states that (1) "the small nerve contains motor fibres which supply the extensor muscles of the limb, and especially the divaricator muscle of the claw"; and (2) "the large nerve contains the motor fibres to the muscles which raise the limb and close the claw". Also the small nerve contains "afferent fibres which cause reflex contraction of the claw through the large nerve which supplies the occlusor muscle", while the large nerve contains "afferent fibres which cause opening of the claw by reflex action through the small nerve which alone supplies the divaricator muscle". These experimental results I have been able to substantiate in the case of the nerves to the chelæ of *Astacus*. I have further found that the small nerve passes to a median centre, the large nerve to an external centre in its ganglion.

In contradistinction to the posterior ventral and dorsal nerves the fibres of the anterior nerves do not appear to directly decussate in any case. Retzius failed to find any decussating fibres;[1] but, as has been pointed out, bilateral connection is made by fibres large or small between the various centres.

Concerning the cell connections of the fibres of the anterior nerves we find that:

1. The large tubular efferent fibres are connected with large nerve cells.

2. The fine fibres, afferent in function, are connected with small nerve cells. The connection with the cells is, as Retzius (*loc. cit.*) has already shown, by lateral branches. The nerve cells, therefore, do not lie in the direct path of the nervous impulses to or from the ganglionic plexus. Exceptions to this appear, however, to occur in the case of the small tubular fibres with thick neurilemma.

INTERNAL CONNECTIONS OF THE POSTERIOR PAIR OF NERVES

The first and most striking facts in connection with the posterior nerves is that they do not pass to two distinct centres, and that a certain number of the fibres, as Retzius also found, do directly decussate.

On entering the ganglion the fine fibres of each nerve take up a position posterior to the large fibres, and form a well-defined single stream which passes to a single mass of fine plexus on the ventral surface and towards the median line (fig. 34, *f.pl.*). This centre, therefore, occupies the same relative position as the internal centre, or centre for the ventral root, of the anterior nerve. Sometimes the fine fibres enter the nerve tissue of the gang-

[1] Retzius, *Biologische Untersuchungen*, Stockholm, 1890.

lion at a point higher up than the large fibres, and arch round to pass to the ventral-lying fine plexus. They may be said generally to take a more or less arched course through the nerve substance of the ganglion and to enter their centre from above. Certain of these fine fibres directly decussate and pass to the centre of the opposite side. This, however, does not always occur.

The large fibres form two sets, some ascend, and after passing for a short distance as external arcuate fibres, plunge into the nerve substance of the ganglion and either

(*a*) Break up in a zone of coarse plexus lying at the junction of the upper and middle third of the ganglion; or

(*b*) They directly decussate, passing to a similar region on the opposite side (fig. 34, *a*).

The second stream of large fibres (fig. 34, *b'*) passes in more ventrally and anteriorly, and communicates with a cell group lying on the ventral surface of the ganglion and extending round the entrance of the posterior nerve (figs. 34 and 35, *b*). Some of the fibres, however, break up directly in a mass of plexus, lying anterior and external to the fine plexus related to the fine fibres of the posterior nerves.

We thus see that the fine fibres or sensory elements of each posterior nerve pass to a single centre, while the large fibres or motor elements are distributed to two regions of coarse plexus. I take it that we may correlate this dual character of the central connections of the motor elements with the fact that the nerve supplies two sets of muscles—the external and internal extensors, differing enormously in their general arrangement, and in the position of their fibres. At the same time these two masses of plexus are intimately connected by unbranched fibres and bridges of plexus.

Retzius (*loc. cit.*) has shown that a certain number of the fibres of the posterior nerves have no connection with the elements of the ganglion, but turn directly backwards to descend in the longitudinal commissure. A similar arrangement, as will be seen, obtains in the case of the posterior dorsal nerves. On the other hand, fibres have not been traced directly from the commissures to the anterior nerves.

Internal Connections of the Posterior Dorsal Nerves

These, as will be remembered, arise from the dorsal and external surface of the inter-ganglionic cord, a variable but short distance posterior to the ganglion. As was pointed out before, they each consist of two parts derived respectively from the ganglion above and the ganglion below. This fact is only demonstrable by sections, since simple dissection merely reveals the fact that they arise a very short distance posterior to, but in very close connection with the second abdominal ganglion. I have examined these nerves in detail in the first five abdominal ganglia. Each nerve is composed of from ten to thirteen tubular fibres, which vary in size from 12 to 13 μ.

On entering the nerve cord, or a little distance before entering it, each nerve divides into two unequally sized roots, of which the smaller, containing generally three fibres, passes down the inter-ganglionic commissure to the ganglion next below. It at first lies on the external dorsal angle of the cord, but soon makes its way obliquely over the external giant fibre[1] to a more median position, and then curves ventrally to run in the external region of the dorsal group of longitudinal fibres. These I propose to call the *descending root*. The remaining fibres, eight to ten in number, turn sharply upwards and form a column of fibres external to the external giant fibre. These form the *ascending root* (fig. 23).

Tracing the whole bundle in its upward course to the second ganglion, it is found to assume a more dorsal position until it overlies the external giant fibre. Almost before any indications of ganglionic structure have appeared in the cord, the most median and largest fibre (a, 35 μ) detaches itself from its fellows and passes obliquely forwards and inwards immediately under the internal giant fibre. It continues its course across the middle line to the opposite half of the ganglion, where it divides into two main branches. One is given off ventrally immediately after the decussation has been completed. It runs downwards and backwards, at first in the dorsal sulcus between the lateral halves of the cord in this region, and then plunging into the nerve substance ends in or rather forms the single process of one of the largest giant cells, situated in the external and anterior part of the posterior division of the ventral plate of nerve cells (fig. 33, a). It will thus be seen that the most internal fibres of each side, after decussating, each give off a branch which runs forwards for a considerable distance before passing to its cell.

The rest of the fibre, or the other branch, continues horizontally across, under the internal giant fibre of the opposite side, to finally break up in a small mass of dorsally situated plexus which overlies the nerve substance between the two giant fibres. This plexus is intimately related to a small group of small nerve cells on its dorsal surface.

Continuing anteriorly, a second fibre (b, 18 μ) next detaches itself from the bundle and runs inwards under the internal giant fibre, then curves downwards and decussates by the dorsal motor commissure, which is largely composed of the decussating motor fibres of the posterior pair of nerves (fig. 34, c). It breaks up in the dorsal coarse plexus of the opposite side.

---

[1] In each lateral half of the inter-ganglionic cord are two large giant fibres which are respectively external dorsal and internal dorsal. They lie completely dorsal to all the other fibres.

Two fibres, one $18\,\mu$ (c) the other $15\,\mu$ (d), are the next to leave. They take different courses. The first arches close round the external giant fibre and runs inwards a short distance towards the median line. It then turns downwards and forwards, running in a sulcus or fissure between two columns of longitudinal fibres which I will call the internal dorsal and median dorsal columns (cf. later). It then appears to divide into two branches, of which one turns inwards towards the median line, and breaks up in the dorsal plexus of its own side, in the region corresponding to that of the opposite side in which fibre (b) is lost.

Fibre (d) leaves the bundle at the same time that fibre (c) does and curves *outwards* over the external giant fibre, then passes forwards and downwards on the external surface of the nerve tissue, and breaks up in the most lateral portion of the dorsal plexus of its own side, and at a level lying between the points of origin of the anterior and posterior nerves. It is the only fibre which passes laterally instead of mesially, and from its superficial and isolated course can be easily traced.

Two fibres (e, $12\,\mu$, and f, $12\,\mu$) leave next and turn inwards and downwards, to curve under the external giant fibre and pass to the same lateral-dorsal region as fibre (d).

The three last-mentioned fibres (d), (e), and (f) thus take the same course

Two fibres (g, $20\,\mu$, and h, $20\,\mu$) alone remain, and they turn abruptly downwards and outwards to curve down ventrally between the entering anterior and posterior nerves, and end in two large unipolar cells of the ventral cell plate on the same side corresponding in position to the cell of the opposite side in which fibre (a) ended. Each gives off a branch to the dorso-lateral plexus as it passes near it. They thus completely resemble fibre (a) in their connections, except in the fact that they do not decussate.

The following table summarises the connections of the descending column of the posterior dorsal nerve on each side.

It will be seen that the fibres fall in three groups defined by distribution and size.

Group 1. Three large fibres which break up in the coarse plexus on most dorsal aspect of the ganglion.

$$\text{Fibre } a \text{ decussates } \dots \quad \dots \; 35\,\mu.$$
$$\left.\begin{array}{c} ,, \quad g \\ ,, \quad h \end{array}\right\} \text{do not decussate} \;\dots\; \left\{\begin{array}{c} 20\,\mu. \\ 20\,\mu. \end{array}\right.$$

Group 2. Three fibres, of which one, the largest, arches above and others below the external giant fibre to pass to the lateral portion of the dorsal plexus.

$$\text{Fibre } d \text{ passes above external giant fibre } \quad 15\,\mu.$$
$$\left.\begin{array}{c} ,, \quad e \\ ,, \quad f \end{array}\right\} \text{pass below external giant fibre} \quad \left\{\begin{array}{c} 12\,\mu. \\ 12\,\mu. \end{array}\right.$$

Group 3. Two fibres which break up in the more lateral portion of the dorsal plexus.

Fibre *b* decussates ...   ...   18 μ.

„   *c* does not decussate...   18 μ.

The table shows that there is a close agreement in size between fibres having the same connections.

THE MOTOR SYSTEM OF NERVES

When we consider together the central origin and mode of distribution of the motor fibres, we see that they present certain features of remarkable interest. Their relations are most clearly shown, because the peculiarities are most exaggerated in the case of the motor system of the flexor muscles just described.

The unit of the system is a large unipolar cell, characterised by the abundance and solidity of its cell-substance, which is loaded with granules of a basophile nature. Like basophile granules generally, they colour more or less intensely with osmic vapour. The cell is enclosed in a nucleated sheath, and suspended by its process in the blood stream of the ganglion.

The single process from this cell runs for a considerable distance without giving off branches. The first branches leave the main process in the ganglion, and break up in the general plexus of that structure. These relations are very clearly shown in the figures illustrating Retzius's work (*loc. cit.*). The cell process then leaves the nerve cord in one or other of the nerves and, on its course to the muscle, it branches *into a great number of fibres which pass to a large mass of muscle fibres.* In the case of the nerves to the very large flexor muscles each myomere (fig. 26) may be innervated by as few as ten nerve fibres, which arise from the same number of nerve cells.

The fibres branch dichotomously, but one of the branches is smaller than the other, while the larger branch is equal in size to the fibre before the branching. This is shown at fig. 12, and, as it occurs in the more peripheral parts of the system, in Haeckel's beautiful drawings (*loc. cit.*). By this method of branching *the transverse sectional area of the unit of the motor system continuously increases as one passes from the nerve cell to the final ending in the muscles.*[1]

The whole structure, which we may call the unit of the motor system, is enclosed in a continuous nucleated sheath which forms the capsule of the nerve cell, and the "tube" of the tubular process.

The substance of the processes is very different from the substance of the cell. The latter is solid and granular, the former is extremely transparent

---

[1] These connections have been traced by Retzius (*loc. cit.*) with the aid of methylene blue; and by myself in sections and dissections.

(compare Haeckel, *loc. cit.* figs. 6, 8, 10, 11 and 12), and consists of two parts:

1. Fine filaments running longitudinally; and suspended in

2. A more or less fluid plasm (Krieger, *loc. cit.*). The junction between cell substance and process substance is extremely abrupt. Similar tubular fibres enclosing filaments are described by Schiefferdecker and Kossel in *Petromyzon*,[1] and the facts at present at our disposal warrant the suggestion that the filaments are the structures which convey the nervous impulses.

The unit of the motor system of *Astacus* thus consists of the following parts:

1. A single nerve cell which, from its histological characters and relation to the blood stream, appears to be a highly metabolic structure; and which is removed by a considerable length of nerve from the direct track of the nervous impulses.

2. A single nerve process from this cell which branches in a characteristic fashion, and consists of a number of filaments, presumably processes of the cell, which are suspended in a plasma.

3. The branches of this process, and, therefore, of the single nerve cell. These are very numerous, are distributed to the plexus of the ganglion and to a very large mass of muscle fibres.

If the prevailing conception of the trophic functions of the central nervous system is correct, we must regard this single nerve cell as the trophic centre for this large mass of muscle fibres.

Though at first sight the motor system of the flexor muscles, consisting, as it does, in each metamere of only a few fibres and cells, seems a simple one, yet a consideration of the arrangement of the flexors themselves shows that we must regard it as a simple contrivance arranged so as to secure a complex result. Figs. 26, 27 and 28 represent dissections of the flexor muscles, and it is there seen that each myomere extends over three metameres. The contraction of the muscles produces flexure of the abdomen, and, if we define the terms "origin" and "insertion" to mean respectively the fixed points from which the muscle pulls and the point of attachment which is moved by the contraction, then we may say that each myomere has three origins and one insertion. Taking the myomere, the main mass of which lies in the fourth abdominal segment, for description, its most anterior origin is from the transverse thickening of the sternum in the second abdominal segment, and it is inserted into the anterior edge of the transverse thickening of the sternum in the fifth abdominal segment (figs. 9, 26 and 28). From the anterior origin to the insertion the muscle runs as a mass of tissue, with a great thickening in the fourth abdominal

---

[1] Schiefferdecker und Kossel, *Gewebelehre d. Menschlichen Körpers*, figs. 129 and 130.

segment, and with an S-shaped curl, arranged in such a way that the fibres in the thickest part of the muscle run almost transversely to the long axis of the animal (figs. 10 a, 10 b, 26 and 28). The other two origins are situated in the fourth abdominal segment and serve to steady the muscle. They are (a) a tendinous attachment to the similar muscle of the opposite side by means of a ventral sheet of fibres springing from the most superficial aspects of the belly of the myomere (fig. 28), and (b) an origin from the pleuron of the same side by a dorsal sheet of fibres (figs. 26, 27 and 28). The ventral sheets of fibres are best developed in connection with the myomeres of the second and fifth abdominal segments.

It is thus clear that we are dealing with a muscular machine of very great complexity, and one the proper working of which must depend upon the co-ordination of the contraction waves in different regions of the large mass of tissue in respect to the time when they occur.

The mode of innervation of the electric apparatus of the *Torpedo* helps us to form a first mental conception of how this correlation is accomplished. In the electric organ the correlation is the simplest possible, namely, the discharge of the individual batteries at the same instant. This, as Wagner has pointed out, is accomplished by the agreement in the length which intervenes between the first branching of the fibre and the final end of each filament. In the case of the flexor muscles of *Astacus* the correlation is a threefold one:

1. A sequence in the time of contraction of the large and distinct masses of fibres which in sections are seen to compose each myomere;

2. The contraction at the same instant of time of the fibres of any one mass, and

3. The serial contraction of the separate myomeres from before backwards.

In case (2) the simultaneous contraction of a large number of fibres receives the simplest explanation, if we suppose that they are innervated by one unit of the motor system, and in this connection I would again draw attention to the constant relation which obtains between the size of the fibres and their morphological relations, and that the distance along the nerve fibre from the central origin of the impulse to where it breaks on the individual muscle fibres is the same. In other words, the explanation would exactly resemble that put forward by Wagner to account for the simultaneous discharge of the batteries of the electric organ. The sequence in the time of contraction of the various fibre masses of the myomere would receive the simplest explanation if we referred it to *differences* in the length of fibre interposed, in the case of each unit of the motor system, between the central origin of the impulses and the muscle fibres.

Lastly, the sequence in the contraction of the different myomeres we might regard as a phenomenon of central origin and to be referred to the time occupied in the transmission of the disturbance from the higher centres down the abdominal nerve cord.

It will be remembered that in describing the motor arrangements of a typical ganglion of *Branchippus* allusion was made to a diffuse motor system connected, not with the purely metameric appendage muscles, but with the longitudinal body muscles which simply send slips to the skin in each segment. And it was there pointed out that this diffuse system consisted of a column of cells on the median dorsal aspect of the cord, from which fibres passed to these muscles. One cannot fail to see in the motor system of the abdominal body muscles of *Astacus* a distinct reproduction of these conditions in *Branchippus*, for the nerves, both to the extensor and flexor muscles of the abdomen, that is, to the body muscles, each arise from two ganglia.

The Inter-ganglionic Regions, or Longitudinal Commissures of the Cord

The longitudinal commissural parts of the cord are composed solely of tubular fibres running parallel to one another and of very various sizes. As has been already noticed, the two lateral halves in the inter-ganglionic regions are clearly marked off from one another by a median vertical septum of connective tissue. There is absolutely no admixture of what may be called ganglionic elements, that is, of nerve cells or plexus, with the commissural fibres; on the contrary, the inter-ganglionic regions correspond, accurately and solely, to the white matter of the spinal cord of Vertebrates (fig. 23). As has been mentioned previously, many of the fibres continue straight through the ganglion to regions below. In the abdominal region, and in sections passing through those parts of the inter-ganglionic cords more removed from the ganglia, a grouping of the fibres into columns is not very obvious, owing to the slight development of the inner sheath. On approaching the ganglia, however, whether from below or from above, a distinct division into columns is very apparent. The fibres in different regions diverge from one another, while at the same time the median fissure widens, and the inner sheath becomes thicker, in order to fill the spaces thus formed, which appear in transverse sections as fissures, or sulci, occupied by an inward extension of the inner sheath (fig. 18). It should be distinctly stated that there is no confusion of the columns of fibres along the whole inter-ganglionic tract; there is no decussation or branching of fibres into this column or that, but, on the contrary, they maintain a parallel course without branching, the different columns simply diverging from one another at the upper or lower limits of the ganglia.

The result of this divergence is that the transverse sectional area of the cord rapidly increases as the ganglion is approached. This increase, however, is also due to another cause, namely, that in each column some of the fibres branch. In one case the fibres were counted in one lateral half of the cord about a quarter of the distance between the first and second abdominal ganglia away from the latter; and again, in a section taken through the region where the divergence of the columns passing to the second ganglion had taken place.

The number of fibres of all sizes found in the hemisection farthest from the ganglion was 613, in the hemisection near the ganglion 815; while in the former case the large fibres numbered 161, and in the latter 225. The division of the fibres may be readily observed in longitudinal sections, or teased preparations, and it is then seen to take place *wholly without the intervention of cells*. Passing on into the ganglion the division of certain of the fibres in each bundle is continued until they form tufts of fine fibres, which, just before they enter the ganglionic tissue, form a plexus, the fibres of which lie mainly in the general direction of the bundle. The transverse or obliquely running filaments also are rather finer, so that, at any rate near where it merges into the parallel bundles of fibres of the inter-ganglionic cord, it may be described as a *nerve plexus, the longitudinal fibres of which are the more prominent.*

In describing the inter-ganglionic region of the ventral chain of *Branchippus* in the first part of this paper, I said that it consisted of a fibrous core of longitudinally arranged fibres with oblique filaments, and a study of the primitive nervous system strongly impressed me with the idea that it is derived from a nerve plexus by the condensation of that plexus along certain lines and in certain places. The connections between the nerve fibres and plexuses described above I regard as suggesting this phylogenetic origin in the case of the nervous system of *Astacus*. In the Zooea of a crab (one of the Paguridæ) which I have examined, the ganglia in the thoracic region are practically continuous. There is no inter-ganglionic cord, but a continuous internal plexus. Here and there in this are delicate strands composed of only a few fibres, and traceable right up to the brain. These represent the great longitudinal columns of the Mammalian cord, and appear to end in the thick sheath of nerve cells surrounding the fibrous core. The longitudinal commissural columns or fibres between ganglion and ganglion have not yet appeared. In other words, the general plexus still performs the functions carried out in the mammalian cord by the root zones of white matter, and probably, in part, by the grey matter itself, and in *Astacus* by the parallel fibres of the inter-ganglionic regions.

As the growth of the Zooea proceeds, the ganglionic masses, the cells and fibrous elements of which are at first quite continuous, are, as it were, pulled apart from one another, to remain connected by an inter-ganglionic zone of parallel fibres.

We may, with Retzius, distinguish three connections for the longitudinal fibres of the ventral cord:

1. Directly with the plexus of the ganglion.

2. Directly with the plexus of the ganglion, but with a T-shaped junction, with cells.

3. With cells whose processes merge in the plexus of the ganglion (fig. 30). The connection through cells is more common in the higher parts of the cord.

Paths of conduction in the cord. In transverse sections through the cord just above or below the ganglion, the fibres in each half of the cord are clearly seen to lie in three main columns—dorsal, median, and ventral. Each of these again is divisible into three regions—internal, median, and external (fig. 18). The ventral columns are related chiefly to the centres of the anterior nerve, the median to the centres of the posterior ventral nerves, and the dorsal to the dorsal nerves. The divisions of the respective columns which abut on the median fissure contain a larger proportion of fibres, which run through the ganglia without change. Thus the great conducting columns, that is, those fibres which correspond in the cord of *Astacus* to the fibres of the pyramidal tracts, and Goll's column in the cord of Vertebrates, form, in the abdominal region, a mass of fibres, wedge-shaped in transverse section, and disposed symmetrically on each side of the median fissure. The base of the wedge is dorsal. Higher up the cord, in the thoracic region, these fibres occupy a still more dorsal position (fig. 24). Further, these fibres are stratified according to their connections, those related to the appendages being most ventral; while of the fibres related to the body muscles, the flexor system is dorsal to the extensor system.

The more lateral and ventral longitudinal fibres of the cord are commissural between one metamere and another.

MINUTE STRUCTURE OF THE FINE PLEXUS

The fine plexus is added to the cord on its ventral side in the abdominal ganglia, and is itself composed of a most complex and dense plexus of filaments which, in teased preparations, and under a high power, sometimes appear faintly moniliform. Teased preparations or sections show that the plexus varies in the fineness of its elements, and in the complexity of their arrangement in different portions. Here and there one sees in the general mass of plexus regions of the most extraordinary delicacy and density (fig. 32). These are often connected with one another by bars of equally dense material, and the masses and bars are embedded in a plexus, the elements of which are larger and more loosely arranged. The appearance is very striking in hæmatoxylin or gold-stained preparations. Into the fine plexus of the ganglion may be traced bundles of fine fibres derived from the nerve bundles, from longitudinal columns, or from the coarse plexus in which the large tubular nerve fibres break up, and each bundle ends in one of these dense masses (fig. 32). The fine plexus is strikingly free from cells of any description. We may thus distinguish in the fine plexus what we may call "centres", each of which forms the immediate termination of a bundle of fine fibrils. Some of these are placed in immediate connection by bars of a similarly dense nature, while all are connected by the general mass of *relatively* less dense plexus.

I have already pointed out that there are regions of the ganglionic plexus to which the fibres of the various nerves may be traced. These we may speak of as the "centres" for the different nerves, and each comprises a coarse plexus of interlacing tubular fibres and a fine plexus. The former structure is merely the expression of the first branching of the large nerve fibres, while to the latter may be traced bundles of fine fibres derived (*a*) directly from the fine nerve fibres, or (*b*) from the coarse plexus. Also by means of serial sections or teasing one can follow a tubular nerve fibre from the inter-ganglionic commissure to where it breaks up into a bundle of fibrils which are lost in the fine plexus. We must therefore look upon this structure as the place where the fibres of the nervous system ultimately communicate with one nother.

Lastly, the nerve fibrils passing to the fine plexus enter it in well-defined bundles which go to histologically distinct regions, and this structural feature we may correlate with the fact that each nerve contains different groups of fibres which supply either different muscles, or regions of the sensory surface supplied by the nerve as a whole, which differ in the fact that stimulation of the one region or the other does not produce quite identical disturbances in the central nervous system.

EXPLANATION OF PLATES VI–IX

Fig. 1. Transverse section through ridge bounding the ventral groove in an inter-ganglionic region. The inter-ganglionic cord also shown. Oc. 4, ob. E, Zeiss, cam. luc.

Fig. 2. Transverse section through mandibular region and just posterior to the mouth, showing the deep post-oral groove lined by sense cells. The long posteriorly directed upper lip is seen in the lower part of the figure. Oc. 4, ob. D, Zeiss, cam. luc.

Fig. 3. Transverse section through the right-hand cord between the circumoral ganglion and the brain. Oc. 2, ob. $\frac{1}{15}$, cam. luc.

Fig. 4. Transverse section through the anterior commissure and just before the exit of the anterior nerve of the right-hand part of the third thoracic ganglion. Oc. 4, ob. E, Zeiss, cam. luc.

Fig. 5. Transverse section of the right-hand half of the same ganglion passing through the exit of the anterior nerve. Oc. 4, ob. E, Zeiss, cam. luc.

Fig. 6. Transverse section of the right-hand half of the same ganglion, and showing the posterior commissure and exit of posterior nerve. Oc. 4, ob. E, cam. luc.

Fig. 7. A portion of a fibre from one of the appendage muscles, showing the termination of a nerve fibre in the protoplasmic sheath. Oc. 4, ob. E, Zeiss, cam. luc.

Fig. 8. A group of ectoderm cells connected with a single (condensing) ganglion cell, which is suspended freely in the body cavity. Oc. 2, ob. $\frac{1}{15}$, cam. luc.

Fig. 9. Dissection of the second abdominal ganglion from the ventral surface. The tissue, which lies on the ventral face of the cord and between it and the posterior sternal arteries has not been disturbed. The course of the anterior nerve on the right side is shown, and the two main branches may be traced, the anterior to the pleuron and more dorsal appendage muscles, the posterior to where it divides into two and curves down into the appendage.

Fig. 10. Dissection of that portion of the posterior nerve of the left side which runs along the ventral aspect of the flexor muscles of the abdomen.

Fig. 10a. Drawing of the same dissection from the side. The posterior nerve is still seen to be running on the surface of the flexor muscles.

Fig. 10b. The same dissection seen from the dorsal side, showing the breaking up of the branches of the posterior nerve into fan-like tufts of fibres, which pass into the extensor muscles.

Fig. 11. Transverse section through one of the anterior nerves of the second abdominal ganglion a short distance away from the ganglion. Osmic vapour, and Flemming's fluid. Right-hand and smaller branch passes into the appendage. Oc. 2, ob. E, Zeiss, cam. luc.

Fig. 12. Large nerve fibre dividing shortly after its exit from the ganglion. Fresh preparation.

Fig. 13. Transverse section through an anterior nerve of the second abdominal ganglion, before it has divided into the two main branches. Oc. 2, ob. E, Zeiss, cam. luc.

Fig. 14. Transverse section through one of the posterior ventral nerves of the second abdominal ganglion close to its exit from that structure. Oc. 2, ob. E, Zeiss, cam. luc.

Fig. 15. Transverse section through the posterior dorsal nerves of the second abdominal ganglion shortly after their exit from the cord. Oc. 2, ob. F, Zeiss, cam. luc.

Fig. 16. Part of a transverse section through cord, showing the exit of the posterior dorsal nerves. Oc. 2, ob. D, Zeiss, cam. luc.

Fig. 17. Section through the skin near the anus, showing the process of a sense cell traversing the cuticle.

Fig. 18. Transverse section through the most anterior portion of the second abdominal ganglion, showing divergence of columns. Oc. 4, ob. A, Zeiss, tube 16·8 cms.

Fig. 19. A piece of the external sheath, or perineurium, of the ganglion isolated by teasing, and viewed from its internal surface. Oc. 2, ob $\frac{1}{15}$, cam. luc.

Fig. 20. Optical section through the external or perineurial sheath of the ganglion. Oc. 2, ob. $\frac{1}{15}$, cam. luc.

Fig. 21. A bundle of fibres isolated by teasing from the inter-ganglionic cord where it enters the ganglion, showing an imperfect sheath of flattened cells. Oc. 2, ob. $\frac{1}{15}$, cam. luc.

Fig. 22. Supporting or "neuroglia" cell from the ganglion. Isolated by teasing. Oc. 2, ob. $\frac{1}{15}$, cam. luc.

Fig. 23. Transverse section through the inter-ganglionic cord just below the second abdominal ganglion, showing the ascending roots of the posterior dorsal nerves. From a micro-photograph.

Fig. 24. Dissection of the ganglion of the chelæ to show the relative positions of the fibres which are going to regions below, and of the fibres passing to and through the substance of the ganglion. Viewed from below.

Fig. 25. Dissection, showing the long double roots of the nerves to the chelæ and walking legs. Viewed from below.

Fig. 26. Dissection, showing the flexor muscles and the origin of the nerves of the second and third abdominal ganglia. The extensor muscles and gut have been removed, and the preparation is viewed from the dorsal surface. Thin sheets of muscle fibres arise from the belly of the great mass of the flexors in each segment, and stretch across to be joined in the middle line and above the nerve cord by tendinous tissue. These are shown intact in Segment II, and cut in Segments V and VI, and completely removed in Segments III and IV.

Fig. 27. Dissection, showing the most superficial aspect of the flexor muscles. The extensor muscles of the right-hand side have been removed.

Fig. 28. Dissection, showing two myomeres of the flexor muscles. On the left, the dorsal muscle sheet, which extends outwards into the pleuron, has been removed. Diagrammatic.

Fig. 29. Dissection of the extensor system. Viewed from above.

Fig. 30. Group of nerve cells from ganglion of chelæ, isolated by teasing from the ending in the ganglion of a bundle of fibres from the longitudinal commissures. Oc. 2, ob. $\frac{1}{15}$.

Fig. 31. Teased preparation of "Punkt Substanz" with two nerve cells. From ganglion of chelæ. Oc. 2, ob. E, Zeiss, cam. luc.

Fig. 32. Section through "Punkt Substanz", showing the entrance of a bundle of fine fibres from the posterior nerve. From section next in series to fig. 34. Oc. 4, ob. $\frac{1}{12}$, cam. luc.

Fig. 33. Section (slightly oblique) through region of entrance of anterior nerves, showing external and median centres, and the fine fibres of dorsal and ventral roots passing to them. From a micro-photograph.

Fig. 34. Section through entrance of posterior ventral nerves. Oc. 4, ob. D, Zeiss.

Fig. 35. Longitudinal vertical section, showing the entrance of the anterior and posterior nerves into the ganglion. (Figs. 32, 33, 34 and 35 are from the serial sections through the second abdominal ganglion.)

Fig. 36. Part of transverse section through the ganglion, showing the coarser and more dorsal ganglionic plexus. The nerve cells, and those of the lateral group, which are connected with the thick-sheathed tubular fibres of the interior nerves. Oc. 2, ob. $\frac{1}{15}$, cam. luc.

Plate VI

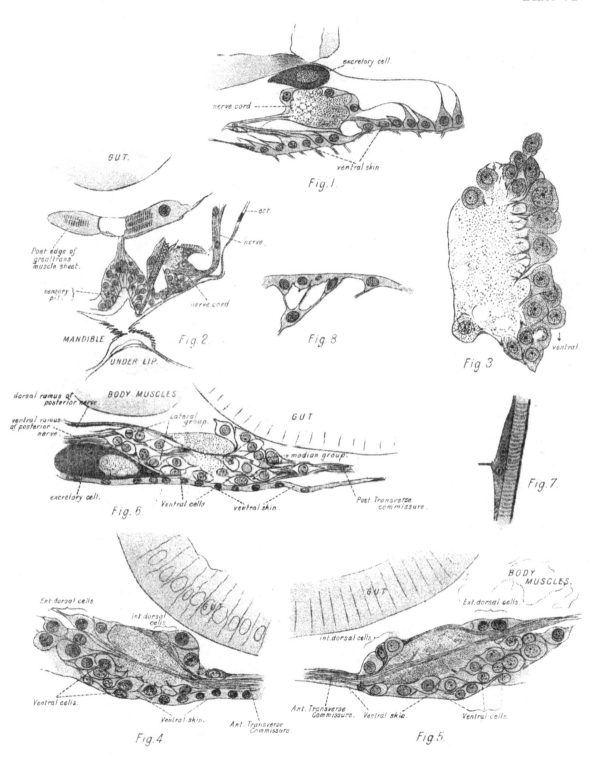

Fig. 1.

Fig. 2.

Fig. 8.

Fig. 3.

Fig. 6.

Fig. 7.

Fig. 4.

Fig. 5.

Plate VII

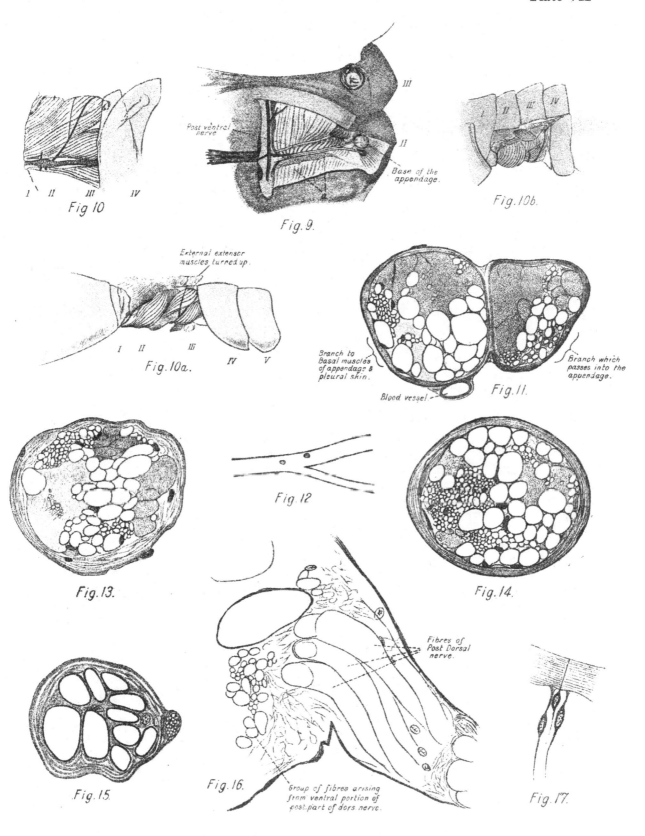

Fig 10

Fig. 9.

Post ventral nerve

III

II

Base of the appendage.

Fig.10b.

External extensor muscles turned up.

I II IIi IV V

Fig. 10a.

Branch to Basal muscles of appendage & pleural skin.

Branch which passes into the appendage.

Blood vessel.

Fig. 11.

Fig.13.

Fig. 12

Fig. 14.

Fig. 15.

Fig. 16.

Group of fibres arising from ventral portion of post. part of dors nerve.

Fibres of Post Dorsal nerve.

Fig. 17.

Plate VIII

Fig 22.

Fig. 18.

Fig. 19.

Fig.21.

Fig.20.

asc. root
of post dorsal
nerve.

Fig. 23

a'

Post vent nerve.

Dorsal root
ant. nerve.

Ventral root
ant. nerve.

b

Fig. 35.

d.c.

v.c.

Fig. 33.

Fig. 31.

c

a

b'

c

b

f. pl.

Fig 34.

large tubular
nerve fibres.

Fig. 32.

Plate IX

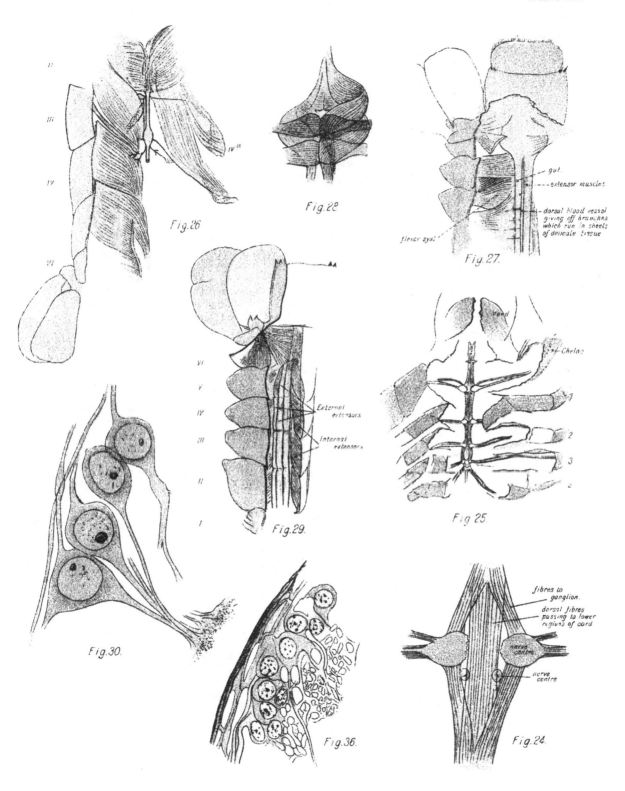

Fig. 26

Fig. 28

Fig. 27.

gut.

extensor muscles

dorsal blood vessel
giving off branches
which run in sheets
of delicate tissue

flexor syst.

Fig. 29.

External
extensors

Internal
extensors

Fig. 25.

hand

Chelae

Fig. 30.

Fig. 36.

fibres to
ganglion.

dorsal fibres
passing to lower
regions of cord

nerve
centre

nerve
centre

Fig. 24.

10

ON THE CHARACTERS AND BEHAVIOUR OF THE WANDERING (MIGRATING) CELLS OF THE FROG, ESPECIALLY IN RELATION TO MICRO-ORGANISMS

With A. A. KANTHACK

[*Phil. Trans.* (1894), B, CLXXXV, 279]

CONTENTS

INTRODUCTORY. (Added September 1893)

[The most salient feature of the wandering cells of the body is their marked increase in numbers in inflammation. Virchow, originally, in his great work, *The Establishment of the Cellular Pathology*, traced the "Rundzellen Infiltration" to a proliferation of connective tissue cells. Later, however, Recklinghausen and Cohnheim showed that the white cells of the blood could penetrate the walls of the blood capillaries, and so pass to an inflamed area. Thus, the first step was taken towards defining the morphological position of the wandering cells as an independent portion of the body.

The fact that the wandering cells ingest discrete particles has long been known. Claus, Schultze, Haeckel, Hoek, and others described the process as occurring in both Vertebrates and Invertebrates. But the genius of Metschnikoff[1] first suggested the importance of this process to the body. He showed that ingestion was, in the case of bacteria and digestible ingesta generally, followed by digestion. This discovery enabled Metschnikoff to attain to some conception of the real significance of the inflammatory process, and the part the wandering cells play therein.

While Metschnikoff was working at this question, startling additions

[1] *Annales de l'Institut Pasteur* (1887), pp. 321–36; *ibid.* (1889), p. 25 *et seq.*; *ibid.* (1889), p. 289 *et seq.*; *ibid.* IV, 65 *et seq.*; *ibid.* IV, 465. Virchow's *Archiv*, XCVII, 177; *ibid.* CVII, 209; *ibid.* CIX, 176; *ibid.* CXIII, 63.

were being made to our knowledge of the life-history of vegetable micro-organisms and the part they play in disease. The enormous importance of his discoveries, as suggesting an explanation of the resistance of animals to the invasion of these bodies, became at once obvious, and Metschnikoff was led to formulate a theory of the immunity of animals in which the resistance to the disease germs is referred solely to the phagocytic activities of the wandering cells.

From this time onwards the wandering cells have been regarded almost solely from the relatively narrow standpoint of their relation to the conflict with pathogenic germs.

The next advance in our knowledge was the direct outcome of the labours of those bacteriologists who refused to accept the phagocytic theory of immunity as a complete solution of that problem.

The discussion of the origin of immunity, like the earlier discussion of the origin of pus corpuscles, led to the enunciation of extreme theories. Those observers who found themselves in opposition to Metschnikoff, went so far as to deny to the wandering cells any direct participation in the processes which attend the development of immunity, and, without attempting to explain how the plasma gained such properties, they gave to the world one or other form of a "humoral theory".

In this way two distinct schools arise, one which attributed all to the phagocytic activity of the cells, another which regarded the fluid plasma as all important.

The result of an attempt to effect a compromise between these extreme views was the formation of a third school, which regards any special activity of the plasma as being due to substances derived from the cells, and its adherents find their support in the work of Wooldridge, of Hankin, and of Buchner.

In order to see how this led to a wider appreciation of the functions of the wandering cells it is necessary to trace briefly the main lines of the discussion.

In 1887, Fodor, Nissen, and others established the fact that the blood serum possesses marked bactericidal powers, and the result of this discovery was the enunciation of a humoral, as opposed to a cellular theory of immunity.

Metschnikoff, and other supporters of the cellular theory, soon proved that this bactericidal power was not possessed by blood in the body, but was developed as a result of post-mortem changes, and, in 1888, Buchner clearly correlated the development of the bactericidal power with the breaking up of the leucocytes in shed blood, and, in 1890, Hankin succeeded in isolating a bacteria-killing substance from lymphatic glands and the spleen. Thus, the idea that some, at any rate, of the wandering cells were concerned in the production of a peculiar substance characterised by its bactericidal power was suggested.

The nature of this substance, and especially the nature of its action on bacteria, whether, for instance, it destroys them or merely hinders their growth, are questions which have engaged the attention of many workers, but which do not concern us here.

Turning to the development of our histological knowledge of the wandering cells, we find that the existence of more than one kind of leucocyte in Vertebrates was recognised at a very early period. But it was not until Ehrlich,[1] in 1878, drew attention to the specific granulation of these bodies, and, with the help of the aniline dyes, provided us with a method of histochemical analysis, whereby the different forms could be readily recognised, that any great advance in the direction of determining differences of function became possible.

The striking histological advances made by Ehrlich and his school were, until quite lately, completely ignored in the discussion which raged around the cellular theory of immunity, and the wandering cells still continued to be spoken of indifferently as phagocytes, with no recognition of the possible existence of diverse forms endowed with diverse functions.

It is difficult to determine when or how the attention of those concerned in the discussion came to be attracted to the granulation of the cells, but it may be traced mainly to the work of Buchner, Hankin, and those others who derived the bactericidal substance from the leucocytes.

Although the observations which we are about to record have, we venture to think, a direct bearing on the theory of immunity, we would ask that they might be regarded, not in this light only, but as some contribution to the more general natural history of the wandering cells.]

On investigating the histology of the body fluids of certain Invertebrates and Vertebrates, we find that animals widely separated in structure and habits possess the same kinds of wandering cell. But we also find (1) that, within the limits of a single group of animals, the simplest forms possess only one kind of wandering cell, while those with greater structural complexity have all the three typical forms sharply and completely distinct from one another; and (2) that during the fœtal period a mammal has only one kind of wandering cell. These facts suggest two ideas; firstly, that a certain fixity of type must be accorded to each kind of wandering cell, that the different forms found in the more complex animals must be regarded as distinct from one another in their development and life-history, even if they be regarded as having a common origin; and, secondly, that, corresponding with this divergence and fixity of type, there must be divergence and fixity of function. We have endeavoured to demonstrate a disparity

[1] Ehrlich, *Farbenanalytische Unters.*, Berlin.

of function comparable to the disparity of form by comparing the behaviour of the different kinds of wandering cells towards various substances when added to the lymph or blood.

Since the wandering cells of the frog retain their vitality for a long time after removal from the body, that animal has mostly been used. The lymph is obviously most serviceable for examination, but the cells of the blood are similar and offer similar phenomena.

Some observations have also been made on *Astacus*, the rat, and the rabbit.

The experiments on the cells when out of the body have been controlled by parallel experiments on the cells while still within the body.

We have to thank many friends for willing help and criticism, but special acknowledgment is due to Miss Greenwood for allowing us to see her preparations, and follow the results of her later and still unpublished work on digestion in Protozoa.

Section I

Methods Employed

In determining the identity of the wandering cells, we have had regard to differences as to shape, whether the cell be resting or amœboid, as to the texture of the cell substance, as to the nuclear type, and as to the presence or absence and the histochemical nature of the cell granules. The shape and appearance of the cells when resting is not more serviceable for their identification than their appearance when active. The manner of emitting pseudopodia and the appearance of the cell when it has thrust out these appendages are markedly different in different cells.

Differences in the texture of cell substance are brought into marked prominence by the use of iodine, and this reagent cannot be too highly praised in this connection.

The nuclear type of the various cells has been studied with the aid of a solution of methyl green, slightly acidulated with acetic acid, and to which a trace of osmic acid has been added.

Nuclear characters are also shown by treatment with an alkaline alcohol-osmic acid solution of methylene blue, which is practically Lœffler's solution with much less methylene blue present and with a trace of osmic acid added. With this solution eosinophile[1] granules remain entirely uncoloured and unchanged. Amphophile granules are stained blue, or rarely a very dull violet when viewed with yellow light. And the basophile granules appear violet with daylight, and brilliant rose with yellow light. Nuclei and microbes are blue with both lights. The substance which produces the

---

[1] Ehrlich, *Farbenanalytische Unters.*, Berlin.

rose-coloured modification of methylene blue does so whether it be present as granules in the cell substance, or dissolved in the surrounding fluid. The reaction also survives with unaltered intensity when the preparation is dried at the temperature of the air and mounted in Canada balsam.

The study of the living cells and their behaviour towards noxious or innocuous substances has been carried out (1) by injecting various substances into the lymph spaces of the frog, and withdrawing drops of lymph for examination at varying intervals of time, and (2) by hanging drops. The hanging drops were suspended on the under side of a coverslip in moist chambers sufficiently large to provide air enough for the needs of a small drop of lymph for about ten hours, without introducing a fresh supply. In this simple way we have been enabled to continuously observe the processes taking place in a drop of lymph or blood after inoculation with microbes, poisons, etc., for periods up to ten hours. Discontinuous observation has been kept up for 40–50 hours. The coverslips used were always carefully cleaned with acid and absolute alcohol, and then sterilised by heat immediately before use.

The study of these drop cultures was controlled by examining lymph taken from the lymph sac and peritoneal cavity of a frog, into which microbes, etc. had been injected. In all cases the most complete accord was found, frequently extending to the element of time.

To study the effects of temperature, the drops were placed on the ordinary metal stage, through which warmed or cooled water was circulating. In these experiments it is essential that the whole chamber and the coverslip should be brought to the requisite temperature before the lymph is added. Otherwise the earlier stages will occur before the temperature has either risen or fallen to the required point.

It is most important to note that, in order to obtain satisfactory results, it is necessary to use only freshly captured frogs.

Section II

Histology of the Frog's Wandering Cells

In the body of the frog three kinds of wandering cells occur. These are (1) the eosinophile cell, (2) the hyaline cell or phagocyte, and (3) the basophile cell with rose-reacting granules. These, together with the red corpuscles and platelets, constitute the sporadic mesoblast of the frog, thus constituting a tissue whose elements, unlike the elements of the other tissues, have no coherence, and but little fixity of place. Like the other tissues of the body, however, this particular tissue increases or decreases in bulk in correspondence with certain bodily needs, the increase in bulk being largely due to the multiplication of the cells, whether eosinophile, hyaline, or rose staining, by binary fission. This occurs freely in the body fluids, and may be watched outside of the body in a hanging drop.

(1) *The eosinophile cell* (Pl. X, figs. 1–3, 10, 14), when resting, is spherical in shape. The more central portion of the cell is occupied by a greater or less number of highly refractive granules, each granule having a yellow-green lustre. These granules were identified by Ehrlich with his α, or eosinophile group.[1] With very high magnification the individual granules sometimes appear to have a short and long axis, and to be slightly spindle-shaped. The central portion of the cell, or endosarc, which contains the granules and the nucleus, is clearly distinguished from the delicate ecto-sarcal layer of very transparent mobile protoplasm, though, in the frog, owing to the smaller size of the cell, the distinction is not so beautifully shown as it is in the larger eosinophile cell of *Astacus*.[2] The nucleus of the eosinophile cell is exceedingly characteristic. It is elongated and bent to a horse-shoe shape. The chromatin filaments are either irregular or radiate from two nodes, situated towards either end of the nucleus. Sometimes the nucleus is trilobed. When proliferation of the eosinophile cells is taking place they are to be found with more than one nucleus. Under appropriate stimuli the eosinophile cell becomes very active. Such stimuli are, normally, of a chemical nature and may be regarded as a change in the surrounding fluid. This may be produced either by clotting, or by the introduction of foreign substances, such as microbic poisons. If the cell is floating freely in the fluid, then the activity is confined to the thrusting out of the ectosarc as short filiform pseudopodia, which radiate from the still spherical endosarc, and do not necessarily result in locomotion. If, however, the cell is in the neighbourhood of a chain of active microbes, then the pseudopodial activity becomes so far modified, that the cell progresses towards the chain. Lastly, when the cell effects actual contact with the microbe the pseudo-podial activity becomes suddenly changed into violent streaming move-ments, which result in the extension of the cell along the chain (figs. 10, 14, 17). So-called indifferent particles of any kind or shape, such as indian ink, in no wise affect the activities of the cell, even though accidental contact occur. Contact with an active microbe, however, not only stimulates the cell to increased movement, but also produces a new activity of a glandular nature. The granules are thrust from the endosarc into the ectosarc, and travel towards that portion of the cell which is in contact with the chains; there they rapidly lose their brilliant refractive nature, shrink in size and disappear (fig. 10 *b*).

Complete discharge of the granules may occur, and then after an interval

---

[1] Ehrlich, "Ueber die specifischen Granulationen des Blutes", *Verhand. d. physiol. Gesellsch. zu Berlin* (1878).

[2] Compare figs. 1 and 5, Pl. VII, *Journ. Physiol.* vol. XIII.

the cell may reform its granules which, at first, are different from the initial eosinophile granule. These processes are described in detail in a later section; enough has been said here to substantiate the statement that in the eosinophile cell we are dealing with one which has both glanular and amœboid properties corresponding to differentiation of the cell's body into a central glandular endosarc, and a peripheral contractile ectosarc.

A change of medium also stimulates cell division. This process we have watched in a drop culture (Exp. I), and in all cases the daughter-cell has been free from granules. In a comparatively short time granules commence to be formed at a point situated rather to one side of the centre of the daughter-cell.

The Re-formation of Granules

The amphophile granule (fig. 17). The glandular activity of the cell does not always result in a complete obliteration of the granule as a histological feature of the cell. Under the influence of poisons, *e.g.* urari, or bacterial products, the granules may be so far altered that they stain with methylene blue as well as with eosin. In other words they become amphophile in the sense in which Ehrlich uses the term. The relation of this change to the complete discharge of the granule can be better discussed after the different phenomena exhibited by the eosinophile cell have been more fully described. The amphophile granule is also produced in another way. The eosinophile cell will, after the complete or partial discharge of its granules, load itself with newly formed ones. These at first appear to be always amphophile, and there is, therefore, a stage in the elaboration of the eosinophilous substance at which this is amphophile.

(2) *The hyaline cell* (figs. 1, 4 and 17) is the only one of the three elements which has been seen by us to manifest the phenomenon of phagocytosis, that is, which ingests and digests discrete particles. This cell is the "mononuclear" cell of Metschnikoff.[1] Since, however, the other cells of the frog are strictly mononuclear, we have thought it wise to adopt a different term. In the resting condition the cell is spherical. The cell substance betrays no differentiation into ectosarcal and endosarcal portions. Therefore, when this cell becomes active, it may exhibit the wildest irregularities of form. There is no special pseudopodial character, the processes may be simple, or branched, they may be extraordinarily long and attenuated, or mere rounded eminences. Lastly, the cell may become excessively flattened until it is reduced to a protoplasmic film. The cell substance is very clear and transparent, and free from specific granules. The nucleus is exceedingly characteristic. It presents, when stained, the appearance of a spherical bladder, formed by a very delicate staining

---

[1] *Leçons sur la Pathologie comparée de l'Inflammation*, Huitième Leçon, p. 131.

membrane, and enclosing a sharply defined spherical nucleolus, which is placed at the centre of the sphere (fig. 4). Fine filaments may often be traced from the nucleolus towards the nuclear capsule, but they are seen with difficulty. When proliferation of these cells occurs nuclei may be found containing two nucleoli (fig. 17). The most prominent functional characteristic of the hyaline cell is its power of ingesting and digesting solid particles. At the same time this power is strictly limited, for the hyaline cell is absolutely incapable of ingesting, or even of effecting contact with, for instance, an intact and virulent anthrax bacillus. The bacillus must first be killed or maimed. In the presence of such substances as coagulated proteid, or dead bacilli, the cell exhibits the following well-marked phenomena. The food particle is ingested by pseudopodial activity in the ordinary way, and at first lies merely embedded in the cell substance. A digestive vacuole is formed round the particle, which floats freely in the vacuolar fluid. The particle, if soluble (e.g. the bacillus or proteid mass), is dissolved, and, lastly, the now empty vacuole closes. In other words, this cell accurately reproduces the ingestive and digestive phenomena of the carnivorous Protozoa.

Though we have not endeavoured to witness the fission of these cells, yet we have found them increase in number in a hanging drop. We have also seen a hyaline cell multiply in a hanging drop by a process of budding, rather than by binary fission. In this case the daughter-cell is about one-third the diameter of the mother-cell. Such young hyaline cells may be seen in hanging drops when leucocytosis, that is to say, an increase in the number of cells, has been induced by the presence, for instance, of some microbe (fig. 7).

The hyaline cell is much more easily killed by change of medium than the other two kinds of cells. It, however, exhibits considerable differences in this respect in different frogs. If the cells be asphyxiated, as, for instance, by putting a coverslip down on to a drop of lymph, they frequently burst in a way precisely similar to the bursting of the "explosive" or hyaline cells of *Astacus*.[1] Rarely a similar fate befalls some of the hyaline cells in a hanging drop, and it is probably due to the formation of a very dense clot. In the relative instability of its cell substance the hyaline cell recalls the similar ingestive cell of *Astacus*. In *Astacus*, however, the cell is so unstable that it has been called the "explosive cell". The hyaline cell of *Astacus* is similarly the phagocyte of that animal.

(3) *The rose-reacting basophile cell* (figs. 4 and 5). This cell, like the similar cell in *Astacus*, is characterised by its relative immobility. We have never seen any signs of active pseudopodial movement. In normal lymph the rose-reacting cells are very few in number (about 2 per cent. of the total

[1] Hardy, *Journ. Physiol.* vol. XIII.

number of cells), and those present are also small as compared with the eosinophile and hyaline cells. As a result of the presence of foreign matter in solution in the plasma, these cells increase both in number and in size, and become highly charged with granules. A similar increase is frequently associated with an œdematous condition of the animal. A comparison between figs. 5 *a* and 5 *b* will render clear the great difference between what may perhaps be called the resting and active condition of the rose-staining cells. The transition from the resting to the active condition will take place in hanging drops. In the resting condition the cells are small, spherical, and have a large spherical nucleus. The cell substance is very scanty, but is charged with granules which are much smaller and very much less refractive than the eosinophile granule. It was in these cells that Ehrlich first described the existence of the fine basophile granule, or δ-granulation.[1] In the latter character, namely, the duller appearance in the living cell, they resemble the similar rose-reacting granules of *Astacus*. The spherical nucleus either stains uniformly, showing no chromatin filaments, or it encloses a central irregular chromatin mass, from which filaments stretch to the nuclear capsule.

The enlarged active cell is usually angular in shape, and the nucleus is then mostly oval. The cell substance is highly charged with granules, which, as in the similar cell of *Astacus*, extend quite to the margin of the cell. There is never any trace of an ectosarc. Not infrequently the cell body contains a vacuole. We have never witnessed the manifestation of any kind of ingestive activity on the part of the rose-reacting cell, and the increase in size and number appears to be related solely to the presence of substances in solution in the lymph. Little is known of the distribution of these cells in the body.

In *Astacus*, as has been previously shown,[2] similar cells inhabit a peculiar tissue, which forms the adventitia of the anterior sternal artery.

In the frog they are normally present in the connective tissues, and in the Mammalia, probably, identical cells are found in the peculiar adventitia of the arteries of the spleen and grouped about the capillaries, especially of the splanchnic area. Dr Sherrington exhibited specimens to the Physiological Society, in May, 1891, which showed very similar cells grouped about the blood vessels of the intestine in cases of cholera.[3]

---

[1] "Beiträge z. Kenntniss d. Granulirten Zellen, etc.", *Verhand. d. physiol. Gesellsch. zu Berlin* (1878–79), No. 8.

[2] Hardy, *Journ. Physiol.* XIII. 177.

[3] Since writing the above we have found that the basophile cell found in the peritoneal cavity of the frog differs from the basophile cells found elsewhere in the body. The differences are of two kinds: (1) the peritoneal cells are larger, and (2) the basophile granules which they contain are larger and are well-defined spheres. The peritoneal cells thus show granules belonging to Ehrlich's group γ.

Giant cells (figs. 11 and 16). These do not normally occur in the frog, but are formed at a certain stage after the introduction of microbes into the lymph. Their formation may be followed in the hanging drop, and they are then seen to be very remarkable bodies, produced by the partial fusion of eosinophile cells or hyaline cells, or both. They are, therefore, of a plasmodial nature. As will be seen later, their formation recalls, in its mode and effects, the temporary conjugation of some Protozoa, and similarly these plasmodia ultimately disintegrate into their constituent cells (see Section V). These plasmodia are produced either by the fusion of (1) eosinophile cells—in this case the fusion is limited to the ectosarc—or (2) hyaline cells, when the fusion of cell substance is, so far as can be seen, complete. Lastly, there is also (3) a plasmodial mass, formed of both eosinophile and hyaline cells. With regard to the third case we are not prepared to say whether the fusion is complete, a single plasmodium resulting, or whether the plasmodium is double, the eosinophile cells being in contact with an inner hyaline plasmodium.

Various observers have noted the fact that the eosinophile cells are relatively more numerous in frogs during winter. Our observations have been confined to the summer months, and we have found that the relative number of the different classes of cells varies in different parts of the body. In the lymph from the subcutaneous lymph spaces the eosinophile variety form from 17 to 25 or 30 per cent. of the total number of cells. The basophile cells form about 2 per cent. In the peritoneal fluid the percentage of eosinophile cells is higher, ranging from 30 to 50 per cent.

The histological structure of the sporadic mesoblast of the frog, excluding the red corpuscles and platelets, which stand on a different footing from the rest, may be summarised as follows:

Normal	I. Cells normally free in the blood and in the lymph.	(a) Eosinophile cells, nucleus horse-shoe shaped or lobed; do not ingest particles; but are motile unicellular glands.
		(b) Hyaline cells, free from specific granulation: nucleus round with central nucleolus. Phagocyte, *i.e.* they possess the power of ingesting and digesting discrete particles.
	II. Cells are few in number and small in normal lymph. Normally present in the lacunar spaces of areolar tissue.	(c) Basophile cells, spherical, with scanty protoplasm when small, angular, rounded or flattened when large, cell substance charged with tiny basophile granules, which give a vivid rose colour with methylene blue. Large oval or round vesicular nucleus, sometimes containing irregular chromatin mass and filaments.
Abnormal	III. Large amœboid cells, vacuolate, with ingesta frequently in the vacuoles, multi-nuclear, very active and phagocytic.	Giant cells formed by fusion of hyaline cells similar to the large phagocytic cell of *Astacus*.
	IV. Small bodies either round and quiescent or amœboid.	Nucleated cells budded off from the eosinophile or hyaline cells.

SECTION III

Leucocytosis

The most constant phenomenon of inflammation in a vascular part is the appearance of an increased number of wandering cells in the tissue outside the blood vessels. Some of these cells appear on the scene by means of migration from the interior of the blood vessels, others find their way thither by migration from neighbouring tissues along lymph paths, but the increase is also largely due to the direct multiplication of the cells by fission on the spot. Whether these cells are attracted by means of "chemotaxis" or not, does not concern us here, and we shall abstain from offering an explanation of the process or cause leading to this collection or infiltration of cells. A question which was of much greater importance to us is the nature of the cells, and the sequence in which these cells appear. On injecting substances like anthrax culture under the skin of the frog, do the three kinds of cells appear simultaneously, or does one class of cell appear before another?

This is a point which, so far, has been neglected by those who have made "chemotaxis" a subject of special investigation. Being satisfied, or assuming, that the attracted leucocytes are phagocytic in property, they have considered it sufficient to prove that, in certain animals, by means of certain substances, or bacilli, leucocytes are attracted, and positive chemotaxis was considered to be of special use to phagocytosis, because it was tacitly understood that the leucocytes attracted to the spot were always active phagocytes. We shall now show that this conception is true in part only, and that the usefulness of the wandering cells in the conflict against injurious substances does not lie wholly in their phagocytic powers.

We have seen that the frog possesses three kinds of wandering cells. Of these, the eosinophile cells are never phagocytic; the hyaline cells, on the other hand, are so. If, therefore, the usefulness of "positive chemotaxis" in the battle against micro-organisms lies solely in the fact that thereby phagocytes are attracted, the hyaline cells should be the ones to appear on the battle-field. We found, however, that the cells which first collect in greatest number (or are attracted) are the eosinophile ones, and that it is not until some time has elapsed that the hyaline cells become evident.

On injecting a fresh culture of anthrax bacilli into the subcutaneous tissue of a pithed frog, and keeping it at the ordinary temperature, which was 12° C. at the time these experiments were performed, and removing drops of lymph with a capillary pipette at intervals of half an hour, it was noticed that the leucocytes which first appeared were the eosinophile ones. They rapidly increased in number until the third or fourth hour after inoculation, and collected in masses around the bacilli. From the third or fourth hour onwards hyaline cells become conspicuous, increasing gradually in number. The eosinophile cells still further increased, and after eighteen or twenty-four hours their number was very great, and often masses of bacilli could be seen surrounded by those cells, and wherever the number of eosinophile cells was greatest, there also the hyaline cells were most conspicuous, and phagocytosis most marked.

The first phenomenon noticed, therefore, after an inoculation with anthrax, is the appearance of eosinophile cells, or, *sit venia verbo*, an eosinophile leucocytosis. Many of these cells certainly find their way to the field of action by migrating from the vessels, others from the neighbourhood, eosinophile cells being always found free, *i.e.* outside the vessels, in the lymph spaces. Undoubtedly also the cells multiply *in situ*. For, as we shall show later, a multiplication of the eosinophile cells may be easily demonstrated on the slide in a hanging drop; again, all the above phenomena may be watched in the amputated leg of a frog, and also if after stripping the skin of a frog's thigh from the muscles, and tying it at both ends so as to convert it into a closed tube, we inoculate this tube with anthrax bacilli, and keep it at the ordinary temperature. "Chemotaxis", therefore, cannot by itself explain this leucocytosis, for besides an attraction, if such exist, we also have an active proliferation of eosinophile cells.

On repeating the above experiments with cultures of *Bacillus pyocyaneus*, or with beer yeast, or even with inorganic irritants, such as nitrate of silver, the same result was always obtained: the cells to appear first were invariably the eosinophile leucocytes, and it was only later that the phagocytes increased at the seat of lesion.

Other methods also were employed to show that the eosinophile cells are the first attracted. On placing small sponges dipped in cultures of anthrax, pyocyaneus, or yeast, or even in a very dilute solution of nitrate of silver, either under the skin or into the peritoneal cavity of a frog, after two to four hours eosinophile cells were almost exclusively found in the meshes of the sponge. Lastly, on placing capillary tubes filled with the same substances under the skin, or into the peritoneal cavity of a frog, after two to four hours the same result was obtained.

It is only after the eosinophile leucocytosis is established that the hyaline cells begin to appear in appreciable quantity, and they now rapidly increase in number. At the same time, however, the eosinophile cells do not cease to increase, so that there is both an eosinophile and a hyaline leucocytosis. The hyaline cells at once become active, and begin to take up the bacilli, and, as we said above, are most active where the eosinophile cells are most numerous. This has been observed so consistently, that even in the absence of better evidence, a correlation between phagocytosis and the eosinophile leucocytosis suggests itself.

The hyaline cells now rapidly increase in number, and after eighteen to twenty-four hours are often very numerous; they may eventually outnumber the eosinophile cells, and they mostly contain bacillary remains in their interior.

When phagocytosis is at its height we notice yet another change. The rose-reacting cells, which under ordinary conditions are but sparse, are now greatly increased, so that at one time (about the eighteenth to twenty-fourth hour) all the three cellular elements are numerous.

We have then the following changes after inoculating a frog with anthrax. At first, the eosinophile cells appear and collect around the bacilli (eosinophile leucocytosis); then the hyaline cells appear (hyaline leucocytosis) and soon show a keen phagocytic action, the eosinophile cells increasing, however, at the same time; and latest of all the rose-reacting cells also increase (rose-reacting leucocytosis). We shall attempt to show later the specific value of each of these groups of cells. One form of leucocytosis merges into and overlaps the other, and it is difficult to separate them from each other by any time limit. This much, however, is certain, that we have never observed phagocytosis with virulent anthrax without previous eosinophile leucocytosis.

If, for some reason or other, the eosinophile cells do not appear, the hyaline cells are apparently powerless. Thus, on warming a frog which has been inoculated with anthrax, the eosinophile leucocytosis is absent, and though phagocytes appear at times in large numbers, the bacilli will thrive, being left unharmed. It cannot be objected that heat "paralyses" the phagocytes, because some of them did contain bacilli, and they will not refuse indian ink or vermilion particles, though exposed to a temperature of from 25 to 30° C.

Again, on narcotising a frog with a mixture of chloroform and ether, and inoculating it, while under the influence of the anæsthetic, with anthrax baccilli, the latter, as Klein and Cowell have shown, will often grow well. In some cases, however, even under these conditions, they will refuse to grow. In the latter cases, the eosinophile leucocytosis is always extremely well marked, in the former, the number of eosinophile cells is very small. Here also it cannot be claimed that the chloroform-ether mixture paralysed the phagocytes, because the latter, at times, were present in fairly large numbers, and, though they left the bacilli untouched, they invariably took up spores.

The same was observed when the circulating blood was replaced by saline solution (0·75 per cent.). A cannula was tied in the conus arteriosus and the vena cava ant., and the NaCl solution was allowed to circulate under a low pressure for two to four hours, and then the frog was inoculated in its subcutaneous tissue. The bacilli grew in most cases, and then the absence of an eosinophile leucocytosis was marked, although the phagocytes were often present in large numbers, containing, in many instances, spores.[1]

No doubt here, chloroform and ether and the artificial circulation cause changes in the chemistry of the tissues, but it must always remain a significant fact that in the absence of eosinophile cells the bacilli thrived excellently. The result of warming a frog for a long time, keeping it for instance at 25° C. for a week or more, is quite different from that of a brief warming. If, at the end of the longer period, the animal be inoculated with anthrax, it is found to be immune as in the normal condition, and now, unlike the case of the brief warming, an abundant eosinophile leucocytosis takes place. Lastly, it may be mentioned here, that in many instances a cluster of bacilli was seen to be closely surrounded by a mass of eosinophile cells, with hardly a phagocyte near. The bacilli were then extremely degenerate and broken up, or forming spores, staining very badly or indifferently with methylene blue.

All these observations tend to show the great important of the eosinophile cells in the conflict against the bacilli or micro-organisms.

Leucocytosis in a hanging drop. The three stages of leucocytosis can also be demonstrated outside the body on the slide, by means of a hanging drop. On removing a drop of lymph from a frog and inoculating it with anthrax, and selecting a suitable spot for examination, it is seen that the eosinophile cells collect in numbers around the cluster of bacilli. In a successful specimen there is a well-marked increase of these cells, and it is possible at times to watch the division of these leucocytes.

Later on, the hyaline cells approach the cluster, so that at this time the mass of bacilli is surrounded by both eosinophile and hyaline cells. The former appear to be present in larger number. On examining the hanging drop 12–18 hours after inoculation, staining it previously with a rapidly fixing solution of methylene blue, a decided increase in the number of rose-reacting cells is noticed. The latter cells are found in all stages of development, from the small round form to the large and very granular one.

On the slide therefore, as within the body, the phenomena are identical, and in each case the inoculation with bacilli is followed first by a collection or aggregation of eosinophile cells, next the hyaline cells appear in appreciable numbers, and the rose-reacting cells also increase at the scene of action.

We have, in the hanging drop, also seen that whenever the initial eosinophile leucocytosis was absent, the hyaline cells did not exert any activity and were not attracted in numbers. On the other hand, whenever the eosinophile cells had collected around the bacilli in large numbers, in all typical cases the hyaline cells streamed towards the bacilli and ingested

[1] Compare footnote to p. 174.

them eagerly. Here, therefore, we again find that there exists a distinct correlation between phagocytosis and eosinophile leucocytosis.

The correlation between the increase in number of the basophile cells and the other kinds of leucocytosis is much more difficult to prove, because it seems that on simply keeping a drop of lymph over night in a moist chamber these cells will increase, though not in the same measure as after a previous inoculation with anthrax.

Again, on inoculating a drop of lymph with a solution of albumen, the rose-reacting leucocytosis is well marked, though an increase in the eosinophile cells may not be observed.

The question naturally arises as to how far the three kinds of leucocytosis are really distinct, as to how far they may occur independently.

There appears to be little doubt that an increase in the numbers of one or other of the three types of cells may occur in the absence of a corresponding increase in the numbers of the remaining kinds.

In œdematous frogs the lymph is usually highly charged with rose-reacting cells, and this may occur with a diminution in the number and vitality of the eosinophile and hyaline cells. Similarly the injection into a lymph sac of finely divided and sterile coagulated proteid, such as is produced by boiling a solution of egg albumen, leads to an enormous increase in the number of hyaline cells, without a corresponding increase in the eosinophile cells. On the other hand, we have never witnessed an eosinophile leucocytosis unaccompanied by a subsequent increase in the numbers of the other cells.

Phagocytosis

Having thus far described the general changes which follow on an inoculation with bacilli, and the order in which the various cellular elements appear at the seat of injection, it remains now to discuss the minute changes and the parts which the different cells play in the conflict. Before doing so we must indicate what we consider to be the proper phenomena connoted by the term "phagocytosis".

M. Greenwood's study of the Protozoa, and particularly of *Amœba*, has made it possible to give to this word a precise meaning.[1] The phenomena which follow the ingestion of a particle by one of the Protozoa are as follows: (1) If the particle is digestible it at first lies embedded in the cell substance, then a vacuole is formed about it so that the food mass now floats freely in a digestive fluid. Solution of the particle is more or less completely effected, and lastly, the vacuole closes, and if there be any insoluble remnant it is extruded. (2) If, however, the ingested body is insoluble the physiological reaction is, as it were, incomplete, and, as in

[1] *Journ. Physiol.* vols. VII and VIII.

the stomach of the highest animals in similar circumstances, there is no secretion of a digestive fluid, and therefore no formation of a vacuole.

Thus the salient feature of the process is the inclusion within the cell's body of discrete particles. If the ingesta be of a nutrient nature, then certain well-defined phenomena follow, namely, the formation of the digestive cavity and the digestion of the fragment. If, on the contrary, the ingesta are insoluble, then the phenomena stop short at the ingestive act. Following these lines we would define phagocytosis as being primarily the inclusion of discrete particles in the body of a cell. This may be followed by the formation of a digestive vacuole and the solution of the particles. This is the complete process of phagocytosis; or the included particles may simply remain embedded in the cell's body, and to this incomplete act the term must also be applied. In brief, phagocytosis implies an intracellular process. Extracellular digestion may occur, but it is not phagocytosis.

Section IV

(1) We shall now proceed to describe the appearances and changes in a hanging drop of frog's lymph, kept on a moist stage without being previously inoculated.

On preparing a hanging drop as described above, and examining it under a high power (Zeiss D or E, Oc. 4) we see that it is rich in cellular elements. These consist of the wandering cells with which we are dealing, together with some red blood corpuscles. The eosinophile cells at first sight usually appear to be most numerous, owing to the striking appearance presented by their refractive granules, the rose-reacting cells are by far the least numerous. It is, however, by no means easy to recognise the latter in the fresh unstained condition; but we have, by means of stained control specimens, convinced ourselves of the great scarcity of these cells at this time.[1]

It may here be remarked that good and healthy lymph clots rapidly, and that this property may serve as a criterion of the excellence of the lymph. It was often found that when our frogs were diseased their lymph refused to clot, and, on the other hand, whenever lymph refused to clot, the changes and phenomena to be described later did not appear.

The drops were placed under the microscope as speedily as possible. It was then noticed that the eosinophile cells at once became active, throwing out pseudopodia. Their movements are rather sluggish, and the change of form is also a slow one. The cell will soon again retract its pseudopodia and become spherical, and then once more throw out pseudopodia. It will thus

[1] Compare the relative numbers of the different cells given at the close of Section II.

alternately change its shape, and cease to do so only when it is apparently dead, being then spheroidal and regular in outline (fig. 3).

The hyaline cells are more numerous, but are less readily seen. As a rule they are very active, and are seen to wander about the field and to throw out long and thin pseudopodia.

It is, as we have just said, extremely difficult to recognise the rose-reacting cells in unstained drops, the regular spherical nucleus surrounded by a granular protoplasm being the only guide. In an ordinary drop they are apparently quite inert. They, however, increase in number and size.

Effect of Inoculating a Hanging Drop with various Substances

(2) By noting the different effects produced by different substances, when added to a hanging drop of frog's lymph, one is soon convinced that these substances might be divided into classes according to whether they affected more particularly one or other of the different cells. In one class would be grouped those substances which are completely unnoticed by the eosinophile cell, while they were readily ingested by the hyaline cell. Indian ink, anthrax spores, and coagulated proteid are instances. Another class would include those bodies which attract both kinds of cells, such as vermilion, which has a slight action on the eosinophile cells, and is readily incepted by the hyaline cells, and yeast cells, which have a very pronounced attraction for the eosinophile cells, and are taken up by the hyaline cells to a relatively limited extent. Yet another class would embrace substances which at first profoundly attract the eosinophile cells, and only after they have been subject to their influence can become the prey of the hyaline cells, such as active growths of *B. anthracis* and *B. filamentosus*. Lastly, there would be the soluble substances, the action of some of which, at any rate, is, like that of egg albumen, mainly limited to the rose-reacting cells. But the completion of such a series can only be contemporaneous with the attainment of a full knowledge of the functions of the various wandering cells, and, for the present, we can only note the differential action of various substances when introduced into the body, or into a hanging drop of lymph.

The effect of some of the substances, which we have so far examined, may now be given in detail.

(*a*) *Inoculation with anthrax spores.* When a drop of lymph is inoculated with anthrax spores obtained from an old agar-agar culture, many of the spores are at once taken up by the hyaline cells present.[1] A single cell may

---

[1] The observations of Ward, on the destructive action of light on anthrax spores, which have been published since the MS. of this paper was completed (*Proc. Roy. Soc.* LIV, 472) make it possible that the spores, which were so readily ingested by the hyaline cells, were already dead. It will be noticed that the spores used by us were obtained from old dry (agar-agar) cultures.

take up several of them. The spores are apparently destroyed by the phagocyte, for, on allowing the specimen to rest at the temperature of the room (15–20° C.), while many of the spores developed into bacilli—a process which can easily be watched under the microscope—these spores were invariably extracellular; in no case did we observe spores, which had been ingested, develop into bacilli. Food vacuoles were, in many cases, observed within the cell around the spore, showing that these bodies were being digested. The hyaline cells seemed to take up the spores immediately without the intervention of the eosinophile cells, there being thus, as we shall show, a great difference between the ingestion of bacilli and the ingestion of spores. It is difficult to say whether or no the eosinophile cells exert any influence on the spores, but this much is certain, that the spores may be taken up and digested by the unaided hyaline cells. Nor did we notice a movement of the eosinophile cells towards a mass of spores collected on the coverglass, like that which occurs in the case of the bacilli.

On the other hand, when the spores had grown out into bacilli, the eosinophile cells were at once seen to move towards them and attack them in the manner we shall describe later when we come to discuss the conflict between the cells and the bacilli.

(b) *Inoculation with indian ink*. The particles are completely unnoticed by the eosinophile cells, but are readily ingested by the hyaline cells.

(c) *Effect of finely divided coagulated proteid*. The results are most interesting, and may be mentioned here, though they are most useful as aiding us to interpret the phenomena of repair.

The coagulated proteid was prepared by dissolving white of egg in tap water, filtering, and then boiling the filtrate. An exceedingly fine precipitate may be thus produced. If this be injected into a lymph sac of a frog, or added to a hanging drop, it is found to induce an excessive leucocytosis, which is almost limited to the hyaline cells. These ingest the exceedingly fine proteid particles, and mass them into balls which superficially resemble a specific granulation, and stain blue with methylene blue. For the present, the important point to notice is that the injection causes a proliferation, or leucocytosis of the hyaline (ingestive) cells.

(d) *Effect of solution of egg albumen*. This has been only partially investigated, and one result will alone be emphasised here. If a sterile solution of egg albumen be injected into a frog, or added to a hanging drop of frog's lymph, it results in an abundant leucocytosis characterised by an increase in the number and size of the rose-reacting cells.

(e) *Inoculation with vermilion*. In the hanging drop we have observed that the phagocytes may take up vermilion particles without the intervention of the eosinophile cells. But we never saw a destruction or digestion of the pigment particles, so that it is quite possible that the hyaline cells simply take up vermilion granules in order to carry them away without destroying them. We can, however, safely conclude from our experiments that the vermilion particles may be taken up by the hyaline cells without the intervention of the eosinophile cells.

These observations were controlled by experiments in the body. Anthrax spores and vermilion particles are rapidly taken up and carried away, without a previous eosinophile leucocytosis at the seat of inoculation. It is often stated that the spores first develop into bacilli and are then taken up by phagocytes. Observations on the living (pithed) frog, however, show that most of the spores are taken up by cells, and at once destroyed. Many of the extracellular spores, however, develop into bacilli and are subsequently destroyed.

Saline solution injected subcutaneously does not lead to any apparent changes, the eosinophile cells are not increased, and the hyaline cells seem to be unaffected.

Section V

The Behaviour of the Cells towards Active Growths of Bacillus anthracis *and* Bacillus filamentosus

We shall now proceed to describe, in detail, the phenomena observed in a hanging drop following on an inoculation of bacilli (and their products). *B. anthracis* and *B. filamentosus*[1], in fresh cultures, readily grow into long chains, which form very convenient objects for continued observations. These bacilli are non-pathogenic, so far as the frog is concerned, but they by no means react as indifferent substances, since when injected under the skin they always lead to a typical inflammation.

It will be seen that the phenomena observed when these bacilli are used are widely different from those noticed when innocuous substances are employed. With very slight differences the appearances are the same, whether anthrax or filamentosus bacilli are employed.

On inoculating a drop with a trace of bouillon culture (the latter should be fresh, preferably not older than twenty-four hours), and observing a suitable chain of bacilli, we find the eosinophile cells which happen to be in the neighbourhood travel towards the chain, performing slow amœboid movements during their progress. Some of the cells apparently come from some distance. This may be an attractive influence exercised on the cells by the bacilli and their products, or it may be the expression merely of the active movements induced by the change of medium. When the cells reach the chain or individual bacilli, they at once become extremely active, extending and retracting their pseudopodia, and a constant streaming of

---

[1] This bacillus has been separated by Klein, who kindly gave us a culture. It grows in beautiful filaments, being, however, non-pathogenic.

the granules may often be observed. The granules, moreover, may be discharged, or at least wholly disappear, so that the same cell at one time appears highly granular, at another almost hyaline. Then suddenly the cell will apply itself along the bacillus and, so to speak, pour itself along it, being in some cases applied to one side of the rod or chain, in others surrounding it, and yet in others fixing itself at the extremity of a bacillus (figs. 10, 13, 14, 15, 17).

At first sight it seems as though these cells had taken up the bacilli, but careful or continued examination will easily prove that this is simply an appearance. The bacilli can, in most cases, be clearly seen to lie on or under the cell, or the cell is, so to speak, folded around the bacillus, and, if from any cause the cell contracts to a sphere, the sphere does not enclose the bacillus but only touches it. Moreover, as we shall show later, the eosinophile cell eventually leaves the bacillus, handing it over to the hyaline cells to be devoured. In fact, almost numberless observations have proved to us the correctness of Metschnikoff's[1] statement that the eosinophile cells are never phagocytic. We have tempted them with charcoal, vermilion, indian ink, with many kinds of bacilli and cocci, with curare and other substances, but they refuse all things alike.

Thus the eosinophile cell comes into apposition with the bacillus and streams along it, losing its previous spherical shape. It is now more or less fusiform, so that we have the appearance shown in figs. 10 *b* and 17. In many cases by focusing one can detect the bacillus and determine with ease that it does not lie *in* the substance of the cell, but in other cases it is impossible to recognise any trace of the bacillus in the area occupied by the cell.

The latter will remain motionless for some time, the only change noticed being a continual fluctuation in the appearance and number of the granules. After a time the cell may again change its shape, becoming once more spherical, soon to assume another shape, and once more to "pour itself out" along the chain.

During this time other cells apply themselves along the chain, passing through the same phases. Others, on the other hand, simply swarm around the chain, so that at one stage the latter is both actively attacked by some eosinophile cells and closely surrounded by others. In the case of the later arrivals the loss of granules is not so complete. This may be said to be an inverse function of the number of cells attacking the chain. The mass of eosinophile cells gradually contracts, so that the result is a "plasmodial" mass hiding the chain (figs. 11, 11 *a*, etc.). It should be stated, however, that some of the eosinophile cells may leave the bacilli soon after they have poured themselves along the chain.

[1] *Leçons sur l'Inflammation*, Paris, 1892, p. 136.

The next phase is the approach of the hyaline cells or phagocytes. Several of these come up and become hidden in the plasmodial mass, which now presents the appearance of an opaque rounded or nodulated mass with separate heaps of granules on the surface (fig. 11 *b*). The contact of a fresh cell, as it comes to lose itself in the mass, acts as a stimulus, and causes active streaming and writhing movements. These become fainter and fainter until they are reawakened in their first intensity by the arrival of another cell. Now gradually one eosinophile cell after another separates from the mass and moves off. Eventually we find that all which is left behind is a plasmodium of the hyaline cells containing fragments of bacilli enclosed in vacuoles, thus showing that an actual digestion is going on (fig. 16 *b*). The chain of bacilli has thus succumbed. The digestion of the bacillary fragments is completed, and finally, at any rate in some cases, the plasmodium of hyaline cells breaks up into separate cells.

This is the description of a typical conflict. We have thus two stages: (1) the attack by the eosinophile cells; (2) the ingestion and digestion by the hyaline cells.

The attack by the eosinophile cells is an active one, and one which must be carefully distinguished from phagocytosis. Phagocytosis it is not, because we have neither ingestion nor digestion. There can be no doubt that the eosinophile cells have a distinctly harmful action on the vitality and growth of the bacilli, for it is seen that in cases where a chain is attacked by a sufficient number of these cells, without the subsequent access of hyaline cells, all growth is suspended. The bacillus apparently dies, or, at any rate, nutritive activity is impaired, because, ceasing to grow, it will either form spores or becomes absolutely unstainable. An actual solution or destruction of the bacillus by the eosinophile cells has not been observed, but a chain will often be seen to be broken up into fragments, or to become diminished in size.

In many specimens, where the number of eosinophile cells was large and the hyaline cells sparse, no growth was noticed anywhere in the drop, but many plasmodial masses were observed, within which the bacilli were hidden.

The eosinophile cells proliferate, cell division being not infrequently observed, and young stages being by no means uncommon. They also seem to possess a high degree of vitality, for eosinophile cells which apparently had done their duty were seen to leave the plasmodial masses or the chain of bacilli, and to direct themselves to another field of action.

At the end of the process the number of rose-reacting cells is greatly increased, cells in all stages often being found in great numbers.

We have, by numerous observations, convinced ourselves that these phenomena are not simply stage phenomena, but that they also occur in the organism. By removing lymph at different intervals from the seat of inoculation of a frog injected with anthrax, all the various stages can be easily obtained.

These experiments then throw new light on phagocytosis, showing that previous to the ingestion and digestion of the bacilli there is an attack by the eosinophile cells, which apparently prepares the bacilli to be taken up by the hyaline cells. That this is actually the case, observations in cases where the bacillus is victorious will show.

In most hanging drops there are always some parts where the bacilli are not destroyed or impaired, but develop typically. Now here there has been either a total absence of an attack by the eosinophile cells, or a very incomplete one, the cells, after a feeble attempt, leaving the bacillus to its own fate. A part of a chain actually in contact with or surrounded by an eosinophile cell, however, seems to be hopelessly beaten and refuses to grow, though rapid growth may occur in the unattacked parts of the same chain. Lastly, we have never seen a hyaline cell take up and digest a bacillus or chain of bacilli that had not previously been attacked by the eosinophile cells.[1]

The following is a detailed description of a few among our many observations:

Experiment I

DROP I

Two drop cultures made from lymph from thigh of a frog.

Hanging drop inoculated with a small quantity of anthrax. A moist chamber was made, as described in Section I, and over this was placed a sterilised coverslip, with a tiny drop of bouillon from an anthrax culture on the under surface. Lymph was then taken from subcutaneous spaces of the thigh with a fine pipette, a drop rapidly added to the anthrax, and the coverslip again inverted over the chamber. The preparation was then as quickly as possible brought on to the stage of a microscope and examined with Zeiss ob. D, Oc. 4. From time to time sketches were made (cf. figs. 11, 11 a and 11 c).

Watched a long chain of bacilli in the centre of the field. Two eosinophile cells came to each end, and one to the middle of the chain. Five cells in all.

As soon as the cells reach the bacillus rapid streaming movements occur, and spread the cell substance along the chain. Rapid discharge of granules.

Half hour. Eosinophile cell at one end divided. Daughter-cell free from granules. It moved away from the chain. Within the next half hour the same cell was seen to divide again. Also one of the cells at the other end of the chain was seen to divide.

Two more eosinophile cells have reached the chain, on which there are now seven cells.

Chain broke in two. Two eosinophile cells are on the north piece, and this was watched.

1 hour. A hyaline cell has now moved to the chain and ingested one end. It has pushed out a long pseudopodium to the other end of the chain. (The chain is slightly bent, fig. 11.)

The pseudopodium is being contracted, thus slowly bending the chain.

_____

[1] This is only true when freshly made cultures of virulent bacilli are used. Compare Section IV, p. 173, and Section VI, on the conflict with yeast torulæ.

The hyaline cell is displacing the eosinophile cells and more completely ingesting the chain; part of the chain still attacked only by eosinophile cells.

A third eosinophile cell rapidly moves towards the mass, and then throws out a delicate fringe of pseudopodia, which touch the mass, finally fusing with it. The contact of the new cell provokes violent streaming movements. Another eosinophile cell fuses with the mass (fig. 11 b).

The mass has now become rounded, the eosinophile granules have been re-formed, and we have four heaps of granules representing the four eosinophile cells on the surface of the mass. The hyaline cell and bacillus are completely hidden.

N.B. The last two eosinophile cells, and all those which join the mass later, do not completely discharge their granules.

2 hours. The whole field of the microscope has now become very full of cells, all of which, both eosinophile and hyaline, are in active amœboid movement.

Churning streaming movements of the mass slacken. A fifth, and then almost immediately a sixth eosinophile cell fuse with the mass, their onset, as before, causing violent streaming movements. Blunt pseudopodia are also thrust out.

Sometimes the mass becomes lobed, and one sees a heap of very numerous, large, and very refractive granules in each lobe. Again all outlines fade, and the spherical mass has a remarkable refractive, curdled appearance, and is very opaque (fig. 11 c).

2 hours 20 minutes. A second hyaline cell has fused with the mass.

2½ hours. Preparation accidentally shifted. Low power put on to find the mass again. This is quite impossible, for the whole drop is now studded with similar masses. There are also a great number of free eosinophile cells. An enormous multiplication of these cells has taken place. They are all heavily laden with granules. No free bacilli.

3 hours. Another mass chosen for watching; a large one (fig. 11 d). Hyaline cells very extended and numerous.

3½ hours. Hyaline cells exceedingly numerous and exceedingly irregular. Some which have wandered into the field are very large, and show vacuoles which contain fragments of bacilli. They are *very* active (fig. 12).

The mass all this while assuming more and more the appearance of a heap of cells. All pseudopodial and streaming movements ceased.

The mass now looks like a heap of eosinophile cells. These are separating from one another and are moving away.

4 hours. Twenty-five minutes later the eosinophile cells are gone, and disclose a large central hyaline cell. This stage carefully drawn with camera lucida (fig. 11 e).

4½ hours. Drop examined, and shows general phagocytosis. Observation discontinued.

Drop II

Contained, in addition to scattered chains, one large mass of anthrax filaments. Cells very much less abundant than in Drop I.

Three eosinophile cells attacked a cluster of anthrax filaments. Owing to relative paucity of cells the greater part of the anthrax in the field remained unattacked.

One cell watched attacked a mass of bacilli, and streamed along a chain showing active movements.

Movements gradually ceased, and the cell contracted into a ball, *which did not enclose the bacillus but remained just touching it.*

After a prolonged period of absolute quiescence the cell again became active, and now moved away from the bacilli. It twice changed its line of movement to avoid bacilli, and finally came to rest in an isolated position.

During the third and succeeding hours the anthrax grew with immense rapidity.

The notes of another drop-culture experiment, in which the phenomena were followed to the final dissolution of the plasmodium of hyaline cells and the close of phagocytosis, are as follows:

Experiment II (figs. 16, 16 *a* and *c*)

Drop culture made with subcutaneous lymph and observation started at 11.45.

11.45. Eosinophile cells moved up to a chain.

A hyaline cell is near the south end of the chain.

There are now six eosinophile cells on the chain.

12.30. The granules are constantly changing. Only four distinct cells now, three having fused.

12.35. The movements of the eosinophile cells have bent the chain.

12.45. Fusing of the eosinophile cells only partial. Can now count seven cells on chain.

A phagocyte at the south end passes into the mass and is soon lost sight of, as the eosinophile cells went all round it, moving *en masse* in a churning manner.

1.0. Another phagocyte came into the field at the north end but wandered off again.

1.15. The eosinophile cells are turning round and round, so that at one time it appears as though there were three or four cells, and then shortly eight or more.

1.40. The eosinophile cells are separating. One already gone. A minute later another eosinophile cell went off with a sudden jerk and then came back.

1.45. Central hyaline cell seen now. A phagocyte in the field has ingested a pigment mass. It is vacuolated and the vacuoles contain fragments of bacilli.

The same eosinophile cell left again and again came back.

2.12. The eosinophile cells are leaving. The hesitating cell gone at last. Only one eosinophile cell left.

2.17. Two hyaline cells have come from outside and fused.

2.35. A fourth hyaline cell has joined.

3.35. The hyaline cells now form a mass with food vacuoles (16 *b*).

5.45. The hyaline cells separating. Digestive vacuoles collapsing.

6.25. Vacuoles collapsed.

As an instance of a case where the bacteria have ultimately triumphed, owing to their being present considerably in excess of the eosinophile cells, we may take the following experiment. In all such cases the conflict may be followed up to a point where the accumulation of bacterial products produces a complete paralysis of the eosinophile cells. Naturally this may occur at any stage in the conflict, and we have witnessed instances where the initial dose, so to speak, has at once paralysed the cells and instances which might aptly be styled drawn battles. In the latter the conflict will endure for hours, the eosinophile cells killing some of the bacteria, while in others the bacteria are so far worsted that they have to resort to spore formation.

Experiment III

DROP CULTURE MADE CONTAINING A LARGE QUANTITY OF *Bacillus filamentosus*

11.15. Scarcely any eosinophile cell with granules. In one case saw the granules shrink in size, leaving vacuoles.

11.30. One long chain in the field stretches up from the bottom of the drop where cells mostly are into an almost cell-free region. Three cells on this chain. When first seen they were quite free from granules. Now they are re-forming them.

11.35. Cells are moving up the chain and others are approaching the lower end.

11.38. The eosinophile cells generally are re-loading themselves with granules.

11.40. The appearance of the hyaline and eosinophile cells in the preparation is very striking (cf. fig. 18). Upper part of the chain still unattacked. A fourth cell is streaming up from the lower end. Another chain in the field, which lies lower down in the drop, is already closely surrounded by 12 granular eosinophile cells.

From this a short branch chain extends upwards which is free from cells.

The cells are all moving very sluggishly.

11.55. The last cell to come on to the chain (No. 4) is hopelessly defeated. It bursts up and only droplets are left.

12.45. No further attack has taken place. Cell No. 3 looks very bad and curdled. The bacilli are not growing. Two cells still on the chain are spherical, as are also the free cells.

1.10. Some of the 12 cells on the deeper-lying chain are still extended. Others are rounded.

2.30. Bacillus growing now where it has been unattacked. The chain which was attacked by 12 cells appears quite dead, but the short chain projecting upwards has increased in length.

Later. Very rapid growth of bacilli. The chain attacked by the 12 cells, however, is completely dead. Of the other chain part was killed, and part which was exposed for only a short time to the attack of the eosinophile cells grew.

(The growth of the bacilli was always determined by making a plan of the chains and numbering the rods.)

The fact that the bacilli are attacked at first exclusively by the eosinophile cells was clearly shown in the following way: About 0·2 cub. cm. of a fresh bouillon culture of anthrax was injected into the peritoneal cavity of a pithed frog. When 20 min. had elapsed the cavity was opened and film preparations made, and stained with eosin and methylene blue. On counting the cells in these preparations, it was found that no less than 82 per cent. of the eosinophile cells were in contact with bacilli. On the other hand, only 2·6 per cent. of the hyaline cells were even in contact with bacilli.

In a second experiment of a similar nature it was found that 42 per cent. of the eosinophile cells were in contact with bacilli, and only 2 per cent. of the hyaline cells.

The cells were counted with Oc. 10, ob. $\frac{1}{12}$ apochromatic oil immersion, by Powell and Lealand.

The phenomena seen in a hanging drop of lymph inoculated with anthrax or *B. filamentosus* show clearly that the intruding organism is always attacked first by the eosinophile cells before the hyaline cells attempt to ingest them or their remains. But we are able to go a step further than this and say that the hyaline cell is incapable of ingesting an anthrax or filamentous bacillus before it has been killed or maimed by the eosinophile cell.

The sequence of events observed in experiments similar to those described in detail would afford strong presumptive evidence for this statement, and we find further support in noticing the different behaviour of the two kinds of cells towards the bacteria and indifferent particles when they are present together in the hanging drop.

We have already seen that the eosinophile cells are not at all phagocytic, and that they are completely unaffected by the presence of indifferent particles in the plasma. On the contrary, the hyaline cells are attracted by such particles as vermilion, indian ink, or coagulated proteid. They ingest them, and, if possible, digest them.

Now, in the drop cultures, we find that the phagocytes are never attracted by the fresh bacilli. They are even repelled by them. We have seen a freshly budded phagocyte come into the near neighbourhood of an anthrax chain (fig. 8) and there exhibit active pseudopodial movements, which neither resulted in locomotion of the cell, nor in effecting contact with the bacillus. After a while the movements slowed and ceased. Then they were renewed and now the cell moved away from the bacillus.

Bacilli and indian ink or vermilion. In a drop of lymph inoculated with bacilli and with indian ink or vermilion, we see strikingly displayed the differential activity of the two kinds of cells. The phagocytes will readily ingest the indifferent particles, while the eosinophile cells with even greater readiness attack the bacilli.

Bacilli and spores. This is only true of the kinetic phase of the life-history of the bacillus. The potential form, the spores, are at once ingested by the phagocytes,[1] and in a hanging drop, inoculated both with bacilli and spores, we witnessed the eosinophile cells attacking the bacilli, while the phagocytes ingested the spores. In illustration of this we may quote the notes of the following experiment:

Experiment IV

A drop of frog's lymph was inoculated with fresh bouillon culture of anthrax, and also with anthrax spores from an old agar culture.

3 p.m. Spores taken up rapidly by hyaline cells without intervention of eosinophile cells—quickly vacuoles are formed, and some of the spores are being digested. At the same time the usual attack of eosinophile cells on the bacilli took place.

7.45. Many extracellular spores have become bacilli.

9.30. Eosinophile cells have attacked the newly developed bacilli and everything is proceeding as before. There appears to have been a proliferation of eosinophile cells.

But to prove the statement that the destruction of anthrax and *B. fila-mentosus* by frog's lymph is due primarily to the eosinophile cell, it was necessary in some way to paralyse the eosinophile cell while leaving the phagocyte unharmed, and then show that under such conditions, namely, in the absence of the eosinophile attack, the phagocyte is powerless and the bacteria grow freely in the lymph. We have been enabled to do this in two ways.

Action of heat. It has already been shown (Section III) that when a frog is warmed to a temperature of 25–35° C. it becomes susceptible to anthrax.

---

[1] Cf. note, p. 174.

We therefore tested the effect of heat on the processes taking place in our drop cultures, and found that the eosinophile cells are completely paralysed by a rise of temperature, and became quite incapable of attacking the bacteria. *At the same time the phagocytes retain their activity and freely ingest indian ink particles or spores present in the same drop.* The thermometer of the warm stage indicated a temperature of 35 to 37·5° C. in our experiments.

Action of urari. Turning to poisons, we obtained a similar differential action with urari. This drug produces a marked alteration in the eosinophile cells: their eosinophile granulation becomes changed into an amphophile granulation. This change occurs both in the body and out of the body with a sufficient dose of the drug. In the body, after a while, recovery takes place, and eosinophile granules are again found. In the same way, owing probably to the elimination of the poison, the neuromuscular mechanism will, at length, recover, and the animal regains the power of movement. Out of the body in the hanging drop the cells appear incapable of recovery and the amphophile granulation persists.

If a drop of lymph, taken from the body of a urarised frog while the amphophile granulation persists, be inoculated with anthrax bacilli, we find that the bacilli are not attacked by the eosinophile cells, and after a time they grow.

Abnormal lymph and bacilli. Lastly, among frogs which have been long kept in confinement in the laboratory and starved, individuals are sometimes found more or less œdematous, and whose lymph contains few eosinophile cells incompletely charged with granules which are sometimes largely amphophile. *B. filamentosus* has been found to multiply in the body of such animals, and, ultimately, to kill them, and both filamentosus and anthrax bacilli will grow freely in drop cultures of lymph taken from such an animal. At the same time hyaline cells are present, and have not lost their phagocytic power. Take, for instance, the following experiment.

Experiment V

A drop of non-clotting lymph from a frog which had been a long time in the laboratory was inoculated with anthrax bacilli and anthrax spores.

The eosinophile cells did not act at all, and therefore the hyaline cells made no attempt whatever to ingest the bacilli.

At the same time the spores were ingested as usual by the hyaline cells.

The bacilli grew rapidly.

Ingested spores were seen to be lying in distinct food vacuoles, and many were undoubtedly digested.

How essential the removal of the dead bacterial remains by ingestive solution on the part of the phagocytes must be in the body, we see when we watch hanging drops made from lymph either abnormally deficient in hyaline cells or in which the hyaline cells are abnormally sensitive to a change of medium. The relative number of the two kinds of cells varies in lymph drawn from different parts of the body. Occasionally in hanging drops made with peritoneal lymph hyaline cells may be almost absent. Under these circumstances the eosinophile attack takes place and the bacilli are killed, but there are no scavengers to clear away the traces of the fray, and the field remains strewn with bacilli, which, if they are not dead, have at any rate completely lost the power of growing in the drop.

The Changes in the Specific Granulation of the Eosinophile Cell

The amphophile granule (fig. 17, *a*). The facts discovered by following the phenomena exhibited in a hanging drop, were supplemented by the examination of such drops, or of lymph taken from the body, when fixed and stained with the methylene blue solution, or with eosin and methylene blue at different intervals after inoculation.[1] The same phenomena, the attack of the eosinophile cell, the formation of plasmodial masses, and the final phagocytosis, are again seen, but new light is thrown on the glandular nature of the activities of the eosinophile cell. In the attack on a bacillus the cell discharges, more or less completely, its granules. The amount of discharge depends on the strength of the stimulus. If, for instance, a considerable number of cells attack a bacillus, then the discharge will be only partial in many of the cells. The process of the discharge of granules was followed with high powers. If a cell attacking bacilli be watched under Oc. 4, ob. $\frac{1}{12}$th, the granules will be seen to travel either singly or in groups of two, three, or four, from the central mass of granules to those parts of the cell which are immediately in contact with the bacillus. There they will be gradually seen to shrink in size until they disappear. This phenomenon, namely, the diminution in size and loss of granules, can be demonstrated in preparations fixed with the methylene blue solution, or in films which have been rapidly fixed by heat, or by corrosive sublimate. As compared with the rate of loss of granules in, for instance, the secreting cells of a salivary gland, the process is very rapid. After the discharge the cell often re-forms its granules. Tracing these events by means of the basic and acid dyes, we find that the re-formed granules are at first amphophile, and only towards the close of the conflict do they become truly eosinophile.

But the amphophile substance is a stage, not only in the construction of an eosinophile granule, but also in its destruction. The action of urari may again be cited, for if this drug be added to lymph, the eosinophile granules alter to amphophile granules, and the change may be followed in a hanging

---

[1] Preparations stained first with eosin and afterwards with methylene blue were made according to Ehrlich's film method.

drop of lymph to which has been added a little urari, and a trace of methylene blue. We have watched the granules in a cell become slightly smaller and less refringent, and then colour with the dye. Similarly, when urari or bouillon containing a considerable quantity of bacterial products is added to the lymph, a change from eosinophile to amphophile granulation rapidly occurs. In the latter case, if the poisons be not present in too great quantity, the eosinophile condition is re-established.

When the conflict with the micro-organisms is carried out swiftly to a successful issue, the changes appear to be as follows:

(i) The initial sluggish eosinophile cell.

(ii) The discharge of the granules, either completely, when the cell effects contact with a bacillus, or so far as to make them amphophile.

(iii) The reconstruction of amphophile matter by those cells which have completely discharged, the amphophile matter being a stage in the elaboration of the eosinophile granule.

(iv) The complete regeneration of the eosinophile granules. And this occurs in what may be spoken of as the plasmodial stage of the conflict.

The change of the granulation of the eosinophile cell from eosinophile to amphophile and back again to eosinophile may be illustrated by the following experiment:

Experiment VI

A series of anthrax drop cultures were made with the lymph from the same frog. These were stained at different intervals with the methylene blue solution.

(a) Stained as soon as possible after the addition of the lymph to the drop of bouillon culture. Eosinophile cells amœboid, granules amphophile or discharged. Many chains already attacked by eosinophile cells. Hyaline cells amœboid. Rose-staining cells present (fig. 17).

(b) After 15 min. Eosinophile cells with eosinophile granules present.

(c) After 30 min. Eosinophile cells present, with perfectly eosinophile granules (i.e. they absolutely refuse to stain). Where a cell is in contact with a bacillus it contains granules which are irregular in shape, are being discharged, and which stain.

(d) After 45 min. Where the eosinophile cells are in contact their granules are amphophile (blue-staining), where they are floating freely they contain eosinophile granules.

(e) After 1 hour. Some eosinophile cells applied to chains are still quite free from granules. Contents of the bacilli no longer stain as a homogeneous blue substance. Disintegration has commenced, and the staining is in patches. Free amphophile cells rare.

(f) After 1¼ hour. Rose-staining cells in marked abundance. Giant eosinophile cells (plasmodia) present, always with heaps of truly eosinophile granules.

The series was continued up to 3½ hours, and the most prominent feature was the increase in the rose-staining cells.

SECTION VI

Action of the Cells on Yeast Torulæ

There are present then, in the body of the frog, two kinds of wandering cells, the glandular eosinophile cell and the ingestive hyaline cell, and the most superficial study of these cells shows that the former is much more stable and can endure much greater changes in its environment than the latter. The eosinophile cells will persist as spherical bodies, with granules intact, for days in a drop of lymph, and for long after the other elements have disintegrated. They may or may not be dead, but absence of movement is the only justification for supposing that life is extinct and not in abeyance. Similarly in *Astacus* we find that the eosinophile cell is very much more stable than the hyaline cell. Correlated with this fact we find that the change of environment due to the presence of bacterial poisons will call forth the most active manifestation of the special functions of the eosinophile cell, while the hyaline cell may be paralysed or destroyed. It is possible, however, that the poisons of all micro-organisms do not possess this discriminating value. It may be that if a sufficiently large number of micro-organisms were taken, they could be placed in a series commencing with those which acted towards the cells like indifferent particles, being ingested by the hyaline cells and unnoticed by the eosinophile cells, and ending with those like *B. anthracis* or *filamentosus*, which provokes the most profound activity of the eosinophile cell while the hyaline cell is incapable of attacking them. Bearing in mind the complex conditions of the conflict, we are forced to conclude that the organisms against which the frog is not immune may be found either at one end or at the other end of the series, or at both.

Yeast, in its most virulent form, represents to a certain extent the middle of the series. Clusters of yeast cells are attacked by the eosinophile cells, just as are anthrax chains, and to these come hyaline cells; in fact, the story is the same as that which has already been related. At the same time hyaline cells will ingest stray yeast cells which have not been attacked by eosinophile cells. The notes of one experiment will illustrate this.

Experiment VII

Bouillon yeast culture of great virulence when tested on rabbits. Inoculated lymph drop with this. Frogs completely unharmed by large doses of this yeast.

1 o'clock. Watched large mass of yeast. Three eosinophile cells came up and were lost in north end of mass.

Saw hyaline cell ingest stray torula.

1.12. The same hyaline cell has now reached north end of mass and is attacking it. After 3 min. it is lost in mass.

1.25. Another eosinophile cell appeared and became lost in south end of mass.

1.30. Another eosinophile cell, followed by hyaline cell, came in at north end and both fused with mass.

All this time the eosinophile cells have been covering the mass all over.

1.35. At north end yeast has lost much of its distinctness. After some hours plasmodium formed, as with anthrax.

9 o'clock. One eosinophile cell left plasmodium.

9.10. Same cell came back and apparently fused with another.

9.15. Another eosinophile cell left.

Discontinued observation, and therefore did not see the general breaking up. Stained the preparation, and found many rose-reacting cells.

The hyaline cell is mostly, but not always, capable of digesting the yeast cells, which it ingests. Similarly it can digest some, but not all the anthrax spores which it ingests, and we may classify the activities of the hyaline cells under two heads:

(1) The ingestion of particles which they can digest, such as coagulated proteid, dead micro-organisms, and certain living micro-organisms or spores. These the hyaline cells eliminate from the body in which they dwell by dissolving them.

(2) The ingestion of particles which they cannot digest, such as indian ink, or dust particles. These they do not eliminate from the body, but carry them to the connective tissues and there hide them away. Instances of this are found in the removal of dust particles by the phagocytes of the lung, and in the fate of indian ink when injected into the body of *Dytiscus*, as shown by Durham.[1]

Section VII

The Rose-reacting Cells

It would appear, from the facts so far set forth, that the fate of the microbe in the frog's lymph depends upon the activity of two kinds of cellular elements. Are we to attribute, therefore, no part in the conflict to the non-amœboid cells which are charged with rose-staining granules? This question is at once answered by the fact that one of the conditions apparently necessary to a successful issue to the conflict between the eosinophile cells and the micro-organisms, whether the conflict occur in the body or out of the body, is an increase in the number, and still more in the size and granulation of these cells. In the lymph of newly captured, healthy frogs, the rose-staining cells are usually very small, and so few in number, that they would be completely overlooked were it not for the brilliant rose tint which their granules take with methylene blue. This condition is shown in figs. 4 and 5 *a*.

The injection of micro-organisms is followed by a continuous increase in the number of these cells, and they now appear as shown in fig. 5 *b*. The

[1] Durham, "On Wandering Cells in Echinoderms", *Quart. Journ. Micr. Sci.* vol. XXXIII.

increase is a striking fact from the second to the fortieth hour. Beyond that time we have not followed them.

To state the same fact in other words, the rose-staining cells are continuously abstracting some substance or substances from the plasma and depositing it within themselves as rose-staining (basophile) granules. For the present we are content to suggest that the substances abstracted are foreign, abnormal substances in solution, such as, for instance, the poisonous products of bacterial activity. Frequently, in watching the processes taking place in a drop culture, one sees the eosinophile cells at first vigorously attacking the bacteria, and then the whole process will come, sometimes almost abruptly, to a standstill, the eosinophile cells contracting and becoming spherical, while the hyaline cells disintegrate. This, we explain, by supposing that the bacterial poison in small doses stimulates the eosinophile cells, while in larger quantity it paralyses and ultimately kills them. If this be true, it is obvious that in the body some mechanism or mechanisms must exist for keeping the amount of bacterial poisons either taken in at once, or produced by the as yet unattacked or unkilled bacilli, below the limit at which it is fatal to the wandering cells or to the body at large. We suggest that this function is, in part, filled by the rose-reacting cells.

The evidence bearing on this question which we have so far obtained may be briefly summarised as follows.

The presence in the plasma of foreign bodies in solution, such as egg albumen or anthrax albumose, leads to an increase in size, number, and granulation of the rose staining cells. This is true both of the frog and of *Astacus*. Also, as has already been stated, if *Daphnia* is exposed to toxic substances, an increased formation of a rose-reacting (amphophile, in this case) substance results.

The alteration in the chemical composition of the plasma of a hanging drop of lymph due to clotting also leads to a similar result. Lastly, we may cite the absence, often noted by us, of rose-reacting cells in lymph in which the bacilli have grown, and this in spite of the fact that these cells are very resistent to the toxic substances. The presence pointed out by Sherrington of similar cells in large numbers round the intestinal blood-vessels in cases of enteric disease suggests that they are there to intercept the toxic substances streaming in from the lumen of the intestines. The increase in the number of these cells in inflammation, in carcinoma, etc., and, generally speaking, in cases normal or pathological, in which there is an abnormal production of normal or abnormal products of metabolism was noticed by Korybutt, Daskiewicz and Ehrlich.[1]

The importance attached to the removal of the bacterial poisons, and the

---

[1] Ehrlich, "Beiträge zur Kenntniss d. granulirten Zellen, etc.", *loc. cit.*

endeavour made by the body to get rid of them, are strikingly shown by *Daphnia* and *Astacus*.

In *Daphnia* the whole blood stream may be watched on the stage of the microscope. The presence of bacterial poisons in the body is then seen to increase the adhesiveness of the corpuscles; they tend to adhere to the walls of the blood spaces, but they are mainly attracted to the excretory organs (the shell glands) around which they cluster in large masses. At the same time the epithelium of the excretory organs becomes more granular, vacuoles appear, and the inner surface of the cells becomes irregular, being pushed out into processes.

In *Astacus* the injection of very large quantities of *B. filamentosus* into the pericardial sinus is followed by the almost complete disappearance of the filaments from the blood in half an hour. Drops of blood taken from the pericardial sinus, and from the ventral sinuses, both abdominal and thoracic, were stained and examined, but prolonged searching failed to reveal more than isolated rods or pairs of rods, and these only at the rarest intervals.

Finally, the animal was bled into some iodine solution, and the cells allowed to sink to the bottom. After twelve hours the sediment was examined, and only one or two cases of eosinophile cells attacking bacilli were found. The bacilli themselves were almost absent. The walls of the blood spaces were then examined and the missing bacilli discovered embedded in plasmodial masses clustering round the green glands, the excretory organs of *Astacus*.

It may be said in passing that the eosinophile attack appeared to precisely resemble that observed in the frog's lymph.

Section VIII

Preliminary Observations on Mammals (Rabbit and Rat)

(1) The peritoneal fluid of these animals is full of cellular elements, these being, just as in the frog, chiefly eosinophile and hyaline cells. On killing the animal (by decapitation), and quickly placing a little of the fluid on a small drop of a bouillon culture of anthrax on a warm stage, the initial attack by the eosinophile cells at once takes place. The latter, in the case of the rat, is pronounced and rapid; while in the rabbit, which is a more susceptible animal, it also takes place, but is less vigorous. It was impossible, with the method employed, to observe more than the initial stage, on account of the difficulty experienced in keeping the cells alive for longer than fifteen minutes. However, the eosinophile cells deported themselves in exactly the same manner as they do in the frog, moving towards the chain of bacilli, and fusing along it.

(2) The following experiments and observations, made on rabbits, will throw some light on the mode of action of the eosinophile cells:

Leber has demonstrated the powerfully solvent action of pus on such substances as copper, gold, and silver, etc. It has been shown that the cellular elements of pure and fresh pus consist of practically nothing else than cells with fine or coarse eosinophile granulation. On repeating some of Leber's experiments, and placing minute pieces of sterilised copper, steel and silver wire into the anterior chamber of a rabbit's eye, under strict aseptic and antiseptic precautions, in all cases a suppuration rapidly ensued, most sudden when copper was used. This pus invariably contained nothing but granular cells. Already, after 24 to 48 hours, the copper was found to be roughened and corroded. We conclude from the observations that the solvent action of pus is due to the eosinophile granulation.

Cellulose, in the shape of sterilised cotton-wool, placed in the anterior chamber under similar precautions, was not affected in the slightest, even after an interval of six weeks. Cellulose is both an extremely indifferent substance and also most resistent.

(3) Others have already demonstrated the presence of a ferment in pus. Rossbach, before us, succeeded in separating an amylolytic ferment from the leucocytes (*Deutsch. Med. Woch.* 1890), while Leber (*Die Entstehung der Entzündung*) found that pus digested fibrin and liquefied gelatine.

Summary

Some of the phenomena described and discussed in §§ II to VII may be summarised as follows:

(1) The three different kinds of wandering cells, the eosinophile cell, the hyaline or non-granular cell, and the basophile rose-reacting cell, proliferate while free in the body fluids. This may be demonstrated in the frog, and has, in the case of other animals, been recorded by ourselves and other workers.

(2) The different kinds of cells multiply independently, so that the numbers of any one kind of cell may vary without a corresponding variation in the numbers of the other cells. There are thus three kinds of leucocytosis, corresponding to the three forms of wandering cells found in lymph or peritoneal fluid.

(3) The three kinds of cells are differently affected by different substances when introduced into the plasma.

(a) Solid substances of the nature of what are commonly called indifferent substances affect only the hyaline cells which ingest the particle. Coagulated proteid and indian ink are examples, as are also anthrax spores.[1]

[1] Cf. note, p. 174.

(*b*) Anthrax and filamentosus bacilli, when first introduced, attract only the eosinophile cells, which kill or maim them by means of a substance derived from their stored eosinophile granules. After the bacilli have been thus acted on they can become the prey of the ingestive hyaline cells.

(*c*) Vermilion and yeast cells stand midway between indifferent substances and these bacilli, and attract both hyaline cells and eosinophile cells. Vermilion only slightly attracts the eosinophile cells. Yeast cells attract them strongly, but also to a certain extent are immediately ingested by the hyaline cells.

(*d*) The rose-reacting cells are increased in number and size by alteration in the chemical composition of the plasma, such as is produced by clotting, or by the introduction of toxic albumose or egg albumen. Hence they are probably chiefly active in maintaining the normal constitution of the plasma so far as dissolved substances are concerned.

(4) The eosinophile cells are highly specialised bodies endowed with the power of movement, in virtue of the possession of a pseudopodial ectosarc, and with glandular powers directed to the production of a bactericidal, or at least antibiotic, substance.

SECTION IX

The facts brought forward in the preceding sections lead to certain conclusions as to the morphological position and physiological attributes to be accorded to the sporadic mesoblast.

Morphology

In the frog, the sporadic mesoblast consists of three kinds of cells, the eosinophile cell, the hyaline cell, and the rose-reacting cell, together with the red corpuscles and platelets. Of the two latter, we, personally, can say nothing from our own knowledge; nor have we any reason, beyond the general facts of their distribution in the body, for placing them in the same morphological group with the three first-named elements.[1] Leaving then the red corpuscles and platelets out of account, we find that the sporadic mesoblast of the frog contains elements precisely similar to those which compose the similar tissue of such a widely divergent animal as *Astacus*. The large phagocyte, derived from the hyaline cell and of temporary existence only, is not only found in both animals, but produced by essentially similar conditions in both. This exact resemblance between animal forms

---

[1] Ziegler, "Die Entstehung des Blutes", *Ber. d. Naturforsch. Gesellsch. zu Freiburg*, Bd. IV, H. 5, as the result of his investigations on the development of the blood, arrives at the conclusion that the red corpuscles are morphologically distinct from the white corpuscles. The former are of intravascular origin, the latter are extravascular at first, but migrate into the blood system.

so diverse is helpful to us, since it gives us good reason for supposing that the three cell elements of the frog have arisen by the differentiation of a primitive homogeneous sporadic mesoblast, that is, one the cell elements of which are all alike.

This question of the comparative morphology of the sporadic mesoblast may, like other morphological questions, be approached in two ways. We may investigate the phylogenetical position of the tissue or its ontogenetical position. In both directions our knowledge is very incomplete. No conclusions as to lines of development of the tissue within the limits of the Chordata can be based on our present knowledge, for in the lowest member of the Chordata which we have examined, the *Ammocœte* larva, all three types of cells are already present. At the same time, so far as we know, there is no exact description of the histology of the wandering cells of the Tunicata or Cephalocorda. In the Crustacea, the conception of the origin of the different wandering cell types from an archetype finds strong phylogenetic support. The group of the Crustacea is remarkable for including within its limits animals of the very simplest and animals of the most complex organisation. The very simple animal *Daphnia* has a sporadic mesoblast composed of only one kind of cell; *Astacus*, on the other hand, representing the most complex members of the group, has three kinds. But the evidence does not rest here. The wandering cells of *Daphnia* are not only of one kind, but also this one cell performs all the functions, and has all the morphological characters of the three cells of *Astacus*. This fact has already been discussed by one of us in earlier papers,[1] but we are able to add a further archaic character of the cell to those given there. In the earlier papers, the specific granulation was justly described as rose-reacting, further investigation has revealed the fact that the rose-reacting substance of *Daphnia* is also amphophile. In other words, even in its specific granulation the wandering cell of *Daphnia* is archaic, it is a rose-reacting, amphophile granulation. We may even pursue the search for the cell element of the undifferentiated sporadic mesoblast a stage further back to the wandering cell of the larval *Daphnia*, while still within the broad pouch of the mother. Then we find an amœboid cell with no granules, but the whole cell substance reacts with aniline dyes, giving "a faint but distinct rose-reaction with methylene blue".

Since tadpoles are not available in summer or autumn, we have not as yet been able to pursue the ontogenetical study of the corpuscles of the frog. Fœtal cats, however, are to be found in all seasons, and in these animals we have so far found only one kind of wandering cell which is without specific granulation, and has a spherical nucleus and scanty cell substance. In the very late fœtus the wandering cells still show no

[1] *Journ. Physiol.* XIII, 184 *et seq.*, and p. 318.

granulation but there are now two kinds of cells present, one with a round nucleus and one with a lobed nucleus. In the adult all these three types are undoubtedly present. Thus we find a certain phylogenetic and ontogenetic support for the statement that the very divergent elements of the sporadic mesoblast have a morphological homogeneity in the fact that they have arisen from a primitive amœboid wandering cell with no specific granulation, by a process of morphological and physiological differentiation akin to that which has, *pari passu*, led to the increasing complexity of the animal generally.

Physiology

The question of the functional significance to be attributed to the sporadic mesoblast also finds a partial solution in the very incomplete series of facts we possess bearing on the physiology of this structure. Undoubtedly the wandering cells are present largely as a protective mechanism to guard against the intrusion of foreign substances, living or non-living, into the organism. But this is probably only a small part, and not the most primitive of their functions. They are also related to the general processes of the body, notably to the bodily nutrition. In *Daphnia* they may be readily watched engaged in the transportation of fatty particles from the alimentary canal to its place of storage.[1] Also any abnormal condition of the perivisceral fluid, or blood, leads to the massing of these cells around the special excretory organs of this animal, and this fact suggests that their activities are partly directed to maintaining the normal constitution of the body fluids so far as the dissolved matters are concerned.

Thus, though exact knowledge is wanting, yet it is abundantly clear that the homogeneous sporadic mesoblast of *Daphnia* is intimately related to the processes of general physiological importance taking place in the alimentary canal and the excretory organs; and there is no sufficient reason to lead us to believe that this connection has been entirely, or to any great extent, lost in the course of the divergence and specialisation of these cells, which has produced the different wandering cell types of the higher animals. On the contrary, it is a matter of common physiological knowledge that the wandering cells are profoundly affected by events occurring in the alimentary canal. Thus, though we are not able to point to any particular physiological process carried out by these cells, still we have sufficient grounds for opposing the supposition that they exist only to protect the body from the invasion of foreign particles, and that appears to us to be a valuable step and a most necessary preliminary to a successful study of their functions.

In endeavouring to indicate the lines along which the specialisation of the sporadic mesoblast has advanced, probably in most animal groups, we may

---

[1] *Journ. Physiol.* XIII, 184 *et seq.*

proceed with perhaps greater sureness. The primitive mobile, ingestive, and glandular cell of *Daphnia* has become in *Astacus* the specially glandular eosinophile cell, the specially mobile and ingestive hyaline cell, and the specially absorptive rose-reacting cell, and what we know of the sporadic mesoblast of the Vertebrates points to its having developed along similar lines.

But both older and later observations[1] on the part played by wandering cells in the formation of scar tissue indicate that an even more extended conception of the relations of the sporadic mesoblast may at some future time become necessary, and the convenient fiction, wherein the blood was regarded as a tissue, like cartilage or connective tissue, only with a fluid matrix, may yet be found to embody a morphological truth.

The relation of the attack of the eosinophile cell to the ingestive act of the hyaline cell. A study of the way in which some of the carnivorous Protozoa capture and ingest their prey throws a clear light on the relations of the peculiar mode in which the eosinophile cell attacks a bacillus to the simple ingestive act of a hyaline cell, or *Amœba*. Further, if all these facts are placed together, they suggest many thoughts on the relation of intracellular to extracellular digestion.

By comparing the various accounts given by Leidy, M. Greenwood, and other observers, of the manner in which an animal like *Amœba* or *Actinosphœrium* captures and ingests its prey, we find that the following processes may be recognised: (1) Contact is effected with the prey and its movements are arrested; (2) then, after it has thus been maimed or killed, the prey is ingested and digested. But the captured infusorian may resist the benumbing influence of the captor, and may, after being exposed for a long time to the *extracellular* attack, as it were, acquire tolerance of it, recover all its powers and escape. The best instances of the killing of prey as a result of mere contact, are furnished by the Suctoria. When a moving animalcule comes into contact with the long stiff, very specialised pseudopodia of these animals, its movements are suddenly arrested. The phenomenon is so striking that Grüber, Maupas, and others, speak of it as a poisoning of the prey by some substance excreted by the captor.

Expressing these facts in terms of the processes occurring in the cell, we have

(1) The contact of the prey stimulating the captor to *excrete* a poison;

(2) The ingestion of the now inert body;

(3) The *secretion* of a digestive fluid which dissolves the ingested prey.

Leidy's account of the capture of a *Urocentrum* by an *Amœba* clearly shows how essentially similar the excretion of the poison is to the secretion

---

[1] Compare Sherrington and Ballance, "On Formation of Scar-tissue", *Journ. Physiol.* x, 550.

of the digestive fluid. His words are, "a second victim of the same kind was included in the fork of a pair of pseudopods, the ends of which were brought into contact so as to imprison the animal in a circle. The latter moved restlessly about within its prison but, after a time, became motionless, and shortly after the ends of the pseudopods which enclosed it fused together...and finally the *Urocentrum* was enclosed."[1] If we now turn to M. Greenwood's[2] carefully detailed account of the phenomena of ingestion in *Amœba,* and notice the close morphological relation between the space "included in the fork of a pair of pseudopods" and the digestive vacuole, the primitive unity of the two processes cannot fail to be seen. In the very specialised Suctoria, the excretion of the poison and the killing of the prey is a specialised and much more perfect process, as is also the very remarkable manner of its ingestion.

Turning now to the eosinophile and hyaline cells, we see that in the former the *extracellular* act has become its special character, and, as in the Suctoria, the perfection of the mechanism for the production of the poison has shortened the latent period of its discharge. The cell is, in short, a unicellular gland, *but* it preserves intact the initial step of the primitive process, it effects contact with its prey; the hyaline cell also effects contact with its prey, but the extracellular discharge is insignificant or absent, the ingestive and *intracellular* act being, on the other hand, complete and rapid.

We have strictly parallel phenomena in the endodermal cells, and this enables us to advance the line of thought a step further. Metschnikoff[3] has given us good reasons for supposing that the primitive endoderm, or nutritive cell, of the Metazoa, carried out its functions in a manner exactly resembling the *Amœba* described by Leidy. But in the Cœlenterates we find an endoderm composed of (1) a gland cell which forms *extracellularly* a digestive fluid, (2) an ingestive cell which produces *intracellularly* a digestive fluid, and (3) an absorbent cell which is correlated with the gland cell and takes up the fluid products of the extracellular digestive process. Viewed in this way the eosinophile cell is analogous, in its activities, to the gland cell of the Cœlenterate endoderm, and, for instance, to the cells lining the fundus of a gastric gland of the Mammalia.[4]

Other instances of an extracellular digestive act are found in:

(1) The attack of *Vampyrella spirogyræ* and *V. pendula* on the Algæ which furnish them with food. This furnishes a most interesting parallel.

[1] Leidy, *Rhizopods of North America.* See the figs. 5 and *ss* on Pl. I.

[2] "On the Digestive Process in some Rhizopods", *Journ. Physiol.* vols. VII and VIII.

[3] Metschnikoff, *Zool. Anz.* (1878), p. 387.

[4] Compare M. Greenwood, "Digestion in Hydra", *Journ. Physiol.* vol. IX.

(2) The digestion of fibrin by the surface cells of the mesenterial filaments as described by Krukenberg.[1]

(3) The hollowing-out of spaces in the first bone spicules by the osteoclasts.

Lastly, we are not without some feeling of the bearing of our observations on the most difficult question of immunity. But further knowledge has only increased our perception of the complexity of the problem. We see that it depends upon all the possible permutations and combinations of the activities of the three wandering cell elements engaged in the conflict, and of that yet entirely unknown factor, the general physiological reaction of the organism to the microbic poisons. In the face of these difficulties we have deemed it wise to stifle the temptation to theorise on the subject, until, with wider knowledge, a greater capacity to cope with the difficulties shall have been attained.

EXPLANATION OF PLATE X

Fig. 1. A group of two hyaline cells and one eosinophile cell fixed with 0·25 per cent. iodine as rapidly as possible after removal from a subcutaneous lymph sac of a frog. The hyaline cells are larger than usual. The vertical line represents the relative length of the major axis of a red corpuscle. Oc. 10, ob. apochr. ½th, Powell and Lealand. Camera lucida.

Fig. 2. Eosinophile cell fixed by heat and stained with eosin and methylene blue. From lymph of healthy frog. Same magnification as fig. 1. Camera lucida.

Fig. 3. Eosinophile cell and a small hyaline cell. Normal frog. Fixed and stained with methylene blue solution. Oc. 4, ob. E, Zeiss. Camera lucida.

Fig. 4. Rose-reacting cell and hyaline cell of normal frog. Fixed and stained with methylene blue solution. Viewed with artificial light. Oc. 4, ob. E, Zeiss. Camera lucida.

Fig. 5. Rose-reacting cells (a) of normal frog, (b) 12 hours after injection of yeast, from the seat of the inoculation. Stained only with methylene blue. Illuminated with yellow light. Same Oculars and Objectives as in fig. 1. Vertical line = length of major axis of red corpuscle.

Fig. 6. Cells from blood of late fœtus of cat. Fixed with heat and stained with eosin and methylene blue. (a) and (b) different forms of white corpuscle, of which (b) is much more abundant, (c) nucleated red corpuscle—very few present. The vertical line represents the diameter of the non-nucleated red corpuscle. Oc. 10, ob. apochr. ½th, Powell and Lealand. Camera lucida.

Fig. 7. Hyaline cell budding. From hanging drop inoculated with anthrax.

Fig. 8. Young hyaline cell showing active movements, but unable to effect contact with anthrax chain. From same field of microscope as fig. 7.

Fig. 9. Hyaline cell dividing. Leucocytosis produced by injection of urari. Acidulated solution of methyl green. Oc. 4, ob. E, Zeiss. Camera lucida.

[1] Krukenberg, "Ueber d. Verdauungsmodus d. Actinien", *Vergleichend-Physiol. Studien*, Heidelberg, 1881.

Fig. 10. Eosinophile cell in neighbourhood of anthrax chain.

Fig. 10 *b*. The same cell shortly after it has effected contact with the chain.

Figs. 11, 11 *a*, 11 *b*, 11 *c*, 11 *d*, 11 *e*. Successive stages in the attack of the cells on a chain of anthrax bacilli in a hanging drop of lymph. 11 *e* drawn with camera lucida. Oc. 4, ob. D, Zeiss. Illustrating Exp. I, p. 179.

Fig. 12. A hyaline cell in the same field as fig. 11 *e*. Not drawn to the same scale. The two sketches illustrate successive phases of this cell's movements.

Fig. 13. An eosinophile cell just about to attack the end of a chain of *B. filamentosus*.

Fig. 14. Eosinophile cell which has just attacked an anthrax chain. Fixed and stained with the methylene blue solution.

Fig. 15. An eosinophile cell with granules completely discharged, attacking anthrax chains. Fixed and stained with the methylene blue solution.

Figs. 16, 16 *a*, 16 *b*, 16 *c*, 16 *d*. Successive stages of conflict with a chain of anthrax, which in fig. 16 is already hidden in the cell mass. Illustrating Exp. II, p. 181.

Fig. 17. The amphophile condition of the eosinophile cells, also an eosinophile cell which has completely lost its granules, and a hyaline cell. See Exp. VI, p. 186.

Fig. 18. A hyaline cell and an eosinophile cell. In the same field of the microscope. See Exp. III, p. 181.

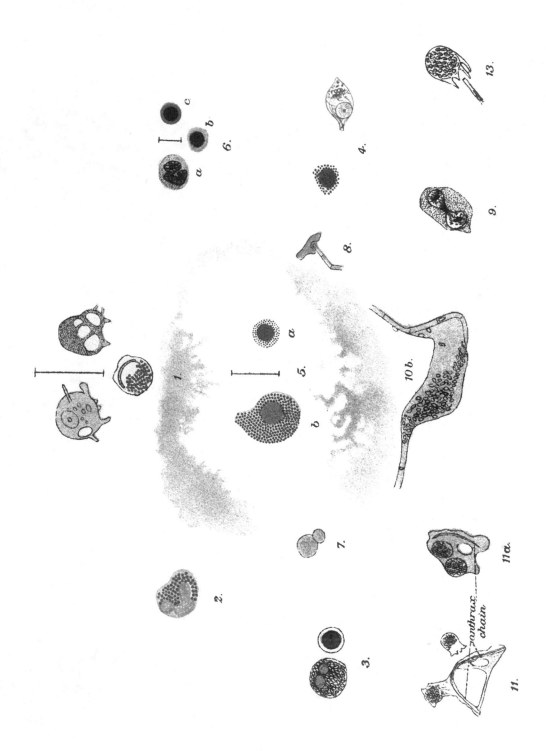

Plate X

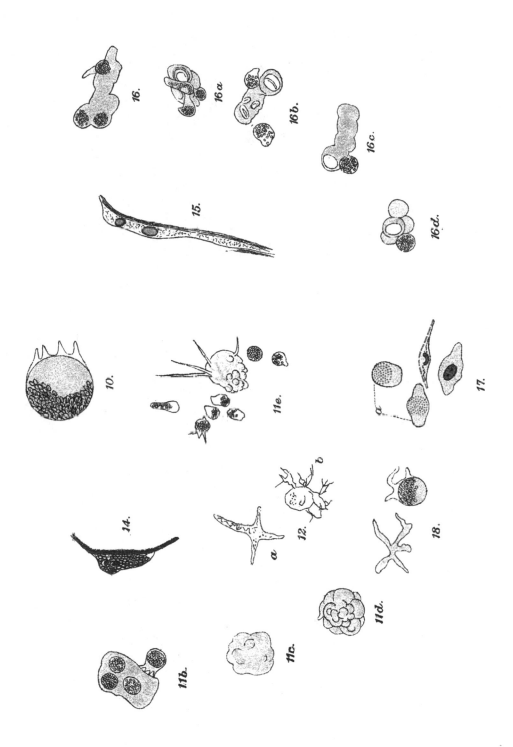

11

NOTE ON THE OXIDISING POWERS OF DIFFERENT REGIONS OF THE SPECTRUM IN RELATION TO THE BACTERICIDAL ACTION OF LIGHT AND AIR

WITH R. F. D'ARCY

[Journ. Physiol. (1894), XVII, 390]

On December 14, 1893, Prof. Marshall Ward[1] communicated to the Royal Society the results of his experiments on the bactericidal action of light. By exposing films sown with anthrax spores to the spectrum of the sun or of an electric arc Prof. Ward was able to show that the bactericidal power of light, a power which it had long been known to possess, is the peculiar property of light of short wave-length. Radiation corresponding to the infra-red, red, orange, and yellow exerted no action, the bactericidal power "begins at the blue end of the green, rises to a maximum as we pass to the violet end of the blue, and diminishes as we proceed in the violet to the ultra-violet regions".

All those who have investigated the antibiotic property of light are agreed that it is manifested only when oxygen is present. Other workers again have shown that ozone and hydrogen peroxide destroy bacteria, and, lastly, Würster[1] proved that when evaporation takes place in direct sunlight "active oxygen" is produced. These results made it possible that the bactericidal action of light and oxygen is due to the production of some oxidising substance at the surface of the fluid, and, if this be the case, it in turn would account for the fact that the action of light is limited to a thin layer next the free surface—that is to the neighbourhood of the region where, according to Würster's results, the "active oxygen" is formed.

In this way we were led to attempt to determine whether the active portion of the spectrum in Ward's experiments might be shown to be that in which the production of "active oxygen" takes place when evaporation occurs in the presence of light, and we have found that such is the case. Our experiments show that, when the spectrum of a powerful arc light is allowed to fall on a moist surface in the presence of a delicate indicator, oxidation occurs and the action commences at the blue end of the green and continues through the blue and violet to the ultra-violet regions. In other

---

[1] *Proc. Roy. Soc.* (Dec. 14, 1893).
[2] Würster, *Ber. d. deut. chem. Gesellschaft*, XIX, 3201.

words, the action is confined to a portion of the spectrum which corresponds to the region of activity in Ward's experiments.

The experiments. On the advice of our friend Dr Ruhemann we made use of tetramethylparaphenyldiamine as a delicate indicator of the presence of "active oxygen". This substance is known commercially as "Würster's Tetra-substance", and it was employed by him to determine the presence of "active oxygen". Paper sensitised with this body is prepared by Dr Schuckardt of Görlitz. The substance itself is an almost colourless solid which forms solutions having an exceedingly ill-defined and very faint yellow tint. In the presence of a trace of, for instance, impure hydrogen peroxide the substance becomes a deep violet. By the use of this very delicate indicator Würster found that "active oxygen" is formed on a moist surface when evaporation takes place in the presence of sunlight. This result we confirmed by moistening strips of sensitised paper with distilled water and placing some in direct sunlight and others in a dark chamber. Those exposed to evaporation and light changed to violet, while those exposed to evaporation only remained unchanged. In order to determine whether the action was due to light or to heat a strip of black paper was laid over each strip of sensitised paper. After about 20 min. exposure to direct sunlight an uncoloured bar was found under the black paper.

Exposure to the spectrum was carried out in a dark room, the illuminant being a powerful electric arc. Unfortunately quartz prisms and lenses were not available. The spectrum as projected on the paper was about an inch in length and very brilliant. In order to secure any result the sensitised paper must be kept thoroughly moist and open to currents of air so that rapid evaporation may take place. The sensitiveness of the paper was tested both before and after each experiment with an exceedingly dilute solution of hydrogen peroxide (one or two drops of a 2 per cent. solution to about 15 c.c. of water).

All our earlier experiments were fruitless owing to the diffusion of the colour through the moist paper. This difficulty was obviated by cutting the paper into thin strips about a millimetre wide, which were arranged on a glass slide vertically, side by side with a slight interval between each, and kept moist by a continuous strip of absorbent paper laid over their upper ends, which was wetted occasionally with a small brush.

In successful cases the violet tint commenced to appear in about half an hour, and was very obvious in from 1 to 2 hours. When well defined it commenced abruptly at the blue end of the green and extended thence into the ultra-violet region.

The apparatus at our disposal was not well adapted for the purpose and it was some months before we obtained results. In the larger number of our experiments the paper showed no coloration. Only a small percentage of the large number of experiments were completely satisfactory in giving a defined patch of colour. In none of these cases did the colour extend over the red end, and no experiment furnished evidence of the production of the colour in the red end.

The Behaviour of the Indicator towards Light

Captain Abney was good enough to criticise this result, and he pointed out that it might be due to the fact that the indicator used absorbed only light of small wave-lengths and therefore would be affected only by the blue half of the spectrum, we therefore determined the absorption of a solution of the tetra-substances made by soaking a number of slips of paper sensitised by Schuckardt in distilled water. Complete reduction was effected with zinc dust and a trace of acetic acid.

In the reduced solution no observable absorption could be detected with a Zeiss spectroscope. With the spectrophotometer the reduced solution gave the same slight absorption as did the glass vessel in which it was contained when filled with distilled water—namely about 3° in the violet.

The oxidised solution gave absorption of the spectrum from about 550 to 620, with two marked bands, or regions of maximal absorption at 565 and 610.

In the spectrophotometer the absorption was equal to a rotation of:

    2° in the Red,
    31° in the Orange-red,
    40° in the Yellow-green of about 560 as matched by the eye with the
           spectrum,
    16° in the Violet.

It therefore seems clear that we may regard the tetra-substance, at any rate so far as our experiments are concerned, as an indicator of the presence of oxidising bodies and not as being itself directly affected by the light. Würster in the paper referred to gives good reasons for believing that the tetra-substance is a definite test for ozone. For instance it is not oxidised by *pure* hydrogen peroxide.

We find therefore that when a spectrum falls upon a moist surface from which evaporation is taking place in the presence of oxygen, there is evidence of the production of some oxidising substance, possibly ozone, in that portion of the surface acted upon by the blue and violet portion of the spectrum.

A comparison of this result with that of the experiments of Prof. Ward leads us to the view that the bactericidal action of light and air is in part at any rate due to the production of some oxidising body or bodies.[1]

[1] Wesbrook (*Journ. Path.* 1894) has shown in an ingenious manner that air is used up when bacilli or their products are destroyed by sunlight.

12

THE WANDERING CELLS OF THE ALIMENTARY CANAL

With F. F. WESBROOK

[Journ. Physiol. (1895), XVIII, 490]

CONTENTS

For the purposes of this research the following animals have been examined: the frog and newt representing Amphibia, the common grass snake representing Reptilia, and among mammals, dogs, cats, and ferrets as carnivorous types, and guinea-pigs, rabbits, oxen and sheep as herbivorous types. In addition to these the rat was used as an animal of omnivorous habit, and the hedgehog as a type of a group, the Insectivora, which in its anatomical characters lies in many ways midway between the Carnivora and the other deciduate Mammalia. Lastly as young animals puppies and kittens were used, and fœtal calves were also examined.

The regions examined were the stomach, chiefly the pyloric end, small and large intestines and rectum. Attention was chiefly directed to the small intestine.

PART I. METHODS

It is absolutely necessary to fix the pieces of gut intended for examination within at most a minute or so after the cessation of the blood flow. Therefore immediately the animals were killed, usually by beheading, short pieces of the gut were removed, cut open, and plunged in the fixing fluid.

The reagents employed for fixation were absolute alcohol and a saturated solution of corrosive sublimate in 0·6 per cent. salt solution. These were employed boiling, at the temperature of the room (16–20° C.) and at 0° C. Absolute alcohol is the only reagent which preserves all the varieties of cells. In the case of this reagent it is the rule that fixation at high temperatures is best for basophile matter, while fixation at medium to low temperatures is best for oxyphile matter. Corrosive sublimate is useful in checking the results obtained with absolute alcohol in respect to the number of oxyphile cells.

Gold chloride and osmic acid vapour were employed occasionally.

In all cases the histological analysis of tissues was made with a variety of stains. Those employed were Ehrlich's neutral mixture, and the Ehrlich-Biondi mixture; the acid stains, eosine, methyl eosine, orange G, acid fuchsin in presence of acetic acid, sodium sulphindigotate, and hæmatoxylin. The acid stains were used in aqueous, glycerine, and alcoholic solutions after the manner described by Kanthack and Hardy.[1] The basic stains employed were methylene blue, thionin, and dahlia in saturated alcoholic solutions, and the intensity of the staining power was controlled by the percentage of alcohol present. A solution in 85 per cent. alcohol is perhaps most generally useful, but by using various strengths of alcohol accurate comparative determination of the affinity of particular substances for the dye may be made.

The basophile cells of the gut are always exceedingly sensitive to the presence of even minute traces of water, therefore in basic staining the percentage of alcohol should not fall below 80. It also follows from this fact that rapid dehydration with absolute alcohol is essential to the preservation of these bodies.

It is necessary to say a few words concerning the effect of temperature on staining. We distinguish two processes in the colouring of a tissue with a dye, one a physical process in which the dye is precipitated probably by surface tension acting in the minute capillary spaces, and another in which a chemical union between the dye and some particular substance occurs. The physical action leads to the diffuse coloration of the tissue, and this precipitated colour can generally be removed with ease by washing with some solvent of the dye. Further it is obvious that the general coloration

[1] *Journ. Physiol.* XVII, 81.

of the tissue by this physical process will be diminished by increasing the solvent power on the dye of the medium in which the dye is applied. Thus, in the case of methylene blue, if a saturated alcoholic solution of this dye be heated it becomes an unsaturated solution since methylene blue is more soluble in hot alcohol than in cold alcohol. Therefore a hot alcoholic solution of methylene blue produces less physical (diffuse) coloration, but on the other hand the chemical processes are intensified and the true dyeing action is more marked. On the other hand with cold solutions the physical process is accentuated and the chemical process diminished or, it may be, arrested.

Turning to the facts of the case it was found that during the cold weather of last winter the temperature of the solutions at times fell to a point at which the chemical processes—namely the dyeing of the basophile granules (dog) with methylene blue—took place with such exceeding slowness that in eight hours only a faint trace of colour was visible here and there. This was because the critical temperature for the union of the granule substance of the basophile cells of the dog and methylene blue is about 13° C. At this temperature however the tissue generally colours deeply.

With a boiling solution of the dye very different results are given—the basophile granules colour very intensely while general coloration is almost absent—if the hot solution is rapidly blotted off the section.

This brings us to the practical issue that basic staining with methylene blue is in most cases best effected by boiling solutions of the dye.

Imbedding was always performed by means of chloroform.

The tissue was passed very gradually from absolute alcohol into chloroform, which was then saturated at a moderate temperature with paraffin, and the chloroform driven off by moderate heat. After the sections were cut they were fixed to coverslips and the paraffin removed by xylol or chloroform.

Morphology and Histology

In the bodies of all the Metazoa so far as we know there exists a tissue composed of isolated units, each of which is a cell endowed with the power of amœboid movement to a greater or less extent. These cells have for their home the fluids of the body and, in the lowest animal types, the cells are all of one kind while in the higher forms they have become differentiated into three kinds; the oxyphile cell bearing discrete granules which stain with acid dyes, the basophile cell bearing discrete granules which stain with basic dyes, and the hyaline cell which does not form discrete granules. These wandering cells are derived from the mesoblast, and to them as a whole the name "sporadic mesoblast" has been given.

So far as is known all the members of the Craniata possess these three types of wandering cells, but in the higher forms—namely in those animals which constitute the Mammalia—the three types are no longer uniform in character throughout the body as they are in the lower Craniata but differ in certain structural details. In these animals the sporadic mesoblast is further differentiated into portions, each composed of three kinds of cells, oxyphile, basophile, and hyaline, and each related to a special part of the body.

In an earlier paper[1] two of these groups were recognised, namely a hæmal group related to the blood system, and a cœlomic group related to the great cœlomic spaces, and, a closely related structure, the peripheral lymph system. This present paper has for its object a consideration of the minute structure and function of those wandering cells which lie in the interspaces of the mucous coat of the gut, and of the relation of these cells to those of the cœlomic and hæmal groups. To these cells of the gut the term "splanchnic" cells might conveniently be given.

Oxyphile cells. The oxyphile cells of the gut in shape are round or irregular and each bears a round or horse-shoe nucleus, resembling in this particular the coarsely granular oxyphile cell.[2] The cell substance is crowded with discrete granules which stain with acid dyes, the size of the granules and the intensity of their affinity for acid dyes varying in different animals.

Thus in Amphibia (frog and newt) and Reptilia (grass snake) the splanchnic oxyphile cells agree in every particular, in the size of the cell, in the nuclear characters, and in the size, highly refringent nature, and staining reactions of the granules with those found generally in the blood, lymph, and connective tissues of the body (cf. Pl. XI, fig. 1 *a* with fig. 1 *b*). They are cells of about 10 μ in diameter with an eccentrically placed nucleus which is in most cases elongated, and bent round to form a more or less complete horse-shoe.

In the Mammalia the splanchnic oxyphile cells differ from those found elsewhere in the size and shape of the cell and in the size and shape of the granules. The nucleus, as in the case of the cold-blooded animals, is mostly horse-shoe in shape though a varying and sometimes large percentage of the cells present in any section show round nuclei.

In type these cells most closely approach the oxyphile cell of the cœlomic group, that is the coarsely granular oxyphile cell, and in the ox, sheep, and rabbit the difference is solely one of size. The same typical crowding of granules, characterised by a high refractive index and an intense affinity for acid dyes, is observed in both, but the splanchnic cells and granules are somewhat smaller (fig. 2). Thus in the rabbit the splanchnic

[1] Kanthack and Hardy, *Journ. Physiol.* XVII, 81.
[2] *Ibid.* p. 86.

oxyphile cell averages 7μ in diameter while the cœlomic oxyphile cell averages 10μ.

In the rat too these splanchnic oxyphile cells closely resemble those of the cœlomic system except in the matter of size but here the difference is greater than in the previously mentioned forms, the splanchnic oxyphile cell averaging $6-6.5\mu$ while the cœlomic oxyphile cell averages 10μ (fig. 3). The splanchnic oxyphile cells of the hedgehog resemble those of the rat in type.

In the Carnivora, ferret, cat and dog, the splanchnic oxyphile cells depart very widely from the types found elsewhere (fig. 4). The granules are small and have a much diminished affinity for acid dyes, in point of fact the granules agree in size and reaction to dyes much more with the finely granular oxyphile cell of the hæmal system than with the coarsely granular oxyphile cell of the extravascular fluids. On the other hand the cell has a rounded or curved nucleus and never shows the characteristic irregular nucleus of the hæmal cell. The granules are smallest and the cell is most extreme in type in the dog and cat, while in the ferret the granules have a very appreciable size and lustre, so that the structural features more closely resemble those found in the splanchnic oxyphile cells of the rat and hedgehog.

It will be noticed that we have pointed to the arched or horse-shoe nucleus of the splanchnic oxyphile cell as an important characteristic, and we have done this because we regard the rounded nucleus of the hyaline cell, the curved elongated nucleus of the cœlomic oxyphile cell, and the irregularly branched nucleus of the hæmal oxyphile cell as being true morphological features, and not as has been suggested, accidental appearances dependent upon the process of fixation.

It has been supposed[1] that the shape of the nucleus in fixed cells is a measure of the activity of the cell. The statement which has been made is twofold, (1) that in the position of rest of these cells both cell body and nucleus are spherical, and (2) that the extent to which the nucleus departs from the spherical shape in preparations is a measure of the ante-mortem activity of movement of the cell. Thus in histological preparations we find the coarsely granular cell for the most part with a more or less curved nucleus, and the finely granular oxyphile cell with a very irregular and lobed nucleus, and, according to the above hypothesis, both curved and lobed nuclei are departures from the shape of the nucleus in the living cell when at rest due to the fact that, in the process of fixation of a mobile cell, the cell substance contracts and is fixed so much more rapidly than the nuclear substance that the latter remains more or less extended in an irregular shape.

[1] Sherrington, *Proc. Roy. Soc.* (Dec. 14, 1893).

This hypothesis appears to us to be opposed to the facts of the case, and to be based on evidence which, strictly speaking, does not touch the question at issue.

In the first place so far as the histological facts are concerned it is inconceivable how, if the shape of the nucleus chiefly depended on such chance factors, it should be so remarkably constant a feature of the different kinds of cells, and so typical in certain animals that the coarsely granular oxyphile cell of the rat can be identified as belonging to that animal eight times out of ten by its nucleus alone.

In the second place the statement that the nucleus of the living cell when at rest is always spherical in shape is apparently based entirely on the observation that when any of these cells are allowed to die slowly the nucleus is found to be spherical.[1]

If we assume that the morphological structure of a cell is the same when it naturally and slowly reaches the condition of non-living matter by the gradual ebb of its store of energy, in other words, when it has reached what may be called the absolute zero of life, as when it is in the condition of rest of the living cell, or the conventional zero of life, then the truth of the hypothesis naturally follows.

But this is a great assumption to make, for even in the condition of rest, that is when rate of change of energy is minimal, the living cell, in virtue of the potentialities involved in the statement that it is "living", constitutes a dynamical field marked by molecular stresses induced by balanced forces; and, until direct evidence to the contrary is forthcoming, we may well believe that those forces maintain the nucleus in a shape which it would not possess if it yielded to those external agencies such as surface tension which, as in the case of an oil globule suspended in water, would otherwise reduce it to the spherical shape.

Further, though we cannot see the nucleus in the healthy living cell, and therefore are not able to appeal to direct observation to settle this question, yet in the case of those cells which bear refringent granules, the disposition of the latter in immobile individuals may help us to form a pretty shrewd guess as to its position and shape by noting the shape of the space unoccupied by granules. In most cases this space is clearly curved in the case of the coarsely granular oxyphile cells (see for instance the photograph reproduced in fig. 1 of Prof. Sherrington's paper), and equally clearly a flattened spheroid is outlined in coarsely granular basophile cells.

Direct observation may be brought to bear by observing the staining of spherical cells in a hanging drop faintly tinged with some dye. If for instance a trace of methylene blue be used, the cells die and stain so slowly as to afford ample time for observation and the making of sketches. A

[1] Sherrington, *Proc. Roy. Soc.* (Dec. 14, 1893), p. 187.

quiescent cell was brought into the field of the microscope *before* the addition of methylene blue. Figs. 15 *a, b* and *c* represent stages in the slow appearance and staining of the nuclei. The drop was from "inflamed" lymph of a frog and the cell happened to be one of the apparently truly multinuclear oxyphile cells sometimes found in that condition.

In discussing this question it is important to bear in mind that the nucleus may play a part in the movements of the cell purely as a mechanical factor. The junction of the nuclear substance and cell substance forms an internal surface which must have a certain directive effect on energy liberated within the cell. In this way the difference in the character of the movements shown by the various wandering cells may be in part referable to differences in the form of the nuclei.

Basophile cells form a very remarkable feature in the mucous coat of the gut. They are especially striking in Carnivora both as regards their number (as many as three to five hundred are present in a single villus of a dog) and the orderly arrangement they exhibit. In Herbivora on the other hand these cells are both less numerous and less ordered.

In all the animals examined the splanchnic basophile cells agree in one feature of peculiar interest, namely the exceeding instability of the granule substance in presence of even small quantities of water. After death water induces a complete disruption of the granules of so marked a nature that it involves the whole cell body, swelling it up to a bladder-like form or even destroying it so completely that no traces can be recognised by subsequent treatment. In the latter case, breaks in the continuity of the tissue appear where the basophile cells previously lay.

It follows from this liability to change in the presence of water that these cells rapidly disintegrate after death. They can only be preserved intact by very rapidly dehydrating small pieces of the gut with absolute alcohol. Corrosive sublimate in aqueous solution fails to preserve them, though sometimes a few imperfect specimens persist.

The instability of the granule substance of these cells is so great that even after being acted on by absolute alcohol for as long as eight and a half months, exposure to water for a few seconds[1] suffices to destroy or alter them in such a way that subsequent treatment with stains completely fails to show their presence. Alcohol, even as high as 50 per cent., only serves to slightly retard the destructive action of water.[2] Therefore staining must be carried on in solutions of dyes with not less than 70 per cent. of alcohol present. Owing probably to the instability of these cells they have, so far as we know, been overlooked by previous observers.

---

[1] By immersing a section fixed on to a coverslip in distilled water.

[2] On damp, muggy days the moisture in the atmosphere or in the blotting-paper used to blot off excess stain sometimes suffices to destroy the granules.

Though nothing is known of the chemical nature of these granules, yet there is reason for believing that they are not simply proteid in nature. Thus even after prolonged exposure to absolute alcohol they are unstable in presence of cold water, and of boiling water; and are even destroyed by a saturated aqueous solution of corrosive sublimate. The destructive action of the saturated solution of corrosive sublimate was determined as follows in the case of the splanchnic basophile cells of the dog, ox and rat. Two successive sections from a ribbon were mounted on coverslips, freed from paraffin by xylol, and brought into absolute alcohol. One section was then plunged for a few seconds into a saturated solution of corrosive sublimate and, after blotting off excess, was then placed in absolute alcohol containing free iodine. The other section was placed directly into the iodine solution, and both were allowed to remain there for some hours. The iodine was then completely removed by prolonged washing with 95 per cent. spirit and the sections stained with methylene blue. The section which had been exposed to the corrosive sublimate solution showed no trace of basophile cells. Control experiments were made to determine whether, for instance, the abrupt transference from alcohol to a watery solution was responsible for the change by, among other methods, first drying the sections before immersion in the watery solution, but, so far as we could determine, the sole cause of the change was the presence of water.

Slight differences in the degree of instability of the granules of the splanchnic basophile cells were found in different animals; those of the rat for instance being a little more resistant than those of the rabbit. The difference is not great but, however slight, it is interesting when we remember the peculiarly great stability, in the face of reagents and treatment, of the granules of the coarsely granular (cœlomic) cell of the rat, and the equally marked instability of those of the rabbit.[1]

The differences between the splanchnic and cœlomic basophile cells of the rats are exceedingly striking both as regards the size and stability of the granules and the size of the cells.[2]

It should be noted that the splanchnic basophile cells occur only in the mucous coat of the gut, and a section treated with water will still show coarsely granular basophile cells of the ordinary type in the submucous and muscular coats.

The splanchnic basophile cells themselves vary much in shape, and in sections of the gut they may appear elongated and flattened or rounded and lobed. They are usually more or less flattened in a plane parallel to the internal surface of the gut and are irregular in shape (fig. 10). The cell

[1] Kanthack and Hardy, *Journ. Physiol.* XVII, 81.
[2] Compare fig. 5a with fig. 5b reproduced from Kanthack and Hardy, *Journ. Physiol.* XVII, 81—both drawings to the same scale.

substance is commonly so crowded with granules as to obscure the nucleus. When that structure can be seen it is ovoid or spherical.

In Amphibia and Reptilia, these cells resemble those found elsewhere both in the size of the cells and of the granules. In Mammalia the splanchnic basophile cells always differ from the coarsely granular (cœlomic) basophile cell in being smaller and possessing usually very much smaller granules. In the rat for instance the difference between the large basophile spherules of the cœlomic basophile cells and the fine dust-like granules of the splanchnic basophile cells is especially striking (figs. 5 and 6).

The arrangement of these cells in the villi of Carnivora is very remarkable. It can be most readily studied by mounting entire villi which have been fixed in absolute alcohol and stained with methylene blue in 80 per cent. alcohol, in Canada balsam. The whole villus then appears closely studded with basophile cells and, in optical sections, one sees that they form a defined layer lying immediately under the epithelium (figs. 12 and 13). Each cell is flattened in the plane of the surface of the villus and is of an irregular shape. Allowing for contraction in preservation it is clear that during life they form a practically continuous layer. We thus find this feature, characteristic of the villi, and of the villi only, namely that a layer of flattened granular cells underlies the epithelium, and to this layer from its histological characters the name *basophile layer of the villi* might fitly be given.

In addition to the basophile cells of this (basophile) layer, others, but only a few, lie scattered in the parenchyma of the villus, and as we proceed downwards into the crypt region, these scattered specimens become more and more numerous so that a considerable number lie between the crypts and in the basal region. In other words the basophile cells of the crypts and basal region do not show such regularity of arrangement, as marks their disposition in the villi.

The cells vary in shape. Those of the basophile layer may be rounded so that they fail to touch one another, or they may be so irregular in shape as to be almost branched and to overlap. The overlapping may be much more considerable than appears on first sight, since it is impossible to determine how far the unstaining cell substance may extend beyond the granule-bearing region, and we are therefore unable to fix the extreme limit of the cells.

Judging from our entire experience of these cells we are inclined to believe that, though they usually have a fixed habit, they perform movements of extension and retraction which result in large alterations of the shape of the cells. Thus the layer may be described as composed of mobile cells, which are laden with basophile granules, and which, in virtue of their mobility, form a fenestrated membrane immediately underlying the epithelium which has the power of altering the size of its perforations. As we shall see later the splanchnic basophile cells under extreme conditions move from place to place.

We have failed to demonstrate the existence of such a well-defined basophile layer in mammals other than Carnivora. In the ox however a similar but very much less perfect structure is seen in some of the villi. Basophile cells are abundant in the mucous coat of the intestine of this animal but they mainly lie dispersed irregularly throughout the parenchyma. The sheep agrees with the ox in possessing a large number of splanchnic basophile cells, but we did not succeed in demonstrating a definite and continuous basophile layer.

In the hedgehog splanchnic basophile cells are numerous but, in the two examples examined by us, they appeared to be specially unstable. In these animals the bulk of the cells lie in the parenchyma, though many occur immediately beneath the epithelium. Those which occupy the latter position are too far removed from one another to allow us to speak of them as a definite basophile layer. The rat resembles the hedgehog except that the cells are relatively less numerous. Lastly in the rabbit the cells are still less numerous in the small intestine, though they are relatively abundant in the cæcum and vermiform appendix.

In the frog, newt and snake basophile cells, resembling in size and appearance those found elsewhere, are found in the gut in fair numbers; they are scattered, sometimes in groups.

Putting these facts together we see that in the Carnivora, especially the dog and ferret, the basophile cells found in the villi are arranged beneath the epithelium in such a way as to form a definite basophile layer; whereas in the non-carnivorous forms they lie scattered throughout the parenchyma. When we compare these non-carnivorous animals one with another we find in some a greater, in others a less proportion of the scattered basophile cells lying immediately under the epithelium, and in this we probably have an expression of the manner in which the arrangement in the Carnivora has been arrived at, namely by the segregation of the basophile cells into a particular region in immediate relation to the epithelium.

The *hyaline cells* found in the mucous coat of the gut are rounded or angular in shape and possess a rounded nucleus. They vary considerably in size, being from 5 to $10\,\mu$ in diameter, and the nuclei show considerable differences in the amount of basophile matter present.

As Heidenhain[1] pointed out the nuclei sometimes stain a deep opaque colour with nuclear stains, or they may show an open network with or without nucleoli (fig. 14).

The dark staining nuclei lie in small cells with scanty cell protoplasm, the open, lighter staining nuclei are larger and occur in larger cells.

[1] *Pflügers Archiv*, Bd. XLIII, Suppl. 1888.

Two extreme conditions of the splanchnic hyaline cells may be recognised, in the first condition the cells present appear to be all of one order, resembling one another in size (about $7\,\mu$) and in the fact that the nuclei all show the open network and do not stain intensely (fig. 14 b). Here and there however are a very few of the smaller cells with deeply stained nuclei. In the second condition cells are present in all the varying stages from small young cells with darkly staining nuclei, so small and with such scanty cell substance as to deserve the term "lymphocyte", to hyaline cells larger than common and reaching to a diameter of $10\,\mu$ (fig. 14). Under these conditions the percentage of small cells with dark nuclei varies markedly in different regions of the gut and is maximal in the neighbourhood of solitary follicles or Peyer's patches. To this point we shall return later. The larger cells in the second condition sometimes manifest phagocytosis and when swollen with ingesta they may be larger than $10\,\mu$ in diameter (fig. 16). We believe that these hypertrophied hyaline cells are identical with the large phagocytes which Metschnikoff includes under the term "macrophage".

In mammals the hyaline cells usually form the larger proportion of the cells present in the mucous coat. Various countings give them from 40 to 70 per cent. In frogs they are sometimes less numerous than the oxyphile cells.

Splanchnic wandering cells of the fœtus. In the course of this work we were fortunate enough to secure two fœtuses from slaughtered cows sufficiently fresh to enable us to examine the condition of the wandering cells of the gut at this period. In both cases the organs were fully formed. Sections of small intestine showed that the villi though prominent structures were very imperfect. The adenoid tissue of the mucous coat was dense and very richly supplied with cells in the basal region. Here obviously rapid growth was taking place. Nearer to the lumen of the gut however, that is to say in the region of the crypts, it was looser and in the villi it was represented only by a network with such large interspaces, occupied by so few loose cells as to give one at first sight the impression that the villi were hollow structures consisting of a wall of epithelium on a basement membrane which enclosed a large central space.

The wandering cells present comprised both oxyphile and basophile individuals, while non-granular free cells were abundant. The oxyphile cells were quite typical but very infrequent (an average of not more than one to an entire transverse section). On the other hand, and this was the remarkable feature of the intestine, basophile cells in all respects resembling the basophile cells of the adult were present in relatively considerable numbers, even in the villi, and were fully charged with granules.

Thus in the fœtal intestine basophile cells are already present and functional.

The cells of the parenchyma of the intestinal mucous coat have been described at some length by Heidenhain.[1] He distinguishes "wandering" cells, "sessile" cells, and "phagocytes", of which the sessile cells represent merely the resting condition of the wandering cells. Unfortunately neither the description in the text nor the figured representation (Taf. III, fig. 17) have enabled us to identify these "sessile" cells. With regard to the phagocytes we are in agreement with Prof. Heidenhain. He regards them as hypertrophied leucocytes, and the Fig. 22 in which he illustrates the stages in their development clearly shows that in his opinion they are derived from the non-granular or in other words the hyaline cell.

By staining with Ehrlich-Biondi's fluid Heidenhain distinguished (1) a small cell with dark staining nucleus and scanty cell substance. This is the lymphocyte or young hyaline cell. (2) A larger cell with clear cell substance and round nucleus. This is the hyaline cell. (3) A cell with round or bent nucleus; the cell substance charged with granules staining purple. This clearly is the oxyphile cell. Lastly Heidenhain distinguishes cells, not always present, with dark intensely staining nuclei and cell substance colouring intensely and evenly purple. These he regards as dying or dead cells—for cells which have been ingested by phagocytes show similar staining reactions. Hoyer,[2] a pupil of Heidenhain, verified this conclusion by showing that if pieces of lymphatic gland were kept at the temperature of the body for some hours after removal from the animal they contained a largely increased number of cells which stained in this way.

Thus Heidenhain distinguished the hyaline and oxyphile cells but overlooked the basophile cells.

PART II. ACTIVITIES OF THE SPLANCHNIC WANDERING CELLS

When we turn from the relatively secure ground of the structural characters of the splanchnic wandering cells and attempt to trace out the nature of their activities we are met with a variety of phenomena which, though fairly simple in themselves, yield curiously contradictory results when an attempt is made to correlate them with obvious processes taking place in the alimentary canal.

What we have to say on this point and the conclusions we offer we regard as being tentative. The reader will readily gather from what follows how contradictory the results may appear on the surface, and how large a series of experiments would have to be conducted to secure anything approaching a complete history of the relation of these splanchnic wandering cells to the processes going on in the alimentary canal.

[1] *Pflügers Archiv*, Bd. XLIII, Suppl. 1888.
[2] *Arch. f. mikr. Anat.* (1889), XXXIV, 208.

Broadly speaking the cells undergo changes (1) in total and relative number, (2) in the size and composition of their granules, and (3) in their distribution, as influenced by their own proper movements, in the mucous coat.

The immediate stimulus which leads to these changes is of a chemical nature and consists of alterations in the chemical composition of the fluid, the lymph, that is to say, which occupies the interstices of the mucous coat.

The chemical composition of this fluid is dependent (1) upon the blood stream in the capillaries of the mucous coat, and (2) upon the absorptive and secretory activity of the lining epithelium.

The nature of the diet affects these factors in two ways. In the first place the chemical character of the substances absorbed from the gut will be modified by changes in the composition of the diet. But in addition to this obvious relation we must bear in mind the fact that the chemical nature of the end products of digestion—*e.g.* leucin, tyrosin, ptomaine—vary in accordance with alterations in the *quantity* of food given even though the quality of the food remain constant.

Taking our experiences as a whole we find that the splanchnic wandering cells are affected by (1) the presence of food in the gut, (2) the absence of food from the gut, and (3) the extent to which the gut is invaded by micro-organisms. The most difficult case to deal with experimentally is the absence of food from the gut, for starvation so frequently leads, especially in herbivorous animals, to a striking increase in the number of micro-organisms present.

A further consideration to be borne in mind is that starvation, when uncomplicated by secondary phenomena, has a twofold effect. Initially it leads to a hypertrophy of such structures as suffer loss of substance during digestion, *e.g.* the loading of the alveolar cells of the pancreas during starvation; while secondly if pushed far enough it will cause a diminution of these same structures as well as of the other structures of the body. An instance of the first action is furnished by the splanchnic oxyphile cells, these in certain pure cases of starvation have been found to increase in numbers and in the loading of individual cells with granules (p. 226).

We will as far as possible deal with the different cells in order in considering the changes which they show under various conditions.

Oxyphile cells. The splanchnic oxyphile cells may occur throughout the entire depth of the mucous coat of mammals from the epithelium, where they lie embedded between the cells, to the peculiar "basal region" which is placed between the bases of the crypts or glands and the muscularis mucosæ; or they may be restricted entirely to this basal region.

So far as these cells are concerned we agree with Heidenhain in regarding the basal region as their place of origin and from this level, in response to certain changes in the constitution of the contents of the lumen of the gut, they move up toward the inner surface.

It is impossible to make a similar statement with regard to either the hyaline or the basophile cells, for these latter never lie wholly in, nor can they be said to have any special relation to, the basal layer of the mucous coat, though both forms occur there together with the splanchnic oxyphile cells.

A comparison of all the mammals examined by us leads us to the conclusion that the moving upwards of the oxyphile cells from the basal layer towards the free surface of the gut is at first associated with an increase in their numbers. If therefore we are right in regarding the basal layer as the place of origin of the oxyphile cells, and in viewing the movement thence towards the lumen as an indication of activity on their part, we may say that, at the onset of a period of activity, the cells migrate towards the lumen of the gut and increase in numbers, and some of them pass into the epithelium and through it into the lumen of the gut (fig. 18).

The appearances seen towards the close of a period of activity would naturally vary according to whether the rate of production of these cells, whatever the machinery for that might be, was or was not maintained, and whether it was less, equal to, or in excess of the rate of destruction of the cells in the epithelium and in the lumen.

If we trace the oxyphile cells into the epithelium we find that they thrust themselves between the bases of the cells and move between the cells towards the lumen. Some of the cells suffer degenerative changes while still within the epithelium (fig. 19). The granules disappear and the nucleus alters. The latter loses all trace of a network and passes into a condition in which it stains evenly. It may also break up into two or three masses. These degenerated nuclei frequently stain a remarkably intense and brilliant green with Ehrlich-Biondi's fluid. Sometimes the nucleus suffers these extreme changes long before the granules have completely disappeared. The appearances presented force one to the conclusion that the oxyphile cells perish in the epithelium and indeed they may shrink there, as though dissolved, and leave cyst-like spaces.

However all the oxyphile cells which enter the epithelium do not perish there, some, an unknown but possibly at times a very large percentage, traverse the epithelium and gain the lumen; others again may re-enter the parenchyma. Degeneration of the cells may occur at any level and in the same section one may find more or less degenerated cells together with individuals which so far as one can see are absolutely intact, in situations ranging from the base to the inner, free surface of the epithelium.

As might be expected free cells in the lumen of the gut are only found in cases of starvation, the presence of digestive juices would no doubt almost instantly destroy them. On the other hand the oxyphile cells are found in the epithelium of animals in full digestion and, in the case of mammals, intact cells have been seen in lowest portions of the lumen of the crypts of Lieberkühn. It appears to us to be clear that immigration of the oxyphile cells into the epithelium and thence into the lumen is a process of constant occurrence, at times so slight as to be barely detectable, at other times so excessive that the epithelium appears to be riddled with these bodies.

The most obvious structural change which the oxyphile cells manifest within the epithelium or in the lumen of the gut is a diminution in the number of the oxyphile granules even to the total disappearance of these structures. This appears in fig. 18, where a small portion of a crypt of Lieberkühn is shown containing oxyphile cells some of which are completely charged with granules, others have only a few, while one has lost its granules. If we limit ourselves strictly to the evidence furnished by stained sections, it is impossible to decide whether this change in the oxyphile cells is or is not necessarily followed by death and complete disintegration.

We have never witnessed any signs of the possession of phagocytic powers by the splanchnic oxyphile cells—in this negative feature they agree with the coarsely granular oxyphile cells.

The oxyphile cells may lie evenly dispersed in the mucous coat, or they may be collected into foci. In the frog the cells very commonly lie in groups, and over each group the epithelium is sometimes astonishingly crowded with immigrating cells. If we ask why immigration should be so marked over certain areas of the epithelium and practically wanting in intermediate regions, absolutely no answer is forthcoming. The cells of the epithelium in a region of migration are naturally displaced to a certain extent by intrusive cells, but the hyaline border may be normal and shows no signs of rupture other than the scanty presence of rounded perforations leading into cyst-like spaces immediately below it from which the oxyphile cells have crept. The epithelium cells themselves, so far as appearances go, exactly resemble those found in regions not infiltrated with intrusive cells.

Parasites in the epithelium appear to have little effect on the oxyphile cells; thus in one frog, and the case is interesting as showing the different behaviour of the hyaline and oxyphile cells, the epithelium of the small intestine was largely occupied by a parasite which occurred with a single nucleus, a fragmented nucleus, or as a heap of spores. Oxyphile cells were practically absent from the gut, but hyaline cells were present in number in the epithelium where they ingested the parasites and, in some cases, obviously digested them.

Mammals resemble the lower forms in the fact that the oxyphile cells do not enter the epithelium indifferently at any point, but a curious and apparently permanent specialisation has taken place, namely that the oxyphile cells migrate especially by way of the crypt epithelium : and further when migration of the oxyphile cells is not very marked, the lower one goes down the crypts the more marked is it until a maximum is reached absolutely at the extreme lowest point of the gland. On the other hand when migration of the oxyphile cells is marked they may be found in the epithelium quite at the apices of the villi. In point of fact the apparent preference of the cells for the crypt epithelium may be very largely due to the fact that it is so much nearer than the epithelium of the villi to the basal layer from which, as we have already stated, the oxyphile cells appear to start, but we do not regard this as the sole cause since oxyphile cells may be present in the villi though absent from the villous epithelium, while at the same time they are present in the epithelium of the crypts. Further, as we have already pointed out, the rare migration of basophile cells follows precisely similar lines. We have never yet seen a basophile cell in the epithelium of a villus.

This limitation of immigrated oxyphile cells to the crypt epithelium is, so far as our experience goes, much more marked in the case of the dog, cat and ferret than in other mammals. It is doubtful whether oxyphile cells find their way into the epithelium of the villi of dogs except under very extreme conditions. Paneth has described the existence of granule-bearing cells at the base of the crypts of Lieberkühn, and we were careful to convince ourselves that these had not been mistaken for intrusive oxyphile cells.[1]

It will be wise here to attempt to give some idea of the number, both absolute and relative, of the oxyphile cells. Comparison is made between sections of similar thickness about $8\,\mu$ thick.

In frogs apparently normal and taken at random from the tank oxyphile cells may be present to the number of three to four hundred in the mucous coat of a transverse section of the small intestine; or only 50–60 in an entire transverse section. When abundant the cells lie in groups and at each focus from 35 to 64 can be counted in a single field of a $\frac{1}{12}$th.

From 20 to 60 or an even higher percentage of the oxyphile cells may lie either in the epithelium or so closely attached to the base of the cells that they remain adherent to the epithelium when that structure is torn away from the subjacent connective tissue.

The above figures are from sections through the upper portion of the small intestine. It is not the purpose of this paper to consider accurately the distribution of the cells throughout the length of the gut, but we may note that this was determined in the case of four frogs, and they agreed in showing a continuous and very marked decrease in the number of oxyphile

[1] See note at the end of this paper. Also figs. 21 and 22.

cells as one passed from the duodenum either towards the rectum or towards the lower end of the œsophagus.

It is difficult to compare the relative number of the three kinds of cells in the frog's intestine. The oxyphile cells are readily seen, the hyaline cells are even more readily overlooked, while the basophile cells are only preserved with difficulty. Careful countings seem to show that the oxyphile cells frequently may form as much as 50 per cent. of the total number of wandering cells present. In the mammals examined the oxyphile cells were found to form from 10 to 50 per cent. of the total number of wandering cells present in the mucous coat.

We will now proceed to the connection of the movements and number of the oxyphile cells with processes taking place in the gut.

The most striking fact we have met so far is that these cells are, if one may use the expression, finally used up either within the limits of the endodermic epithelium or even in the lumen of the gut itself. The movement of the cells from the basal layer, through the parenchyma to the epithelium, we rank as a phenomenon falling under the heading of chemiotaxis.

It is certain that the oxyphile cells perform some work either as such, or in virtue of the products of their disintegration, when in the epithelium and also when in the lumen of the gut, but we have no evidence to offer as to the nature of this work. It is also equally certain that they are "used up" in the performance of certain functions while still within the parenchyma—if proof be needed of this it is found in the fact that ingestion of effete oxyphile cells on the part of the larger hyaline cells occurs within the limits of the parenchyma. Here again evidence sufficient to determine the part the oxyphile cells play is at present lacking.

What our experience seems to show is that the movement of the oxyphile cells into the epithelium is increased (1) by the presence of an unwonted number of micro-organisms in the gut, (2) by the onset of starvation, (3) by the onset of a period of digestion when it follows a short period of starvation. The difficulty in arriving at certain conclusions has already been dealt with. As we have pointed out during a period of starvation micro-organisms usually increase in the gut and one cannot then determine how far the changes observed are due to the absence of food and how far to the micro-organisms. But this is only an example of one side of the difficulty for it is equally difficult to obtain a measure of the activity of the cells. It is obvious that when the activity of the cells results in migration and destruction the total number present is of no use to us. It merely represents the balance for the time being between the rate of production and the rate of destruction.

This may be illustrated by reference to the hyaline cells. In starving rabbits these cells were found in excessive numbers in the epithelium of the small intestine, but the total number present in the parenchyma was very obviously below the normal. In other words here for the time being expenditure was far in excess of income.

An estimate of the number of the cells in the epithelium might be regarded as an accurate criterion but this again is useless unless we assume that the rate of progress through or from, and the rate of destruction in, the epithelium do not vary. To take an instance. In rabbits and rats deprived of food and prevented from consuming their own fæces, in two days the stomach was occupied by incredible masses of inwandered cells, yet on examination the living epithelium was found to contain a ridiculously inadequate number of intrusive cells. In point of fact it was difficult to be certain that they were more numerous than in the normal animal, although there is always a paucity of intrusive cells in the lining epithelium of the stomach. In view of these difficulties we have taken the percentage of the total number of cells present in the entire mucous coat which is formed by those lying within the epithelium as the best guide to the migratory activity, and, in the case of the oxyphile cells of mammals, we have been guided in our estimate also by the position of the cells in the mucous coat, and by the extent to which they are or are not limited to the basal layer.

The action of micro-organisms when present in the gut on the oxyphile cells was, it seemed to us, prettily illustrated by experiments on frogs. Diluted cultures (broth) of anthrax bacilli and cholera vibrios were injected into the intestines of six pithed frogs. The intestines of three of the animals were removed and fixed after $5\frac{1}{2}$ hours, and from the remaining three, 19 hours after the operation. In one case a length of intestine was removed from a recently killed frog, and after having some cholera culture injected into its lumen, it was hung up in a moist chamber. In this and in all the experiments the result was the same, namely a gradual diminution of the oxyphile cells present, with at the same time a continued increase of the proportion of the total number of these cells present in the epithelium, until the number there formed 100 per cent. of the whole; and finally, a total disappearance of oxyphile cells from the gut.

In attempting to determine the relation of the oxyphile cells to periods of feeding and of hunger we were met with the difficulty that, in spite of the utmost care directed to keeping the animals clean and to prevent them eating their own fæces, the micro-organisms increased in the small intestine and penetrated even as high as the duodenum. We believe that our failures in this respect were largely due to the fact that we used the rabbit and the rat for the purpose, our experience going to show that in animals of purely carnivorous habit the small intestine remains free from any obvious number of micro-organisms during at any rate short periods of starvation.

Whatever may be the true effect of starvation on the oxyphile cells there can be no question but that when it is complicated by the presence of micro-organisms in the gut, a large percentage are found in the epithelium. In digesting mammals and frogs the oxyphile cells are more numerous in the duodenum and decrease in number as we proceed down the small intestine. In starving animals whose intestines have become invaded by micro-organisms, this condition is usually altered, the oxyphile cells are more numerous at the lower end of the small intestine or in the large intestine, than in the duodenum, and this is in keeping with the fact that the invasion by micro-organisms takes place from below upwards.

Those cases of starvation in which the gut remained free from detritus agreed in the fact that the oxyphile cells increased in numbers to a great extent. They were abundant in the basal layer and extended thence right to the tips of the villi. On the other hand a comparison with the condition of the cells in other animals fed on the same diet, and under the same conditions, but killed when in full digestion, revealed the fact that the migration of the oxyphile cells was very much less in the starving than in the fed animals. In other words the accumulation of oxyphile cells in the case of starving animals appears to be due to a diminution of the drain on the cells which occurs in feeding animals, especially on a flesh diet, and in starving animals when increase occurs in the micro-organisms present.

The conditions which we have met with often seem curiously contradictory, but taking them together this seems to be the most satisfactory interpretation, and it is further borne out by the fact that when accumulations of wandering cells appear in the gut as a result of starvation they are exclusively of the hyaline type.

In his experiments on feeding and starving animals Heidenhain[1] found that the oxyphile granules were larger in the former; and he illustrates the two conditions of the cells in a starving and a full-fed dog on his Pl. IV, figs. 27, 28. Our experience goes to show that these relations do not always hold and indeed in very clear and pure cases of starvation, that is free from any excess of micro-organisms in the gut, the reverse condition may occur, the granules being larger in the starving animal. This was noticed for instance in the experiment further detailed on p. 226; and in connection with this and like experiments we have been led to think that the partial hypertrophy of the oxyphile cells and granules was due to a diminished drain. It is a phenomenon of hunger, as it modifies this tissue in a thoroughly well-nourished animal.

The results of an attempt to determine experimentally the effects of a flesh diet are given on p. 225. The most marked oxyphile granulation met with by us among dogs, when the granules were large and the cells very

[1] *Pflügers Archiv*, Bd. XLVIII, Suppl. 1888.

numerous, occurred in a portion of small intestine from a fat bitch. The mucous coat however was injured by the presence of worms, and migration was particularly marked, the number of oxyphile cells in the crypts of Lieberkühn being very striking. Fig. 18 is drawn from this animal.

Hyaline cells. The hyaline cells, unlike the oxyphile cells, are distributed evenly throughout the parenchyma, and are always present, though in variable numbers, in the epithelium. We have already pointed out that the hyaline cells vary largely in character. The cells differ in size, and the nucleus may be either very full of basophile material so that they stain as dark bodies, or they may be relatively free from basophile matter showing only the open nuclear network.

The most obvious phenomena displayed by the splanchnic hyaline cells are migration into and through the epithelium and phagocytosis. The latter may be manifested both in the parenchyma, in the epithelium, and, under favourable conditions, in the lumen of the gut. They also display changes, too obscure to be dealt with at present, in the staining reaction of their nuclei and cell substance. Practically the discoverable activities of the hyaline cells are limited to migration and phagocytosis, and the variations in the nuclei ("dark" or "open") and in the size of the cells are as Heidenhain pointed out correlated with the general activity of the cells as displayed in these two ways.

Heidenhain notices that when migration is unusually marked (that is as determined by the number of cells present in the epithelium), a large proportion of hyaline cells with small dark nuclei are found. As a matter of fact this is only a part of the change, for whereas when migration is slight the cells are all of about the same size and all have open nuclei, when migration is marked variations on both sides of this mean position occur, small cells with "dark" nuclei being present and so with all intermediate conditions to cells considerably larger than common with nuclei containing only a very loose network. In other words, not only is the tissue occupied by numbers of the youngest cells but growth of individual cells progresses under the particular stimulus, whatever it may be, which gives rise to the condition of greater activity, to a point beyond that reached in the relatively inactive state.

From Heidenhain's description and figures one gathers that only the smaller hyaline cells were found in the epithelium. As a matter of fact this is not so, the cells in the epithelium, probably as a result of the staining of the tissue about them, appear much smaller than they really are—actual measurement shows that the smallest cell does not occur in the epithelium but that both intermediate and larger ones migrate, the intermediate ones being apparently more active in this respect than the larger ones.

Hyaline cells migrate chiefly into the epithelium of the villi. In Carnivora the distinction between hyaline and oxyphile cells in respect to their place of migration is sometimes exceedingly sharp, the former being confined to the villous, the latter to the crypt epithelium. Hyaline cells, as a result of their ingestive activity, may contain intrusive bodies (parasites), micro-organisms, other wandering cells, *e.g.* oxyphile cells,[1] and amorphous debris probably derived from one or other of the above.

So far as can be seen there is no necessary connection between the migratory and phagocytic activities of the hyaline cells. Thus migration may be excessive and yet one may be unable to detect any trace of phago-cytosis, and, on the other hand, the number of cells within the epithelium may be by no means excessive when instances of ingestive activity occur both within the epithelium and in the parenchyma.

In point of fact the ingestion of solid particles is an act infrequent in its occurrence and of minor importance among the processes carried out by these cells. It has been shown[2] that the hyaline cells of the lymph of the frog will remove from the lymph plasma foreign substance dissolved there. The particular substance employed in the experiment was methylene blue and it was found precipitated as solid masses within vacuoles in the cell substance.

Similarly, by feeding with peptonate of iron, we have been able to show that the splanchnic hyaline cells remove from the interstitial lymph of the villi substances which are in solution there.

Absorption of iron by splanchnic wandering cells. Macallum[3] found that when the "peptonate" of iron is given to animals (guinea-pigs) the metal may be detected not only in the epithelium of the villi but also in wandering cells which crowd to the tips of the villi and there load themselves. It seemed to be probable that the hyaline cells and these only were the iron carriers. We accordingly repeated the experiments, using Denayer's peptonate of iron. Guinea-pigs and rats were fed with this for one to four days, and then killed. The intestines were preserved in absolute alcohol and sections were tested for iron both with ammonium sulphide and with hydrochloric acid and potassium ferrocyanide, while other sections were stained according to the methods already described. It was found that clean staining of the oxyphile granules was possible after the application of the ferrocyanide test to sections. Very beautiful preparations can be made in this way. After the application of the iron test, the section should be washed fairly quickly in distilled water and then very thoroughly in several changes of re-distilled spirit (95 per cent.). They are then lightly stained with hæmatoxylin, washed in tap water, stained with very dilute aqueous

---

[1] Fig. 9, compare also Hardy and Lim Boon Keng, *Journ. Physiol.* (1894), xv, 361.
[2] Hardy and Lim Boon Keng, *Journ. Physiol.* (1894), xv, 361.
[3] *Journ. Physiol.* (1894), xvi, 268.

eosine, dehydrated, cleared in cedar wood oil, and mounted in balsam. Bismarck brown is perhaps a more effective nuclear stain than hæmatoxylin. In these various ways we demonstrated the presence of iron in large numbers of wandering cells in the villi and in the spleen. *These iron-holding cells were in all cases hyaline cells* (figs. 16 and 17).

It seems to us that in this we have a clear case in which the hyaline cells perform important work directed to the removal of dissolved substances from the fluid in which they are bathed. This result is important since it at once widens our conception of the activities which may be ascribed to these structures—they are not only as phagocytes concerned in the removal of solid particles from the fluids of the body, but we may also conceive them as actively and directly controlling the chemical constitution of those fluids by removing dissolved constituents.

The presence of iron in quantities above the normal sometimes induces a marked immigration of oxyphile cells into the epithelium of the deeper portion of the crypts. The oxyphile cells not only pass into the epithelium but continue through and are found in the lumen. This fact is the more remarkable and inexplicable when we remember that there is no evidence that the iron salt is absorbed by the lining epithelium of the crypts.

Basophile cell. In many respects this is the most interesting cell in the splanchnic series. Unlike the other forms the basophile cell does not lie free in the interspaces of the mucous coat but is attached to the supporting framework of the adenoid tissue. We were for a long time under the impression that these splanchnic basophile cells were not in the strictest sense wandering cells, we supposed that their movements were limited to changes of shape of the cell body and did not carry the cell from place to place. This idea however must be given up since, in two cases—the first of a starving rat in which the lower part of the small intestine contained very large numbers of bacilli, the second of rats fed for $2\frac{1}{2}$ months on fresh flesh from the butcher—undoubted basophile cells were found thrust between the cells of the endodermic epithelium.

These two cases stand alone and we must regard the migration of the basophile cells as being an event of very rare occurrence. The fixed nature of these cells is shown by the fact that, though the numbers present do vary in different animals, even in different individuals of the same species, yet we have never met with any increase or decrease in number sufficient to warrant us in thinking that they commonly vary very greatly.

The striking changes which the basophile cells exhibit are limited to changes in their granules. The cells are markedly granular in well-nourished animals and become less granular during starvation though the latter change is not readily brought about and it is difficult to produce extreme exhaustion of the granules (fig. 8).[1]

If however the condition of starvation be complicated with increase in the micro-organisms of the gut, or in any case where micro-organisms are abundant, the basophile cells are much distended with granules which are often much more stable than those found normally (compare fig. 7 with fig. 5).

The hypertrophy of basophile cells which commonly occurs when abnormal chemical substances are present, *e.g.* during inflammation, is an old observation of Korybutt-Daskiewicz and Ehrlich, and in a paper dealing with the wandering cells of the frog, Kanthack and Hardy suggest that the basophile cells remove certain substances, probably either of the nature of foreign substances, or the more extreme products of the general metabolism, which may be present in solution in the lymph or blood plasma. The facts which we have stated concerning the splanchnic basophile cells lend support to this view, that is so far as it confers on the basophile cells important chemical activities. The position of the remarkable basophile layer of Carnivora is such that the fluid elaborated by the activity of the endodermic epithelium from the contents of the gut must flow past it and be exposed to the action of the basophile cells before it reaches the blood vessels and lymphatics. Similarly the hypertrophy of the granulation when micro-organisms are very abundant in the gut points in the same direction.

As we have said we have never met with any very striking hypertrophy of the splanchnic basophile tissue as a whole but Dr Sherrington has described cells, apparently basophile in nature, in the intestines of cholera patients as being present in numbers apparently far above any observed by us, and Mr Hankin writes to one of us that at the time Dr Sherrington demonstrated his preparations the disposition and appearance of the cells strongly impressed him with the fact that they were present in order to absorb the microbic poison streaming in from the lumen of the gut.

The further question of the exact nature of the action of these cells on the chemical substances brought to them by the cells of the epithelium is one we do not propose to discuss at any length here. The knowledge we possess of the chemical processes taking place in the walls of the gut points to their being mainly synthetic in character, as an instance we have the condensation of peptone to the level of more complex proteids; and we will leave the question of the part played by the basophile cells with the statement that evidence has been accumulating for some time past showing that the basophile granule is a centre of synthetic activity.

[1] Small though they sometimes are these granules show signs of being complex in substance. Strictly speaking the change above mentioned is limited to that portion of the granule which fixes the basic dye.

The presence of basophile cells in a fœtus at a period of life when the gut is not occupied in the digestion of food, does not necessarily conflict with the view that, at a later time, they take part in the manipulation of the absorbed products of digestion. In the broadening of our physiological ideas we have been led of late to a much less narrow view of function and to a clearer recognition of the interdependence of the chemical processes which are carried out in the various organs. Each part of the body as we now know is a blood gland and in the gut for instance the blood and lymph is exposed to chemical action ·probably chiefly synthetic in character, by which products of metabolism in other tissues suffer change. Such processes will go on even before digestive activity is manifested and in the splanchnic basophile cells of the fœtus we may see a part of the mechanism involved.

The differences between carnivorous and other forms in respect to the arrangement and number of the basophile cells and in the character of the oxyphile granulation led us to attempt to determine whether these differences had any special relation to a flesh diet. Rats were chosen for the experiment mainly because we had examined a greater number of these animals than of any other species and could therefore better judge what might be called the normal condition. Four tame rats were isolated, in a large wire cage kept scrupulously clean, and fed for $2\frac{1}{4}$ months on fresh butcher's meat freed from fat and obvious masses of connective tissue. One rat died after about a fortnight, the others remained very healthy and active to the end of the experiment. After $2\frac{1}{4}$ months one rat was starved for $3\frac{1}{2}$ days and then the three were killed by beheading, and portions of intestine were fixed in absolute alcohol cold and boiling, and corrosive sublimate both cold and boiling. Apart from the special differences between the starving and fed examples which will be dealt with later there were obvious changes in the condition of the splanchnic wandering cells, all in the direction of an approximation to the carnivorous type. The oxyphile cells did not depart from those found in rats generally either in size, or numbers, or in the size of the granules. On the other hand they differed widely from normal rats and approached the carnivorous type in two respects, (1) the granules were scarcely preserved by absolute alcohol, and (2) they did not stain very readily. The basophile cells were present in more than normal numbers especially in the mucous coat of the stomach, they were however scattered in the parenchyma and there was absolutely no formation of a definite basophile layer. But though there was no formation of a basophile layer an unparalleled state of affairs was found, for basophile cells in considerable numbers were in the epithelium, thrust between the cells and either spherical, irregular, or elongated in shape. These intrusive basophile cells were remarkably abundant in the walls of the upper portion of the pyloric glands, in the small intestine they occurred in the walls of the crypts.

It was very clear that the basophile granule had altered considerably, especially in its chemical composition. The granules were smaller than in the splanchnic basophile cells of normal rats, indeed one might describe them as being powdery. The change in the composition of the granule was striking. The most satisfactory preservative of splanchnic basophile granules is boiling absolute alcohol, but these rats were quite unique in our experience since boiling absolute completely destroyed all trace of the basophile granules though they were preserved by absolute alcohol at 19° C. The difference in chemical composition was further shown by the fact that union between the granule substance and methylene blue could only be brought about by high temperature, and in a solution saturated with the dye at the boiling-point. On the other hand once the combination was effected the stain was intense and very characteristic, the granules showing a bright rose colour.

So far as this initial experiment goes, it points to the conclusion that the peculiar condition of the oxyphile and especially of the basophile cells in the gut of Carnivora is a direct result, an acquired feature, impressed on them by the nature of their diet. This is rendered almost certain so far as the basophile cells are concerned by the fact that in three-day-old puppies, though basophile cells were present in considerable numbers, we were unable to demonstrate the presence of a complete basophile layer.

The differences in this experiment between the starving and full-fed digesting rats were that the oxyphile cells were more numerous and particularly full of granules, while the basophile cells possessed an exceedingly scanty granulation in the starving animal. On the other hand there was little or no migration of the oxyphile cells in the starving animal, the increased number of the cells therefore appeared to be due to a diminished drain on this tissue. It should be remembered in this connection that the animal was exceedingly well nourished up to the onset of the short period of starvation.

PART III. ORIGIN OF THE SPLANCHNIC WANDERING CELLS

Oxyphile cells. It is clear from what we have already said that the splanchnic oxyphile cells of Mammalia betray in their structural features a close affinity to the cœlomic oxyphile cells. In Herbivora, Rodents and Insectivora the differences between the two are very small and it is not until we come to the Carnivora that any marked divergence appears. This suggests that the splanchnic oxyphile cells are closely connected with, or are a specialised portion of, the cœlomic cells, and we accordingly find in the frog that a focus of proliferation of oxyphile cells may supply both the small intestine and the peritoneal cavity (fig. 20).

The walls of the body spaces—pleural, pericardial and peritoneal—of all animals contain areas, usually related to lymphatic capillaries and to blood vessels, which are crowded with wandering cells. In some areas the wandering cells are oxyphile, other areas again, notably about the diaphragm, contain only vast numbers of basophile cells. These areas are probably foci for the proliferation of particular kinds of wandering cells, and if this be the case they are comparable in this respect to the nodules of lymphoid tissue in spleen, or lymphatic gland. We have met with such foci in rabbits, guinea-pigs, rats and frogs—in all animals in short in which they have been looked for—and Klein noticed proliferating areas in his study of the cœlomic spaces as portions of the lymphatic system.[1] In the frog a focus of oxyphile cells exists in the connective tissue which accompanies the hepato-pancreatic duct, and in sections taken through the small intestine at the point where the hepatic duct opens into it one sees that this focus of oxyphile cells extends into the walls of the gut, and continues a short distance upwards and downwards, gradually thinning out until it merges into the scattered oxyphile cells of the gut walls (fig. 20). In other words the splanchnic and cœlomic oxyphile cells in this region clearly have a common origin.

Hyaline cells. There is no special level in the mucous coat from which the hyaline cells may be said to originate.

In the solitary follicles and Peyer's patches however we find within the reticulum hyaline cells of various sizes, the larger number being of the smallest kind, and in cases of inflammation of the gut (enteritis) we have found the mucous coat near these structures flooded with an amazing number of hyaline cells. The lymphatic glands of the small intestine thus form an obvious source of hyaline cells. On the other hand the distribution of these structures is very irregular and large tracts of the intestine are free from them. Further, when, in the condition mentioned above, the lymph glands are flooding the mucous coat with hyaline cells only a part, and that possibly a small part, remain there, the rest passing into the lacteal vessels in such numbers that they appear in sections distended with packed masses of hyaline cells.

These cells which are so drained off are mainly small cells with dark nuclei—in other words what we recognise as the young form of the cell.

[1] Klein, *Anatomy of the Lymphatic System,* i, 1873.

General Conclusions

Although this enquiry has led to only very partial and tentative results so far as the elucidation of the function of the three forms of wandering cells present in the gut is concerned, yet it has established certain solid facts of structure, notably in determining the presence and peculiar arrangement of the splanchnic basophile cells, and in establishing the solidarity which exists between the characters of the wandering cells of the gut and those found in the cœlomic spaces, the peripheral lymph system, and the blood system, in respect to the presence there of the three great types, oxyphile, basophile and hyaline cells.

It has been claimed for the wandering cells that they form a distinct tissue in the body which has undergone a peculiar development converting it from a system of free cells all possessed of similar characters to one characterised at first by being composed of three kinds of cells,[1] while later in its history it becomes specialised in different portions of the body, in the blood, the great extravascular spaces, and, as we now see, in the walls of gut. In each of these places the three kinds of cells are present, though the cells of any one place differ from those found elsewhere sufficiently to enable us to recognise them as being distinct. The magnitude of these structural differences in the case of the gut of Carnivora is brought home to us by the fact that when Heidenhain submitted his preparations of the intestine to Ehrlich the latter was unable to rank the cells there present (those which contained granules staining with acid dyes, *i.e.* the oxyphile cells) with those which he had classified from other parts of the body.

Those views receive fresh support from this work not only from the facts of structure set forth in the preceding pages but also from such scanty light as has been thrown on the activity of the splanchnic cells, for we have seen throughout how distinct are the activities displayed by the oxyphile, the hyaline and the basophile cell. The oxyphile cells for instance may be very numerous without, so far as one can see, any immediate reference to the number of hyaline cells; the oxyphile cells too chiefly wander into the crypt epithelium, while the hyaline cells chiefly wander into that of the villi. Again in the case of the absorption of iron we saw that the hyaline cells alone charged themselves with the iron compounds.

The absorption of iron by the hyaline cells to which reference is made has, as has been pointed out, an interesting significance, since it widens our conceptions of the activities of these structures. We now know them to ingest solid particles and also to absorb into vacuoles in their cell substance and precipitate there matters previously in solution in the fluids which bathe them.

[1] Hardy, *Journ. Physiol.* vol. XIII. Kanthack and Hardy, *Trans. Roy. Soc.* (1893).

Finally, if we turn to the specialisation of the sporadic mesoblast in different parts of the body the structures found in the gut suggest certain pregnant reflections.

The gut of the Mammalia with the exception of the œsophagus and extreme portion of the rectum contains, in its mucous coat, a sheath of lymphoid tissue crowded with wandering cells. Placed here and there in this sheath are foci of proliferation specially related to the lymph stream and apparently peculiarly the seat of origin of hyaline cells. These are the solitary follicles and Peyer's patches. No such development of lymphoid tissue is found in the gut either of Amphibia or of Reptilia, nor are wandering cells so constantly present in large numbers. How then does this large development of lymphoid tissue and of wandering cells in the wall of the gut in Mammalia fit with what we know of the development of the lymphatic system in Vertebrates?

We have already alluded to the fact that there appears to be a specialisation of the wandering cells in different regions of the body. A comparison of the condition of the lymphatic system and lymphoid tissue of the body as a whole in different Vertebrates presents this fact in a new light.

In the less-specialised lymphatic system of Amphibia and Reptilia the peripheral part consists largely of irregular spaces rather than of defined vessels, and lymphoid tissue is not distributed along their course in a manner comparable to the lymphoid masses which lie as lymphatic glands on the course of the lymph vessels of Mammalia. In the gut, as we have seen, the amount of lymphoid tissue is exceedingly small, and the contrast in this respect between a section of the small intestine of a frog or snake and that of a mammal is very great, and the spleen of a frog does not contain masses of lymphoid tissue such as form the malpighian bodies of the spleen of Mammalia.

In these groups of animals in place of the widespread development of lymphoid tissue in lymphatic glands, in the gut wall and the solitary follicles and Peyer's patches of the gut, etc., there is a concentration of this tissue in one organ, namely the thymus gland.

In the Mammalia a very different condition is found. The lymphatic vessels now ramify as specialised tubes to the most remote parts of tissues, and the lymphoid tissue instead of being gathered into one mass is scattered about the body in masses having special relations to the lymph stream from definite areas of the body.

The significance of this arrangement is obvious when we consider the effect of localised inflammatory lesions. If an infection of some part of the body take place, such for instance as the leg, the effects of the lesion may be traced along the lymphatic vessels of the thigh as far as the lymphatic glands of the groin, but there it abruptly ends. The afferent vessels of those glands bring lymph in quantities above the normal, laden, probably with noxious substances in solution in the plasma, certainly with dead and dying wandering cells and even with microbes. With this disorganised lymph the glands deal, eliminating its poisons and destroying its effete corpuscles, and so long as the glands are capable of coping with the difficulty so long will the lesion remain localised.[1]

In the processes which go on during inflammation we probably see merely a gross exaggeration of events occurring under normal conditions. Each group of glands in the body is related to a definite group of tissues in that it receives the lymph flow from those tissues: and we must suppose that the numberless events to which the body is exposed affect the delicate balance of the chemical process, so that from time to time metabolites appear which so far depart from the common either in quantity or quality as to deserve the title "abnormal". The effect of excessive exercise may be cited as a case in point.

On such bodies, if we may trust the phenomena of disease, lymphatic glands act so as to preserve the mean composition of the lymph.[2] In the spleen again we have tissue possessing broadly the same histological characters related to the blood stream as lymphatic glands are related to the lymph stream. The processes which we know to occur in the spleen resemble those which are so obvious in inflamed lymphatic glands. In the spleen effete solid matter, such as red and white corpuscles, and bacteria are eliminated from the blood and there too the plasma suffers chemical changes.

An instance of the action of the spleen on the blood plasma is furnished by animals fed with peptonate of iron—in these the spleen will be found to contain an incredible amount of arrested iron held in part by wandering cells entangled in the reticulum, and apparently in part by the cells of the reticulum itself. Unless we make the rash assumption that the iron is carried from the intestine wholly by wandering cells—an assumption which scarcely agrees with the fact of the rapid absorption of iron by the stomach and the paucity of leucocytes there—then some of this very large amount found overloading the spleen must have been removed from the blood plasma.

If we turn now to that portion of the lymphoid tissue and wandering cells which forms the special subject of this paper we see that the great development of this tissue in the gut of Mammalia is only a part of the developmental process which has perfected the peripheral vessels of the

---

[1] Cf. Hoyer, *Arch. f. mikr. Anat.* (1889), XXXIV, 208.

[2] Hofmeister (*Arch. für exp. Path. u. Pharmak.* (1885), XIX) starting from a consideration of the fate of peptone in the body arrives at a similar view of the utility of the chemical processes carried out in lymphatic glands.

lymphatic system and placed on their course masses of lymphoid tissue each having a special functional relation to the tissues drained by its afferent lymph vessels.

And as these masses of lymphoid tissue modify and control the histological structure and chemical composition of that peculiar overflow of the tissues and vascular system called lymph, so the lymphoid tissue of the gut with its contained wandering cells modifies the composition of that special lymph which owes its composition chiefly to the activity of the endodermic epithelium.

Note on changes in the cells at the base of the crypts of Lieberkühn in feeding and hungry animals. The presence of large granules in the cells at the extreme base of the crypts of Lieberkühn was first noticed by Paneth.[1] The possession of these bodies seems to be a peculiar character of these basal cells, marking them off from the rest of the epithelium lining the crypts.

In the course of our work on the wandering cells our attention was called to changes in the extent of the granularity of Paneth's cells and a comparison of various preparations showed that those from well-fed animals agreed in possessing a scanty granulation (fig. 22), while those from hungry animals agreed in possessing numerous and large granules (fig. 21). This change was observed in rats, and it would appear to show that the granules, like those of the salivary glands, pancreas, and other digestive glands, suffer loss during digestion.

EXPLANATION OF PLATE XI

Figs. 1, 2, 3, 4, 5, 6, 7, 8, 9, 11, 14, 16, 17, 18, 19, 21 and 22 are drawn with camera lucida and Ocular 4, Zeiss Objective $\frac{1}{12}$th, Leitz, T.L. 170. Same scale as in plate Kanthack and Hardy, *Journ. Physiol.* XVII, 86.

Figs. 10, 12, 13 are camera lucida drawings with Ocular 4, Objective D, Zeiss.

Fig. 20 is camera lucida drawing, Ocular 2, Objective A.

Fig. 15 *a*, *b* and *c* are sketches from Ocular 4, Objective $\frac{1}{12}$th.

Figs. 1, 2, 3 and 4 show oxyphile cells from mucous coat side by side with oxyphile cells from peritoneal cavity, or blood.

Fig. 1. Frog. (*a*) Oxyphile cell from mucous coat of intestine, (*b*) oxyphile cell from peritoneal cavity.

Fig. 2. Rabbit. (*a*) Oxyphile cell from small intestine, (*b*) coarsely-granular oxyphile cell from omentum.

Fig. 3. Rat. (*a*) Oxyphile cell from small intestine, (*b*) coarsely-granular oxyphile cell from peritoneal fluid.

Fig. 4. Dog. (*a*) and (*b*) Oxyphile cells from small intestine, (*c*) coarsely-granular oxyphile cell from blood.

[1] *Arch. f. mikr. Anat.* (1888), XXXI, 112.

Fig. 5. Rat. Basophile cells from mucous coat of small intestine.

Fig. 6. Rat. Basophile cell from peritoneal fluid.

Fig. 7. Rat. Basophile cell from small intestine with micro-organisms very abundant in lumen. Animal starved but ate its own fæces.

Fig. 8. Rat. Basophile cell from small intestine, $3\frac{1}{2}$ days' starvation, gut healthy.

Fig. 9. Ferret. Basophile cell from group shown in fig. 10. Animal digesting.

Fig. 10. Ferret. Tangential section through apex of villus.

Fig. 11. Dog. Almost completely disrupted basophile cell. Small intestine. Compare with *Journ. Physiol.* XVII, Pl. 11, fig. 13.

Fig. 12. Dog. Optical section, longitudinal through villus of dog showing layer of basophile cells at base of epithelium. Eosine and methylene blue.

Fig. 13. Dog. Optical section, transverse through villus of dog showing the basophile layer.

Fig. 14. Dog. Hyaline cells, various sizes.

Fig. 15. (a) Living oxyphile cell in hanging drop of inflamed lymph. Frog. Trace of methylene blue added. Successive stages in staining of nucleus shown in (b) and (c).

Fig. 16. Guinea-pig. Hyaline cell holding an ingested oxyphile cell, and absorbed iron as droplets in its cell substance. Animal fed with Denayer's peptonate of iron. Absolute alcohol. Potassium ferrocyanide and hydrochloric acid. Methyl-eosine and methylene blue. Before the application of the iron test the droplets appear bright yellow.

Fig. 17. Guinea-pig. Hyaline cell holding iron.

Fig. 18. Dog. Portion of crypt of Lieberkühn about three-quarters down showing wandering of oxyphile cells into lumen. Corrosive sublimate, methyl-eosine and methylene blue.

Fig. 19. Dog. Portion of epithelium of crypt of Lieberkühn showing degenerated oxyphile cells. Corrosive sublimate, methyl-eosine and methylene blue.

Fig. 20. Frog. Near junction of hepato-pancreatic duct with intestine: (a) focus of oxyphile cells which continues into gut wall and into peritoneal membrane, (b) pancreas.

Fig. 21. Rat. Cells at base of crypt of Lieberkühn in starving animal ($3\frac{1}{2}$ days), (a) oxyphile cell.

Fig. 22. Rat. The same cells in full-fed animal.

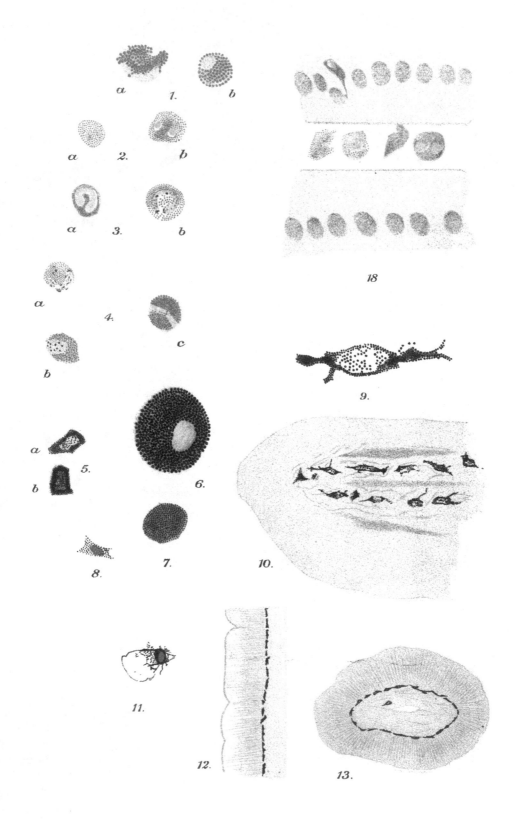

Plate XI

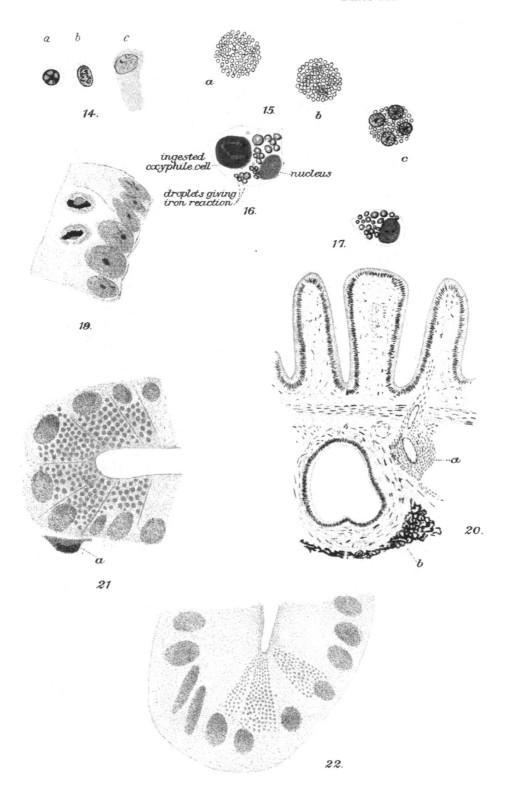

14.

15.

a

b

c

ingested
oxyphile cell

nucleus

droplets giving
iron reaction

16.

17.

19.

20.

21.

a

22.

13

FURTHER OBSERVATIONS UPON THE ACTION OF THE OXYPHIL AND HYALINE CELLS OF FROG'S LYMPH UPON BACILLI

[*Journ. Physiol.* (1898), XXIII, 359]

The observations which form the subject-matter of this paper were made in the winter of 1895–6. They constitute a direct continuation of earlier researches which the author made on the same subject, in the first instance in collaboration with the late Prof. Kanthack.[1] In this paper no new view is advocated, but perhaps more convincing proof is brought forward of the view that the oxyphil cells possess the power of injuring bacilli although they are not phagocytes.

The immediate interest which the experiments have for the author centres in the fact that they offer strong evidence of the secretory nature of the oxyphil granules. These granules, like the secretory granules of the frog's pancreas, form the nodal points of the network which can be demonstrated in the cells after fixation.[2] The fact that granules which are of the nature of non-living stored products form part of the network of fixed specimens seems to me to make it impossible to believe that this structure constitutes the contractile portion of the cell, as some authors urge, or indeed is anything more than an artifact.

The experiments consisted in observing the events which occurred when bacteria were placed in drops of frog's peritoneal fluid suspended in a moist chamber. A simultaneous record of temperature was made from a thermometer placed on the stage of the microscope.

Answers were sought to three questions: (i) Do the oxyphil cells possess the power of killing or maiming bacteria? (ii) If they do so, then what is the manner of their action? (iii) Do oxyphil cells change to hyaline cells—or is there at any time any confusion of the identity of the two kinds? To the solution of the first question twenty-six experiments were devoted, and in each of these chains of bacilli on to which oxyphil cells crawled, and chains which were free from cells were measured. Fresh broth cultures, 30 hours old, of *B. filamentosus* were used. These experiments furnish their own control observations, seeing that chains in contact and chains not in

---

[1] Kanthack and Hardy, *Phil. Trans.* B. (1894), p. 279. See also *Journ. Physiol.* (1892), XIII, 309; (1894), XV, 361; (1895), XVII, 81; (1895), XVIII, 490.

[2] This point I propose to deal with in a paper shortly to be published.

contact with cells were measured in each case, and except in three cases, in the same field of the microscope. The results, however, were also con-

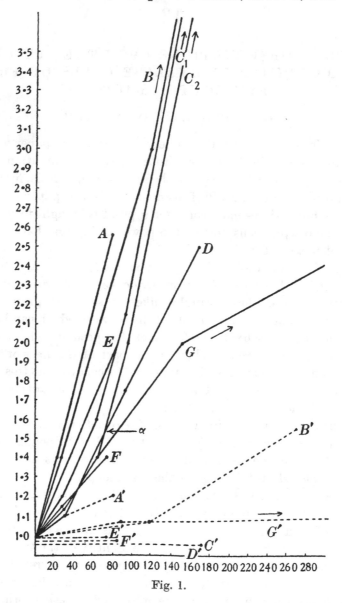

Fig. 1.

trolled by a series of twelve experiments in which measurements were made of the rate of growth of the bacilli in drops of centrifugalised serum of frog's blood.

Observations calculated to afford answers to the second and third questions were episodical in the case of the first group of twenty-six experiments; they, however, formed the chief aim of a second group of nine experiments performed with very virulent *B. anthracis*. It is important to notice that the lymph was obtained from winter frogs, in which the wandering cells have much less vitality as compared with the same cells from the summer frogs.

The general effect of the contact of a coarsely granular oxyphil cell with a growing chain is to retard its growth or to kill it. This is most conclusively shown by the measurements. In Fig. 1 are shown the curves[1] of growth in seven successive experiments chosen at random from the series. *A...G* are the curves for the chains on to which oxyphil cells did not move, while *A'...G'* show the growth of the chains with which cells were in contact for a longer or a shorter time. The latter show no growth or slow growth. The details of one experiment will serve to illustrate them all.

Exp. A. Two chains in the same field were measured. One was free from cells—this is chain *A*: two oxyphil cells moved along the other chain (*A'*), one from each end: they did not extend along the whole length of the chain, and after 10 min. the cells began to move away from the chain. The chain therefore was not exposed to the action of the cells for very long, or along its entire length.

Time min.	A μ	A' μ	Temp. ° C.
0	70	100	20
16	100	110	21
83	180	125	20

The change of direction in the curves of experiments *B* and *G* are due to alterations of temperature. The curve bends upwards in the former case, owing to a rise of 3°, and downwards in the latter case, owing to the fall of temperature in the laboratory on a cold winter night.

In the experiments which afford the data for the curves in Fig. 1 the chains which were measured in each experiment lay on the same field of the microscope, the Zeiss objective D and the ocular 4 being used. Thus the distance which separated the chains on to which oxyphil cells moved from those which throughout were free from cells was small; in experiment A it was $\pm 80\mu$.

Fig. 2 gives the curves of growth of seven chains widely separated in different parts of the drop; the position of each chain was marked by the vernier scale on the movable stage. Chains 1–4 were not at any time in contact with cells: oxyphil cells moved on to chains 5–7. The chains were measured every 30 min.

---

[1] In all the figures the curves are drawn to the same scale. In the ordinates 5 mm. represents an increase in length of 0·1 per cent.; in the abscissæ 5 mm. is equal to 10 min. of time.

It might be urged that the oxyphil cells only attack or crawl on to chains which are dead or dying. This is extremely unlikely since the control experiments prove conclusively that when *B. filamentosus* is transferred from broth to frog's serum it suffers little or no injury.

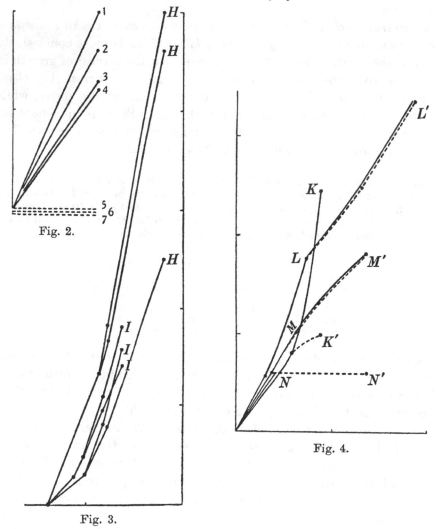

Fig. 2.

Fig. 3.

Fig. 4.

Fig. 3 gives the curves of growth of chains transferred from broth culture to the fresh centrifugalised serum. The measurements in two experiments only—the first two—are represented. In each case three chains were measured.

Any doubt which might still exist as to whether an oxyphil cell was able to injure or kill an actively growing chain is removed by the observation made many times of retardation, extending to complete stoppage, of growth in chains which were actively growing before the cells moved on to them.

Fig. 4 illustrates this point. It shows the curves of four chains chosen from different experiments. The number was restricted to four to avoid confusion in the figure, and the examples were selected so as to show complete and partial cessation of growth and to illustrate the effect of temperature on the conflict.

Curve K, K' is a singularly conclusive example. At the beginning of the experiment the chain was quite free from cells. It had a sharp bend or kink, and measured from one end to the kink 11μ, from the kink to the other end $13{\cdot}5\mu$. After about 50 min. an oxyphil cell rapidly crawled on to the part $13{\cdot}5\mu$ long, without, however, touching the portion beyond the kink. The curve as far as the bifurcation represents the growth of the entire chain up to the time when the cell moved on to this one end; beyond this the ascending portion is the curve for the part which was not touched by the cell, while the curve K' represents the growth of the other portion after the cell had moved over it.

In the other curves the portions to L, M, N give the growth when the chain was free from cells.

Curve 0, Fig. 6, illustrates the retardation due to the attack of a second oxyphil cell (see p. 239).

The deflection in these curves which follows contact of the chains with an oxyphil cell varies in amount according to the temperature, and according to the ratio of the length of the part which is at some time in contact with the cell to the total length of the chain and to the rate at which the cells move over the portion large, or small, which they may be said to attack.

Chain L, for instance, was 40μ long when a single oxyphil cell crawled on to it. A little later a second oxyphil cell joined the first, and in the second hour and a half of the experiment these two cells had moved slowly over about one-half of the length of the chain. Thus the attack, as we may for convenience call the process, was sluggish and incomplete and the retardation proportionately slow. The action of the cells on the bacillus was diminished by the temperature, which was high ($21{\cdot}5°$ C.), for, as we shall see later, the action of oxyphil cells upon *B. filamentosus* is hindered when the temperature is between 20 and 25° C.

Chain N was 220μ long when it was attacked. But no less than nine oxyphil cells moved on to it and rapidly traversed the whole length. The temperature was $18{\cdot}5°$ C.

What these various curves mean is shown by Fig. 5, which shows the appearance of the field as seen with a low power, and the aspect of the same field 16 hours later. Chain α was attacked by the only oxyphil cell in the field; it did not grow. Chain β, which was originally a continuation of chain α, grew rapidly. The first drawing represents the field, not at the

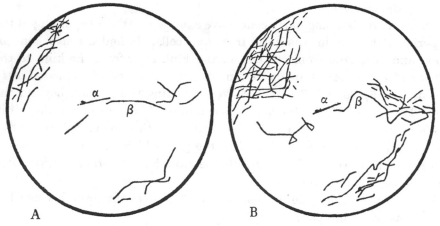

commencement of the experiment, but at the close of the 6th hour of observation; the second drawing represents the appearance at the close of the 18th hour. The oxyphil cell died but remained attached to one end of chain α.

Fig. 5.

The general result of the twenty-six experiments may be stated as follows: of the chains which were attacked by oxyphil cells,

25 did not grow at all,
18 showed slight growth,
1 showed free growth.

By "slight growth" I mean that the rate was much less than that of the chains which were not at any time in contact with cells. Slight growth does not imply that that part of the chain which was actually traversed by an oxyphil cell continued to grow. In most cases it means that only a fraction of the length of the chain was at some time in contact with the cell. In one case the rate of growth of the chain was not changed. In this case contact with the bacillus was not followed by any visible change in the cell; nor was there any of that production of slime which is described in a later section.

The effect of changes of temperature upon the conflict would probably vary for different bacilli. My conclusions relate solely to *B. filamentosus*, and to the action of the cells when in hanging drops. There would appear to be a critical temperature situated at or near 19·5° C. Below this the cells are able to and readily do kill the bacilli; above this they retard the growth only, and at 25–26° C. (the highest temperature in my experiments) the chain recovers from the retardation.[1]

On the other hand, *if the chain has been attacked by an oxyphil cell at the*

[1] A parallel to this is found in the fact that frogs are said to lose their immunity to *B. anthracis* when warmed to 25° C.

lower temperature (15–19° C.), *a subsequent rise of temperature will not lead to growth.*

In the different curves given in the figure, the temperatures were as follows:

$$\left.\begin{matrix} A\text{–}A' \\ B\text{–}B' \end{matrix}\right\} 21\text{–}22°, \qquad \left.\begin{matrix} C\text{–}C' \\ \text{to} \\ F\text{–}F' \end{matrix}\right\} 16\text{–}19°,$$

$$1\text{–}7 \left\{\begin{matrix} 19° \text{ at the beginning raised} \\ \text{at the close to } 23°, \end{matrix}\right. \qquad \left.\begin{matrix} K\text{–}K' \\ L \end{matrix}\right\} 21°, \qquad \left.\begin{matrix} M \\ N \end{matrix}\right\} 18\cdot5\text{–}19°$$

Fig. 6 illustrates the recovery of the chain when the temperature is high. The lines indicate the rates of growth of the chains which were measured in one particular experiment. Ten chains were measured. Of these two were not at any time in contact with cells, six were completely traversed by oxyphil cells. The temperature at the commencement was about the critical point but no exact record was made; it was rapidly raised to 24·4°, and at the close was 25·2°. Curve 8 gives the mean rate of growth of the chains which were not touched by cells, curve 9 gives the mean rate of those which were attacked, and curve 0 is the complete curve for one particular chain among the latter group. It is this curve which merits attention.

Fig. 6.

The chain 0 was 17 μ long at first. An oxyphil cell rapidly extended along it, but after a short time it ceased to move and appeared to be dead. Thirty minutes later a second oxyphil cell moved on to the chain. This produced a marked retardation of growth; this cell remained attached to the chain throughout, but in spite of that fact the rate of growth returned to the normal.

Action of the hyaline cells. When the enfeebled cells from these winter frogs are placed in a hanging drop the hyaline cells rarely possess sufficient vitality to retain their activity in presence of the substances in solution introduced with the broth culture. On five occasions, however, an opportunity was found of observing the effect which contact with their cells has upon the growth of the chains. In three of these instances the hyaline cell extended itself along the chain and did nothing more—this did not alter the rate of growth at all. In the remaining two cases the hyaline cell extended itself along the chain and then developed vacuoles within its substance which were traversed by the chain. In these cases the growth of the chain was retarded, and in the one case when the chain was small enough to come wholly within the sphere of action of the hyaline cell its growth was ultimately suspended and the chain was broken up in the vacuoles.

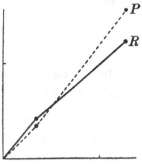

Fig. 7.

In Fig. 7 the curves of growth of two chains are given, chain R was free from cells; while a hyaline cell extended itself over about half the length of chain P. The cell did not develop vacuoles. Chain D (Fig. 1) furnished another example. From point α in the curve onwards about one-half of the chain was covered by a hyaline cell.

These few experiments are not sufficiently numerous to stand alone, but they offer excellent confirmation of the conclusion arrived at by Kanthack and Hardy[1] that the action of the oxyphil cell upon the bacilli is solely an extracellular act, while the action of the hyaline cells is purely an intracellular, a vacuolar activity. To put it into a phrase—*the attack of the hyaline cells to be effective must be vacuolate.*

Mode of action of the oxyphil cells; the character of the attack. Kanthack and Hardy[1] described the loss of the oxyphil granules when the cells move on to bacilli, and then came to the conclusion that the cells were amœboid unicellular glands, the secretory granules being the characteristic oxyphil granules. The same conclusion was reaffirmed by Hardy and Wesbrook[2] on other grounds. It is impossible to obtain such decisive proof of the truth of this observation as can be given of the result of the action of the cells upon the bacilli. It needs many hours of exhaustive stimulation of the nerves of a salivary gland to produce a condition of activity which is readily demonstrable in sections. Production goes hand in hand with loss, and it is only under exceptional strain that a gland cell loses the majority of its stored secretion masses. In these hanging drops the cells rarely retain sufficient vitality to enable them to discharge a great quantity of the granules. The result, however, of these experiments was to confirm in my mind the conclusion that the action of the oxyphil cells on the bacilli is exerted at the expense of the granules.

It is very difficult to count the number of granules in a cell. I made the attempt on some occasions. It is impossible to count the granules in the normal "resting" cell, they are too crowded and numerous. Film preparations would fix the number at between 40 and 80. The number of granules present in cells which had been applied for some time to bacilli were as follows:

[1] *Phil. Trans.* B (1894), p. 279.   [2] *Journ. Physiol.* (1895), XVIII, 490.

	Before	After	Number of granules
B. filamentosus	Crowded	90 min.	±13
B. anthracis	,,	50 ,,	±25
,,	,,	33 ,,	"A few"
,,	,,	—	6 faint and a small heap
,,	,,	36 ,,	6 dull looking
,,	,,	13 ,,	Cannot detect any
,,	,,	28 ,,	Not more than 15
,,	,,	40 ,,	Not a single granule
,,	,,	52 ,,	7*

* Examined with an apochrometer objective and various eyepieces, and with various illuminations, this cell was found to show: seven small refractive granules, one blot about the size of two granules, the rest of the cell substance showed dull round areas.

It is quite certain that this loss of granules is in part only apparent. Observation with an apochrometer lens, moderate magnification (± 1000 diameters) and careful illumination enables one to trace the changes in the granules due to the stimulus which the cell receives from the bacillus. The granules lose their high refractive index; they become dulled and larger, and one can then see them as dull round areas in the cell substance. The loss of granule substance in the part of the cell which is in contact with the chain is shown at the upper end of the cell in Fig. 9 which was drawn with a camera lucida, oc. 8, ob. 2 mm., apochr. Zeiss after fixation with formaline vapour.

After a prolonged and active attack upon bacilli the oxyphil cells are unquestionably diminished in size. This observation was made on many occasions. One experiment I devoted to an attempt to obtain a numerical measure of the change. Four hanging drops of lymph were made, to two of which there was added a small quantity of a fresh culture of *B. fila-mentosus*. The four drops were placed in the dark for 2 hours and were then fixed with the vapour of formaline. Fifty oxyphil cells were then measured in the control drops, and fifty in the inoculated drops. The mean diameters were found to be, for the former $11 \cdot 5 \mu$, for the latter $10 \cdot 9 \mu$. This represents a very large change of volume.

As an oxyphil cell crawls over a chain it leaves behind it a slime, which frequently contains fragments of still unchanged oxyphil granules. When the attack is very vigorous the slime is so abundant as to be readily seen. So far as my observations go a chain which has a coating of slime which can be detected under the microscope never grows. In one case I saw the actual production of the slime. It is the seventh cell in the column in the table above, and it was extended and moving along a chain of anthrax. At the hind end of the cell little blebs were formed at the surface, each of which had a dull appearance. They ran together to form the slime.

The formation of the slime is not a consequence of, nor is it necessarily followed by, any diminution in the vitality of the cell. The movements after slime formation may indeed be more vigorous than before. It is, however, possible that such cells do die as a result of the process, though hanging drops fail to afford evidence of this. When the inflammatory process occurs within the body in the later stages a large part of the activity of the hyaline cells is devoted to the ingestion of oxyphil cells.[1] This process indeed may be carried to quite extraordinary lengths,[2] oxyphil cells which are apparently quite healthy being ingested.

The effect of the attack of an oxyphil cell upon the aspect of the bacillus is interesting. A living growing *B. filamentosus* or *B. anthracis* under a high power has a dull translucent appearance, and shows no trace of structure beyond the brighter transverse lines which mark the junction of the segments of the chain. The effect of the attack may be limited to an increase in the refractive index of the material, which may be slight or may result in the development of a double contour. Fig. 8 shows the end of a chain of *B. filamentosus* before and after the attack of a cell. The coating of slime holding the remains of three oxyphil granules is very obvious. The effect of the attack sometimes is to make the contents

a   b

Fig. 8.

of the rodlets appear curdled and discontinuous; and they seem to shrink away from the cell wall.

The production of a bactericidal slime by the oxyphil cells recalls the extrusion of a slime on to the external surface of the entomostracan *Daphnia* when it is stimulated electrically or chemically. This slime unquestionably serves to keep the carapace free from ectoparasites;[3] and confers on the animal the external cleanliness which is so striking a characteristic of this species as compared with other freshwater entomostraca.

Do the oxyphil cells or hyaline cells develop the one into the other? I should say most emphatically that they do not. The reasons are twofold. Prolonged observation of the events occurring in many hundreds of hanging drops has failed to furnish a single case of loss of identity, and oxyphil cells have been seen to divide and produce daughter-cells of like kind, while hyaline cells have likewise been seen to divide and to produce daughter hyaline cells. The cells always differ in the type of amœboid movement, and in the refractive index of the cell substance. Anyone who has witnessed the formation of the curiously active plasmodia, each of which includes both oxyphil and hyaline cells, and has seen such a plasmodium disperse after a period of active existence, enduring perhaps for 5 or 6 hours, into the same

[1] In mammals this occurs to a very great extent in the lymphatic glands which first receive the lymph from the inflamed area.

[2] Cf. Hardy and Keng, *Journ. Physiol.* (1894), xv, 361.

[3] Hardy, *Journ. Physiol.* (1892), xiii, 309.

number of hyaline cells and of oxyphil cells as originally joined together
to make it, will not readily believe that cells with such persistent identity
and diverse function are merely developmental states of one another. And
to my mind reasons of this kind, founded upon observations which directly
touch the life of the cells, transcend an indefinite number of observations
of what are vaguely styled "transitional forms" in preparations from fixed
material. Wandering cells are specially prone to suffer change in the process
of fixation, and, according to my experience, and the experience of those
who demonstrate these cells to classes, an undiscoverable variation in the
mode of procedure will produce alterations which will furnish an extensive
series of transitional forms.

It should be remembered that this question, like many other histological
questions, cannot be put to the test of a decisive quantitative experiment.
In this, as in other similar questions, however, the general knowledge of
tissue change in animals indicates pretty clearly the one of several alternatives which is the most probable solution. In this particular case one
may say that all the evidence available goes to show that specialised tissue
cells cannot manifest marked change of function. The many cases of
discontinuity of function in a structure, such as occurs in e.g. ossification,
or in the change from the gut of the caterpillar to that of the imago, are
accompanied by a destruction of pre-existent cells, and the production of
new cells from latent embryonic cells which are fitted to carry out the new
functions. Thus general considerations drawn from conditions in which, if
anywhere, tissue cells might be expected to manifest the phenomena of
metaplasia, place the balance of probability very much in favour of the
view that the very much differentiated wandering cells of the higher
members of any animal series such as the Crustacea or the Vertebrata are
permanently distinct: and the assertion to the contrary demands therefore
a correspondingly convincing amount of experimental evidence.

This concludes the chief part of the paper. Certain minor points remain
to be dealt with.

Formation of plasmodia. I have again observed this just as Kanthack
and Hardy described it. Oxyphil cells and hyaline cells come together to
form a mass which shows no trace of limits between cell and cell, and
manifests confused writhing movements without progression. This plasmodial phase is temporary only in Vertebrates, but it is always present in
Nematodes.[1]

Character of the movements of the oxyphil cells. When a cell is wandering
through the serum amid the invisible fibrils of fibrin pseudopodia are
protruded at the anterior end which may be filiform, so that there is as it

[1] Cf. Nasanow for observations on wandering-cell tissue in *Ascaris. Arch. de
Parasitologie* (1898), I, 170.

were a fringe of waving processes, or thick and blunt. When the cells are in contact with bacilli the movements however are often very singular. They are:

(1) Streaming: this is like a flowing of the cell substance over the chain.

(2) Peristaltic: this does not lead to progression. The extended cell thickens at one end and the thickening travels as a wave slowly to the other end, where it is as it were reflected and so travels to and fro.

(3) Pulsating: the whole cell body rhythmically expands and contracts in some one plane.

(4) Bending movements: the cell, elongated along a chain, will bend and then straighten rhythmically and regularly. The extension appears to be due to the elastic recoil of the bent bacilli.

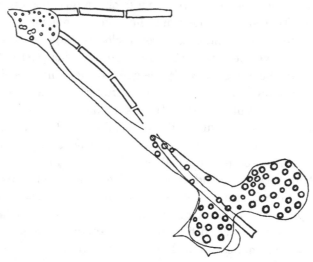

Fig. 9. Mag. 1000 diam., oc. 8, ob. apoch. 2 mm. Zeiss. Camera lucida. Fixed with formaline vapour. Note shrunken granules at the north end of the cell.

The last movement may be compared to another action. A long thin pseudopodium, which may be 20–30 μ long, is thrust out from a cell attached to one end of a chain to the other end of the same chain. The pseudopodium is then retracted slowly, bending the chain at the same time: at a certain stage the force of the elastic recoil of the chain proves too great and the chain slowly extends, drawing with it the pseudopodium. The action is rhythmic and it offers marked periods of activity and fatigue. It may end in the breaking of the chain, or in the pseudopodium being torn from its attachment to the chain.

The movement of the granules in a cell offers the clearest evidence that they are not connected to one another by bars or threads.[1] They move

---

[1] This point will be discussed more fully in a later paper.

quite independently of one another, sometimes singly, sometimes in small groups from one part of the cell to another. For instance, in such a cell as is shown in Fig. 9 single granules will travel from the body of the cell to the peripheral mass at the extremity of the pseudopodium.

"Waiting" of the hyaline cells. In a freshly made hanging drop both oxyphil cells and hyaline cells are in the ordinary amœboid phase. Very commonly, or most commonly, the latter after a few minutes cease to move, but remain extended in the most irregular shapes; in some cases they extend as thin plates which become more and more tenuous until they are barely visible. After a longer or shorter interval the cells resume their amœboid movements. Nothing in the actions of these cells gives one so vivid a sense of the great difference between the oxyphil cells and the hyaline cells, and no description will succeed in conveying an adequate idea of the singularity of the phenomenon. When an oxyphil cell is motionless it is spherical, or only slightly irregular. Hyaline cells which show no signs of movement for an hour or more will exhibit the wildest irregularity of form. They may expand as thin plates, or they may thrust forth delicate pseudopodia which extend over more than one field, or even traverse the drop from top to bottom. If the surface tension of the cell differs from the surface tension of the fluid, then these irregular shapes imply the existence of strains in the cell substance which it endures in consequence of either a temporary loss of its fluid characters, or of the continuous expenditure of energy.

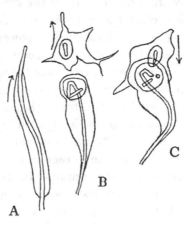

Fig. 10.

Fission of a hyaline cell while it is ingesting a chain. The bacillus was *B. filamentosus*, and the event is singular as showing that a cell can divide even when actually ingesting a foreign body. The details are as follows:

Time 0 min. Fig. 10 A: hyaline cell crawling north over a short chain.

,, 5 ,, Still moving north.

,, 13 ,, Cell is bending the north end of the chain and bringing it into a vacuole.

,, 50 ,, Fig. 10 B. The cell has divided: and a daughter-cell has moved north. It contains a vacuole enclosing a small length of the chain.

,, 58 ,, The daughter-cell has moved south and over the other daughter-cell. (Fig. 10 C.)

,, 74 ,, The first daughter-cell has moved past the second daughter-cell on to the south part of the chain.

The notes and drawings are very definite, and I do not think it possible that the advent of a second cell between the 13th and 50th minute was overlooked.

The source of the bactericidal power of the body. Certain points in this series of observations have a special bearing on this question. In the years between 1885 and 1890 two views arose, one, which is associated with the name of Metschnikoff, referred the bactericidal power of the body to the phagocytic powers of the wandering cells; the other, attributable in the first instance to the discovery by Fodor and Nissen of the bactericidal power of serum, regarded it as due to the chemical constitution of the body fluids. The supporters of the cellular theory, however, affirmed that the bactericidal power of blood serum is developed after the removal of the blood from the body; and the work of Buchner and Hankin went to show that it arises as a result of the solution in the serum of substances set free by the breaking up of the white corpuscles. In this way the humoral and the cellular hypotheses seemed in a fair way to be reconciled. One could picture the bactericidal power of the normal animal as being due to an internal secretion, the secretory agents being the wandering cells.

The observations recorded in this paper appear to me to make it difficult to accept such an explanation of the high immunity to *B. filamentosus* possessed by the frog. They show throughout that the retardation of the growth of chains is affected by actual contact with cells. A chain which is only 1 or 2μ distant from active oxyphil cells will grow as readily as chains which lie in regions completely free from these cells. And if oxyphil cells come into contact with the whole of a chain with the exception of the two terminal rodlets, as was the case in one instance, these terminal rodlets appear to grow as freely as any others in the drop. In other words, there is no evidence of the production in these hanging drops of a soluble diffusible bactericidal substance, the antibiotic agent is a colloidal slime which remains adherent to the bacilli. Even a fairly large dose of a culture does not lead to the destruction and solution of the bodies of the cells; and *B. filamentosus* grows freely from spores or from already germinating plants in centrifugalised frog's serum.

Now if we put aside all general theories of immunity and confine the discussion to an analysis of the complete immunity which the body of a frog possesses for *B. filamentosus* we have the following considerations:

The immunity may depend upon the fact that the chemical composition of the normal plasma of the blood or lymph is such as to render it either merely unsuitable as a nutrient medium, or actually harmful to these bacilli. The free growth in the freshly separated serum gives a general indication that this natural chemical immunity is small or absent.

The immunity is certainly in part due to that destructive action of the cells upon the bacilli which, as we have seen, demands actual contact.

Immunity to natural infection by any chance spores or rods of *B. filamentosus* which might succeed in passing whatever barriers a mucous surface interposes can be referred chiefly, if not entirely, to this direct cellular action.

Is there therefore produced at any time a temporary chemical immunity due to the solution in the body fluids of substances derived from the wandering cells? The most probable answer is yes. This would be the second, or if one includes the mechanisms of the mucous surface, the third line of defence. It can be shown in hanging drops that if the concentration of the bacilli, or more properly of the medium in which they have been living, be made very great it leads to a destruction of wandering cells, which is accompanied, in the case of the oxyphil cells, by a partial or complete solution of their granules. The facts can be demonstrated by placing side by side a drop from an active culture, preferably of *B. anthracis*, and a drop of lymph, and bringing their edges into contact. In the region of junction one can see that the great change in the constitution of the medium due to the advent of the constituents of the broth and the products of the activity of the bacilli kills a large number of the oxyphil cells. They become spherical and are seen to contain only a few granules, as compared with cells in the lymph remote from the bacilli, or only curdled masses produced by the swelling and fusion of such of the granules as remain. Others, however, become motionless but still retain all their granules. A few retain their activity so far as to move on to chains of bacilli, when they rapidly extend, and may there show singularly complete loss of granules. Now in the living animal this high concentration of abnormal chemical substances is produced artificially when a considerable quantity of a culture is injected into a lymph sac, and naturally when, either from mechanical injury or otherwise, necrotic tissue offers a shelter to invading bacilli.

SUMMARY

The chief points noted in this paper are the following:

1. Actual measurements show that contact with an oxyphil cell of frog's lymph retards or stops the growth of a chain of *B. filamentosus*.

2. The action of the cells (upon *B. filamentosus*) is generally determined by temperature. Below 19° C. the cells commonly completely arrest growth: between 20 and 25° C. the growth is only retarded.

3. The cells exert the action by coating the chain with a slime which is derived from the oxyphil granules.

4. Contact with a hyaline cell does not necessarily have any effect whatever upon the rate of growth of *B. filamentosus*. If the bacillus is enclosed within vacuoles developed in the cell substance, then retardation of growth occurs.

14

ON THE STRUCTURE OF CELL PROTOPLASM

PART I. THE STRUCTURE PRODUCED IN A CELL BY FIXA-
TIVE AND POST-MORTEM CHANGE. THE STRUCTURE OF COL-
LOIDAL MATTER AND THE MECHANISM OF SETTING AND OF
COAGULATION.

[*Journ. Physiol.* (1899), XXIV, 158]

CONTENTS

Since Brücke made the deduction that the complex phenomena of the life
of the cell can only be exhibited by structured matter nearly forty years
have elapsed. During that period a great part of the facts and theories of
modern physiology have been discovered and formulated; and during that

time more than fifty workers have endeavoured to convert Brücke's deduction into an induction founded upon the direct investigation of the structure of the cell substance. It is not going beyond the facts to say that in spite of their labours not one single point has been placed beyond the domain of ardent controversy; almost nothing has been added to the settled body of knowledge.

At the present moment the living cell protoplasm[1] is regarded by many as being composed of two substances, one of which is disposed as a contractile net according to some, as a relatively rigid framework according to others, or as free filaments. Other workers again regard cell protoplasm as being built up of a more solid material, and of a more fluid material which occupies the minute spaces or vacuoles which are hollowed out in the former. Still others view it as a homogeneous jelly holding granules. Lastly there are still those who deny the truth to all these views and maintain that the living cell protoplasm is homogeneous in so far that it does not manifest the relatively coarse structure which these theories ascribe to it. Its peculiar and transcendent qualities are according to them associated with molecular and not molar structure.

In this paper I endeavour to show that, if one puts on one side the confusion which has arisen from the inheritance of two nomenclatures of the parts of the cell, the one due to the botanist Hanstein, and the other started by the zoologist Kupffer, this lack of consonance in the views held as to the structure of the cell protoplasm is traceable in the main to the fact that they are largely based on details of structure visible both in fresh and in fixed cells which are the result of the physical changes which the living substance undergoes in the act of dying, or at the hand of fixatives.

Flemming[2] was perhaps the first to draw special attention to the fact that the appearance of structure in the cell protoplasm might be due to the action of fixatives or to post-mortem changes. He showed that a fine net structure is produced when the sap of a vegetable cell is fixed with osmic acid—this net therefore is an artifact due to the action of the fixing reagent.

The possible origin of structure in the dead cell substance which is not present in the living state, and which therefore is artifact for the purpose of the present discussion, appears to me to be threefold: (i) the rearrangement of the constituents of the cell body which is due to and indeed con-

---

[1] Throughout this paper the words "cell substance" are used to indicate the whole of the cell exclusive of the nucleus; and the words "cell protoplasm" to indicate the cell substance with the exception of secretory granules, glycogen or fatty masses, or other stored (paraplastic) products.

[2] *Zellsubstanz, Kern, und Zelltheilung* (Leipzig, 1882), p. 51.

stitutes the actual process of dying—these are the sub-mortem[1] changes; (ii) the changes in the stored products of the cell activities which may occur sub-mortem, or only post-mortem as a result of the action of reagents; for instance, the swelling of mucin granules and consequent distortion of the cell protoplasm, or the solution of the oxyphil granules of the coarsely granular blood cells of the crayfish; (iii) the action of fixing reagents in coagulating the cell substance.

Of these three possible sources of artifact the last mentioned is dealt with first. The subject-matter of the paper therefore falls into three sections, which deal with (i) the nature of the structure produced in solid and liquid colloids by the action of fixatives, (ii) the production of structure by sub-mortem and post-mortem changes in the animal cell, and (iii) the application of the facts and principles set forth in the first two sections to the interpretation of structure seen in the cell substance (α) when dead or dying but unfixed, (β) after fixation.

I take this opportunity, at the close of the introductory section, to express my sincere thanks to my friend Mr Neville, of Sidney Sussex College, for the unwearied patience with which he has aided me in these investigations. I have received both suggestions and criticism from him; indeed without his help, so generously given, the sections on the nature of the colloidal state could not have been written.

I. The Nature of the Changes produced in Colloids by Fixatives

It is, I think, one of the most remarkable facts in the history of biological science that the urgency and priority of this question should have appealed to so few minds. Yet the urgency lies patent to the most superficial consideration. It is notorious that the various fixing reagents are coagulants of organic colloids and that they produce precipitates which have a certain figure or structure. It can also readily be shown, as will appear more fully in the sequel, that the figure varies, other things being equal, according to the reagent used. It is therefore cause for suspicion when one finds that particular structures which are indubitably present in preparations are only found in cells fixed with certain reagents, used either alone, or in particular formulæ. Altmann demonstrates his granules by the aid of an intensely acid and oxidising mixture, while Martin Heidenhain is dependent almost exclusively upon corrosive sublimate.

Though many writers deal episodically with the action of fixing reagents, the matter has received special attention at the hands of, so far as I know,

---

[1] This phrase conveniently indicates that broad zone which intervenes between the completely living and the completely dead—the per iod μεταξὺ τοῦ θανάτου.

only four workers. These are the three botanists Berthold[1] (1886), Fr. Schwarz[2] (1887) and Fischer[3] (1894–5) and the zoologist Bütschli[4] (1892). These workers are singular in the fact that they attacked the problem from without; they endeavoured to determine the nature of the action of fixing reagents on cell substance by control experiments made on material such as solutions of gelatine or peptone, the physical homogeneity of which can be taken for granted.

Berthold added little to what had been previously said by Flemming. He criticised the net which Schmitz described in *Bryopsis* and *Saprolegnia* preserved with picric acid and roundly styles it an artificial product. He supports this assertion by experiments which prove that when egg-white is precipitated it takes a most beautiful "framework" structure. In the next year Fr. Schwarz in the course of a long treatise on the morphological and chemical composition of protoplasm, stated that the examination of living and fixed vegetable cells led him to the conclusion that the cell protoplasm does not contain a "preformed" net or framework (Gerüst). This conclusion he supports by experiments on solutions of gelatine, peptone and dried egg albumen, which showed that when these solutions are fixed by various reagents, they exhibit all the various appearances, granular, fibrillar and fine and coarse nets, which are to be seen in fixed cell substance.

Fischer attacked the subject quite independently, and added two important observations to those made by Berthold and Fr. Schwarz.

The object of his experiments in the first instance was to determine whether Altmann's granules are or are not artifacts due to the action of the reagents. The result of his experiments proved, however, so striking as to lead him to condemn as artifacts not only Altmann's granules, but also the "tingible bodies" of Flemming, and possibly the whole of those appearances in the nucleus and cell body grouped by Flemming under the heading "chromatolysis".

Fischer used solutions of peptone (2–10 per cent.), hæmoglobin, nuclein, and various proteids such as serum albumin, paraglobulin, etc., either alone or mixed. He found, like Berthold and Schwarz, that the precipitates produced by various fixing reagents (osmic acid, chromic acid, Flemming's and Altmann's mixtures, etc.) had a certain figure—granular or net-like. He also showed that this is true, even when the solution is enclosed in thin solid membranes. Small pieces of elder pith were infiltrated with 2–10 per cent. peptone solution, and then fixed in 1 per cent. osmic acid,

---

[1] *Studien über Protoplasmamechanik* (Leipzig, 1886), p. 62.
[2] Cohn's *Beiträge zur Biol. d. Pflanzen* (1887), v, 1.
[3] *Anat. Anzeiger* (1894), IX, 678; (1895), X, 769.
[4] *Protoplasm and Microscopic Foams* (London, 1894).

or in Altmann's mixture. Sections were made, and they showed the contents of the pith spaces arranged so as to offer a picture remarkably similar to that shown by sections of an actual cell. "In the middle was a nucleus-like body, from all sides of which there stretched to the cell wall beautiful delicate threads which anastomosed with one another, and were composed of small and large granules." The second important point established by Fischer was the fact that when mixtures of colloids are used, such as peptone and serum albumin, the constituents could be "differentiated" by fixing and staining. In the case quoted, after fixation and staining by Altmann's method, the serum albumen was found to form a matrix holding peptone granules.

The importance of this last observation can scarcely be overrated. Solutions of peptone and serum albumen are miscible in all degrees, so there can be no question of the completeness with which the two substances are intermixed at the commencement of the experiment. Therefore it is clear that the fixing agent not only separates both dissolved substances from the water, but also separates the one from the other, so that a structure is obtained differentiated both in form and in staining reaction which bears no resemblance whatever, and gives no clue to the relation of the three constituents, water, albumen and peptone, before the fixing agent acts.

Bütschli's observations are especially interesting, not from the interpretation he places on them, because in this he has, I think, fallen into error, but from his clear recognition of the extreme importance of the facts. In the text of the book he is disposed to deal somewhat summarily with the view that the appearance in fixed cell substance is an artifact. Some investigations which he carried out himself led him completely to alter this attitude, and the changed view finds expression in the appendix. There he confesses that coagulated white of egg or commercial gelatine show all the appearances which are presented by fixed cell protoplasm; those appearances which Bütschli interprets as indications of a very fine emulsion, or foam, structure. Ardent advocate as he is for structure in cell protoplasm, Bütschli is compelled to admit that this fact deprives observations upon fixed cells of any evidential value. But, he says, the view held by Berthold, Schwarz and Kölliker, that the structures which are alleged to be present in protoplasm are really simply due to processes of coagulation of this kind, is untenable "in view of the numerous cases in which the structures are distinctly demonstrable in living protoplasm". The value of the last-mentioned evidence has, it seems to me, been very much overrated. This point, however, I hope to take up in a future paper.

My own work leads me to think that the action of fixing agents may be viewed in a much more comprehensive way.

I would start the discussion with no statement as to the nature of cell protoplasm other than that it is, as Dujardin described it, "glutinous". Now this glutinous character is a special characteristic of that state of matter to which Graham applied the word "colloidal". If then we amplify Dujardin's adjective so far as to define living matter as matter either in or not far removed from the colloidal phase, we shall be within the limits of absolutely assured knowledge. But living matter does not lose its colloidal characters when it dies slowly, or when it is killed quickly by the action of some fixing reagent. The colloid substance of a cell does not become crystalloid as a result of the action of, *e.g.*, mercuric chloride. But, though it does not become crystalloid, the action of the fixing reagent is such as to produce an insoluble modification of the colloidal matter. Now the question which concerns us is, by what internal rearrangement of the solid and liquid constituents is this modification brought about?

My own experiments furnish the following general answer to that question. In the formation of an insoluble modification of a colloid from a soluble form, there is a separation of the solid from the liquid, so that the particles of the former adhere to form a framework which holds the liquid in its interstices.

This statement holds without modification whether the initial stage, that is the soluble colloid, be entirely fluid (colloidal solution), entirely solid (jelly), or a mixture of the two; or whether the physical change is or is not accompanied by chemical change. Lastly, the figure which the framework offers depends primarily upon the concentration of the initial stage, *i.e.* the proportion of solid to liquid, and upon the nature of the fixing agent which is used, and upon the particular colloid employed, and to a less extent upon the temperature during and subsequent to fixation, and the nature and concentration of the crystalloid bodies not immediately concerned in fixation, which may be present during the process.

Among fixatives I would include heat, and, in addition to this, I used osmic acid, as a 1 per cent. solution, and vapour; potassium bichromate, 1 per cent.; corrosive sublimate solution saturated in 0·6 per cent. sodium chloride; formaline and potassium sulphocyanate. This list offers examples of fixing agents of both the slowly and the quickly acting types.

Two very different experimental methods were used to determine the condition of the material—these were direct microscopical examination, and what may be called pressure experiments.

The first consisted in the examination of films, preparations made by teasing, and sections prepared with the freezing microtome, and, after embedding in paraffin, with the rocking microtome. The preparations were examined in air, water, alcohol, or canada balsam. For extreme magnification, I used an admirable Zeiss apochromatic objective with angular aperture 1·40 and a focal length of 2 mm. Sections cut in paraffin were usually cut with only one tooth of a rocking microtome. Such sections vary in thickness from 0·6 to 1 μ.

In a fair number of cases, notably after osmic vapour, sections even thinner than this were made—that is to say, the wheel of the microtome was moved less than one tooth.

The pressure experiments need a word of introduction. It occurred to me that if, as microscopical examination seemed to show, all insoluble gels[1] have an open sponge structure, then one ought to be able to express from them the fluid contained in their meshes by comparatively small pressures, which should bear some very general relation to the size of the meshes.

The statement that the pressure necessary to express the fluid varies in some inverse measure of the size of the meshes of the sponge follows from Poiseuille's law for the outflow from capillary tubes. Guerout made a similar use of this law to obtain a first approximate measure of the diameter of the pores in bladder, gold-beater's skin, and parchment paper.[2] Pressure experiments also enable us to discriminate between an open sponge structure and a vesicular or honeycomb structure. The pressure required to separate the fluid being greater in the latter than in the former case, when the dimensions of the structures are approximately the same.

The pressure required to separate fluid and solid from a solid structure containing both will depend upon the nature of the solid, and the size, shape and relative disposition of the fluid-holding spaces. Information on these points was desired for the purpose of these experiments. The pressure, however, also depends upon other relations, namely, the presence or absence of a continuous external limiting membrane of impervious solid, the viscosity of the fluid, and the cohesion between fluid and solid. The influence of any limiting solid membrane would be very difficult to estimate.[3] It may, however, be eliminated by cutting the mass into pieces. The viscosity of the expressed fluid did not vary in the different cases sufficiently to contribute much to the difference of pressure.

Actual experiment showed that fixation of any colloidal mass involves a change in the relations of solid and liquid, so that the latter can be more readily expressed. Thus a hydrogel containing 13 per cent.[4] pure gelatine at a temperature of 15° C. will endure a pressure of 400 lb. to the square inch—this is the maximal pressure I was able to employ—without separation of fluid. After fixation with formaline or corrosive sublimate, the solid and fluid are so much separated from one another as to permit of the latter being squeezed out, as from a sponge, with simple hand-pressure.

[1] Graham's nomenclature is as follows: The fluid state, colloidal solution, is the "sol", the solid state the "gel". The fluid constituent is indicated by a prefix. Thus an aqueous solution of gelatine is a "hydrosol", and on setting it becomes a "hydrogel". There are etherogels, sulphogels, etc., which contain respectively ether and sulphuric acid.   [2] *Ladenburg*, III, 299.

[3] Cf. Osborne Reynolds on "Dilatancy", *Nature* (1885–6), XXXIII, 430.

[4] Here and elsewhere this means grams per 100 c.c.

The phrase "insoluble gels" needs definition. Gels, such as a jelly of agar, of silica, gelatine, and celloidin are mixtures of solid and liquid. Of these mixtures I distinguish two classes which can be discriminated by the action of changes of temperature. They may be styled the gelatine class and the silica class after the best known examples; members of the former class are rendered more fluid by a rise of temperature; while members of the latter class become less fluid under the influence of the same change. Thus hydrogels of gelatine are mobilised by heat, they absorb water at a more rapid rate, and at a certain point, determined in the main by the proportion of solid and liquid, they become hydrosols. A hydrogel of silicic acid, of many metallic hydrates and sulphates, or of gelatine after it has been exposed to the action of formaline, loses its power of holding water when the temperature is raised. To use Graham's terms, the "clot" shrinks and "serum" is expressed.

For the purpose of a purely histological treatise, and without going too deeply into the physical problems involved, these gels of the second class may be defined as those in which the solid constituent is insoluble in the fluid constituent, and this is what is meant by the words "insoluble gel" when they are used in the text. The words also imply that the gels in question are not solid solutions. This is proved by the fact that the curves of the rate of evaporation, and of the relation of the vapour pressure to concentration, are bending lines and not straight lines.[1]

Physicists, notably van Bemmelen, have arrived at the conclusion that certain of the properties of gels of silicic acid, cupric hydrate, etc., which belong to my insoluble class, can be best explained by supposing that they have a sponge structure, being built up of a more solid framework holding fluid in the interstices. My own investigations have shown that this view can be confirmed by the microscope—the framework can be detected with, in some cases, even low magnification.

The great value of the pressure experiments lies in the fact that they supplement and verify microscopical observations, in one particular wherein the latter is notoriously weak. Sections of cells or of more simple colloidal masses can furnish positive evidence of the existence of a framework; but they yield only unsatisfactory negative evidence as to the condition and nature of the material which occupies the meshes, unless indeed that substance can be differentiated by stains. In by far the majority of cases it is impossible to stain matter in the meshes of the net: this however would scarcely serve as an adequate reason for believing that the meshes are occupied by a simple fluid, the "serum" as Graham styled it of the coagulated mass, seeing that the cell histologist speaks freely of the existence in the cell of "non-staining" substances. Pressure experiments therefore serve most opportunely in affording proof that "non-staining" substances means for the most part simple spaces occupied only by a fluid which varies according to the menstruum in which the coagulated colloid mass or the tissue may chance to be lying.[2]

---

[1] Spring and Lucion, *Zeits. f. anorg. Chem.* (1892), II, 195. Tamman, *Zeits. f. phys. Chem.* (1893), x, 263. v. Bemmelen, *Zeits. f. anorg. Chem.* (1893), v, 467 and (1896), XIII, 233.

[2] One becomes peculiarly sceptical as to the existence of colloid masses which cannot be infiltrated with some dye when one remembers that, as Graham and Voightländer (*Zeits. f. phys. Chem.* (1889), III, 316 have shown, salts diffuse through gels with so little hindrance that the rate is the same as in pure water.

It will simplify the method of treatment if I state at the outset the significance to be attached to the remark wherever made that: it was found impossible to demonstrate the existence of any material in the meshes of a net. This means that the following separate methods failed to drive stain into any substance in the meshes: the iron-hæmatoxylin method, used without any washing out soever; staining with saturated solution of acid and basic dyes at various temperatures; and evaporation to dryness in such solutions with or without subsequent heating.

Fixation of colloidal solutions. *White of egg* was broken up thoroughly, diluted and filtered. Solutions varying in concentration were made from the filtrate either by partial evaporation, or by evaporation to dryness and resolution. Tiny droplets, about 1–2 mm. in diameter, held in loops of silk thread were fixed in various ways and sections cut. Larger drops and masses of the solution in moulds of porous paper of different sizes were also fixed in the same reagents. The following stains were used: iron-hæmatoxylin alone and followed by osmic, methylene blue, eosine, and Ehrlich-Biondi's triple stain.

The following facts appeared to be established. 13 per cent. solution fixed with corrosive sublimate sections made after embedding in paraffin were found to show a sponge or net structure, the aspect of which is given in Pl. XII, fig. 1. It proved to be impossible to stain any substance in the meshes of the net. They contained simple fluid, of which direct hand-pressure[1] or a centrifugal machine expressed a quantity equal to ± 60 per cent. by volume of the entire gel.

This net structure, visible under the microscope, might be due not to the action of the fixative but to heat shrinkage. This is not the case. It is formed in the process of fixation. The net can be demonstrated in films of the solution fixed by immersion in sublimate, in preparation from the fixed droplets made by teasing, and in sections cut from the gel immediately after fixing and washing in water by the freezing microtome.

The effect of a rise of temperature upon the fixed colloid is complicated. If the heat is applied while the coagulum is in its original fluid, or in water, it behaves as insoluble gels commonly do—it shrinks. The rise of temperature leads to a closer aggregation of the solid particles—a synæresis, as Graham called it. When the original fluid has been replaced by such a fluid as xylol a rise of temperature produces a very small effect, but the process of dehydration itself causes a shrinkage amounting to ± 30 per cent. of the original volume.[2] The initial coagulation, or fixing, is not necessarily or perhaps usually accompanied by a change of volume. This is an agree-

---

[1] That is, pressure applied by a glass rod to the substance when in a glass vessel.

[2] It should be remembered that a change of 30 per cent. in cubic capacity implies a very much smaller diminution in any linear measurement.

ment with the fact first established by Graham that an insoluble gel when first formed has the same volume as the solution from which it is produced.

This statement holds good not only of simple colloids but also of such tissues as were experimented with, and the figures for shrinkage are about the same. Pancreas and liver of the frog were examined. I subjoin one example.

A strip of liver was cut out with fine scissors and when suspended it was found to measure 11 by ± 2 mm.

Suspended in the sat. solution of sublimate in 0·6 sodium chloride.

After   2 hours length = 11 mm.
After 24 hours length = 11 mm. } no diminution of length.

Brought into 95 per cent. spirit, containing a trace of iodine: after 12 hours length = 9 mm.   = decrease in length of 18 per cent.

Further 24 hours in 95 per cent. spirit = 9 mm.
20 hours in cedar oil   = 9 mm. } no further shortening.

If the decrease were the same, namely 18 per cent., in all directions, then the diminution in volume would be about 50 per cent.

The average linear shrinkage for strips of liver and pancreas was found to be nothing on fixing, and about 20 per cent. on dehydration.

We may set these facts, the absence of any change of volume when the insoluble gel is formed, and the diminution of volume which follows when it is infiltrated with various fluids, or warmed, beside Graham's account[1] of the volume change of a gel of silicic acid—the type of insoluble gels—when its water is replaced by other fluids.

When a hydrosol of silica changes into a hydrogel there is no change of volume. But when the water in the sponge of the hydrogel is replaced by other fluids there is a decrease in volume. Alcohol and glycerine produce only a slight diminution of volume; sulphuric acid, however, leads to a diminution equal to one-fifth or one-sixth of the original volume.

I have determined the existence of a solid framework having an open net structure in the following gels: in white of egg coagulated by corrosive sublimate, heat, potassium bichromate, or a trace of potassium sulphocyanate; in a hydrogel of silicic acid, and in gelatine coagulated by sublimate, ammonium bichromate, or formaline; in a gel of celloidine produced by the action of chloroform on an ether-absolute solution, and lastly in common black indiarubber.

The last-mentioned case is interesting. Indiarubber is produced by the coagulation of an organic fluid, and it behaves like an insoluble colloid— it shrinks on warming and extends on cooling. It is therefore interesting to find that sections of black indiarubber cut with the freezing microtome show the characteristic structure, namely a fine open sponge.

[1] *Journ. Chem. Soc.* (1864), II, 318.

The sponge structure of these insoluble gels, which is here demonstrated by direct observation, has been inferred by physicists from their examination of the relations of fluid and solid in such gels. Thus, to quote only the chief worker in the field, van Bemmelen,[1] who constructed curves of the removal and absorption of water for various gels, comes to the conclusion that the process of coagulation by which these gels are formed consists essentially of a separation of solid from liquid; the solid particles then hang together to form a framework which encloses the separated solvent.

In other words, and this is what I wish to insist upon here, the very essence of the process of fixation is the separation of solid from liquid and the formation thereby of a structure which may have had no counterpart whatever before fixation occurred.

I turn now to consider the various factors which modify the configuration of this structure.

The influence of the nature of the fixative, and the nature and concentration of the colloidal solid upon the final configuration. The relations which fall under this heading are fairly simple for solutions of egg-white, but very complex for solutions of gelatine.

Egg-white. The figure is always that of an open net with spherical masses at the nodal points except when the fixative is osmic vapour. The size of the nodal spherules and dimension of the meshes of the net vary according to the nature of the fixative and the concentration of the egg-white solution.

The procedure in all cases was as follows. The droplets of the solution of egg-white were suspended in the fixative in loops of silk. They were exposed to the sublimate solution for 20 hours; to osmic vapour for 12 hours in the dark; to potassium bichromate, a 1 per cent. solution, for 21 days. To fix by heat the droplets were held for a moment until opaque white in a jet of steam. After fixation the droplets were washed in distilled water for about 6 hours, and dehydrated by changes of spirit, 15, 30, 40, 50 per cent., and so on; and embedded in paraffin from cedar oil.

Fixation by potassium sulphocyanate needs special mention. Meusel in 1886[2] described the coagulative action of this salt in a brief paper which does little more than record the experiments. He found that when 8 gr. of the dry salt are added to 30 c.c. of white of egg or serum the latter sets to a firm jelly in the course of a few days. As the original paper is bare of detail I subjoin an account of one of my own experiments.

[1] *Berichte d. deut. chem. Gesell.* (1880), XIII, 1467; *Journ. f. prakt. Chem.* (1881), XXIII, 324 and 379; *ibid.* (1892), XLVI, 497; *Zeits. f. anorg. Chem.* (1893), V, 467; *ibid.* (1896), XIII, 233.

[2] Dr Ed. Meusel, *Die Quellkraft d. Rhodanate* (1886).

White of egg beaten and centrifugalised: solids in clear fluid determined and found to be 13 g. in 100 c.c.; to 30 c.c. of the clear fluid 8 g. KCyS were added; the salt dissolved with great rapidity and without any clouding; placed in a moist chamber—temp. ± 16° C.

24 hours still fluid and clear.
48 hours becoming viscid and slightly cloudy.
72 hours much more viscid and still more cloudy.
96 hours set to an opaque white jelly; not very firm.
Later jelly quite firm and is contracting slowly with expression of a clear serum.

The jelly is insoluble in distilled water even after the sulphocyanate has been allowed to diffuse out into several changes of distilled water.

I found, however, that it is not necessary to employ such a large quantity of the salt; coagulation can be effected by a small trace only if the temperature be raised. Thus: 5 c.c. clear solution of egg-white holding 10 per cent. solids was warmed to 45° C. and a minute crystal of KCyS added. In $1\frac{1}{2}$ hours it had set completely to a slightly opaque jelly.

Films were also fixed by traces of the salt in a warm moist chamber. In all cases a delicate net structure was demonstrable. These details are needed, since these gels were used to investigate the action of ordinary fixing reagents on already formed "insoluble" gels.

The effect of the nature of the fixative is shown by the following table, which gives the average value for the length of a diagonal of the mesh when the amount of coagulable solid in the original solution was ± 13 gr. in 100 c.c.:

Osmic vapour	0·5–0·7 μ
Steam	1 μ
Potassium sulphocyanate	1·0 μ
Potassium bichromate	1·3 μ
Corrosive sublimate	1·7 μ

It will be seen from the marked difference between the figure for sublimate and that of the rest that in the fixation of proteids this reagent has, as Henneguy remarks, "a certain brutality of action" which unfits it for use in the study of the structure of protoplasm.[1]

The configuration after fixation by osmic vapour appears to be different from that produced by the other reagents. It looks to be that of a number of vesicles hollowed out of a continuous solid mass and therefore not communicating with one another. It is, however, very difficult to be certain on this point.

The effect of the concentration of the solution of the egg-white when it

[1] *Leçons sur la cellule*, pp. 42 and 61. Henneguy is so impressed by this "brutality" as to lead him to reject the observations of the important school of Louvain (Carnoy) owing to the fact that the workers rely almost exclusively on this fixative (p. 42). He makes the interesting remark that when one looks through the plates in *La Cellule* one is struck by "l'uniformité de structure que revêtiraient les cellules les plus diverses", and he asks "si celle-ci n'est point due à l'uniformité de la méthode employée".

is above the minimum which is necessary for the production of a continuous mass in the fixed state is restricted to an alteration in the size of the meshes. Figs. 1, 2 and 3 show the appearance of the net produced by sublimate in solutions of egg-white containing respectively 13, 30 and 60 gr. solid per 100 c.c. The microscopical analysis of the figure in the last case is a very doubtful matter. When the microscope is pushed to these limits its powers are more potent in magnifying the personal equation of the observer than in really elucidating details of structure, seeing that the structures under observation become commensurate with the diffraction areas. However, it is certain that there are discontinuities—the doubt is limited to whether the open net structure is still retained, or whether the spaces are discontinuous in the form of separate vesicles.

If the concentration of the colloidal solid is very small, a granular precipitate may be produced. The particular concentration necessary for the production of a continuous net varies with the rate of fixation, the absence of mechanical disturbance, and, as we shall see later, the concentration and nature of any electrolytes which may be present. The continuous net and the precipitate are not discontinuous states. They pass into one another with variations in concentration.

With egg-white when the concentration is 10 gr. per 100 c.c. and the fixative potassium sulphocyanate the figure is a continuous open net. When the concentration is lowered to 6 gr. per 100 c.c. a double structure is visible, the typical fine net is no longer continuous throughout the entire mass. Clefts and spaces appear in it so that sections have the aspect of a very coarse sponge the bars of which are built out of the typical fine net. With still further dilution the fine net is formed in completely discontinuous patches which constitute a "flaky" precipitate. This flaky precipitate therefore is formed as it were by the rupture of the coarse framework mentioned above. Finally, with yet greater dilution, the fine net is not formed even in patches, but in its place there appears a suspension of discrete spherical granules.

Gelatine. The figure in this case changes not only in dimensions but in character. If gelatine of medium concentration, say 7–15 gr. solid to the 100 c.c., be fixed by alcohol, or sublimate, instead of forming an open net, it takes the form of a continuous solid hollowed out by vesicles about 7μ in diameter (fig. 6) which as a rule become deformed to polyhedra by mutual pressure. If it be fixed by formaline, however, an open net is produced; the formaline, however, must be in excess and be allowed to act on the gelatine for about 16 hours. The open net figure may be formed with alcohol or sublimate, but in that case, supposing the reagent to be present in large excess, the percentage of gelatine in grams per 100 c.c. must be less than 5. The figure then consists of nodal spherules joined by bars as when white of egg is fixed (fig. 5). The dimensions of the meshes and spaces are as follows:

Fixative sublimate: sat. sol. in 0·6 per cent. NaCl

Gr. gelatine per 100 c.c.	Figure	Dimensions		
4	Open net	Diagonal of mesh		2μ
10	Honeycomb	Diameter of sphere		7μ
25	,,	,,	,,	3μ
50	,,	,,	,,	$2\cdot5\mu$

Figs. 5–8 illustrate these cases.

The point at which the figure changes from the open net to the honeycomb depends not only upon the concentration of the colloidal solid but also upon the proportion of alcohol or sublimate in the changing system. I have not followed this point in detail though it is of considerable theoretical importance. The following instance will serve to prove the general truth of the statement:

$\dfrac{\text{vol. alcohol}}{\text{vol. H}_2\text{O}}$	Gr. gelatine per 100 c.c. necessary to effect the change from the open net to the honeycomb
1	±20
∞	4–5

These involved relations depend upon the fact that sublimate and alcohol are largely miscible with gelatine in presence of water, and they thus form with the water and gelatine a ternary system of partially miscible substances. The sublimate and alcohol do not destroy the gelatine-water system as does, *e.g.*, formaline, they only modify it. Thus one volume of a 50 per cent. solution of gelatine mixes completely with one volume of a saturated solution of sublimate, forming a system which, like the original gelatine-water system, is heat-reversible, being fluid at above 50° C. and a transparent jelly below.

We must therefore distinguish two cases: (i) the production of a true insoluble modification, and therefore the complete destruction of the original system, as when formaline acts fully upon gelatine, or when sublimate acts upon egg-white; and (ii) the modification of the original system so that though altered it still remains within the class of soluble colloids—becoming fluid on heating in presence of a certain quantity of water, and solid on cooling. The final figure appears to be in the former case always that of an open net if the concentration of the colloidal solid be above a certain minimum; in the latter case when the fixative is in excess it is a honeycomb unless the concentration of the colloidal solid be very low.

Fixation of a soluble gel. Fixatives produce the same general effect, namely, a separation of a more fluid from a more solid part, whether the soluble colloid is exposed to them in the solid state or the fluid state. This was determined by comparing the results of fixation when both the mixture of gelatine and water and the fixing solution were above the temperature of melting of the former, and when they were below that temperature. The internal change in the jelly is manifested clearly by the effects of pressure. A hydrogel of gelatine holding ± 13 gr. solid in 100 c.c. will endure a pressure of ± 400 lb. to the square inch without separation of fluid. Fixation with formaline produces little superficial change. The jelly becomes firmer but retains its transparency. Its internal structure is, however, changed to such an extent that fluid can be expressed by hand-pressure alone.

The absence of any obvious optical changes during the process of fixation is remarkably evident in the fixation of an ether-alcohol solution of celloidin by chloroform. A clear transparent solid is produced. The meshes, however, of this gel are so coarse that the greater part of the contained fluid can be wrung out with the fingers as though the mass were a common sponge.

The Theory of the Changes which are Produced by Fixatives[1]

The facts which are stated empirically in the foregoing sections are susceptible of rational treatment to a limited extent. Put briefly the facts show that when a soluble colloid is fixed by the action of a fixing reagent it acquires a comparatively coarse structure in the process which differs wholly or in part from the structure of the soluble colloid. It is possible to make a general statement of the relation between the initial and final states in this change.

On p. 254 I drew attention to the fact that colloidal mixtures[2] differ from one another in their physical characters, and I distinguished between two classes. To this point it is necessary to return.

Fluid mixtures of water and, *e.g.*, agar, gelatine, proteid, silicic acid, aluminium hydrate, or antimony hydrosulphide; or of absolute alcohol and celloidin form binary systems which are capable of passing into the solid state known as the gel. In some cases the change of state follows on a simple fall or a simple rise of temperature, in other cases it occurs only when the primary system is modified by the introduction of some reagent, and in the process the colloidal solid may suffer chemical change. No matter how it is accomplished the change to the gel state enables us to place the

---

[1] The purely physical conceptions, and the experimental results which are briefly stated in this section, will be more fully stated elsewhere.

[2] Colloidal mixtures can be classified but not individuals, for the same individual will form a mixture of one order with one solvent, of another order with another solvent.

systems in two groups; in the first the change of state is reversible by altering the sign of the temperature change—*e.g.* gelatine or agar and water; in the second the change of state is irreversible, permanent molecular aggregates being formed. We will consider first the former, the heat-reversible systems, or soluble colloids as they are called in the preceding pages.

Heat-reversible colloidal mixtures. Up to the present nothing has been known and but little even conjectured as to the relation of such mixtures as gelatine-water, or agar and water to other states of matter. I have, I think, obtained experimental proof that they may be regarded as special cases of partially miscible fluids.

Agar-agar and water was the mixture experimented with. It was prepared free from diffusible bodies. By small pressure a jelly holding 1 gr. agar in 100 c.c. may be separated into two parts: (*a*) a solution of water in agar, (*b*) a solution of agar in water. The former is at 15° C. a solid solution. These two solutions form a binary system of partially miscible solutions—that is to say they are completely miscible above a certain temperature, in which state a homogeneous fluid is formed which is a solution of agar in water. Below a certain temperature the homogeneous liquid divides into a pair of liquids which are separated by a surface across which diffusion occurs. Such a pair of liquids have been called conjugate liquids.[1]

With still falling temperature one phase of the pair—the solution of water in agar—becomes a solid solution, the solid phase forming first as a membrane at the surface. This change is the cause of the change in the whole system whereby it sets to a jelly. This agar-water system therefore resembles a mixture of for instance phenol and water. Above 80° C. phenol and water are miscible in all proportions; below that temperature if more than 71 or less than 76 per cent. of phenol is present the mixture forms a pair of conjugates, one a solution of phenol in water, the other a solution of water in phenol. What is known of these mixtures is almost entirely due to the work of Alexejeff.[2] The chief properties which have been determined for partially miscible liquids are that the composition of the two phases of the conjugate is independent of the total composition of the entire mixture, but is determined by temperature. These two relations hold for agar-water. In a mixture holding 3·5 per cent. (grams agar per 100 c.c.), at 15° C. the solution of agar in water phase contained 0·45 per cent. agar; while in a mixture holding 1 per cent. agar at the same temperature the same phase also had 0·45 per cent. agar. On the other hand the percentage composition of the two phases was found to vary with the

[1] Neville, *Science Progress* (1895), IV, 77.
[2] *Ann. der Phys. u. Chem.* N.F. XXVIII, 305.

temperature. Thus, taking the same phase of a mixture holding 2·5 per cent. agar the figures obtained were

°C.	%
±1	0·35
10·5	0·42
32	1·00
42	1·66

The difference between the agar-water system and any partially miscible fluids hitherto investigated lies in the contour of the surface of contact of the conjugates. In phenol-water, ether-water, or any such systems the surface is a plane—that is to say, it is minimal. In the agar-water system it is curved and discontinuous, so that when the conjugates form, one phase separates as droplets immersed in the other. It will, I think, be possible to refer this peculiar feature to the very great difference in the mobility of the molecules of the two constituents of the mixture—the agar and the water. Owing to this feature the system attains true equilibrium very slowly indeed—in the case of, e.g., gelatine holding 2·5 per cent. solids only after many months.

Apart from this difference a mixture of agar and water closely resembles a mixture of salicylic acid and water.[1] Salicylic acid has a high melting-point (151° C.), but when heated under water to 100° C. it melts, forming a fluid hydrate which becomes miscible with water at 106° C. On cooling this mixture it separates into conjugates. Thus this substance, like agar, is solid at ordinary temperatures but forms a fluid hydrate which in turn forms a partially miscible system with water.

The actual configuration of these soluble jellies can be studied with even low powers of the microscope by using ternary instead of binary mixtures. Alcohol, pure gelatine,[2] and water form such a ternary mixture. If 13·5 gr. gelatine are mixed with 50 c.c. H$_2$O and 50 c.c. absolute alcohol, one has a mixture which is homogeneous at about 17° C., but which forms a pair of conjugates below this temperature. The changes on cooling as seen on the stage of the microscope are these—when the temperature reaches 17° C. small droplets appear in the homogeneous fluid which attain a maximal size of 3 μ. The droplets are at first separate and fluid; owing to their presence the viscosity of the mixture very much increases. As temperature falls the droplets adhere to one another, forming a framework which at 12° C. is composed of a solid solution composed of spherical masses hanging together to form anastomosing threads. The droplets slightly deform each other where they touch. At this stage the entire mass forms a jelly. At 14° C. the distribution of gelatine in the two phases

---

[1] Alexejeff, Ann. d. Phys. u. Chem. N.F. xxviii, 305.

[2] That is gelatine freed from diffusible substances by washing with distilled water for some months.

is as follows: 100 c.c. of the droplets hold 18 gr. gelatine; 100 c.c. of the nexus fluid hold 5·5 gr. gelatine.

Ternary mixtures have been studied only by Duclaux.[1] They are formed most readily by converting a pair of immiscible fluids into partially miscible fluids by the addition of a third substance, which dissolves both. His figures, as Ostwald points out[2], show that these mixtures behave in a simple manner resembling a binary system if the solubility of the two immiscible substances in the common solvent is approximately the same. If the solubility is very different, then the relations are complex. A mixture of benzene, acetic acid and water is a system of this kind. The common solvent is the acetic acid, but whereas water and acetic acid readily mix with rise of temperature, benzene and acetic acid only mix freely at temperature above 15° C. In this case the distribution of the constituents in the conjugates varies widely with variations in the composition of the whole mass.

Gelatine, water and alcohol form a ternary mixture of this type. The common solvent is water—it mixes freely and in all proportions with alcohol with liberation of heat: it is only completely miscible with gelatine at temperatures above ± 40° C. Therefore one finds that the distribution of the constituents in the two phases is much altered by changes in their relative proportions in the entire mass. So far I have only studied the effect of changes in the amount of gelatine on the distribution of this substance. Stated in grams in 100 c.c. the figures are as follows:

	Mixture	Droplets	Nexus fluid
$T = 15°$ {	6·7	17	2
	13·5	18	5·5
	36·5	8·5	±40

These figures illustrate a remarkable property of these mixtures which appears to hold whether they are binary or ternary mixtures. In the particular case under consideration the gelatine is chiefly contained in the droplets up to a mixture holding 13·5 per cent.; somewhere between this and 36·5 per cent. the system becomes inverted[3]—the droplets now containing the smaller quantity of gelatine. Thus if the mixture with the lower concentration is heated until it is homogeneous and then allowed to cool it separates into droplets containing a large amount of gelatine which at a slightly lower temperature become solid solutions of water and alcohol in gelatine. Therefore in the jelly stage the framework is a solid

[1] *Ann. Chem. Phys.* (1876), (5) II, 264.

[2] *Lehrbuch d. allgem. Chemie*, 2nd ed. (1891), I, 821.

[3] That is the major part of the gelatine changes from the concave to the convex side of the surface of separation.

net holding fluid in its interstices. In the case of the mixture with a higher gelatine content the configuration of the jelly is the exact converse. As cooling goes on in this case droplets separate with a small gelatine content, and it is the nexus fluid which passes into the stage of solid solution. The droplets in the latter case therefore are quite regular vesicles $10\,\mu$ in diameter.

There are therefore theoretically three states in which these mixtures of water and solid forming the class of heat-reversible or soluble colloids can be, namely, (1) a homogeneous fluid; (2) a heterogeneous fluid, having the form of an emulsion of conjugate, as opposed to immiscible, fluids; (3) a pair of conjugates, one a fluid, the other a solid solution. In the third state the configuration as already pointed out differs remarkably according to concentration. The reaction of such a system to external agencies varies according to the state in which it happens to be; in the third state for instance, not only is there the interdiffusion between the two phases but also capillary phenomena are manifested. Mixtures of agar and water, gelatine and water, certain carbohydrates and water, such as in Lyle's golden syrup, are in one or the other state according to the relative amounts of the constituents present and the temperature. The conjugate state is reached at ordinary temperature in the two first-named mixtures, even when the concentration of water is very great; in the last case, however, the concentration of water must be small and the temperature low.

It is clear that of these states one, the solid or gel state, is of secondary importance—that is to say, it is dependent upon the formation of a membrane at the surface of separation of the conjugates, or upon the passage of one phase into the state of solid solution. This point may be made clear by an experiment which was suggested by my friend Mr Neville. Ether and water form a pair of conjugates which are separated by a plane surface. If they are violently shaken an emulsion is produced which is, however, unstable and passes rapidly into the stable state in which the surface of contact is minimal. The emulsive stage may, however, be made permanent so that a jelly-like mass of adherent droplets is produced by the introduction into the system of a substance whose solubility in the two phases is very different, and which is a solid at ordinary temperatures. Thus if water is saturated with iodine and then shaken with ether a permanent emulsion is formed, the droplets of which adhere to form a jelly-like mass. On warming, the droplets fuse and the jelly stage vanishes. The phenomenon is due to the formation of a tenacious film of concentrated iodine at the surface of the droplets.

The phenomenon is related to the fact that iodine is solid at ordinary temperatures. Bromine is fluid at ordinary temperatures. It has, however, about the same relative solubility in the two phases as has iodine. In spite of this, however, an ether-water-bromine mixture made in the same way will not form a permanent emulsion, nor will it make a jelly. The iodine-water-ether mixture manifests a striking property of soluble colloids, namely that it forms very permanent foams. One might also point out in passing that the vapour-pressure of the iodine would be quite abnormally low. It would be that of the minute amount dissolved in the nexus fluid.

The formation of a surface film on droplets was observed by M. Traube[1] in a variety of cases, by Lehmann,[2] and by myself in the case of ternary mixtures holding gelatine.

These facts and conclusions show that the class of heat-reversible colloidal mixtures may possibly include members which at no concentration form gels, though at certain temperatures and concentrations they may form conjugates.

One may perhaps regard the proteid-water system, or rather the egg-white-water system, as being one of these for the following reasons. So long as the changes of temperature are kept within those limits which do not destroy the system by producing chemical change, the changes of viscosity are of the same sign as those of aqueous solutions of agar or gelatine, and of the opposite sign to those of a solution of silica—that is to say, irreversible molecular aggregates are not formed on raising the temperature. The change in viscosity produced by varying the concentration of the egg-white differs from that produced by changes in the concentration in true solution. The only data are those furnished by Bottazzi.[3] His figures do not stand very close together but they show that the viscosity of proteid solutions does not change continuously with changes in concentration as it does in crystalloid solutions, but it changes *per saltum*. That is to say, there is a critical concentration marked by a sudden and extraordinarily great increase in viscosity above the values for lower concentrations. This cannot be due to change in the size of the molecular aggregates, for that would proceed continuously with change in concentration. It would however be produced by change from a homogeneous fluid to a pair of conjugates with a very large surface of contact.[4]

Colloidal mixtures which form irreversible molecular aggregates when they pass into the gel state. These mixtures have been examined by a number

[1] *Arch. Anat. u. Physiol.* (1867), pp. 83, 129.
[2] *Molekularphysik* (Leipzig, 1888), i, 508.
[3] *Arch. Italienne de Biol.* (1898), xxix, 401.
[4] To safeguard myself from misconception I must point out that the viscosity of colloidal solutions is probably of different origin from that of the viscosity of crystalloid solutions. A clue to its nature is found in the experiments of Frankenheim (*Journ. f. prakt. Chem.* (1851), liv, 433), on the influence of suspended particles upon fluid properties. I hope to discuss this and kindred points elsewhere.

of physicists. In the fluid state the relations of the solid and fluid are very far removed from those found in true solution, so far indeed as to render conclusions as to, *e.g.*, molecular weights, derived from the application of laws which are known to be true only for crystalloid solution in its most perfect form, that is when concentration is very small, worthless to the verge of ridicule. These colloidal solutions cannot be discriminated from non-settling suspensions of solid particles in a fluid medium such as those of gold in water which were prepared by Faraday, and are still preserved at the Royal Institution. The characteristic phenomena which they manifest are the following.

The molecular aggregates which form the dispersed particle of solid becomes larger by a process of clumping as a result of the presence of electrolytes—that is of salts; and the process is aided by a rise of temperature. The action of electrolytes in increasing the size of the molecular aggregates was worked out by Picton and Linder[1] with the aid of various optical methods including the microscope. If the concentration of the electrolyte is sufficient, however, the internal changes proceed beyond the simple increase in the size of the molecular aggregates to the actual solidification of the entire mass—that is to the point of coagulation of the mixture. It is obvious therefore that there exists some connection between the size of the particles and the process of coagulation, and Graham I believe somewhere suggests it. One obvious inference from the facts is that when the molecular aggregates attain a certain size the fluid condition is no longer possible; this would follow immediately from Graham's observation that actual coagulation is preceded by a continuous increase in the viscosity of the liquid. I have been able to follow the entire change from the hydrosol to the hydrogel under the microscope, so that it is possible now to give a connected account of the mechanism of coagulation. Before proceeding to this I may note in passing that the similarity of these colloidal solutions to simple suspensions, such as, *e.g.*, China clay in water, is borne out by the fact that electrolytes and temperature act on both alike, in making the discrete particles of solid larger by a process of aggregation.[2]

The generic action of salts as coagulants, as distinguished from any specific chemical action, was first discovered by Graham, and has since been thoroughly investigated by Schulze,[3] Prost,[4] and Picton and Linder.[5] Putting the facts and conclusions reached by these workers together and

---

[1] *Trans. Chem. Soc.* (1895), LXVII, 63.
[2] Cf. Bodländer, *Neues Jahrbuch f. Mineralogie* (1893), II, 147.
[3] *Journ. f. prakt. Chem.* (1882), XXV, 431; *ibid.* (1883), XXVII, 320.
[4] *Bull. d. l'Acad. Roy. d. Sci. d. Belg.* (1887), Ser. 3, XIV, 312.
[5] *Trans. Chem. Soc.* (1895), LXVII, 63.

adding my own observations the conditions which determine coagulation may be stated as follows:

(1) The point at which coagulation occurs is determined by the concentration of the solid—*e.g.* silica in the colloidal mixture, the temperature, and the molecular concentration[1] and nature of the electrolytes which are present.

(2) The concentration necessary for coagulation is lowered by a rise of temperature, or by an electrolyte.

(3) The coagulative energy of electrolytes as measured by the number of gram-equivalents per litre necessary to produce coagulation is determined almost solely by the nature of the metal of the salt; and among the metals themselves it is determined by the valency of the metal. The coagulative energy of metals of different valency is approximately in the ratio

$$V' : V'' : V''' = x : x^2 : x^3.$$

(4) If two electrolytes are present then the action is additive if the metals are of the same valency, and subtractive if of different valency. Therefore there is interference between, *e.g.*, salts of the form $R'Ac^n$ and $R''Ac^n$—the one "inhibits" the other.

These relations therefore are exceedingly definite and seem to hold for all colloidal mixtures which form irreversible aggregates; and they differ as widely as possible from the relations which govern alteration in the conditions of colloidal mixtures of the heat-reversible type.[2]

Those colloidal mixtures which form irreversible molecular aggregates in the process of forming a gel may unhesitatingly be placed in the class we are considering together with the type examples—silicic acid and certain metallic hydrosulphides, or hydrates and water.

Such a mixture is the one which is produced by boiling a filtered solution of egg-white in water containing ± 1 gr. solids per 100 c.c. Haycraft and Duggen[3] first pointed out that if white of egg be diluted by the addition of about 8 volumes of water and boiled it does not coagulate but merely "becomes a little milky". If calcium or barium chloride or sulphate of magnesium be added to the fluid after boiling it readily coagulates, forming a gel either at the temperature of the air or at some higher temperature determined chiefly by the amount of the salt added.[4] I have reinvestigated these phenomena and I find that this heat-modified solution of egg-white forms irreversible molecular aggregates which may or may not proceed to the stage of coagulation under those conditions which characterise the

[1] That is gram-molecules per litre.
[2] Cf. for example Pascheles, *Pflügers Archiv* (1898), LXXI, 333.
[3] *Brit. Med. Journ.* (1890), p. 167.
[4] Ringer, *Journ. Physiol.* (1891), XII, 378.

colloidal mixture of the silica-water class. The mixture coagulates when the molecular concentration of the electrolytes which are present reaches a certain point which, for a given temperature, is determined by the valency of the metal. It is in this particular case that I have followed the events in coagulation from the process of formation of larger and larger aggregates, at which stage Picton and Linder left it, to the actual formation of a firm gel. The events are easily seen under the microscope because the refractive index of the aggregates is very different from that of the fluid in which they lie and they rapidly grow to a relatively large size ($0\cdot75$–$1\,\mu$). The simplest method is the following. A thin thread of silk is dipped in some solution of a salt (*e.g.* a 2 per cent. solution of calcium chloride) and laid on a slide, over this is placed a coverslip, and the boiled solution of egg-white is allowed to fill the space. The coagulum may be watched as it gradually forms round the silk thread. A magnification of 500 diameters is sufficient. The fluid at first is free from visible particles. We know however from Picton and Linder's work that the heat by chemically altering the dissolved proteid has by a process of desolution, to use Lord Rayleigh's word, produced a suspension of particles having an average diameter commensurable with the mean wave-length of light. Under the influence of electrolytes these particles aggregate to larger and larger masses, so that one first sees near the silk thread a fine cloud, the particles in which grow in size until they form spherules having a maximum diameter of $0\cdot75$–$1\,\mu$. They are now seen to be arranged in patterns forming an open net with regular polygonal meshes having diagonals as long as $6\,\mu$. The threads of the net are formed of contiguous spherules. This stage, however, is not one of equilibrium—the net shrinks, the meshes become smaller and the spherules apparently shift their points of attachment until, in place of being bounded by threads composed of several spherules, the image has the appearance of the typical fine net with spherules at the nodal points joined by tiny threads. Whether these joining threads or bars have a real existence, or whether they are purely optical and the spherules actually touch one another, it is impossible to say at present. One important point struck me, namely that when the particles are large enough to be clearly visible with a magnification of 500 diameters they do not show Brownian movement—in other words they are probably already in some way linked to one another.

This then is the process of coagulation. To complete what can be said concerning the mechanism, a few words would need to be devoted to the nature of the directive force which aggregates the particle to visible spherules and then links these together in a definite figure. This matter, however, will be dealt with in a paper on coagulation produced by electricity.

Action of reagents upon heat-reversible colloidal mixtures. This can best be considered by dealing first with gelatine. Mercuric chloride and alcohol have the same action. The latter, as we have already seen, forms a ternary mixture with gelatine and water which, at certain concentrations and within certain limits of temperature forms a pair of conjugate fluids. Sublimate, gelatine and water also form mixtures which are fluid above certain temperatures and which on cooling form a pair of conjugate fluids, one of which on still further cooling becomes a solid solution. To give an example, one part of a 50 per cent. solution of gelatine mixed with twice its volume of a saturated aqueous solution of mercuric chloride is fluid above 50° C., and as the temperature falls it undergoes the changes described above.

The figure of the conjugate fluids differs according to the concentration of the gelatine in the mixture. If the mixture of gelatine and water contains more than 4 per cent. of the former the droplets are fluid; and the figure in the gel stage is that of a solid mass with vesicles hollowed out of it. If the percentage is less the figure is the obverse—an open net of solid solution is formed having fluid in its spaces. The configuration is not destroyed when the water is replaced by alcohol and the sublimate dissolved out by iodine.

If the sublimate is dissolved out, as, e.g., by an aqueous solution of sodium chloride, the coarse figure reverts to the much finer structure of the gelatine-water system.

Formaline, at any rate when in excess, has a different action. The gelatine is altered so that it becomes permanently and absolutely insoluble in water. The jelly which it now forms behaves like a jelly of silicic acid and water—it shrinks on warming. Its physical characters are unaltered by complete removal of any free formaline and exposure to distilled water for 10 months. The structure of this gel is that of an open net holding fluid in its interstices. I have found this figure in all the true insoluble gels which undergo shrinkage on warming which I have examined.

These are thus two extreme cases. In the first the primary system is modified in degree but not in kind by the introduction of a third substance. It remains a heat-reversible system. In configuration the sole change in the particular cases examined was a decrease in the surface of contact of the conjugates. In the second case the primary system is destroyed and a new one formed belonging to another group of colloidal mixtures. With this change an entirely new configuration appears.

All the fixing agents examined act on egg-white as formaline does upon gelatine, with the possible exception of osmic acid in minimal amounts. Therefore an irreversible change is produced, and a true insoluble phase having the typical open net structure, even when the concentration is pushed as high as 60 per cent. solids, is the result.

The entire process, as we have seen, can be completely followed in the case of heat fixation, by using dilute solutions, containing only 1·0–1·5 gr. of coagulable solids per 100 c.c.

Two stages are then clearly recognisable; in the first there is formed by the action of temperature upon the egg-white a colloidal mixture of the silicic acid and water type, the dissolved colloidal matter undergoing a process of desolution in the change; in the second the new system so produced, following its own laws, forms irreversible molecular aggregates which may or may not link themselves together to form the open network of solid according to the concentration of the colloidal solid, the concentration and nature of the electrolytes present, and the temperature. This statement constitutes a first approximate conception of the physical events in fixation when an insoluble modification is produced.

The conclusion that two types of structure are met with in heat-soluble gels can fortunately be freed from the doubt which would always attach to it, if it depended solely upon the interpretation of appearances seen with high magnification. The transition from the gel built up of adherent droplets of solid solution—the solid-within type—to the gel built of a continuous solid frame with enclosed droplets of fluid solution, or the solid-without type, can be followed in ternary systems with lenses of a focal length long enough to eliminate diffraction errors. And, secondly, the inversion of the system is marked in the pressure experiments by a great rise in the pressure necessary to separate the conjugates, so that from pounds on the square inch it rises to hundredweights. If, however, the solid-without system is mechanically destroyed, the fluid conjugate can be separated by comparatively slight pressure. Thus the fluid conjugate is separated by a centrifugal machine if a layer of fine sand is placed above the gel. The sand particles are driven through the gel, and in their passage they cut open the walls of the fluid droplets.

For the same reason, namely the ease with which fluid can be expressed from it, we may rest absolutely certain that the figure of the true insoluble gel, the gel of silicic acid, of egg-white fixed by heat, sublimate, etc., is that of an open net. On the other hand, the detailed configuration of this net is arrived at solely by inference from an image, the dimensions of the parts of which are in some cases of the same order as the diffraction phenomena of a lens of 2 mm. focal length—that is of the lens used to produce the image.

The effect of fixing reagents upon a pre-existent net of insoluble colloid. This point was examined at some length. An insoluble gel was made from 13 and 6·5 per cent. of solutions of egg-white with the aid of potassium sulphocyanate. Excess salt was allowed to diffuse out into distilled water, and the final product was a firm white jelly in which a net could be demonstrated by sections cut with the freezing microtome. Portions of this jelly were treated, just as though they were bits of tissue, with 1 per cent. ammonium bichromate, osmic vapour, or the saturate solution of corrosive sublimate. Sections were cut from paraffin.

The results of the various experiments were the following. The pre-existent net is not markedly altered by treatment with fixing reagents. The meshes were distinctly smaller, but the figure was otherwise unchanged. The meshes were smallest in the specimens which had been treated with the vapour of osmic acid. That, however, I regarded as the effect of slight drying and consequent shrinkage of the mass.

Influence of solid particles when present in a colloidal solution before fixation. This appears to me to be a particularly important point from its practical bearing. Carmine of various grades of fineness was prepared by suspending ordinary carmine in distilled water, and separating the precipitates which fell during the third and fourth hours, and the second and third 24 hours. The carmine so separated was added in varying quantities to solutions of white of egg, which were then fixed with corrosive sublimate and embedded in paraffin. Examination of sections seemed to establish the following points:

(1) The carmine grains occupy nodal points in the net and modify within wide limits the size of the meshes and therefore also the thickness of the bars.

(2) The larger the grains and, up to certain limits, the fewer, the wider are the meshes and the thicker the bars. The grains of carmine commonly occurred in patches between which one had regions fairly free from solid particles. A study of such sections showed that when the grains were larger than the mesh in the grain-free parts of the colloid—that is to say the mesh formed in the absence of solid particles—the effect of a group of grains was to cause the mesh to widen out very much. If the grains are sufficiently numerous then each nodal point contains, or is formed by, one. If, however, they are too few and scattered the normal net is formed, but opens out here and there to lodge a single grain, or a group of grains on its nodal points (fig. 2). So we may conclude that the grains modify the size of the meshes and the thickness of the bars by their number and their size.

Carmine grains correspond in their insoluble solid nature with the granule of the secretory cells of for instance the frog's pancreas and the coarsely granular blood cells of the same animal; and not with such soluble, swellable granules as those of mucin secretory cells, or of the coarsely granular blood cells of the crayfish.

Fixation under stress. In thinking over the nature of the evidence which supports the view that the cell-protoplasm has a fibrillar or reticular structure I came to the conclusion that the spindle and aster of dividing cells furnished the strongest proof. It occurred to me, however, that the fibrous appearance seen in cells under these conditions might be due not to a pre-existent *structure*, but to pre-existent stresses which would fashion the figure of the net, so that the latter would be, as it were, a coarse diagram of a dynamic phase of the cell history. I was not then aware that Bütschli had actually shown that a fibrillate appearance can be seen in the threads of coagulated albumen which are formed when "filtered albumen is sprinkled by means of a paint-brush into a drop of picro-sulphuric-osmic acid on a slide", and that he had come to the conclusion that "the fibrillar structure had already been produced before the coagulation by the drawing out of the fluid white of egg into threads".[1]

My own experiments differ slightly from those of Bütschli. They included the fixation of films of solution when stretched, and of a solution immediately after it had been mechanically sheared. Films of soap and egg-white of fair thickness can be made to stretch across a ring of cork. When the film was made a tiny drop of mercury was rolled on to it and the film was then fixed by alcohol in the former or sublimate in the latter case. In other cases stretched films were made by pouring a colloid solution over a drop of mercury on the slide. A fibrillar appearance radiating from the centre of application of the stress can be seen in either case after fixation. Another method is to draw a small quantity of colloidal solution along a slide with, *e.g.*, the point of a needle or a glass rod and then to fix it at once. In favourable cases the appearance of fibrillæ is so striking that they look as if one might isolate them by teasing.

The phenomenon is due to the well-known "after-action" phenomena of colloids; which in turn are due to the fact that they are capable of retaining a shear strain. When colloidal fluids are sheared they become doubly refractive, and in this way Maxwell[2] demonstrates the existence of a shear strain in Canada balsam. The "rate of relaxation", as Maxwell called it, of balsam is, however, so great that the double refraction does not survive this shearing motion. Klocke[3] demonstrated double refraction due to strain in a way which especially touches our subject. He stretched thin films of gelatine over tinfoil frames; then, as the films dried, they tended to contract but were unable to do so owing to the rigid frames. The films therefore became strained in drying, and when dry they were found to be doubly refractive. The more viscid the colloid the more marked is the "survival" shear as one might call it. This is obvious, since the phenomenon depends upon internal friction. Therefore, as Kundt[4] showed

[1] *Protoplasm and Microscopic Foams*, p. 218.   [2] *Proc. Roy. Soc.* (1873), XXII, 46.
[3] *Neues Jahrbuch f. Mineralogie* (1881), II; or Lehmann's *Molecular Physics*, I, 533.
[4] *Ann. Phys. Chem.* (1881), XIII, 110.

it, as too elusive to be demonstrated at all in pure fluids. Soft wax, on the contrary, manifests the phenomenon to an amazing extent.[1] Now we may look upon the undoubted fact that an internal strain is as it were visible after coagulation of a colloid mass in the fibrillar aspect and disposition of the fibrils in the direction of the lines of force in two ways.

We may say that shearing a colloidal mass, fluid or solid, actually does produce internal heterogeneity, or simply structure, which is fixed by the process of coagulation. Or we may say that, though the strain does unquestionably dispose the molecules in some definite linear pattern, as is clearly proved by the phenomenon of double refraction, yet the pattern is not identical with that found in the coagulated colloid, though it certainly does condition the rearrangement of solid and liquid in such a way as to produce that pattern.

Whichever be true it is clear that neither view affords any support whatever to the hypothesis that the cell substance is composed of two different kinds of material. The internal structure developed as a result of strain may appear in a mass previously homogeneous and has nothing in common with, for instance, filar and interfilar substances, or with spongioplasm and hyaloplasm, as those words are used by the histologist.

The more purely physical aspect of this last question is of interest. In the first place, if we define a fluid by an inversion of Maxwell's definition of a solid as a mass which is incapable of maintaining its shape when exposed to forces which are not equal in all directions, then it is clear that colloidal solutions depart from the definition and approach the solid state just so far as their molecules have fixed positions in the mass and are therefore capable of manifesting "survival" shear. That view of the liquid state first set forth by Poisson and developed by Maxwell which has been revived lately by Poynting offers an explanation of the phenomenon. According to it a liquid may be "in the main a solid structure inasmuch as the molecules cohere and resist strain of any kind. But the molecules have so much energy that...the solid structure is continually breaking down and renewing itself. If we impose a shear strain on the structure, the strain will of course disappear with the structure in which it is produced."[2] Colloidal mixtures of fluid and solid depart from this fluid condition inasmuch as the strain does persist for a certain period; and this would imply an equivalent survival of the structure in which it was produced.

The principle involved in this discussion may be applied to an observation made by Engelmann. Threads of cell substance, such as for instance

---

[1] Kohlrausch, quoted by Lehmann, *Molecular Physics*, I, 533. Speaking of the after-movements of a wax cylinder he says, "Ich kenne wenig so überraschende Vorgänge, wie diese freiwilligegen Bewegungsänderungen eines leblosen Körpers."

[2] *Phil. Mag.* (1896), XLII, 289.

the fine pseudopodia of Foraminifera, frequently manifest a double streaming; the direction of the stream on one side of the thread being outwards, on the other inwards. Engelmann showed that the thread under these conditions is doubly refractive and, in the light of what has been said, we may regard this as due to the shear strain imposed upon the substance at, and on each side of, the surface of separation of the two streams.

The effect of cold upon colloids. This investigation yielded results of some interest to the histologist. If a portion of a gel containing 1·5 per cent. approximately pure gelatine be cooled while on the stage of the microscope it shows no change which can be detected by a magnification of 450 diameters until the temperature has fallen to 1° C., when exceedingly minute droplets about $0·5\mu$ in diameter appear in its substance. At about $-1°$ C. there is a very rapid change from this condition to a honeycomb or sponge-like arrangement.[1] The solid matter disposes itself in bars or lamellæ having a very constant thickness of about 2μ. The meshes of the structure are rounded and have an internal diameter of $20–30\mu$. Fluid can readily be expressed from such a sponge, and it was found to contain 0·1 per cent. of solid matter. I was unable to satisfy myself as to the nature of this small amount of solid. At a slightly lower temperature the fluid within the spaces deposits a mass of ice crystals. The gel is transparent, or at any rate translucent, until the ice crystals appear. There are thus three stages of the solid; the ordinary gel which exists between 26 and 1° C.; the condition with minute droplets and the honeycomb with fluid-containing spaces 1 to $-1°$ C., and the stage with ice crystals from $-1°$ C. downwards. The changes in the second stage are so rapid that I succeeded in recognising the beginning and the end only. There is therefore as it were a critical point about zero in the case of the gelatine, or about $-5·5°$ C. in the case of 2 per cent. agar when the gel assumes the structural features of an "insoluble" gel. The reverse changes differ very much in the temperatures at which they occur. The quite unimportant point, namely, the melting of the ice crystals, occurs at the same temperature as their formation, and the transparent or translucent spongy gel holding fluid is reformed. This retains its sponge-like character until a temperature of about 26° C. is reached in the case of gelatine, and $\pm 75°$ C. in the case of agar, when it passes almost abruptly into the fluid state.[2]

[1] Temperatures were more carefully determined for 2 per cent. agar. The sponge appears at $-5·75°$ C. Even 10 min. exposure to $-4·5°$ or $-0·5°$ C. produces only slight clouding due to formation of evenly dispersed droplets each about $0·5\mu$ in diameter.

[2] The description in the text may be condensed into this statement. When soluble gels are cooled below a certain critical temperature the new condition imposed on them is no longer simply reversible by a small rise of temperature above the point of change. The conditions of the change therefore recall the phenomena of hysteresis.

If the cooling is carried only so far as to cause the appearance of the small droplets and the temperature is then allowed to rise, the droplets, unlike the sponge structure, gradually shrink and disappear and the gel once more becomes glass-like and optically homogeneous.

If the melted gelatine is very rapidly cooled to the critical point the first stage of the solid condition is completely cut out. Thus, if a small drop of 1·5 per cent. gelatine solution at 90° C. be placed on a glass plate cooled to − 15° C. the edges at once become solid owing to the formation of a sponge the bars of which project into the central still fluid portion.

The spongy gel formed by the action of cold therefore is stable over a large range of temperature, and within this range it behaves in some respects like an insoluble gel, in that it undergoes spontaneous shrinkage with expression of fluid, and the shrinkage, the synæresis, is hastened by a rise of temperature up to a certain point. There is therefore true cold fixation. Cold causes a separation of solid and liquid; and this leads to the formation of a structure which persists during a considerable rise of temperature.

The significance of these facts may be shown by applying them to a special case. They make it clear that the expression of Kuhne's muscle plasma from muscle fibres which have been frozen is no evidence of the presence of a distinct fluid constituent in the living muscle substance.

Optical homogeneity of very thin films after fixation. An observation which at first sight appears to offer conclusive proof of the hypothesis that the living cell substance is composed of two distinct portions was made by Schäfer on leucocytes fixed by heat. He found that the thin pseudopodial expansions differed from the rest of the cell body in showing no net and in being resistant to stains. He therefore urges that they are composed of different material—the hyaloplasm—which has flowed out of the spongioplasmic framework of the cell body.

The observation will not bear this interpretation, since thin films of a simple colloid solution, such as white of egg, show the same resistance to stains and are apparently homogeneous after fixation by heat. These unstained homogeneous films are produced at the edge of a layer of colloid and they are commonly separated by an abrupt line from the thicker part which manifests a net and stains deeply with hæmatoxylin, or other dyes.

These edge films do not owe their characters to drying since they are seen at the junction line of white of egg and olive oil when the former is coagulated, when the colloid solution is fixed while actually flowing over the surface, and also within the colloid mass itself in thin films which are formed between contiguous air-bubbles.

They may readily be made by dipping the corner of two coverslips held in the forceps into a solution of white of egg and at once fixing by momentary immersion in a steam jet. The preparation may then be examined in water, alcohol, or, after staining, in balsam. If the fixation was quickly done the colloid solution will have penetrated only a few millimetres from the corner; it was therefore fixed while still flowing. Fig. 10 is a drawing of an edge film; it shows the abrupt transition from the non-staining to the staining portions. Fig. 9 shows the optically structureless film at the junction of two air bubbles.

To my mind the phenomenon deprives Schäfer's observation of any great weight. The fact that thin pseudopodial films are clear and unstaining ceases to be conclusive evidence that they are composed of material which is permanently different from the staining mass which constitutes the cell body.

The phenomenon is probably related to the fact that when colloidal mixtures separate to form conjugates a thin layer next the surface remains homogeneous. The depth of this layer appears to bear some definite numerical relation to the diameter of the droplets.

II. STRUCTURE PRODUCED BY SUB-MORTEM OR POST-MORTEM CHANGE IN THE CELL

Fixation and clotting. In the foregoing sections the process of fixation is considered much as if it was the same as clotting. This has been done because the redistribution of solid and liquid is the same in both cases. In the clotting of, *e.g.*, blood plasma irreversible molecular aggregates are formed by a process of desolution and the aggregates become grouped together to form a solid sponge traversing the fluid.

If now we discard the distinction between clotting and coagulation and retain only one word, then the following methods of production of an insoluble gel can be recognised:

(i) *Coagulation without concurrent chemical change.* An aqueous solution of silicic acid will set to a gel on mere lapse of time, or at once if powdered graphite be added to it. A solution of celloidine in a mixture of ether and absolute alcohol coagulates when exposed to the vapour of chloroform.

(ii) *Coagulation with concurrent chemical change.* There are three modes by which this may be brought about: (*a*) by heat, (*b*) by ferments, and (*c*) by chemical reagents. In the last class falls the process of "fixation"; and somewhere in the three classes will fall the coagulation which the whole or a portion of living matter undergoes when it dies.

Therefore, and this is the practical outcome of these considerations, a relatively solid framework may be formed in the cell substance, either (1) by what we may call the spontaneous coagulation of the death change, or (2) by the action of reagents in fixation. If a continuous framework is to be formed then, as our experiments on simple colloids show, the process of desolution must separate a considerable proportion of insoluble matter. If only a small quantity is separated a granular deposit will result.

In the process of fixation, until proof to the contrary is forthcoming, we must recognise the following possibilities. The separation of practically the whole of the solid matter as a coarse framework by such a reagent as corrosive sublimate; or a discriminative action whereby, as Fisscher's experiments show (cp. p. 251), part may be separated as a framework, part as granules.

Structure due to the presence of, or to post-mortem changes in, secretion masses. The production of a false appearance of structure in cell protoplasm by "secretion masses", as Flemming styled them, has been dealt with by histologists in special cases only.

Flemming writes that before one can take the appearances seen in preparations as true "vital structure" it is necessary to be very vigilant lest things which have no resemblance to such living structures are taken for them. The first case which he chooses as an instance of things which have no relation to living structure is the net seen in gland cells, for these cells hold (1) cell substance, (2) secretion masses in "secretion vacuoles", and "it is very possible that the cell protoplasm between the secretion vacuoles are pressed out into such thin masses (Bälkchen) that they become the same as or equivalent to the threads in cells of other kinds".[1]

Flemming, however, at the time he wrote could scarcely appreciate how fruitful a source of artifact these secretion masses may be: this wider knowledge follows from Langley's work upon mucous salivary glands published in 1886—that is to say, four years after Flemming wrote.

In no structure is a more beautiful and more definite "net" observable than in the secretory cells of the alveoli of these glands, yet Langley was able to show that this net only appears as a result of a post-mortem swelling of the secretory granules which distends the cell protoplasm.

Secretion masses, still to use Flemming's words, not only produce deformation of the cell protoplasm by post-mortem change in themselves, but, as in the case of the net which Kupffer described in the liver cells of winter frogs, the mere presence of these masses, glycogenic in this case, implies that the whole of the cell protoplasm is thrown into a structure or framework to hold them. In this case one can hardly speak of the secretion masses, with which in this connection one must group simple vacuoles, as sources of artifact; yet they need mention here since in many cases the net or sponge structure which writers have described as occurring *within* the cell protoplasm is as a matter of fact something quite different, namely a frame formed of the whole of this substance, and not from a part only. A single instance may be taken to exemplify this.

[1] *Zellsubstanz, Kern und Zelltheilung* (Leipzig, 1882), p. 60.

Probably no cells have furnished more observations which have been used to support the hypothesis of the existence of a net or sponge-work in protoplasm than the blood corpuscles of *Astacus*. Frommann, Heitzmann, Leydig and Griesbach may be quoted amongst others who have described the "Gerüst" in these cells. Yet it is certain that the sponge-work figured by these authors is the very obvious and beautiful sponge-work which in almost all preparations of the blood cells of *Astacus* is produced by the disappearance of the soluble and apparently semi-fluid oxyphil granules contained in the corpuscles during life.

There can be no doubt on this point. It is possible to watch at leisure the solution of the granules and the formation of the sponge-work, the spaces of which are readily distorted by pressure, and so to obtain appearances truthfully figured by Griesbach.[1]

The most striking instance of structure in cells charged with secretion masses is furnished by the secretory cells of mucous glands.

Opinion is still divided on the question whether the very definite and delicate network so often seen in these cells includes within itself the whole of the cell protoplasm, and is due therefore simply to the distension and compression of the protoplasm by the swelling of the secretion granules, or whether it is a true net present in the living cell and formed of only part of the cell protoplasm. I endeavoured to come to some conclusion by a prolonged examination of the orbital glands of kittens and puppies.

The minute structure of an alveolar cell of the orbital glands. Glands from young animals were used because it was found impossible to cut very thin sections from glands of adult animals.

The material was preserved in three ways: (1) with absolute alcohol, (2) with the vapour of osmic acid, (3) with chromic acid.

Appearances after fixation with absolute alcohol. Small pieces of gland were blotted thoroughly to remove adherent blood and lymph, and then shaken quietly in a considerable bulk of absolute alcohol for about 10 min., after which they were removed to a fresh quantity of the fixative. After 24 hours the tissue was examined and found not to be hardened; it softened readily in solutions containing less alcohol, or in water.

Study of sections cut with the free hand. These are not infrequently exceedingly thin near the edges, offering sections 2μ and less in thickness. The sections were examined in absolute alcohol, and in xylol when unstained; they were then irrigated in succession with 90 per cent. alcohol containing a large amount of methylene blue, with absolute alcohol, and with xylol. Tested as carefully as possible neither the xylol nor staining fluid was found to alter the disposition of the parts. Observations were made with the Zeiss apochromatic lens, aperture 1/40, focal length 2 mm.

[1] *Pflügers Archiv* (1891), L, 473.

The granules were found in different stages of preservation. In some cells distinct granules were seen, in other cells the granules had vanished, and one had irregular flakes or masses.

The granules and the flakes or masses stained an intense opaque blue with the methylene blue. The cell protoplasm at the base of the cell coloured a vivid green. Probably in no case had the absolute alcohol prevented all swelling of the granules. Even when quite distinct they always measured $2\,\mu$ in diameter; while Langley fixes the size at $1\text{--}1\cdot 5\,\mu$.

The configuration of the cell protoplasm was beautifully shown in these preparations. At the edge of the preparation exceedingly thin sections of cells were found from out of which the blue staining substance had fallen. One then had without doubt a honeycomb, the spaces of which did not appear to communicate with one another (fig. 11). The substance of the honeycomb portion of the cell was continuous with and in no sense different from the cell protoplasm which formed the base and sides of the cell: clearly the honeycomb structure is simply the expression of the fact that the cell protoplasm was hollowed out to hold the secretory granules: the spaces have been slightly enlarged by the partial swelling of the granules.

The configuration of the cell protoplasm in the cells which are occupied by intact granules cannot be directly determined, since the granules colour so intensely with the methylene blue as to overshadow completely any substance between them. It can be discovered, however, by taking advantage of the fact that there is a certain critical point in the process of inhibition of water, when the staining properties of cell protoplasm and granules is reversed.

If a cell containing distinct granules is kept in view while the preparation is being irrigated with weaker and weaker alcohol, lightly charged with methylene blue, one witnesses the following changes. A more or less isolated cell at the edge of the section should be chosen for observation since it can expand almost without obstruction. When the percentage of alcohol has fallen to about 60, the whole cell rapidly expands to twice its volume,[1] and this is seen to be due to a swelling of the granules which, in spite of the increase in size, still remain distinct.[2] Coincident with the increase in size the granules become less deeply coloured, and their tint changes from an opaque blue to a transparent purple, while at the same time the cell protoplasm becomes more coloured and its tint changes from green to blue. One can now see that the blue-stained basal cell protoplasm is continuous with the substance of a honeycomb in the polygonal spaces of which lie the

[1] Langley gives an account of the swelling with dilute alcohol. *Journ. Physiol.* (1889), x, 433.

[2] The changes in volume were computed from several linear measurements. I may therefore remind the reader that doubling the volume of a sphere means roughly an increase in a diameter from 3 to 4.

granules. With a further fall in the percentage of alcohol the cell continues to expand in size to many times its original volume; the granules become less and less coloured until one sees only the blue-stained cell protoplasm, now expanded into exceedingly thin plates. If the percentage of alcohol has not fallen below 40, one can produce contraction of the cell by again raising the concentration of the alcohol until its volume is almost what it was at the beginning. The granules, however, do not recover the property of dyeing deeply with the methylene blue (fig. 12). One can therefore stain the cell protoplasm and so show that the spaces in the honeycomb agree in size with that of the granules before swelling occurred.

If the percentage of alcohol falls below 40, to be more accurate when it approaches 30, the distension of the cell becomes so great as to rupture the cell protoplasm. Raising the percentage of alcohol naturally will not now restore the honeycomb.

If the percentage is kept above this destructive lower limit a cell can be swollen and contracted and swollen again an indefinite number of times without modification of the configuration at any particular stage. At no time does the material between the granules differ in staining from the cell protoplasm round the nucleus, and it always appears to be a simple continuation of the whole of that protoplasm.

It is particularly noticeable how the granules and cell protoplasm expand and contract together; and it is clear that the size of a cell in a preparation is no indication of the extent to which swelling may have occurred during fixation.

Portions of the glands were transferred to xylol after 24 hours in absolute alcohol, and embedded in paraffin. This process, though it could not conceivably have given rise to swelling, was found to have profoundly altered the physical character of the tissue; in this way the granules retained the power of swelling though to a much less extent, while the cell protoplasm was completely fixed and inextensible. The change is of advantage since it enables us to determine with certainty the configuration of the cell protoplasm in those cells which contain intact granules, especially as one could use sections of $1\,\mu$ and less in thickness. These thin sections were floated with 95 per cent. spirit on to the coverslip; and the paraffin removed by xylol. The same cell and the same alveolus were measured at various stages while the preparation was being irrigated with 80, 75, 60, 55 per cent. and so on down to 30 per cent. alcohol. Neither alveoli nor individual cells showed any increase in size along any axis; on the other hand, measurement of the diameters of individual granules gave the following average results—in absolute $2\,\mu$, in 50 per cent. $2\cdot7\,\mu$, in 30 per cent. $3\cdot3\,\mu$. In these thin sections, since the thickness is less than that of a single granule, these bodies are not restrained from swelling by the bars of cell substance: they can expand at right angles to the plane of the section and over the bars.

Now in these thin sections from paraffin it was found that the cell protoplasm could be stained with methylene blue while the granules remained almost uncoloured when the percentage of alcohol fell below 40. One then has a green-blue stained cell protoplasm at the base of the cell which continues into a continuous network. Seeing that the whole cell has not changed its dimensions we may take the meshes as being unaltered in size by the irrigation. Actual measurement of meshes and of the granules before irrigation, and counting both meshes and granules, proves (1) that there are the same number of meshes as granules, and (2) that the granules must incompletely fill the meshes when they are unswollen.

That the net is the section of a honeycomb is proved by this consideration. The section is less in thickness than half the average diameter of a mesh. If it were a true net composed of threads a section so thin would show broken lines joining the nodes; instead of that one always found continuous lines.

The study of these absolute fixed glands appears to me to prove that the typical net of the mucous secretory cell is the optical or actual section of the honeycombed cell protoplasm distended by the swelling of the secretory granules, and that it is not a net within, but a net composed of the whole of, the cell protoplasm.

These conclusions were confirmed by the study of specially thin sections from glands hardened in chromic acid which were stained in various ways, notably with iron-hæmatoxylin.

It is difficult to raise a histological question above the region of individual judgment to the level of secure proof. In the case we are considering this fortunately can be done, by comparing the configuration of cells fixed with osmic vapour with that found in absolute or chromic fixed cells in which the granules have partly or wholly swollen. It can be clearly established that when the granules are completely preserved the configuration is totally different to the honeycomb already described; it becomes the conventional net of fixed colloids which has the granules at the nodal points; such a net as is formed when egg-white holding carmine granules is fixed.

Appearances after fixation with the vapour of osmic acid. Special precautions were taken to prevent swelling of the granules. Very small pieces of the tissue, about 2 mm. across, were suspended in the vapour for 8, 24, or 96 hours in the dark. They were transferred either to absolute alcohol, and then to xylol, or direct to xylol, and embedded in paraffin. Sections were floated on the coverslip with 95 per cent. alcohol. They were examined unstained in xylol, in absolute alcohol and in lower alcohols; or after staining with methylene blue saturated in 90 per cent. alcohol, or with iron-hæmatoxylin. The latter method was of necessity modified in order to avoid the injurious action of water. In place of the solutions recommended by Martin Heidenhain I used a saturated solution of the ammonio-ferric alum in 80 per cent. ethylic alcohol,[1] and a strong solution of Grübler's purest hæmatoxylin in the same strength of alcohol.

After 8 hours' exposure to the fixative it was found that the granules still retained the power of absorbing water; but it had almost or completely vanished after 24 hours' exposure.

The examination of unstained preparations in absolute and xylol shows that even with the small pieces of tissue which were used the granules are not preserved throughout. In by far the greater number of the cells they were quite distinct, in a few, however, they had run together; in some cells the granules had "smeared" in places, four to a dozen or more having stuck together, while over the rest of the cell they were completely separate. One could therefore trace in the most complete way the change in the configuration of the cell protoplasm induced by the smearing of the granules.

The configuration of the cell substance when the granules were perfectly preserved was found to be that of a sponge-work of threads holding the granules on the nodal points. The form of the net varies according to whether the granules are very close set or not. In the former case the threads run simply from granule to granule, and iron-hæmatoxylin preparations of sections about 0.5μ thick (less than one tooth) convey the impression that each granule is enclosed in a thin shell of cell protoplasm from which bars run to the similar shell on neighbouring granules. Where the granules are less closely set the cell protoplasm forms a net with nodal points of its own substance which stretches between the granules. A reference to the camera lucida drawings reproduced in figs. 13, 14 and 15 will obviate further description.

The meshes of the sponge are simple fluid-holding spaces. This was proved in two ways.

(1) By microscopical examination. After 24 hours' exposure to osmic vapour the configuration was found to be unaltered by irrigation with aqueous fluids. After many failures, I succeeded on several occasions in keeping the same alveolus in the field of the microscope, with the large Abbe camera lucida always in position, for periods up to 10 days, during which time the alveolus was stained with iron-hæmatoxylin by the alcoholic method, then again by the aqueous method, and finally it was irrigated with a saturated solution of eosine and methyl eosine in water. Drawings were made at intervals, and a series from one experiment are

[1] Here and elsewhere in the paper by the word alcohol I mean ethylic alcohol, and I distinguish between solutions of this body and solutions of commercial spirit.

reproduced in figs. 14a–d. In other cases the alveolus was irrigated with a 20 per cent. solution of acid fuchsine in water saturated with aniline oil, and various saturated solutions of basic dyes were tried. In short, by every method which suggested itself to me I tried to stain material in the meshes. If there be any present, then it must have these negative characters: it is invisible in fluids whose refractive index varies from that of absolute alcohol to that of Canada balsam; it cannot be stained by eosine, methyl eosine, acid fuchsine, iron-hæmatoxylin, methylene blue and osmic acid.

(2) By direct measurement of the volume of the fluid-holding interspaces in the tissue. This was done in the following way. A number of the tiny pieces of gland were removed from the absolute alcohol in which they had been placed after fixation, and surface-dried by being rolled about on fine filter-paper. Their volume was then determined by measuring the amount of absolute alcohol they displaced. They were then again dried by being rolled on the filter-paper for as nearly as possible the same length of time, and at once compressed by screw pressure between two glass plates. The fluid which exuded was collected on dried and weighed filter-paper, which was at once enclosed in a weighed airtight capsule, and the whole weighed. The fluid was colourless, and on evaporation left no solid residue; it was, therefore, assumed to be absolute alcohol of the specific gravity of that used to preserve the gland. The specific gravity being arrived at in this way, the volume was calculated from the weight. I was not able to collect all the expressed fluid: part remained and was visible between the small masses of gland. I was only able to collect the overflow. There was also loss at the various stages from evaporation. The volume of the quantity collected represented 46 per cent. of the total volume of the bits before compression. Now examination of sections appeared to show conclusively that the interalveolar clefts and the lumina taken together could not represent anything approaching 50 per cent. of the volume of the mass; part, and probably the greater part, of the fluid must have been expressed from the sponge-work of the alveolar cells.

The study of material which has been exposed to the fixative vapour for only 8 hours affords the strongest confirmation of the account which has been given of the configuration of the cell substance. Unstained sections examined in absolute alcohol show the granules distinctly, but it is difficult to determine the disposition of the cell substance, owing to the diffraction halo round the granules. The disposition of the granules can, however, be exactly planned, and one has (1) cells with distinct granules only, (2) cells with granules for the most part distinct, but here and there in heaps forming a mass having a curdled aspect, (3) cells wholly occupied by an irresolvable curdled mass. On irrigation with xylol the cell substance becomes visible; it is now found that the separate granules are joined by threads, while the heaps of granules, and the cells in which the granules have completely run together, show the characteristic honeycomb net. Staining with iron-hæmatoxylin and methylene blue, while still in the field of the microscope, does not alter the picture. Irrigation with lower alcohols causes the granules to swell; without, however, producing a honeycomb net in regions where the granules had been completely separate at the commencement. Figs. 13 *a* and *b* show the same cells as they appeared in 95 per cent. spirit and in 30 per cent. spirit.

It might be urged that the net structure which is characteristic of osmic-vapour preparations might be produced by the granules first swelling so as to touch, and then shrinking, leaving, however, a thread joining the granules. Langley actually saw granules do this. This supposition is untenable for the following reasons. The granules stain a deep opaque blue with the alcoholic methylene blue, while the basal cell protoplasm and the threads stain a true translucent green. Flooding the field with light makes the contrast in the staining reaction very obvious. Further, the threads of the network may be seen to be continuous with the basal cell protoplasm. Lastly, if these threads are formed from granule substance, what has become of the cell protoplasm? As we have seen, the thinnest section and the most drastic staining fails to show any substance other than granules, and the threads which link them together, or which, in more open spaces, form an independent net.

The study of this material which has been only lightly fixed by the vapour offers still further confirmation of the statement that the meshes of the sponge-work are simple fluid-holding spaces. When a preparation is irrigated with lower alcohols the granules may be seen to expand in the plane of the section, the threads of cell protoplasm are displaced by this so that they appear as bits irregularly placed and wedged between the swollen granules. This process can be followed on the slide by irrigating with lower alcohols and then staining with methylene blue or with iron-hæmatoxylin. As one follows the process it is evident that the granules expand freely without having to displace solid matter other than the threads, and these are readily moved.

III. The Interpretation of Structure seen in Fixed and Fresh Cell Substance

The preceding pages constitute a general criticism on the inferences which have been drawn from the study of fixed and fresh cell substance. They may be reduced to three propositions:

(i) That a regular geometrical figure may be conferred on the cell protoplasm by the presence in it of secretion masses and especially by post-mortem swelling of secretion masses.

(ii) That, so far as can be judged by control experiments with colloid masses of known character, the radical changes in the physical characters of the cell substance produced by fixation are of such a nature as necessarily to produce a structure.

(iii) That the process of dyeing without fixation, since it appears to consist in all cases of a coagulation of some part of the cell substance, must also produce structure not present in full life.

The study of the action of fixatives upon colloidal solution and upon soluble gels cannot be applied to the elucidation of the effect of fixation upon living cells until we know whether the living substance reacts as a soluble colloid.

If it can be shown that the configuration of the cell protoplasm after fixation depends upon the fixative used, and upon the size and physical characters of granules which may be present, the conclusion is rendered probable.

The following tissues were examined: pancreas, mucous coat of duodenum and stomach of frog, gut of *Oniscus*, red marrow and orbital gland of kittens and puppies, and peritoneal fluid of frog, and blood of the crayfish. Very small bits of tissue were taken and separated from adherent connective tissue as much as possible, since the presence of this substance makes it difficult to obtain sections of less than $1\,\mu$ in thickness. Corrosive sublimate saturated in 0·6 per cent. sodium chloride and the vapour of 2 per cent. osmic acid were used as fixatives. The tissue was exposed to their action for from 2 to 4 hours; then washed in 0·6 per cent. salt solution for about 15 min., dehydrated by very gradually raising the percentage of alcohol, and embedded in paraffin from cedar oil. Sections varying from $\pm 0·5$ to $\pm 3\,\mu$ thick were used.

On the whole living cell substance does react to fixatives just as does solid or fluid soluble colloid. The facts may be summarised as follows:

Gland cells. When the granules neither swell nor dissolve, corrosive sublimate throws the cell substance into an open net. The figure of the net depends upon the number and size of the granules. Figs. 16 *a* and *b* are from different regions of the same cell of the pancreas of a frog separated by only a short space. Fig. 17 is from another cell.

The vapour of osmic acid produces a much finer structure. Where there are no granules the structure appears to be a fine honeycomb and not a net. Where granules occur the honeycomb opens out, the spaces becoming larger. Figs. 18 and 19 representing the structure of the mesenteron cells of *Oniscus* after sublimate and osmic vapour respectively will illustrate this. The difference is partly due no doubt to the general shrinkage which accompanies fixation by osmic vapour. A rough numerical estimate of this was obtained in the following way. Two specimens of *Oniscus* of approximately the same size were chosen, and kept without food for 5 days. The intestine of the one was exposed to the vapour of 2 per cent. osmic acid for 2 hours, that of the other was placed in the sublimate solution for the same period. They were then embedded in paraffin by the same process and sections were cut as nearly as possible at right angles to the length of the gut.

The long and short axes of a number of nuclei in each gut were measured and the mean of the measurements were taken as the diameter of a sphere, the volume of which was calculated. It was found that the ratio of the mean volumes of a nucleus—roughly estimated in this way—in the osmic-vapour preparation to the mean volume in the sublimate preparation was 7 : 11. To put these figures in another way, if the percentage of solid in the sublimate fixed cells were 10, then it would be about 16 in the osmic-vapour fixed cells.

In osmic-vapour preparations of the pancreas of the frog an interesting point was noticed which showed the effect on the configuration of the cell protoplasm of the rapidity of fixation and of shrinkage. In cells quite at the surface of a bit of tissue the spherical discontinuities in the cell protoplasm where it was fairly free from granules were so excessively minute as to be perceptible only in sections not thicker than $\pm 0.6\,\mu$ which were deeply stained with iron-hæmatoxylin. Passing more toward the centre of the mass of tissue one found the discontinuities in the cell substance larger and more obvious. This effect of the rapidity of fixation demands for its elucidation a knowledge of the volume of the cells in the various conditions. This I do not possess. The facts, however, may fairly be set beside the account which Mikosch gives of the changes in the configuration of living cell substance which occur as it is dying.[1] He observed the epidermic cells of various plants—chiefly *Sedum telephium*. The tissue was mounted in water or sugar solution and when first viewed the cell substance appeared to consist of a completely homogeneous matrix which embedded small refractive granules commonly in rows like a string of pearls; some, however, were grouped in pairs, others separate. After 20–30 min. minute vacuoles not more than $1\,\mu$ in diameter appeared in the cell substance. They gradually increased in size so that a net appearance was produced with rounded and polygonal meshes. The granules before mentioned have their arrangement destroyed by this change; they now lie at the nodal points of the net, and the appearance is similar to that usually figured by Bütschli. The vacuoles continue to increase in size, and they burst into one another until finally a structureless mass is produced which encloses still persisting vacuoles.

The same conclusion, namely, that the configuration of the cell proto-

[1] *Verhand. d. Gesell. deutsch. Naturforsch. u. Ärzte* (1895), Th. II, 179.

plasm is determined by the fixative is most clearly shown in the case of the secretory cells of the orbital gland. The facts have already been described and nothing need be added here.

In the case of the giant cells of red marrow the configuration of this cell protoplasm after fixation with corrosive sublimate is such as to show in sections a net, the average diagonal of a mesh being about 0.6–0.7μ. After fixation with osmic vapour the average measurements are at least one-fifth less.

Special mention might be made here of a criticism based upon the structure of oxyphil wandering cells after fixation. On the ground that the granules in these cells form the nodal points of the network Gulland[1] claimed that they could not be of the nature of secretory granules, as had been urged by Hankin, Kanthack, and the writer. If these granules really are secretory granules they would, according to Gulland, occupy the position of paraplastic matter—namely the meshes of the net. The nature of this criticism serves to bring into prominence the shifting artificial nature of the structure in fixed cells, for, as we have seen, the secretory granules of the alveolar cells of the frog's pancreas form the nodal points of the network, while the secretory granules of the orbital gland lie sometimes on the net, sometimes in the meshes, according to the nature of the fixative.

Effect of the thickness of the section upon the image. The very great difference which exists between the image offered by sections which do not vary very much in thickness will scarcely be realised except by the aid of actual observation.

In a section approximately 1μ in thickness through a gut of *Oniscus* which has been fixed with sublimate the cell substance shows as an open structure like the skeleton of a dead leaf. In sections say 4μ and upwards in thickness the appearance presented is that of a coarsely punctate material.

The point was studied in detail in cells of the red marrow fixed with sublimate. The lenses used were the same Zeiss apochromatic with the ocular 18 (*i.e.* × 2250 diameters). Sections of three grades of thickness were compared; the thinnest were about the thickness of the visual field, the next were between two and three visual fields in thickness.[2] The thickest sections were too thick for this method of estimation, but they were roughly twice as thick as the medium sections. Now, in the thinnest sections one has a clear open mesh such as is shown in fig. 20. As the sections increase in thickness one has more and more decidedly the impression that the net at any particular focal plane lies embedded in a faint

[1] *Journ. Physiol.* (1896), xix, 385.
[2] Absolute measurements in μ represent the relation less accurately. It is easy to convince oneself, both by reasoning and practice, that the possible error in absolute measurement is fully equal to the depth of a visual field.

continuous grey ground substance. This impression is due to the haze from the focal planes above and below the one actually in focus. Figs. 20*a*, *b* and *c* are from a cell in a section of medium thickness. The section of the cell was a little more than two visual fields in depth. Fig. 20*a* is a freehand drawing of the upper visual field. Fig. 20*b* a similar drawing of the lower field. Fig. 20*c* is a drawing made with the camera lucida, the focal level being that of the upper visual field.

In addition to the confusion in the image which results from the fog so to speak of material out of focus, there is a possible source of error in what might be called false differentiation by double staining. This may be explained by citing an instance. In many or most cases the bars of a net differ in thickness in different parts of the cell. If a section is stained with iron-hæmatoxylin and the process of washing out be used, the dye is first removed from these finer structures. On double staining with say eosine the fine bars from which the hæmatoxylin has been removed dye red and one then has a net black in some places, red in others. The whole process can be followed in the field of the microscope.

The foregoing paragraphs may be summarised as follows: The study of the behaviour of certain gland cells and of the cells of red marrow, of frog's lymph and of the intestine of the wood-louse, leads me to conclude that the cell protoplasm reacts to corrosive sublimate and osmic vapour in the same way as does a soluble colloid to a reagent which converts it into an insoluble colloid. I hold therefore that there is no evidence that the structure discoverable in the cell substance of these cells after fixation has any counterpart in the cell when living. A large part of it is an artifact. The profound difference in the minute structure of a secretory cell of a mucous gland according to the reagent which is used to fix it would, it seems to me, almost suffice to substantiate this statement in the absence of other evidence. The framework which is visible in fixed cells contains within itself all the solids of the cell; it is produced by the action of the fixing reagent in converting the $\pm$ 10 per cent. of solids in the living cell substance into an insoluble state. The meshes of the framework therefore are mere interstices occupied by alcohol, xylol, or balsam as the case may be. This conclusion is borne out not only by microscopical analysis and pressure experiments, but also by argument from the physical characters of colloidal matter in the soluble and insoluble states.

Though it is impossible to deal with questions of such magnitude in anything other than a suggestive manner it would be unfair to omit pointing out the lengths to which criticism of this kind carries one. I have confined myself in this work to a consideration of those views which ascribe definite coarse structures to the cell protoplasm: but the dead or fixed cell body also holds other structures—the nucleus and the attraction sphere, each of which represents something which is present in the living cell. The same difficulty exists in determining the structure of these bodies during life as exists in the case of the general cell protoplasm—it is the difficulty of deciding to what extent structure demonstrable in the fresh or fixed state is the product of the chemical and physical changes which constitute the death change, or which may be due to the action of fixatives.

In discussing the view of Berthold and Fr. Schwarz according to which the reticular structure of fixed cell protoplasm is an artifact Bütschli appeals to the structure which various workers have described in "living protoplasm". These observations satisfy him that criticism such as was advanced by Berthold and Schwarz and as appears in the pages of this paper is false and "requires no further refutation". It is valueless in face of the fact that net or fibrillar structures "are frequently to be observed quite plainly in the living condition and therefore cannot be any artificially produced appearances of precipitation or coagulation".

To this point of view there are two fundamental objections. The first may stand in the form of a question. As Bütschli has himself contributed to show, fixatives do produce structures in colloids; is not therefore the statement that the structure seen in the fixed cell agrees with that seen in the unfixed cell singularly suspicious? What has become of the structure which must be produced by those complete redistributions of solid and liquid and those chemical changes which are the very essence of the process called fixation? In the second place an examination of the original memoirs convinces me that it is very doubtful whether the structure in question has been observed in actually normal living cells. The discussion of this point, together with an account of some attempts to repeat the more noteworthy observations upon cells regarded as living will form the subject-matter of the second part of this paper.

SUMMARY

1. A study of the action of reagents upon colloidal matter shows that when an insoluble modification is formed there is a separation of solid particles which are large molecular aggregates, and that these become linked together to form a comparatively coarse solid framework having the form of an open net which holds fluid in its meshes.

2. In some cases, however, the reagent is partially miscible with the colloidal mixture. In this case the latter is modified in degree but remains the same in kind. The action of corrosive sublimate upon gelatine is an instance.

3. The general statement is, however, possible that reagents which have any action at all confer a structure upon the colloidal matter which

differs in most cases in kind, in some cases in degree, from the initial structure. Hence it is inferred that the structure seen in cells after fixation is due to an unknown extent to the action of the fixing reagents.

4. The structure of dead matter which was once living may also be referred to the coagulation (clotting) phenomena of death, as well as to post-mortem change. These points were specially examined in the case of the secretory cells of a mucous gland.

5. Incidentally certain investigations which are still in progress as to the phenomena of the colloid state are touched upon. The existence of two classes of colloidal mixtures is noted and the chief features of each are described.

6. The mechanism of coagulation, and of setting, receives special discussion.

EXPLANATION OF PLATE XII

Egg-white. Sections 0·6–1·0 μ thick cut from paraffin; stained with iron-hæmatoxylin. Magnified 1500 diameters; camera lucida.

Fig. 1. Solids per 100 c.c. 13 gr. Fixative sublimate.

Fig. 2. Solids per 100 c.c. 30 gr. Fixative sublimate. *a. a.* Carmine grains.

Fig. 3. Solids per 100 c.c. 60 gr. Fixative sublimate.

Fig. 4. Solids per 100 c.c. 13 gr. Fixative potassium sulphocyanate.

Gelatine. Film preparations fixed with sublimate. Mag. 1500 diameters; camera lucida.

Fig. 5. Solids per 100 c.c. 4 gr.

Fig. 6. Solids per 100 c.c. 10 gr.

Fig. 7. Solids per 100 c.c. 25 gr.

Fig. 8. Solids per 100 c.c. 50 gr.

Fig. 9. Egg albumen. Solids 13 gr. per 100 c.c. Film fixed by steam while flowing between two coverslips—to show abrupt transition from net region to optically homogeneous film. *a, a'*, interior of two contiguous air-bubbles. Mag. 450 diameters; camera lucida.

Fig. 10. Gelatine droplet fixed while flowing over coverslip with sublimate. Stained with iron-hæmatoxylin. Note unstaining optically homogeneous edge film. Mag. 1000 diameters; camera lucida.

Orbital gland. Kitten.

Fig. 11. Fixative absolute alcohol 20 hours. Freehand sections stained with methylene blue in 90 per cent. alcohol and examined in xylol. Framework a brilliant green, shrunken remains of mucous granules a dull opaque blue. Mag. 1500 diameters; camera lucida.

Fig. 12. (*a*) Net in 50 per cent. spirit; intact granules swollen. *Not* camera lucida.
   (*b*) Same net, after dehydration in xylol; methylene blue.

Plate XII

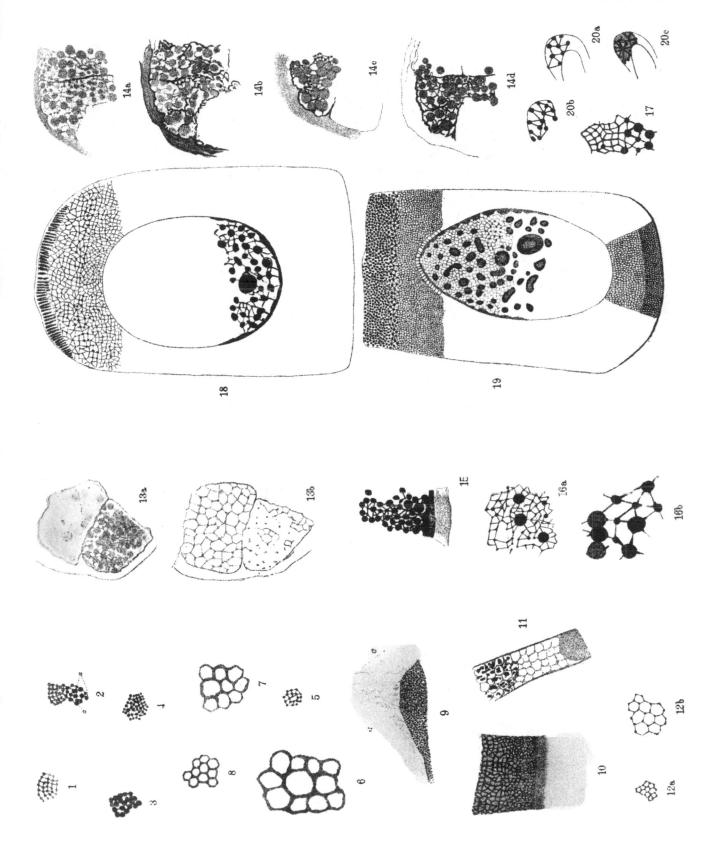

Fig. 13. Fixative osmic vapour 8 hours. Section 2 teeth ($\pm 1\cdot4\,\mu$) cut in paraffin. Mag. 1000 diameters; camera lucida. (a) Appearance after removal of paraffin and mounting in 95 per cent. spirit. (b) After irrigation with lower strengths of spirit down to 30 per cent.; drawn in 30 per cent. Section unstained.

Fig. 14. Fixative osmic vapour 24 hours. Mag. 1000 diameters; camera lucida. Section 1 tooth ($\pm 0\cdot7\,\mu$). Same portion of the alveolus was drawn in each case. (a) section unstained mounted in absolute alcohol after removal of paraffin. Stained with iron-hæmatoxylin while on stage of microscope, brought back into xylol, drawing (b) made. Staining light. Restained with iron-hæmatoxylin— ferric alum 2 per cent. 13 hours; strong hæmatoxylin 24 hours; staining very intense. Brought into absolute, drawing (c) made; irrigated with xylol and drawing (d) made. Illumination moderate, the condenser being achromatic. The change in the image due to the refractive index of the mounting medium is noteworthy. In (c) resolution was difficult owing to the intensity of the stain.

Fig. 15. Orbital gland. Puppy. Section 2 teeth ($\pm 1\cdot4\,\mu$). Fixative osmic vapour 24 hours. Stained with methylene blue saturated in 80 per cent. alcohol. Mag. 1000 diameters. Drawn when mounted in absolute; illumination of field varied, but mostly bright; very careful study with camera lucida of part of a cell.

Fig. 16. Pancreas. Frog. Fixative sublimate. Section 1 tooth ($0\cdot7\,\mu$). (a) and (b) were in the same cell, (a) being almost continuous with (b). Mag. 2250 diameters. Probably not made with camera lucida.

Fig. 17. Pancreas. Frog. Fixative sublimate. Section 1 tooth ($\pm 0\cdot7\,\mu$). Drawing was not made with the camera lucida. Mag. 2250 diameters.

Gut of Oniscus. Animals starved for 6 days.

Fig. 18. Fixative sublimate 2 hours; no exposure to water; dehydrated in alcohol with a trace of iodine. Section 2 teeth ($1\cdot4\,\mu$). Stain, iron-hæmatoxylin. Camera lucida.

Fig. 19. Fixative osmic vapour 2 hours. Section 1 tooth ($0\cdot7\,\mu$). Stain, iron-hæmatoxylin. Camera lucida.

Fig. 20. Oxyphil cell of red marrow. Fixative sublimate. Section between 2 and 3 focal planes thick. Illumination moderate. Mag. 2250 diameters.

> (a) Freehand drawing of upper focal plane.
> (b) Freehand drawing of next deeper plane.
> (c) Combining these two planes and adding a third.

15

ON THE COAGULATION OF PROTEID BY ELECTRICITY

[Journ. Physiol. (1899), XXIV, 288]

The material used in the experiments which form the subject-matter of this paper was the slightly opalescent fluid which is formed when white of egg is mixed with eight or nine times its volume of distilled water, filtered, and boiled. The proteid matter in solution is changed by the heat so that, according to Starke, the greater part presents all the characters of alkali albumen.[1] In its physical characters the solution is indistinguishable from a solution of silicic acid. It belongs to the silicic acid and water type of colloidal mixtures.[2] It contains, in addition to inorganic matter, organic matter which may be separated into two kinds by the passage of a constant current. One of these is the alkali-albumen-like body: it is present as a fine suspension of discrete particles the size of which may be increased by adding any electrolyte. These particles have a limited mobility in the solvent which is dependent solely upon the alkali present. Therefore, if the percentage of alkali is gradually lowered by dialysis against distilled water, or by slowly allowing free acid to diffuse into the fluid, demobilisation occurs with production of a coagulum. This may be called the specific relation of the substance to free acid or alkali as opposed to the generic relation which, in common with all colloidal solutions of the suspension type, it bears to electrolytes. Electrolytes when added to the solution cause an increase in the size of the colloid particles. Different electrolytes vary in the power they possess: three stages in the process have been chosen for measuring this; the development of a clearly visible opalescence[3]; the formation of the hydrogel;[4] and the synæresis of the hydrogel.[5] The results in the first two cases are absolutely concordant. They have been stated in various ways but I propose the following as the simplest method of indicating the relation.

[1] J. Starke, ref. in *Maly's Berichte* (1897), XXVII, 19.

[2] *Journ. Physiol.* (1899), XXIV, 174.

[3] Schulze, "Antimony and Arsenic hydrosulphide", *Journ. f. prakt. Chem.* (1882), XXV, 431; *ibid.* (1883), XXVII, 320.

[4] Prost, "Cadmium sulphide", *Bull. d. l'Acad. Roy. d. Sci. d. Belg.* (1887), Ser. 3, XIV, 312; Picton and Linder, "Antimony hydrosulphide", *Journ. Chem. Soc.* (1895), LXVII, 63.

[5] Spring and Lucion, "Copper hydrate", *Zeits. f. anorg. Chem.* (1892), II, 195.

If we define the coagulative power of a substance as the inverse of the concentration in gram-molecules per litre necessary to convert a given hydrosol into a hydrogel, then we can speak of the specific molecular coagulative power of a salt. If we denote this by k and the number of molecules per litre by m, then for two salts,

$$\frac{k}{k'} = \frac{m'}{m}.$$

The value of k has been found to vary for different salts according to the valency of the metal when the hydrogel produced is not liquefiable by heat. The actual numbers may be obtained directly or by calculation from the papers quoted in the footnotes. Picton and Linder give the following figures: if the value for salts of trivalent metals be taken as 1000, then the value for salts $R''Ac$, $R'Ac$ will be respectively 100 and ± 2.

In the case of colloidal sulphides of arsenic, antimony or cadmium the salts of hydrogen—that is to say free acids—have the coagulative power proper to salts of univalent metals. In this respect the proteid solution is different as we have already noticed. Otherwise my experiments show that the action of neutral salts as coagulants agrees in this case with the law of the valency stated above.

The action of free acid or alkali varies according to the condition of the material. The original opalescent fluid is alkaline in reaction: on dialysing this against distilled water it at length coagulates, and the coagulum may be broken up and suspended in distilled water without solution. On the addition of acid a dispersion[1] of the flakes of coagulum occurs with production of an opalescent fluid having an acid reaction. It therefore is necessary to distinguish three conditions which are according to the reaction with litmus (i) when the fluid is alkaline, (ii) when it is acid, (iii) when it is neutral. The electrical reactions show that the proteid substance is electro-positive or in an alkaline phase when the fluid has an acid reaction, and it is electro-negative or in the acid phase when the fluid has an alkaline reaction. This more exact nomenclature is, however, likely to be confusing if it is used before the actual experiments are described and discussed. I therefore propose to distinguish the three conditions according to the reaction of the fluid to litmus paper.

The similarity of this substance to a solution of silicic acid in water is complete. In the neutral condition it coagulates spontaneously unless the solution be very dilute. The coagulum is liquefied by an exceedingly minute trace of free alkali. One part by weight of caustic soda dissolved in 10,000 of water will liquefy 200 parts by weight of silicic acid in 60 min. at 100° C.[2] The proteid coagulum is perhaps even more sensitive to free

[1] The following pages show that it is inadvisable to use the word "solution" here.
[2] Graham, *Trans. Chem. Soc.* (1864), II, 318.

alkali. If a pin point is just dipped in a strong solution of caustic soda, shaken, and then washed in 5 c.c. of distilled water, an alkaline solution is produced which will rapidly liquefy the coagulum at the boiling-point. Finally the coagulum, like the hydrogel of silicic acid, and unlike the hydrogel of gelatine, is not liquefied by a rise of temperature, but it shrinks with expression of "serum". The coagulum differs from a hydrogel of silicic acid in the fact that it is liquefiable by traces of acid as well as by traces of alkali. The coagulation of a solution of silicic acid is retarded or prevented by free acid although the process of coagulation is not reversed by acid, therefore the difference is probably one of degree rather than of kind. Still, in the ease with which it is liquefied by free alkali, the proteid coagulum reacts as an acid coagulum such as the hydrogel of silicic acid, and in the ease with which it is liquefied by free acid it resembles a basic coagulum such as the hydrogel of alumina or peroxide of iron.

The action of the acid or the alkali in liquefying these various hydrogels cannot be regarded as a simple chemical action such as the formation of a soluble salt, a silicate, or albuminate as the case may be, since the quantity of free acid or alkali necessary to produce the change is so excessively minute.[1]

The minor part of the organic matter present in the boiled fluid is readily soluble and forms a fine film on drying. Supposing the separation which is effected by the electric current to be complete, then the ratio by weight of this matter to the alkali-albumen-like matter would be about 1 to 12.

The Action of a Constant Current

The current was supplied by storage cells at 105 volts. Platinum electrodes were used and the electromotive force was varied from 8, 13, to 105 volts in various experiments. The resistance was always very great. It was not thought necessary to measure it, but the current passing was always less than 0·000001 ampère. The cell employed had the shape of a U-tube with vertical limbs. The bend of the U was narrowed in order to make it easier to withdraw the contents from the limbs without mixing. The narrow and wide parts were joined by funnel-shaped regions each 14 mm. long. The transverse diameter of the limbs was 11 mm. The bending narrowed part was 80 mm. long and had a bore of 3–3·5 mm. It was found that this was wide enough to allow of an axial flow sufficient to neutralise the electrical endosmotic action so that, except when the tube became blocked, no alteration of level in the two limbs was produced by the potential gradient. Important endosmotic action was, however, noticed in the course of the

---

[1] In the parallel case of silicic acid Thomsen's determination of the heat evolved, when different quantities of alkali are added, shows that there is no definite point of neutralisation.

experiments when a coagulum formed so as to block the tube. In one experiment the discharge from a Holz electrical machine was tried. In this case the voltage, as measured by the length of the spark in air, was about 25,000. In this experiment a tube having throughout a transverse diameter of 13 mm. was used.

The fluid has an alkaline reaction. The effect of the passage of a constant current is the formation of an opaque white coagulum about the anode. The density of the coagulum is dependent upon the electromotive force: with 105 volts it is yellowish with transmitted light and of a tough almost rubber-like consistency as though the particles had been impacted with considerable force. When the coagulum forms as it sometimes does in the neck of the narrow part of the tube nearest to the + electrode it is of the dense yellow type, even though the electromotive force be only 13 volts; in the narrow part of the tube the lines of force which join the electrodes are closer together and therefore the intensity of the electric field is greater. The formation of the coagulum about the anode is accompanied by a diminution in the opalescence of the fluid in the cathode limb.

These changes in the aspect of the contents of the U-tube are due to an actual movement of the proteid particles through the liquid from the cathode to the anode. This was determined by estimating the concentration of the solids in the two limbs after a current had been allowed to run for a certain time.

The following figures give the results in one experiment. I may note in passing that all the figures quoted in this paper are comparable, since the same cell, the same electrodes, the same electromotive force, and the same solution of egg-white were used to obtain them.

Exp. A. White of egg beaten thoroughly, to one volume eight volumes of distilled water were added: filtered and filtrate boiled. Placed in dialysing trays in shallow layers a few millimetres deep and dialysed against distilled water in large volume, which was renewed every 24 hours for 6 days. At the close of the third day the fluid was removed and brought to the boiling-point for a few seconds in order to check any bacterial action. The cups were thoroughly rinsed with distilled water. The fluid was again boiled at the close of the sixth day. The optical characters of the fluid were not modified by boiling. The total solids were now found to be 1·4 gr. per 100 c.c. This fluid was used in experiments A, B, and C.

Some of the fluid was placed in a cell, E.M.F. = 105 volts, current < 0·000001 ampère, $T = 17°$ C. After 24 hours a dense coagulum had formed about the anode; the fluid was at the same level in both limbs. The coagulum in the anode limb was thoroughly broken up and the solids in each limb determined.

Limb	Gr. per 100 c.c.
+	2·5
−	0·2

The solid matter from the cathode limb formed a film of a much finer texture than the coarse-grained film produced by the dried solid of the anode limb.

If the coagulum which forms round the anode be broken up with a glass rod so as to convert it into a fine milky suspension it is found that the particles have changed their electrical characters, so that they now move in the reverse direction, namely towards the cathode. There is thus a primary movement of the proteid particles in the direction of the negative stream, and a secondary movement in the direction of the positive stream. The secondary movement is not manifested unless the coagulum be broken up because of the purely mechanical resistance. With a lower electrical pressure, say 8–13 volts, the coagulum is loose enough to allow of the secondary movement, so that after 48 hours a membranous coagulum appears at the commencement of the narrow part of the tube on the side of the cathode—the flocculent coagulum as it were jams at this point on its reverse journey. In this way the proteid particles may be driven out of both limbs and impacted in a dense coagulum occupying a mean position.

The force which drives the proteid particles must be considerable. If one end of the narrow part of the tube be blocked by a plug of glass wool driven home as hard as possible the movement is not arrested though it is retarded. Now a pressure of about an atmosphere does not drive the particles through a similar plug of glass wool in the base of a funnel. The retardation is shown by the fact that with the same voltage—105—the amount of solids in the cathode limb was reduced only from 1·4 to 1·0 gr. per 100 c.c. in 60 hours.

The experiments which I performed furnish the grounds for a qualitative statement only of the influence of the conductivity of the fluid upon the movement of the particles. A comparison of the results obtained with undialysed and dialysed solutions shows that in the former the rate of accumulation in the anode limb is less and the percentage of solids there does not rise to so high a level. This difference is in part due to the fact that the secondary or reverse movement is established sooner.

So far no mention has been made of any movement of the finely divided organic matter which is found in the cathode limb after the proteid particles have passed over to the anode limb. When a plug of glass wool is present in the narrow part of the U-tube the secondary movement forms a coagulum on the anode side which is dense, translucent, and yellow by transmitted light. On the cathode side, however, a much smaller amount of an entirely different coagulum appears. It forms a thin membrane only about 1 mm. thick, of an opaque Chinese white appearance. Sometimes (volts 105) the force of the secondary movement drives the yellow translucent coagulum through the glass wool so that it reaches the cathode side. In this case the opaque white coagulum abuts with a sharp line of junction on the yellow coagulum.

This white coagulum which is so scanty in amount is, I take it, formed by a movement, primary or secondary as the case may be, of the very finely divided organic matter which remains in the clear fluid of the cathode limb after the current has been flowing for say 24 hours.

When an impacted coagulum is formed in the narrowed part of the tube endosmotic phenomena are manifested in a change in the level of the water in the limbs. The fluid rises in the cathode limb so that the flow is from the anode to the cathode. To sum up therefore—in a fluid having an alkaline reaction the proteid particles move from the cathode to the anode, that is in the direction of the negative stream; the water moves from the anode to the cathode, that is to say in the direction of the positive stream. The proteid particles after completing their transit to the coagulum on the anode, or being in the anode region for some time, become altered so that they now tend to reverse their movement and to travel with the positive stream.

The discharge from a Holz machine in passing through the alkaline fluid also produces a coagulum about the anode. In this case the electricity was led into and from the fluid by two platinum wires. The voltage was $\pm 25,000$. A tiny coagulum formed slowly on the point of the positive wire, and in about 30 min. it grew to the size of a small pin's head.

The fluid has an acid reaction. The movements of the proteid particles and of the water are the reverse of those described. The particles now move with the positive stream from the anode to the cathode, the water moves with the negative stream. No attempt was made to demonstrate any secondary reverse movement of the particles.

Exp. D. To a part of the fluid in Exp. A which had been dialysed for 6 days, hydrochloric acid was added until it became opaque white. Acetic acid was then added until the fluid became again translucent. It was then dialysed in a very thin layer against a large volume of distilled water for 24 hours: the distilled water was twice changed. Current run for 24 hours: E.M.F. = 105 volts: current < 0·000001 ampère.

Limb	Solids in gr. per 100 c.c.
+	0·625
−	2·20

In this case also endosmotic action is obvious when a coagulum blocks the tube. The water, however, now flows from the cathode to the anode. Therefore, in both acid and alkaline fluids the movements of the water and of the particles are in opposite directions.

Fluid is neutral. The coagulum which forms when dialysis is pushed far enough was broken up very thoroughly and suspended in distilled water. There is now so little movement of the particles under the influence of a current that it is difficult to detect, and what movement there is, is due to the fact that the material is not absolutely neutral. Thus, if the coagulum be formed by dialysing an acid fluid the particles when suspended in distilled water will move to the cathode.

Exp. C. Remainder of acid fluid used in Exp. B dialysed until it formed a clear hydrogel. This was broken up in distilled water by shaking so that a milky fluid resulted. Some was exposed for 24 hours to the current, the electromotive force being again 105 volts. Fluid in both limbs still milky, though the particles have settled somewhat and to an equal extent in both limbs. This settling is due simply to gravity, seeing that it occurs at the same rate in a tube when no current is passing. The narrow part of the cell where the electrical density is greatest was, however, swept quite free from particles, and a small nodular mass of yellow coagulum had gathered round the cathode.

The movements of the particles and the water seem to be readily explained on the lines of Quincke's theory of electric endosmose. When a current passes through a fluid in a capillary tube there is movement of the fluid. The capillary tube in question may be a single capillary tube of, for instance, glass, or it may be the capillary spaces in a permeable membrane of china clay, or such organic membranes as frog's skin, cat's lung, etc., as in Engelmann's experiments. The movement of the fluid is almost always in the direction of the positive stream; it is, however, determined by the nature of the wall of the capillary tube and the nature of the liquid. Thus water in glass moves with the positive stream, turpentine in glass moves with the negative stream. Similarly solid particles, or minute gas-bubbles, are driven through a fluid by the passage of a current, the movement in this case is almost always in the direction of the negative stream.

An excellent account of the whole subject is given in Wiedemann, *Electricität* (1893), I, 982–1023, ed. 2. In addition to this a special discussion of the mathematical theory is to be found in a paper by Prof. Lamb in the *Brit. Assoc. Report* for 1887, p. 495; and Picton and Linder have described the movement of colloidal particles under the influence of a current in the *Journ. Chem. Soc.* (1897), LXXI, 568.

The explanation of electrical endosmose and of the movement of particles suspended in a fluid through which a current is passing which was furnished by Quincke is based upon the assumption that there is a discontinuity of potential at the surface between the liquid and the solid particles or the wall of the capillary tube; there is thus developed a double layer of positive and negative electricity. The direction of the flow of the fluid or of the movement of the particles therefore is an indication of the nature of the charge which is carried by the one or the other. Thus since water in a glass tube flows with the positive stream the charge it carries at the surface will be positive, that of the glass negative.

The application of Quincke's theory to the experimental facts set forth in the first part of this paper yields very interesting results. When the fluid is alkaline the proteid particles move with the negative stream and therefore they carry a negative charge, while the water moves with the positive stream and therefore carries a positive charge. When the fluid is acid, however, the proteid particles carry a positive charge and they therefore are positive to the water. When the fluid is neutral there is little or no difference of potential between the water and the particles—that is to say, the water and the particles form an electrically homogeneous mass.

The proteid particles therefore have this interesting property that their electrical characters are conferred upon them by the nature of the reaction, acid or alkaline, of the fluid. If the latter is alkaline the particles become electro-negative; and *vice versa*.

The secondary or reverse movement of the proteid particles must on Quincke's theory be a result of a change in their electrical characters as referred to the water, so that whereas they may have been for instance negative to the water, they become positive to the water after a long or short sojourn in the neighbourhood of the anode. This change may I think be referred to the electrolytic action of the current upon the electrolytes present. Owing to this an electro-negative particle which moved to the anode would there reach an acid region in which its character would change from the electro-negative to the electro-positive character. This view is supported by the observation that the reverse movement is more pronounced the higher the concentration of electrolytes and therefore the greater the concentration of acid or alkali about the electrodes. In an undialysed solution the particles do not as a rule reach the electrodes to which they are travelling; they are arrested and form a coagulum at some intermediate point the position of which with regard to the electrodes is determined by the fall in potential in the cell.

The movement of the fluid may be explained as follows. When the tube remains patent throughout the experiment the level remains the same in both limbs. When a coagulum forms so as to block the tube the fluid flows with the positive stream if it is positive to the proteid particles; with the negative stream if it is negative to the latter. Quincke and von Helmholz refer the movement of particles in a fluid which is conveying a current to the contact difference of potential between them, whereby the particle and the fluid are urged in opposite directions. If the particles are free the result will be that they will move through the mass of the fluid; if, however, they become fixed, as happens when they aggregate to a clot which becomes jammed in the tube, the fluid is urged past the particles, though of course in a direction opposite to that of the free particle.[1] Therefore when the particles are electro-positive, as is the case when the fluid is acid, the

---

[1] The statement in the text is purely a qualitative one. I have therefore omitted any mention of the movement of the water in the narrow part of the cell which will result from the difference of potential between it and the glass wall.

water moves with the negative stream, a direction which is most unusual in the electrical endosmose of this fluid.

If the proteid in the boiled solution of egg-white be, as Starke affirms, a representative of the derived albumens, then the experiments perhaps throw a little light on the relation of the proteid to the acid or base in this class.

The proteid molecules in distilled water form large aggregates so that a coarse suspension or a coagulum is the result. A minute quantity of acid or base will disperse these aggregates so that a much finer suspension is produced; and the particles at the same time assume the electrical characters of basic or acid particles respectively. The proteid molecules seem therefore to act as basic or acid particles according to the circumstances in which they find themselves. This conclusion is justified by the relation which Picton and Linder have established, namely, that the direction of the movement of colloidal particles under the influence of an electric current is determined by their chemical nature. Thus basic ferric hydrate moves with the positive stream, acid arsenious sulphide moves with the negative stream.

The oxides of silicon and iron which form colloidal hydrates also act as acid, and as base. Thus from silicon we can get a silicate of potassium, or a sulphide, or tetrachloride of silicon. Similarly there are ferrous or ferric salts, and ferrites and ferrates in which the iron oxide forms the acid ion.

The experiments suggest certain ideas concerning the nature of the directive force which determines the aggregation and arrangement of the colloid particles in the formation of a hydrogel. Graham attributes it to a quality of "idio-attraction" possessed by colloid molecules. Many colloidal solutions are, however, stable unless an electrolyte be added, and in this fact a clue to the nature of the force may perhaps be found.

In speaking of the precipitating action of crystalloids upon colloids, Hofmeister referred the property to the "water-attracting power" of the former, and the phrase has become current in physiological literature. It may I think be put aside once and for all. It is vague and meaningless in that it cannot be expressed in terms of any definite stoichiometrical quantity. Also any possible meaning it can bear must be shared by non-electrolytes, and these apparently possess no aggregating action upon colloid particles. The boiled egg-white solution, for instance, is freely miscible with alcohol.

I endeavoured to arrive at the nature of the action of electrolytes by assuming that if the electrolyte acts by modifying the characters of the solvent it must be possible to find some value for one of their properties such as osmotic pressure, surface tension, viscosity, etc., which is approximately constant for all those solutions which are of the strength which is just necessary to produce coagulation. I calculated the concentration of Schulze's equicoagulative solutions[1] in gram-equivalents and used these figures as the basis for further calculations. No agreement in any values was found for equicoagulative solutions.

The following table shows the values for the product of the number of gram-molecules and the coefficient of osmotic pressure (mi) for chlorides.

Equicoagulative solutions of chlorides of	Gram-equivs. per litre	Calculated value of mi
Am	0·090	0·16
K	0·098	0·18
Na	0·07	0·13
Mg	0·002	0·003
Ba	0·003	0·004
Ca	0·004	0·006
Al	0·0002	0·0003

As little agreement was found in the viscosity of equicoagulative solutions. Thus for the solutions of potassium and barium chloride Wagner's constant gave the respective values 0·0947 and 0·00367. Lastly, a comparison of the way in which surface tension of the free surface of solutions of various salts varies with the valency of the metal shows a complete lack of agreement with coagulative action. Thus for chlorides we have:

	R'	R''	R'''
Surface tension*	1	1·7	4
Coagulative power	1	30	1650

* The value for the constant K and the formula given by Dorsay, *Phil. Mag.* (1897), ser. 5, XLIV, 369, were used.

Also the amount of the salts of bi- or trivalent metals necessary to produce coagulation is so small as to produce negligible changes in the surface tension of the free surface, though it is of course possible that by electrical action the ions may largely alter the surface tension between the colloid particles and the water. The amount of zinc sulphate necessary to produce coagulation of the solution of arsenious sulphide used by Schulze was 0·0037 gram-equivalents per litre. Dorsay by actual measurement with the ripple method found the tension of the free surface of a solution holding 0·0199 gram-equivalents to be 73·17. This value is below that which he obtained for distilled water, namely, 73·27. The conclusion I would draw from these results is that an electrolyte does not act as a coagulant by modifying in any simple manner the stoichiometrical characters of the solvent. The same conclusion indeed might be drawn at once from the effect of valency upon coagulative power. Stoichiometrical values vary generally with valency according to the ratio $x : 2x : 3x$ and not according to the square and the cube as coagulative power approximately does.

[1] *Journ. f. prakt. Chem.* (1882), xxv, 431.

The dependence of coagulative power upon electrical conductivity is, however, very clear. Non-electrolytes are almost if not quite without any coagulative power, that is unless they act chemically. Feeble electrolytes possess an equally feeble power. Cane sugar, absolute alcohol, glycerine, chloral hydrate, boracic acid, benzoic acid, salicylic acid, and tartaric acid are miscible in all degrees with a solution of arsenious sulphide. The conductivity of an electrolyte increases with a rise of temperature, therefore at high temperatures tartaric, boracic, and glacial acetic acids coagulate antimony trisulphide.[1] If Linder and Picton's measurements of the coagulative powers of various acids are set beside the values for the specific molecular conductivity of solutions of the acids holding 1 gr. per litre, the same relation is clearly shown.[2]

Acid	Value of K referred to Al_2Cl_6 as unity	Sp. mol. conductivity when 1 gram-equiv. $= 1000$ c.c.
Hydrobromic Hydriodic Hydrochloric Nitric	0·001	2950
Sulphuric	0·0006	1935
Oxalic	0·0005	578
Phosphoric	0·00007	230

On the ground of this relation Bödlander[3] suggested that the aggregating or clumping action of electrolytes might be electrical in nature. This view indeed follows naturally from the assumption that the ions carry static charges. I constructed a table giving the molecular conductivities of Schulze's equicoagulative solutions for colloidal arsenious sulphide. The values for neutral salts stand remarkably close together, so that in this particular case equicoagulative solutions of these salts are those in which the molecule carries approximately the same electric charge. The difficulty which stands in the way of any simple electrical explanation is the fact that the volume density of the electricity in the interior of the water due to the static charges on the ions must vary so widely in equicoagulative solutions. Thus comparing solutions of potassium and aluminium chloride Schulze's figures would give for the volume densities respectively $\frac{\rho}{\rho'} = \pm \frac{300}{1}$. The coagulative power of electrolytes cannot depend simply on the volume density of electricity in the liquid. For equimolecular solutions the value of this density would, by Faraday's law, be proportional to the valency, and the rapid increase of coagulative power which is observed as the valency rises would be impossible.

[1] Schulze, *Journ. f. prakt. Chem.* (1883), XXVII, 320.
[2] *Journ. Chem. Society* (1895), LXVII, 63.
[3] *Neues Jahrbuch f. Mineralogie* (1893), II, 147.

I owe to the kindness of Mr Whetham a suggestion which appears to remove this difficulty and, in conjunction with the experimental facts described in this paper, seems to afford a basis for a hypothesis which explains the relation of specific molecular coagulative power to valency. It is as follows.

The numbers which represent the specific molecular coagulative power for salts of metals of different valency suggest the ratio $R' : R'' : R''' = x : x^2 : x^3$ or some similar relation.

Now if we assume that in order to produce a certain state of aggregation of a given group of colloidal particles the presence of a critical amount of charge equal to that on a trivalent ion is necessary, and that no such aggregation occurs in presence of a smaller charge, it will need the conjunction of two divalent and of three univalent ions to produce the same result, for the charge on an ion is proportional to its valency.

In a solution where ions are moving freely, the probability that an ion is at any instant within reach of any fixed point is, putting certainty equal to unity, represented by a fraction proportional to the ratio between the volume occupied by the spheres of influence of the ions to the whole volume of the solution, and may be written as AC, where A is a constant and C the concentration of the solution. The chance that two ions should be present together is the product of their separate chances, that is $(AC)^2$, and the chance that three ions should be in conjunction with the colloidal particle is $(AC)^3$.

If we have three solutions whose ions are trivalent, divalent, and monovalent respectively, the probability that one of the first-named ions should be within its radius of action of a given point is AC_1, the probability of two of the second kind being within reach together is $A^2C_2^2$, and the probability of three of the third kind being present together is $A^3C_3^3$, the constant being assumed to be the same in each case within the order of accuracy required.

Now in order that these three solutions should have equicoagulative power the frequency with which the necessary conjunctions occur must be the same in the three solutions. We shall then have

$$AC_1 = A^2C_2^2 = A^3C_3^3 = \text{a constant} = B;$$

therefore
$$C_1 = \frac{B}{A}, \qquad C_2 = \frac{\sqrt{B}}{A}, \qquad C_3 = \frac{\sqrt[3]{B}}{A}$$

or
$$C_1 : C_2 : C_3 = B : \sqrt{B} : \sqrt[3]{B}.$$

Since B is a fraction less than unity, this gives the kind of relation observed, the concentration for trivalent solutions being much smaller than that for an equicoagulative univalent solution.

This result is only approximate; the constants A will not be strictly identical in the three cases; moreover, since the two divalent ions really give a charge greater than that on one trivalent ion the equicoagulative concentration will, in the case of the divalent solution, be less than that deduced above, and it is probable that a nearer approximation to the truth would be given by

$$C_1 : C_2 : C_3 = B : \sqrt[3]{B} : \sqrt[3]{B} = B^3 : B^2 : B.$$

Let us take an example. By choosing a proper value for B, in the case of our first relation $0 \cdot 0000156$, we get $\sqrt[3]{B} = 0 \cdot 025$, and the ratio $C_1 : C_3 = 1 : 1600$. This gives $\sqrt{B} = 0 \cdot 00395$, making C_2 proportional to 253, so that we should have

$$R' : R'' : R''' = 1 : 253 : 1600.$$

But by our second relation, if we adjust B to give the $1 : 1600$ ratio for $C_1 : C_3$, we must take $B = 0 \cdot 025$, when B^2 becomes $0 \cdot 000625$ and B^3 $0 \cdot 0000156$, so that

$$C_1 : C_2 : C_3 = R' : R'' : R''' = 1 : 40 : 1600,$$

a relation agreeing within the limits of error with that given on p. 303 for chlorides, viz. 1 : 30 : 1650.

It seems likely, therefore, that this hypothesis may explain the relation between valency and coagulative power.

The fact that the proteid molecules can have their electrical characters altered suggests a reflection concerning the classification of colloidal mixtures. It can be based upon the relation to temperature or the relation to electrolytes. The former distinguishes the well-defined groups which form heat-reversible, and those which form irreversible or insoluble hydro-gels.[1] From the latter we get a division into hydrosols in which the power of an electrolyte to effect demobilisation depends upon the basic ion; and hydrosols in which it depends upon the acid ion. In all those cases in which the latter relation has been determined the two methods of classification coincide. The number of cases is, however, very small. As they are widely scattered in different journals I have brought them together into the following table:

Substance	Electrical characters	Chemical character	Nature of hydrogel	Prepotent ion
Antimony sulphide	No obs.	Acid	Insoluble	Metal (Schulze)
Arsenic sulphide	Negative (Picton and Linder)	Acid	Insoluble	Metal (Schulze)
Cadmium sulphide	No obs.	Acid	Insoluble	Metal (Prost)
Alkali albumen	Negative (Hardy)	Acid	Insoluble	Metal (Hardy)
Gelatine	Doubtful (Hardy)	?	Heat soluble	Acid (Pascheles)*

* *Pflügers Archiv* (1898), LXXI, 333.

Now in this table we see that in those cases in which the metal ion has been shown to be prepotent the colloid particles react to a constant current or to reagents like acid particles. This may be a coincidence, or it may mean that it is the chemical character of the colloidal matter which determines whether the basic or acid ion is prepotent in modifying the spatial relation of the particles in the fluid. The latter view is supported by the following facts. Silicic acid is electro-negative (Picton and Linder) and it is coagulated by salts but not by free acids (Graham). Ferric hydrate is basic and electro-positive (Picton and Linder), and it is coagulated by alkalis but not by acids, with the exception of sulphuric acid. Hydrate of alumina acts both as acid or base, it is coagulated by acids and alkalis as well as by neutral salts. Spring and Lucion[2] examined the influence of some sulphates and chlorides upon the rate of synæresis (dehydration) of the hydrogel of cupric hydrate. The numbers do not help us, however, since they differ completely from those found when a hydrosol is used. The figures agree with the common stoichiometrical relation in which the effect is directly proportional to the molecular and ionic concentration.

[1] *Journ. Physiol.* (1899), XXIV, 174.   [2] *Zeits. f. anorg. Chem.* (1892), II, 195.

If we assume that the action of an electrolyte is governed by the chemical nature of the colloid particles, then there is no reason why the two classifications indicated at the beginning of this section should coincide and the reversibility or non-reversibility of the hydrogel remain as the only fundamental difference, since it must depend upon a radical difference in the relations of the colloidal solid to the fluid.

The point is of great importance as a guide in the obscure region one is endeavouring to penetrate. In order to throw some light upon it I endeavoured to determine the movement of gelatine in water under the influence of a constant current. If that could be determined then the relation detected by Picton and Linder between the direction of the movement and the acid or basic nature of the substance could be applied. The experiments did not give decisive indications. The action of the current is obscured by the remarkable influence it exerts upon the separation of the solution into conjugates. Spherules $2–6\,\mu$ in diameter with a high gelatine content separate even though the original solution is too dilute to permit of setting. These are repelled from both anode and cathode, so that they aggregate to form a loose jelly in some midway position.

Summary

Under the influence of a constant current the particles of proteid in a boiled solution of egg-white move with the negative stream if the reaction of the fluid is alkaline; with the positive stream if the reaction is acid. The particles under this directive action of the current aggregate to form a coagulum. The physical characters of the material are discussed, and the action of the current is explained and developed on the lines of Quincke's theory of electric endosmose.

A PRELIMINARY INVESTIGATION OF THE CON-
DITIONS WHICH DETERMINE THE STABILITY
OF IRREVERSIBLE HYDROSOLS

[*Proc. Roy. Soc.* (1900), LXVI, 110]

It has long been held that a large number of colloidal solutions are related
to or identical with suspensions of solid matter in a fluid in which the
particles of solid are so small as to settle at an infinitely slow rate. Such
solutions are the colloidal solutions of metals and of sulphides such as those
of antimony, arsenic and cadmium. Such solutions belong to the class of
irreversible colloidal mixtures. A rise of temperature assists the process of
coagulation or precipitation;[1] but neither a further rise nor a fall of temperature will cause the reformation of the hydrosol. On this ground they
may provisionally be classed with such colloidal solutions as those of
silica, ferric hydrate, alumina, etc., and with the modification of the
albumen of white of egg which is produced by heating an aqueous solution
to the boiling-point. I also add to the class, for reasons to be developed in
the following pages, the suspension of mastic in water which is produced by
adding a dilute alcoholic solution of the gum to water.

Looked at from the point of view of the phase rule, the equilibrium in
these hydrosols, if they really consist of minute solid particles dispersed in
a fluid, is not necessarily between the solid particle and water, but between
the solid particle and a solution of the particular solid in water. The
hydrosol of gum mastic gives off a vapour of the gum of a density sufficient
to affect the olfactory organs, and, therefore, the water must contain a
definite quantity in solution. Similarly, as it is probable that no substance
is completely insoluble, we may assume that in all the examples a portion
of the solid is in true solution in the fluid. As the solid which is not in true
solution is dispersed in particles whose diameter is, as a rule, very much
smaller than the mean wave-length of light, it follows that the surface of
contact between solid and fluid is very great for unit mass of the former.
The opportunity for evaporation and condensation of the solid matter of the
particles afforded by the immense surface of contact is so very great that,
although only an immeasurably minute quantity of the solid may be in
true solution at any one time, this quantity, minute though it be, is probably
an important factor in determining the equilibrium between solid and fluid.

---

[1] Elsewhere (*Journ. Physiol.* (1899), XXIV, 172) I have shown that precipitation and
coagulation are not discontinuous processes. Coagulation gives way to precipitation
when the concentration of the solid phase falls below a certain amount.

It is necessary to keep such considerations as these in mind in view of the readiness with which these mixtures have been regarded as simple suspensions[1] in which the only relation between solid and fluid is a mechanical one. These hydrosols are, as a matter of fact, singularly stable when pure. They can, for instance, be concentrated by boiling to a remarkable extent, and their stability depends upon complex relations between fluid and solid, which give the former, so to speak, a definite hold over the latter.

Mode of preparation of the different solutions. The hydrosol of gold was prepared by adding a couple of drops of a solution of phosphorus in ether to about a litre of a very dilute solution of gold chloride. The fine ruby-coloured fluid which was formed was dialysed against distilled water[2] for 14 days, and then concentrated by boiling. The hydrosol of silicic acid was prepared by acting on soluble glass with excess of hydrochloric acid, and dialysing the product. A hydrosol of ferric hydrate was prepared by prolonged dialysis of the solution in ferric chloride.

The hydrosol of gum mastic was prepared by adding a very dilute solution of the gum in alcohol to distilled water. It was dialysed for 14 days against distilled water. The hydrosol of heat-modified egg-white was prepared by dissolving white of egg in nine times its volume of distilled water, filtering and boiling. The result should be a brilliant fluid which scatters blue light. Surface action, however, plays an extraordinary part. If the solution is boiled in a test-tube a milky fluid is formed and a film of proteid is left on the glass; a second quantity boiled in the same test-tube comes out less milky, until, when the proteid film is sufficiently thick to eliminate all action by the glass, the solution after boiling contains the proteid dispersed as particles so small that they scatter pure blue light. After preparation the hydrosol was dialysed against distilled water for some days.

Behaviour of the hydrosols in an electric field. It has long been known that the particles in these colloidal solutions move in an electric field. Zgismondy[3] found that the gold in colloidal solutions moves against the current. Picton and Linder[4] established the important fact that the direction of movement of the particles, as compared with the direction of the current, depends upon their chemical nature. I have shown that the heat-modified proteid is remarkable in that its direction of movement is determined by the reaction acid, or alkaline of the fluid in which it is suspended.[5] An

---

[1] Cf., for instance, Stoeckl and Vanino, *Zeits. f. phys. Chem.* (1899), xxx, 98; also Ostwald, *Lehrbuch.*

[2] In working with these colloidal solutions it is very necessary to use distilled water freed from dissolved carbonic acid.

[3] *Lieb. Ann.* ccci, 29.   [4] *Journ. Chem. Soc.* (1897), lxx, 568.

[5] W. B. Hardy, "The coagulation of proteid by electricity", *Journ. Physiol.* (1899), xxiv, 288.

immeasurably minute amount of free alkali causes the proteid particles to move against the stream, while in presence of an equally minute amount of free acid the particles move with the stream. In the one case, therefore, the particles are electro-negative, in the other they are electro-positive.

Since one can take a hydrosol in which the particles are electro-negative and, by the addition of free acid, decrease their negativity and ultimately make them electro-positive, it is clear that there exists some point at which the particles and the fluid in which they are immersed are iso-electric.

This iso-electric point is found to be one of great importance. As it is neared, the stability of the hydrosol diminishes until, at the iso-electric point, it vanishes, and coagulation or precipitation occurs, the one or the other according to whether the concentration of the proteid is high or low, and whether the iso-electric point is reached slowly or quickly, and without or with mechanical agitation.

This conclusion can be verified experimentally in many ways. If a coagulum or precipitate of the proteid particles made either by the addition of a neutral salt, or by the addition of acid or alkalis, be thoroughly washed, made into a fine mud in an agate mortar, and suspended in water in a U-tube, it rapidly subsides. The establishment of an electric field having a potential gradient of 100 volts in 10 cm. has no influence on the level of water or precipitate in 48 hours. If, now, the smallest possible amount of caustic soda or acetic acid be added, the proteid will commence to move, so that in 20 hours the precipitate will rise in one or other limb until it nearly touches the platinum electrode.

Speaking generally, the hydrosol of ferric hydrate is stable only in the absence of free acids or alkalis or neutral salts. The hydrosol of heat-modified proteid is stable only in presence of free acid or alkali. The hydrosol of gum mastic is readily precipitated by acids, but is stable in presence of any concentration of monovalent alkalis. The general conditions of stability of these various hydrosols, therefore, are very different, yet they agree in manifesting the same important relation between the iso-electric point and the point of precipitation as is shown by the hydrosol of proteid.

In the hydrosol of ferric hydrate the particles are markedly electro-positive. A dilute hydrosol is coagulated by citric acid when the concentration of the latter reaches 1 gram-mol. in 4,000,000 c.c. No matter how small the concentration of the ferric hydrate, the hydrosol becomes cloudy and settles. The rate of settling is, however, slow, being about 1 cm. an hour. In an electric field, having the form of a U-tube, the particles always settle slightly faster from the negative electrode—the acceleration due to the electric field being about 5 mm. an hour. The suspended particles of ferric hydrate show, therefore, an exceedingly slight movement in a direction *opposite* to that which they manifest when in colloidal solution.

In the latter condition they are markedly electro-positive; in the former they are exceedingly faintly electro-negative. An exceedingly faint electro-negative character is also conferred upon the ferric hydrate when the hydrosol is coagulated by ammonia, 1 gram-mol. of the latter being present in 100,000 c.c.

If a fresh gel of silica is broken up in distilled water and carefully washed to free it from still uncoagulated silica, and from impurities, it is completely iso-electric with the water. It becomes markedly electro-negative, however, on the addition of the minutest trace of free alkali.

Gum mastic precipitated from a dilute hydrosol by adding barium chloride until the concentration is 1 gram-mol. in 600,000 c.c. is found to be iso-electric with the fluid. It is markedly electro-negative when in colloidal solution.

Picton and Linder have shown that the particles in these hydrosols gradually grow in size as the coagulation or precipitation point is neared.[1] It might, therefore, be urged that, as the movement of the particles in the electric field is, on Quincke's theory of electric endosmose, due to surface action, the fact that they do not move when in simple suspension as opposed to colloidal solution may be due to the diminution of the impelling force acting on a given volume.[2] This is, however, negatived by the character of the experiments. The addition of a minute amount of free alkali to a mass of particles of coagulated silica which have settled to form a "mud" cannot alter the size of these relatively very large masses to any appreciable extent. And since in the case of ferric hydrate and proteid, the sign of the charge which the particles carry in the electric field is different on each side of the actual point of precipitation, that point must of necessity be an iso-electric point.

If the stability of the hydrosol is dependent upon a difference in electrical potential between the solid particles and the fluid, then one would expect that for at any rate a short distance from the iso-electric point, the stability would vary simultaneously with the variation in the difference of potential. The experimental investigation of this question is beset by many difficulties. At present I know of no way of approaching the iso-electric point other than by the addition of salts, acids or alkalis. One may, therefore, approach the point by the addition of, say, acid or alkali, and use a salt to measure the stability of the system, as in the experiment described later. In such experiments, however, the colloid particles are immersed in a complicated system of three components, the conditions of equilibrium of which cannot

---

[1] *Journ. Chem. Soc.* (1892), LXI, 148.

[2] As a matter of fact, Lamb finds that the velocity of a particle is independent of its size or shape, provided that its dimensions are large compared with the slip, so perhaps the objection scarcely needs discussion. Lamb, *Brit. Assoc. Report* (1887), p. 502.

be arrived at from existing data. The conditions could be simplified by using, say, KHO or H_2SO_4 to approach the iso-electric point, and K_2SO_4 as a measure of the change of stability. A series of determinations with different systems of this kind may afford the requisite measurements.

A direct and conclusive proof that stability does decrease as the iso-electric point is approached was, however, obtained in two ways. The iso-electric point can be approached in the case of the hydrosol of proteid by the withdrawal of either the free acid or the free alkali, as the case may be. As it is neared, the proteid particles increase in size, so that instead of scattering blue light, they scatter white light; thus the surface of contact of fluid and solid gradually diminishes as the point is neared. The second experiment, though not a quantitative one, is very convincing. A hydrosol of gum mastic dialysed as pure as possible is not destroyed by mechanical agitation even when long continued. If, however, a salt is added in an amount so small that it just fails to coagulate the hydrosol, the latter is rendered so unstable that it is destroyed by shaking.

Experiments were made to determine whether the particles actually carry a charge. An electric field which was practically uniform was made by using flat electrodes of the same size, which were placed parallel to one another at the ends of a straight tube. The particles were found to move in all parts of the field; they therefore carry a definite charge which, according to Quincke's theory of the movement of particles in an electric field, would be a surface charge, each particle being surrounded by a double layer of electricity.

Action of salts. The power possessed by salts of destroying colloidal solutions was noticed by Graham. The subject was, however, first accurately investigated by H. Schulze.[1] He showed that the power which various salts possess of precipitating a hydrosol of sulphide of arsenic is related to the valency of the metal, while the valency of the acid has little influence. The increase in the precipitating or coagulating power produced by increase in valency is very great. If coagulative power be defined as the inverse of the concentration in gram-molecules per litre necessary to convert a given hydrosol into a hydrogel, then from Schulze's measurements the coagulative power of metals of different valency is:

$$R' : R'' : R''' = 1 : 30 : 1650.$$

Schulze's conclusions were verified by Prost,[2] who used sulphide of cadmium, and Picton and Linder, who used the sulphide of antimony.[3] The last-named workers added the important fact that a small portion of the coagulating salt is decomposed, the metal being entangled in the coagulum.

[1] *Journ. f. prakt. Chemie* (1882), xxv, 431.
[2] *Bull. de l'Acad. Roy. de Sci. de Belg.* (1887), ser. 3, xiv, 312.
[3] *Journ. Chem. Soc.* (1895), lxvii, 63.

The measurements which I have made with various colloidal solutions both confirm Schulze's results, and bring out a new relation, which may be stated as follows:

The coagulative power of a salt is determined by the valency of one of its ions. This prepotent ion is either the negative or the positive ion, according to whether the colloidal particles move down or up the potential gradient. The coagulating ion is always of the opposite electrical sign to the particle.

The salts employed to determine this point were the sulphates of aluminium, copper, magnesium, potassium and sodium; the chlorides of copper, barium, calcium and sodium, and the nitrate of cadmium. Solutions containing 1 gram-mol. in 2000 c.c. were prepared.

The experiments may be summarised as follows:

Silica dialysed free from chlorides, electro-negative

Temperature 16° C. Concentration of coagulating salt 1 gram-mol. in 120,000 c.c.

Coagulated at once	In 10 min.	In 2 hours	In 24 hours	Still fluid
$Al_2(SO_4)_3$	$CuSO_4$ $CuCl_2$ $Cd(NO_3)_2$ $BaCl_2$	$MgSO_4$	K_2SO_4 Na_2SO_4	$NaCl$ Control

This illustrates many experiments.

Proteid in presence of trace of alkali, electro-negative

Temperature 16° C. Coagulating salt 1 gram-mol. in 80,000 c.c.

Coagulated at once	On slightly warming	Did not coagulate
$Al_2(SO_4)_3$ $Cd(NO_3)_2$ $CuSO_4$ $CuCl_2$	$MgSO_4$ $BaCl_2$ $CaCl_2$	Na_2SO_4 K_2SO_4 $NaCl$

Proteid in presence of trace of acetic acid, electro-positive

Coagulated instantly	No effect
$Al_2(SO_4)_3$ $CuSO_4$ K_2SO_4 Na_2SO_4 $MgSO_4$	$CuCl_2$ $Cd(NO_3)_2$ $BaCl_2$ $NaCl$

Mastic, dialysed, neutral, electro-negative

Temperature 16° C. Concentration of coagulating salt 1 gram-mol. in 50,000 c.c.

Coagulates at once	No coagulation
$Al_2(SO_4)_3$ $CuSO_4$ $CuCl_2$ $Cd(NO_3)_2$ $MgSO_4$ $BaCl_2$	K_2SO_4 Na_2SO_4 $NaCl$

Ferric hydrate, dialysed, neutral, electro-positive

Temperature 16° C. Coagulating salt 1 gram-mol. in 100,000 c.c.

Coagulates at once	Does not coagulate
$Al_2(SO_4)_3$	$CuCl_2$
$CuSO_4$	$Cd(NO_3)_2$
$MgSO_4$	$NaCl$
K_2SO_4	$BaCl_2$
Na_2SO_4	

Gold, dialysed for 14 *days against distilled water, very faintly acid,*
electro-negative

Temperature 16° C. Coagulating salt 1 gram-mol. in 200,000 c.c.

Red changes to blue* instantly	No change
$Al_2(SO_4)_3$	$NaCl$
$CuSO_4$	Na_2SO_4
$CuCl_2$	K_2SO_4
$Cd(NO_3)_2$	
$MgSO_4$	
$BaCl_2$	

* The relation of the colours of hydrosols of gold to the size of the particles has been investigated by Stoeckl and Vanino (*Zeit. f. phys. Chem.* (1899), xxx, 98). The change from red to blue indicates an increase in the size of the particles.

Only one comment on these experiments is needed. Solutions of $Al_2(SO_4)_3$, $Cd(NO_3)_2$, $CuCl_2$, and $CuSO_4$ are acid to litmus, while $MgSO_4$ and $BaCl_2$ are neutral to litmus, but acid to phenolphthalein. This acidity has a disturbing action in some cases—the system acts not only as a neutral salt, but also as a free acid. Thus the hydrosol of proteid when brought very near to the point of precipitation by dialysis is more sensitive to the more acid than to the less acid salts of the bivalent metals. The effect of the acid or basic reaction of the salt on the hydrosol is as a rule small compared with the effect of the metal ion. Thus the stability of a hydrosol of electro-positive proteid is increased by free acid, yet the acid salts find their proper place in the scale of valency. Again, ferric hydrate is coagulated by nitric acid when the concentration reaches 1 gram-mol. in 2500 c.c.; yet the cadmium salt of this acid is not much more potent than the "neutral" salts $MgSO_4$, $BaCl_2$.

Temperature 16° C. Concentration necessary to coagulate ferric hydrate

Salt	
K_2SO_4	1 gram-mol. in 4,000,000 c.c.
$MgSO_4$	,, 4,000,000 ,,
$BaCl_2$	,, 10,000 ,,
$NaCl$	,, 30,000 ,,
$Cd(NO_3)_2$	,, 50,000 ,,

The extraordinary rise in coagulative power with an increase in valency, which was observed by Schulze, Prost, and Picton and Linder, holds in all cases. In order to measure it for ferric hydrate, I used Schulze's method, in which a drop of the hydrosol is allowed to fall into a large volume of the solution of the salt. A number of experiments were made until the concentration of the salt was found which just sufficed to coagulate the drop. In the case of gold and mastic the process was reversed, the salt solution being added drop by drop to a measured quantity of the hydrosol. I append the results:

Temperature 40° C. Gum mastic, neutral.

Salt	
$BaCl_2$	1 gram-mol. in 86,000 c.c.
NaCl	,, 8,000 ,,
$MgSO_4$	,, 68,000 ,,
K_2SO_4	,, 8,000 ,,

Temperature 16° C. Gold, very faintly acid.

NaCl	1 gram-mol. in 72,000 c.c.
$BaCl_2$	,, 500,000 ,,
K_2SO_4	,, 75,000 ,,

The figures for ferric hydrate have already been given. It has been pointed out that if specific molecular coagulative power be defined as the inverse of the volume occupied by 1 gram-mol. of a substance when it just suffices to bring about coagulation, then this value (K) varies with the valency of the active ion approximately according to the square and cube:

$$R' : R'' : R''' = K : K^2 : K^3.$$

The relation really is not as simple as this; it is complicated by the change which the specific molecular conductivity of a salt undergoes with change in concentration. The theoretical considerations have been dealt with elsewhere.[1] For convenience of description, however, I will call this relation the relation of the square and cube.

Action of acids and alkalis. The values for K furnished by these substances show relations to valency even more interesting than that found with salts. As in the case of salts, their action is entirely dependent upon the electric properties of the colloid particles.

When the colloid particles are electro-negative, alkalis either do not cause precipitation at any concentration, or if they do cause precipitation the value of K does not vary in any simple way with variations in valency.

When the particles are electro-positive, K increases with valency, but the relation of the square and cube does not hold. Instead, one finds that K varies directly with the chemical activity of the solution.

[1] Hardy and Whetham, *Journ. Physiol.* (1899), XXIV, 288, and Whetham, *Phil. Mag.* (November, 1899).

Acids have the reverse relations. When the particles are electro-negative, the value of K varies directly with the chemical activity of the solution; while if these particles are electro-positive, acids either have no precipitating power, or if K has any value, then (in the particular case measured) the value varies with valency according to the square and cube.

The various measurements are brought together in the following table. The specific conductivities were calculated from the British Association tables.[1]

Coagulating solution

Hydrosol		Concentration necessary to produce coagulation		Specific conductivity of coagulating solution at $18° \times 10^{13}$	Temperature °C.	
		1 gram-mol. in c.c.	Gram-equiv. per litre			
Mastic, electro-negative	Ammonia	0	∞	—	16–100	
	NaOH	0	∞	—	16–100	
	KOH	0	∞	—	16–100	
	$Ba(OH)_2$	40,800	0·048	100	16	
	H_3PO_4	194,400	0·015	13·9	17	
	Acetic acid	1,360	0·7	12·6	17	
	HCl	260,000	0·004	14·5	17	
	HNO_3	260,000	0·004	14·3	17	
	H_2SO_4	460,000	0·004	13·2	17	
	Oxalic acid	220,000	0·009	14·4	17	
	$BaCl_2$	86,000	0·022	[20]	40	The values for specific conductivity are given for a temperature different from that of observation, but they serve to show the general relation
	$MgSO_4$	68,000	0·028	[18]	40	
	NaCl	8,000	0·12	[110]	40	
	K_2SO_4	8,000	0·24	[250]	40	
Gold, electro-negative	Ammonia	0	∞	—	17–100	
	NaOH	11,800	0·08	152	17	
	KOH	10,800	0·09	189	17	
	$Ba(OH)_2$	12,000	0·16	—	100	Solution saturated at 17° C. has no action
	$Ca(OH)_2$	—	—	—	17	No action when saturated
	HCl	123,000	0·008	29	17	
	H_2SO_4	238,544	0·0084	26	17	
	NaCl	72,000	0·013	13	17	
	$BaCl_2$	500,000	0·004	4·4	17	
	K_2SO_4	75,000	0·026	28	17	
Ferric hydrate, electro-positive	KOH	1,000,000	0·001	2·2	16	
	$Ba(OH)_2$	2,000,000	0·001	2·3	16	
	HCl	1,800	0·5	1650	16	
	HNO_3	2,000	0·5	1589	16	
	H_2SO_4	1,000,000	0·002	6·8	16	
	Oxalic acid	1,000,000	0·002	3·4	16	
	Citric acid	4,000,000	0·0007	[0·7]	16	Specific conductivity by analogy with similar acids will not be greater than the value given
	K_2SO_4	3,200,000	0·0006	0·77	16	
	$MgSO_4$	4,000,000	0·0005	0·5	16	
	$BaCl_2$	6,000	0·3	255	16	
	NaCl	20,000	0·5	28	16	

[1] T. C. Fitzpatrick, "The electro-chemical properties of aqueous solutions", *Brit. Assoc. Report* (1893).

The figures in the fourth column are very remarkable. When the particles are electro-negative, equicoagulative solutions of acids agree in their electric conductivity within the limits of experimental error. The same relation is clearly shown if one takes the measurements which Picton and Linder made of the power possessed by acids of coagulating the hydrosol of arsenious sulphide.

Acid	Value of K referred to Al_2Cl_6 as unity	Sp. mol. conductivity when 1 gram-equiv. = 1000 c.c.
HBr HI HCl HNO₃	0·001	2950
H₂SO₄	0·0006	1935
Oxalic	0·0005	578
H₃PO₄	0·00007	230

When, however, the particles are electro-positive, the conductivity of equicoagulative strengths of acids varies to a remarkable extent.

Acids	H′	H″	H‴
Mastic, electro-negative	12·6	14·4	13·9
Ferric hydrate, electro-positive	1650	6·8	0·7

Now specific conductivity (C) has the relation

$$C = n\alpha\,(u + v),$$

where α is the fraction of the total number of molecules (n) which are dissociated at any one moment, and $u + v$ is the sum of the velocities of the two ions. The factor $u + v$ plays an important part, as will be seen by comparing the values for $n\alpha$ in equicoagulative solutions of acids with slowly moving ions with those with rapidly moving ions:

	$n\alpha$
H₃PO₄	0·01
Acetic acid	0·07
HCl H₂SO₄ HNO₃	0·004

This, however, is probably partly due to the fact that owing to the manner in which the coagulative power was measured, time has practically a constant small value. The values for n might, perhaps, be different if the duration of the experiments were prolonged indefinitely.

The important point, however, which calls for notice is that the function $\alpha\,(u + v)$ is a numerical measure of the chemical activity of the substance at a given concentration, so that we reach the important conclusion that *the concentration of acids necessary to coagulate electro-negative colloid particles, and of alkalis necessary to coagulate electro-positive particles, is determined by the laws which govern ordinary chemical equilibrium.*

In the case of the action of salts on these hydrosols, the relation is not so simple. K does not vary directly with $\alpha\,(u+v)$, but contains a factor which is approximately squared or cubed by a change from a monovalent to di- or trivalent ions. The relation can therefore be best expressed as

$$K = n\alpha\,(u+v)\,A^{x},$$

where x is positive and increases rapidly with an increase in the valency of the ion whose electric charge is of the opposite sign to that on the particles.

I should interpret these relations by the suggestion that in the former the acid or alkali alters the difference of potential at the surface of the particles by altering the character of the fluid, and in that way modifies the stability of the hydrosol; in the latter the active ions of the salt act directly upon the solid particles, or, perhaps, on the charge which these carry, and thus play a part which is, perhaps, generally similar to the action of ions when they furnish nuclei for the condensation of vapour. Picton and Linder have shown that the active ions are actually entangled in, and form part of, the coagulum.[1]

The former relation may profitably be placed beside Brühl's conclusions that the action exerted by a fluid upon the substance dissolved in it is determined by the chemical characters of the former, as well as of the latter. He has shown that the molecular refraction, the dielectric coefficient and the power possessed by the fluid of dissociating or chemically changing the molecules of the substance dissolved in it are measured by the unsatisfied valency, or, to use another phrase, the residual energy of its molecules.

The action of acids or alkalis on a hydrosol, the particles of which are of the opposite electrical sign, seems to be compounded of these two actions. The acid or alkali may act as a salt, and exhibit the characteristic relation between K and the valency of the ion of the opposite electrical sign. An instance is furnished by the action of various acids on ferric hydrate. Or the acid or alkali by *increasing* the difference of potential between the fluid and the solid particles may increase the stability of the hydrosol. This is markedly manifested by the increased stability given to the hydrosol of gum mastic by the addition of univalent alkalis. In the action of barium hydrate on this hydrosol, the segregating action of the metal ion overcomes the action exerted by the reagent in virtue of its alkalinity, the result is that the coagulative concentration of the alkali $Ba(OH)_2$ gives a value for K which is less than that given by salts of bivalent metals, and the specific conductivity of the solution is of the same order as that of the coagulating concentration of salts of univalent metals. Against these suggestions,

---

[1] *Journ. Chem. Soc.* (1895), LXVII, 63.

however, must be set the anomalous relations of the various alkalis to the hydrosol of gold.

Action of a salt in presence of varying amounts of acid or alkali. This was measured for one salt only, potassium sulphate, the colloidal solution being gold. The figures are as follows:

Temperature 16° C.

	Concentration 1 gram-mol. in c.c.	Concentration of the salt necessary to produce blue tint. 1 gram-mol. K_2SO_4 in c.c.
Acetic acid	1,087	0
	16,000	324,000
	66,000	64,000
	330,000	50,000
	(neutral)*	28,500
Ammonia	113,333	10,000
	22,666	9,000
	4,900	20,000
	2,450	24,000
	980	20,000
	200	Large amount of salt needed
	100	Salt unable to act when saturated at 16° C. or at 100° C.

* Except for a faint acid reaction of the gold solution, due probably to a trace of phosphoric acid.

These results are shown in Fig. 1. Ammonia alone will not aggregate the particles of gold. Up to a certain point, however, it decreases the stability of the system.

The conclusions can be summarised as follows: The irreversible hydrosols which have been investigated are systems composed of solid particles dispersed through a solution of the substance of the solids in the water.

The stability of the system is related to the contact difference of potential which exists between the solid and the fluid phases, and which forms round each solid particle a double electrical layer. Such double electric layers round particles of any kind immersed in a fluid would resist any movement of the particles through the fluid, because, as Dorn's experiments show, electric work is done in displacing the particles.[1] The effect would be the same as if the viscosity of the fluid was increased.[2]

The stability of the system may be destroyed by altering the difference of potential. Free acid, added to a hydrosol in which the particles are negative to pure water, will diminish the relative difference of potential of the water. In this case the reagent acts directly on the water, and the coagulative activity of unit mass of the substance varies directly with its chemical activity when dissolved in water. The same relation seems to hold

[1] *Wied. Ann.* (1880), x, 70.
[2] This mode of stating the result I owe to Prof. J. J. Thomson, and I gladly acknowledge his kindness in discussing this and kindred points with me.

when free alkali is added to a hydrosol in which the particles are electro-positive.

The stability of the system may also be destroyed by induction, the active agents being free ions carrying a static charge.[1] In this case the action may be said to be on the particles, or rather on the electric layers immediately around them, and the active ions are those whose electric sign is the opposite of that of the charge on the surface of the particles. In

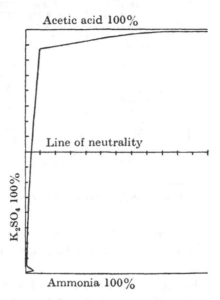

Acetic acid 100%

Line of neutrality

K$_2$SO$_4$ 100%

Ammonia 100%

Fig. 1. Action of potassium sulphate in presence of varying amounts of acetic acid or ammonia upon the hydrosol of gold. The abscissæ represent the volume of water which holds 1 gram-mol. of the salt. The positive ordinates represent the reciprocals of the volume which holds 1 gram-mol. of the acid, and the negative ordinates the reciprocals of the volume which holds 1 gram-mol. of the alkali. Each division = 50,000 c.c.

this case coagulative power does not vary directly with variations in chemical activity. It rises exceedingly rapidly with a rise in the valency of the active ion, so that the relation

$$I' : I'' : I''' = n : n^2 : n^3$$

is approximately satisfied.

Picton and Linder have shown that when the concentration of the salt is insufficient to completely destroy the system, it is not wholly without action. A fresh point of equilibrium between solid and fluid is reached by an increase in the size of the particles and therefore a diminution in the extent and curvature of the surface of contact. The fact is of importance, because it introduces us to the possibility that the reagent may affect the

---

[1] Whetham, *Phil. Mag.* (November, 1899).

size of the particles by altering the equilibrium between the part of the solid in solution and the part in suspension. Double electric layers round each particle are, according to Thomson, separated by a region of "uncompleted chemical combination" between the components.[1] The density of the field round the particles in hydrosols will therefore be a measure of the velocity of the solution and condensation between the particle and the liquid, and therefore the factor which determines whether the particles will on the whole grow, diminish, or remain stationary in size.

When acids or alkalis are added to hydrosols holding particles of the opposite electric sign to themselves, the simplest relation seems to be that univalent acids or alkalis increase the stability; bivalent acids or alkalis decrease it.

The view advanced in this paper implies that each particle in a hydrosol is surrounded by a zone in which the components are in a condition of chemical instability. According to Rayleigh,[2] such a zone is of finite thickness, and deep enough to contain several molecules. We therefore have in these hydrosols two phases, separated by a layer of extraordinarily large extent, which possesses considerable chemical energy. This, it seems to me, suggests an explanation of the catalytic powers so markedly manifested by hydrosols.

[1] *Discharge of Electricity through Gases* (Scribner, 1898), p. 24.
[2] Thomson, *loc. cit.* p. 26; Rayleigh, *Phil. Mag.* (1892), XXXIII, 468.

17

ON THE MECHANISM OF GELATION IN REVERSIBLE COLLOIDAL SYSTEMS

[Proc. Roy. Soc. (1900), LXVI, 95]

Speaking generally, colloidal matter occurs in three conditions:

(1) As fluid mixture, colloidal solutions, or sols, as Graham called them;

(2) Solid mixtures of fluid and solid, the gels; and

(3) Solids, such as dry silica or dry glass.

The property of forming gels is not possessed by all those mixtures which have been classed as colloids. Some only form slimes, which even to the point of actual drying retain the fluid property of flowing. Serum albumen and water is an instance.

Those which form gels fall into two well-defined classes, according to whether the change from the sol to the gel is, or is not, reversible by a reversal of the conditions which produce it. Silica and water may be taken as the type of the latter, gelatine and water of the former. When a hydrosol of silica forms a hydrogel, the latter is "insoluble". To this class belong hydrosols of metallic hydrosulphides and oxides. A hydrosol of gelatine sets to a hydrogel by lowering the temperature; the process is, however, reversed when the temperature is again raised. As the inner mechanism of the gelation of the hydrosols must differ in the two cases, since in the one irreversible, in the other reversible molecular aggregates are formed, I propose to distinguish the processes by different names. The production of an insoluble gel I will call "coagulation", of a soluble gel "setting". This nomenclature is in accordance with general usage.

Temperature is the most potent factor in determining whether a mixture which forms reversible gels is in the sol or gel state. There is also a limiting concentration of the solid below which the gel state is impossible at any temperature.

"Setting", as a rule, follows on a fall of temperature. Caseine, the chief proteid of milk, furnishes, I believe, the only known exception. In the presence of a small quantity of free alkali it forms a hydrosol. When a small quantity of a solution of calcium chloride or nitrate is added to this, a mixture is produced which forms a hydrogel on warming, and which reforms the hydrosol on again cooling.[1]

---

[1] Sydney Ringer, *Journ. Physiol.* (1890), XI, 464.

PART I. REVERSIBLE COLLOIDS

Systems containing two or three components occur, that is, binary or ternary mixtures. The binary system, agar-and-water, was studied at considerable length; and initial experiments were made with the ternary systems, gelatine-water-alcohol, gelatine-water-mercuric chloride, and agar-water-alcohol. These mixtures are homogeneous when heated, but, on cooling, there occurs a division into two fluid phases. In the binary systems and in the ternary system agar-water-alcohol, the conjugate phases have approximately the same refractive index. In the ternary system, gelatine-water-alcohol, the refractive index of the one phase differs so much from that of the other as to permit of direct microscopical investigation of the form of the surface which separates the phases. For this reason I propose to treat this ternary system first.

The Ternary System—Gelatine-water-alcohol

When 13·5 gr. of dry gelatine are dissolved in 100 c.c. of a mixture of equal volumes of absolute alcohol and water, a system is produced which is clear and homogeneous at temperatures above 20° C. As the temperature falls below this limit a clouding occurs, which I find to be due to the appearance of fluid droplets which gradually increase in size until they measure 3μ. On cooling further these fluid droplets become solid, and they begin to adhere to one another. In this way a framework is built up, composed of spherical masses hanging together in linear rows which anastomose with one another. The framework, therefore, is an open structure, which holds the fluid phase in its interstices. The macroscopical result of the change is the conversion, with falling temperature, of the fluid into a loose gel. The droplets can be readily separated from the interstitial fluid by the help of a centrifugal machine.

The phenomena above described undoubtedly depend upon the separation of a homogeneous mixture into two phases owing to a fall of temperature. Each phase contains water, alcohol and gelatine, and the system may be described as a conjugate composed of a fluid solution of gelatine in a mixture of water and alcohol, and a solid solution of water with a trace of alcohol in gelatine. Both phases, however, are fluid within a small range of temperature. The surface of separation of the two phases is curved, and at first discontinuous, and owing to the small size of the droplets it is very large.

When the gel is heated the two phases again mix to form a clear fluid. Owing, however, to the fact that the droplets adhere to one another, they tend to fuse as temperature rises, so as to form irregular masses of viscous fluid, which are separated from the other phase by a surface of no particular shape. The irregular form of this surface, and the ease with which it is modified by any chance slight currents, show that at this stage the surface tension between the two phases must be exceedingly slight. In order to simplify the description I shall call that phase which separates as small spheres the "*internal phase*".

The concentration of the gelatine in the mixture exerts a very remarkable influence upon the configuration of the hydrogel. When it is present in large quantity the internal phase is less viscous and of smaller gelatine content than the external phase, and on cooling it is the external phase which becomes a solid solution. The effect of increasing the proportion of gelatine above a certain amount is therefore very striking—it, so to speak, turns the system inside out, so that the gel is composed of a continuous framework of solid solution, out of which are hollowed spherical spaces filled with fluid. The general mechanical properties of the gel, built on this plan, naturally differ very much from those of a gel with a small proportion of gelatine, which consists of an open framework of solid holding fluid in its interstices.

A mixture of gelatine, water and alcohol is a ternary mixture which resembles a mixture of benzene, acetic acid and water. In each there are two immiscible substances and a common solvent. The immiscible substances are gelatine and alcohol in the one case, and benzene and water in the other, while the common solvent is water in the former and acetic acid in the latter case. In both systems the solubility of the immiscible substances in the common solvent varies widely. Thus acetic acid and water, and water and alcohol, mix readily with rise of temperature; while acetic acid and benzene and water and gelatine mix freely only when the temperature is above 15° C. in the former case, and above 40° C. in the latter case. Duclaux's researches[1] show that in ternary mixtures having this last characteristic the distribution of the constituents in the two phases varies widely with variations in the composition of the whole mass.

These different characters are illustrated by the following figures, which give the amount of gelatine present in grams per 100 c.c. They are, however, only approximate for the solid phase, owing to the difficulty in separating it completely from the fluid phase.

Temperature 15° C.

Total mixture	Internal phase	External phase
6·7	17·0	2·0
13·5	18·0	5·5
36·5	8·5	40·0

The temperature at which the internal and external phases in this ternary system mix was found to be altered by altering the ratio of the

[1] *Ann. de Chim. et de Phys.* (1876), sér. 5, VII, 264.

masses of the components. Increasing the proportion of either of the two immiscible components, alcohol or gelatine, was found to raise the temperature, while an increase in the proportion of the common solvent water was found to lower it.

The curvature of the surface which separates the phases was found not to be constant for a given mixture. The internal phase formed droplets which were large or small according to whether the mixture was cooled slowly or rapidly. Thus with a mixture containing 13·5 gr. gelatine per 100 c.c. the droplets (of solid solution) were very regularly $3\,\mu$ in diameter when about 20 c.c. was allowed to cool slowly in air. Cooled rapidly, however, in an ether spray, the droplets were so minute as barely to be visible with a magnification of 400 diameters. The effect of the rate of cooling is the same when mixtures with a large gelatine content are used, and when, therefore, the internal phase is a fluid solution at ordinary temperatures. When cooling is very rapid the droplets are excessively minute; when it is slow they may be as large as $10\,\mu$ in diameter (gelatine 36·5 per cent. of the mixture). One can therefore make the general statement that *the more slowly the division into two phases occurs the smaller and less curved is the surface of separation.*

The effect upon the structure of the rate at which a fresh condition is imposed upon the system is manifested in a very striking way when an already formed gel is cooled. The experiments upon the effect of temperature on the composition of the two phases in the case of the hydrogel of agar show that when heat is added to or taken away from the system the balance of the phases is altered, water, and perhaps agar, passing from the one to the other. It might be expected that this would take place solely by the passage of material across the surface which separates the two phases. The study of the ternary mixtures, however, makes it clear that a new approximate equilibrium may be reached in two distinct ways.

When a portion of the hydrogel of gelatine-water-alcohol is cooled slowly from 16 to, say, 3 or 4° C., one can see with the microscope no change beyond an alteration in the size of the droplets already present, that is to say, the fresh (approximate) equilibrium is attained by exchange across the surface which separates the phases. But if the cooling is rapid, say a fall of 10° C. in a few minutes, a secondary system of small droplets appears.

In all the mixtures which I examined these were formed in the external phase. Thus, when the concentration of gelatine in the whole mass was low, it was the fluid phase which underwent a division into secondary phases; when it was high, it was the solid phase. To put this fact in a general way, one can say that *when the hydrogel is exposed to a rapid fall of temperature the phase which lies on the convex side of the surface of separation undergoes division into two secondary phases.*[1] When the temperature is again allowed to rise these secondary phases fuse before there is any obvious change in the relation of the primary phases.

When once formed the phases have considerable stability. If the droplets are composed of a solid solution one may, by the addition of water, cause them to increase to relatively vast dimensions without their being destroyed, as they increase in size their refractive index approximates more and more to that of the external phase until finally they are lost sight of. The addition of alcohol, however, once more brings them into view and causes them to shrink. Owing to this stability once a configuration is established one has to far overstep the conditions of its formation in order to destroy it. This would account for the remarkable hysteresis observed in reversible gels. Thus a 10 per cent. solution of gelatine in water sets at 21° C. and melts again at 29·6° C., and solutions of agar in water set at temperatures about 35° C. and melt at temperatures about 90° C. Similarly with the ternary mixtures. In one holding about 35 per cent. gelatine, the internal and external phases separate at 20° C., but they mix again only at 65° C. When water is added to a ternary mixture so as considerably to swell the droplets the system is unstable, and the two phases mix at once when it is mechanically agitated.

The properties of the ternary system, alcohol, gelatine and water, are the following:

(i) Below a certain temperature it exists in two phases separated by a well-defined surface. The temperature at which the separation occurs depends upon the relative proportion of the components in the mixture. Increasing the proportion of gelatine raises it; as does also an increase in the proportion of alcohol. An increase in the proportion of the common solvent, water, however, lowers the temperature at which the biphasic character develops.

(ii) Both phases are at first fluid; with further fall in temperature one becomes solid.

(iii) The surface of separation is curved and discontinuous. In some cases, strictly as a secondary change, the discontinuous masses of the internal phase become continuous with one another.

(iv) The more slowly the two phases are established the less is the surface which separates them both in extent and in curvature.

(v) The solid solution phase is formed sometimes on the concave, sometimes on the convex side of the surface of separation. The former happens when the proportion of gelatine is small, the latter when it is large.

It follows from the last (v) of these properties that a hydrogel may be

[1] The formation of the secondary phases therefore occurs in that one of the primary phases which is under the lower hydrostatic pressure.

built on two very different plans. It may consist of a solid mass containing spherical fluid droplets, or of solid droplets which, by hanging one to the other, form a framework in the spaces of which fluid is held. These two types present important mechanical peculiarities. The former is firm and elastic, and it maintains its structural integrity even under high pressure. The latter is much more brittle, and manifests a tendency to spontaneous shrinking, which is due to a continuous increase in the surface of contact or possibly union between droplet and droplet. These gels with an open solid framework therefore specially manifest that property of spontaneous shrinkage to which Graham applied the term "synæresis".

In the building of a hydrogel of the second type two distinct events occur. The first is the separation of droplets, which rapidly become solid; the second is the linking of these droplets together to a pattern so that they build a framework throughout the fluid phase. The first is the separation of a homogeneous mass into two phases; the second is a phenomenon akin to the grouping of particles which are suspended in a fluid. It is probable that these two events are not directly connected with one another.

Binary Mixtures (Agar-water)

I know of no binary reversible system in which the optical characters of the two phases differ sufficiently to permit of direct microscopical investigation of the surface of separation. It is, however, easy to prove that in such a system as agar and water the property of gel building is dependent upon the appearance of two phases.

The agar which was used was prepared from commercial agar as follows. The strips were suspended in a large volume of distilled water for 24 hours; the water was then drained off, and a large part of the water absorbed by the strips was squeezed out by a powerful press. The strips were again suspended in distilled water, and again drained and squeezed after 48 hours. This washing with distilled water was continued for some weeks. The strips were then melted and the hydrosol filtered, and the filtrate allowed to set. The clear hydrogel so obtained was sliced and suspended for a further period of weeks in many changes of distilled water. In this way a colourless gel was obtained free from all foreign diffusible bodies. It was not found necessary to take precautions against micro-organisms. With the removal of the salts the agar ceased to afford them a suitable nidus.

Effects of Pressure upon the Hydrogel of Agar

When gels containing 1–3 per cent. solids[1] are broken up and slightly squeezed by hand a fluid exudes. In order to collect this fluid a screw press

---

[1] By this is meant 1–3 gr. per 100 gr.

was made use of. The gel was cut into pieces, which were wrapped in fine cotton canvas which had been completely freed from soluble substances by treatment for months with hot and cold distilled water. The packet was then pressed in a screw press, and the large yield of fluid collected. When the fluid ceased to flow, the solid which remained in the canvas was removed to a stoppered vessel.

The fluid was found to be a solution of agar. This was proved by evaporating some after it had been thrice filtered to a small bulk, when it was found to set to a typical clear gel. The results of the study of the ternary systems give sufficient grounds for defining the expressed fluid as a solution of agar in water, and the solid which remains in the canvas a solution of water in agar. The effect of the composition of the gel and of temperature upon the distribution of the water and the agar in these two phases was determined. The percentage composition was arrived at by drying a known weight of each, and assuming that the residue was entirely composed of agar. The results which were obtained lie far outside of any error which could have been introduced by this assumption when one considers the pains which were taken to free the gels from foreign bodies. Further, in every case an examination of the dry residue was made in order to prove that it was composed of matter capable of forming with water a typical agar gel.

Experimental Difficulties

The method used to separate the two phases, though at first sight crude, was found to be the most effective. The great error to be avoided is the blocking of the canvas pores by a mass of the solid phase, so that, instead of the true fluid phase, one really expresses fluid which has been forced through a membrane (pressure filtered). Owing to this error, a press, which I had specially made, and in which the piston drove the gel directly down on to a disc of canvas, proved quite useless. Very great force was necessary to express a fluid which was found to be almost pure water.

To succeed, it is necessary to avoid direct or great pressure. The masses of gel are loosely placed in a long canvas packet, which is then deformed by pressing the ends together. The pressure necessary to yield abundant fluid is now quite small, for the solid framework of the gel is destroyed by being rubbed against the canvas, and is reduced to fine particles, while the fluid easily makes its way through the coarse pores of the canvas. Raising the pressure always expresses a fluid poor in agar, while with slight to moderate pressure the concentration of the expressed fluid, as tested by determining the solids in the yield at different stages, remained fairly constant, but always with a slight decrease as time went on.

The expressed fluid was filtered before the solids were estimated: this was found to lower the amount of solid to a very slight extent.

The following figures illustrate the variation in the composition of the fluid as the gel becomes more completely expressed:

Successive equal quantities of expressed fluid contain dry agar in 100 c.c. Temp. = 14° C.

Exp. I	Exp. II
0·12	0·11
0·14	0·12
0·1	0·09

The mechanical pressure used to separate the phases will modify their composition by deforming the surface of separation. This error cannot be estimated.

The Influence of the Ratio of the Masses of the Two Components upon the Composition of the Phases

Two portions of a fairly concentrated gel were taken. To one part water was added to dilute it, and both were then heated to 100° C., and equal portions of each were poured into two glass-stoppered cylindrical vessels of identical shape, make and size. The two vessels were then set aside to cool.

After 48 hours samples were cut from different levels in each gel, and used to determine the percentage composition.

Five hundred and eighty grams of each of the gels were then expressed. The results were as follows:

Temp. ° C.	Agar in 100 gr. of the gel gr.	Expressed fluid		Solid solution	
		Volume c.c.	Agar %	Volume c.c.	Agar %
18	1·1	440	0·1	140	4·7
	3·3	230	0·14	350	5·6
In another experiment:					
15	1·6	—	0·12	—	—
	2·2	—	0·14	—	—

Thus an increase of the concentration of agar in the mixture produces an increase in the concentration of the agar in both phases. An explanation of this relation is suggested, and discussed later.

Effect of Temperature upon the Composition of the Phases

This was determined by running a large mass of the hydrosol into a number of glass vessels of the same shape and size. Each vessel held 600 c.c. of the hydrosol. They were close stoppered and allowed to cool to the room temperature. After 48 hours they were placed in chambers of known temperature, where they were kept for 5–7 days before the contents were subjected to pressure. In these experiments, as is obvious, the internal changes are those which follow on raising or lowering the temperature of the hydrogel from the air temperature. In other experiments the hydrosol was cooled down to, but not below, the temperature of observation. This distinction is important, because it was found that the composition of the phases varied for a given temperature according to whether that temperature was the lowest of a descending series or the highest of an ascending series. This is shown clearly in the two curves AB, AB' (Fig. 1). The arrows indicate the direction, ascending or descending, of the changes of temperature.

No. I. Agar content of mixture 1·6 per cent.

	Temp. ° C.	Agar in expressed fluid %	Agar in solid %
Ascending series	14	0·14	—
	33	0·29	—
	50	0·80	—
Descending series	14	0·14	—
	33	1·10	—

No. II. Agar content of mixture 2·23 per cent.

	Temp. ° C.	Agar in expressed fluid %	Agar in solid %
Ascending series	13	0·12	4·7
	36	0·25	5·0
Descending series	5	0·09	3·0
	13	0·12	4·7
	36	0·47	3·2

Putting on one side for a moment the different effect of an ascending or a descending temperature change, these experiments show that (1) a hydrogel or agar is a structure form of a more solid part and a fluid, and (2) each of these two phases is a mixture of agar and water, (3) the composition of the phases is dependent to a lesser degree upon the ratio of agar to water in the entire mass, to a greater degree upon the temperature.

While recognising as fully as possible that only an approximation to the actual composition of the two phases at different temperatures is obtained by these experiments, it is obvious that they afford reliable information on two points. These are, firstly, the marked "lagging" action or passive resistance to change offered by the system agar-water. The difference in composition of the phases according to whether any given temperature lies in an ascending or a descending series shows how slow the system is in reaching final equilibrium.[1] Secondly, the experimental results seem to me to indicate pretty clearly the general form of a part of the concentration temperature curve. I give the curve as it appears from the figures in Experiment II. AB and CD are the curves for the system—solution of agar in water, solution of water in agar, and vapour. If they correctly represent the general form of the curve, then, by the theorem of Le Chatelier, it follows that the change from the system solution of water in agar and vapour to the system solution of water in agar, solution of agar in water, and vapour will be accompanied by a liberation of heat when the change takes place along the isotherms from 5 to $\pm 20°$ C., and by an absorption of heat when the change is along the isotherms ± 20–$35°$ C., while the change from the system solution of agar in water and vapour to the system of two solutions and vapour will always be accompanied by absorption of heat.

I have not established this deduction experimentally, but it finds a considerable amount of support in the following facts. When water is allowed to dissolve in pure dry agar at 14° C., a considerable amount of heat is given off. 1 c.c. of dry agar in coarse powder added to 10 c.c. of water gives a rise of more than 6° C.,[2] while control experiments with carefully dried finest graphite or sand gave a rise of temperature of 0·15 and 0·17° C. respectively. Wiedemann and Lüdeking[3] also found that when dried gelatine absorbs water heat is liberated, but when gelatine saturated with water is dissolved in water heat is absorbed.

I have verified the general form of the curve AB in a way which eliminates all the errors due to the expression of the fluid phase from the gel. A cylindrical column of gel 15 cm. high was divided by two vertical cuts at right angles into four equal pieces. Four stoppered glass vessels were taken of the same size and shape, and in each one of the pieces was placed and just covered with water. Two of the bottles were kept at 14° C. for a week, and two at 44° C.; the water in both was found to have dissolved some of the agar, and to contain per 100 gr. of the solution 0·50 and 0·12 gr. of dry agar respectively.

[1] The systems salicylic acid and water, and thorium sulphate and water are perhaps comparable cases. The former readily yields two fluid phases which, however, are throughout in labile equilibrium (Bancroft, *The Phase Rule*, p. 105). In the case of the latter system supersaturated solutions can be obtained over a wide range of temperatures, and even in presence of the stable hydrates it is often hours or days before equilibrium is reached (Bancroft, *loc. cit.*, p. 54, or Roozebrom, *Zeits. f. phys. Chem.* (1890), v, 198). The lagging action in colloids, which is so markedly shown by van Bemmelen's researches into the effect of time on the hydrogel of silicic acid, ceases to be extraordinary when one remembers that one of the phases is a solid. Gels reach equilibrium much more rapidly than does, for instance, a bar of metal in which the reaction velocity is so slow that final equilibrium may never be reached.

[2] The mercury in the Beckmann thermometer was driven beyond the scale into the upper reservoir.

[3] *Ann. der Phys. u. Chem.* N.F. (1885), xxv, 145.

The curves AB, DC continued upwards will meet at some point which marks the consolute temperature for agar and water. I have attempted to fix this point by observing the changes in the intensity of the beam of polarised light scattered normal to the ray when parallel light is passed through a gradually cooling hydrosol. The observations, which are still in an initial stage, have so far failed to fix the point.

The study of ternary systems under the microscope makes it probable that as the curves AB, DC are continued upward they reach a point beyond which the equilibrium is no longer between a fluid solution and a solid solution, but between two fluid solutions.

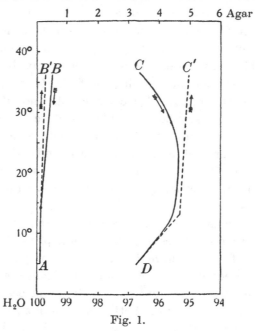

Fig. 1.

The first worker to regard gelation as being due to the formation of two phases, one fluid and the other solid, was van Bemmelen.[1] He has given a suggestive discussion of the formation and structure of gels, based chiefly upon the manner in which amorphous material is precipitated from a solution, and he is led to the conclusion that "coagulation or the precipitation of a gel from a solution seems to be a similar phenomenon; a desolution (Entmischung) which forms, not two layers completely separated from one another, but

"1. A framework of a material which is in a more or less transitional state between fluid and solid, and which presents those special properties to which the term colloid is applied.

"2. A fluid which is enclosed within this framework."

[1] *Zeits. f. anorg. Chem.* (1898), XVIII, 20.

Van Bemmelen, however, does not consider that these two parts can be considered as two phases in the sense of the phase rule, since there is no sharp line between them,[1] and he therefore concludes that the phase rule cannot be employed to elucidate the phenomena. This opinion is based upon a study of the equilibrium between the water content of various gels and the vapour pressure, so patent and thorough as to give it very great weight. The curves of the equilibrium points are gradually bending lines if the dehydration of the gels is sufficiently slow, but if dehydration is relatively rapid there is a sudden change of direction (Fig. 2) when the water content is very much diminished (1 to $2H_2O$ to $1SiO_2$).

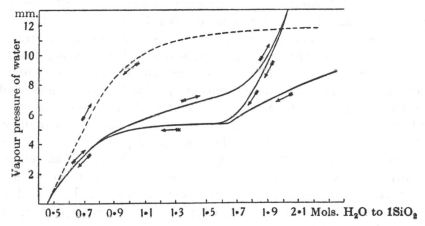

Fig. 2. (Reproduced from van Bemmelen, *Zeit. f. Anorg. Chem.* (1896), XIII, 233). Equilibrium between a hydrogel of silica and water vapour. Curve ---- the rate of removal of water very slow. In curve —— much more rapid. The arrows indicate whether the curve shows the removal or the reabsorption of water.

It is possible that the form of these curves does not necessarily depend upon the absence of a clear separation between the fluid and the solid portions of the gel. When one considers how small is the mutual solubility of silica and water and how slight therefore the influence which a given mass of silica is likely to exert upon the vapour pressure of even a relatively small mass of water, it is probable that the form of the curve is determined more by the operation of secondary influences, such as capillary tension, which depend on the structure of the gel, than upon the direct interaction of silica and water. Capillary tension would tend to lower the vapour pressure[2] with which the gel is in equilibrium to a greater and greater extent as the spaces in the solid framework of the gel became smaller and smaller with

[1] *Zeits. f. anorg. Chem.* (1898), XVIII, 121.
[2] The vapour pressure which van Bemmelen measured is that of the free surface of the gel. It is analogous to that at the open ends of a number of capillary tubes filled with fluid.

the decrease in the water content. The tendency to reduce the surface energy at the surface of separation of fluid and solid to a minimum, which manifests itself in the spontaneous shrinkage of some of these hydrogels, would act so as to raise the vapour pressure with which the gel is in equilibrium, but the operation of this factor would diminish as the surface was diminished by decrease of the water content. These two forces operating simultaneously would alone produce the characteristic gradual diminution in the vapour pressure of the gel as the fluid component is diminished. The break in the direction of the curve when dehydration has been relatively rapid and is nearly complete, is what must occur when the capillary spaces in the framework become commensurate with the masses (small spheres for instance) of solid out of which the framework is built, and when, therefore, any further diminution in the capillary spaces involves deformation of those masses, unless the removal of water is so slow that the very slow rate of readjustment in the solid phase is not exceeded. Lastly, the very limited powers of reabsorption of fluid by completely dried irreversible gels would, on this view, again not necessarily represent a reformation of the phases, that is to say, a real interaction between silica and water, but the refilling of capillary spaces by water due to the excessive capillary tension of these very minute capillaries. The capacity for reabsorption would therefore be diminished by any agent which facilitates the annealing of the dried gel and so destroys the capillary interspaces. Such an agent is heat, and van Bemmelen found that brief heating to red heat destroyed the reabsorptive powers of the gel of silica.[1]

There is one binary system in which gelation is an irreversible process (*i.e.* coagulation) which can be readily studied under the microscope. The hydrosol is a ternary system composed of water, a minute trace of free acid or alkali, and the modification of egg albumen which is produced by heating it to 100° C. Coagulation occurs when the free acid or alkali is removed. As the coagulation point is neared the proteid particles in the hydrosol increase in size, so that spheres $0.75-1\mu$ in diameter are formed. These become arranged in rows which anastomose so that an open net with regular polygonal meshes is formed.[2] In this case the process of gel building is the same as that which can be followed so easily in ternary mixtures, and in both cases a definite surface separates the phases. It is probable that the hydrogel of silica is formed in the same way, since Picton and Linder have shown by optical tests that, as the point of coagulation is approached, larger and larger particles of silica form in the hydrosol.[3] These particles

[1] *Zeits. f. anorg. Chem.* (1896), XIII, 289.

[2] The process is described in detail in an earlier paper by the author in the *Journ. Physiol.* (1899), XXIV, 182, and the information which the microscope affords as to the manner in which irreversible gels are built is discussed there.

[3] *Journ. Chem. Soc.* (1892), LXI, 148.

may be solid solutions of water in silica, or they may be large molecular aggregates of silica free from water. I incline to think that the latter is the more probable assumption, since, if they were solutions, it is difficult to see why the process should be irreversible.

In the case of the reversible systems agar-water, or gelatine-water-alcohol, the particles seem to be of the nature of solid solutions.

The system agar and water consists of two components, and, therefore, a nonvariant system should be defined by four coexistent phases. Since the gel stage consists of three phases, namely two solutions and a vapour phase, it should be a monovariant[1] system. That is to say, the composition of the phases should be fixed by fixing either the temperature or the pressure of the vapour phase. The experiments show that this is not the case. The composition of the fluid and solid phases is not constant for a given temperature. This result might be regarded as being due to the passive resistance to change in the system which is introduced by the formation of a solid phase. On this view if the velocity of the reaction were known, the phases would be fixed if the element of time were introduced and accorded a definite finite value. This is the method which Bancroft suggests for dealing with such cases;[2] it is, however, possible that there are really more than $n + 2$ independent variables, so that the hydrogel is not a monovariant system. In an ordinary system the independent variables are the components (n), temperature and pressure. Agar-water, however, is a system with two components, temperature and two pressures. This follows from the fact that the surface which separates the fluid and solid phases is curved. In point of fact the system is most closely represented by a system of two solutions separated by a membrane which is permeable by only one of the components, for while water will readily pass the surface of separation, agar, having the heavy immobile molecule characteristic of such organic bodies, will be almost unable to do so. Hence, if time be considered finite and small, the surface may practically be considered to be permeable by only one component. As Bancroft[2] points out, in a system of two solutions separated by semipermeable membrane, there are two pressures and there will be $n + 3$ phases in a nonvariant system when $n =$ the number of components. The hydrogel is a system of three phases and, therefore, on this view, to fix the composition, it would be necessary to fix the temperature and one pressure. This relation would probably be true if the curvature of the surface of separation could be fixed. This, however, is not the case, and in order to fix the composition of the phases it would be necessary either to fix the temperature and both pressures, that of the internal as well as of the external phase; or to fix the temperature, one pressure and the

---

[1] That is to say, a system having one degree of freedom.
[2] *The Phase Rule*, p. 234.

form of the surface. Practically we can only fix the temperature and the pressure of the external phase. I have succeeded in obtaining two phases separated by a plane surface by cooling a hydrosol slowly in an electric field. This method may prove suitable if the system is able to recover from the forces operating during its formation. The method of taking known weights of dry agar and water and keeping them at constant pressure and temperature until equilibrium is obtained is simple, but unfortunately there is the fallacy that the dry agar is a preformed system. The structure of the hydrogel from which it is reproduced is not destroyed by drying, and the system tends to reform itself on the old lines by the filling of the original capillary spaces.

To sum up these remarks, we may describe the hydrogel of agar as a system of three phases, a solid, a fluid and a vapour phase. The equilibrium is determined by the chemical potential of the components in the various phases, by two pressures, and by temperature. Other operating variables are capillary tension and the energy of the surface between the fluid and solid phases. I have made no measurements to determine how soon the system reaches equilibrium, but the analogous system, gelatine and water, attains to a constant melting point twenty-four hours after the formation of the hydrogel.[1]

---

[1] *Gelatinöse Lösungen* (van der Heide, München, 1897).

18

THE ACTION OF SALTS OF RADIUM UPON GLOBULINS

[*Journ. Physiol.* (1903), XXIX P]

Two solutions of globulin from ox serum were used, one electro-positive made by adding acetic acid, the other electro-negative made by adding ammonia.

They were exposed in shallow cells one wall of which was made of thin mica to the radiations from 50 mg. of pure radium bromide enclosed in a capsule covered with mica, so that two sheets of mica were interposed between the radium and the solution. No action took place in 1 hour.

The globulin was then exposed as naked drops separated by 3 mm. of air from the radium salt, with suitable controls shielded from the radium.

In the + solution the opalescence rapidly diminished—that is to say, solution became more complete. The − solution was turned to a jelly, at first transparent, rapidly becoming opaque. The action was complete in about 3 min.

Radium like other radioactive bodies gives off matter in three states: (1) an emanation having the mobility of a heavy gas, (2) positively charged particles of little penetrating power and relatively large size, (3) ultra-material negatively charged particles of a size much smaller than atoms. A mica plate will screen off 1 and 2, therefore the globulin solutions are unaffected by the ultramaterial negative particles. The rate of action and conditions of the experiment make it unlikely that the emanation was the active agent, though, owing to its nature, intense solvent or coagulating power may safely be predicated of it. The action observed was almost certainly due to the + particles. These are of material dimensions.

Globulin systems therefore seem to be completely transparent to the ultramaterial electrons and so too probably are the living tissues, since the physiological influences of the discharges from radium seem to be limited to a superficial layer a few millimetres deep.

The experiments were carried out by the kindness of Sir William Crookes in his laboratory and with his assistance.

19

ON THE SENSATION OF LIGHT PRODUCED BY RADIUM RAYS AND ITS RELATION TO THE VISUAL PURPLE

With H. K. ANDERSON

[*Proc. Roy. Soc.* (1903), LXXII, 393]

It is now well known that when a few milligrammes of a salt of radium are brought near the head in the dark a sensation of diffuse light is produced. We have examined this phenomenon with the object (1) of determining the place of origin of the sensation; (2) of identifying the particular rays which cause it.

The rays from radium falling upon the skin produce no sensation, and we failed in evoking sensations of sound, smell, or taste by their agency. The only response immediately traceable to them seems to be this one of diffuse luminosity.

It may be described as an appearance of diffuse light of steady intensity disposed in the external space in front of the head and filling that space fairly or quite uniformly. If the radium, covered of course with some opaque screen such as, for instance, black cardboard, to cut off the pale light which it emits, be held in front of one eye, one notices that the intensity of the glow is considerably reduced by closing the eyelid. When the eye is open it is possible in a very general way to locate the radium from the fact that the sensation is strongest when the axis of vision is directed to it and diminishes when the head is turned to one side. The sense of direction arises solely from variations in the intensity of the glow and not from variations in its quality.

When the eye is closed the sense of locality is completely lost. This is due to the fact, to be dealt with more fully presently, that the glow is due to the β and γ rays, and that the eyelid is peculiarly opaque to the former, stopping apparently the whole of them. The γ rays, on the other hand, having a very great penetrating power, pass almost equally well through the eyelid or through the bones and other tissues forming the orbit. There is, therefore, no possibility of a differential screening action when once the eyelids are closed and the β rays stopped, with the result that all sense of the direction of the source of light is lost.

The sensation of light is purely of retinal origin. It is not due to a response to the rays on the part of the optic nerve, the optic tract, or the brain itself. We were led to this conclusion in the following way. When the crystals of radium bromide are spread out so as to form a layer about 1 mm. deep, covering a circle of about 5 mm. diameter, the glow is very intense when the flat face of the layer is presented to the eye,[1] and practically vanishes when the layer is turned edgeways. This must be due to the fact that the volume of the stream of rays is roughly proportional to the surface of the mass of the salt, and that the stream is densest in a plane normal to the surface. The effect of rapidly rotating the little plate of crystals before the eye is very striking; it is as though a series of blinding flashes of light were thrown on to the retina.

If now the radium be moved about over the surface of the head, one notices that for the sensation of light to be present the axis of greatest ray density must cut one or other eyeball, and that the further the radium is removed from an eyeball the weaker is the sensation.

This statement can be illustrated by the following case. When the radium is held with the face of the crystals opposite the middle of the forehead, there is at most a luminosity so feeble as to be detected with difficulty; at the same time no other sensations are evoked. If, however, the plate of crystals be rotated downwards and outwards so as to face towards one eyeball, a strong light is at once seen with that eye.

Experimenting in this way, we convinced ourselves that no sensations are directly produced by the rays reaching the brain substance,[2] and that the origin of the sensation of light lies solely in the eyes themselves.

On passing directly from daylight into a dark room one's eyes are at first almost completely insensitive to the rays, and sensitiveness grows only slowly. In the evening, after exposure for some hours to yellow artificial light, the sensation of light is felt almost immediately tho ordinary sources of light are extinguished. Radium vision, if we may so call it, will continue to increase for a full hour in the dark—in other words, it is a phenomenon of the dark-adapted eye.

Radium vision, therefore, resembles the response to light of low intensity, and as a great many facts go to show that this is connected with the visual purple, we examined the effect of the rays upon that substance. To our surprise we were unable to detect any bleaching action whatever on the visual purple of the eye of the frog or rabbit.

The method we adopted was to keep the animal in the dark for some hours, to kill it and dissect out the retinæ in sodium light, and to expose one

---

[1] To avoid repetition, it must be understood that the crystals are always covered with opaque paper or cardboard, so as to exclude the actual light rays.

[2] This statement is of course limited to the stream of rays from 50 mg. of radium bromide. It is possible that a more powerful stream might evoke direct response from the brain.

retina to the rays from 50 mg. of pure radium bromide in a moist dark chamber, while the other retina remained in a similar dark chamber as a control.

The retinæ, when removed, were spread on to thin sheets of mica and placed over the radium at about 3 mm. distance, in such a way that either the mica alone was between the radium and the retina, or the mica and a layer of opaque black paper to screen off the light rays. In a few cases the retina was so placed that nothing more than about 2 mm. of air separated it from the radium. The time of exposure was, as a rule, 20 hours. In no case was there any difference in tint between the exposed and the unexposed retinæ, while both control and exposed retinæ were found to bleach in full daylight in a few seconds.

It will be noticed that in those experiments in which the light rays from the radium were allowed to fall on the retina, no bleaching could be detected. This does not prove that they were wholly without action since, as Kühne showed, even the excised retina has a feeble power of reproducing the purple. The experiments prove merely that the α, β, and γ rays, either alone or with the light rays, fail to overcome this recuperative power sufficiently to effect a detectable decrease in the quantity of pigment in 20 hours.

When one considers the density and activity of the stream of "invisible" rays at 3 mm. distance from 50 mm. of radium bromide, we may fairly conclude that these rays have no action on the pigment. This conclusion led us to consider whether the sensation of light which is caused by the invisible rays is, indeed, due to the direct response of the retina. When one considers how limited in extent is that portion of the spectrum to which the retina responds, it would be a matter of surprise if so specialised an instrument were found to respond to rays as different from light waves as are the radium rays. The evidence we have been able to obtain points to the conclusion that the retina probably does not respond directly to the radium rays, but to light rays, which are given out by the tissues of the eyeball— chiefly the cornea and lens—when they are traversed by the β and γ rays.

The fresh lens of a sheep, ox, or rabbit was found to glow strongly when exposed to the rays. The cornea and vitreous humour also glow, but to a less extent, and the retina itself gives a strong glow. The sclerotic, on the other hand, glows very slightly. The glow of the lens alone is so striking as fully to account for the sensation of light produced by the rays.

This quality of emitting light under the influence of the rays is not at all peculiar to the tissues of the eyeball. The skin glows, as may readily be seen by bringing the hand near to radium covered by opaque paper. Fat and muscle glow strongly.

The fact that direct stimulation of the retina contributes at most but little to the sensation of light comes out clearly when the nature of the rays which cause the sensation and their power of penetrating the eyeball are examined. From this investigation the α rays were always excluded. They possess so little penetrating power that they are arrested by the thickness of opaque paper which is necessary in order to screen off the light rays.

The β and γ rays, therefore, remain for consideration, and they both contribute to the production of the sensation of light.

The following experiment proves this:

Fifty milligrammes of radium bromide, covered with opaque paper, were placed at the bottom of a thick-walled lead tube 6 cm. in length, on each side of which were the poles of an electric magnet giving a field of more than 5000 lines over nearly the entire length of the lead tube. The physical effect of this field would be to divert the β rays into the lead. The optical effect was very striking. When the field was established while the eye was applied to the tube, the glow suddenly diminished to what appeared to be about a fifth or less of its previous intensity. The change resembles the sudden turning down of a gas light. The persistent or residual glow must be that due to the γ rays.

The effect of the γ rays can be demonstrated by screening off the β rays with lead screens. A plate of lead 2·3 mm. thick was found to reduce the intensity of the glow very much. Such a plate cuts off the β rays. The further addition of four more such screens, thus increasing the thickness of the lead to 11·5 mm., scarcely diminishes the remaining glow, for they arrest only a small fraction of the γ rays. In these screen experiments it is, of course, necessary to keep the distance between the radium and the eye constant. This was effected by a simple apparatus designed for the purpose which does not need description.

The glow is still quite strong when 4 cm. of lead are interposed between the eye and the radium (50 mg.), but it is faint with more than 5 cm.

Similarly the fluorescence of the excised lens of the eye which is excited by those rays which traverse lead 2·3 mm. in thickness is not sensibly diminished by adding two more such screens, thereby increasing the thickness to 6·9 mm.

So far as the β rays are concerned, it is certain that they cause a sensation of light solely by inducing fluorescence of the tissues of the eyeball in front of the retina. Taking the eyeball of the sheep as corresponding in size fairly well with the human eyeball, it was found by measurement with the electroscope that the cornea, lens, and front half of the vitreous humour taken together had the same screening action as 6·6 mm. of lead. That is to say, they arrested all the β rays and a measurable fraction of the γ rays. Therefore the β rays do not penetrate to the retina, except perhaps in small quantity to the most anterior portion.

The part played by the γ rays is not so obvious. It is certain that they can excite the excised lens to become luminous, since a lens separated from the radium by a centimetre of lead emits light enough to be visible. The fluorescence is, however, exceedingly faint, much less for instance than that excited in a piece of glass of about the same size. On the other hand, the sensation of light evoked by those rays which traverse even 4 cm. of lead is quite considerable, and no one can fail to be struck with the fact that the fluorescence of an excised lens and the sensation of light produced by the γ rays are very dissimilar values. It is, therefore, possible that the γ rays produce the sensation of light in part only when they actually strike the retina. Here again, however, the response need not be directly to the γ rays but to light waves started in the retina itself, for, as has been already pointed out, the excised retina glows strongly when brought near radium.

The following experiment helps to prove that the γ rays produce the sensation of light in great part when they actually strike the retina. The radium was covered with 5 mm. of lead to cut off the β rays, and a vertical plate of lead 40 mm. deep and 2 mm. thick was moved backwards and forwards over the radium so as to form a bar-like screen 40 mm. thick and 2 mm. wide between it and the eye. The effect produced was that of a bar of shadow moving across the glow.

Now the γ rays are not refrangible, and therefore the vertical lead screen would cast a partial shadow its own breadth upon the lens and retina. The effect upon the lens would be to diminish its fluorescence, but as the light waves are given off in all directions from the lens, there would not be a definite shadow on the retina. Therefore this bar-like shadow must be due to the shadow on the retina itself.

Reference was made in passing to the remarkable stopping power of the eyelid. The human eyelid seems to be capable of arresting all the β rays. With the eyelid closed the insertion of a lead screen 9·2 mm. thick makes no[1] difference in the intensity of the sensation of light. This, however, is true only when the eyelids are tightly closed. If they are stretched by raising the eyebrows, and their thickness thereby reduced, the interposition of the lead screen produces a detectable decrease.

Some measurements of the screening power of various tissues were made with a Curie electroscope. The numbers give the rate of leak in divisions of

---

[1] Many trials with six different individuals resulted in a strong balance of opinion in favour of the screen producing no diminution. Only one person thought he could detect a lessening of the glow.

the scale per second. The tissues were from a freshly killed rabbit. 5 mg. of radium bromide were used, enclosed in a lead capsule having an aperture 4 mm. in diameter on one side.

Lead, 0·2 mm. thick	25·0
„ 0·4 „	18·0
„ 2·3 „	11·3
Skin of back, 0·8 mm. thick	20·5
„ (shaved), 0·8 mm. thick ...	22·0
Eyelid, 0·7 mm. thick	17·2
Nictitating membrane, 0·8 mm. thick ...	28·6
Iris	28·6
Lens, 7·5 mm. thick	14·0
Sclerotic ⎫	
Choroid ⎬	23·0
Retina ⎭	
Sclerotic <0·5 mm. thick	36·3
Entire eyeball	11·5
Muscle (skeletal), 3·0 mm. thick	19·0

These figures show a remarkable difference in stopping power between the eyelid and the nictitating membrane, in favour of the former, which equals 0·4 mm. of lead. The screening power of the eyelid is also noticeably greater than that of the skin of the back, although the latter is the thicker.

20

ON THE OXIDISING ACTION OF THE RAYS FROM RADIUM BROMIDE AS SHOWN BY THE DECOMPOSITION OF IODOFORM

WITH MISS E. G. WILLCOCK

[*Proc. Roy. Soc.* (1903), LXXII, 200]

In the course of certain experiments one of us noticed that a solution of crystals of pure iodoform in chloroform rapidly became purple. The colour change is due to the liberation of iodine, and the purple solution readily gives the starch test, and is decolorised by thiosulphate. This decomposition of iodoform occurs in a variety of solvents, namely, in chloroform, benzene, carbon bisulphide, carbon tetrachloride, pyridene, amyl alcohol, and ethylic alcohol. In alcohol the change is shown only by a deepening of the original yellow tint of the solution to brown; iodine when dissolved in alcohol having a yellow-brown tint.

As the reaction itself seems not to have been described, a few words may be devoted to it before passing to the main point—namely, the influence of the rays from radium upon this chemical change.

The liberation of the iodine needs the presence of oxygen—though exceedingly minute amounts are sufficient—and some form of radiant energy. When oxygen is washed out by a stream of CO_2, or the vapour of the solvent, no change takes place; the solution of iodoform retains its faint yellow tint in full daylight.

If a minute quantity of oxygen be left behind, the colour deepens in daylight to a brownish yellow, but iodine is not liberated—the solution will not give the starch test. That is to say, there is an intermediate stage of chemical change which is reached in presence of minute amounts of oxygen, and which falls short of the actual liberation of iodine.

If a pair of platinum electrodes be dipped in a solution which is in process of changing to purple, and a field (± 4 volts per centimetre) be established, a heavy, oily, colourless liquid slowly drips from each electrode. The nature of this liquid has not yet been determined, but from its specific gravity probably it is methylene iodide.

In the complete absence of any radiant energy, and in presence of abundant oxygen, the solution of iodoform undergoes no change at ordinary temperatures. When heated to near the boiling-point, however, the solutions change even in the dark.

Salts, when present, have a remarkable influence on the reaction, although they can hardly be said to be soluble in the reagents employed. For instance, if a solution of iodoform in benzene be divided into two parts, and to one part solid sodium chloride be added, and then both be heated, the one with the salt decomposes much more rapidly. Compared in this way, it was found that

$NaCl$; KCl; KNO_3; $Pb(NO_3)_2$; $Ba(NO_3)_2$; $BaCl_2$ accelerate.

K_2SO_4; $CaCO_3$; $BaSO_4$; $MgCO_3$ retard.

It is remarkable that the salts which were tried should so group themselves that those with univalent acids accelerate, those with bivalent acids retard.

The influence of salts appears to be purely a case of surface action. If the salt first be heated for a few minutes with two changes of the solvent (benzene), it entirely loses its power.[1]

Probably owing to obscure catalytic action of this kind one finds that in certain apparently clean test-tubes the very sensitive solution of iodoform in chloroform changes to purple even in the dark. The catalysing power of such a tube, however, is very rapidly exhausted.

Probably also owing to catalytic action some samples of iodoform decompose when dissolved in chloroform even in complete darkness. The impurity which brings about this apparently spontaneous change can be distilled off by suspending the sample in water and boiling for a considerable time. The first distillate condenses as a red liquid, when this ceases to come over the distillate will be found to be approximately or quite stable. The impurity can be got rid of more effectively by recrystallising from ethylic alcohol.

The chemical feature on which we wish to lay most stress is that for the liberation of iodine oxygen is needed—it is in all probability due to an oxidation, and, like many oxidation processes, it is carried on, under ordinary circumstances and at ordinary temperatures, only in the presence of light.

The reaction, on the one hand, is a delicate test for the presence of oxygen, on the other, a convenient method for measuring the chemical activity of various rays. We found the trace of oxygen which remains after CO_2 has been bubbled through the chloroform, and over the iodoform for one hour, sufficient to produce a decisive change of tint.

The beautiful purple colour which the liberated iodine makes in solvents other than alcohol lends itself readily to measurements which may be made

---

[1] According to Würster (*Ber. d. Deut. Chem. Gesellschaft*, XIX, 3201), finely powdered bodies in general occlude "active" oxygen. The two carbonates, however, were the most finely powdered of the salts used.

by choosing some solution of iodine in chloroform as a standard colour and matching the fluids under examination with it.

The reaction, when once started, continues for a time in absolute darkness and then ceases. Thus, if light be allowed to play upon a tube so as to produce, say, a faint purple tint, and the tube then be removed to the dark, the faint purple tint will deepen to a certain extent. On renewed exposure to light the action recommences.

Action of the Radiations from Radium

We used 5 mg. of pure radium bromide, supplied by Buchler and Co., of Brunswick, and we found that a solution of iodoform in chloroform was turned deep purple by simply resting the test-tube containing it on a plate of mica covering the radium salt. That is to say, the active rays penetrate mica and glass.

They also penetrate cardboard. Tubes containing a solution of iodoform in chloroform were enclosed in a box of black cardboard, and they remained unchanged for 60 hours. On placing the box over the radium salt the tubes became purple in about 10 min.

For the following reasons we believe that the active rays from radium are entirely different from the active rays of light.

The active rays of sunlight are completely arrested by an opaque layer of lamp-black deposited over a test-tube, by black cardboard, by aluminium, or, in short, by any substance opaque to visible light rays. The active radium rays traverse lamp-black, black cloth, or cardboard, and aluminium sheet 1 mm. thick, without any measurable loss.

An ordinary yellow gas-light was found to emit active rays in quantity sufficient to change iodoform dissolved in chloroform at 1 ft. distant in a few minutes, even when the test-tube was jacketed with water in order to prevent any heating. There is, therefore, no reason to believe that the activity associated with light is different from the ordinary chemical activity of light.

The radium rays which produce the change were identified by measuring the effect of screens upon the time necessary to produce a standard depth of purple in 1 c.c. of a standard solution of iodoform in chloroform.

A comparison of radium unscreened and screened so as completely to intercept the α rays failed to show any action on the part of these rays.

Attention was then turned to the more penetrating β and γ rays. A corked test-tube was suspended at a constant distance from the radium (approximately 3 mm.), the same test-tube being used throughout. 1 c.c. of the standard solution was used for each measurement. In these measurements the wall of the test-tube was always present as a screen.

Time necessary to read the standard colour:

		Minutes			Means
1. Radium uncovered		13	12	13	12·6
2. Thin screen of mica		11	12·5	12	11·8
3. „ „ and sheet of writing paper		12	—	—	12·0
4. Glass, ±0·5 mm.		15	14·5	14·5	14·8
5. Aluminium, ±1 mm. thick		15·5	15·5	15·5	15·5
6. Lead plate, ±2 mm. thick		Between 200 and 250			225·0
7.* Four lead plates, each ±2 mm. thick		Less than 1000			<1000

* In this case the distance between the radium and the solution was, of course, increased in order to make room for the screens.

Prof. Rutherford was good enough roughly to measure the stopping power of the screens actually used. Nos. 2, 3, 4, and 5 stopped all the α rays. No. 6 stopped 80 per cent. of the β rays, and allowed the γ rays to pass. No. 7 stopped practically all the β rays, and allowed only γ rays to pass.

The obvious conclusion from these figures is that the action is mainly due to the β rays—that is to say, to the stream of negative electrons. On the other hand, the fact that action is not arrested by as many as four of the lead screens makes it certain that the very penetrating γ rays also are chemically active.

As the γ rays are said to be the same as the Röntgen rays—that is to say, ethereal pulses—the action of the latter was tried by exposing tubes of iodoform dissolved in chloroform, which were enclosed in light-tight cardboard boxes. The Röntgen rays were found to be active, the solutions were purple at the end of 15 min.

An exact comparison of the relative activity of light, radium rays and Röntgen rays cannot be attempted, but the experiments prove that light is the most active. The difference appears to be very great. The profound and often lethal physiological action of radium rays must therefore, for the present, be looked upon as being due to their power of penetration rather than to the fact that they exert any novel or very intense action. They reach parts which are shielded by a cuticle very impervious to light waves. Viewed in this way the pigmentation of the human skin found in tropical races, and in those exposed to sunlight, may be regarded as an increased protection to the internal structures which acts by increasing the opacity of the epidermis.

One of us has already shown that the α rays profoundly modify the *physical* state of colloidal solutions.[1] If the colloid particles be electrically negative, the α rays act as coagulants; if the colloid particles be electrically positive they act as solvents, that is to say, the rays decrease the average size of the particles.

As a provisional basis for investigating the physiological action of

[1] *Journ. Physiol.* (1903), XXIX, xxix.

radium rays we may therefore regard the α rays as altering the physical state of the living matter, the β and γ rays as altering the chemical processes, especially perhaps the oxidation processes of the tissues.

It may be well to mention briefly the instances of chemical decomposition produced by radium rays which have been described up to the present.

Berthelot[1] gives the following cases. Iodic acid is decomposed with liberation of iodine by rays from radium and by light. Unlike the liberation of iodine from iodoform the change proceeds very slowly, free iodine being present only after 14 days' exposure. Nitric acid forms nitrous fumes when acted upon by the radium rays or by light.

Becquerel[2] mentions the case of mercuric chloride which in presence of oxalic acid is decomposed by light rays, and by radium rays.

[1] *Comptes Rendus* (1901), cxxxiii, 659.
[2] *Ibid.* (1901), cxxxiii, 709.

21

COLLOIDAL SOLUTION. THE GLOBULINS

[*Journ. Physiol.* (1905), XXXIII, 251]

CONTENTS

Colloidal solutions differ in their relation to small concentrations of salts, some, such as the hydrosols of metals, of silica and alumina, etc., are precipitated, others appear not to be changed, such as, for instance, hydrosols of albumen or gelatine, others again depend for stability upon the presence of salts. The chief and perhaps the only true example of the last class is the species of proteid known as globulins.

About five years ago I commenced the investigation of this last class with the object of gaining further insight into the mechanism of colloidal solution, and the results were communicated to the Physiological Society in 1903, to the British Association in 1904, and a statement as complete as time would allow to the Royal Society as the Croonian Lecture of 1905.

In the communication to the Physiological Society I attempted to express the relation between globulins and salts, acids or alkalis in a purely physical way, which may be briefly recapitulated as follows. The globulin is dispersed in the solvent as particles which are the colloid particles and which are so large as to form an internal phase. These particles at any one moment contain within themselves an excess of the most penetrating and fastest moving ion present, and they therefore have the electric charge of this ion. Therefore, in presence of acid they contain an excess of the hydrogen ion and they are charged positively, in presence of alkali they contain an excess of the hydroxyl ion and are therefore charged negatively, in presence of neutral salts the excess ionic concentration in the colloid particles will be so slight as to be unrecognisable and the particles therefore are uncharged: they are in point of fact related equally to both ions.

This purely physical hypothesis is attractive from its simplicity, it covers most of the facts, and it is identical with the theory of the colloidal state advanced by Perrin[1] as a result of certain most interesting investigations upon the origin of contact difference of potential. The hypothesis fails in two respects: (1) it does not adequately recognise the chemical phenomena involved, such, for instance, as the influence of the selective affinity of metals for either hydrions or hydroxyl ions in the formation of metallic hydrosols,[2] or the connection between the colloidal state and the hydrolytic reaction between water and soaps, or water and proteid-acid or proteid-alkali compounds. (2) It does not cover those cases of colloidal solution where there is no potential difference between the colloid particles and the fluid in which they are immersed. In globulin salt solutions and solutions of gelatine there is no trace of drift of the colloid in an electric field if the possibility of electrolytic decomposition of the organic colloid be excluded as in the method described later. Lastly the theory can be disproved by direct experimental evidence. If the relation between the colloid and the fluid depends solely upon the relative velocity of the ions present in the latter as I at first thought, and as Perrin thinks, then in a solution of globulin by a salt such as LiCl or LiBr, in which the ionic velocities are in the ratio of about 1 to 2, the globulin should show a negative charge due to the faster anion. I failed to find indications of even the feeblest charge provided secondary electrolytic decomposition were excluded. This same experimental fact stands in the way of the acceptance of Perrin's theory of the relation of contact difference of potential to ionic velocities since the ratio of the velocities of the ions in acids or alkalis is only of the order 1 to 4 or 1 to 5.

In the following paper I have tried to deal with the phenomena of colloidal solutions in the special case of proteids from a frankly chemical standpoint. The chief difficulty in the way of a chemical theory of colloidal solution is the apparent need for postulating the existence of continuously

---

[1] *Journ. Chim. Phys.* (1904), II, 61; (1905), III, 50.

[2] Linder and Picton, *Journ. Chem. Soc.* (1897), LXXI, 568; Burton, *Phil. Mag.* (not yet published).

varying chemical compounds, or what van Bemmelen calls absorption compounds. The need to my mind is more apparent than real. The absence of transition points, the smoothing of the curves, is probably merely an expression of the inertia of the colloidal system due to the presence of electrified surfaces and to the large molecules involved. True equilibrium values for matter in the colloidal state, that is, values which have the same magnitude however the condition is approached, have not so far as I am aware yet been obtained in any one single instance. The form of the temperature curve for hydrosols of agar or gelatine depends entirely on whether the particular temperature is approached by adding heat or by subtracting heat, and upon the rate of addition or subtraction of heat.[1] The reaction velocity of colloidal matter may be of the order of that of matter in the solid state so that, as van't Hoff says of the hydrates of calcium sulphate, the time values are of the magnitude of geological epochs. Owing to the large size and low mobility of the molecules of colloids any surface once formed in a fluid plays to a great extent the part of a semi-permeable membrane, e.g. the surface of the colloid particles, and if that surface becomes strongly electrified it acquires a special stability. The system then as a whole is in equilibrium with *two* pressures.[2] In this way the instable state of spheroidal subdivision in which a new fluid phase first of all appears and which gives place rapidly to the crystalline state in crystalloid solutions[3] becomes subpermanent in colloidal solutions.

The globulin referred to except where express statement to the contrary is made is that precipitated from diluted ox serum by acetic acid. Four states can be recognised in both the solid condition and in solution, globulin itself, compounds with acid or with alkali, and compounds with neutral salts. Following old usage I propose to call the compound with acid acid globulin, those with alkali alkali globulin, those with salts salt globulin.

Globulin itself is insoluble in water, though it forms filterable non-settling suspensions which are solutions of very low grade.[4] Acid globulin

[1] Hardy, *Proc. Roy. Soc.* (1900), LXVI, 95.

[2] *Ibid.*

[3] Lehmann, *Molecularphysik*, I, 730; Bancroft, *The Phase Rule* (1897), p. 242; and especially Garnett, *Trans. Roy. Soc.* (1904), A, CCIII, 385.

[4] Defined exactly, a solution is a homogeneous mixture or a uniform phase of continuously varying composition. In this exact sense a hydrosol is not a solution but a mixture of matter in two states more or less clearly distinguishable since it consists of a fluid which has dispersed through it large particles, the colloid particles which are not of the same order of magnitude as the molecules of the fluid and which are separated from the fluid by a surface, the nature and properties of which may almost be said to define the colloidal state.

Colloidal solution therefore is a state of partial solution, and it passes, as Picton and Linder showed, by insensible gradation into true solution. Its exact position in the phenomena of the liquid state may be indicated as follows.

When a homogeneous liquid separates into two states, say a liquid and crystals, the and alkali globulin can be separated in the solid state by dissolving globulin with minimal amount of acid or alkali and evaporating to dryness *in vacuo* over sulphuric acid and caustic potash.

When hydrochloric acid is used the dried HCl globulin is found on analysis by Carius' method to contain all the chlorine used to dissolve the globulin. HCl globulin therefore is stable *in vacuo* in presence of solid KHO, and the acid may be regarded as being in true combination. Acid globulin dissolves in water.

Alkaline globulin can be separated in the solid state from, *e.g.*, solution in ammonia by drying *in vacuo* over sulphuric acid. When redissolved the solid shows the same order of molecular conductivity as the solution from which it was dried out.

Both acid and alkali globulins ionise in solution. That is to say in solution the globulin is electrically charged and takes part in any electric transport, and the velocity can readily be measured by the boundary method. Therefore one must recognise "ionic" globulin as a species, but reasons will be given for regarding the ions as peculiar and special to the colloidal state. By reason of their ionisation solutions of acid and alkali globulin are good conductors.

Globulin therefore is an amphoteric substance and its acid function is much stronger than its basic function. As an acid it is strong enough to form salts readily with bases so weak as aniline,[1] glycocol, and urea; acting as a base it forms salts with weak acids, such as acetic and boracic acid, which are very unstable in presence of water.

The precipitation of globulin by salts. Globulin solutions can be precipitated by neutral salts, and in this respect they exhibit very characteristic relations. Each salt acts as a precipitant at low concentration, and at high concentration. Between these two it acts as a solvent. The first precipitation occurs only when acid or alkaline globulin is present, and it is similar to

new state appears first as minute droplets which represent an instable state and which changes to the stable crystalline state in a longer or shorter time according to the relative magnitude of the force of crystallisation and the surface tension (Garnett, *Trans. Roy. Soc.* (1904), A, CCIII, 385).

Substances which do not readily crystallise, whose molecules therefore have feeble polarity, permit the instable emulsion state to become subpermanent, and this is the state known as colloidal solution. All cases of colloidal solution, however, are not necessarily states of incomplete crystallisation, though they are states of incomplete separation of a new phase.

To include these states Picton and Linder introduced the conception of grades of solution, the grade being high or low according to the degree of dispersion of the solvee in the solvent.

[1] Of a weak base of limited solubility such as aniline it is almost more accurate to say that it is dissolved by globulin, than that the globulin is dissolved by the aniline. Aniline will not combine with anhydrous globulin.

the precipitation of hydrosols by small concentrations of salt in that the colloid particles are electrically charged. The second precipitation is a separation of solid globulin from a solution of salt globulin, and it is the precipitation of the colloid from a hydrosol in which the colloid particles are completely uncharged.

Thus solutions of acid or alkali globulin are precipitated by the addition of neutral salts, further addition brings about re-solution always in the case of alkali globulin, sometimes in the case of acid globulin. Still more salt brings about reprecipitation. In the first solution the globulin is ionic, in the second solution it is not ionic.

The mechanism of these two precipitations, in spite of views to the contrary,[1] I take to be distinct.

The precipitation of colloidal solutions by salts has been explained in two ways as being due (1) to a condensation of the electrically charged colloid particles to larger and heavier masses owing to the nuclear action of those salt ions which carry a charge of the opposite sign to that on the colloid particles, or (2) to a dehydrating action of the salts. In case one, as Picton and Linder first showed, the precipitate carries with it a fraction of the active ions. The solid phase, therefore, is a compound between the colloid and one of the salt ions, of the nature of an absorption compound, and the nuclear action of the salt ions is due to chemical reaction which is peculiar only in the chemically indefinite nature of the resulting compound. As Picton and Linder showed long since the compound formed between the colloid and the ion of the salt can be decomposed in quite an ordinary way, the salt ion being replaceable by an equivalent weight of another ion. It is wrong to speak of the action of salt on a colloid particle as "electrical combination" as though it were something quite apart. It is "electrical combination" only in the sense that reaction between electrolytes is electrical.

Now if the action between salt and colloid is chemical and identical for instance with the selective reaction in dyeing which discriminates between acid and basic dyes, it would be represented by a generalised equation of the form:

$$C'H^{\cdot} + B^{\cdot}S' = CB + HS.$$

That is to say, the colloid functions as an acid and by the law of mass action, the compound $C'B^{\cdot}$ is formed because of its insolubility.

Consider the hydrosol of a metal such as platinum. The colloid particles are negatively charged, they are anionic in character, and the charge is due to a reaction between the metal and water at the moment of formation of the hydrosol whereby the hydride PtH is formed which ionises in the sense

$$PtH \rightleftharpoons Pt' + H^{\cdot}.$$

[1] Pauli, *Beit. f. exp. Physiol. u. Pharmak.* (1905), VI, 233.

The chief number of the ions Pt′ are in the form of masses so large as to have the properties of matter in mass, they are not of molecular dimensions and they form an internal phase.[1] Ionisation is a phenomenon of the surface of these masses only. It confers on the particle its electric charge, and it is in this case the "incomplete chemical combination" which Lord Rayleigh regards as the source of contact differences of potential to which I referred in an earlier paper.[2] In reactions therefore an electro-positive colloid is a weak alkali, *e.g.* hydrosol of ferric hydroxide, and electro-negative colloid a weak acid, *e.g.* silica.

This view of the source and nature of the charge on the particle of a hydrosol, as it seems to me, is proved by the researches of Burton,[3] who finds clear evidence of selective chemical reaction when hydrosols of metals are formed in various media.

Though one may speak of the colloid particles as being ionic in nature they are sharply distinct from true ions in the fact that they are not of the same order of magnitude as are the molecules of the solvent, the electric charge which they can carry is not a definite multiple of a fixed quantity, and one cannot ascribe to them a valency, and their electrical relations are those which underlie the phenomenon of electrical endosmose. To such ionic masses I would give the name "pseudo-ions", and I propose to treat globulin solution from the standpoint of a hypothesis of "pseudo-ions".

Amongst crystalloids there are substances which exhibit electrical phenomena in solution, the electrolytes, and substances which do not exhibit electrical phenomena, the non-electrolytes. Electrolytes are so called because the conditions of solution are such that the charged units of the solvee can carry equal amounts of + and − electricity and discharge on to the electrodes. It occurred to me that in non-electrolytic solution the particles of the solvee might be charged all of the same sign, and the condition such as to make discharge on to the electrode impossible. I was therefore led to look for instances of electrical convection in non-electrolytic solution and I satisfied myself that the solvee may be absolutely iso-electric with the solvent, but I failed to find a single such case when the solvent was water. Fields up to 500 volts per centimetre were used, and in aqueous solutions such bodies as urea, glycerine, glycocol, cane-sugar, and iodoform moved. From their chemical constitution and relations the first five are weak bases or weak acids, and they move as such. Iodoform undergoes decomposition.[4] In a non-ionising solvent such as chloroform solutions of azobenzene showed absolutely no drift. They are completely iso-electric.[5]

[1] Hardy, *Proc. Roy. Soc.* (1900), LXV, 110.
[2] *Ibid.* p. 66.   [3] *Phil. Mag.* (not yet published).
[4] Hardy and Willcock, *Proc. Roy. Soc.* (1903), LXXII, 200.
[5] The experimental method is given in the appendix.

These facts are paralleled amongst colloids. One has hydrosols such as those of metals, of acid and alkali globulin, and of acid and alkali albumen, of silica, etc., in which the solvee is highly charged. One has relatively iso-electric hydrosols such as those of glycogen and starch, and hydrosols in which the colloid particles are completely uncharged such as solutions of salt globulin or gelatine.[1]

On the theory already stated the charge on the colloid particle will be due to its acid or alkaline nature.[2] Thus cane-sugar acts in chemical combination as an acid, and in aqueous solution it moves as an anion. Similarly its chemical relatives, the colloids glycogen and starch, move in a field as though they were anionic.

The theory of precipitation of colloidal solution by salts which has been considered so far, clearly is applicable only to cases when the colloid solution is electrically active. The precipitation of electrically inactive hydrosols by neutral salts needs some quite different explanation, and Billitzer[3] and Pauli,[4] in a series of most interesting papers on the precipitation of proteids, are wrong in attempting to bring all cases of precipitation by electrolytes within the one theory.

The precipitation of electrically inactive hydrosols can be adequately explained on the lines suggested by Spiro[5] as a separation into two phases, one a solid phase rich in the colloid, poor in salt and water, the other a fluid phase, rich in water and salt, poor in the colloid, the action being exactly similar to the salting out of alcohol from a mixture of alcohol and water by the addition of magnesium carbonate.

Electrical precipitation is distinguished from salting out in the same way as is the precipitation of calcium from a solution of the hydroxide by potassium sulphate to the precipitation of potassium sulphate by calcium sulphate as the double salt $K_2SO_4CaSO_44H_2O$.[6] In the one case the precipitant is decomposed, in the other case it is not. The precipitation of electrically active hydrosols is distinguished also by the small concentration of electrolytes necessary to produce the change, whereas a high concentration is necessary to salt out.

According to Hofmeister[7] the precipitating action of salts is due to their "water-depriving" action. Clearly this view is not applicable to the cases where salts act as precipitants at very low concentration. It is, however, a

[1] The drift measured by Whitney and Blake (*Amer. Chem. Soc.* (1904), XXVI, 1339) is due to partial electrolytic decomposition of the gelatine.
[2] Picton and Linder, *Journ. Chem. Soc.* (1897), LXXI, 568.
[3] *Wied. Ann.* (1903), II, 902; *Zeits. f. phys. Chem.* (1903), XLV, 307; (1905), LI, 129
[4] *Beit. f. exp. Physiol. u. Pathol.* (1901), III, 225; (1903), V, 27; (1905), VI, 233.
[5] *Ibid.* (1902), IV, 300.
[6] Ditte, *Compt. Rend.* (1904), p. 1266.
[7] *Arch. f. exp. Physiol. u. Pharmak.* (1888), XXIV, 246; (1889), XXV, 1.

convenient way of considering salting out, the separation of the precipitate being due to the appearance of a lower hydrate of globulin just as the concentration of added salts determines by their dehydrating influence the particular hydrate of $MgSO_4$ which separates from a solution of the salt.[1]

Cases of precipitation in which both actions concur obviously are possible, as when for instance an electrically active colloid by reaction with one of the ions of a salt is converted into an electrically inactive colloid which is separated as a solid phase only when the salt concentration reaches a certain point. In this case, owing to the feeble chemical potential of the colloid, the first reaction by the law of mass action can take place only when the salt concentration reaches the point when the new compound is thrown out of solution. Precipitation of proteids by hydrolysed salts, such as ammonium sulphate, sodium acetate, and the salts of the heavy metals perhaps, are examples.

The simple salting out of proteids, so far as the imperfect knowledge which is available goes, is similar to the separation of a solid phase when ammonium sulphate is added continuously to a solution of potassium sulphate. The solid phase is a continuously varying solid solution of the two salts, therefore the composition of both fluid and solid phases alters continuously with addition of either component. There are no breaks in the curve. Both Pauli's papers and Osborne's[2] recent paper on globulins give indications, however, that when salts of the heavy metals are used the concentration curve shows breaks which would correspond to a change in the nature of the solid phase, such as is seen for instance when lead iodide and potassium iodide in solution are in equilibrium with a solid phase. The latter according to the composition of the fluid phase is lead iodide, or potassium iodide, or a double salt.[3]

In an earlier paper[4] I used the adjective "reversible" to distinguish certain colloidal systems such as gelatine water from systems such as silica water, because in the former case gelation occurs with a fall of temperature, and is reversed by a rise of temperature. The word "reversible" was unfortunate in the first instance, since the last thing one would say of colloidal systems is that they undergo changes of state which are reversible in the strict physical sense of the word, but no other word equally convenient suggests itself. Pauli has applied the word in quite another way, and unless the point is cleared up confusion will arise.

When a new state of matter is formed in a system it may be said to be

[1] Van't Hoff, Meyerhoffen, and Smith, *Preuss. Akad. Wiss. Berlin* (1901), XLI, 1034.

[2] *Amer. Journ. Physiol.* (1905).

[3] Schreinemakers, *Zeits. f. phys. Chem.* (1893), x, 467.

[4] *Proc. Roy. Soc.* (1900), LXVI, 95.

irreversible with respect to any one of the factors which produced it, which are the relative masses of the components, temperature, and pressure. A solid phase for instance may become again miscible with a fluid phase owing to a rise or fall of temperature, in which case the state is reversible with respect to temperature. I used the word reversible originally only in this case. The solid phase may disappear when any one of the components is added, or removed, as when globulin is redissolved by the addition of water, or of salt. Pauli describes precipitates of proteids as being reversible or irreversible solely with reference to one component—water, though amongst the cases which he has studied are instances of irreversibility with respect to water, reversibility with respect to either of the other components salt and proteid: and the work of others from Alex Schmidt downwards shows that many of his cases of irreversible precipitates may be expected to be reversible with respect to temperature.

Preparation of the globulin. Except where it was otherwise stated globulin from ox serum prepared as follows was used.

The serum was centrifuged to free it from particles, diluted tenfold, and acidulated with acetic acid. The copious precipitate was purified by being twice dissolved in alkali (usually ammonia) and precipitated by acid (usually acetic) or *vice versa.* To free it from excess of adherent salts it was suspended in water centrifuged off, resuspended, collected on a hardened filter and washed with water. Sometimes prolonged dialysis was resorted to. Toluol and camphor were tried as disinfectants, and the dialysis was carried out in closed vessels. Water saturated with the former seems to be without action. Camphor acts on alkali globulin, producing an insoluble modification; it seems to be without action on acid globulin. Where it was specially important to follow the washing out of the salts, hydrochloric acid was used in the preparation in place of acetic as a precipitant, and the wash-water tested for chloride.

Acid and Alkali Globulin

The point of solution by acids or alkalis is indeterminate. In the change from the mixture of globulin suspended in water to a brilliant transparent solution, or backwards from a brilliant solution to a suspension, a continuous series of events occur which make it impossible to say where solution is complete.

The reverse change, clear solution towards suspension, may be produced by dialysis. When the acid or alkali is monovalent the solution gradually changes towards whiteness until it becomes opaque. At the same time no precipitate falls and, in the absence of any disturbing factor, dialysis alone will not precipitate the globulin.

For example—some globulin was dissolved in acetic acid to a clear transparent solution; concentration of acid, 0·006 equivs. per litre. It was dialysed in a closed vessel against four successive lots of boiled distilled water which in each case were allowed to remain until equilibrium between dialysate and dialysee had been approximately reached.

	Acidity of dialysate gr. equivs. per litre	Condition of the dialysee
0	0·006	Limpid transparent solution.
I	0·0023	Transparent: light opalescence.
II	0·00095	Translucent: white opalescence, smoky amber by transmitted light.
III	0·00025	Opaque white: objects still visible through thin layers (0·5 cm.) if strongly illuminated.
IV	0·00012	Completely opaque milky fluid, still no precipitation.

No. IV endured without precipitation for 20 days, when the equilibrium was destroyed by moulds appearing and the globulin was thrown down *en masse*.

Similar phenomena are seen when strong acids such as HCl, or alkalies such as KOH, NaOH, or NH₄OH are used.

It follows from the above that the grade of solution which a given quantity of solvent will produce depends primarily upon the ratio mass of globulin to mass of acid or alkali. The effect of the degree of dilution, that is to say of the mass of the third component, water (freed from gases), on the grade of the solution is too slight to be detected when the solution is one of alkali globulin, and it is almost negligible for solutions of globulin in strong acids. Thus the continuous addition or subtraction of either component produces a continuous series of changes with no break corresponding to the separation of a second phase. As a special case take the removal of the component water by evaporation; at no point is there a separation into two phases, the components are miscible in all degrees. Dry globulin, dried *in vacuo* over sulphuric acid and caustic potash, swells to a jelly in presence of exceedingly dilute alkali or strong acid, and the jelly is miscible with the fluid in all degrees.

As the ratio of the mass of acid or alkali to the globulin is reduced and the grade of the solution diminished its stability also is diminished, a smaller and smaller quantity of acid or alkali, as the case may be, or neutral salt being needed to destroy it.

The statement that no second phase separates needs qualification. The change from transparency to opacity means the appearance within the fluid of particles of the order of at least 10^{-6} cm. in size and which therefore are defined by a surface. They are not of molecular dimensions and they constitute a phase which separates in a special way in that it never aggregates to form a separate layer. It has long been known that in the separation of a new phase from a fluid the first stage is a labile state, in which the new phase appears as minute spheres condensed on to some nucleus; this is true even when each sphere will ultimately give rise to a crystal. The characteristic of the globulins, and probably of very many hydrosols, is that the labile stage of crystalloid solution becomes subpermanent, the condition being in all probability one of false equilibrium.

Looked at in this way the change of state due to the continuous addition or subtraction of either acid or alkali or globulin is a continuous change in the composition of *two* phases, a fluid phase poor in globulin, and an internal fluid or solid phase rich in globulin. Both phases vary continuously, as do for instance the fluid and solid phases when either ammonium or potassium sulphate is added continuously to a system composed of both salts and water. The solid phase which separates is a continuously changing mixture of the two salts to which the name solid solution has been given.

Now it is possible to speak of the solution of acid in alkali globulin as being composed of two continuously varying solutions or of two continuously varying compounds, and the latter is the better because it takes account of a number of purely chemical phenomena such as reaction to indicators, hydrolytic reaction with the water, or electrolytic dissociation, which occur.

The system differs from a mixture of the gases ammonia and hydrochloric acid, since though the latter in the solid state combine to form NH_3HCl the gas system composed of reacting weights $(NH_3 + HCl)$ does not contain the compound $NH_3.HCl$. Acid and alkaline globulins show by neutralisation and other phenomena that throughout there is true combination, though when the compound is between globulin and a weak acid the condition of free globulin and free acid, similar to $NH_{3gas} + HCl_{gas}$, is realised at great dilution.

In the system two salts and water when a solid phase appears the equilibrium over a certain range of concentration is usually between a fluid phase and a definite double salt (*e.g.* potassium and silver nitrates, calcium sulphate and potassium sulphate, silver chloride and sodium chloride), and the two phases do not vary continuously. When globulin is dissolved by the bibasic acid, sulphuric acid, a similar discontinuity is found. Removal of the acid by dialysis leads to the formation of a solid phase which separates completely as a precipitate and which appears to be a special insoluble compound of acid and globulin. After washing with water it is found to be insoluble by neutral salts, that is to say the special globulin feature is lost. It is, however, soluble by dilute alkali, and the globulin can then be recovered by neutralisation with its original properties unchanged.

The foregoing considerations make it clear that to define the globulin system two curves are needed, one for the changes of composition of the external fluid, the other for the internal non-settling or settling phase, the two curves being those to which Roozeboom gave the names respectively of "liquidus" and "solidus".[1]

[1] Cf. Heycock and Neville, "On the constitution of the copper-tin alloys," *Phil. Trans.* (1903), A, CCII, R. 1.

The solvent powers of various acids and alkalis. In order to measure the different solvent power of reagents two methods were employed, namely the addition of minimal amounts of acid or alkali to a suspension, which after standing for some time was centrifuged to get rid of the excess proteid, or the matching of solutions, taking care to allow for the depth of the layer, and to make the final comparison at the same concentration of the proteid.

In the second case one measures the amount of solvent needed to produce an arbitrary grade of solution at volume V. The first case is more complex, since it includes the distribution of acid and proteid between a fluid and a precipitate when the reacting masses are the same. The grade of solution in this case may differ very widely, being perhaps opaque white in the one case and transparent opalescent in the other.

By the second method some 250 measurements were made at temperatures from 2 to 30° and at concentrations from 0·28 to 4·18 dry globulin per 100 c.c. with nine different specimens of globulin.

If the mean value for the amount of acid required to dissolve 1 gr. of dry globulin is given in terms of $HCl = 1$, we have:

HCl	1·0	H_2SO_4	1·91	Citric	3
HNO_3	0·995	Tartaric	1·994	H_3PO_4	2·9
$CHCl_2COOH$	1·0	Oxalic	1·9	H_3BoO_3	Very great excess
CCl_3COOH	1·0				
$CH_2ClCOOH$	1·05				
$HCOOH$	1·25				
CH_3COOH	5·2				
CH_2CH_3COOH	7·56				

The values for strong monobasic acids are approximately the same within the limits of error. The first four acids are strong acids, and the value is that of HCl. Monochloracetic acid is rather less than half the strength of these, but the slightly diminished solvent action is within the limits of error. In the last three, which are less than a tenth the strength, the fall in solvent power is decided. As the dilutions employed were of the order of 0·005 normal all the acids are strong acids with the exception of the fatty acids.

	μ_v	$\dfrac{0.006\,N}{\mu_\infty}$	t	$\dfrac{\mu}{\mu_\infty}$
HCl	3654	3672	18	
HNO_3	3636	3665	18	>0·99
CCl_3COOH	3570	3560	25	
$CHCl_2COOH$	3300	3584	25	0·92
$CH_2ClCOOH$	1550	3603	25	0·43
$HCOOH$	673	3740	25	0·180
CH_3COOH	207	3610	25	0·057
CH_2CH_3COOH	193	3560	25	0·054
H_2SO_4	3190	3552	18	0·898
Oxalic	1680	3500	18	0·480
Tartaric	850	3500	18	0·243

For the strong acids the solvent power is a molecular function,

$$HCl = H_2SO_4 = H_3PO_4 ;$$

and if solution takes place through salt formation the salt is GHCl, GH_2SO_4, or GH_3PO_4.

This result agrees with that of Bugarsky and Liebermann[1] for combination of egg albumen or albumose with HCl, and as they point out, it resembles the combination of NH_3 and HCl to form NH_4Cl.

This type of combination in which molecules combine together without replacement is characteristic of the amido acids which react with other acids as basic anhydrides,

$$CH_2\begin{cases} NH_2 \\ CO_2H \end{cases} + HCl = CH_2\begin{cases} NH_2HCl \\ CO_2H \end{cases}.$$

The values for solution by alkaline bases show the same interesting relation. Taking the value for NaOH as unity we have

KOH	1
NaOH	1
NH_4OH	0·98
$Ba(OH)_2$	2·008

Aniline
Urea } dissolve; relation not measured quantitatively.

Here again the molecule of $Ba(OH)_2$ has the same solvent power as the molecule of KOH, NaOH, and NH_4OH. But the series differs from the acid series in the fact that the base NH_4OH, which is about as weak as acetic acid, has the same solvent power as the strong bases.

Aniline in aqueous solution readily dissolves globulin, urea only with difficulty and at high concentration. It can be compared to boracic acid, which also dissolves globulin but only in saturated or nearly saturated solutions. Picric acid is a good solvent of globulin, but a secondary change rapidly sets in which leads to precipitation, the precipitate being resoluble in excess alkali. Whatever the strength of acid or base (urea and picric acid alone were not tried) the effect of solution is that the globulin becomes electrically active and moves in a field as though true salt formation had occurred. That is to say, dissolved by acid it is positive (GS $\rightleftarrows$ G· + S′), dissolved by alkali it is negative (GB $\rightleftarrows$ G′ + B·). It behaves like an amphoteric electrolyte.

Gravimetric determination of the solubility of globulin was made for HCl, H_2SO_4, and H_3PO_4 in order if possible to confirm the molecular relation described above, $G = HCl = H_2SO_4 = H_3PO_4$. The conditions are complex owing to the presence of the solid phase. The grade of solution is not the same, that of hydrochloric acid being much higher (more transparent,

---

[1] *Pflügers Arch.* (1898), LXXII, 51.

less opalescent) than the others. The mass of globulin retained in solution by equivalent weights of the acids depends not only upon the concentration of the acid but also upon the concentration of the globulin. It is not sufficient to define it as being in excess. There must be a partition of the acid between the globulin in solution and the globulin in the precipitate, and one is measuring the distribution of the globulin between two solutions, the one poor, the other rich in proteid, which are separated under the stress of centrifugal force. The latter factor cannot be neglected, as van Calcar and Lobry de Bruyn[1] have shown that the composition of solutions of ordinary crystalline bodies such as sodium sulphate is altered by centrifuging. The four groups of measurements which were made agree in showing an approximation to the molecular solvent power already noticed.

Globulin 2·721 gr. per 100 c.c.

Vol. in lit. of 1 equiv. of acid	HCl	H_2SO_4
1333	0·0084	—
1000	0·0186	0·0074
500	0·0775	0·0183
383	0·0939	—
328·6	0·0979	—
250	—	0·0479

The figures show the grammes of dry globulin in the solution per 1 c.c. $N/100$ acid added.

A similar comparison of HCl and H_3PO_4 when equal amounts of $N/100$ HCl and $3N/100$ H_3PO_4 were run in gave:

1 c.c. $N/100$ HCl $= 0·0245$ gr.   1 c.c. $3N/100$ $H_3PO_4 = 0·0254$ gr.

The differences in the solvent power of the various acids and alkalis can be explained by assuming that the proteid compounds undergo hydrolytic dissociation in the way so well discussed by Osborne[2] for caseinogen.

According to the well-known theory of hydrolysis the fraction hydrolysed will be greater the weaker the acid, and will be diminished by an excess of acid, the excess needed being smaller the stronger the acid. Therefore, in order to reduce opalescence to a minimum, the acid required is the amount necessary to combine with the globulin plus the excess necessary to depress hydrolytic dissociation. Call these amounts respectively m and n, then for a strong acid $m + n$ will not differ largely from m in value, for a weak acid $m + n$ will be much larger than m, and this is what the figures show. In the case of an acid so weak as boracic acid n is so large compared with m as to lower the solvent power to a very low level.

The effect of volume on the amount of acid necessary to dissolve also can be explained by the theory of hydrolysis.

[1] *Rec. Trav. Chim. Leiden* (1903), xxiii, 218.
[2] *Journ. Physiol.* (1901), xxvii, 398.

I take as the unit the gram equivalent $\times 10^{-5}$ and express in this unit the amount of acid necessary to dissolve 1 gram dry globulin:

Conc. proteid gr. per 100 c.c.	H$\bar{A}$	HCl
0·28	183	23
1·46	56	15
4·18	40	13

On the theory of hydrolytic dissociation the fact that the weak alkali NH_4OH has the same solvent power as KOH and NaOH can only mean either that the ammonium proteid compound is not ionised or that globulin is a much stronger acid than base. As we shall see, the electric conductivity shows that NH_4OH globulin conducts rather better than NaOH globulin, therefore the proteid must be a stronger acid than base, and this conclusion is supported by other observations.

In comparing the solvent power of acids and alkalis comparison is only possible for the strong acids.

Acids are always feebler solvents than alkalis.

The actual figures are (as mean values) in the same units (equivs. $\times 10^{-5}$ to 1 gr. proteid):

$$\left.\begin{array}{l} NaOH, KOH \\ NH_4OH \\ C_2H_3NH_3OH \end{array}\right\} 10 \qquad \left.\begin{array}{l} HCl, HNO_3 \\ CCl_3COOH \\ CHCl_2COOH \\ CH_2ClCOOH \end{array}\right\} 18$$

There can be, I think, little doubt that the true ratio is 10/20, since the method of comparison being an optical one, the volume correction would increase the difference actually observed.

The degree of hydrolysis, that is the fraction of a salt split into free acid and free base by interaction with water, usually increases with rise of temperature. Osborne[1] used this temperature relation to support the view that the opalescence of certain compounds of caseinogen is due to liberation of the proteid by hydrolysis. The opalescence of caseinogen solutions increases with a rise of temperature, as it should do according to theory. But the opalescence of solutions of globulin diminishes with rise of temperature, even when the solvent is a weak acid. The amount of acid required to produce a given grade of solution therefore *falls* with a rise of temperature.

The change of state from 2 to 40° is not very great, and probably complex. If it were due to simple change in the degree of hydrolysis the opalescence should return on cooling. This it does either very slightly or not at all.

Globulin in combining with molecules and not equivalents of acids

[1] *Journ. Physiol.* (1901), XXVII, 398.

behaves like the amido acids; *i.e.* it acts like an anhydrous base. But judging from the solubility data it reacts also with the molecule of a base, *i.e.*

$$G = NaOH = Ba(OH)_2.$$

That is, it reacts as though it were an acid anhydride analogous to CO_2 in the hypothetical equation:

$$CO_2 + NaOH = NaHCO_3.$$

So far as I know no instances of such a reaction have ever been detected. The behaviour of globulins with indicators furnishes another suggestion.

Acid or alkali globulin and indicators. Freshly precipitated and washed globulin or globulin dried *in vacuo* in presence of H_2SO_4 and KOH reacts acid to thoroughly dialysed litmus, gives no colour with phenolphthalein, and only very slightly depresses the orange tint of methyl orange. The acid reaction with litmus might be due to CO_2, or to other acids held by the proteid.

To test this globulin, dissolved in NH_4OH and precipitated by HCl three times, was washed at $5°$ C. until the wash water was free from chlorine, and the proteid when dissolved in large excess of HNO_3 gave no clouding with silver nitrate. It was then dried and kept *in vacuo* over H_2SO_4 and KHO for some weeks. Distilled water free from CO_2 coloured with litmus was turned red on addition of the solid *in vacuo*. A solution of $Ba(OH)_2$ was made neutral to phenolphthalein by it *in vacuo*.

A strong alkali such as NaOH even in $N/100$ solution gives a fairly sharp neutral point with globulin, two drops excess giving decisive colour with phenolphthalein.

Methyl orange behaves in a specially interesting way. In presence of a suspension of globulin it is orange, on addition of $N/100$ HCl or H_2SO_4 drop by drop it turns at first bright canary, that is it sways over to the *alkaline* side, further addition of acid finally gives pink.

The relation of the point of maximum insolubility of the globulin to the reaction with indicators can be shown diagrammatically.

In the diagram, solution means the point at which opalescence is minimal. Now it is remarkable that in order just to neutralise globulin to phenolphthalein with the strong bases KHO or NaOH the amount of alkali needed is approximately twice that required to dissolve the proteid. On the other hand, the point of solution and of neutrality to the same indicator coincide for $Ba(OH)_2$. Therefore for the monovalent alkalis there are two values, one the amount necessary for solution (of the order of 10×10^{-5} equivs. per 1 gr. globulin), the other of the order 20×10^{-5} necessary to produce excess OH ions.

On the hypothesis of salt formation these facts suggest that globulin has at least two replaceable hydrogens, and that its acid salts of sodium or potassium are soluble, while its acid salt of barium is relatively insoluble, the neutral salt being soluble and neutral to litmus and phenolphthalein.

With a weak base, such as ammonia, the point of solution remains the same, but the point of neutrality to phenolphthalein is much more indeterminate but clearly beyond that for the strong bases.

The evidence which indicators furnish of the combination of globulin with alkalis being due to replaceable hydrogen is strengthened by other experiments.

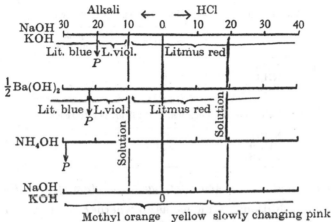

P = Colour appears with phenolphthalein
Scale: gramme equivalents to 1 gramme dry globulin

In our ignorance of the effect of dilution upon solution of the colloid type one is chary of ascribing much weight to coincidences between values obtained for molecular conductivity and those given by ordinary solutions. If, however, colloidal solution be an emulsion, one is reassured by the remarkable fact that in partially miscible liquids near the critical point, though internal friction and some optical properties show abnormalities, electric conductivity data are normal.[1] Now alkali globulin compounds ionise freely, as is shown by the fact that the conductivity of the weak base NH_4OH is raised by the addition of globulin (Table, p. 372). Therefore it is a remarkable coincidence that the values for $\mu_{32 \times 32} - \mu_{32}$ for NaOH just neutralised with globulin gives Ostwald's characteristic value for a dibasic acid.

Two determinations at 18° C. gave respectively

$\mu_{32 \times 32}$	μ_{32}	Diff.
55·3	34·4	20·9

But, as Säckur[2] points out, with such low values for μ the simple difference cannot be used, but comparison must be made between the ratio

[1] Friedländer, *Zeits. f. phys. Chem.* (1901), XXXVIII, 385.
[2] *Zeits. f. phys. Chem.* (1903), XLI, 672.

of the differences to the value at ν_∞. By this method, where the values of $\dfrac{\mu_1 - \mu_2}{\mu_1}$ are compared, globulin comes out as five basic.

Whatever the significance of the molecular relation between globulins and alkalis it is fundamental so far as proteids are concerned, since it is exhibited also by acid and alkali albumen derived from egg-white and freed from electrolytes by prolonged dialysis, using the same units, equivalents $\times 10^{-5}$ needed for 1 gr. dry proteid.

Acid albumen from egg-white (0·475 gr. per 100 c.c.).

	Solution	Faint pink with phenolphthalein
NaOH	27·3	57
NH$_4$OH	29	100
Ba(OH)$_2$	57	65

Alkali albumen from egg-white (0·363 gr. per 100 c.c.).

NaOH	36
NH$_4$OH	37
Ba(OH)$_2$	77

Measurements of the Acid and Basic Function of Globulin

In order to get additional evidence as to whether globulin is a stronger acid than base I measured the degree of hydrolytic dissociation of the HCl and of the NaOH compound by the velocity of catalysis of cane-sugar and of methyl acetate respectively.

Inversion of cane sugar. A suspension of ox-globulin containing 6·296 gr. dry globulin per 100 c.c. was used and the following solutions made:

    (i) Globulin, HCl, cane-sugar.

    (ii) Water, HCl, cane-sugar.

    (iii) Water with globulin ash dissolved in it, HCl, cane-sugar.

    (iv) Globulin, HCl, and water in place of the cane-sugar solution.

Acid and globulin were present in ratio 1 gr. globulin $= 9\cdot53 \times 10^{-5}$ equivs. acid.

Concentration of HCl 0·001837 equiv. per litre or $N/544\cdot4$,

    ,,   ,, globulin 19·275 gr. per litre,

    ,,   ,, sugar 8·2 per cent.

Solution (iii) was a control to determine the effect of the salts of the globulin on the velocity; 30 c.c. of the suspension was evaporated to dryness, carefully ashed and the ashes taken up with 30 c.c. water. This was used in making up solution (iii) in the same volume as the actual suspension in numbers (i) and (iv).

The four solutions were kept at 20 in closed flasks, the disinfectant being the hydrochloric acid present, and a trace of toluol.

At the outset the amount of $N/100$ NaOH necessary to produce maximal precipitation of the proteid was determined, and the proteid was removed in this way from some of (i) and (iv) to get initial readings.

After 140 hours solution (iv), *i.e.* the control, was precipitated by $N/100$ NaOH, and the precipitated proteid filtered off. The filtrate gave practically no rotation.

The proteid therefore was removed from solution (i) in the same manner by the addition of $N/100$ NaOH: the same volume of NaOH solution being added to numbers (ii) and (iii).

It was found that the amount of alkali needed to precipitate the proteid increased remarkably in the flasks with sugar present (i) but remained unchanged in the flasks with no sugar (iv).

In the former it was abnormally *low* at first, complete precipitation being produced by alkali equal to one-third the acid added, but after 140 hours the amount of alkali necessary slightly exceeded the amount of acid originally added to dissolve.

	a_0	a_n	A	Constant
(i)	$+11 \cdot 039$	$+10 \cdot 815$	$-3 \cdot 753$	$79 \cdot 4 \times 10^{-8}$
(ii)	,,	$+10 \cdot 215$	,,	$296 \ \times 10^{-8}$
(iii)	,,	$+10 \cdot 725$	,,	$111 \cdot 7 \times 10^{-8}$

This gives two values for the amount of "free" acid according to whether comparison is made with HCl alone, or with HCl and globulin salt ash. The former gives $26 \cdot 83$ per cent. of the HCl free, the latter $71 \cdot 1$ per cent. As the change of state of the proteid in presence of acid and sugar introduces an unknown error, these figures can be taken only as indicating in a general way that the proteid is a weak base and that therefore a large fraction of the compounds with acids undergoes hydrolytic dissociation in aqueous solution.

Catalysis of methyl acetate. Globulin purified by twice being dissolved in ammonia, and precipitated by HCl, then washed at $5°$ until filtrate gave no chloride with silver nitrate: dried *in vacuo* over H_2SO_4 and KOH.

Made into suspension with distilled water and kept as far as possible free from CO_2.

Three previous attempts to measure the constant showed that to obtain readable values and to take advantage of its antiseptic action it was necessary to have the methyl acetate present in a molecular concentration in excess of the NaOH.

The catalysis was carried out in a number of flasks at 20, and in order to obtain the constant for NaOH alone the same relative concentrations of alkali and ester were employed as in the globulin holding flasks.

Conc. globulin 1 litre $= 25 \cdot 76$ gr.

,,   NaOH $= 0 \cdot 0044$ normal,

1 gr. globulin $= 17 \cdot 127 \times 10^{-5}$ equivs. NaOH.

The NaOH used was freshly prepared from metallic sodium, and the globulin was dissolved in it to a point just short of giving colour with phenolphthalein, but in excess of the amount needed to dissolve (1 gr. $= 10 \times 10^{-5}$ equivs. NaOH). That is to say the point is as near as possible to the hypothetical neutral salt GNa_2.

Conc. methyl acetate $0 \cdot 02395$ gram-mol. per litre.

The fundamental equation for the velocity of the decomposition of the ester is

$$-dC = KC (C+s) dt^1,$$

where C is the concentration of the alkali and $(C+s)$ the concentration of the ester.

The solution of this equation given by Arrhenius and ordinarily used is an approximation which is valid only when s is very small compared to C. In my measurements s is large compared with C, and therefore it was necessary to use the general logarithmic solution,

$$K = \frac{1}{s} \times \frac{1}{t_n - t_1} \log_e \left(\frac{C_n + s}{C_1 + s} \times \frac{C_1}{C_n} \right).$$

I obtained a mean value for K for NaOH of $9 \cdot 9$ at $20°$, Hantzsch[2] measured this constant at $25°$ and found $10 \cdot 2$.

---

[1] Arrhenius, *Zeits. f. phys. Chem.* (1887), I, 115.

[2] *Ber. d. d. chem. Ges.* (1899), XXXVII, 3, 3066.

The constant for the NaOH proteid was calculated by the equation given by Shield:[1]

$$K = \frac{\dfrac{c_2}{c-c_2} \log_e \dfrac{c_2-x_0}{c_2-x_1} - \dfrac{c}{c-c_2} \log_e \dfrac{c-x_0}{c-x_1}}{K\,(t_1-t_0)}.$$

K is the constant for NaOH, c conc. of ester, c_2 the conc. of the NaOH proteid is reckoned as the normality of the NaOH present, x is the amount of ester decomposed in time (t_1-t_0) reckoned in minutes. Concentrations are given in hundredths of a gram-molecule per litre.

For the purpose of calculation the above equation is more useful in the form with Briggs' instead of natural logarithms:

$$K = \frac{c}{0 \cdot 4342945 \times K \times (c_2-c)} \left[\frac{\dfrac{c_2}{c} \log_{10} \dfrac{c_2-x_1}{c_2-x_0} - \log_{10} \dfrac{c-x_1}{c-x_0}}{t_1-t_0} \right].$$

Titrations were carried out with $0 \cdot 01$ N Ba(OH)$_2$ and phenolphthalein. That is to say one directly measured the amount of acetic acid produced by decomposition of the ester (methyl acetate + water = acetic acid + methyl alcohol) as the difference between titre 1 and the following titres in succession:

t	$c_2 - x$	x	$c - x$	K
1490	0·382	0·058	2·337	388×10^{-9}
2665	0·404	0·036	2·359	333×10^{-9}
4024	0·390	0·050	2·345	399×10^{-9}
5644	0·355	0·085	2·310	$[853 \times 10^{-9}]$

The last value at $t = 5644$ obvious bacterial decomposition had set in. The ester is a good antiseptic, and until it is largely decomposed bacteria do not seem to make headway. Considering the difficulty in judging the end-point of the titration owing to the opalescence of the solution the agreement in the values for K are satisfactory.

From the mean value of K one finds that $0 \cdot 288$ per cent. of the NaOH present is in the free form, that is to say, only $0 \cdot 288$ per cent. of the globulin NaOH compound is decomposed by the water.

The amount of NaOH present was, as I have said, in excess of the amount necessary to dissolve the proteid and near to the hypothetical "neutral" salt, therefore it is interesting to see that the degree of hydrolysis agrees with that of Na$_2$HPO$_3$ when Shield's value for the latter is corrected for dilution by the effect of volume on the degree of hydrolysis in other salts.

If this value is correct therefore globulin acts as an acid of a strength not far inferior to phosphoric acid.

These results show that acids react with globulins to form soluble compounds in molecular and not in equivalent ratios, but that the relation is obscured in the case of weak acids owing to the high degree of hydrolytic dissociation which occurs.

Alkalis also react to form soluble compounds in molecular and not in equivalent ratios, and owing to the stronger acid function of globulin the relation is less obscured by hydrolytic dissociation, so weak a base as ammonia not being displaced in the series.

The simplest interpretation of these relations would be that globulin compounds are:

$$\text{G} + \text{HS} = \text{GHS} \quad \text{or} \quad \text{G} + \text{BOH} = \text{GBOH}.$$

[1] *Phil. Mag.* (1893), (5), xxxv, 365.

The first is similar to the manner in which acids combine with the amido-acids to form salts

$$RNH_2.COOH + HS = RNH_2COOH.HS.$$

No analogy can be found for the latter equation.

In this difficulty the reaction with indicators suggests that globulin has a series of replaceable hydrogens and that it can therefore form acid salts, the acid salt of barium being relatively insoluble. This would account for the facts and yet leave open the possibility of combination by replaceable hydrogen,

$$GH_n + NaOH = GH_{n-1}Na + HOH.$$

The Electric Conductivity of Solutions of Acid and Alkali Globulin

Measurements of conductivity were made, except where it is otherwise stated, at 18°. Platinum terminals coated with platinum black in the ordinary way are not suitable for work with proteids; the platinum itself, or substances absorbed by it, cause changes in the proteid such as to raise the conductivity. This source of error is avoided by heating the coated terminals to red heat as recommended by Whetham, the platinum black being thereby at once changed to a fine grey deposit.

The units employed in the following pages are ohms and reciprocal ohms, and specific molecular conductivity or μ, following the plan in the British Association Tables, is given in c.g.s. units $\times 10^{11}$. K is used to designate specific conductivity, R specific resistance.

A suspension of a globulin always has a conductivity which, though low, is large enough relative to the conductivity of solution in acids or alkali to form an important correction. As an instance of the relative values to be dealt with the following will suffice.

A suspension holding 7·737 gr. globulin per 100 c.c.

Part was dried and charred, the char extracted with water, the char ashed at low red heat and the ash taken up with water so that the final volume of electrolytes of char and ash was that of the original suspension.

$$18° \qquad K_{suspension} \qquad 31 \times 10^{-6} \qquad K_{ash} \qquad 20 \times 10^{-6}$$

Various alkalis as 0·1 solution were added to dissolve the globulin,

$$1 \text{ gr. globulin} = 10·34 \times 10^{-5} \text{ equivs.},$$

and gave the following values corrected for the conductivity of the water used $[3 \times 10^{-6}]$:

K_{NaOH}	$288·5 \times 10^{-6}$	$\mu\ 389$	$\mu_{corr.}\ 352$
K_{NH_4OH}	$390 \quad \times 10^{-6}$	$\mu\ 526$	$\mu_{corr.}\ 489$
K_{HCl}	$409·7 \times 10^{-6}$	$\mu\ 553$	$\mu_{corr.}\ 515$
$K_{H\overline{A}}$	$183·4 \times 10^{-6}$	$\mu\ 283$	$\mu_{corr.}\ 243$

μ is the value obtained directly from the specific conductivity; it is K/c.

$\mu_{corr.}$ is the value corrected by subtracting $K_{suspension}$, or

$$\frac{K - K_{suspension}}{c}, \qquad c = 0·00741.$$

The correction for the conductivity of the globulin alone cannot be neglected, it is doubtful how it should be employed. If $K_{\text{suspension}}$ represents undecomposed proteid compound, either the observed conductivity or the sum $(K + K_{\text{suspension}})$ would be nearest to the true value; if it be due to absorbed salts, then observed conductivity minus the conductivity of the suspension is probably nearest to the true value. I have chosen the latter alternative and the values for μ given are $\dfrac{K - K_{\text{suspension}}}{c}$ except when it is otherwise stated.

In determining the effect of volume upon conductivity, therefore, two sets of measurements are used, one of the globulin compound at certain degrees of dilution, the other of the suspension at the same dilutions. The corrected conductivity is the difference.

HCl globulin. 1 gr. globulin $= 9 \cdot 318 \times 10^{-5}$ equivs. acid. 18°.

ν is the volume in litres of 1 gr. equiv.

ν	107·2	[135]	216·9	433·8	867·6	1735	3470
μ	439	[515]	596	705·6	834	991	1167

NaOH globulin. 1 gr. globulin $= 9 \cdot 75 \times 10^{-5}$ equivs. alkali. 18°.

ν	[32]	152·2	304·36	608·7	1217	2435	4869
μ	[320]	375·6	408·8	463	523·7	592	748

NaOH globulin. 1 gr. globulin $= 18 \times 10^{-5}$.

ν	32	64	128	256	1024
μ	334	401	451	484	549

The values in brackets do not belong to the series, they were made at different dates and on different samples of globulin and in both the ratio globulin to acid was different:

$[\mu_{135}]$ HCl proteid. 1 gr. globulin $= 10 \cdot 34 \times 10^{-5}$ equivs.

$[\mu_{32}]$ NaOH ,, ,, ,, $= 10 \ \ \times 10^{-5}$ equivs.

μ_{32} is difficult to obtain. The globulin solution at this concentration, and indeed up to and beyond ν_{107}, is as viscid as stiff treacle. Accurate measurements of volume therefore are practically impossible. A comparison of weights and volumes showed that the method adopted might give a maximal error of 1·8 per cent.

These values are plotted in curves (Fig. 1).

They are completely uncorrected for variations of viscosity—therefore it is remarkable that the curve of NaOH proteid should be of quite ordinary slope. The curves for the three phosphates of sodium are plotted alongside for comparison.

The table on p. 379 shows that in these solutions viscosity rises rapidly about $\nu = 110$. Botazzi[1] found that at a certain concentration of casein (8 per cent.) and egg albumin ($\pm$ 10 per cent.) viscosity rises to an "enormous value". It is therefore remarkable that the curves should show so even a sweep. The fact is in agreement with the observation of Friedländer[2]

[1] *Arch. Ital. de Biol.* (1898), XXIX, 420.

[2] *Zeits. f. phys. Chem.* (1901), XXXVIII, 385.

that in partially miscible liquids near the critical point conductivity is
normal, viscosity abnormal.

The chief point in a comparison of the curves for μ-volume is that the
molecular conductivity increases with increase of volume much more for
HCl globulin than for NaOH globulin owing to the weaker basic function of
the globulin. The molecular conductivity of a hydrolysed salt is given by the
well-known equation

$$M_\nu = (1-x)\,\mu_\nu + x\mu_{\text{HCl : NaOH}},$$

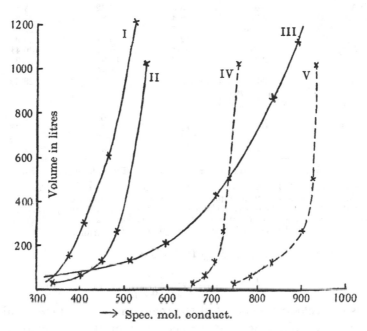

Fig. 1. Ordinates, the volume in litres occupied by 1 equivalent. Abscissæ, the
specific molecular conductivity (μ) in c.g.s. units $\times 10^{13}$ at 18°. I, NaOH globulin,
1 gr. globulin $= 9.75 \times 10^{-5}$ equivs. II, NaOH globulin, 1 gr. $= 18 \times 10^{-5}$ equivs. III,
HCl globulin, 1 gr. $= 9.32 \times 10^{-5}$ equivs. IV, NaH_2PO_4. V, Na_2HPO_4.

in which x is the fraction hydrolysed, μ_ν the molecular conductivity of the
undissociated salt, and $\mu_{\text{HCl : NaOH}}$ the molecular conductivity of the free acid
or base. Therefore as x increases with dilution and $\mu_{\text{HCl:NaOH}}$ is always
greater than μ_ν owing to the high ionic velocity of H˙ and OH, it follows
that M_ν must increase with dilution the more rapidly the greater the degree
of hydrolysis.

The slope of the curve for NaOH proteid is incompatible with any but a
slight degree of hydrolysis: that for HCl proteid might signify a considerable
degree of hydrolysis.

A comparison of the effect of globulin upon the conductivity of various

24-2

acids and alkalis, in other words a comparison of the value for $\dfrac{K_{salt}}{K_{acid}}$ or

$\dfrac{K_{salt}}{K_{alkali}}$, very clearly points to a stronger acid than basic function.

	$\dfrac{K_{sa.}}{K_{ac.}}$ or $\dfrac{\mu_{sa.}}{\mu_{ac.}}$	1 equiv. = ν = litre	gr. prot. per litre	1 gr. = equiv. $\times 10^{-5}$
$\dfrac{H_3BoO_3}{3}$	0·9	0·72	4·78	29050·0*
C_2H_5COOH	0·84	151·7	4·54	145·2
HA	0·8	225·6	16·72	23·49
$CH_2ClCOOH$	0·336	436·4	10·81	21·2
$CHCl_2COOH$	0·188	438·7	10·81	21·09
CCl_3COOH	0·191	442·7	10·81	20·9
HNO_3	0·24	453·4	10·81	20·41
HCl	0·24	454·7	10·81	20·3
H_2SO_4	0·18	273·2	8·96	40·8
H_3PO_4	0·22	289·7	12·05	28·63
	$\dfrac{K_{sa.}}{K_{alk.}}$ or $\dfrac{\mu_{sa.}}{\mu_{alk.}}$			
NH_4OH	2·3	1047	9·6	9·8
NaOH	0·29	1217·4	8·41	9·75
,,	0·21	304·36	33·67	9·75
,,	0·257	612·7	14·57	11·2

K_{salt} and K_{acid} are uncorrected for conductivity of the globulin suspension.

* This cannot be regarded as a measure of the solvent power of H_3BoO_3. The proteid was in large excess and the greater part precipitated on standing. A rough estimate would be 1 gr. proteid needs 100,000 equivs. to dissolve. The acid was twice purified by recrystallisation. The first column gives the relative specific conductivity of the solution of acid or alkali globulin (K_{salt}) to that of the same concentration of the acid or alkali alone ($K_{ac.:alk.}$).

The fundamental equation of hydrolysis is

$$K = K_{HOH} \frac{K_{salt}}{K_{alk.}\,K_{acid}},$$

where K is the velocity constant of the reaction salt + water = alkali + acid and K_{salt}, $K_{alk.}$, K_{acid} are the respective ionisation constants in the Guldberg-Waage equation, $KC_{AB} = C_A \times C_B$.

Therefore the degree of hydrolysis varies directly with the degree of ionisation of the salt (globulin HS or globulin BOH) and inversely with the degree of ionisation of the acid or alkali (HCl, NaOH, and globulin).

Assuming the salts globulin HS and globulin BOH to be ionised to about equal amounts in all cases, the degree of hydrolysis, that is to say the fraction x of HS and BOH in the free state, will be greater the weaker HS or BOH.

Since in the equation

globulin HS + HOH = globulin HOH + HS,

the chief conducting species will be HS and globulin HS, the molecular conductivity of the other two being negligible, it follows that the greater the degree of hydrolysis the more closely must the molecular conductivity approach that of the pure acid or pure base at the same concentration. Therefore the weaker the acid or base combined with the globulin the more closely should the value for μ_{salt} agree with $\mu_{acid:alkali}$, and the table shows this to be the case throughout the acid series in which, owing to the globulin being a very weak base, the fraction x hydrolysed is always considerable, and with such weak acids as the first three must approach unity.

The features in the alkaline series again can only be explained by supposing that the globulin salt ionises freely and therefore is a good conductor, and secondly that it is not hydrolysed to any great extent even when combined with the weak base ammonium.

In the equation $M_\nu = (1-x)\,\mu_{salt} + x\mu_{acid:alkali}$

it is clear that if $\mu_{acid\ or\ alkali}$ is less than μ_{salt} the value of M_ν will be reduced by hydrolytic splitting. This is the case with those salts of ammonia which ionise freely and hence are good conductors. μ_{NH_4OH} is low at all dilutions, the base being a weak one.

Now the table shows that M_ν is considerably larger than μ_ν. Therefore the ammonium globulin compound must ionise freely, and it must be hydrolysed to a relatively small extent, and this can be the case only if the globulin functions as an acid of considerable strength.

If NH_4OH globulin ionises freely and hydrolyses slightly, its molecular conductivity should exceed that of the NaOH globulin owing to the ionic velocity of $NH_4^{\cdot}$ being greater than that of $Na^{\cdot}$. A comparison made with the same globulin under identical conditions proved this to be the case.

$18^{\circ},\ \mu = \dfrac{K}{c},\ \mu_{corr.} = \dfrac{K_{observed} - K_{suspension}}{c},\ c -$ gram equiv. NaOH or NH_4OH per litre.

ν	NaOH		NH_4OH		HCl	
	μ	$\mu_{corr.}$	μ	$\mu_{corr.}$	μ	$\mu_{corr.}$
135	389	352	526	489	553	515
251	422	384	575	534	644	606
468	463	407	—	—	715	673

The most striking fact, however, is the difference in the values for HA globulin and NH_4OH globulin.

Acetic acid and ammonia have about the same molecular conduction. Therefore, if these globulin compounds behaved in the same way in solution, in the equation for hydrolysis

$M_\nu = (1-x)\,\mu_{salt} + x\mu_{acid:alkali}$

the relative values of μ_{salt} and $\mu_{acid:alkali}$ would be not far distant. The chief difference would be in μ_{salt} owing to the difference in the ionic velocities of acetion and ammonium ion, their respective values at infinite dilution at 18° being 36 and 67.

Later on we shall see reason to give a velocity of 15 to the globulin ion; therefore, assuming the degree of ionisation to be approximately the same, the ratio would be

$$\mu_{\text{NH}_4\text{OH globulin}} : \mu_{\text{H}\overline{\text{A}}\text{ globulin}} = 820 : 510 = 0.63.$$

In both cases, owing to the small conductivity of NH_4OH and $\text{H}\overline{\text{A}}$, μ_{salt} must be greater than $\mu_{\text{acid : alkali}}$ if the salt ionises freely (as the figures for NH_4OH globulin show it must do) although the velocity of the globulin ion is somewhat low.

Therefore, the greater the degree of hydrolysis in $\text{H}\overline{\text{A}}$ globulin as compared with NH_4OH globulin, the smaller will be the relative value of the molecular conductivity of the former. For unhydrolysed salt the probable value of $\dfrac{\mu_{\text{H}\overline{\text{A}}\text{ globulin}}}{\mu_{\text{NH}_4\text{OH globulin}}}$ is as we have seen 0.64. The best values obtained give 0.3 for $\mu_{\text{uncorrected}}$ and 0.27 for $\mu_{\text{corrected}}$. And whereas the addition of globulin raises the conductivity of solution of ammonia very markedly (i.e. M_ν is $> \mu_{\text{alkali}}$), the conductivity of acetic acid is lowered ($M_\nu < \mu_{\text{acid}}$) except in certain cases due to impurities, when it is only very slightly raised. No fact which I have met with so impresses me as this high conductivity of NH_4OH globulin. The solution presumably is colloidal, it has a high viscosity as judged by the time of transpiration through a capillary tube, it is opalescent, and the globulin can be recovered from it by precipitation unchanged. Indeed the alkali I have most freely used in the purification of globulin has been ammonia. Yet the globulin ammonia compound has a high conductivity. How high and how close to the molecular conductivity of an ordinary ammonium salt can be seen from the following figures.

Ostwald[1] showed that ionic velocity falls with increase in the number of atoms in the ion. He also showed that as the ion increased in size the further addition of atoms produced less and less effect. The lowest value obtained by him was 29 for an ion of 28 atoms.

The globulin ion contains many hundreds of atoms, but direct measurement proves that its velocity is not excessively low, and in the ammonium compound it must lie between 10 and 20 at 18. Take it as 15. The velocity of $\text{NH}_4^{\cdot}$ is 67, therefore, in our units,

$$\mu_\infty = v + \nu = 150 + 670 = 820.$$

Now the values actually obtained for NH_4OH globulin with different globulins and at different times were:

ν	μ	$\mu_{\text{corr.}}$
135	526	489
251	575	534
1047	696	
1940	827	

[1] *Zeits. f. phys. Chem.* (1888), XI, 840.

At $\nu = 1940$ ionisation is usually nearly complete for any ordinary salt and the agreement with the calculated and observed value of μ is very striking.

It is too good—so good as to be suspicious—therefore an attempt was made to determine the magnitude of certain probable errors due to viscosity or electrolytic impurities.

As all the ordinary salts of ammonia conduct much better than the alkali itself, any impurity which combined with the ammonia to form a salt, or even was set free by it, would raise the apparent value of μ. In passing I may notice that exactly the same holds for acetic acid whose conductivity is not raised by globulin. I therefore took a solution of starch of a conductivity higher than that of the suspension of globulin and compared its action on the conductivity of ammonia. I also determined the effect of the ash of globulin on the conductivity of ammonia.

In order to avoid loss by vaporisation of the salt the method adopted was as follows: A given volume of suspension was dried down at 100, charred at a temperature considerably below red heat and the black carbon extracted with hot water, the carbon was then ashed completely at dull red heat, the process occupying 24 hours, and the ash taken up with water and added to the previous extract. The combined solution was brought to the same volume as the original solution. The final solution I call "ash".

The starch solution used was made from finest potato starch washed by decantation. It gave practically no proteid reactions. It was converted into a paste with boiling water, the "solution" being thick and opalescent.

Conductivity values are given $\times 10^6$.

Globulin suspended in water. 7·74 gr. dry globulin in 100 c.c

K suspension 30·87. K water 3.

Starch solution. 1·74 gr. solid in 100 c.c.

K starch 38·6.

K "ash" 20.

These solutions were titrated with $0\cdot1\,N$ HCl, $H\bar{A}$, NH_4OH, and NaOH, the last being freshly made from metallic sodium, so that the final concentration was $0\cdot0074\,N$ or $\nu = 135$, $c = 0\cdot0074$. At this stage the globulin was dissolved to a transparent opalescent solution, as viscid as a heavy oil, by the alkalis, the solutions in acid were however white and opaque.

1 gr. dry globulin $= 10\cdot34 \times 10^{-5}$ equivs.

Control solutions of the same concentration ($\nu = 135$) were made by adding the $0\cdot1\,N$ solution to water.

Values of μ: those on the left corrected only for the conductivity of the water, those on the right ($\mu_{\text{corr.}}$) corrected for the conductivity of the "ash", starch, or globulin respectively $\left(\dfrac{K_{\text{obs.}} - K_{\text{susp.}}}{c} \right)$.

	Water	Ash		Starch		Globulin	
	μ	μ	$\mu_{\text{corr.}}$	μ	$\mu_{\text{corr.}}$	μ	$\mu_{\text{corr.}}$
$H\bar{A}$	172·8	191	164	178	131	247	210
HCl	3516	3554	3535	3301	3255	553	515
NH_4OH	122·7	155	123·5	158	111	526	489
NaOH	1961	1892	1865	1350	1301	389	352

How far are these numbers normal? Taking the globulin ion as having a velocity of 15, then we have:

$$\text{NH}_4\text{OH globulin } v + \nu = 15 + 67 = 82,$$
$$\text{NaOH } \quad \text{,,} \quad \text{,,} = 15 + 45 = 60,$$

$\frac{60}{82} = 0.73$ and $489 \times 0.73 = 357$, which is in close agreement with 352, the value actually found. Similarly the ratio for NaOH globulin/HCl globulin should be 0.7 and $515 \times 0.7 = 360$, which again does not depart widely from the number found. The HCl value is too high, and that would be accounted for by the greater degree of hydrolysis, but the degree of hydrolysis is depressed here by the excess globulin present, beyond the amount dissolved by the acid. Calculated in the same way $\mu_{\text{H}\overline{\text{A}} \text{ globulin}}$ should be ± 300. The actual value is therefore low, and this is accounted for by the greater degree of hydrolysis replacing the globulin $\text{H}\overline{\text{A}}$ compound by badly conducting acetic acid.

The effect of the "ash" on conductivity is remarkably small, and practically nil in the case of NH_4OH. The effect of the starch solution also is surprisingly small. When the conductivity of the starch solution itself is allowed for the effect always is to depress conductivity. But the depression is small compared with the high viscosity and it varies in the different cases, which proves that it cannot be any simple function of the viscosity of the liquid.

Conductivity is depressed in the various cases as follows:

	$\dfrac{\mu_{\text{starch}}}{\mu_{\text{water}}}$	100α
$\text{H}\overline{\text{A}}$	0.758	—
HCl	0.926	4.3
NH_4OH	0.904	—
NaOH	0.678	19.3

I have calculated Arrhenius' α-function[1] for the extreme cases and the result appears in the last column. The values found by Arrhenius for non-colloid non-electrolytes, and various electrolytes range between 1.84 and 2.95. It is clear therefore that the high viscosity does diminish conductivity but its effect is surprisingly small seeing that the starch solution had a viscidity too high to be measured by any ordinary viscometer.

The Viscosity of Solutions of Acid and Alkali Globulins

Sackür[2] refers the low conductivity of solutions of caseinogen compounds to their high viscosity. This explanation cannot be accepted without qualification for reasons already given and chiefly because the drift of the ionic proteid when directly observed is found to be remarkably rapid.

[1] *Zeits. f. phys. Chem.* IX, 487.   [2] *Ibid.* (1903), XLI, 672.

The observations already recorded on the effect of starch upon conductivity show that what is usually measured as the viscosity of a colloidal solution is not identical with or perhaps closely related to the internal molecular friction to which the ions are subject. I ventured to point out in 1899[1] that the high viscosity of colloidal solution is probably of complex origin and that it is very imperfectly analysed by ordinary methods of measurement. If colloidal solutions are heterogeneous states of matter the viscosity as measured by the flow in capillary tubes will be in part due to the influence of the friction of internal surfaces on the fluidity—a point which, as I then drew attention to, was discussed as early as 1851 by Frankenheim.[2] Transpiration methods measure fluidity, and this will be simply related to the true molecular internal friction only in homogeneous fluids.

Recently the point has been investigated by Garrett working under Prof. Quincke,[3] by the two classical experimental methods in which the time of transpiration through a capillary tube, and the decrement in the period of oscillation of a disc immersed in the fluid are taken as measures of viscosity. In the case of colloidal solutions the two methods gave different values for the viscosity owing, as Garrett points out, to the different incidence in the two cases of the various factors which modify the fluidity.

The fluidity of a non-homogeneous liquid will depend upon (1) the internal friction of each of the several states, (2) the surface friction of the internal surfaces, and (3) the surface tension of the internal surfaces. To these I would add a fourth factor, the density of the electric charge if the internal surfaces be electrified. This will modify the surface tension and it will decrease the fluidity owing to the fact that electric work must be done in deforming the solution.

Electric conductivity will depend upon the internal friction of each of the several states, upon the distribution of the electrolytes between the different states, and upon the coefficient of ionisation in each of the states and in the surface film bounding the states.[4] Therefore changes in gross fluidity, and that is what is measured, need not agree with changes in electric conductivity in magnitude though they usually will in direction. For instance the fluidity of gels is infinitely small, but Graham[5] and Voightländer[6] have shown that the velocity of diffusion of electrolytes in them is approximately

---

[1] *Journ. Physiol.* (1899), XXIV, 180.

[2] *Journ. f. prak. Chem.* (1851), LIV, 433.

[3] *Ueber den Viskosität u. d. Zusammenhang einigen colloid. Lösungen* (Inaug. Dissert. Heidelberg, 1903).

[4] Molecular conductivity and chemical potential are greater in the surface layer of a solution. Reinold and Rücker, *Phil. Trans.* (1893), A, CLXXXIV, 505. W. Spring, *Zeits. f. phys. Chem.* (1889), IV, 658.

[5] *Phil. Trans.* (1861), CLI, 183.   [6] *Zeits. f. phys. Chem.* (1889), III, 36.

the same as in water, and Whetham[1] has shown by direct measurement that the mobility of ions in agar agar jelly is very little less than in ordinary solutions.

The fluidity of a solution of gelatine decreases with lapse of time. But Garrett found that while the logarithmic decrement of a disc oscillating in the solution increased from 0·2502 to 0·4049 the electric conductivity remained constant.

For globulins I have found the following relations when the fluidity is measured by the transpiration method:

	Fluidity	Conductivity
HCl globulin	Rises with time	Remains constant
NaOH globulin	Falls with time	Falls with time

Time here means intervals up to 24 hours.

Bayliss[2] finds that in tryptic digestion electric conductivity does not change concurrently with change in fluidity, and instances might be multiplied further.

The absence of any simple relation between viscosity and electric conductivity was noticed by Reyher in the case of solutions of non-colloidal organic bodies, but there is a fair general agreement in the values.[3]

Measurements of viscosity yield some unexpected evidence as to the relative acid and basic functions of globulin.

On the view that acid or alkali casein solutions are solutions of proteid salts there would be present if hydrolysis took place the following molecular species, expressed in ordinary notation:

$$\text{BS,} \quad \text{B}^{\cdot}+\text{S}', \quad \text{BOH,} \quad \text{B}^{\cdot}+\text{OH}', \quad \text{HS,} \quad \text{H}^{\cdot}+\text{S}',$$

and Sackür[4] asks to which of these can the high viscosity of solutions of casein salts be attributed.

By a train of ingenious argument he fixes upon the proteid ion as the chief factor. As the casein acted always as acid in the cases considered by him, the high viscosity is due chiefly to S'.

This conclusion can be tested with extraordinary ease and certainty in the case of globulins since their solution in neutral salts contains no ionic, no electrically charged globulin, as is shown by the absence of all movement in an electric field, while solutions in dilute acids and alkalis always are "ionic". Therefore, if Sackür's view be correct, the former solutions should have relatively a quite low viscosity.

[1] *Phil. Trans.* (1895), A, CLXXXVI, 507.
[2] "Kinetics of Tryptic Action", *Arch. d. Sci. Biol.* XI, 261.
[3] *Zeits. f. phys. Chem.* (1888), II, 743.
[4] *Zeits. f. phys. Chem.* IX, 487.

Some measurements of viscosity made to test this point gave striking results. The specific gravity of each solution was determined for Series III, and the values found to differ only slightly; therefore, as the formula is a linear one, viscosity may be taken to vary directly as the transpiration times.

Pressure 29 mm. mercury. Temp. 19°.

Conc. of acid, alkali or salt		Gr. equivs. of acid, alkali or salt to 1 gr. globulin × 10^5	Transpiration time for the same volume in seconds	State of the solution
		Series I. 0·62 gr. of globulin in 100 c.c.		
NH₃	0·0005 N	8	57·1	Transparent whitey blue
	0·0008	13	58·1	„ brilliant blue
	0·002	32	58·7	„ „ „
	0·01	161	57·5	„ „ „
NaOH	0·002	32	58·7	„ „ „
HA̅	0·002	32	63·3	White opalescent
	0·005	81	56·6	Transparent
	0·015	242	58·1	„ .
NaCl	0·15	2419	58	
		Series II. 2·13 gr. of globulin in 100 c.c.		
NH₃	0·006	28	234	Transparent
	0·03	141	143·7	„
HA̅	0·012	56	130·5	Opaque
	0·03	141	134·3	Whitish transparent
NaCl	0·28	1315	107·2	Transparent
		Series III. 7·59 gr. of globulin in 100 c.c.		
NaOH	0·0084	11	3292	(a) Transparent slight opalescence
HCl	0·0084	11	753	White but non-settling
½MgSO₄	0·32	423	226	Matches (a)
½MgSO₄	0·32	No globulin	60	
	Water		48·5	

These figures show that the viscosity of solutions of alkali globulin and of acid globulin is much higher than that of solutions of salt globulin of the same strength and of the same grade. Therefore the presence of ionic globulin in solution raises the viscosity very much. In Series III, in which the proteid content is of the same order as the proteid content of blood plasma, the difference is surprising.

Dilution diminishes the difference between the various solutions, as is seen on comparing Series I, II and III.

When the opalescence of a solution is diminished by increasing the concentration of the solvent the viscosity is increased to a maximum; beyond this further addition of the solvent diminished the viscosity (Series I and II, NH₃).

The viscosity of acid globulin is less than that of alkali globulin even when the grade of solution is approximately the same.

The relation to opalescence shows that viscosity is diminished as hydrolytic dissociation increases. Reyher[1] found a similar relation between hydrolysis and viscosity in the case of soaps, and Sackür in the case of caseinogen salts. Therefore one finds an interesting and somewhat unexpected confirmation of the conclusion already arrived at that the acid function of globulin is much stronger than its basic function in the fact that an equivalent quantity of alkali will produce a more viscid solution of globulin than an equivalent quantity of acid.

Excess of alkali diminishes the viscosity at both high and low concentrations, though the effect is decisive only for the latter. This is no doubt partly due to the free alkali or acid depressing the degree of ionisation of the globulin salt, and so far it confirms Sackür's view that the high viscosity is due to the proteid ions, but, in view of the fact that ammonia globulin conducts better than ammonia itself, and that ammonia is a feeble base, the action is too marked to admit of this simple explanation. Taken in conjunction with the extraordinary rise in viscosity produced by increasing concentration it points to the presence in the solutions of complex ions which attain great size at high concentration and have then the properties of matter in mass. With their appearance the fluid ceases to be truly homogeneous and has the characteristic low fluidity of the heterogeneous state. Excess of alkali diminishes the size of these complex ionic states, and owing to their absence in dilute solution the proteid ion though present has no special influence on the viscosity (cf. Series I).

GLOBULINS AS PSEUDO-ACIDS OR PSEUDO-BASES

Hantzsch[2] has drawn a distinction between true acids and true bases and pseudo-acids and pseudo-bases. Pseudo-acids do not contain a replaceable hydrogen but they can change to an isomeride of a salt-forming type, and pseudo-bases are substances which by an isomeric change can act as bases of the ammonium hydroxide type. In each case the substance itself is neutral in solution, but by a change in the molecular structure it may act as an acid or alkali as the case may be and form salts. The chief characteristic of these substances is that, owing to the change of molecular state, the process of salt formation is gradual, there is a measurable latent period. Cohnheim and Krieger[3] suggest that proteids may be pseudo-acids and pseudo-bases in virtue of a labile atomic grouping. The suggestion is tempting, but I can find no justification for it in the case of globulins. I failed completely to detect any latency of the kind described by Hantzsch

[1] *Zeits. f. phys. Chem.* (1888), II, 743.

[2] Hantzsch, *Ber. d. deut. chem. Ges.* (1899), XXII, 306 and 575; (1899), XXXII, 607, 3109 and 3066; (1902), XXXV, 210. Kauffmann, *Zeits. f. phys. Chem.* (1904), XLVII, 618.

[3] *Zeits. f. Biol.* (1900), XL, 95.

in the reactions with acids or alkalis either with indicators, or by measurements of electrical conductivity. Amongst other indications of reactions with pseudo-acids or pseudo-bases is the existence of an abnormally large positive temperature coefficient of molecular conductivity. I measured the temperature coefficients for acetic acid globulin, and for NaOH globulin, and found them to be of normal magnitude

$$\text{H}\bar{\text{A}} \text{ globulin } 0 \cdot 5\text{–}23 \cdot 3° \quad \frac{1}{K_{23 \cdot 3}} \left(\frac{\delta K_{23 \cdot 3}}{\delta t} \right) = \cdot 0242,$$

$$\text{NaOH globulin } 0 \cdot 5\text{–}20 \cdot 5° \quad \frac{1}{K_{20 \cdot 5}} \left(\frac{\delta K_{20 \cdot 5}}{\delta t} \right) = \cdot 0214,$$

and neither system showed any appreciable lag.

One property which Hantzsch claims for pseudo-acids is manifested by globulins, namely in their inability to react with a dry base. Dry globulin is quite insoluble in and unchanged by dry aniline.

Direct Measurement of the Specific Velocity of Ionic Globulin

In order to measure the movement of the globulin in an electric field I tried first the method already used by me[1] of analysing the contents of each limb of a U-tube through which a current had passed, but it was soon discarded in favour of the "boundary" method which was used by Whetham to measure the rate of movement of coloured ions.[2]

Essentially the method consists in using two solutions, one of slightly greater specific gravity than the other, so that when carefully poured the one over the other a definite surface of separation is formed. One of the two solutions contains a salt with a coloured ion. By passing a current this ion can be made to move and its velocity and direction determined by the movement of the boundary of the coloured zone.

In observations on globulins the lower layer was the opalescent solution of the globulin in acid or alkali as the case might be, and the upper layer a clear solution of the same electrolyte, and the movement of the boundary between opalescent and clear regions is the measure of the movement of the globulin.

The theory of the boundary has been dealt with by Whetham (*loc. cit.*) and by Orme Masson.[3]

Very briefly stated it is as follows:

$$C = A \frac{n}{\eta} (U + V),$$

---

[1] *Journ. Physiol.* (1899), XXIV, 288.
[2] *Phil. Trans.* (1893), A, CLXXXIV, 337.
[3] *Phil. Trans.* (1899), A, CXCII, 331.

where C is the current, A the transverse sectional area of the tube, n the number of gram equivalents of the electrolyte per litre, $1/\eta$ the charge carried by each monad ion, and U and V the average velocity of the cation and anion respectively.

Now the velocity of any ion varies directly with the fall of potential per unit of length (π), and the average velocity of the ions for a given value of π decreases as the concentration of the electrolyte expressed in gram equivalents per litre increases. The factor which expresses this last relation has the same value as that which expresses the relation of specific molecular conductivity to concentration, and it is called the coefficient of ionisation (ϵ).

Therefore
$$C = A\,\frac{n}{\eta}\,\pi\epsilon\,(u+v),$$

where u and v are the actual velocities of the ions when free, and not the average velocities.

The principle of the method is due in the first instance to Lodge,[1] who measured the rate of the hydrogen ion by means of a jelly containing phenolphthalein with enough alkali to give a colour. The tube connected two vessels containing sulphuric acid, so that when a current was passed the hydrogen ions entered the tube, and as they moved decolorised the phenolphthalein. Whetham[2] modified the method by using a vertical tube containing two salt solutions of slightly different density, the lighter one floating on the heavier one. The heavier and lower solution contained a coloured ion and the rate of this ion was determined by measuring the rate of movement of the surface separating the coloured from the uncoloured solutions.

Picton and Linder[3] observed the movement of colloids in an electric field in a tube near the ends of which were two platinum electrodes. Whitney and Blake[4] have followed this method in their measurements of the velocity of colloids. The method is open to the objection that the colloid is concentrated by the field, it is driven into a smaller volume. This and secondary electrolytic decomposition I take it account for the complicated movements which these observers describe.

If the solution under observation is bounded by two fluid surfaces there is no compression of the colloid, and one reaps the further advantage of getting readings in pairs in which the most common source of error— namely electrification of the boundary surface—appears of opposite sign in the members of each pair. The chief advantage however lies in the fact that the sensitive colloid is removed from the electrodes by a column of fluid

[1] *Brit. Assoc. Report* (1886).
[2] *Phil. Trans.* (1893), A, CLXXXIV, 343.
[3] *Journ. Chem. Soc.* (1897), LXXI, 568.
[4] *Amer. Chem. Soc.* (1904), XXVI, 1339.

about 5 cm. long so that, provided each experiment be not too prolonged, the risk of secondary electrolytic decomposition is obviated.

The only method by which one can secure two fluid surfaces is by using a U-tube (Fig. 2), and having the liquid which contains the ion to be measured as the lower liquid. In order to avoid mixing, this liquid must be introduced from below and allowed to flow under the liquid which is to form the upper layers into which the electrodes will dip. The tube I had made is shown in the accompanying figure, and I find that a similar form for use in the direct measurement of ionic velocities has been suggested by Nernst. On each limb of the U-tube a millimetre scale is etched.

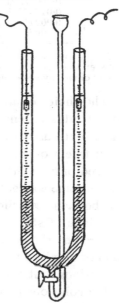

Fig. 2.

The apparatus is used as follows. It is first washed out with the fluid which is to form the upper layers, drained, the tap turned and the U-tube filled with the fluid to a standard level. The second fluid—the solution of globulin—is then allowed to run slowly under the first fluid until the level in each tube stands at a standard mark—the 2 cm. mark in my experiments. The tube is then placed in an upright position in a bath at constant temperature, the electrodes lowered to a certain point in each limb and the apparatus is ready.

In order to avoid convection currents which would destroy the boundaries it is first necessary to bring the fluids, the tube and the bath to the same temperature.

Electrodes of very large surface were used, each consisting of a strip of platinum foil coated with grey platinum deposit 1 cm. broad and 4 cm. long rolled on itself.

The E.M.F. between the electrodes was measured by a Siemens-Halske millivoltmeter.

In order that a boundary shall remain sharp during the passage of a current it is necessary (1) that a specifically slower ion should follow one specifically faster, and (2) that the specific resistance of the fluid on each side the boundary should be the same. If these conditions are not fulfilled the boundary becomes indeterminate. The explanation was furnished by Whetham,[1] and it may be illustrated by this simple example. Take the case of a difference in specific resistance, and let the current pass across the boundary from the liquid of lower to the liquid of higher resistance. By Ohm's law the potential slope will be greater on the distal (as defined by the direction of the current) than on the proximal side of the boundary. Any straggling ion which falls behind therefore will find itself in an area of lower potential gradient and it will be still further slowed; any ion which gets in advance of the boundary will find itself in a region of greater potential slope and it will be still more hurried forwards.

When there are two boundary layers, one preceding, one following, the theoretical conditions obviously are fulfilled when both boundaries remain sharp and move at the same rate. This is a very precise and accurate test

[1] Whetham, *Phil. Trans.* (1893), A, CLXXXIV, 343.

and when it is satisfied also on reversal of the current the readings may be taken as thoroughly trustworthy.

Lower layer methylene blue, 1 gram-mol. in 144·93 litres. Temp. 13°. The movement of the boundaries actually observed in 10 min. is given in millimetres.

	−	+	Mean
0·0476 N, MgSO$_4$ in both layers	4 ↑	4 ↓	4
0·0244 N, „ „ „	4	4	4
0·0012 N, „ „ „	6	2	4
0·000 N, „ „ „	22	1	11·5
0·0476 N, K$_2$SO$_4$ „ „	4	4	4
0·002 N, „ upper layer only	20	3	11·5
0·009 N, „ „ „ „	4	2	3

The figures show the effect on the movement of the boundaries of diminishing the difference between the resistance of the layers by adding salt to the upper layer or to both layers. Mean values obviously are of use as giving an approximation only when the difference is small.

The choice of an upper layer for globulins is very limited seeing that all electrolytes act energetically. It must always be an aqueous solution of the electrolyte used to dissolve the globulin. A second difficulty arises in the adjustment of the resistances. A solution of acetic acid has its conductivity altered only slightly by the presence in it of globulin, and an upper layer of the same concentration of the acid in water was found to give values for the boundaries which agreed within the limits of error of the readings. But solutions of strong acids and alkalis have their conductivity very much lowered by the addition of globulin. Adjustment of the aqueous solution of acid or alkali to the same resistance by dilution did not seem to satisfy the conditions, but the most perfect adjustment was obtained by dialysis.

For solutions in alkali toluol is the best disinfectant, for solutions in acid either toluol or camphor.[1] The dialysing membrane was ordinary parchment coated with formalised gelatine, and dialysis was carried out in glass vessels with the stoppers luted with soft paraffin.

The final equilibrium is a remarkable one. Both dialysee and dialysate stand at the same level—that is to say there is no difference in the osmotic pressure. This agrees with Weymouth Reid's observation,[2] but with such prolonged dialysis I always found that the dialysate though perfectly limpid and free from opalescence gave a just detectable xanthoproteic reaction. The electric conductivity is identical. Therefore the product of the ionic concentration (supposing the ions to have the same valency in both fluids) into the mean ionic velocity must be the same for both dialysee and dialysate. It seems as though the proteid were of no account in the equilibrium. The globulin remains globulin throughout. That is to say on precipitation it is resoluble in a neutral salt. It is not changed to derived albumin. The theory of this equilibrium is considered later.

[1] Camphor changes a solution of alkali globulin into an insoluble gel.
[2] *Journ. Physiol.* (1904), XXXI, 438.

The velocities are tabulated below:

	Concentration	$\times 10^{-5}$ equiv. to 1 gr. globulin	Temp.	Specific velocity cm. sec. $\times 10^{-5}$
H$\overline{\text{A}}$ globulin	0·0004 to 0·0100 N	17–80	13·2	19·78
			18	23
H$\overline{\text{A}}$ globulin dialysed in successive stages	Character of the solution	Transparent	18	22·5
		Transparent, highly opalescent	18	22·9
		White opaque	18	19·4
HCl globulin dialysed in successive stages	Character of the solution	Transparent	18	11·5
		Transparent, highly opalescent	18	9·0
NaOH globulin dialysed	Character of the solution	Transparent opalescent	18	7·66
Not so good as the foregoing				
H$_2$SO$_4$ globulin	0·00048 to 0·0045 N		18	18·5
H$_3$PO$_4$ globulin	0·0092 N		18	23
For comparison with the above; measured in the same manner				
(AgCl)$_x$Cl$_y$	—		18	57
Phase $\frac{\text{water}}{\text{phenol}}$:	The emulsion of phase $\frac{\text{water}}{\text{phenol}}$ in phase $\frac{\text{phenol}}{\text{water}}$, near critical temperature		Velocity of the former ± 105 at 9·5	
[Ag$^\bullet$	$\nu = \infty$		18	58]

The accuracy of these values may be tested by calculating from those for the H$\overline{\text{A}}$ globulin the temperature coefficient. Expressed as a percentage of the velocity at 18° it is 2·6 per cent. per degree. The coefficient for electric conductivity of H$\overline{\text{A}}$ globulin I find to be 2·69 per cent. of the conductivity at 18°. These values are fairly close to Thorpe and Rodgers' determination of the change in the viscosity of water with temperature. Over the same range their figures give 2·58 per cent. per degree of the value at 18°.

Two points are noticeable:

(1) The specific velocity of the globulin in H$\overline{\text{A}}$ globulin is constant over a wide range of concentration both of the acid and the globulin. Over wide limits the ratio acid/globulin does not affect it. As the ratio acid/globulin rises, the solution changes from milk-white or non-settling suspension stage to transparency. Over the greater part of this range the velocity is constant, while from the appearance of the fluid the colloid particles are altering in size. Therefore over this range the velocity of the globulin particles must be independent of their size and the quantity of electricity which each carries cannot be constant. The particles, since they refract light, must be of much more than molecular dimensions. Each is bounded

H

by a surface, and from Helmholtz' theory of electric endosmose, since the velocity is constant, the density of the charge per unit of surface is also constant. The total quantity of electricity on each must therefore be directly in proportion to the surface or

$$Q = \sigma . 4\pi r^2.$$

The fact that all the colloid particles have the same specific velocity is remarkable since Zsigmondy and others have shown that the particles in colloidal solutions are not all of the same size. It agrees with Quincke's observation that the velocity of particles suspended in a fluid in a field is independent of their size, and it proves that the globulin particles cannot be simple ions, for the charge on an ion is a fixed quantity. Nor, as has been suggested, can the charge be due to a haphazard association of the colloid particles with ions which may be present. *It is due to a real electrolytic dissociation at the surface of the particles, the degree of dissociation depends upon the same factors which in the interaction between solvee and solvent determine the degree of dissociation of different electrolytes, and it fixes the density of the charge, while the amount of electricity carried per unit of weight is the product of the density and the area of the surface between the particles and the fluid in which they float.*

The stability of the system will vary with the density of the charge because amongst other things this measures the degree of interaction between solvent and solvee (the solvee here being taken to include the two other components, water and acid or alkali in the case of globulins), but it is not the only factor which operates since in solutions of salt globulins and salt alkali globulins the colloid is entirely uncharged.

Electrolytic dissociation is not the only thing which defines the degree of miscibility in ordinary salt solutions. K_2SO_4 and Na_2SO_4 do not differ in this respect, but the upper limit of miscibility with water is very much higher for the latter, and the reason is that the latter can form molecular compounds with the solvee while the former, for reasons which are entirely unknown, cannot.

Similarly one has electrolytic dissociation as the factor which chiefly defines miscibility in solutions of acid and alkali globulins, and in metallic hydrosols; while the capacity for forming molecular compounds with the molecules of the solvee (salt and water) is the factor which determines miscibility in the case of the electrically inactive solutions of salt globulins, while both factors operate probably almost equally in the case of feebly electrical colloidal solutions such as those of starch and glycogen.

It is interesting to consider for a moment the possible relation between the density of the charge on the surface which separates the internal phase (colloid particles) from the external phase, if, as is not unlikely, the density of the charge defines the shape of the surface and fixes the limits within which the degree of curvature can change without destroying the stability of the system. Below a certain grade of solution which lies well within the region of "white opaque" solution the specific velocity falls. Therefore when the colloid particles become very large, that is when the curvature is diminished very much, the velocity and therefore the density of the charge is decreased, the limiting state being that of a plane uncharged surface. If this argument be correct, the potential difference between the phases phenol/water and water/phenol is zero and the charge which can readily be demonstrated (cf. the table, p. 385) when the two phases have been shaken together to form an emulsion is due to the deformation of this surface which disturbs the equilibrium by placing the two phases under unequal pressures due to surface tension.

The second remarkable feature is the abnormally high value of the specific velocity of the globulin when combined with acetic acid or phosphoric acid, both as compared with known ionic velocities, and as compared with the value when combined with strong acids and bases. According to the table the specific velocity is greater in acid globulin than in alkaline globulin, and rises the weaker the acid. Therefore, on the theory of hydrolytic splitting of the globulin, the greater the degree of hydrolysis the higher the specific velocity of the globulin.

The other relation, namely that to known ionic velocities, must be considered with this.

The velocity of the globulin ion can be approximately calculated from the conductivity data by Ostwald's method.[1] For Na globulin the result is in round numbers 10 at $V = \infty$, 7·66 therefore is the order of magnitude one might expect with incomplete ionisation. Reckoned by the number of atoms the specific velocity of a proteid ion cannot be more than 10, and should be much less, and the values actually observed for HCl and NaOH globulin show that Ostwald's law of the decrease in ionic velocity with increase in the size of the ion ceases to hold beyond a certain size of ion. The values for HCl proteid and NaOH proteid therefore agree very well with the view that the globulin is in the ionic state, and they are low as compared with the velocity of movement when conditioned by electric endosmose.

The other values, however, are altogether outside the limits which could be reached by ions of such magnitude as proteid ions must be, and the theory of hydrolytic splitting furnishes an explanation.

According to it the globulin would be chiefly in the ionic state in NaOH globulin, chiefly in the state of anhydride or hydrate or of ionic complexes of compounds with lowered acid content (cf. p. 396) in $\overline{HA}$ globulin, while

---

[1] *Zeits. f. phys. Chem.* (1886), II, 840.

between these extremes would be the HCl globulin. Therefore the velocity observed in NaOH and HCl globulin, which it must be remembered is the mean velocity of all the globulin, no matter what its state may be, would be chiefly due to the globulin ion. It would be an ionic velocity and it actually is of the expected magnitude of an ionic velocity.

The velocity of the globulin in $\overline{HA}$ globulin will be mainly that of the free or partly free globulin to which the opalescence is due and which therefore is not in a state of molecular subdivision. It is present in masses so large as to be defined by a surface. Such masses, according to the view with which we started, would have a surface charge owing to ionisation at their surface, they are pseudions of the form $G_n G_x' + Na$, where n is large compared with x, and in their case the quantity of electricity would be a function of surface and not a fixed quantity.

Now the order of magnitude of the velocity of pseudions can be arrived at by measurements from known cases.

I. To $0.1N$ NaCl I added a solution of silver nitrate until it became highly opalescent. This formed the lower layer in the cell, the upper layer being sodium chloride adjusted by trial and error to approximately the same conductivity.

Both boundaries moved at the same rate, and the specific velocity at 18° was found to be 57×10^{-5} cm. per second.

This value, by what is probably a coincidence, agrees almost exactly with the speed of $Ag^{\cdot}$ at infinite dilution; that is with the maximal velocity of a rapid ion, $Ag^{\cdot}$ having the value 58.

Clearly therefore the large pseudions $(AgCl)_x Cl_y'$, which are so large as to refract white light, do not conform to Ostwald's law of the decrease in specific velocity with increase in the number of atoms.

II. A solution of pure sodium stearate from Kahlbaum was dialysed for 10 days against five changes of water. It became highly opalescent owing to the formation of acid salts. The dialysate formed the upper layers and the specific velocity of the boundaries, *i.e.* of the opalescence, was 50 at 18°. Again the value is much too high to be in agreement with Ostwald's values.

With these parallel cases as a guide we may conclude that the globulin which is turned out of combination by hydrolysis will show a higher specific velocity than "ionic" globulin in spite of the fact that it is very slightly dissociated. In other words the specific velocity is high, but the quantity of electricity carried per unit of mass is low, and this is the feature of a hydrosol.

The entire proteid moves with the same average velocity, which is therefore neither that of the fastest nor that of the slowest ions or pseudions. In ordinary solutions if U be the absolute velocity of an ion when entirely free, and α the ratio of the period in which it is in conjunction with the related ion and therefore not moving to the period during which it is free, then the observed velocity will be $\alpha\,(U)$, which is a measure of the mean free path of the ion.

In a globulin solution the globulin according to the above theory exists as free ions, as globulin combined in the molecular state, or as pseudions. Let the absolute velocity of the free ion be U, of the pseudion U_{coll}. If x be the fraction hydrolysed, then of the total mass x is the fraction in the state U_{coll}, therefore the observed velocity will be

$$\frac{xU_{coll}\,[\alpha\,(1-x)\,U]}{U_{coll}\times U}.$$

The equilibrium between dialysee and dialysate for globulin compounds with monovalent acids and alkalis is very remarkable. There is agreement within the limits of measurement in osmotic pressure,[1] and therefore in the composition of the gas phase, and in specific electric conductivity.

Assuming that the molecules furnish the same number of ions, this means that in unit volume of dialysee and dialysate there are the same number of molecules which contribute to the osmotic pressure and the same number of molecules which are electrically active at any one moment. If one set of values cease to agree, so also do the other. Thus in some cases when owing to slight bacterial decomposition the osmotic pressure of the dialysee remains above that of the dialysate, the electric conductivity also remains different, but *below* that of the dialysate in the few cases which I have examined.

These relations are explicable only on the view that the globulin solution is biphasic, the one phase being present as a number of separate spheres, and continuously varying with variations in the composition of the external phase.

In the equilibrium of dialysis there are four phases, one gas phase, internal and external phase of the dialysee, and the dialysate. If there be any tension on the surface of the internal phase there are two pressures, the lower one of the gas phase with which both the dialysate and the external phase of the dialysee are in equilibrium, and the higher one of the internal phase. The osmotic equilibrium is between the dialysate and the external phase of the dialysee, the agreement in electric conductivity is an equilibrium between the dialysate and both phases of the dialysee, since the internal phase carries a charge on its surface and therefore contributes to the electric transport. Considered in this way it seems at first sight that there must be a difference of pressure between dialysate and the external phase of the dialysee due to the globulin in the latter, and this may well be the case, the value being altogether too low for detection by experiment.

[1] Waymouth Reid, *Journ. Physiol.* (1904), XXXI, 438.

GENERAL THEORETICAL CONSIDERATION OF THE STATE OF
ALKALI AND ACID GLOBULIN IN SOLUTION

Attention was drawn to the possibility of direct chemical combination between proteids and acids or alkalis by the study of the action of the digestive juices.

Proteid was found to "mask" a portion of free acid or alkali. A few attempts have been made in recent years to investigate the nature and extent of the masking action by the newer methods of physical chemistry.

Sjöqvist,[1] working in close touch with Arrhenius, and later Sackür,[2] applied the method of electric conductivity, and Bugarsky and Liebermann[3] measured the electromotive force of concentration cells containing acid or alkali in presence of proteid, and from the values so obtained they calculated the concentration of the OH or H ions: they measured also the change in the freezing-point due to the addition of various amounts of proteid to $0.05 N$ acid and alkali. Cohnheim[4] measured the free acid by the velocity of the catalysis of cane-sugar. Paal,[5] Spiro and Pemsel,[6] Cohnheim and Krieger,[7] Erb,[8] and Osborne[9] used a more ordinary chemical method, namely the precipitation of the proteid-acid, or proteid-alkali compound and the determination of the quantity of acid held in the precipitate.

The methods used have been varied, and all these workers, with the exception of Spiro and Pemsel, conclude that proteids combine to form salts with both acids and bases, they themselves being possessed of both basic and acid properties. Such bodies are known as amphoteric electro lytes, and from the nature of the case they must necessarily be weak acids and bases, since the presence in the same molecule of acid and basic groups diminishes the importance of both. Sjöqvist, who dealt with proteids as bases only, found them to be weak bases, and therefore, like all weak bases, their salts in aqueous solution are much hydrolysed. Bugarsky and Liebermann showed that proteids also are weak acids, the compounds with alkalis hydrolysing freely.

From the measurements of these workers it is possible to estimate the acid and basic functions. In the only native proteid used by them, egg albumen, they appear to be almost exactly balanced. Sjöqvist calculated the basic dissociation constant for egg albumen and found it to lie between the values for aspartic acid and urea. It is a base about 11 times weaker

[1] Skan. Arch. f. Physiol. (1895), v, 277.
[2] Zeits. f. phys. Chem. (1903), XLI, 672.
[3] Pflügers Arch. (1898), LXXII, 51.   [4] Zeits. f. Biol. (1896), XXXIII, 489.
[5] Ber. d. deut. chem. Gesell. (1892), XXV, 1202.
[6] Arch. f. physiol. Chem. (1898–9), XXVI, 233.
[7] Zeits. f. Biol. (1900), XL, 95   [8] Ibid. (1901), XLI, 309.
[9] Journ. Physiol. (1901), XXVII, 398.

than asparagin, 6 times weaker than aspartic acid, but decidedly stronger than urea.

Proteid salts in aqueous solution are fairly good conductors, that is to say, they ionise. Bugarsky and Liebermann conclude that the degree of ionisation is not large, but they speak of true proteid ions. When reacting with acid, for instance HCl, the resulting salt has the form BHCl, and the ions are BH˙Cl′.

The proteid therefore combines like ammonia:

$$NH_4Cl \rightleftarrows NH_3H^{\cdot} + Cl',$$
$$BHCl \rightleftarrows BH^{\cdot} + Cl'.$$

All these workers are agreed that proteids form salts which behave quite normally. They ionise in aqueous solution, forming a true proteid ion. They react with water according to the ordinary equation of hydrolysis $BS + HOH = BOH + HS$. The only marked feature is the relatively low value for specific molecular conductivity, and this according to Sackür is due not to an abnormality in the salt itself but to the high viscosity of the solution.

If the standpoint of these various authors is accepted we are bound to consider the question whether proteids in aqueous solutions are colloids at all.

Since Graham's time true colloids as a class have been distinguished by the feebleness of their chemical reaction. But according to the authors quoted proteids react directly and fully with acids and alkalis, the reactions being sufficiently definite to admit of the calculation of molecular and combining weights.[1] That there is, however, a theoretical difficulty not to be passed over so lightly has been recognised in the analogous case of soaps.

Krafft[2] urged the colloidal nature of these bodies on the ground that they gelatinise and do not lower the vapour pressure in concentrated solution.

Kahlenberg and Schreiner,[3] however, using the method of electric conductivity, found soaps to be salts of ordinary type undergoing ionisation and hydrolysis in the ordinary way. They therefore refuse to regard soaps as colloids, on the ground that matter in the colloidal state does not react in this simple chemical way.

Smits,[4] from direct determinations of vapour tension, proves the accuracy of Krafft's contention that the vapour pressure becomes that of pure water when the concentration of soap is sufficiently high; soap, therefore, at these concentrations cannot be in true solution, it must be in colloidal solution. Kahlenberg and Schreiner, however, found even at concentrations when by the vapour-pressure method the colloidal state must

[1] Sjöqvist, Bugarsky and Liebermann, Sackür, *loc. cit.*

[2] *Ber. d. deut. chem. Gesell.* (1896), XXIX, 2, 1334, and earlier papers.

[3] *Zeits. f. phys. Chem.* (1898), XXVII, 552.   [4] *Ibid.* (1903), XLV, 608.

be fully established, soaps still are good conductors and therefore must be ionised. Smits regards this result as due to absorbed electrolytic impurities which act as the conductors, but this view is hardly tenable, for the impurities if sufficient to give normal values of molecular conductivity for the soap would also affect the vapour pressure.

Probably the controversy may be in part referred to our ignorance of the true interpretation of conductivity data derived from concentrated solution. A similar discrepancy between vapour-pressure measurements and conductivity measurements appears in the work of Jones and Getman[1] on concentrated solution of chlorides, sulphates and nitrates of various metals, and the explanation which they offer probably has a direct bearing on our subject.

The position with regard to the nature of soap solutions is exactly paralleled by the case of proteids. Reid[2] has shown that proteids in solution do not alter the vapour pressure, on the other hand in presence of acids or alkalis these solutions are good conductors, and as we have already seen the conduction cannot be referred either to the free acid or alkali, or to absorbed electrolytic impurities.

Further, the proteid in presence of acid or alkali is electrically charged, and therefore so far is "ionic", and, as may be proved by indicators or by measurements of the rate of catalysis of methyl acetate or cane-sugar, it neutralises the acid or alkali. These and other facts already discussed point to salt formation.

Proteids therefore show relatively vigorous chemical action, and as a class they gelatinise with difficulty.[3] In what respect then are they colloids?

The colloidal nature of their solutions appears mainly in the presence of a marked time element in the reaction. Kohlrausch found on dilution of sodium silicate, or on adding caustic soda to an already diluted solution, that the electric conductivity reaches a constant value only after a long time. Similarly van Bemmelen found that when $SnCl_4$ is added to water tin oxide is liberated owing to hydrolytic decomposition and is dissolved in the HCl, but the relation between acid and oxide is not one of chemical equivalence. The tin oxide slowly changes its state to one of true colloidal solution, the "colloid particles" retaining absorbed acid. Similarly with globulin solution I have found that viscosity, electrical conductivity, sometimes the one, sometimes the other, sometimes both, continue to change for days. And the amount of globulin held in solution by a given quantity of salt is not constant for a given temperature and concentration but varies according to the previous history of the system; e.g. according to

[1] *Amer. Chem. Journ.* (1904), XXXI, 303.
[2] *Journ. Physiol.* (1904), XXXI, 438.
[3] Alkali globulin at high concentration sometimes gelatinises like a soap.

whether the particular volume relations have been reached by dilution of a more concentrated solution, or by direct solution of a suspension of globulin.

Enough has been said to show the need for caution in regarding the reaction between proteids and acids or bases as one of simple salt formation. The possibility of the phenomena belonging to the intricate borderland of absorption combination must not be lost sight of.

It must be borne in mind that the possibility and existence of definite chemical compounds between proteids and basic or acid radicles is not here called in question. The isolation of a score of such bodies would help little or nothing towards understanding the mechanism of solution of proteids. The existence of definite compounds of silica helps us not one whit to understand why silicic acid in aqueous solution has no definite heat of neutralisation, or why in presence of alkali it needs days to reach electrical equilibrium. And from the point of view of the dynamics of the body fluids what we need to know is the state of the proteid in aqueous solution.

The possibility of the relation between proteids and electrolytes in aqueous solution being that of "absorption" needs consideration, in view of the fact that the ratio of the reacting masses by weight is from 300 to 1000 to 1; and van Bemmelen[1] has shown that 1 molecule of colloidal Fe_2O_3 can absorb 0·31 molecule of, for instance, $Ba(OH)_2$.

Absorption compounds are marked by the indefiniteness of the reacting masses. The same indefiniteness is very clearly seen in certain relations of globulins. Dialysis of solutions of alkali or acid globulin proves that the quantity of acid or alkali necessary to dissolve unit weight of globulin is quite indeterminate, though the "grade" of solution, i.e. the degree of dispersion of the globulin, does depend upon the mass ratio acid or alkali/globulin. Dialysis causes the solution to become more and more opalescent, but no precipitate settles out, and the entire mass of proteid retains its electric charge, the specific ionic velocity being almost unaltered.

Conversely indefinitely minute concentrations of acids or alkalis will convert a globulin precipitate into a solution of low grade just as a minute amount of alkali will dissolve ("peptonise" as Graham called it) an indefinitely large quantity of silica, the solution being of proportionately low grade.

There is, therefore, no point in the solution of globulin by acids or alkalis at which it may be said to be complete. Continuous addition of reagents produces a continuous and parallel increase in the dispersion of the globulin, so that, starting from a solution which is opalescent and opaque, that is to say which is optically heterogeneous, one ends with a solution which is transparent and is, relatively speaking, optically homogeneous. Similar

---

[1] van Bemmelen, *Zeits. f. anorg. Chem.* (1903), XXXVI, 380.

variations in the grade of the solution were described by Picton and Linder,[1] who were the first to draw attention to this important point, in hydrosols of sulphides and of silica, and Siedentopf and Zsigmondy[2] detected them by the method of ultra-microscopical vision in metallic hydrosols.

If globulin is dissolved by acids and alkalis owing to chemical compounds of the nature of globulin salts being formed, it is clear from the facts of dialysis and of solution, and from measurements of the specific ionic velocity of the globulin, that the mass of alkali or acid needed to bring about solution of 1 gr. of globulin, and to confer on it its maximal specific ionic velocity, is quite indeterminate, but the mass required to produce a given grade of solution, that is to produce a given degree of dispersion, is fixed and measurable.

If the globulin salt be $G_n(HS)_y$, and $G_n(BOH)_y$ respectively, the ratio n/y, the combining ratio, that is, can vary continuously, and this I take it is the essential characteristic which van Bemmelen claims for absorption compounds.

A theory based on simple salt formation, therefore, will not cover the facts, it fails to explain what I take to be the central fact, namely the indefinite nature of the alkali or acid globulin compounds. It breaks down in another respect.

The molecular conductivity μ of a salt which behaves normally in solution is the sum of the mean velocity of the two ions,

$$\mu = U + V,$$

and the mean velocity is a value which can be directly measured by Lodge's[3] boundary method. It is low when the degree of ionisation is low.

Acetic acid ionises only slightly, its salts as a rule ionise well. Therefore the molecular conductivity of the salts stand as a rule much higher than that of the acid. But the molecular conductivity of acetic acid globulin is lower than that of the acid. Sackür and others explain these low values for the molecular conductivity of proteid compounds by supposing that the high viscosity of the solutions lowers the ionic velocity, and that the salts ionise freely. The explanation is contradicted by fact. Direct observation shows that the specific velocity of the proteid, the mean ratio of drift of the proteid in the electric convection, is not low but abnormally high. The low electrical capacity of acetic acid globulin in solution therefore cannot be explained by hydrolytic dissociation, low degree of ionisation, or high internal friction. The only other way which suggests itself is by the formation in solution of ionic complexes of large size and low electrical capacity per unit of weight.

[1] *Journ. Chem. Soc.* (1895), LXVII, 63.   [2] *Ann. d. Phys.* (1903), (4), x, 1.
[3] *Brit. Assoc. Report* (1886), p. 389. "Mean velocity" is not the velocity of the free on, but the average speed including periods of freedom and of conjunction.

The relation observed between concentration and viscosity, and the effect of excess alkali upon viscosity, also points to the existence of ionic complexes of a size so large as to form masses of matter separate from the general mass of the fluid.

Walker[1] found the molecular conductivity of solution of the acetates of asparagin, asparaginic acid and glycocoll to be lower at certain dilutions than that of acetic acid itself, and he suggested as the only explanation that molecular complexes are formed. Arrhenius[2] proposed another explanation based upon changes in the degree of ionisation, which would however directly conflict in the case of acetic globulin with the observed rate of drift of the proteid. Since these papers the formation of ionic complexes as a factor in reducing molecular conductivity has been freely recognised. Acid and alkali globulin in solution form such complexes, some of which are so large as to cease to have the properties of molecules. They form internal systems each separated from the general mass of fluid by a surface.

The acid or alkali must, I think, be held to be truly combined with the globulin, because it is wholly or partly neutralised, the proteid assumes an anionic, or cationic character as the case may be, and there are evidences of normal hydrolytic phenomena. The compounds may be called absorption compounds, but the further development of the theory shows that the distinction is one of name only.

Recent work on complex ions, chiefly by the method of concentration cells, has extended the possibility of reaction by chemical equivalents to a point where it seems to be undistinguishable from absorption combination. I propose to consider the nature of globulin compounds in the light of these recent investigations.

At the present moment the conception of the colloidal solution, which finds greatest favour, is that it differs from ordinary solution, because the solution is not uniformly distributed throughout the solvent in a state of molecular dispersion, but is for the most part gathered into masses which may be either solid particles, as in metallic hydrosols, or fluid spheres, in which case the "colloid particles" are themselves a solution within a solution.

In an earlier paper[3] I ventured to describe colloid solution as being essentially biphasic, and to apply the terms internal phase and external phase to the general mass of the fluid and to the colloid particles respectively. In certain cases, as may easily be seen under the microscope, three phases may be present—external, internal$_1$ and internal$_2$, as when gelatine alcohol-water is over-cooled. Now the conception which I wish to introduce here is that the internal phase may form not necessarily because

[1] *Zeits. f. phys. Chem.* (1889), IV, 319.
[2] *Ibid.* (1890), V, 16.   [3] *Proc. Roy. Soc.* (1900), LXVI, 95.

the solvent is supersaturated with respect to the solvee as a whole, but because it is supersaturated with respect only to one of the ions of the latter.

Consider the case of the double salts of silver or mercury and the halogens. The work of Bödlander and Fittig,[1] of Sherrill,[2] and of Bödlander and Eberlein[3] with others shows that when for instance AgI is dissolved by KI a double salt is formed, which ionises forming

$$[(AgI)_n K^{\cdot} + I'], \quad [Ag^{\cdot} + I'], \quad [K^{\cdot} + I''],$$

all these species being in equilibrium. An increase in the ratio AgI/KI in the solution causes the value of n to rise and the complex ion to increase in size.

I have examined the system $(AgCl)_n (NaCl)_y$ which represents the class, and one cannot fail to be struck with the many analogies it offers to solutions of globulins. As the relative concentration of AgCl is increased the solution becomes gradually more opalescent, rising to complete opacity. Up to this point the solution is stable in clean vessels. The material which causes the opacity is ionic in character, that is to say it moves in a field with a normal ionic velocity, the value I found by actual measurement at 18° being 57×10^{-5} cm. per second. The constituent which causes the opalescence further is always positive in sign, clearly therefore it represents an overgrowth of the complex ion $(AgCl)_n Na^{\cdot}$ to dimensions so huge as to refract white light.

Now an ion of this size will very largely cease to have the property of matter in a molecular state, it is matter in mass defined by a surface. It moves in a field according to the laws of electrical endosmose—that is to say the total charge it carries is not fixed as it is for all true ions but the *density* of the charge is constant. This would account for the high specific velocity observed by me which is almost exactly that of Ag′ at infinite dilution, whereas, as Ostwald showed, ionic velocity should decrease with increase in the size of the ion. A case such as this is, I think, best described by speaking of the fluid as being supersaturated with a complex ion, the instability of which decreases with increase in the ratio n/y.

Colloidal solutions, as we have seen, do not always contain colloid particles carrying a charge. There are, therefore, two types of hydrosol, electrically active and electrically inactive. On the view enunciated above in electrically active sols the colloid particles, *i.e.* the internal phase, is formed owing to supersaturation by an ionic species; in the latter it is due to supersaturation by an unionised species, that is by the molecule in the chemical sense. Therefore in those colloid solutions which are electrolytic the electric transport is due in part to the drift of charged particles so

[1] *Zeits. f. phys. Chem.* (1902), XXXIX, 597.

[2] *Zeits. f. phys. Chem.* (1903), XLIII, 707; (1904), XLVII, 103.

[3] *Zeits. f. anorg. Chem.* (1904), XXXIX, 2, 197.

large as to be defined by a surface and to have the properties of matter in mass. These are the "pseudions" or "colloid ions". Each differs from a true ion in the fact that owing to its large size as compared with the molecules of the solvent it cannot partake of the molecular movement of the latter. It has *to a greater or less extent* the properties of matter in mass. It will contribute little or nothing to the osmotic pressure—nothing if it be so large as to cease completely to act as a "molecule" of the general fluid system. It will then have as little effect as have the particles of an emulsion

Its relation to the true molecular species of the solvent will be peculiar in that, though the relation will be that of an average state, of a continuous interchange, it might be expressed statically by regarding the molecules of the surface layer of the pseudion as being in a state of incomplete ionisation, or orientation, so that the double electrical layer is formed which is postulated in the theory of electrical endosmose.

If a colloidal solution consists of two phases and comprises two components, then for a given temperature the vapour pressure must be independent of the concentration. In the case of soaps this point is reached, according to Krafft and Smits, only when the concentration is above a certain minimum [at 0·18 mol. per litre], when it is constant and equal to that of pure water within the limits of accuracy of measurement. Below this concentration the vapour pressure rises with increasing dilution (Smits). In this region of varying vapour pressure the solution cannot be biphasic since it is bivariant, and the components are two in number, water and soap. Are these dilute solutions not colloidal at all, and is the change to the colloidal state an abrupt one (as Krafft holds) due to the separation of a second fluid phase?

From the analogy of other cases an abrupt change to the colloidal state is not probable. In the many instances investigated by van Bemmelen changes of state are always gradual. On the other hand Smits' observations show that the pressure-concentration curve for soap must have the general form shown in *ABCD* (Fig. 3), and this is remarkably similar to the curve for hydrobromic acid and water.[1] *LMO* is the curve for the solution hydrobromic acid/water, at *O* a second fluid phase appears, and the system becomes monovariant, *OR* being the characteristic curve of constant pressure.

The difficulty may be avoided by the hypothesis that over the region of changing vapour pressure the internal phase is not fully separated. It is as it were growing by increase in the size of the particles, which, however, are still so small as to have to only a limited extent the properties of matter in mass. In other words, with increasing concentration the colloidal particles

[1] Bancroft, *The Phase Rule* (1897), p. 96.

are completely separated only gradually, just as van Bemmelen found them to appear gradually in the case of SnO or SnO_2 by mere lapse of time[1].

The conception of an imperfectly separated phase is not altogether visionary. Mixtures of isobutyric acid and water form two fluid phases. Friedländer[2] has studied these mixtures and he describes an appearance of opalescence which is not diminished by lapse of time over a range of 10° above the critical temperature for the complete separation of the phases. Konowalow[3] describes a similar stable opalescence in mixtures of pentane and dichloracetic acid. Over the range of temperature in which the opalescence is stable the vapour pressure does vary with the composition

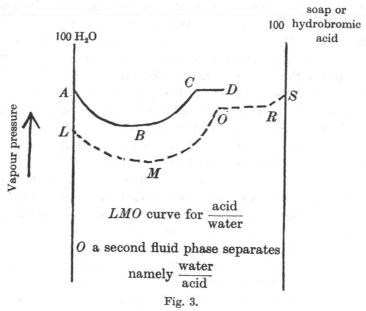

Fig. 3.

of the mixture, but the variation is much smaller than it would be if there were complete homogeneity. There is thus an incomplete separation of a second fluid phase, and the separation remains incomplete over a wide range of temperature (10°) and composition.

This conception of a heterogeneous fluid implies that in the external phase there will be some of the material in true molecular solution, which would pass through a porcelain filter while the internal phase would be arrested. A solution of pure sodium stearate strong enough to set to a jelly will yield a filtrate which contains a trace of soap, the filtrate from a clear filtered solution of NaOH globulin similarly contained 0·024 per cent. of solids which gave the proteid and phosphorus reactions.

[1] *Zeits. f. anorg. Chem.* (1903), xxxvi, 380.
[2] *Zeits. f. phys. Chem.* (1901), xxxviii, 385.
[3] *Ann. d. Phys.* (1903), (5), 12, 1160.

SALT GLOBULINS

The special interest of these bodies lies in the fact that colloidal solutions as a class are precipitated, not preserved, by the addition of salts. Solutions of salt globulins, however, as their low viscosity and the decisive and clear way in which they can be precipitated shows, are much less colloidal than solutions of acid globulin or especially alkali globulin. Alkali globulins have a high viscosity and I have twice when endeavouring to measure electric conductivity at high concentration found the solution set to a firm jelly like a soap.

The solution of an insoluble body by an added neutral salt is not an exceptional phenomenon. Silver chloride is dissolved by sodium chloride, silver cyanide by potassium iodide, and calcium carbonate is dissolved by a wide range of salts.[1] Where the relations have been investigated solution is found to be due to the formation of molecular compounds, and the quantitative and qualitative relations between globulin and neutral salts can be adequately and simply explained in the same way. Salt globulins therefore recall the double compounds which amido acids make with salts, while alkali and acid globulins recall the simple salt compounds which these bodies make when they react with free acid or free alkali.

The saturation of a salt solution with globulin alters the molecular conductivity of the former only slightly:

NaCl	$0.16 N$	Conductivity depressed		2.4%
	0.11	,,	,,	2.3
	0.09	,,	,,	2.1
MgSO$_4$	0.3	,,	,,	1.4

These values are very small, and they need to be diminished still more for the increase in the internal friction produced by the globulin. Salt globulin, however, has a low viscosity, and in the light of the previous discussion this correction is probably negligible.

These low values show that the bulk of the salt is not attached to the globulin, 98–99 per cent. being free.

Salt globulins in solution form no ionic globulin—or at any rate the concentration is such as to prevent its presence being recognised. Salt globulins therefore may be taken to be unstable in presence of water, stable only when their dissociation is completely suppressed by excess of salt. As only 1 or 2 per cent. of the salt present is combined with the globulin, the solvent power of salts—the weight of globulin which one equivalent can keep in solution—will therefore be affected numerically only to an insignificant extent by variations in the combining weight of the salt and the globulin. What one actually measures is the concentration of salt necessary

---

[1] Cameron and Seidell, *Journ. Phys. Chem.* (1902), vi, 50.

to prevent dissociation of the salt globulin. Owing to the extreme instability of the salt globulin this value is dependent upon a number of values which are obscure by reason of their small magnitude.

A comparison of the relative solvent powers of strong and medium alkalis, where by reason of the stability of the salt in aqueous solution we come nearest to the true combining ratios, and the solvent powers of salts, give us a measure of the depression of electric conductivity which would be produced by the presence of the globulin, if the combining ratios of salt and globulin, and alkali and globulin were the same. The salt globulin is neglected as it does not contribute to the electric transport.

To dissolve 1 gr. of dry globulin it needs 10×10^{-5} equivalent of alkali, and from 1000 to 2000 of salt. The calculated lowering of conductivity in the latter case therefore would be between 0·5 and 1 per cent. The higher values observed are this theoretical value plus an increment due to increased viscosity.

The compound of silver chloride and ammonia which has been studied by Bödlander and Fittig[1] is analogous. The compound ionises in solution according to

$$AgCl \cdot 2NH_3 \rightleftarrows Ag\,(NH_3)\dot{}_2 + Cl',$$

and also a minute fraction according to

$$AgCl \cdot 2NH_3 \rightleftarrows Ag\dot{} + Cl' + 2NH_3.$$

The ions of the first equation are stable, of those in the second equation the ions $Ag\dot{}$ and Cl' can coexist only at very low concentration. Therefore the system as a whole is stable only when NH_3 is present in excess sufficient to depress ionisation according to the second equation to the critical saturation point of $Ag\dot{} + Cl'$.

The stability of ionic globulin in presence of salt seems to be nil, and this is the critical value which defines the system salt + salt-globulin.

When a salt solution is saturated with globulin, there being no excess, the system therefore being monophasic, the solvent power is found to vary for different classes of salts but, on the whole, to rise with a rise in valency. This means that as a class the salt globulins are less easily decomposed by water the higher the valency of the salt.

The increase in solvent power with a rise in the valency is not peculiar to salt-globulin solutions. It is seen in the solution of slightly soluble salts by other salts.

The method adopted was either just to saturate a given volume of salt solution by running in a measured amount of a suspension of globulin in water, or the reverse. What one measures is the relative mass of salt and globulin at the concentration when the salt globulin compound just ceases to be stable.

[1] *Zeits. f. phys. Chem.* (1902), xxxix, 597.

	Globulin in 100 c.c. suspension in gr.					
	0·19	0·426	0·426	0·55	0·95	0·95
KCl	1	1	1	1	1	1
NaCl	1	1	1	1	1	1
KBr	1·4	1·37	—	—	1·2	—
NaBr	1·5	1·37	—	—	1·3	—
KNO_3	1·5	1·5	1·5	1·5	1·5	—
$NaNO_3$	1·5	1·5	1·5	1·5	1·5	—
$AmNO_3$	1·5	1·5	1·5	1·5	1·5	—
$Ca(NO_3)_2$	1·8	—	1·6	—	1·5	—
$Mg(NO_3)_2$	1·8	—	1·7	—	1·5	—
$BaCl_2$	1·8	—	—	—	1·5	—
$MgCl_2$	1·85	—	1·6	—	1·5	1·52
$CaCl_2$	1·8	—	1·6	—	1·5	1·52
Na_2SO_4	1·99	2	1·99	—	1·86	1·89
K_2SO_4	1·99	—	1·99	—	1·86	1·89
$MgSO_4$	1·87	1·8	1·83	1·73	1·56	1·6
Mg succinate	2·4	2	1·79	—	1·56	—
$AmSO_4$	—	—	—	—	1·96	—
K oxalate	—	—	—	—	2·1	—
Am oxalate	—	—	—	—	1·96	—
Na citrate	—	—	—	—	3·3	—

The figures give the weight of dry globulin held in solution by 1 equivalent of salt in equilibrium with the CO_2 of the air, the values for KCl and NaCl being taken as unity.

The following relative values were obtained from the series of papers by Cameron, Seidell, Atherton and Breazale.[1] The numbers give equivalents of $CaSO_4$, $CaCO_3$, or $MgCO_3$ dissolved by 1 equivalent of salt:

Concentration of solvent	Solvent	1 equivalent dissolves
1·5 N	KNO_3 $NaNO_3$ $MgKO_3$	0·7 $CaSO_4$ 1·4 ,,
And chlorides "rather less than" KNO_3 or $NaNO_3$:		
1·0	NaCl Na_2SO_4 Na_2CO_3	0·006 $MgCO_3$ 0·011 ,, 0·012 ,,
1·0	NaCl Na_2SO_4	0·0014 $CaCO_3$ 0·0030 ,,

The valency rule is not absolute. It does not always hold for the fluid phase when a solid phase is present.

Normal salt was saturated with edestin crystals and filtered at 65°. Cooled slowly (some hours) to 22° and filtered from the crop of crystals at 22°. The edestin had been crystallised out by cold from a solution of NaCl edestin saturated at 70°; crystals washed with dilute alcohol.

Composition of the fluid in equilibrium with the crystals at 22°.

	Salt %	Equivs. per litre	Proteid %	1 gr. equiv. of salt = gr. proteid
NaCl	5·97 5·97	1·03	1·33 1·20	12·4
K_2SO_4	7·64 7·8	0·885	0·69 0·68	7·7

[1] *Journ. Phys. Chem.* (1902), VI, 50; (1903), VII, 578; (1904), VIII, 335 and 493.

The table shows that solvent power is determined also by the nature of the anion, the order being $Cl < Br < NO_3$. This is explained by the difference in the capacity of ions for forming complexes such as those under consideration (GBS). Bödlander and Eberlein[1] find in respect to this property the order is: $Cl < Br < CNS < I < CN$.

The concentration of globulin is determined not only by the concentration and nature of the salt but also by the initial concentration of globulin. The composition of the fluid phase is not defined by saying that the globulin is in excess since more than one solid phase may be present or separate out of the nature of various hydrates of globulin, or various molecular states of globulin, or of various compounds of globulin and salt. As in other three-component systems it is not sufficient to say that the fluid is in equilibrium with a solid phase, it is necessary to define the particular solid phase. For instance in the case of calcium sulphate, which has been studied by van 't Hoff,[2] the solid phase in equilibrium with the fluid phase may be insoluble anhydrous calcium sulphate, soluble anhydrous calcium sulphate, or various hydrates such as $CaSO_4 . \frac{1}{2}H_2O$ and $CaSO_4 . 2H_2O$, and the composition of the fluid phase varies according to the particular solid phase with which it is in equilibrium, therefore, as van 't Hoff points out, the system as a whole has the inertia of the solid state.[3]

The effect of the concentration of ox globulin upon the solvent power of the salt is manifested very readily when salt solution is saturated by running into it suspensions of various strengths of globulin in water.

Initial concentration of salt, in each case normal. Saturated with globulin by running in suspensions of different concentrations.

I. Concentration of suspension in gr. dry ox globulin in 100 c.c.
II. Weight of proteid in gr. dissolved by 1 equiv. of salt.

NaCl	I.	0·19	0·426	0·55	0·94	1·26
		0·34	0·77	1	1·73	2·3
	II.	14·5	26	31	55	79
		0·47	0·84	1	1·77	2·5
K_2SO_4	II.	22·3	42	55	92	99
		0·4	0·77	1	1·7	1·8

The lower row of figures gives the ratio for each case. It shows a close parallelism between the initial concentration of the globulin in the suspen-

[1] *Zeits. f. anorg. Chem.* (1904), XXXIX, 2, 197.

[2] *Loc. cit.*

[3] See also a series of papers by van 't Hoff on the formation of oceanic deposits. An interesting case also is the effect upon the curve of the solubility of calcium sulphate in solution of potassium sulphate of the nature of the solid phase when the latter is ordinary calcium sulphate or syngenite. Cameron and Breazale, *Journ. Phys. Chem.* (1904), VIII, 334   See also Ditte, *Compt. Rend.* XCIV, 1266. At certain concentrations of KCl and $CaSO_4$ the composition of the fluid phase is determined by the solid phase which is the double salt $K_2SO_4CaSO_4 . 4H_2O$.

sion and the concentration of globulin in the fluid phase. Over a certain range the numbers agree closely.

When I first detected the effect of the mass of the globulin upon the composition of the fluid phase the possibility of the presence of an unsuspected component became obvious. In order to eliminate this factor as far as possible I satisfied myself that the same relation occurs when pure crystals of edestin washed with 20 per cent. alcohol are used.

An example from this globulin will illustrate the facts.

Crystals of edestin were suspended in water freed from carbonic acid. Two suspensions were made containing respectively 13·16 and 4·39 gr. of crystals per 100 c.c.

5 c.c. of each suspension was mixed with 10 c.c. of $\frac{1}{3}N$ NaCl, in test-tubes tightly closed with paraffined corks and set on a mechanical device which ran the fluid from one end of the tube to the other about four times a minute. This was done to secure complete equilibrium, but later measurements of the rate of solution of edestin proved the precaution to be wholly unnecessary. After 20 hours the proteid in the upper layer of each tube was determined. The experiment was carried out in duplicate.

$$\text{Salt } N/4\cdot5. \text{ Proteid} \begin{cases} \text{I.} & 4\cdot39 \text{ gr. per 100 c.c.} \\ \text{II.} & \tfrac{1}{3}\times4\cdot39 \text{ ,, \quad ,,} \end{cases}$$

$$\text{Fluid phase proteid} \begin{cases} \text{I.} & \begin{cases} 0\cdot45 \text{ gr. per 100 c.c.} \\ 0\cdot46 \text{ ,, \quad ,, \quad ,,} \end{cases} \\ \text{II.} & \begin{cases} 0\cdot16 \text{ ,, \quad ,, \quad ,,} \\ 0\cdot18 \text{ ,, \quad ,, \quad ,,} \end{cases} \end{cases}$$

The undissolved residue in each tube was edestin, being entirely soluble in 10 per cent. NaCl.

The composition of the fluid phase depends therefore upon each component, water, salt and globulin, and as the variation seems to be continuous the solid phase must be a continuously varying compound or solid solution containing all three.

The influence of the concentration of globulin in the system is probably in part related to the fact that a suspension of globulin in water differs from a suspension of, for instance, sand in this respect—that it is partly in solution and undergoes a true dissociation. The part dissolved may be of the nature of an impurity attached to the precipitate, but no matter how carefully prepared, ox globulin suspended in water raises the conductivity of the water and the electrolyte, which is dissociated from the globulin, is so much a part of it as to be carried down again when the latter is centrifuged off. The globulin therefore is *ab initio* a solid solution or compound of low grade which dissociates in presence of water, the degree of dissociation increasing with dilution. Globulin reacts acid to litmus even when it has been dissolved and precipitated ten times with volatile acid ($H\overline{A}$ or HCl) and alkali (NH_3), washed on a filter-paper at nearly zero, and dried *in vacuo* over H_2SO_4 and KHO. The reaction of such a globulin in gas-free water when tested *in vacuo* with litmus is decidedly acid.

The following table shows that the degree of dissociation of the globulin precipitate increases with dilution.

Specific Conductivity of a Suspension of Globulin

V = vol. in c.c. of 1 gr. globulin.
C = gr. globulin per litre.
K = specific conductivity in reciprocal ohms at 18°.

V	C	$K \times 10^{-6}$	$K/C \times 10^{-9}$
35·89	27·86	22·9	824
143·5	6·96	6·01	863
287·12	3·48	4·64	1333
574·24	1·74	3·25	1867

Salt globulins are so unstable and their solution so sensitive as to render the carbonic acid of the air an important component. This can be shown in many ways, but most simply by measuring the degree of dilution necessary to precipitate a solution of salt globulin with ordinary distilled water and with the same water freed from CO_2 by the passage of CO_2-free air, and from other gases by boiling.

$N/6$ NaCl was saturated with ox globulin and centrifuged. Divided into two lots and diluted to $N/15·6$ with (i) ordinary distilled water, and (ii) gas-free water respectively, and the precipitate centrifuged away as rapidly as possible. Proteid content of the upper layers: (i) 0·93 per cent., (ii) 0·76 per cent.

A portion of distilled water was freed from CO_2 by long-continued bubbling through it of pure air, and from other gases by being boiled and rapidly cooled just before use.

Two burettes were filled, the one with gas-free water, the other with ordinary water, and determinations were made of the degree of dilution with the two samples of water necessary to precipitate salt globulin from clear saturated solutions. Ox globulin used 1 c.c. salt globulin solution:

	c.c. CO_2 water	c.c. gas-free water
KNO$_3$ globulin	4 quite opaque	10 no change
MgNO$_3$,,	4 opaque	4 very faint opalescence in 20 min., deepening to partial opacity
Na$_2$SO$_4$,,	2·4 faint opalescence, rapidly deepening to opacity	2·4 no change; 10 min. later faint opalescence
NaCl ,,	1·6 opaque at once	1·6 remains clear for 10 min., then becomes cloudy 15 min., not yet opaque

The slow clouding of solutions diluted with gas-free water is due to absorbed gas, and with care it may be seen to start from the free surface.

A thick suspension of ox globulin precipitated from a solution in MgSO$_4$ by dialysis was diluted to thirteen volumes (1) with gas-free water, (2) with CO_2 water, and run into 1 c.c. of a molar salt solution until a faint cloud appeared. Globulin in each suspension 0·426 gr. per 100 c.c.

Salt	c.c. suspension almost free from gases	c.c. suspension saturated with gases
NaCl	6·5	4·5
K$_2$SO$_4$	7·9	7·0
KNO$_3$	8·0	6·8

The NaCl globulin is the more easily decomposed by acid (H_3CO_4). Even more striking is the action of CO_2 in precipitating saturated solutions of salt edestin in gas-free water.

The solvent power of two salts acting simultaneously is the sum of their solvent powers when acting separately. The point is interesting, since, as Picton and Linder showed, when two salts together act as precipitants of colloids the precipitating power is the sum of their actions when measured separately only when the valency of the active ion is the same. When ions of different valency act simultaneously they inhibit one another. The solvent power of pairs of salts was found to be exactly the sum of the solvent power of each for the following pairs:

KCl	KCl	KCl	NaCl	KB_r
NaCl	K_2SO_4	$MgSO_4$	K_2SO_4	Na_2SO_4

Inertia. Once a solid phase is formed it dominates the whole system by reason of the characteristic inertia of the solid state. I have often been struck with the fact that when silica is precipitated from solutions by salt the precipitate is at first resoluble and the change of state readily reversed. The precipitate, however, rapidly assumes its characteristic insoluble resistant nature. The gel of silica in contact with water, as van Bemmelen showed, continues to change for years always in the direction of a more complete separation from the fluid. In these cases the equilibrium is labile and the composition of the fluid phase will alter continuously with the alteration in the solid phase. Van 't Hoff points out the difficulties which arise in the study of the equilibrium states of calcium sulphate and water owing to the inertia of the solid phase, which may be one of various hydrates or one of two different molecular states of anhydrous calcium sulphate. The anhydrous form which separates first is a soluble calcium sulphate, and this slowly changes to insoluble calcium sulphate.

One can trace similar changes in globulin precipitates, which may be of the nature of differences in the degree of hydration of the globulin, of differences in the molecular state of the globulin similar to the distinction between soluble anhydrous and insoluble anhydrous calcium sulphate, or of differences in the amount of salt, acid or alkali combined with the globulin.

Increased stability of the solid phase of globulin due to contact with water. Globulin precipitated from a solution of acid or alkaline globulin by neutralisation and the precipitate is redissolved by salt. The amount of salt necessary to effect resolution rises and apparently reaches a maximum in 120 min., and the rate of solution slows with time. The increase in inertia is well seen in the following curve which represents solution of a globulin precipitate which had been kept in dialysing bags in contact with boiled water for 3 days.

Below the line where no solid phase can be separated by the centrifuge the curves were mapped by estimating the fraction of globulin in the total solid phase, above that line the change was followed by careful comparison of the change in the degree of opalescence. Curve $OABCD$ represents in order solution in KOH, precipitation and resolution by HCl. Curve $OXYCZ$ represents solution in HCl, precipitation and resolution by KOH. In both cases the curve starts from a suspension of ox globulin containing in 100 c.c. 1·34 gr. dry globulin, and the figures to right and left of the line OO give respectively the number of c.c. of $0.01 N$ KOH or HCl added to 4 c.c. of this suspension.

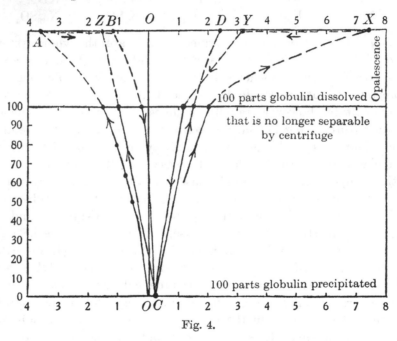

Fig. 4.

When globulin is dissolved by salt in presence of a small amount of acid (HCl) the end-point depends upon the way in which the salt is added. If it be added slowly the first effect of the salt is to precipitate the globulin, and this precipitate acts as a drag on the system so that final solution is only attained by excess of salt. If the salt solution be added rapidly with agitation the initial precipitate has not time to form and clear solution is obtained with less salt.

Two special instances of inertia are worth mention.

When the concentration of salt in the entire system is the same the concentration of globulin in the fluid phase is greater when the salt is added to a thick suspension of globulin than when the concentration of salt is reached by diluting a more concentrated solution of salt globulin free from solid.

Thus $2N$ NaCl was added to a suspension of globulin so that the final concentration was $0.12N$ and $0.09N$ respectively. And portions of $1.8N$ salt saturated with globulin were diluted with water to $0.12N$ and $0.09N$ with the result that precipitation of globulin took place. Therefore in the final state in each case excess solid globulin was present.

The various solutions were centrifuged and the globulin in solution determined with the following result:

Salt	Direct solution %	Diluted from 1·8 normal %
$0.12N$	4·46	2·16
0·09	2·776	0·35

The explanation is, I think, simple. If there were no inertia the state produced by solution and that produced by precipitation should coincide so long as excess globulin is present, and the state would be one in which salt, water, and globulin are distributed through both phases. But when solid globulin is dissolved the system has to come into equilibrium with residual solid globulin which will, owing to the inertia of the solid state, change to the salt globulin state slowly, while in precipitation on dilution the first-formed precipitate is a salt globulin in equilibrium with a certain relative mass of water. The lag in the one case is from one extreme condition of the solid phase, in the other it is from the other extreme. Similar cases are found in systems much less complex than colloidal systems, and they led van't Hoff to a generalisation which connects the degree of inertia with the valency of the ions.[1]

An interesting example of inertia in respect to temperature is furnished by the compounds of edestin with potassium sulphate and sodium chloride.

When crystals of edestin are alternately melted by heat and separated by cold in sealed tubes under saturated solutions of K_2SO_4 edestin and NaCl edestin respectively, it was found that with a *rising* temperature the crystals in the NaCl tube melt slightly later than those in the K_2SO_4 tube, with a *falling* temperature they reappear very much earlier.

Warmed to 15°, the crystals in both tubes fuse, those in the NaCl tube lagging very slightly; cooled to 11°, the whole mass in the NaCl tube at once recrystallises, while that in the K_2SO_4 tube is still fluid ten minutes later.

The composition of the fluid phases from which crystals separated was:

	Proteid %	Salt equiv. in 100 gr.	Water % by weight
K_2SO_4	44·7	0·011	55
NaCl	45·8	0·015	54·4

From these two solutions it would be possible to construct a device which would resemble the thermal sense organs in indicating either a rising temperature or a falling temperature.

[1] *Archiv. Néerlandaise* (1901), VI, 471.

Looked at from the point of view of the Phase Rule most of the measurements made are of consolute points for the globulin-rich and globulin-poor phases, the consolute point being defined by opalescence due to an internal phase being minimal. Like ordinary three-component systems which form phases of continuously varying composition the composition of the consolute depends upon the initial ratio of the components, *e.g.* water and globulin. In the phase diagram the curve for the conjugates is nowhere parallel to the base-line globulin water and therefore it is cut at different points according to the position on that base line of the initial system.

THE EQUILIBRIUM BETWEEN ACID OR ALKALI GLOBULIN AND SALTS

It will be convenient to use BOH for alkali, HS for acid, and G for globulin, the latter symbol, however, is not taken to mean always the same reacting weight.

There are four components, BOH [or HS], G, BS, and HOH, and the equilibrium of these four component systems which are of practical interest are (1) the precipitation of solutions of alkali or acid globulin by low concentrations of salts, (2) the precipitation of those solutions on neutralisation, *e.g.* GHS + BOH = G + BS + HOH. In the second case the right-hand side of the equation which represents the condition when the solid phase forms contains the salt BS, and this is the agent which brings about the precipitation. Dialysis, for instance, in which salt is not formed does not lead to precipitation, at any rate from solutions where the acid or alkali combined with the globulin liberates only univalent ions.

Precipitation by neutralisation therefore is merely a special case of the precipitation of alkali or acid globulin by salts. The stability of the solution is diminished by reducing the ratio acid or alkali/globulin until the concentration of salt is sufficient to destroy it.

The solid phase which settles out may be alkali globulin or acid globulin of very low alkali or acid content, or globulin. This is the only possible explanation of the following facts. The solid phase is completely immiscible with the fluid phase, but it is usually only imperfectly separable from water. When the precipitate is dialysed or washed by decantation a stage is frequently reached at which solution of a low grade is reformed, the solution being filterable and stable on a centrifuge. The proteid has resumed its ionic character and the direction of movement shows that if it has been precipitated from alkali globulin it moves to the anode, if it has been precipitated from acid globulin it moves to the cathode. Between these two conditions of the precipitates there is a point where globulin is precipitated which does not resume its ionic character in presence of any volume of water.

Salts as precipitants. When salts at low dilution precipitate hydrosols a simple and remarkable relation exists. The precipitating action of the salt is due to one only of its ions, always that which carries an electric charge of a sign opposite to that of the colloid particle, and the power of the salt is much greater the higher the valency of the ion.

This valency rule can be detected in the precipitation of acid or alkali globulin by salts, but it is obscured to a greater or less extent by the fact that the salts themselves can form soluble compounds with globulins.

If transparent solutions of acid or alkali globulin are precipitated by salts it will easily be seen that the salts are arranged so far as precipitating power is concerned according to the ionic valency rule stated above, but it is equally obvious that some salts precipitate a larger fraction of the globulin and give a heavier and coarser precipitate than others.

Putting this complication on one side attention was fixed solely upon the point of maximal precipitation for any particular salt, and four salts were examined in some detail, the acid and alkali being HCl and NaOH respectively.

The method used was to add a normal solution of the salt to the solution of globulin until the point of maximal precipitation was reached. The mass of globulin was constant throughout. The volume, that is the concentration, of the globulin varied. The following curves (Fig. 5) give the relative concentrations of acid or alkali, and salt which correspond to the most complete separation of a solid phase. The curves for salt acid globulin are much more extended than for salt alkali globulin. The range of concentration of alkali over which a salt can precipitate is so small as to reduce the curves practically to short almost vertical lines, the slope of which is almost indeterminable.

The curves were followed downwards into the region of opaque opalescent solution as far as possible, as the horizontal lines which mark the level of clear solution show.

The curves for salt acid globulin are extensive enough to permit of some conclusion being drawn. At a concentration of acid when the globulin is completely dissolved the valency law holds, the electric sign of the proteid is + and the salts are:

$$K_2SO_4 = MgSO_4 > MgNO_3 \text{ and } KNO_3.$$

As the acid concentration rises an entirely fresh relationship is seen— that due to the solvent powers of the salts, those powerful solvents the sulphates ceasing to have any precipitating power, the nitrates continue to be good precipitants beyond the range followed.

At 0·002 to 0·003 HCl the concentration of K_2SO_4 or $MgSO_4$ needed to *precipitate* nearly reaches infinity, while at 0·23 $MgNO_3$ the concentration of acid needed to *dissolve* reaches infinity, and for 0·23 KNO_3 is rapidly rising.

The remarkable change of direction in the curve MgNO$_3$.HCl on the analogy of similar cases indicates a change in the constitution of the solid phase which is formed.

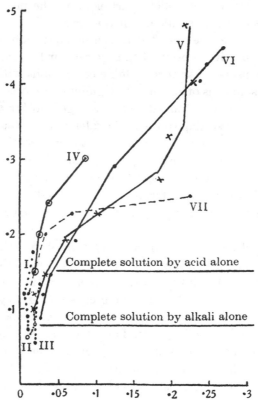

Fig. 5. Ordinates equivalents of HCl or NaOH × 100 per litre. Abscissæ equivalents of salt per litre. Dotted curves are salt alkali. I, MgNO$_3$.NaOH; II, MgSO$_4$.NaOH; III, K$_2$SO$_4$.NaOH; IV, K$_2$SO$_4$.HCl; V, MgNO$_3$.HCl; VI, KNO$_3$.HCl; VII, MgSO$_4$.HCl.

The curves for K$_2$SO$_4$.HCl and MgSO$_4$.HCl cannot be prolonged upwards, for beyond their upper ends a solid phase cannot exist. The curves for MgNO$_3$.HCl and KNO$_3$.HCl can be continued beyond the point to which they are actually mapped, that is to say these salts continue to precipitate with still higher concentration of acid.

Salts as solvents in presence of acid or alkali. I measured the converse relation for two salts, namely the effect of varying amounts of HCl and NaOH upon the amount of salt necessary to dissolve the same weight of globulin.

The salts were K$_2$SO$_4$ and MgSO$_4$ (Fig. 6). The ordinates give c.c. of normal salt, the abscissæ to the right of the line *OO* represent c.c. of 0·01 *N* HCl, to

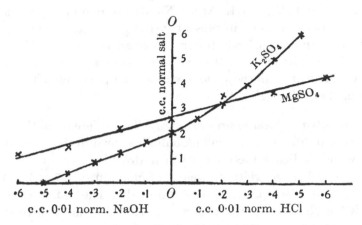

Fig. 6. Ordinates are c.c. normal salt, abscissæ to right, c.c. $0.01N$ HCl, to left $0.01N$ NaOH (for K_2SO_4 curve) or KHO (for $MgSO_4$ curve), in each case added to 5 c.c. of a globulin suspension. OO is line of salt alone.

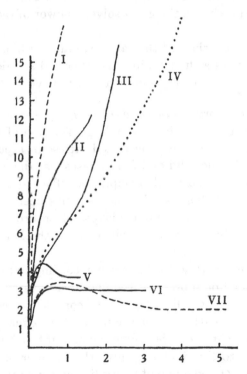

Fig. 7. The curves show the amount of acid (I, II, III, IV) or alkali (V, VI, VII) needed to dissolve to transparent solution 0.107 gr. dry globulin in presence of varying amount of salt.

the left c.c. of $0.01 N$ NaOH. At 0.5 NaOH the mass of K_2SO_4 needed to dissolve the mass of globulin present (0.107 gr.) is zero. The ordinate OO measures the amount of salt needed to dissolve when neither acid nor alkali is added. The curve is the line which separates the system with no solid phase present from the system with a solid phase present. The former lies above, the latter below the curve.

Acid or alkali as solvent in presence of salt. The difference in the relation of salt to acid globulin and to alkali globulin is shown in the following curves (Fig. 7), which indicate the quantity of hydrochloric acid or caustic soda needed to dissolve an arbitrary weight of globulin (0.107 gr.) to the same grade of solution—in presence of certain salts. The ordinates are c.c. of $0.01 N$ HCl or NaOH, and the abscissæ are c.c. of normal salt to 10 c.c. of a suspension containing 0.107 gr. dry globulin.

The salt was added to the suspension first and it produced partial or complete solution. The acid or alkali was then run in, the globulin being reprecipitated by the first few drops. The curve for $Mg(NO_3)_2.NaOH$ is too short to be plotted owing to the great solvent power of the salt in presence of alkali.

The curves show (1) that all the salts examined inhibit solution by the acid, and all the salts at first inhibit and then aid solution by the alkali; (2) that each salt has its own peculiar form of curve.

Acids and alkali as precipitants and solvents in presence of salts. In order that the precipitate should include a large fraction of the globulin the relative concentrations need to be adjusted with some accuracy. The distribution of globulin in the solid and fluid phases was determined by centrifuging and estimating the total precipitated. This gives the state of the system under the centrifugal stress, and as it has been shown that centrifugal force will concentrate the outer layers of an ordinary salt solution[1] this state must not be regarded as coincident with that produced by the much smaller force of gravity.

The following curves (Fig. 8) are drawn according to the recognised convention—the upper line represents the one-phase system, all the globulin being dissolved, the lower line represents complete precipitation of the globulin. The apex of each curve is the point of maximal precipitation. To the left of the line OO the numbers give the amount of NaOH added in c.c. of $0.01 N$ solution, to the right they give the number of c.c. of 0.01 HCl. Therefore the line OO is the line of neutrality so far as the NaOH and HCl alone are concerned. The lower O is a suspension of 0.1 gr. of dry globulin in 10 c.c. It was completely dissolved by alkali (OA), the curve being drawn

---

[1] van Calcar and Lobry de Bruyn, *Rec. Trav. Chim. Leiden* (1903), xxviii, 218.

as a simple straight line because the points intermediate between O and A were not mapped. The solution of NaOH·globulin was precipitated by HCl $0\cdot01\,N$, and the curve AX does not reach the line of complete precipitation. The precipitate was redissolved by HCl (XA). One c.c. of normal KNO_3 was added to 10 c.c. of the suspension and the curve BYB mapped, and $3\cdot1$ c.c. $N\ KNO_3$ gave the curve CZC.

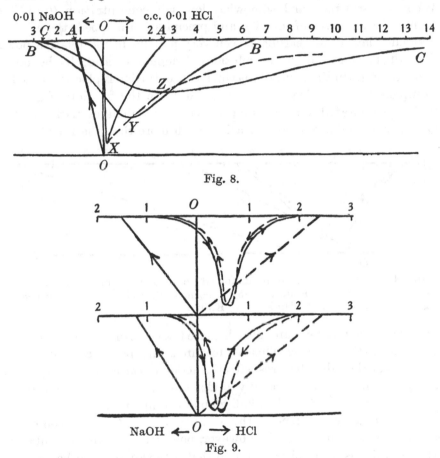

Fig. 8.

Fig. 9.

These curves show that the effect of increasing salt (KNO_3) concentration is to shift the point of maximal precipitation over to the acid side, and to render resolution by acid more and more difficult. At the same time the fraction of the globulin which can be precipitated becomes less and less, the trough of the curve gradually filling up, the apex moving along the dotted line.

The salt which is produced when a solution of acid or alkali globulin is precipitated by alkali or acid similarly shifts the point of maximal precipitation over to the acid side. In Fig. 9 two sets of curves are given. In the lower, starting at O, the following events occur in order in the curves: solution by NaOH, precipitation and resolution by HCl, and, in the dotted curve, reprecipitation and resolution by NaOH. The second precipitation occurs further in the acid region. The other curve traces the same series of events with the same suspension after 20 c.c. $0·02N$ NaCl had been added to 100 c.c. of the suspension.

When an arbitrary and somewhat low salt concentration ($0·13N$) is chosen and the curves of solution and precipitation mapped, the effect of the valency law stated earlier is seen very prettily in the relative power which salts possess of inhibiting solutions of globulin by acid. In the follow-ing curves (Fig. 10) the point of maximal concentration is placed on the line of complete precipitation as a convention, since the fraction of globulin actually precipitated was not determined. The sulphates with bivalent negative ions inhibit solution by acids much more than the nitrates with

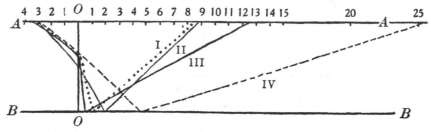

Fig. 10. AA, 100 per cent. globulin dissolved. BB, 100 per cent. precipitated. OO, no acid or alkali. At the beginning 1·5 c.c. normal salt added to 10 c.c. suspension. I, $Mg(NO_3)_2$; II, KNO_3; III, $MgSO_4$; IV, K_2SO_4.

monovalent negative ions, and also displace the point of maximal precipita-tion further to the acid side. This last feature is unexpected and interesting. It shows that the salt interferes with the action of the acid in decomposing alkali globulin, a relation readily understood on the hypothesis of the existence in solution of a salt alkali globulin compound.

These various cases render the complexity of the four-component systems very obvious; there is, however, one feature of fundamental im-portance which is never obscured, and that is the antagonism between the solvent actions of salts and acids, and the additive nature of the combined solvent action of salts and alkali. This feature arises, I believe, from the fact that acid globulin is insoluble by salt. Salts will combine with globulin or with alkali globulin to form soluble compounds; they will not so combine with acid globulin.

If this be true, then, when acid globulin is precipitated and redissolved by salt, the acid must be displaced and the globulin redissolved, not as salt-acid globulin but as salt globulin. The displacement of the acid can readily be followed by methyl orange. When HCl globulin is precipitated by a salt this indicator shows that free acid is liberated.[1] But I failed completely to detect the liberation of free alkali when salt is added to a solution of alkali globulin. This suggests that precipitation of acid globulin is partly a definite chemical replacement of acid by salt,

$$GHS + BS = GBS + HS.$$

The fact that alkalis slightly assist solution in salts while acids very generally depress it suggests an interesting possibility, namely, that in the compounds GHS and GBS the acid HS and the salt BS are united to the molecule of G in the same way, so that they compete with one another, while in GB and GHS the base and salt are united to different parts of the molecule.

On this view the possible equations would be

$$GB + BS \rightleftharpoons GB.BS,$$
$$G + BS \rightleftharpoons G.BS,$$
$$GHS + BS \rightleftharpoons GBS + HS.$$

And the impossible equation is

$$GBS + HS \rightleftharpoons GBS.HS.$$

Analogous case of amido acids. These equations can be developed further on the analogy of the amido acids, if the globulin be taken to have a molecular structure in general agreement with that of Fischer's polypeptides.

The structural formula for an amido acid is

According to Bayer's oxonium theory the upper oxygen of the carboxyl group has two unsatisfied valencies, oxygen being tetravalent $O\equiv$. In the molecule it acts as $O=$. The nitrogen atom has two unsatisfied valencies. In the molecule it is $N=$, while its upper limit of valency is $N\equiv$. Therefore there are two places where other atoms can be linked on without replacement.

Consider the special case of amido-acetic acid. The following compounds are known:

(i) The salts with alkalis formed by replacement of hydrogen:

i.e. potassium amino acetate.

[1] The following salts were tried and in each case free acid was detected, KNO_3, $CaCl_2$, and $MgSO_4$.

(ii) The salts with acids:

$$R \underset{C \overset{O}{\underset{OH}{}}}{\overset{\overset{H_2}{\|}}{N \overset{H}{\underset{Cl}{}}}} \qquad \text{or} \qquad R \underset{C \overset{O}{\underset{OH}{}}}{\overset{\overset{H_2}{\|}}{N \overset{H}{\underset{HSO_4}{}}}}$$

In this class the linkage is due to a change from $N\!=\!$ to $N\!\equiv\!$.

The hydrated form which theoretically exists in water would also probably have the formula

$$R \underset{C \overset{O}{\underset{OH}{}}}{\overset{\overset{H_2}{\|}}{N \overset{H}{\underset{OH}{}}}}$$

(iii) Double salts such as $CH_3C \overset{NH_2}{\underset{COOH}{}} . NaCl$.

Which of the two possible linkages holds the NaCl? Clearly it might be the bivalent $O\!=\!$ on the analogy of the hyper-acid acetate of potassium.

$$R \overset{O \overset{K}{}}{\underset{OH}{\underset{}{}}} CH_3COO$$

But the probability is very much in favour of the linkage being by $N\!\equiv\!$, since $N\!\equiv\!$ is much more stable than $O\!\equiv\!$.

Therefore the probable formula is

$$R \underset{C \overset{O}{\underset{OH}{}}}{\overset{\overset{H_2}{\|}}{N \overset{Na}{\underset{Cl}{}}}}$$

That is to say the amido acid can link on an acid or a neutral salt *but not both*.

Therefore the possible equations are:

$$R \overset{NH_2}{\underset{COOH}{}} + HS \rightleftharpoons R \overset{NH_2.HS}{\underset{COOH}{}}$$

$$R \overset{NH_2}{\underset{COOH}{}} + BOH \rightleftharpoons R \overset{NH_2}{\underset{COB}{}} + HOH$$

$$R \overset{NH_2}{\underset{COOH}{}} + BS \rightleftharpoons R \overset{NH_2BS}{\underset{COOH}{}}$$

$$R \overset{NH_2HS}{\underset{COOH}{}} + BOH \rightleftharpoons R \overset{NH_2HOH}{\underset{COOH}{}} + BS$$

$$R \overset{NH_2BS}{\underset{COOH}{}} + HS \rightleftharpoons R \overset{NH_2HS}{\underset{COOH}{}} + BS$$

$$R \overset{NH_2}{\underset{COB}{}} + BS \rightleftharpoons R \overset{NH_2BS}{\underset{COB}{}}$$

APPENDIX I

The apparatus used to determine the absence of any drift in a field, and therefore of any charge on the molecules of the solvee, was as follows:

AB is a glass tube ground at the ends, on which is a shoulder at E, so that it can be lowered into the U-tube which it nearly fits. The lower part is graduated in millimetres, and filled with a gel of celloidin permeated only by pure chloroform. The gel was prepared by filling the tube to C with a solution of celloidin in ether and alcohol and setting it to a gel by immersion in chloroform. The ether and alcohol were removed by immersion in the dark for weeks in many changes of chloroform. The bend of the U-tube holds mercury, and the space BC a solution of azobenzene in chloroform. The space AC holds a similar solution. Electrodes dip into AC and into the mercury and a field of 130 volts per cm. was maintained for six hours; after which the tube AC was removed and the azobenzene found to have diffused 9 mm. into the celloidin gel from each end. It has therefore moved at equal velocities in the field in a direction from anode to cathode, and from cathode to anode. There is no difference to indicate a drift of charged molecules.

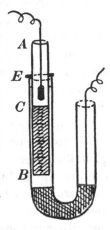

Fig. 11.

In other experiments two plates of solid azobenzene, each 1 mm. thick, were kept at differences of potential the gradient between them being 500 volts per cm. They were then immersed in chloroform and found to dissolve at the same rate.

APPENDIX II

The Relation of the Globulin to Serum Proteid

Ionic globulin is absent from serum, therefore alkali globulin is absent. But salt-alkali globulin which does not liberate ionic globulin in solution may be present and stable, in spite of the low salt content, owing to the great stability of this substance in solution. Dialysis brings about partial precipitation by liberating alkali globulin, which in turn is precipitated by the free salt present, or by carbonic acid. Further decomposition and precipitation occurs when a stronger acid is added.

This hypothesis is simple, and can be extended very easily to embrace the methods of precipitation. It is also in agreement with general opinion, which regards serum as a mixture of certain proteids in solution.

It fails, however, to explain why serum is not precipitated by simple addition of acid, for acid globulin is extremely unstable in presence of salts,

H   27

and on the whole the balance of probability is against it and in favour of there being in serum some (possibly one) complex proteid which breaks down readily into fractions whose composition and properties depend upon the degree of dilution and the reagents used.

The facts of the case are these. The proteids of serum are electrically inactive. Neither the whole nor any fraction moves in a field. It is not possible to detect a trace of "ionic" proteid. Dialysis or dilution disturbs the equilibrium, and "ionic" globulin appears, and can be swept out of the general mass of proteid as an opalescent cloud before dialysis has been pushed to the point where precipitation occurs. The development of even minute quantities of "ionic" globulin can be detected in this way. The direction of the movement is towards the anode, therefore the globulin which appears is the anion of alkali globulin.

As dialysis or dilution proceeds the concentration of the liberated alkali globulin increases, the increase being shown by an increase in the density of the anionic globulin cloud which emerges from the serum in the electric field and when the concentration reaches a certain point precipitation of globulin or of an alkaline globulin having a lowered alkali content occurs, due either to the action of carbonic acid or to the relative concentrations of alkali globulin and salt becoming such that the former is partly precipitated, just as a solution of pure alkali globulin can be more or less completely precipitated by the addition of small amounts of salt. What is known of globulins reduces the possibilities to these two, and as a matter of fact the action of carbonic acid seems to be the more potent. Dialysis in closed vessels against water freed from carbonic acid decidedly diminishes precipitation.

Be this as it may, when dialysis ceases to precipitate one is left with serum which has lost only a small portion of its proteids[1] and which still contains globulin in abundance. Further precipitation of this residual globulin is brought about by the addition of any acid, but even the most cautious acidulation fails to precipitate the whole, a second dialysis giving a still further yield.

Serum which has been dialysed against water with very low carbonic acid content until it ceases to give any precipitate, but which can still give with acid a large yield of globulin, is in a most interesting condition. The whole remaining proteid is now ionic. It moves towards the anode quite uniformly, therefore it behaves as a whole as the proteid ion of an alkali proteid compound.

Now what does this mean? The presence of other proteids does not interfere with the movement of ionic proteid. This point cannot be too much

[1] Hammarsten's values are not for serum, but for serum which has been first neutralised by acid.

insisted on. It lies at the root of the evidence. Therefore one starts with proteid which behaves uniformly and is electrically inactive, one ends with proteid which behaves uniformly and is electrically active, and in the final stage there is no evidence of the presence of more than one kind of proteid ion. But this residue of uniform character still contains a globulin fraction. It can be split by saturation with a neutral salt or by acidulation into fractions differing in properties according to the mode of separation.

Serum proteid in unchanged serum is electrically inactive, but it is possible to make it as a whole electrically active.

When in the cell used for these experiments which is described on p. 384 the upper layer of fluid is a solution of $1/600 N$ acetic acid and the lower layer is serum and a current is passed for 24 hours, the serum proteid becomes slowly charged in such a way that it is repelled from both poles. Therefore in the anodic limb it becomes charged positively, in the cathodic limb negatively. The result of the double repulsion is that the proteid is condensed into a hard plate of rubber-like consistency just midway between the electrodes.[1]

The phenomena would be explained on the theory which has been outlined in this way. In the anodic limb an excess of hydroxyl ions are liberated owing to the electric convection, and these, reacting with the serum proteid, convert it into cationic proteid. In the cathodic limb the converse reaction with hydrions occurs and is possible owing to the amphoteric nature of proteids. The entire mass of serum proteid is thus thrown into the ionic state and in this condition moves with uniform motion, the one half as cationic proteid, the other half as anionic proteid. The electric current, the most subtle of analysers, detects only one substance, and this substance owing to its amphoteric nature can exist in either the cationic, or the anionic state.

The concentrated hard rubber-like mass of proteid obtained in this way cannot be discriminated from serum proteid. It is easily soluble in distilled water, and in dilute salt solutions. From the solution in water a globulin fraction can be precipitated by acid, or by saturation with magnesium sulphate. From the solution in dilute salt solution a globulin fraction can be precipitated by saturation with salt (NaCl or $MgSO_4$). The filtrate after removal of the globulin by saturation contains a proteid which is precipitated by acetic acid. In this respect it is identical with serum albumin.

The plate of concentrated proteid contains the serum phosphorus but only a fraction of the serum pigment.

The position then is that in serum one has proteid which can be thrown

[1] Whitney and Blake describe a similar double repulsion in the case of gelatine (*Amer. Chem. Soc.* (1904), XXVI, 1339).

into the ionic state and which then moves in a field as a single substance. From it an electrically active fraction, namely globulin, can be split off, and the proteid thereby becomes electrically heterogeneous. When the fraction is removed by dialysis and by acid the remaining fraction is again electrically homogeneous, but it is now ionic.

Now the globulin fraction has an abiding characteristic. In all its solutions its molecular state is so gross as to cause the molecules to be arrested by a porous pot. They will not pass such a filter, even under pressure. In this it is sharply distinct from the parent serum proteid, which is readily filterable. If globulin be present as such in serum it is not only non-ionic, but the agent which dissolves it must be something more than alkali and salt since either alone or together they will not produce so high a grade of solution.

The difference in the molecular state of globulin when once separated, and the electrical homogeneity of serum proteid and of the fraction (still capable of further subdivision by salting out) which remains after the alkaline globulin fraction which most readily appears has been removed, suggests that serum proteid is a complex unit. If such a unit exist it is not saturated with globulin. Fresh ox serum has an extraordinary power of dissolving globulin, it will take up almost its own volume of the thick cake at the bottom of a centrifuge tube; and in ox serum so saturated there is not a trace of alkali globulin nor of any ionic proteid.

The question of the relationship of globulin and serum proteid can be approached from the chemical side.

Globulin contains phosphorus, two analyses by Carius' method giving 0·07 and 0·08 per cent. by weight. In the analyses of blood globulins which I can find, this fact is nowhere mentioned, though Porges and Spiro[1] speak of calcium phosphate as a constituent of globulin precipitates, though the manner of its identification is not given. If calcium and phosphate were to be found in the ash in the requisite proportions it is certain that calcium phosphate does not occur in the intact globulin, for the phosphorus-holding portion is not dissolved by 48 hours' digestion at 40° with 50 per cent. acetic acid.

Ox serum globulin was precipitated by $\frac{1}{2}$ saturation with $AmSO_4$, and the precipitate purified by being dissolved and reprecipitated seven times. The preparation contains phosphorus.

Some of this preparation was dissolved and coagulated by boiling, and the fine coagulum suspended in boiling 30 per cent. acetic acid, and then digested for 48 hours with 50 per cent. acetic at 40°. The proteid was filtered off and the filtrate found to give a good xanthoproteic reaction. The undissolved proteid was washed on a filter for 2 days with hot water. It still contained phosphorus, apparently in undiminished amount.

[1] *Beit. z. chem. Physiol. u. Path.* (1902), III, 277.

The phosphorus is not from entangled lecithin.

Ox globulin precipitated by acetic acid was coagulated by heat, digested for 48 hours at 40° with 50 per cent. acetic, and washed on a filter with hot water until all the acid was removed. It now contained phosphorus. It was extracted with alcohol, dried and extracted with two changes of ether in a Soxhlet apparatus for 5 days, and it still contained phosphorus.

The property of serum globulins which first caught my attention in this investigation was their phosphorus content and the close association of the characteristic insolubility of the globulin with the presence of this element.

All the globulin fractions which can be separated from serum contain phosphorus in the first instance, most of them can be fractionated by secondary treatment into a phosphorus-holding proteid which is precipitable by dialysis and a phosphorus-free proteid which is soluble in water.

Consider first two fractions, the one separated from ox serum diluted to 10–15 volumes by acetic acid, the other separated from ox serum by saturation with magnesium sulphate. The first is purified by three times solution in ammonia, and precipitation by acetic acid. The second was purified by solution and precipitation by the salt until the whole of the proteid came down on saturation. The magnesium sulphate fraction, or shortly, the salt fraction, is, as Hammarsten showed, a much larger fraction of the whole serum proteid (63 per cent.), the acid fraction is only 17 per cent.

The phosphorus content of these two fractions is very different, being for the acid fraction 0·07–0·08 per cent. P, in the salt fraction 0·009 per cent. P.[1]

The acid fraction will not yield a fraction soluble in water or free from phosphorus by any simple treatment such as partial precipitation by acids or alkalis, by dialysis, or treatment with salts. I have only two sets of estimations of phosphorus bearing on this point, namely of fractions precipitated from solutions in very dilute caustic soda by acetic acid. The first fraction to come down contained 0·068 per cent. P, the second fraction 0·071 per cent., both fractions being prepared for analysis by washing with alcohol and ether.

The salt fraction, as Burckhardt[2] first showed, readily splits on dialysis into a water soluble and a water insoluble portion. The former contains phosphorus, the latter is free from phosphorus, is soluble in water, but is precipitated by saturation with neutral salt.

Consider another method of fractionating serum proteid. Ox serum is

[1] The analyses were carried out for me by Mr Hall, assistant at the University Chemical Laboratory, by the bomb method. The amount of phosphorus in the salt fraction is too small for accurate estimation. Mr Hall gives me the above figure as the best available approximation.

[2] Arch. f. exp. Path. u. Pharmak. (1883), XVI, 322.

dialysed until precipitation ceases [P. i], the residue is precipitated by addition of acetic acid [P. ii], the filtrate is precipitated by saturation with magnesium sulphate [P. iii]. P. i, P. ii, and P. iii all contain phosphorus. Between P. ii and P. iii a fraction also can be interposed, namely by dialysis. I do not know whether this causes the final salted out fraction to be free from phosphorus.

P. i will not yield a phosphorus-free or water-soluble moiety, P. ii readily does so. Dialysis of solutions of the proteid in either dilute acetic acid, or sodium chloride, leads to the separation of a precipitate which contains all the phosphorus, the residue is a water-soluble phosphorus-free proteid which is precipitated from its solutions by saturation with magnesium sulphate. From the former fraction I failed by subsequent treatment to separate water-soluble proteid. The latter fraction is a proteid of somewhat singular properties. It is not precipitated by a stream of carbonic acid gas, nor by dilute acids. It is precipitated by very dilute alkali and the precipitate is resoluble on the addition of salts. It is entirely and completely precipitated by saturation with magnesium sulphate at 30°. It is not precipitated by 60 per cent. alcohol, though the slightest trace of salt at once and completely precipitates. It is coagulated by heat. It is very readily changed to derived albumen, as for instance by simple evaporation of the solution in air.

The two fractions into which P. ii splits differ not only in relation to phosphorus, but also in relation to the presence in the molecule of a sugar-yielding moiety. I succeeded in preparing osazone from the fraction which is insoluble in water and contains phosphorus. I failed in three attempts with the soluble phosphorus-free fraction. P. iii gives a very poor yield at best of insoluble proteid.

This examination of the various fractions is uncompleted and I am engaged in filling in the gaps, but the knowledge so far gained suggests that serum proteid will yield a phosphorus holding proteid which is insoluble in water and has all the properties associated since Schmidt's time with the word globulin. It corresponds to the euglobulin of later writers.[1]

It can be split off in association with varying amounts of a soluble proteid which corresponds probably to the pseudoglobulin of these writers.

When isolated it forms compounds with acids and alkalis which are remarkable for the indefiniteness of the combining weights, therefore euglobulin itself may be merely a step in a series of proteid compounds, the magnesium sulphate fraction being at once extreme, and the so-called nucleo-proteid of Pekelharing, which has all the characters of a globulin, at the other.

[1] Marcus, *Zeits. f. physiol. Chem.* (1899), XXVIII, 559; Porges and Spiro, *Beit. z. chem. Physiol. u. Pathol.* (1902), III, 277; and others.

This is a very obvious possibility, it is therefore pertinent to ask why euglobulin is dealt with in this paper as a chemical unit? Because in the first place it combines as a whole with bases so weak as aniline and urea, and with acids so weak as boracic, and in the second place when combined with any acid or any base the whole of the proteid becomes ionic, not merely a highly phosphated acid fraction, and the whole in its ionic state shares uniformly in the electric transport.

Summary

This paper deals with the behaviour of globulins to acids, alkalis and salts, and the properties of the solutions considered as cases of colloidal solution.

In the three component systems water, globulin and monovalent acid or monovalent alkali, or neutral salt, when three phases are present they are vapour and two continuously varying states, one rich in globulin the other poor in globulin. The former may separate as a precipitate or it may remain dispersed throughout the latter as an internal phase which is not separable by gravity or centrifugal force, in which case it confers on the fluid a greater or less degree of opalescence. Both these phases vary continuously in composition with variation in the relative masses of globulin and acid, alkali, or salt. Such continuously varying states might be called solutions, solid or fluid, of the one component in the other, the precipitate for instance being a continuously varying solid solution of globulin, water, and salt, acid or alkali, like the continuously varying solid solution of ammonium and potassium sulphate which separates from a solution of both salts. The relations however are best described as continuously varying compounds of globulin, since there are signs in the behaviour with indicators, and in the electrical character of the solution, and in the replacement of acid by neutral salt, of undoubted chemical interaction.

1. *Solution by acids or alkalis.* (Electro-positive or electro-negative globulin.) The facts can best be explained by assuming that salt formation occurs. This agrees with the view of all previous workers on the subject. Globulins therefore have both an acid and a basic function—they are amphoteric electrolytes.

The globulin salts ionise in solution; therefore, in an electric field the entire mass of proteid moves. They also hydrolyse, but the hydrolysis offers special features resembling those which Jordis has pointed out in the case of sodium silicate, and Chevreul in the case of soaps. In both cases hyper-acid salts are formed. By dialysis the degree of hyper-acidity or hyper-basicity can be raised, but, as Jordis finds in the case of silica, with continuously increasing difficulty. With the rise in the degree of hyper-acidity or hyper-basicity, the "grade" of solution diminishes, but, in the case of globulins, a precipitation point is not reached with monovalent acids or alkalis. Electrically active solutions of globulin cannot be precipitated by dialysis, nor at any stage do they cease to be electrically active.

If G be used to denote globulin, the equation of hydrolysis or dialysis would be:

$$x\text{GHS} + y\text{HOH} = (\text{GHOH})_y(\text{GHS})_{x-y} + y\text{HS},$$
or
$$x\text{GB} + y\text{HOH} = (\text{GH})_y(\text{GB})_{x-y} + y\text{BOH}.$$

In dialysis the ratio x/y varies continuously.

Clearly a relation of this kind agrees with van Bemmelen's definition of absorption compounds as chemical combination with variable composition.

The relative solvent power of various acids and alkalis has been measured. For strong and medium acids solvent power is measured by the number of gramme molecules present, not by the number of gramme equivalents:

$$\text{HCl} = \text{H}_2\text{SO}_4 = \text{H}_3\text{PO}_4.$$

Very weak acids have a lower solvent power

$$\text{HCl} = 5\overline{\text{HA}} = \pm\,30{,}000\,\text{H}_3\text{BoO}_3.$$

Comparison was made between the amounts of acid necessary to reach the same grade of solution.

The relations are explained by the very weak basic function of globulin. Therefore, salts with weak acids are much hydrolysed, and to reach the same grade of solution an excess of acid is needed in order to lower the degree of hydrolysis. With alkalis the weak alkali NH_4OH dissolves as well as the strong alkalis, owing to the fact that globulin acts as an acid of considerable strength.

The acid and basic functions were measured by the well-known methods —the catalysis of cane sugar and of methyl acetate—and the acid function found to be much the greater.

Data derived from measurements of electric conductivity, and from the behaviour with indicators, support the view that salts are formed, and that the globulin acts much more strongly as an acid than as a base.

2. *Direct measurement of the specific velocity of globulin "ions"* was carried out by the boundary method.

As the basic function of a globulin is weaker than its acid function, the salts GHS (globulin + acid) will be hydrolysed more than the salts GB (globulin + base). Therefore, in the equation of hydrolysis given above, y/x will be larger for the former than the latter. Similarly, comparing salts with $\overline{\text{HA}}$ and with HCl, y/x will be greater for the former.

With increase in the value of y/x the size of the molecule increases also.

By dialysis, the size can be increased until it is large enough to diffract white light.

The large molecules are ionised, that is to say, they take part in the electric transport. And their specific velocity is exceptionally high. "Ions" of this order of magnitude have the properties of matter in mass. Each is defined by a surface and moves under the influence of a surface contact difference of potential. With their appearance the fluid would cease to be homogeneous. It would have internal surfaces. I propose to call such ions colloidal ions, or pseudo-ions. Their specific velocity, within wide limits, is independent of their size, and is controlled by the laws of electrical endosmose.

These conclusions are borne out by the boundary measurement and by the electrical conductivity of colloidal solutions.

For instance: theoretically the proportion of pseudo-ions in the following solutions of globulin should be:

$$\text{In } \overline{\text{HA}} > \text{in HCl, } > \text{in NaOH,}$$

and the specific velocities at 18° were found to be 23×10^{-5}, $11 \cdot 5 \times 10^{-5}$, $7 \cdot 6 \times 10^{-5}$.

By Ostwald's law of the relation of specific ionic velocity to the number of atoms in an ion the value 23×10^{-5} is an utterly impossible one for an ordinary ion of the magnitude of the proteid molecule.

3. *Viscosity.* The viscosity of a globulin solution diminishes rapidly with the increase in the concentration of the ions, and more especially with the increase in the concentration of the true globulin ions as compared with the pseudo-ions.

4. *Solution by neutral salts.* Globulins are dissolved by neutral salts owing to the formation of molecular compounds (G.BS). These compounds are readily decomposed by water with liberation of the insoluble globulin (G.BS + HOH = G.HOH + BS). Therefore they are stable only in presence of a large excess of salt. Hence the solvent power of salts is from 200 to 500 times less than that of acids or alkalis.

A double salt of the form $AB.CD$ ionises according to

$$AB.CD \rightleftharpoons AB.C + D.$$

I have never succeeded in detecting any trace of such ionisation in the case of globulin and neutral salts; possibly owing to the extreme instability of the ions.

Owing to the instability of the dissociation products the globulin can be precipitated from its solution with neutral salts by simple dilution. No degree of dilution will precipitate globulin from solution with acids and alkalis.

The compounds of globulin and alkalis (GB) are more readily dissolved by neutral salts than is simple globulin. Compounds of globulins and acids are insoluble by neutral salts, being decomposed with liberation of the acid.

If on the analogy of an amido acid a globulin combines owing to the presence of the NH_2 and the COOH group, then it forms salts with alkalis by replacement of hydrogen, and with acids by the change of the trivalent nitrogen of the amido group to the pentavalent form.

There are two possible places where the neutral salt might link on—the unsatisfied valencies of the nitrogen, or the unsatisfied valencies of the upper O of the COOH group. According to the oxonium theory, O has a maximum valency $O\!\equiv\!$, but $O\!\equiv\!$ is much less stable than $N\!\equiv\!$; therefore one would expect the linkage to be by the amido group:

And this would account for the fact that the acid can be turned out of the combination with globulin by an excess of neutral salt, but alkali cannot be.

5. *The relation of globulin to the serum proteid or proteids* is discussed and the complete absence of "ionic" globulin from serum is noted. The probability of globulin being formed owing to the decomposition of a complex proteid present in serum is urged.

22

COLLOIDAL SOLUTION. THE GLOBULIN SYSTEM

[Journ. Physiol. (1903), XXIX, 26 *P*]

The experiment which is shown illustrates how by the use of a simple apparatus devised by Nernst measurements for calculating the rate of movement of a globulin in an electric field may be obtained, and, generally, the electrical state of the globulin may be determined. The globulin which is used was precipitated from diluted ox serum by acetic acid and purified by resolution and precipitation. It has been dissolved by the addition of $1/50 N$ acetic acid and now forms the lower layer in the cell; the upper layer is a solution of acetic acid of the same concentration. When the field is established the opalescent proteid is seen to move with uniform velocity from the anode towards the cathode. It is therefore electro-positive.

The result of these and other experiments may be grouped under two headings—qualitative results, quantitative results.

Qualitative results. Globulins form three systems when in solution: (1) in presence of acids when they are electro-positive, (2) in presence of alkalis when they are electro-negative, and (3) in presence of neutral salts when they are electrically inactive. The third system therefore introduces us to an entirely new type of colloidal solution—one in which the colloid particles are uncharged.

The amount of neutral salt reckoned in equivalents necessary to produce the same degree of solution of 1 gr. of dry globulin varies for different salts, and for the same salt in an exceedingly complicated way. Salts, however, have a much smaller solvent power than acids or alkalis (such figures as 500 to 1 will roughly express the relation).

The three systems are incompatible—that is to say the addition of electro-positive to electro-negative globulin, or of either to electrically inert globulin, produces partial or complete precipitation; and, when the conditions are such that an interface exists between any two of the systems, a membrane is formed which, on Traube's law, would sever completely *all* the components of one system from *all* the components of the other. Similarly, a neutral salt added to electrically active globulin decreases the charge on the particles and, if in sufficient quantity, will destroy it and so cause precipitation. Salts act according to the law which I have shown to govern the case when neutral salts are added to electrically active colloidal solutions, namely—the active ion is of the opposite electric sign to that of the colloid particles. The coagulative energy of a salt is an exponential function of the valency of the active ion, so that it might be written $C = k^y$, where k is a constant.

The proteid of blood serum is electrically inactive—it will not move in a field. It can, however, be made to move by appropriate treatment, and the character of the movement is such as could hardly be exhibited by a mixture of various colloids—in other words the experiments suggest that only one proteid is present in serum and not several proteids.

Quantitative. The velocity of electro-positive globulin in a field of unit potential gradient has a singularly constant value—it is at $13 \cdot 20 \times 10^{-5}$ cm. per second. This is of the order of the velocity of bivalent ions such as the calcium ion in a solution of $0 \cdot 1$ equivalent of salt per litre. The globulin particles vary in size according to the number of equivalents of acid present: when the number is large enough (80×10^{-5} equivalents of acetic acid to 1 gr. of dry globulin) the particles are small enough to diffract a blue light; as the number of equivalents falls the diffracted light becomes whiter until an opaque dead white fluid is produced. These stages can be passed through by gradually adding the electrolyte, or by dialysing it away, and in all the stages down to as low as 10×10^{-5} equivalents of acetic acid (much less in the case of stronger acids or alkalis, *e.g.* 1×10^{-5} or less of HCl to 1 gr. dry proteid) there is at each stage a perfect colloidal solution—that is to say no precipitation occurs on standing even though the particles be large enough to diffract pure white light.

From the blue transparent stage down to the opaque white stage the velocity of the proteid particles in a field of unit potential gradient is the same. Therefore the velocity is independent of the size of the particles, and as by Stokes' theorem the resistance to the movement of small spheres through a viscous fluid varies directly with the radius, it follows that the electric charge on the proteid particles must vary directly with the radius. When the concentration of the globulin is very low (< 5 gr. per litre) the velocity *falls*. Therefore dilution produces an effect the opposite to what it does on electrolytes, where it always increases the velocity of the ions.

Throughout all these relations between acid or alkali and proteid it is obvious that there are no definite points of equilibrium corresponding to definite chemical compounds—to chlorides, acetates, proteates, etc.—just as there is no definite heat of neutralisation of silicic acid by alkali. The effect of the globulin upon the conductivity of electrolytes brings into prominence the same indefiniteness. Globulin added to a solution of electrolyte diminishes its conductivity as follows:

HCl	diminished	75 %
HĀ	,,	10 ,,
NaOH	,,	65 ,,
Neutral salts	,,	2 ,,

These values are independent of the proportion of acid to globulin within the limits from a transparent blue solution down to a white opaque solution.

These various constants seem to differ for globulins from different animals. Their values are least for the dog, highest for the pig, and about the same for sheep and ox.

The properties of globulins in solution seem to justify the following view. They are not embraced by the theorem of definite and multiple proportions. Therefore they are conditioned by purely chemical forces only in a subsidiary way. A precipitate of globulin is to be conceived not as composed of molecular aggregates but of particles of gel. I have shown elsewhere that gelation and precipitation of colloidal solutions are continuous processes. These particles of gel when suspended in a fluid containing ions are penetrated by those ions. Let the fundamental assumption be that the higher the specific velocity of an ion the more readily will it become entangled within the colloid particle. Then as H and OH ions have by far the highest specific velocity, the colloid particle will entangle an excess of H ions in acid and thereby acquire a + charge, and of OH ions in alkali and thereby acquire a − charge. These charges will decrease the surface energy of the particle and thereby lead to changes in their average size.

The diminution in the conductivity of acids or alkalis is due to the loading of the H or OH ions in this way, in much the same way that electrons travelling through certain gases load themselves with impurities. Or, to make the more general comparison, the ion has somewhat the relation to the colloid particle that the electron has to the atom.

In presence of a neutral salt with ions of approximately the same velocity, the globulin particles are loaded with both ions. The particles therefore have no resultant charge, though the conductivity of the salt is slightly diminished. In this case what might be called the energy of solution is obtained by decreasing the kinetic energy of the system $RAc \rightleftharpoons R + Ac$.

23

ON GLOBULINS

(Being the Croonian Lecture delivered May 25, 1905)

[*Proc. Roy. Soc.* (1907), B, LXXIX, 413]

Globulins are a class of proteids which occur in both animal and vegetable tissues. They are peculiar in the complexity of their relations to electrolytes. Insoluble in water, they are soluble in low concentrations of acids, alkalis, or neutral salts. In presence of acids the globulin is electro-positive, in presence of alkalis it is electro-negative, in presence of neutral salts it is electrically neutral. Electrically active globulin (*i.e.* dissolved by acids or alkalis) is precipitated by minute amounts of neutral salts; also, no matter what its electrical state may be, or how dissolved, globulins are precipitated by neutral salts near the saturation point of the latter. The problem I propose to consider is their diversified relation to electrolytes.

Connected with this problem is another, namely, the relation of solutions of globulins to colloidal solution. Do they form hydrosols at all, and, if so, to what extent? Krafft urged the colloidal nature of soap solutions, because, within certain limits of temperature and concentration, they gelatinise, and have the vapour pressure of pure water. Kahlenberg and Schreiner, however, regard soap solutions as being crystalloid in character, because over the whole range of concentration the soap is a good electrolyte—it ionises and undergoes hydrolytic splitting, like other salts of a weak acid and strong base. Smits, again, by measurement of the vapour pressure over a wide range of concentration, is convinced that above a critical concentration the soap passes wholly into the colloidal state.

Exactly the same points arise in connection with proteid. Waymouth Reid has shown that the proteids in a solution exert no measurable influence upon the vapour pressure. On the other hand, solutions of globulins are relatively good conductors over the whole range of concentration through which it is possible to follow them. The dilution curve shows no break indicating a general change of state, even when concentration is pushed to the point where fluidity almost vanishes.

Again, colloids, as a class, are chemically inert. But globulins react actively with acids or alkalis to neutralise them. Globulin solutions, even at extreme concentration, form syrups and not true gels, except, perhaps, under special circumstances. Clearly, therefore, it is pertinent to ask whether globulins form colloidal solutions.

The globulin used throughout this research is that which is precipitated from ox serum by dilution and slight acidification.

Solution by acids or alkalis. Electro-positive or electro-negative globulin. When acid or alkali is gradually added to a suspension of globulin in water, the opaque suspension gradually changes to a clear transparent fluid. Between the opaque suspension, from which the globulin settles on standing, to the stage of limpid transparency there is no break. The most minute addition of acid or alkali (in the absence of salts) converts the suspension into a non-settling "solution" of low grade, which is opaque white from the large size of the particles of globulin dispersed throughout it. Further addition of acid or alkali raises the grade of solution. The globulin particles become smaller and smaller until complete transparency is reached.

The process can be reversed by dialysis, save that, at the lowest grade attainable, no precipitate settles in the absence of disturbing factors.

With the first addition of acid or alkali, and at the last stage of dialysis, the globulin is electrically active, so that it moves in an electric field, the direction of movement being that which one would expect if it combined with the acid or the alkali to form a salt, and the specific velocity is quite, or nearly, at its maximum value. A rise in the grade of solution is, on the whole, associated with a fall in the specific velocity, never in my experience with a rise in this value. The electric charge on the surface of the globulin particles therefore appears to reach maximal density when the solutions are still of exceedingly low grade.

These facts can best be explained by assuming that true salt formation occurs. This agrees with the view of all previous workers on the interaction of proteids with acids or alkalis (Sjöqvist, Bugarsky and Liebermann, and Cohnheim). Since globulins combine either with acids or bases, they have both an acid and a basic function—they are amphoteric electrolytes.

The globulin salts ionise in solution; therefore, in an electric field the entire mass of proteid moves. They also hydrolyse, but the hydrolysis offers special features resembling those which Jordis has pointed out in the case of sodium silicate, and Chevreul in the case of soaps. In both of these cases hyper-acid salts are formed, while hyper-acid salts are formed on dialysis of a solution of globulin by alkali, and hyper-basic salts on dialysis of a solution of globulin by acid.

By dialysis the degree of hyper-acidity or hyper-basicity can be raised, but, as Jordis finds in the case of silica, with continuously increasing difficulty. With the rise in the degree of hyper-acidity or hyper-basicity, the "grade" of solution diminishes, but, in the case of globulins, precipitation does not occur. Electrically active solutions of globulin cannot be precipitated by dialysis, nor at any stage do they cease to be electrically active.

If G be used to denote globulin, the equation of hydrolysis or dialysis would be:

$$x\text{GHS} + y\text{HOH} = (\text{GHOH})_y(\text{GHS})_{x-y} + y\text{HS},$$

or
$$x\text{GB} + y\text{HOH} = (\text{GH})_y(\text{GB})_{x-y} + y\text{BOH}.$$

In dialysis, the ratio x/y varies continuously. It is as indeterminate as the ratio between the combining salts has been found to be by Bödlander, Abegg, Sherrill, and others in the double salts of mercury and silver.

Clearly a relation of this kind agrees with van Bemmelen's definition of absorption compounds as chemical combination with variable composition.

Removal of water from a solution of acid or alkali globulin does not produce precipitation, and the dried gummy residue reabsorbs water and passes slowly again into a state of solution. Acid and alkali globulin, therefore, form, with water, solutions which have the feature characteristic of colloidal solutions, in that there are no saturation points.

In order to compare the solvent power of different acids or alkalis, it is necessary to fix upon some arbitrary point, such as the point of minimal opalescence. By the use of a system of controls and proper illumination this point furnishes very concordant values.

Measured in this way, it appears that for strong and medium acids solvent power is measured by the number of gramme molecules present, not by the number of gramme equivalents:

$$\text{HCl} = \text{H}_2\text{SO}_4 = \text{H}_3\text{PO}_4.$$

Very weak acids have a lower solvent power:

$$\text{HCl} = 5\text{H}\bar{\text{A}} = \pm\, 30{,}000\text{H}_3\text{BoO}_3.$$

These relations are explained by the very weak basic function of globulin. Salts with weak acids are much hydrolysed, and to reach the same grade of solution an excess of acid is needed in order to lower the degree of hydrolysis.

With alkalis, the weak alkali NH_4OH dissolves as well as the strong alkalis, owing to the fact that globulin acts as an acid of considerable strength.

The acid and basic functions were measured by the well-known methods —the catalysis of cane-sugar and of methyl acetate—and the acid function found to be much the greater.

The molecular relation noticed above recalls the salts of amido acids:

$$\text{R}\!\!<\!\!^{\text{NH}_2}_{\text{COOH}} + \text{HCl} = \text{R}\!\!<\!\!^{\text{NH}_2\text{HCl}}_{\text{COOH}}\,; \qquad \text{R}\!\!<\!\!^{\text{NH}_2}_{\text{COOH}} + \text{H}_2\text{SO}_4 = \text{R}\!\!<\!\!^{\text{NH}_2\text{H}_2\text{SO}_4}_{\text{COOH}}$$

Data derived from measurements of electric conductivity, and from the behaviour with indicators, support the view that true salts are formed, and that the globulin acts much more strongly as an acid than as a base.

Direct measurement of the specific velocity of globulin "ions". This was carried out by the boundary method.

As the basic function of a globulin is weaker than its acid function, the salts GHS (globulin + acid) will be hydrolysed more than the salts GB (globulin + base). Therefore, in the equation of hydrolysis given above, y/x will be larger for the former than the latter. Similarly, comparing salts with $H\bar{A}$ and with HCl, y/x will be greater for the former.

Now with increase in the value of y/x the size of the proteid particles increases also. By dialysis, the size can be increased until the particles diffract white light. The growth in the case of solutions of sodium silicate can be traced by the appearance of molecular states incapable of passing through a parchment membrane. In a similar way, by continuous addition of AgCl to a solution of NaCl, molecular states are produced large enough to diffract white light.

In each of these cases these large molecules are ionic, that is to say, they take part in the electric transport. And their specific velocity is exceptionally high; that of $(AgCl)_x Cl_y$ being 57×10^{-5} at $18°$, whereas Ag at infinite dilution is only 58.

"Ions" of this order of magnitude have the properties of matter in mass. Each is defined by a surface and moves under the influence of a surface contact difference of potential. With their appearance the fluid ceases to be homogeneous. It has internal surfaces. I propose to call such ions colloidal ions, or pseudo-ions. Their specific velocity is high, and, within wide limits, is independent of their size, and is controlled by the laws of electrical endosmose.

These conclusions are borne out by the boundary measurements and by the electrical conductivity of colloidal solutions.

For instance: theoretically the proportion of pseudo-ions in the following solutions of globulin should be:

$$\text{In } H\bar{A} > \text{in HCl}, > \text{in NaOH},$$

and the specific velocities at $18°$ were found to be 23×10^{-5}, $11 \cdot 5 \times 10^{-5}$, $7 \cdot 6 \times 10^{-5}$. By Ostwald's law of the relation of specific ionic velocity to the number of atoms in an ion, the value 23×10^{-5} is an utterly impossible one for an ordinary ion of the magnitude of the proteid molecule.

Solution by neutral salts. Globulins are dissolved by neutral salts owing to the formation of molecular compounds (G.BS). These compounds are readily decomposed by water with liberation of insoluble globulin (G.BS + HOH = G.HOH + BS). Therefore they are stable only in presence of a large excess of salt. Hence the solvent power of salts is from 200 to 500 times less than that of acids or alkalis. Hence, also, the presence of the globulin lowers the electric conductivity of the salt to only a small extent (2–6 per cent.).

H   28

A double salt of the form $AB.CD$ ionises according to

$$AB.CD \rightleftharpoons AB.C + D.$$

I have never succeeded in detecting any trace of such ionisation in the case of globulin and neutral salts; possibly owing to the extreme instability of the ions. The proteid does not move in an electric field, and it seems to take no part in the electric transport.

Owing to the insolubility of the dissociation products, the globulin can be precipitated from its solution with neutral salts by simple dilution. No degree of dilution will precipitate globulin from solution in dilute acid or alkali.

The compounds of globulin and alkalis (GB) are more readily dissolved by neutral salts than is simple globulin. Compounds of globulins and acids are insoluble by neutral salts, being decomposed with liberation of the acid.

If on the analogy of an amido acid a globulin combines owing to the presence of the NH_2 and the COOH group, then it forms salts with alkalis by replacement of hydrogen, and with acids by the change of the trivalent nitrogen of the amido group to the pentavalent form:

Where, then, does the neutral salt link on?

There are two possible places—the unsatisfied valencies of the nitrogen, or the unsatisfied valencies of the upper O of the COOH group. According to the oxonium theory, O has a maximum valency $O\equiv$, but $O\equiv$ is much less stable than $N\equiv$; therefore, one would expect the linkage to be by the amido group:

And this would account for the fact that the acid can be turned out of the combination with globulin by an excess of neutral salt, but the alkali cannot be.

It is easy to prove by the boundary method that the globulin, when dissolved by salts, takes no part in the electric transport. Dissociation of the salt-globulin compound by the water may be regarded as being practically completely suppressed by the excess salt present. This being the case, the diminution of the electric conductivity of the salt by the globulin may be used as a measure of the fraction of the salt actually combined with it.

In the following table a few values are given. Comparison is made between solutions containing the same amount of salt per litre; but in the one case the solution is saturated with globulin, in the other case it is a simple solution of the salt in water. No correction is made for the diminution of the molecular conductivity of the salt owing to replacement of a portion of the water by globulin, since it would amount to less than 0·01 per cent. No. III is of the nature of a control, to prove that electrolytic impurities adherent to the purified globulin may be neglected. The salt solution was made up with a dialysate in equilibrium with the suspension of globulin.

Serum centrifuged, diluted to 10 vols., globulin precipitated by $H\overline{A}$. The precipitate suspended in a large volume of distilled water and collected by the centrifuge, and re-suspended in water freed from gases by boiling.

Specific conductivity of the water employed 6×10^{-6} rec. ohm, 18°.

I. Ox globulin suspended in water. 100 c.c. = 4·08 gr. dry globulin. Temperature 18°:

Salt	Normality	Specific conductivity, salt globulin $\times 10^5$	Specific conductivity, salt water, $\times 10^5$	$\dfrac{K \text{ salt globulin}}{K \text{ salt water}}$
K_2SO_4	0·1716	1445	1539	0·938
,,	0·1679	1411	1506	0·937
NaCl	0·27	1850	1950	0·949

Specific resistance of the globulin suspension $3\cdot1 \times 10^{-5}$.

II. Sheep globulin. 100 c.c. = 4·4 gr.:

NaCl	0·2659	2145	2280	0·938
,,	0·2659	2150	2278	0·943

The measurements in II were duplicated in order to test the degree of accuracy of the measurements of volume.

III. Ox globulin dialysed until in equilibrium with water saturated with toluol:

$MgSO_4$	0·11	517	543*	0·95

In this case the salt solution (*) was made up with the dialysate, not with water.

[*Note.* In a recent paper[1] it is stated that globulin in solution does not alter the conductivity of the salt used to dissolve it. The conductivity of the salt "is the same as that of a similar solution in pure water". This statement is directly contrary to my own observations. It leads me to add a few words as to the precautions which are necessary in dealing with proteid solutions.

Electrodes coated in the ordinary way with platinum black are quite useless. The proteid in their immediate neighbourhood is changed very rapidly and very drastically. An instance selected at random will show the magnitude of the effect.

[1] Mellanby, *Journ. Physiol.* (1905), XXXIII, 354.

18°. Globulin dissolved by dilute NaOH. 3·3 per cent. globulin. Readings at intervals of 2 min. approximately:

2560 ohms.	2530 ohms.
2542 ,,	2522 ,,
2537 ,,	2538 ,,

The rise in the last observation followed upon slight shaking of the cell to wash the changed portion of the solution away from the electrodes. Platinum grey electrodes, prepared as Whetham directs, give no trouble. Readings with globulin solutions of all kinds remain quite constant after being for 48 hours and longer in contact with them.

Even with platinum grey electrodes, however, it is necessary so to adjust the cell that the observed resistance does not fall under about 500 ohms. With a lower resistance, and therefore a larger current, the readings become unsteady and too low.]

On the assumption that the whole of the drop in electric conductivity is due to association of the salt with the globulin, the above figures show that, at the particular concentration and temperature, 1 gr. of dry globulin combines with 33×10^{-5} gr. equivalent of NaCl and 26×10^{-5} gr. equivalent of K_2SO_4, while to dissolve 1 gr. of dry globulin 10×10^{-5} gr. equivalent of alkali or 18–38 equivalents of acid are needed. The figures for salt and acid are of the same order of magnitude, and this is what might be expected if salt and acid combine in the same way with the proteid.

APPENDIX

(*Added December* 1906)

Viscosity of solutions of globulins. The measurements of this value, very briefly touched upon in the lecture, have been amplified. They show the following interesting features:

In dilute solutions of globulins the viscosity, as measured by the rate of flow through a capillary tube, is of the same order of magnitude for each of the three types—acid, alkali, and salt globulin.

With increasing concentration the viscosity increases, but the increase is much greater for alkaline globulin than for acid globulin, and for acid globulin than for salt globulin. The difference is very striking; thus, at a concentration of 7·59 gr. globulin per litre, the transpiration times are in the ratio—

Water	...	...	...	1
$MgSO_4$ globulin	...	...	4·66
HCl	,,	...	...	15·5
NaOH	,,	...	...	67·9

In solutions of salt globulin, the globulin particles do not carry electricity; "ionic" proteid is completely absent, therefore the higher viscosity of acid and alkaline globulin solutions is connected with the presence of "ionic" proteid, that is, of very large molecules carrying a charge.

Since globulin acts much more strongly as an acid than as a base, there will be in similar solutions of HCl globulin and NaOH globulin a greater concentration of "ionic" proteid in the latter. Here, again, increased concentration of "ionic" proteid goes with increased viscosity.

These conclusions agree with the observations of Reyher[1] on the viscosity of solutions of soap, and of Sackür[2] on solutions of the proteid caseinogen. It would, however, be rash to fix upon the simple proteid or soap ion as the agent. The effect of dilution, and of varying proportions of acid or alkali, make it possible that the high viscosity is due to the presence of the molecular complexes, which carry a surface charge, and which I have called "colloidal" ions, or pseudo-ions. Thus, further addition of small quantities of ammonia, beyond the amount necessary to dissolve the globulin to a transparent solution, *decreases* the viscosity. Ammonia is a very weak base, which, in solution, has a specific molecular conductivity much less than that of its compound with globulin. One cannot, therefore, refer the diminished viscosity to diminished ionisation of the ammonia-globulin compound, but the optical properties of the solution show that slight excess of ammonia does diminish the size of the particles of the globulin. It raises the "grade" of the solution.

A comparison of the different types of solution of globulin raises, in a special way, the question of the meaning to be attached to the phrase "colloidal solution". Mixtures of very diverse character have been classed as colloidal, some because they form jellies, others because they are physically heterogeneous, others because the osmotic pressure is exceedingly low, others, again, because the electric conductivity is abnormally low. Of these, perhaps, the only constant attribute is physical heterogeneity. Solutions which present any of the other distinctive features of the colloidal state always contain particles which are large enough to scatter light.

Amongst the attributes of colloidal solutions is the possession of an abnormally high viscosity. Judged in this respect, the salt-globulin solutions are only slightly colloidal as compared with the solutions of alkali or acid globulin. They differ also from acid and alkali globulin in the relatively greater definiteness of the relations between the components. Dilution of a solution of salt globulin, for instance, brings about a sharp separation of solid and liquid. No such clean saturation points are found with the other two systems. The definiteness of the relations makes it possible to compare the system—neutral salt, water, globulin—with other similar three-component systems.

Data sufficient for the purpose can be obtained from the measurement of the solubility of globulins in aqueous salt solutions made by Osborne and

[1] *Zeits. f. phys. Chem.* (1888), II, 743.   [2] *Ibid.* (1903), XLI, 672.

Harris,[1] and by Mellanby.[2] The curves which these authors give fail to show the true relations, since only two co-ordinates are used, namely for the salt and the globulin; the third component, water, is neglected. For the following curves, trilinear co-ordinates are used in the manner recommended by Stokes.

Fig. 1 shows curves plotted to scale from values derived from Osborne and Harris.[3] S is pure salt, W pure water, G pure globulin (edestin). The points on the curves, fixed by actual experiment, are shown in the usual

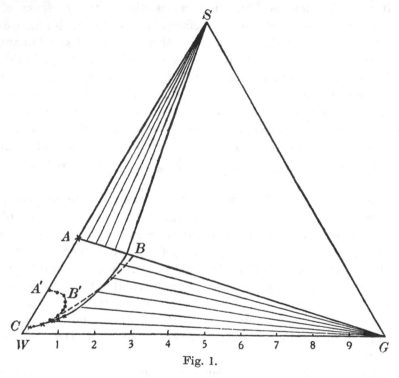

Fig. 1.

way by dots or crosses. $A'B'C$ is the curve for Na_2SO_4, ABC the curve for NaCl. The broken line extending to just beyond B is the probable form of the curve for H_2SO_4. The diagram is drawn as an isotherm for NaCl, water, edestin, the areas for the other salts being left out. The context is not very clear on the point, but the temperature of observation seems to have been 20°.

Each curve clearly consists of two distinct parts, AB, which is in equilibrium with solid salt, and BC in equilibrium with globulin crystals. The whole surface, therefore, consists of certain areas: ABC, all points within which represent a homogeneous state, namely, a fluid solution of

[1] *Amer. Journ. Physiol.* (1905), vol. xiv.
[2] *Journ. Physiol.* (1905), xxxiii, 354.   [3] *Loc. cit.*

globulin, salt, and water; *ASB*, which is an area of heterogeneous states. Each point in it represents a mixture which cannot exist as a uniform state, but divides into two phases, one a fluid solution on the curve *AB*, the other a solid which is pure salt. *WBG*, again, is an area of mixture which cannot form a uniform state, but which divides into a fluid solution on *CB*, and into globulin crystals. *SBG*, each point within which represents a mixture which divides into a fluid having the composition of *B* and solid salt and solid globulin.

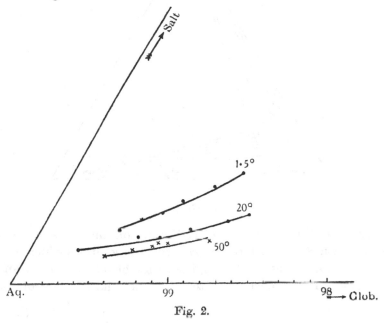

Fig. 2.

In Fig. 2, portions of three isotherms for NaCl, water, serum globulin are plotted to scale, the data being derived from Mellanby's paper. They cover only a very small fraction of the whole curve, *ABC*, but they are sufficient to show, (1) that the curve encloses an area which is placed on the line *WS*, as in the preceding case. This follows from the form of the curve, and from the fact that if sufficient salt be added, a point is reached on the line *WS* where salt solution is in equilibrium with solid globulin; (2) that the area enclosed increases with a rise of temperature.

This last fact, together with some observations of my own, enable us to approximate roughly to an isotherm for edestin at, say, 30°. When edestin is present in sufficient quantity, on warming a further amount passes into solution, so that we have a series of solutions in equilibrium with crystals. Beyond a certain temperature the crystals fuse, and we then have two fluid layers in equilibrium, the lower layer (as I have found by actual analysis) containing all three components.

Fig. 3 is a diagram of an isotherm at a higher temperature; it shows the area ABC enlarged, and it shows a new area, ROP, due to the appearance of a second series of solutions (the melted crystals). Of the shape of this

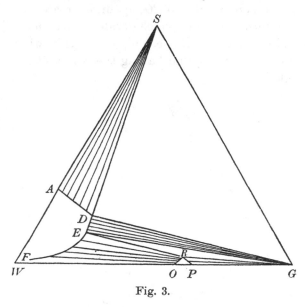

Fig. 3.

area we know nothing. The curve ABC now contains three parts, AD in equilibrium with S, DE with G, and EF with the solutions which lie on OR.

Fig. 1 does not represent the conditions for blood globulin at any temperature, since, in presence of water alone, a solid solution of water in

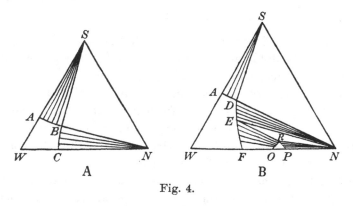

Fig. 4.

globulin is formed. The tie lines from the curve BC, therefore, instead of focusing at G, end on a curve of unknown slope, which starts from about the middle of the line WG. In other words, the isotherms of blood globulins, over the region where we have any information to guide us, are of the form shown in Fig. 4.

This is as far as the known facts carry us. It now remains to compare the system with a similar one. Such a one is found in water, sodium chloride, and succinic nitrile, which has been studied by Schreinemakers.[1] In both there is a pair of immiscible bodies, salt and globulin, or salt and succinic nitrile, each of which is partially miscible with the third substance, water. Fig. 4 A and B are two isotherms reproduced from Schreinemakers' paper. It is obvious that, in all essential features, they resemble the isotherms for globulin-salt-water already given.

It is clear, from this comparison and from the low viscosity, that globulin-salt-water is a system showing few abnormal features. It manifests decisive points of equilibrium, which resemble those in a comparable and purely crystalloid system. The other two globulin systems are as decisively abnormal, especially in the high viscosity, and the absence of definite transition points, and it is significant that, regarded from the purely physical standpoint, the essential difference should be the presence in the latter of large molecular aggregates, each carrying an electric double layer.

The presence of internal electrified surfaces adequately accounts for the high degree of inertia which obscures the transition points. What is the source of the electrification? In all cases in which it can be traced, the charge is due to molecular interaction of the type classed as chemical, and of that particular chemical type associated with the decomposition of neutral electricity, to which the name "ionisation" has been given. Burton[2] clearly shows that metals form hydrosols because they react with the solvent to form hydrides or hydroxides. Acid globulin, alkaline globulin, and soaps are salts which form colloidal solutions because one of the radicles is of large size, is almost or quite immiscible with the solvent, and therefore readily forms molecular complexes which have the electric sign proper to the radicle when dissociated from its fellow in the salt molecule.

Each colloid particle undoubtedly carries on its surface a charge, since, without exception, they move in a uniform electric field. But in the system formed by each particle there cannot be any free electricity, any resultant charge. If there were, the system would, by Earnshaw's theorem, be unstable, and settling would occur the more rapidly the greater the charge. Experiment shows the opposite to be the case. The charge on each particle, therefore, must be bound by an equal and opposite charge on the liquid face opposed to it. The condition of double electric layers is therefore realised.

It has been shown by Waymouth Reid[3] that in a solution of proteid no osmotic pressure can be traced to the proteid, and I find that when a solution of globulin is dialysed against water until equilibrium is reached, the dialysee and dialysate have exactly the same electric conductivity.

[1] *Zeits. f. phys. Chem.* (1897), XXIII, 417.
[2] *Phil. Mag.* (April 1906).   [3] *Journ. Physiol.* (1904), XXXI, 438.

These facts, added to others already stated, afford overwhelming evidence that in these solutions the chief portion of the proteid is, in effect, removed from solution by being gathered into masses of more than molecular dimensions which are separated from the rest of the solvent by a surface. The masses of probably hydrated proteid thus form an internal phase. In colloidal salts such as soaps, solutions of globulin in dilute acid or alkali, solutions of caseinogen or of acid or alkali albumen, in which the internal phase is bounded by double electric layers formed by ionic interaction with the external phase, the stability of the system with reference to forces such as gravity will depend, in part, upon the potential difference between the two faces round each particle. The electric double layers will contribute to stability, since any movement of a particle through the fluid will develop free electricity in quantity proportional to the potential difference between the layers.[1]

Salts in small amount diminish the stability of these colloidal systems and bring about concentration; and salts, as has been shown experimentally by Wiedeman, Perrin,[2] Burton,[3] and others, diminish the potential difference in the condenser system round each particle.

The contribution which the electrification of the internal surfaces makes to mechanical stability is seen in the high viscosity of these solutions as compared with solutions of globulin by salts which resemble them in containing large molecules, but differ in the total absence of electric layers from these molecules.

Solutions of globulins in salts introduce us to an entirely different aspect of colloidal solution. Here, again, the proteid contributes nothing to the osmotic pressure,[4] and, therefore, must be withdrawn from true molecular solution; but as all ionic interaction is suppressed by the large excess of salt present, there are no double electric layers formed, and in their absence viscosity is low and stable equilibrium conditions are rapidly reached.

The origin of the mechanical stability of this system, the nature of the influence which maintains the even dispersion of the proteid throughout the fluid, is difficult to explain, since any factor to which we can appeal seems also to involve the exertion of a finite osmotic pressure by the proteid.

If the distribution of the energy be considered, we arrive at the following: Let the state of thorough mixing of the components be established, and let a movement under the influence of any mechanical force take place, so that the proteid becomes more concentrated in some region. In this region,

[1] Helmholtz, *Wied. Ann.* (1879), VII, 337.
[2] *Journ. de Chim. Phys.* (1904), II, 61; (1905), III, 50.
[3] *Phil. Mag.* (November 1906).
[4] Waymouth Reid, *Journ. Physiol.* (1904), XXXI, 438.

owing to the concentration of the colloidal salt, the degree of hydrolysis and of ionisation will be diminished, and the available energy thereby increased.

The equation of hydrolysis which Arrhenius[1] gives shows that hydrolysis and ionisation diminish faster than concentration increases; therefore, with unequal distribution of the proteid the gain in available energy $(E - \tau\phi)$ will be greater in the regions of higher concentration than the loss in the regions of diminished concentration. The system as a whole, therefore, would gain in available energy, and the change from uniform distribution to unequal distribution could not take place unless work were done on the system.

[1] *Zeits. f. phys. Chem.* (1890), v, 16.

24

THE PHYSICAL BASIS OF LIFE

[*Science Progress* (1906), No. 2]

In a famous lay sermon on the Physical Basis of Life, written nine years after the publication of *The Origin of Species*, Huxley writes as follows:

> When hydrogen and oxygen are mixed in a certain proportion and an electric spark is passed through them, they disappear, and a quantity of water, equal in weight to the sum of their weights, appears in their place. There is not the slightest parity between the passive and active powers of the water and those of the oxygen and hydrogen which have given rise to it. At 32°, and far below that temperature, oxygen and hydrogen are elastic gaseous bodies whose particles tend to rush away from one another with great force. Water at the same temperature is a strong though brittle solid, whose particles tend to cohere into definite geometrical shapes....
>
> Nevertheless, we call these and many other strange phenomena the properties of the water, and we do not hesitate to believe that in some way or another they result from the properties of the component elements of water. We do not assume that a something called "aquosity" entered into and took possession of the oxide of hydrogen as soon as it was formed, and then guided the aqueous particles to their places on the facets of the crystal or amongst the leaflets of the hoar frost. On the contrary, we live in the hope and in the faith that by the advance of molecular physics we shall by-and-by be able to see our way as clearly from the constituents of water to the properties of water as we are now able to deduce the operations of a watch from the form of its parts and the manner in which they are put together....
>
> If the properties of water may be properly said to result from the nature and disposition of its component molecules, I can find no intelligible ground for refusing to say that the properties of protoplasm result from the nature and disposition of its molecules.
>
> But I bid you beware that, in accepting these conclusions, you are placing your feet on the first rung of a ladder which, in most people's estimation, is the reverse of Jacob's, and leads to the antipodes of heaven. It may seem a small thing to admit that the dull vital actions of a fungus or a foraminifer are the properties of their protoplasm, and are the direct results of the nature of the matter of which they are composed. But if, as I have endeavoured to prove to you, their protoplasm is essentially identical with, and most readily converted into, that of any animal, I can discover no logical halting-place between the admission that such is the case and the further concession that all vital action may, with equal propriety, be said to be the result of the molecular forces of the protoplasm which displays it. And if so, it must be true, in the same sense and to the same extent, that the thoughts to which I am now giving utterance, and your thoughts regarding them, are the expression of molecular changes in that matter of life which is the source of our other vital phenomena.

This uncompromising, virile attitude towards the most difficult and stupendous problem of science is characteristic both of the man and of the time. Huxley wrote in 1868 at the zenith of a period of strenuous intellectual life without doubt unsurpassed in the history of the world. The strong new wine of scientific discovery was running in men's veins.

A mere chronological table of the chief scientific events shows how fast was the growth. In the 'forties the labours of Joule provided a basis for the conception of the conservation of energy which at a step unified all the sciences. In the 'forties, too, the unification of the biological sciences was begun by the recognition of the cell as the unit of all life, and of the glutinous sarcode as its physical basis, and was crowned by the publication of *The Origin of Species* in 1859, which gave force and authority to the older doctrine of the continuous development and progression of life.

The spirit of the age was one of conflict, and men's minds were tuned by it to Pisgah-like visions of the country to be conquered. The ideal of the new learning was the unity of all knowledge, its quest the establishment of a scheme of things animate and inanimate which should show them, linked together, without break, in orderly progress from the simple to the complex, from the lower to the higher, and its duty the warfare against a piecemeal and partial outlook of separate creation and catastrophic change. For the new learning no one did battle more strenuously than did Huxley.

The doctrine of the unity of knowledge and experience is not an easy one; it is justified even now rather by the steady trend of science than by its completed demonstrations. Knowledge may be seen to be growing from the sides of many a chasm like the two arms of a cantilever, and we believe that human industry and the human intellect will one day complete a bridge across which all may pass in safety. But in the meantime there are grave signs in the scientific and semi-scientific literature of the day of a growing impatience with the rate of progress, which, on the one hand, ensures ready and uncritical acceptance of the crude attempts of an amateur biologist to make living matter, and, on the other, breeds a feebler purpose which seeks an unhealthy opiate in "vitalism" or some other "ism" of like nature.

Nearly forty years of vigorous scientific work have elapsed since Huxley wrote, and it is still possible for the vitalist to assert that no single vital process can be completely expressed in terms of physics and chemistry, that is, of motion and of matter. The biologist is reproached, for instance, with the undoubted fact that the power which a living cell has of selecting certain chemical substances and of rejecting others cannot yet be explained by, and indeed in some ways seems to contradict, the known laws of molecular physics.

To this reproach I would reply after the fashion of Socrates, and with the same purpose, by a question.

Here are two pairs of gases, one of hydrogen and oxygen, the other of hydrogen and chlorine. I burn the members of each pair together, and from the one pair I get water, a fluid odourless, innocuous, and of relatively slight chemical activity, while from the other I get hydrochloric acid gas, acrid, poisonous, and of the highest chemical activity. Now, the molecules of those three gases have certain inalienable properties, an invariable weight, a fixed capacity for electricity. They perform movements the harmonic periods of which are so fixed that apparent departures from them have been used to detect and measure the velocity of approach of a star towards the earth. I ask the chemist or molecular physicist to explain the amazingly divergent properties of the compound in terms of the properties of the component gases. I ask him to do what over-hasty people, forgetful of the extreme youth, the paucity in years of human knowledge, ask the biologist to do with respect to living matter, and the reply is that the question is unanswerable.

It cannot be sufficiently insisted upon that in many regions not the simplest more advance has been made towards a material explanation of vital phenomena than towards a solution of the simple question why one pair of gases should combine to form a fluid, while another pair combine to form a gas.

An unanswerable question concerning the elements of natural knowledge is a sharp reminder of our ignorance, and such a reminder is needed to curb the spiritual arrogance which in our time has brought this greatest of all mysteries, the relation of living to non-living matter, to the temples of vulgar credulity, and has prostituted it to the purposes of common charlatans and impostors.

Since Huxley wrote, our knowledge of the physical basis of life has developed in many directions. The properties of secretion and absorption, of contractility and irritability, have been studied in great detail. The classical fields of physiology, the detailed investigation of form, and the anatomy of function have been continuously worked. But the greatest advance has come in the domain of chemical physiology. Ten years ago this was a scientific No Man's Land, despised by the pure chemist and traversed only in a distrustful, amateurish way by the physiologist. Now that is changed; on the one hand a race of physiologists has sprung up who are at the same time expert chemists, on the other one sees a pure chemist, Emil Fischer, of Berlin, bending all the resources of his great laboratory in men and materials to the central chemical problem of living matter, the chemical structure of proteid.

The few pages at my disposal would not hold even a *catalogue raisonnée* of the new departures. Therefore it is necessary to select a few problems, and for purposes of contrast I choose not the new but the old, which were agitated half a century ago.

At the outset, however, it is necessary to state certain elementary facts— that there is a unit of living matter called the cell, which everywhere and in all places has recognisably the same structure; and that all forms of life are divisible into two divisions: those in which the individual and the cell are coterminous—the simple-celled forms; and those more complex and larger types in which the individual is a cell complex—the multicellular forms. The former are probably the more numerous, but they escape notice by reason of their small size, which is imposed upon them by a law, well-nigh without exception, which must strike very deeply into the nature of living matter—namely, that no single unit, no cell, that is, can increase to more than microscopical dimensions. When it reaches the limit of size it becomes unstable, a field of force of a peculiar and special nature is formed within it, and by this field of force the cell is presently rent in twain.

The basis of this curious limitation of size is not far to seek. Living matter is composed of very large molecules, and substances so built possess certain special properties which mark them off from simpler substances. To them the name of colloids is given, after the type of the class the jellies. Now, a jelly is a curious half-way house between the solid and the liquid states. Like a solid, it is capable of retaining differences of state, it is rarely of uniform character throughout. The rate of relaxation, as Clerk Maxwell called it, of jellies is slow, much slower than that of simple liquids, much faster than that of true solids. Combined with this characteristic inertia, however, is a degree of molecular mobility sufficient for chemical changes of great velocity. A jelly is in this way a meeting-place of extremes, and this it is which enables the colloidal state to manifest life.

Consider now a small free cell, an infusorian swimming in a wayside pool. It displays many activities, it digests in this region of its living body, it maintains a store of starch in that region, in the movements of its parts there is diversity. Both its chemical and physical characters betoken a complexity which show it to be not a homogeneous droplet, but, in spite of its minute size (less than $\frac{1}{100}$th inch), to be heterogeneous. It has a structure, an architecture, the coarser features of which we can decipher with the aid of the microscope.

It is only in the colloidal state that we could have within so small a space so great a diversity of matter, and such differences of chemical potential as must exist to support the multifarious activities of the living cell, combined with the molecular mobility necessary to give chemical change free play. At the same time this capacity for maintaining differences of state imposes limitations, one of which is that of size. Large molecules can move in the substance of the cell scarcely at all. Therefore, when the size exceeds a certain critical limit, the dynamical balance fails, and internal strains appear of a magnitude great enough to tear the cell apart. On this blending of opposites, on the curious combination of inertia and chemical mobility in the colloid state, is reared the whole fabric of the dynamics of living matter.

Each living cell is a machine; it breathes, taking in oxygen; it feeds, and the food is burnt by the oxygen to chemically simpler bodies. The living cell, like the gas engine, can tap the stores of chemical energy—and, like the gas engine, it is an internal combustion engine. Now in a power station where electricity is being produced to run a score of trams, there is a steady hum or drone, the varying pitch of which marks the speed of the engine. To the engineer in charge, from long habit, that varying sound speaks of events happening in remote parts of the system. A glance at the clock, and he will tell you that the sound is falling because the engine is adjusting itself to the increased load due to such and such a tram breasting such and such a hill. In the same way, watching the movement of a living cell under the microscope, if we were sufficiently skilled we could refer the continual change in the rate and direction of its movement to temporary inequalities of temperature, of lighting or of chemical composition, etc., in the water in which it lives. If we resort to experiment, the effects are obvious: an electric shock causes the irregular *Amœba* to come to rest as a sphere, a trace of acid slows its movements, of alkali accelerates them.

These things—the electric shock, the acid, or the alkali—are what the biologist calls "stimuli", and by varying their nature or intensity he can control the activity of living matter to a very remarkable extent.

Let us return for a moment to the *Amœba*. We watch it crawling amid sand, fragments of decayed leaves, and living diatoms, and we notice that of the particles which it eats some are nutritious food, some are innutritious and absolutely useless. But we also notice that there is a decided balance in favour of the nutritious particle. Like Autolycus, it is a picker-up of unconsidered trifles, guided by a decided preference for things useful to itself. Therefore, the tiny animal manifests discrimination or choice—imperfect, no doubt, but clearly recognisable. And the choice is beneficial; it contains an element of purpose.

Watch an *Amœba* long enough, and it will be seen to divide into two, and these again into two, so forming successive generations the individuals of which resemble one another. This labile, creeping fragment of jelly has recognisable form, and zoologists classify the various forms in so many species, each of which breeds true. They manifest that property of living matter called heredity.

Lastly, the individual *Amœba* is in incessant movement, and with each successive generation there is growth of individuals and increase in the mass of living matter. Now, to these three features, *choice and purpose*, *growth*, and *heredity*, I propose to confine myself, and I will consider them separately and in the order named.

THE FACULTY OF CHOICE

In one of Jules Verne's books, which at one time or another held most of us in thrall, there is an account of a submarine vessel which for a long time was conjectured to be some mighty marine monster. Now, I want you to put yourself in a similar position with respect to a model Whitehead torpedo, to consider yourself as meeting one of these for the first time and studying its movements under the impression that it is a living being. The torpedo must be without its charge, otherwise your experiments would come to an abrupt end; for I wish you to consider yourselves as inquiring why this curious beast should always swim at the same depth. Push it down, pull it to the surface, it would presently be swimming again as many inches below the surface as before, and you would say to yourselves, Why on earth does it *choose* to swim there?

Instead of a model torpedo, here in a drop of fluid are countless thousands of the most minute forms of life, each actively darting hither and thither, each so small that 5000 would make only one large *Amœba*. Into that drop you introduce two fine capillaries, the one filled with very dilute acid, the other with very dilute alkali, and in no very long time you will find that the Vibrios have collected in a mass at one or other of the tubes—probably the acid tube. If you followed ordinary usage, you would express the result again in terms of choice by saying that the animals are attracted to the acid and are repelled from the alkali.

Both cases illustrate the difficulty in freeing the imagination from the tyranny of the counters it employs. The first case, that of the torpedo, has served its purpose as an illustration, and it interests us no longer. Let us see whether a purely mechanical conception will explain the second.

The acid and alkali diffusing out of the tubes destroy the uniformity of the water, so that, starting from one tube and moving to the other, one passes through gradually diminishing acidity, through neutrality, to a region of gradually increasing alkalinity, which reaches a maximum at the orifice of the other tube. The medium between the tubes, therefore, is accurately graded in composition.

Now let us see the effect of a trace of acid and alkali upon these Vibrios. It is not always the same; it depends upon the particular forms we are examining. I choose the case where acid slows the movements, alkali increases them. Each individual *Vibrio*, as we watch it, is seen to move in an erratic and irregular orbit, so erratic that we can consider it as completely irregular.

The problem now becomes a simple mathematical one. Given a number of particles, each moving in an irregular orbit, and uniformly distributed throughout a homogeneous medium. The medium ceases to be homogeneous and is changed so that in one region the mean velocity of each of the particles is augmented, in another it is diminished. What will be the effect upon the distribution of the particles? The answer is, that they will collect where the motion is slowest.

We now try the experiment, and we find that the Vibrios do collect where their motion is most slowed—namely, in the region of maximal acidity. And they do not swim directly there; they, as it were, settle out in that region, as the hypothesis demands.

The influence of a chemically heterogeneous medium upon the free cells living in it is called "chemiotaxis". I have analysed a simple case, but it would take a session's lectures to follow out the application of the principle to biological problems. It explains great regions of disease, it has even been applied to the workings of the nervous system. At one sweep it embraces the directive effects of the surrounding medium upon the movements of free cells, in the waters of the earth and in the bodies of animals and of plants.

The choice of food particles, the discrimination manifested by *Amœba*, is the chemiotactic response of its irregularly flowing protoplasm to the chemical atmosphere, if I may so put it, of the food particle. At the Mint a chance collection of sovereigns are presented to a certain machine, and it sorts them into those of full weight and those of short weight. A chance collection of particles are presented to *Amœba*, and it sorts them very imperfectly into those which modify the incessant streaming of its protoplasm so that they become engulphed, and those which do not so modify the streaming. The element of choice, or, as we may now put it, the directive influence of the surroundings, is much less perfect here than in the case of the Vibrios, because the oscillations—the movements of the protoplasm on which it operates—are both less in extent and much more regular than are the movements of the Vibrios.

In the choice of food particles, and in the sorting of the coins, there is an end to be served. Looked at from this point of view, chemiotaxis sometimes presents novel features. *Amœba proteus*, large and slow-moving, frequently captures an active ciliate called *Colpidium*. Observers describe the capture as being due to the *Colpidium* swimming as though attracted into the pseudopodial jaws, whence it makes no efforts to escape. Here the element of purpose, looked at from the standpoint of *Colpidium*, is that of a Christmas ox marching to the kitchen to be converted into beef-steaks.

The directive effect of the medium upon a free cell is usually more complex than in the case considered. *Opalina* is a large ciliate which in a uniform medium swims straight forward, owing to the movement of the cilia or vibratile hairs which cover its surface. The movement in this case starts at the front end of the animal and sweeps back as a wave like the wave over a cornfield. In a heterogeneous medium the movement starts excentrically, the waves sweep obliquely down the animal, and the direction of motion changes. The net result again is the same—the animal ceases to be distributed evenly when the water ceases to be of uniform composition.

The next example raises the question of choice to a higher level. It brings into the response of the animal its previous history. We will take the simplest case, as it is offered by *Opalina*. This animal is parasitic in the intestine of the frog, and it thrives in a very slightly acid medium. But its attraction to acid is not an inalienable quality. Glut it with acid, soak it in dilute acid for an hour, and it now collects in a region of alkali; bathe it for an hour in very dilute alkali, and its chemiotactic response is once more changed—it collects about the acid.

The mechanism underlying this change of response must be patent to every chemist. There are many substances whose chemical and physical characters are completely reversed by change from a trace of acid to a trace of alkali, or *vice versa*. Amongst these substances, and markedly possessed of this character, are the chemical substances, called proteids, of which all living matter is composed. The varied response to acid or alkali may unquestionably be traced in the first instance to the directive influence of the amphoteric proteid on the surface energy of the animal and upon the train of chemical events in its interior.

A parallel differential response is furnished by *Stentor*—a large, trumpet-shaped animalcule—which fixes itself by its foot to some solid object. Touched on one side by a fine glass hair very lightly, it bends towards the hair; touched more heavily, it bends *away*. Therefore there is a touch of a certain strength which produces no response. To a series of touches regularly repeated it gives the following responses. At first it simply bends away, then it contracts right down on to its foot; if that does not get rid of the irritant, it looses its hold and swims away.

These responses have been analysed with great care in order to elucidate the underlying mechanism. I can stop only to point out the element of purpose. In order to get rid of an irritation, a certain movement is tried; it fails, another movement is tried; it fails, and a third movement is tried.

Now, it must be clearly understood that organisation will account for these phenomena. Quite as remarkable a series of responses, each in turn designed to get rid of an irritant, can be obtained from a frog, which has been deprived of its brain, and therefore presumably lacks both consciousness and intelligence. And step by step, as organisation advances, the response gains in complexity, until the human imagination is unable to unravel the chain of cause and effect. But the biologist is cognisant of no break in the series from the choice of a *Vibrio*, which can be analysed algebraically, to the choice of a child between two toys.

The faculty, clearly seen in the case of *Stentor*, of storing impressions, so that the response to any particular stimulus is in part conditioned by the stimuli which have preceded it, is a familiar property of living matter, and also of matter in the colloidal state. The molecular state of a jelly is not fixed by the conditions of the moment. Just as a piece of wrought iron has properties different from those of cast iron, so the circumstances which attend the making of a jelly—temperature, concentration, and the like—confer on it an internal structure which controls its properties for years to come. Each jelly, therefore, has an individuality due to the record which it bears of its past.

Take another case. A vertical rod of wax is bent, first north, then south, then east, then west, and so on. Left to itself, it will quietly work out these movements in the reverse order. It bends first west, then east, then south, then north, and so on. The molecular structure of the wax is such as to preserve a record not only of the fact that it has been moved, but also of the number, direction, and order of the several movements.

The Faculty of Growth

If it be true, as some chemists think, that in the process of oxidation there are always two processes more or less concurrent—the first one of synthesis, in which bodies of increased chemical complexity are formed by the union of the oxygen and the combustible substance; the second one of analysis, which supervenes only when the synthetic products reach a degree of complexity where they are unstable at the particular temperature and pressure—then, considered in a general way, the processes of assimilation and growth of living matter are exceptional only in the prominence and permanence of the synthetic stage. The living cell, on this view, is like the flame in being an oxygen vortex; it is unlike it in the extraordinary latency or delay in the advent of the analytic processes.

The peculiar feature of the living cell, however, considered as a machine, lies in the fact that, of the total amount of energy which it acquires, a fraction is retained and devoted to the increase of its own substance. In other words, it grows. After a while it divides, and the daughter-cells are like itself, so that there is not only the power of increasing the bulk of living matter by growth, but also a directive faculty called heredity, which constrains the new living matter, made from non-living matter, into the pattern of the old. The problems of growth and multiplication can be reduced to their simplest terms only in the case of minute forms like *Amœba*, each of which is at once a single cell and an individual. Each individual amongst the higher forms of life is built of countless cells, all of which, with one or two exceptions, are predestined to death. The exceptions —the true immortals—are chosen from the germ cells. When, however, an

Amœba multiplies, it divides bodily into two, and by this simple process a new generation is formed. Clearly, as Weissmann first pointed out, in such cases death intrudes only in the guise of accident.

The conditions of life of these simplest forms are by no means simple. Make an infusion of hay in boiling water, and let it cool. In the course of a day or so it will be found to be swarming with rod-like bacteria (*Bacterium subtilis*), engaged in feeding on the organic matter dissolved out of the hay. A few days later numbers of an actively mobile slipper-shaped animal, called *Paramœcium caudatum*, make their appearance to actively swallow and digest the bacteria; and so the round goes on. The bacteria and the Paramœcia alike have developed from wind-carried spores. Therefore in the natural life of these creatures are periods of physiological activity alternating with periods when life seems to be completely dormant— periods which follow one another according to no regular sequence, but in consequence of chance rainfall and of drought, when the inhabitants of the dried-up pool are caught up and carried away as dust.

Watch any chance collection of Paramœcia, and individuals will be seen not only to divide, but occasionally to fuse. Two individuals swim together, adhere closely, and effect an extensive interchange of substance. This is the process of conjugation: it is the first beginning of sexual reproduction. It is followed by increased physiological activity, increased rate of growth and of multiplication. If we could follow this mating process fully, if our imagination could grasp the events which lead up to it and the effects which follow, we should see in it the response of life to the flux of cosmical energy, just as the oscillations of a particle in Brownian movement are the response to the flux of electrical potential. This is no careless phrase: it is sober truth, for the air currents carry and mix spores from far-distant places which have, therefore, had different life-histories. They have lived in waters draining different kinds of soil, and therefore chemically different, the balance of sunshine and shade has been different, and the wind capriciously sows these spores from north, south, east, and west in the pot of hay tea. There they become active, and mate in conjugation, but not fortuitously. Guided by chemiotaxis, unlikes meet and fuse, just as do unlike cells when an ovum and a spermatozoon fuse, and the fusion of a pair of unlike individuals results in a thorough reorganisation, a fresh make-up, of the living matter of each. The continuance of the race depends upon the change of environment, upon the alternation of periods of activity with periods of dormancy, and upon the fusion of unlike individuals; and for these the sequence of natural phenomena, of summer and winter, of sunshine and shade, provides.

Now, the problem of growth is this. Suppose we eliminate these factors; suppose we isolate a pure strain of Paramœcia and keep them abundantly supplied with food, will the race continue to flourish and grow indefinitely, or will it attain old age and die off? The problem is far-reaching. It touches the simple questions of function, of digestion and assimilation, on the one side; while on the other it is concerned with the limitations of heredity in moulding successive generations after the common type. Three workers have attacked it with conspicuous success, Maupas,[1] Calkin,[2] and Woodworth,[3] and in each case the experimental method was the same.

Maupas was the first. He isolated individual Paramœcia under normal and healthy conditions—namely, in hay infusion containing the bacteria on which they feed, which was changed daily. Each individual was the starting-point in a sequence of generations, there being, on the average, two generations in three days. The rate of division was recorded, and the records furnished the basis for a curve of vitality.

The experiment established two points, the first being the presence of fluctuations of vitality of fairly regular character—"rhythms", they have been called. The curve alternately rises and falls, and each complete "rhythm"—a rise and a fall, that is—lasts about a month. The second is that the curve, as a whole, steadily falls, each successive rise in vitality is a little less than its predecessor, each depression a little lower, until—about the 170th generation—the period of old age, of senile decay, is reached, and the race dies out.

There the matter was allowed to rest, until fresh experiments were prompted by a remarkable observation made by Prof. Loeb. He found that the unfertilised eggs of sea-urchins could be made to develop by immersing them for a few hours in sea-water containing a higher percentage of salt than ordinary. If the eggs could be artificially aroused, why not the senile Paramœcia? So argued Prof. Calkin. Therefore, when the period of decay had arrived and the individuals were dying off rapidly, he tried placing them in various infusions. Vegetable infusions were without effect, but infusions of animal tissues, and particularly beef extract, gave the required result. The rate of growth and of reproduction reached the normal level, and death ceased. Senile decay had given way to artificial rejuvenation. Instead of 170 generations being the limit, by stimulation in the periods of depression Calkin succeeded in carrying a race to the 740th generation, and Woodworth to the 860th generation, when the individuals were still healthy and fully active. The living matter of these cells without doubt is potentially immortal!

Consider for a moment what incredible chemical activity and stability of character these figures imply. If it was possible to preserve alive all the

[1] *Arch. d. Zool. Exp.* (1889) (2), VII.
[2] *Arch. f. Protistenkunde* (1902), vol. I.
[3] *Journ. of Exp. Zool.* (1905), vol. II.

individuals, then at the 900th generation we should have a number which would need a row of some hundreds of figures to express. The parent cell would have produced the 900th power of 2 individuals like itself. The increase in the bulk of active living matter which would have been formed from non-living had there been space enough and food enough is not less wonderful. At the 350th generation it would have the dimensions of a sphere larger than the known universe![1] And the surface of the sphere would be growing outwards at the rate of miles a second. Nor is this all, for in addition to the enormous chemical activity implied by a rate of growth which would, if unchecked, produce a mass of living matter larger than the known universe in less than two years, there has been throughout continuous expenditure of energy on incessant and active movement. These animals have been watched continuously for 5 days, and throughout that time they were ceaselessly moving!

The recurring periods of depression show that in the living machine repair is not complete, and that after a time it will, if left to itself, cease working. With the condition of ill-repair there is associated a feature of singular interest. Woodworth specially draws attention to the fact that in the periods of depressed vitality the transmission of characters is imperfect. The moulding power of heredity fails, and many "monsters" are born.

Rejuvenescence can be brought about by a great variety of media, by extracts of muscle, of brain, of pancreas, by simple salts, by alcohol even. It is not food, but a marked and abrupt change of state that is needed—a stimulant, in fact—and boef extract produces its effect not *quâ* food, but as a stimulant pure and simple. Senile decay is due to monotony, under the influence of which the vital potential wears out!

The action of alcohol is remarkable. It was added to the water in which the animals lived so that they were always immersed in a solution of 1 part of spirit in 5000 to 10,000 of water. In the effect produced there is the touch of nature which makes the whole world kin. The periods of depression were wiped out. The curve of vitality no longer showed the ominous recurrent depressions. At the same time the rate of growth and division—that is to say, the physiological activity—was increased by as much as 30 per cent.

Something of the same effect is produced by strychnine, but there is a remarkable and significant difference in the fundamental action of the two drugs, for whereas the beneficial effect of alcohol endures after the drug ceases to be administered, that of strychnine does not. Alcohol, as Calkin says, in spite of the prodigiously increased rate of living, "exacts no physiological usury", it is beneficial in its after effects. Strychnine is harmful in its after effects; the onset of decay and death is hastened.

[1] I owe this rough calculation to my friend Mr Punnett.

What significance are we to attach to artificial rejuvenescence? There are two possibilities. The chemical agent employed may either add something which is missing or diminished in the chemical make-up of the protoplasm, or it may restore a physical state. The former implies that the chemistry of the growth process is imperfect: the process of converting non-living to living matter is subject to inaccuracies—inconceivably small it is true, since they need to be magnified to the 170th power of 2 before they destroy the working of the machine, but cumulative from generation to generation. I incline to think that senile decay is due not so much to such a chemical insufficiency as to the wearing out of a physical state, of a "potential".

Consider a special case. Thirty minutes' immersion of an individual *Paramœcium* in very dilute solution (1 part of salt in 1000) of potassium phosphate was found to restore vitality, and the effect persisted for 282 generations. Now, in this case the restoration and maintenance of the "vital potential", as Calkin calls it, cannot be due to the presence in the individuals of a trace of the salt, for each generation would halve the amount, so that as early as the twentieth generation less than a millionth part would be left for each individual. One is therefore driven to believe that the salt acts by restoring a state which, in the absence of natural or artificial rejuvenescence, wears out in about 170 generations.

The continual flux of energy and of matter which seems to be necessary to the maintenance of life implies a high degree of molecular mobility. It is possible that living matter, like all other forms of matter, tends to come into equilibrium with its surroundings, and to attain a condition of too great stability. To restore it the living substance needs stimulating at intervals, just as a coherer needs tapping after each electric wave has passed, in order to restore its particles to the non-conducting position. These are vague possibilities, but physical science furnishes a case so suggestively akin to artificial rejuvenescence as to merit description.

Matter is composed of molecules which in the liquid or solid state are attracted to one another by forces of prodigious power. Each molecule in the interior of a mass is pulled on the average equally in all directions. But consider the surface layer, a film only a few molecules deep. There the intermolecular forces are necessarily to a great extent unbalanced, with the result that this surface film acts something like a stretched elastic skin—it tries always to compress the mass to the smallest possible dimension. This, however, is not the only feature of the surface layer. It is also the layer which is in contact with adjacent masses of matter, gas, liquid, or solid, as the case may be.

Now, masses of matter which do not mix when in contact, and which therefore are defined by a surface of separation, are rarely, perhaps never without influence upon one another. Interaction of the surface layer takes place, so that the balance of molecular forces is modified, incomplete chemical reactions occur, and a condition of molecular stress is produced which, amongst other things, is manifested by the development of electrical charges.

These molecular events on surfaces are very potent; they can produce effects which are impossible and even inconceivable in matter in bulk. It is, for instance, not only possible, but probable, that in the surface layers the conditions may sometimes be such as to associate decrease of volume with decrease of pressure, a relation so subversive of ordinary experience as to be unthinkable. In the surface layer a gas may be condensed to the liquid state when far above its critical temperature and below its critical pressure. Chemical changes occur or are suspended under conditions of temperature and pressure totally unlike those controlling the same changes in masses of matter. Concentration, electric conductivity, all physical properties in fact, become abnormal, therefore when the surface energy forms a large fraction of the total molecular energy, as in films, or fluid in fine capillaries, ordinary chemical or physical knowledge fails us.

There is no lack of evidence to prove that the lifelike characteristics of colloidal matter, its capacity for storing impressions, the elusiveness of its chemical and physical states, are due to the fact that an exceptionally large fraction of its energy is in the form of surface energy.

There is also direct and unmistakable evidence in the nature of the effect of various salts upon the heart-beat, and in the optical characters of thin films, that living matter also contains a very large proportion of surface energy per unit of mass, and the curious and extreme physical and chemical powers which it manifests are without doubt largely due to this cause. Now it is just in experiments on surface energy that one finds a case analogous to the effect of the salt in bringing about rejuvenescence of senile protoplasm, or in awaking the dormant powers of an unfertilised egg.

It has been shown recently by a French physicist, M. Perrin,[1] that by the use of minute amounts of salts one can give to the surface energy of a solid a certain direction—one can fix in the surface layer certain qualities which, for instance, define the electric properties of the surface. The effect once produced, no amount of washing will undo it; the salt can be removed, the effect remains. So far as we know, in the absence of active chemical intervention it will endure for all time, always exerting a directive influence upon the molecular events in its neighbourhood. In these experiments there is, it seems to me, a real clue to the nature of the phenomena of rejuvenescence.

[1] *Journ. d. Chem. Phys.* ii, 61, and iii, 50.

HEREDITY

On the earth are some half-million different species of animals and plants, each of which breeds true in virtue of what we call heredity. Each species therefore represents a strain or line of descent of living matter always growing, dividing, and increasing in mass, like the little Paramœcia we have already considered, each striving to occupy the whole earth, and restrained in the attempt only by the accident of death.

The strains of living matter are separated from one another by a wide gulf which we do not know how to bridge. Change of state seems to be without effect. Continuous supplies of the richest food will not convert a strain of dwarfs into giants. In the solemn words of the Burial Service, "All flesh is not the same flesh; but there is one kind of flesh of men, another flesh of beasts, another of fishes, and another of birds."

The nature of these differences in the kinds of living matter and their mastery so that we may be able to control them is without doubt the most difficult and the most important problem which science has attempted to solve: most difficult because it deals with a form of matter much more complex than any which the chemist or physicist so far has considered; more important because on the solution of this problem depends the possibility of removing practical medicine, politics, and morality from the domain of empiricism and tradition to that of rational co-ordinate knowledge.

To speak of a strain breeding true is a bald way of describing a force so potent as heredity, so impish in its eccentricities. On the Antarctic ice there abound a race of birds called penguins. They have never seen a tree since they first were penguins; they do not fly, for their wings have been reduced to small flat paddles with which they swim. The bird cannot tuck its head under its wing, because the wings are too shrunken; but still, in mute worship and touching fidelity to its forebears of thousands of years ago, when it composes itself to sleep each individual bends round its head and tucks the tip of the beak—it is all it can do, poor thing!—under the dwarfed wing.

This lingering instinct, this obsession by the great past, is like a whale dreaming of the green fields in which his forefathers browsed! Now, each individual penguin starts life as a microscopic fragment of living matter, a single cell, so wonderfully compounded, so cunningly devised, as to enshrine without loss all the diverse qualities and powers which the word "penguin" connotes, down to the trivial detail I have described! There is little wonder that the naturalists of half a century ago gave the problem up in despair. There is cause for wonder and for congratulation that, impelled by the divine dipsomania for research, knowledge has moved so far as to make a beginning in the assaults.

Given a fulcrum, anything can be moved. The necessary fulcrum was found when attention was directed not, as in Huxley's time, to the more obvious resemblances between the different kinds of protoplasm, but to the less obvious differences. The microscope for the most part fails us here, in the first place because the discrimination between different kinds of matter by the agency of sight is possible only when there are associated differences in optical properties, and when there is the possibility of getting a clear image. Now, living matter is singularly free from definite optical differences, and it has the optical characters of ground glass. Therefore, the ultimate refinements of microscopic vision are for the most part wasted upon it. The dead cell exhibits remarkable structural details, but in the act of death there is of necessity a redistribution of matter which obscures and defaces the finer details of the real living structure, and replaces them by structure which is formed in the process of dying. For the material basis of the difference in the strains of living matter we have to look below the limits of microscopic vision, below the limits even of the living molecule, to the chemical molecule of which that living matter is built up.

The nearest chemical approach to living matter is the proteid, the chemical substance of which all protoplasm is, water excepted, chiefly composed. And, the fulcrum I spoke of, or, better, the thought which loosed the fetters of imagination, was the appreciation of the significance of the fact that proteids chemically are not all alike, and that the strains of living matter differ from one another in the kinds of proteid of which they are built up—that is to say, in their ultimate chemical constitution.

All proteids are not the same proteids; there are proteids of men, others of beasts, others of fishes, and others of birds!

The nature of the differences leads us to a real picture of the underlying differences between the kinds of protoplasm. The tide of thought of the older observers was fettered by the fact that all proteids have about the same atomic composition. The biologist of to-day owes his emancipation to the chemical discovery that the properties of a complex substance are defined not so much by the kind of atoms or number of atoms of which it is built, as by the arrangement of those atoms in space.

Here is a simple and startling case. The molecules of two chemical substances, benzonitrile and phenylisocyanide, are composed of seven atoms of carbon, five of hydrogen, and one of nitrogen:

Benzonitrile   Phenylisocyanide

There is a small difference in the arrangement of these atoms which is illustrated by the diagram. Now, what are the properties of these two substances? They are as unlike as possible. The first is a harmless fluid with an aromatic smell of bitter almonds. The second is very poisonous and offensive.

A vivid impression in regard to the odour of the isocyanides may be produced by the following experiment. In a test-tube bring together a *little* chloroform, aniline, and alcoholic potash. The reaction takes place at once. *It is better to perform the experiment out of doors and in such a place that the tube with its contents can be thrown away without molesting any one.*

In the building of a complex molecule one has atoms gathered together to form groups, these to form larger groups, and the whole structure is arranged on a fundamental plan or style, like, for instance, the ring of carbon atoms in the two substances just mentioned. There is, therefore, a molecular architecture, and, as in ordinary architecture, there are differences of style and of general plan, Gothic, Norman, etc., with endless variety in detail. Amongst the recognised styles of molecular architecture is the proteid style, and the qualities common to all forms of life are based ultimately upon the essential features of that style, while the differences between one kind of living matter and another are the expression of the differences in detail—the omission of this group, the addition of that.

Some of the atomic groups which find a place in the proteid molecule are readily recognisable by chemical tests—one of these groups occurs as a separate chemical substance called *Tryptophane*. It shows a vivid purple colour with sulphuric acid and reduced oxalic acid. Here are solutions of two proteids, one from maize seeds, the other from the white of egg, the one of the vegetable strain, the other of the animal strain. The former lacks, the latter possesses this group.

In order to represent the great varieties of living matter the proteid molecules must be capable of very many variations of structure. That is, after all, mainly a question of size—the larger it is the greater the possibility of variations in detail; and as the molecule of proteid seems to contain from ten to thirty thousand atoms, whereas the most complex molecule known to the organic chemist contains less than a hundred, there is no lack in this respect.

Proteids unquestionably are the material basis of life, but when isolated after the death of the cell they are not living. They are chemically stable bodies. They show no signs of the characteristic chemical flux. It is therefore conjectured on experimental grounds that the living molecule is built up of proteid molecules, that it is so complex, so huge, as to include as units of its structure even such large molecules as these. But when such very large molecules enter into chemical combination with one another, whether by reason of the great magnitude of the masses of matter in each in relation to the magnitude of the directive forces, or because the molecules themselves, owing to their great size, to a certain extent cease to be molecules at all in the physical sense, and possess the properties of matter in mass, it is at any rate certain that in their chemical combinations they cease to follow the law of definite combining weights which is the basis of chemistry. The quantity of the substance A which will combine with a fixed quantity of the substance B is determined not only by the chemical nature of A and of B, but also by the chance conditions of temperature and concentration of the moment. This class of chemical compounds is within limits continuously adjustable to changes in its surroundings, while at the same time it resists those changes by reason of its inertia. Here is a real adumbration in non-living matter of the chemical flux which is the abiding characteristic of the matter of life.

The biologist speaks of these molecular complexes as *molecules*, and in that he is wrong in so far as the word implies a *defined* structure, a chemical unit. The biogen, or chemical unit of living matter, is not a fixed unit like the molecule of dead proteid; it is an average state. That we know from the chemical phenomena of living matter.

Why should this be? Consider what must happen if you make the atomic building much larger than it already is in the molecule of dead proteid. You already have a molecule so large as to be liable to fracture on mere mechanical agitation. A molecule composed of fifty proteid molecules would cease to be a molecule in the physical sense: it would be matter in mass, defined by a surface; it would break up the waves of light, so radiant energy would profoundly affect it.

In a mass so large, a portion of the energy would of necessity be in a borderland between what we call osmotic energy and surface energy, the fraction in the one state or the other being determined from moment to moment by the changing relations with the enveloping matter. If the chemical structure was such as to produce a shape other than a sphere, surface energy would tend to produce chemical rearrangements, and the opposing play of these forces might result in oscillations of form which would reflect the irregular flux of cosmical forces just as does the particle in Brownian movement. The chemical relations of such a mass would be defined in the first instance by the surface layer, but any simple chemical event on the surface would be likely to fire a train of events leading to an eruption like a sunspot on the sun.

It is not, I think, difficult on these lines to conceive of a substance the chemical units of which could maintain themselves only in virtue of a continual flux of matter and energy—only, that is, as an average state; but it is certain that to develop the hypothesis we need what has not, so far as I know, yet been begun—namely, a kinetic theory of those intramolecular relations of atoms which are statically expressed by the geometrical methods of stereo-chemistry. The living cell, like a gas engine at work, is a chemical vortex, and there is no hope of analysing the motions of its parts so long as we are limited to statical methods.

In the history of the study of heredity there is a note of tragedy. In the early days of last century Lamarck began the revolt against the dogma of the immutability of species, which culminated in 1859 in the publication of *The Origin of Species*. Between Lamarck and Darwin, however, stand a scanty band of men forgotten by all but a few specialists, who strove by experiments in cross-fertilisation to pierce the mystery of heredity. Amongst them, and the last of the line, was a monk of the Abbey of Brünn, one Gregor Mendel, who in 1865 communicated to the Brünn History Society the results of eight years devoted to experiments with peas, under the modest title of *Experiments in Plant Hybridisation*.

The fate of Darwin's work is known to every one: how "it was considered a decidedly dangerous book by old ladies of both sexes", and how, "overflowing the narrow bounds of scientific circles, it divided with Italy and the Volunteers the attention of general society." The fate of Father Mendel's work was different. For the rest of the century it lay completely forgotten and buried in the annals of the little local society. But when it was rediscovered in 1900 by Prof. de Vries, of Amsterdam, it was at once realised by the very few competent to judge that the pursuit of a hobby in the abbey garden had led to a theory of the nature and workings of heredity so clear and complete as to leave to others only the application of principles and the amplification of details.

To find an achievement parallel to Mendel's, in the difficulty of the problem attacked and the all-embracing nature of the solution reached, one has to turn to Willard Gibbs's clean sweep of the domain of chemical equilibrium. But the author of the Phase Rule lived to see the work rediscovered—again by a Professor of the University of Amsterdam—and become the inspiration of a cloud of workers in all lands. The Mendelian laws of heredity, established twenty years earlier, are only now beginning to bear fruit, twenty years after Mendel's death.

The magnitude of Mendel's achievement can be appreciated by calling to mind the acute intellects which have been foiled by the problem. For a century the study of heredity has remained a repellent mass of statistics, with scarcely more discernible order than might be found in any chance collection of facts; and of the would-be student it might be said, "Quæsivit cælo lumen ingemuitque reperta." And for half of that century there has lain hidden a solution of the riddle which brings these facts into an order so straightforward that a child might learn it.

We should have nothing to do with the Mendelian laws here were it not that they have given singular meaning and interest to certain details of cell structure which before were a mere collection of unintelligent facts. To take things in their proper sequence I will first state the laws of inheritance so far as they concern us, and then consider the structural characters which seem to be their material basis.

The first Mendelian principle which concerns us is this: that what is transmitted from generation to generation may be analysed into certain qualities or characters—constant characters, as Mendel calls them—each of which is a unit in heredity, each of which, therefore, is capable of independent transmission. Thus in peas are length of stem, character of inflorescence, colour of seeds, flavour, and so on. Underlying these characters—each of which is capable of being picked out or put back by a breeder, forming a substrate on which they are erected—there would seem to be a basal character which is inalienable, and which the breeder cannot, at present at any rate, touch. Thus, in the case of peas, what is of necessity transitted is the fundamental qualities of "plant" as opposed to "animal", and of "pea" as opposed to other plants. To proceed in Mr Bateson's words:

These [unit] qualities or characters whose transmission in heredity is examined are found to be distributed among the germ cells, or gametes, as they are called, according to a definite system. This system is such that these characters are treated by the cell divisions (from which the gametes result) as existing in pairs, each member of a pair being alternative to the other in the composition of the germ. Now, as every zygote—that is, any ordinary animal or plant—is formed by the union of two gametes [in the process of sexual fertilisation], it may either be made by the union of two gametes bearing similar members of any pair, say two blacks or two whites, . . . or the gametes from which it originates may be bearers of the dissimilar characters, say a black and a white. [In the first case, no matter what its parents or their pedigrees may have been, the zygote breeds true indefinitely, unless some fresh variation occurs.]

If, however, the zygote be gametrically cross-bred, its gametes [or germ cells] in their formation separate the pair of characters again, so that each gamete contains only one character of each pair. At least one cell division in the process of gametogenesis is therefore a differentiating or segregating division, out of which each gamete comes sensibly pure in respect of the unit characters it carries, exactly as if it had not been formed by a cross-bred zygote at all.

For our purposes this may be reduced to three propositions: (1) that inheritance consists in the transmission of independent characters, of which each race or species possesses a definite number; (2) that these characters form pairs of opposites or alternatives; (3) that in the formation of the germ cells these characters are sorted out and distributed so that no germ cell carries both members of a pair. Can any material basis be found for these? To this we will now turn.

Five years ago it is doubtful whether there existed in the whole domain of science such a charnel-house of dead facts as in that of the science of cell structure. Thirty years of active study of animal and plant cells prepared for microscopical examination in various ways had resulted in the accumulation of a multitude of details respecting the structure of the cell nucleus and of the extraordinary way in which it behaves in cell division, and especially in those cell divisions which produce the germ cell. It was known that from the characteristic substance of the nucleus—which stains very deeply with aniline dyes, and hence is called chromatin—a continuous thread is spun as the first step in cell division, and that this thread of chromatin splits across into rods called chromosomes, each of which again splits, this time not across, but lengthwise, so as to form two "daughter" chromosomes, which, under the influence of a peculiar field of force formed in the substance of the cell, move away from one another and gather at the opposite poles of the spindle-shaped field, there to fuse and form the two nuclei of the "daughter" cells.

A further very significant and curious fact was known—namely, that the number of chromosomes formed in the process is not a chance one, but, in the first place, it is always an even number, and, in the second place, each species of animal or plant has a characteristic number. In the division of the cells of the human body, for instance, there are formed thirty-two chromosomes. But to these and many other similar facts no significance could be attached, beyond the obvious one that the nuclear substance is not divided grossly to form a new cell generation, but distributed by a complex and minutely detailed process of subdivision and segregation.

Not only were the facts of nuclear division without significance, but the presence of the nucleus itself seemed to be meaningless. The contractility of the muscle cell, the conductivity of the nerve cell, the chemical activities of the gland cell, reside in the cell body, and not, save perhaps in the last case, at all in the nucleus. Throughout cell life it lies to all appearance an inert mass, which becomes active only in the process of cell division. And yet actual experiments on enucleated fragments of Protozoon cells and on the nerve cells of higher forms had proved that in the absence of the nucleus the cell body cannot live. True, it carries on all the life functions for a while, but it seems to have lost with the nucleus the power of growth and of repair.

The last six years have witnessed the rehabilitation of the nucleus, and biologists now see in it the seat of that influence which directs the formative process by which living matter is produced from non-living matter, and controls the distribution of characters in heredity.

The actual agent in the latter process seems to be the chromosome, and the material basis of the limitation in the number of characters transmitted from generation to generation, and of their definiteness lies in the restriction of the number of chromosomes to a definite number for each species of animal or plant. The chromosomes are not fragments of nuclear substances of accidental composition, nor are they all alike. On the contrary, the probability is that they are unlike, and possessed of a high degree of individuality.

The material process which underlies the segregation of characters in the germ cell and the fusion of characters in pure and cross-bred zygotes can also be followed in the peculiar features of the cell divisions which form the male and female gametes.

As I have already said, each species has a characteristic number of chromosomes, but in the cell divisions which form ova or spermatozoa, ovule or pollen grain, this number is halved, so that each spermatozoon or ovum receives only half the proper number. In this sense, therefore, the germ cell is only half a cell. When two germ cells fuse in the act of fertilisation, the full number of chromosomes is restored. Thus, to choose an instance, the full number of chromosomes which make up a nucleus in a cell of the human body, no matter where it be placed, is 32; in the formation of the spermatozoon or ovum, however, there is a redistribution of chromatin, so that each receives only 16 chromosomes. When a spermatozoon fuses with an ovum, a zygote with the full number, 32, is formed.

The chromosomes therefore are the elements, the organs, as it were, of heredity. They have individuality, the limitations of which are not yet known. Each bears a unit character or a group of unit characters. The evidence for the individuality of the chromosomes is very remarkable.

Fundulus and *Menidia* are two fishes belonging to separate orders. Each has 36 chromosomes, but the chromosomes of the former are so much longer than those of the latter as to be readily distinguished. Moenkhaus[1] crossed these two forms, and traced the fusion of the long and short chromosomes in the formation of the hybrid zygote. But when the zygote prepared to divide, the paternal and maternal elements segregated and formed two groups of chromosomes, the one of long and the other of short chromosomes, and in each segment division the paternal long and the maternal short chromosomes reappeared and acted independently. Another case has been furnished by Wilson.[2] In certain groups of insects there is among the chromosomes of the male cells one distinguished by its small size. The total number of chromosomes, instead of being even, is odd: there are thirteen, and this small chromosome is the thirteenth. It is, in point of fact, only half a chromosome, therefore when each of the others divides into two, it does not divide, but passes bodily to one or other of the two new cells. In this way two different kinds of spermatozoa are formed—those

[1] *Amer. J. Anat.* (1904), vol. III.   [2] *Journ. Exp. Zool.* (1906), vol. III.

which possess the odd half chromosome, and those in which it is missing—and they are formed in equal numbers. Now, ova fertilised by the former grow into females, those which are fertilised by the latter grow into males. Therefore this particular chromosome is the carrier of the sex character.

I have stated the theory of the mechanism of heredity as it seems to be developing. A word of caution, however, is necessary. It is quite possible that we are attaching too much importance to the chromosome simply because, owing to the affinity of its substance for dyes, we can follow it in the phases of cell history. The rest of the nucleus and cell body does not happen to show such constant affinities, and therefore the sense of sight yields no evidence as to their action in cell division. Yet, so far as we know, the same detailed processes of synthesis and analysis which we can follow in the chromatin substance may divide the units of the rest of the cell in cell division, and guide the half of each unit to its allotted place in the architecture of the new cells.

The observations of Conklin[1] upon a curious Ascidian egg makes this even probable. The body of this egg is built of five kinds of protoplasm recognisably different to sight during life. These are (1) deep yellow, (2) light yellow, (3) light grey, (4) slate grey, (5) clear transparent. Each of these has a separate history: the deep yellow protoplasm makes the muscular system, the light grey the brain, the clear transparent the skin, and so on. This egg therefore is a mosaic, an architecture of different kinds of living matter, which we can detect and follow owing to associated optical differences. Had these been absent, we should have known as little of the architecture of this egg as we know of that of eggs in general.

The independent transmission of characters, and the presence in the germ cells of different kinds of living matter, are indisputable. They lead us, however, to a riddle which I leave to my readers to solve as they will. We are driven to believe that in the material make-up of any race there are several kinds of living matter which cannot be changed the one into the other, and of which some will mix, others will not or cannot mix. These materials, bricks, as it were, in the building, are transmitted from generation to generation by the agency of the germ cells, which therefore are heterogeneous structures.[2] Now, the doctrine of the direct transmission of the various living substances employed in the make-up of the individual lands us in this difficulty. The fertilised egg has all the material necessary for the make-up, therefore it can, and does, develop into an adult. The generative cells also possess amongst them all the necessary material. Therefore

---

[1] *Journ. Exp. Zool.* (1905), vol. II.

[2] The beginnings of the science of their architecture is to be found in the last report of Mr Bateson and Mr Punnett to the Committee of the Royal Society on Evolution.

amongst the earlier generations of cells produced by the growth and division of the fertilised ovum that cell or those cells which will form the generative organs must contain all the substances. But direct experiment contradicts this conclusion. Possibly in the very first cleavage, certainly in the second cleavage of the egg, there is a distribution of material amongst the two or four cells such that each one lacks something in the general make-up, and therefore can, and will, grow only to an imperfect monster if isolated. But one of those four incomplete cells will give rise, amongst other things, to the generative organs, each cell of which, in the first instance, is complete. Therefore, as we may "neither confound the persons nor divide the substance", we seem to be in a region of incomprehensibles.

"Just as that normal truth to type", says Bateson, "which we call heredity is in its simplest elements only an expression of that qualitative symmetry characteristic of all non-differentiating cell divisions, so is genetic variation the expression of a qualitative asymmetry beginning in gameto-genesis [the genesis, that is, of the germ cells]. Variation is a novel cell division.... What is the cause of variation?" Cross-breeding—that is, the union of unlike germ cells—may modify the character units. So, too, apparently may the long-continued *absence* of cross-breeding. It has been noticed in the cycles of a pure strain of *Paramœcium* that the periods of depressed vitality are also periods when the directive force of heredity is weakened. The individuals of successive generations show great departures from the normal type, and monsters are of frequent occurrence. With the lowered rate of growth, the lowered "vitality", as we call it, for want of a more precise word, there is associated a lowered degree of fixity of type.

25

ELECTROLYTES AND COLLOIDS.
THE PHYSICAL STATE OF GLUTEN

With Prof. T. B. WOOD

[Proc. Roy. Soc. (1909), B, LXXXI, 38]

Gluten, as ordinarily prepared by washing wheat flour in tap water, forms a coherent stringy mass insoluble in water. It consists essentially of a mixture of two proteins, gliadin and glutenin, but even when very thoroughly washed it always includes some starch. Gliadin, which forms rather more than half of the total protein, is soluble in dilute alcohol, and gives to the gluten its peculiar physical properties.

The power which dough possesses of retaining the gas formed during fermentation is due to the tenacity and ductility of gluten.[1] Therefore, the property of forming a light and well-shaped loaf, which is so variable a feature of different flours, is determined by the amount and the physical state of the contained gluten.

The physical state of gluten, like that of other colloids, is conditioned by the electrolytes which are present. Gluten washed out of flour with distilled water obviously is more friable and less tenacious than gluten washed out with tap water which contains salts. It is this influence of electrolytes upon the physical state of gluten which we propose to discuss.

Gluten is peculiarly sensitive to low concentrations of acid or alkali. A tenacious ductile mass suspended in a large volume of, for instance, $0.0001 N$ acid, begins almost at once to show signs of disintegration, and is at once dispersed by slight movement to form a stable opaque colloidal solution or hydrosol.

Action of acids. This action was investigated quantitatively by suspending a small mass of gluten on a bent glass rod in a beaker containing 120 c.c. of a solution of acid of known strength, and noting the concentration at which cohesion was so far reduced as to allow the protein to fall off the rod and disperse in a cloudy "solution". It was found that while very dilute acid causes dispersion, a solution of a strong acid above a certain concentration maintains the cohesion. Gluten, therefore, is coherent in distilled water, and in strong acids above a certain critical concentration. A weak acid, such as acetic acid, brings about dispersion up to as high as twice normal, the highest concentration tried. Inspection of a series of beakers with

[1] Wood, *Journ. Agric. Sci.* II, pt. 2, p. 139, and pt. 3, p. 267.

concentrations of any strong acid from zero to the critical point makes it clear that, starting from the lowest concentration, dispersion increases to a maximum and then falls to zero at the critical point. In other words, the power of destroying the cohesion and dispersing the gluten as a cloud varies with the concentration of the acid, so that the relation can be shown by a curve. The form of the curve will be seen later.

The dispersion of the gluten is not due to a change in the protein molecule of the nature, for instance, of hydrolysis, since it can be recovered as a tenacious stringy precipitate by neutralising the acid or by the addition of salt.

The following table gives the mean of several determinations of the concentration at which the gluten retains its coherence. The exact point is the concentration at which gluten just breaks under its own weight when suspended in the solution of acid; and the results obtained in different experiments are fairly consistent. It is remarkable that there should be no simple relation between the observed concentrations and the strengths of acid used as measured by electric conductivity. The conductivity of the solutions after the gluten had been immersed in them was measured, and the results are given in the second column of figures, the value of the sulphuric acid solution being taken as unity:

TABLE I

Acid	Normality of critical concentration	Relative conductivity
H_2SO_4	0·017	1·0
Camphorsulphonic	0·02	1·59
HNO_3	0·03	1·9
HCl	0·05	3·8
Oxalic	0·15	3·8
H_3PO_4	2·00	—

Action of distilled water. Gluten breaks up when washed very thoroughly in many changes of ordinary distilled water. The distilled water used was acid to litmus owing to the presence of carbonic acid; and the dispersion of the protein is due to this acidity, since (1) it is precipitated by the addition of a trace of alkali, and (2) the protein when dispersed is electropositively charged—that is to say, it displays the characteristic relation of protein to acid.

The influence of salts. Salt in small concentration precipitates a hydrosol of gluten whether it be formed by acid or by alkali. Therefore, salts lessen the power which acids or alkalis possess of destroying the cohesion of gluten, and, in sufficient concentration, completely neutralise it. The concentration of salt necessary completely to nullify the dispersive power of particular acids was investigated in the manner already described, namely, by suspending approximately equal pieces of gluten in varying concentrations of acid and salt, and noting the point at which cohesion was so far reduced as to allow the protein to flow off the rod. The relations appear in the following curves (Fig. 1), which show that for all strong acids and for all salts the concentration of the latter needed to balance the former increases to a maximum as the concentration of acid increases, and then declines to zero at the point where the acid alone is sufficient to maintain

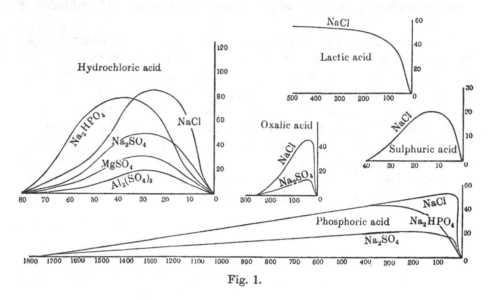

Fig. 1.

cohesion. The curves all agree, therefore, in showing that, measured by the concentration of salt needed to prevent dispersion, the dispersive power of an acid increases with increasing concentration, and then falls until the critical concentration is reached, where dispersive action is *nil*.

These curves are so characteristic that they afford a means of testing a point of general theoretical interest. One great class of colloidal solutions, the aqueous solutions of characteristically insoluble bodies such as metals, some proteins, sulphides, and gums, are characterised by the fact that round each particle of the solute there is an electric double layer, and on the potential difference between which the stability of the solution depends. Coagulation or precipitation of such a solution is approximately coincident with the reduction of the potential difference to zero, the most complete coagulation, *i.e.* mechanically the densest and most coherent coagulum, being formed at the isoelectric point.[1]

[1] Hardy, *Proc. Roy. Soc.* (1900), LXVI, 110; Picton and Linder, *Trans. Chem. Soc.* (1905–6), LXXXVII; Perrin, *Journ. de chim. Phys.* (1904), II, 601; (1905), III, 50.

On this view the formation of the hydrosol of gluten is due to the development of electric charges round the particles of the protein owing to chemical interaction between the protein, the acid or alkali and the water; and the tenacity, ductility, and water-content of a solid mass of moist gluten depends upon the total or partial disappearance of these electric double layers, and the reappearance of what is otherwise obscured by them, namely, the adhesion of "idio attraction", as Graham called it, of the colloid particles for each other, which makes them cohere when they come together.

It is possible to put this hypothesis to the proof. We can measure the potential difference between the water face and the protein face of each particle in the hydrosol of gluten by determining the rate of transport of the particles in a uniform electric field. The method adopted has been described by one of us.[1] Briefly it consists in the use of a graduated U-tube, the opalescent hydrosol is introduced as the lower layer, the upper layer in each limb being a clear solution of the same electrical resistance. Electrodes are immersed in the upper layer, a field established, and the rate of movement of the boundaries between upper and lower layers observed.

The hydrosol was prepared either by washing gluten in distilled water containing carbonic acid, a process which occupied at least two days, or in a few hours by washing in a few changes of $0\cdot0001N$ sulphuric acid. It was freed from all starch by centrifuging. To successive lots of the hydrosol, acid was added in varying amounts, and water when necessary, so that the concentration of protein was constant, while the concentration of acid varied. Finally the resistance was measured, and a fluid to form the upper layer was prepared either by adding the same acid to water or by adding sodium chloride. Hydrochloric, sulphuric, and acetic acids were used, and the results were in all cases the same. The figures for hydrochloric acid are plotted in the following curve, the ordinates being specific conductivity of the solution $\times 10^{-6}$, the abscissæ the specific velocity in centimetres per second for unit potential gradient $\times 10^{-}$ (Fig. 2).

The curve agrees in form with those already given for the effect of salt upon cohesion, and we may therefore conclude that acids, and by inference alkalis also, destroy the cohesion of gluten by forming double electric layers round the particles, and that the potential difference between these layers rises with increasing concentration of acid to a maximum, and then falls.

Action of alkalis. The action of alkali in destroying the cohesion of gluten is essentially similar to that of acid, except that the electric sign is reversed. In a hydrosol of gluten formed by carbonic acid or any other acid the protein is charged positively; when formed by any alkali it is charged negatively.

[1] Hardy, *Journ. Physiol.* (1905), XXXIII, 251.

It is interesting to note that, when alkali is added, it not only neutralises any acid present, but also reacts directly with the protein as though the latter were itself an acid. The alkali, therefore, disappears as such; it is, in point of fact, neutralised by the protein with the formation of new ions. For instance, in a particular hydrosol formed by carbonic acid, the particles of gluten were charged positively, and had a specific velocity 46×10^{-6} cm. per second. Sodium hydroxide was added in quantity sufficient to make the entire solution contain $N/1600$ of NaOH. The fluid was not alkaline to phenolphthalein, but in spite of this the protein was now charged negatively and had a specific velocity of 23×10^{-6} cm. per second.

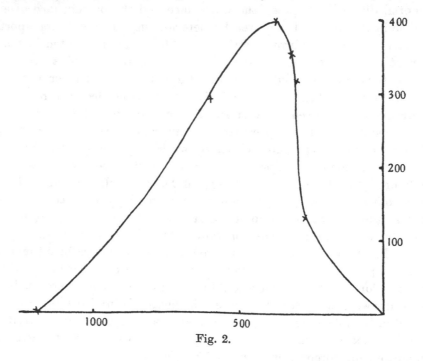

Fig. 2.

Approximately pure gliadin, dissolved in 70 per cent. alcohol, shows relations to acid and alkali the same as those described for gluten. Dropped into 98 per cent. alcohol or distilled water, it forms an opalescence, and is then electro-positive; in presence of $N/4160$ of NaOH it is electro-negative.

Conclusions. The experimental results seem to prove beyond question that the physical state of gluten—that is to say, the degree of coherence or dispersion as a hydrosol—is determined by the potential difference between the particles of the protein and the fluid.

The development of such a potential difference between colloid particles and fluid has been accounted for in two ways. The first, which may be described as a purely physical hypothesis, ascribes it to differences in the speed of the ions of electrolytes present. The colloid particles at any moment contain within themselves an excess of the most penetrating and rapidly moving ions present, and they therefore have the charge of that ion. In presence of acid they will have the charge of the hydrogen ion, in the presence of alkali that of the hydroxyl ion. This hypothesis was advanced by one of us to explain the properties of certain proteins of the globulin class when in solution.[1] It was also advanced independently by Perrin to explain the electrical properties of colloidal solutions in general.[2]

The second hypothesis is frankly chemical in nature, and, as applied to proteins, it may be put as follows: The protein molecule contains H and OH groups. Proteins, therefore, as a class are, like their chemical allies the amino acids, amphoteric electrolytes. They react with acids and alkalis to form salts, but the reactions are not precise, an indefinite number of salts of the form $(B)_n BHA$ being formed where the value of n is determined by conditions of temperature, and concentration, and of inertia due to electrification of internal surfaces within the solution.

The salt so formed is ionised by the water. Positive or negative ions, as the case may be, leave the protein face to enter the water face, and form an electric layer there, while the protein face is left charged respectively negative or positive.[3] On this view, in the particular case under consideration, the decrease and final disappearance of the potential difference which occurs when the concentration of acid rises above a certain value would be due to a suppression of the feeble ionisation by the excess of acid.

The first view seems to be incompatible with certain experimental facts— such, for instance, as the fact that salts such as LiCl or LiBr, the velocities of whose ions are in the ratio of about 1 to 2, do not confer any change on proteins, nor, as Perrin noticed, do they produce any contact difference of potential between a water and a solid wetted by it. It also ignores the purely chemical nature of the conditions which govern the formation of colloidal solutions of metals.[4]

[1] Hardy, *Journ. Physiol.* (1903), XXIX, xxvii.
[2] Perrin, *Journ. de chim. Phys.* (1904), II, 601; (1905), III, 50.
[3] Hardy, *Journ. Physiol.* (1905), XXXIII, 251; *Proc. Roy. Soc.* (1907), LXXIX, 413.
[4] Burton, *Phil. Mag.* (1906), XI, 425.

26

ELECTROLYTIC COLLOIDS

[Gedenkboek—van Bemmelen (1910)]

The final classification of colloidal states will be reached only when a final and satisfying interpretation of the phenomena of the state has been arrived at. In the meantime various classifications are in use each of which has its advantages. Such are the distinction into *lyophil* and *lyophobe* colloids according to an assumed difference in the relation between solvent and solute, or into *suspensoide* and *emulsoide* according to whether the dispersed phase is solid or liquid.

I wish to consider here a classification based upon the view that colloidal solutions are merely special cases of ordinary molecular solution, which are peculiar solely in a wide departure from thermodynamic reversibility due to the operation of a large internal frictional constraint which springs from the limited mobility of one of the constituents.

This immobile constituent may be an ion or it may be a neutral molecule. It is, for instance, an anion in a solution of soap in water, an anion or a cation in solutions of a protein salt in water, or the neutral gelatine molecule in gelatine solutions.

Owing to its immobility, or relative immiscibility, the constituent to the presence of which the peculiar colloidal features of the solution are due, behaves towards any internal surface which may be formed as though the latter were a membrane partially or wholly impermeable by it. It is precisely on this impermeability that the frictional constraint before mentioned is founded.

When a mixture of phenol and water is cooled to the point where a second fluid phase rich in phenol appears, the latter separates in droplets which only slowly coalesce to form a separate layer of fluid. While the emulsion state persists electric conductivity, viscosity, and other physical properties are abnormal and indicate that the system is not thermodynamically reversible. This is due to the fact that the change of internal energy with change of temperature is not solely a function of the making or unmaking of those molecular associations in which the act of solution consists. A portion is involved in electrical and surface tension changes, and it is to the operation of these last-mentioned factors that the departure from reversibility is due.

The droplets of the phenol-rich phase are at a lower potential than the external phase, and they move with a very high velocity in an electric field.[1] The electrification may be traced to the different resistance to migration across the common surface of the phases experienced by the large anion of phenol and the hydrogen ion, and the sign of the charge implies that, possibly owing to the higher hydrostatic pressure on the internal phase, the internal pressure of phenol in that phase is higher than it is in the external phase. If the difference in mobility of the anion and cation were much greater than they are in the case we are considering, the operation of surface forces would be greater and the emulsion stage more permanent. The system would in point of fact become colloidal.

The expression for the velocity of formation of a new phase is in its simplest form:

$$v = \frac{F}{A\alpha} \frac{\Theta_0 - \Theta}{\Theta_0},$$

where F is the latent heat of the reverse change, A the force needed to give unit velocity to a gram-molecule of one constituent through the other, and α the depth of the layer at the surface of the new phase through which this molecule may be supposed to be urged by the force A.

In order to apply this equation to the colloidal state the factor A would need to be expanded so as to include: (1) the special constraint on the movement of the constituent of low mobility, usually or always that one with a very high atomic weight such as the protein or starch molecule, or the anion of a soap, and (2) the constraint upon other mobile constituents due to, e.g., electrostatic forces between them and the constituent of low mobility. In this extended form the expression for the force required to overcome the internal friction of the system would be the expression characteristic of colloidal solution as distinct from ordinary solution.

When the relatively immobile constituent in a colloidal solution is an ion, electrolytic phenomena will occur at internal surfaces. Such cases form special developments of ordinary electrolytic solution and may be classed as *electrolytic colloids*. When the immobile constituent is a neutral substance, we have a second class related to ordinary non-electrolytic solution and therefore to be called *non-electrolytic colloids*. As examples of the former class may be taken metals, soaps, hydrosulphides, and certain proteins in water; and of the second agar or gelatine in water, or celloidin in ether and alcohol.

Electrolytic phenomena may occur in solution of the non-electrolytic class, but in the first place it is doubtful whether they are due to the interaction of solute and solvent or to traces of electrolytes,[2] and in the second place they do not play a large part in determining the state of the solution.

[1] The velocity in unit field is about 105×10^{-5} cm. per second.
[2] For instance Dhere and Gorgolewski, *Comptes Rendus* (1910), CL, 934.

It has so far been taken for granted that a potential difference between the colloid particles (internal phase) and the fluid (external phase) is due to ionisation at the common surface. As other views have been expressed as to the source of this contact difference of potential the point needs detailed discussion.

Consider first metal sols, produced by sparking metal electrodes within some fluid. The sign of the charge on the metal particles is very easily found by determining the direction of the drift in an electric field. The solution tension theory of Nernst would appear to be peculiarly applicable to these solutions. How far will it explain their electric properties?

The theory may be briefly stated as follows:

A metal plate immersed in a fluid gives off ions into solution which are positively charged. The plate therefore is left with a negative charge upon its surface which is opposed by a positive charge on the fluid face due to a layer of positive ions there.

Metals, however, vary in relation to any particular fluid as to the concentration or tension of these ions with which they are in equilibrium. They fall broadly into two classes, those whose ionic tension is high, such as zinc, iron and lead, and those whose ionic tension is low, namely the "noble" metals silver, platinum and gold. The former give off positively charged ions into a dilute solution of their own salts and therefore become themselves negatively charged; the latter, owing to their low tension, receive positively charged ions from solutions and therefore themselves become positively charged.

If the colloid particles consisted simply of particles of metal torn off the electrodes by the arc they would, according to this theory, be charged negatively and the potential difference would be greater in the case of non-noble metals. The exact contrary was found by Burton. The non-noble metals are + the noble metals − in sign, and the potential difference between any noble metal and water was about twice that between any non-noble metal and water.[1]

The facts can be explained by assuming that the non-noble metals zinc, lead, bismuth, copper or iron react with the water to form a skin of hydroxide which ionises, OH-ions passing into the water leaving the metal charged positively. The noble metals platinum, gold, and silver combine with the hydrogen to form hydrides which also ionise the particles becoming negatively charged.

This view is supported by the fact that when these same metals are sparked under methyl or ethyl alcohol which contain a replaceable hydroxyl group the noble metals will not form colloidal solutions, while the non-noble metals form solutions in which the particles always carry a positive

[1] *Phil. Mag.* (1906), (6), xi, 446.

charge. The contrary occurs when ethyl malonate, which has replaceable hydrogen, is used instead of alcohol. Then only the noble metals will form solutions and the particles are always negative. The contact differences of potential between the phases therefore is for the noble metals due to a skin reaction of the form

$$MH \rightarrow M' + H\cdot,$$

and for the non-noble metals of the form

$$MOH \rightarrow M\cdot + OH'.$$

A remarkable feature of these solutions in water is that each of the two classes of metals, noble and non-noble, has its own characteristic potential as the following figures, taken from Burton's paper show:

Metal	Φ in volts
Platinum	$-0\cdot031$
Gold	$-0\cdot033$
Silver	$-0\cdot036$
Lead	$+0\cdot018$
Bismuth	$+0\cdot017$

This fact may perhaps be explained as follows:

The distance any ion will travel away from the particle is a function of two things, the velocity of the ion and the electrostatic force which finally brings it to rest. As the mass of the water is by comparison always infinitely great, we may perhaps consider the expulsive force the same for all ions and put in place of velocity the specific resistance to the passage of an ion through water which is proportional to the specific mobility at infinite dilution. The force which brings the ions to rest at an average distance d from the surface of the particle is

$$F = K \frac{2}{3} \frac{\Phi^2}{d^4},$$

where Φ is the contact potential difference.

$$\therefore \quad K \frac{2}{3} \frac{\Phi^2}{d^4} = m \left(\frac{2v}{3}\right)^2 \frac{1}{2d},$$

where m is the mass of the ion and v its velocity.

$$\therefore \quad \frac{\Phi_H}{\Phi_{OH}} = \frac{v_H d_H}{v_{OH} d_{OH}} \sqrt{\frac{m_H}{m_{OH}}}.$$

By similar considerations $\quad d = \frac{1}{K} \frac{2e^2}{mv^2}.$

$$\therefore \quad \frac{d_H}{d_{OH}} = \frac{m_{OH} v_{OH}^2}{m_H v_H^2},$$

$$\frac{\Phi_H}{\Phi_{OH}} = \frac{v_{OH}}{v_H} \sqrt{\frac{m_{OH}}{m_H}}.$$

This equation would give as the ratio of the potential differences of the two classes 2·2, while Burton's figures give as an average value 1·91.

Soaps and serum globulin form colloidal solutions in water in which the colloidal features depend upon the relative immobility and immiscibility of one of the ions. In both, owing to hydrolytic dissociation, an insoluble species tends to form and in fact does so far separate out as to constitute an internal phase. The conditions have been fully discussed elsewhere.[1] It will suffice here to reproduce the characteristic equations of the solution:

$$x\mathrm{CHS} + y\mathrm{HOH} = (\mathrm{CHOH})_y(\mathrm{CHS})_{x-y} + y\mathrm{HS}$$

or
$$x\mathrm{CB} + y\mathrm{HOH} = (\mathrm{CH})_y(\mathrm{CB})_{x-y} + y\mathrm{BOH}.$$

The internal phase contains, with water, the complex

$$(\mathrm{CHOH})_y(\mathrm{CHS})_{x-y} \text{ or } (\mathrm{CH})_y(\mathrm{CB})_{x-y},$$

which liberates ions into the external phase just as does the phenol-rich phase previously mentioned and so becomes charged + or −

$$(\mathrm{CHOH})_y(\mathrm{CSH})_{x-y} \rightarrow (\mathrm{CHOH})_y(\mathrm{CH})_{x-y}^+ + \mathrm{S}_{x-y}^-,$$
$$(\mathrm{CH})_y(\mathrm{CB})_{x-y} \rightarrow (\mathrm{CH})_y(\mathrm{C})_{x-y}^- + (\mathrm{B})_{x-y}^+.$$

In the form in which the equations are put the ratio x/y must have a small value.

By the help of the ultramicroscope it may be seen that the colloid particles in these solutions are not all of the same size yet they all move with the same velocity in an electric field. The density of the surface charge must therefore be the same for all. These particles therefore are ionic but differ from true ions in carrying a variable quantity of electricity. For this reason I have suggested elsewhere the name of pseudo ion for them, but *colloidal ion* is perhaps more suitable.

Another explanation of the difference of potential between the phases in colloidal solutions has been put forward by Perrin.[2] The study of the electric endosmose through diaphragms made of various, non-conducting materials led him to formulate two laws:

(1) That contact difference of potential exists only between a solid and an ionising liquid.

(2) That, in the absence of polyvalent ions, non-metallic substances are positive to fluids which are acid, negative to fluids which are basic.

Put in another form the second law affirms that the solid is positively charged in presence of excess H-ions, and negative in presence of excess OH-ions in the fluid. These ions therefore are prepotent, and this Perrin ascribes to their smaller size whence their centres can approach the solid surface more nearly than those of other ions. At any moment therefore

---

[1] *Journ. Physiol.* (1905), XXXIII, 251.
[2] *Journ. d. chim. Phys.* (1904), II, 601; (1905), III, 50.

there will be under the influence of diffusion an excess of positive electricity if H-ions are in excess, or of negative electricity if OH-ions are in excess, in the skin of liquid immediately touching the solid.

There is an obvious contradiction between this theory and Burton's results. Methyl and ethyl alcohol contain only OH-ions as a possible source of contact difference of potential in the sense of Perrin's laws. The colloid particles therefore should be negative. They are, as we have seen, invariably positive. Similarly H-ions only, and no OH-ions can be present in ethyl malonate. The metal particles in this fluid should be positive. They are always negative. The most interesting discrepancy, however, is between the theory and Perrin's own admirable data. The table on p. 625[1] gives the flow of fluid per minute due to a field of 10 volts per cm., through diaphragms of various materials. The flow varies in direction and rate in such a way as to show that, broadly speaking, the solid is positive when the fluid is acid, and negative when it is alkaline. It is easy from the figures to fix upon the point in the varying reaction of the water at which the potential of solid and liquid are identical—the iso-electric point. The results appear in the following table, but before turning to them, it will be well to consider the probable position of the iso-electric point according to Perrin's theory.

Pure water contains H- and OH- in equal concentration. Therefore, since H is the faster and smaller of the two a solid in contact with pure water should be charged positively. The table, however, shows that out of ten solids examined, eight were negative to water, and in two cases remarkably high concentrations of acid were needed to destroy the charge namely $N/30$ HCl for cellulose and $N/90$ HCl for carborundum. When it is remembered that these fields between solid and liquid are completely discharged by some salts in concentrations as low as $N/1000$ these values are seen to be very high. As no mention is made in the paper of precautions to exclude carbonic acid, and as all ordinary distilled water contains sufficient to turn litmus red, the iso-electric point is probably more on the acid side than the table shows.

Gelatine	iso-electric in	$N/100$ HCl
Carborundum	,,	$N/90$ HCl
Salol	,,	$N/500$ HCl
Sulphur	,,	$N/500$ HCl
Cellulose	,,	$N/30$ HCl
Boracic acid	,,	N/x HCl
BaSO$_4$	,,	$N/900$ KOH
AgCl	,,	$N/2000$ HCl
Naphthalene	,,	$N/5000$ HCl
Al$_2$O$_3$	,,	$N/900$ NaOH

[1] Perrin, *loc. cit.*

The fact that insoluble solids are almost always negative to water (that is water containing carbonic acid) was, if I remember rightly, emphasised by Quincke. Two of the substances are positive to water, namely Al_2O_3 and AgCl. Aluminium readily forms a basic hydroxide which would liberate negative ions and so become itself positively charged, and traces of a soluble chloride would explain the positive charge of silver chloride. Bödlander[1] and others have shown that silver chloride readily forms double salts with other chlorides which in solution liberate chlorine ions, the residue remaining as a complex ion of the form $(AgCl)^nNa^+$, where n may be very large.[2]

Perrin's observations seem to me to prove two things: (1) that there is a true contact difference of potential between some solids and pure water, and (2) that the field is profoundly modified by H- or OH-ions, and they are valuable precisely because they bring these two relations into sharp contrast.

In many cases, and in all those included in the class of electrolytic colloids, the potential difference is due to ions passing from the solid to the fluid, or from the internal phase to the external phase in colloidal solution.

Invariably when the chemical nature of a substance is such as to give it acid or basic properties, the sign of the charge it carries in colloidal solution is that which it would have if it parted with H- or OH-ions to the fluid. For instance, the pure acid of congo red forms a colloidal solution in pure water with negatively charged particles

$$SH \rightarrow S^-_{coll.} + H^+.[3]$$

H- and OH-ions do two things: they may discharge the field between the phases, and they may in certain cases reverse the sign of the charge on the particle or increase the existing charge. The discharging and the charging effects should be considered separately, though we shall find reasons for thinking them to be due to the same operation in certain cases.

In analysing the influence upon contact potential of H- and OH-ions, that is to say of acids of alkalis when added to the fluid, the explanation will differ according to whether the colloidal substance, or the solid in cases of electric endosmose, is or is not itself an amphoteric electrolyte.

Proteins, for instance, form salts with acids and with bases, the reactions with acids and alkalis being those of simple neutralisation and chemical combination:

$$\text{globulin} + HCl = \text{globulin } H^+_{coll.} + Cl^-.$$

[1] *Zeits. f. anorg. Chem.* (1904), xxxix, 197.
[2] *Journ. Physiol.* (1905), xxxiii, 303.
[3] Bayliss, *Proc. Roy. Soc.* (1909).

On adding alkali the salt is at first decomposed and the insoluble globulin precipitated, more alkali redissolves it and the electric sign of the internal phase is now reversed:

$$\text{globulin} + KOH = \text{globulin}^-_{\text{coll.}} + K^+ + HOH.$$

Aluminium and silicon form amphoteric hydroxides, that of the former having an excess of basic function, that of the latter excess acid function. The neutralisation and reversal of the charges on solid alumina in contact with water is therefore explicable on purely chemical lines.

The position of the iso-electric point in the case of amphoteric substances might, if the disturbing influence of traces of electrolytes could be eliminated, be taken as a measure of the ratio acid function/basic function.

An instructive case is that of gliaden, a protein of wheat. It is insoluble in water but forms colloidal solutions of low grade in presence of acid or alkali, the acid or alkali being partly or wholly neutralised. In these solutions the protein is quite normally cationic or anionic as the case may be. But when the concentration of either acid or alkali reaches about $0.03 N$ the protein is precipitated. It is also, of course, precipitated at the neutral point. Passing from a suspension of the protein in water to the precipitation point on either side, the potential difference between solute and solvent at first increases and then decreases in a regular curve.[1] The relation can be explained by supposing that the first quantity of acid or alkali added combines with the protein to form a salt which hydrolyses according to the formula given earlier for such cases, and that any slight excess either of acid or alkali depresses the ionisation of the colloidal complex and so diminishes and finally destroys the electric field between the phases.

When the solute has no amphoteric properties the action is different. Colloidal copper is positive according to the formula

$$CuOH \rightarrow Cu^+ + OH^-.$$

A trace of alkali depresses the feeble ionisation and precipitates the solution. Acid forms soluble copper salts.

Colloidal gold behaves according to the formula

$$AuH \rightarrow Au^- + H^+.$$

A trace of acid here depresses ionisation and precipitates. Alkali has no precipitating action.

Glass furnishes a more complicated case. It is charged negatively in contact with water, and since it contains silicates of low solubility its equation may be written

$$Si_n Na + H_2O = Si_n^- + H^+ + NaOH.$$

---

[1] Wood and Hardy, *Proc. Roy. Soc.* (1908).

The first quantity of acid added to the water fails to destroy the charge because it reacts with the liberated alkali, but excess acid should destroy the charge by suppressing the ionisation of the excess silicic acid left on the surface of the glass.

Where the view enunciated here seems to fail one is tempted to impute the blame to inadequate chemical knowledge, especially to ignorance of the field these skin reactions may cover. The only safe guide is to refuse to assume the possibility of a skin reaction which does not under some conditions occur in matter in bulk. Carborundum furnishes an instance. It is, according to Perrin, negative to water, and this, according to the view advocated in these pages, implies acidic properties. The existence of graphitic acids makes this possible.

If for the moment we assume that carborundum is negative because it is acidic, the fact that a trace of acid suppresses the potential difference would be due to the depression of the ionisation. But further addition of acid confers a positive potential on the solid as Perrin showed. This brings us at once to the cardinal difficulty, namely, to explain how acid or alkali can confer a charge where chemical reaction of the nature of salt formation seems out of question.

It may be said at once that the present state of our knowledge does not furnish an answer. All that can be done is to point out the relation between the influence of H- and OH-ions and other ions.

The property of discharging the condenser systems at the surface between solid and fluid, or between external and internal phases of colloidal solutions, is common to all ions whose electric charge is of the same sign as that of the phase in which they are dissolved. The higher the valency of the ion the lower is the concentration needed to effect the discharge, so that, if the discharging power be reckoned as the reciprocal of concentration, the relation is

$$R' : R'' : R''' = 1 : x : x^2.*$$

Ions whose valency is three or higher than three not only discharge the condenser system, but if the concentration is high enough will recharge the surface in the opposite sense, so that the phase in which they are dissolved now has a contact charge of the opposite sign to that of the ion.† The position of H- and OH-ions in the series therefore is exceptional only in the fact that, although univalent, they possess about the same efficiency

---

* Whetham, *Phil. Mag.* (1899), (5) XLVIII, 474.

† *Phil. Mag.* (1909), p. 583. For example: colloidal copper, the copper particles are positive, the velocity of migration in unit field at 18° being 30.4×10^{-5}. K_3PO_4 lowers the velocity until at about $17 \times 10^{-3} N$ the iso-electric point is reached. Beyond this concentration the copper particles migrate but in the opposite direction. They are now negative.

as trivalent ions, and is explained if the action of an ion be regarded as a function of mobility as well as of valency.

Burton has measured the changes in the potential difference between the particles and the fluid produced by adding salts in metal sols. When the active ion is trivalent, the sign of the charge on the metal is inverted when the concentration of added salt exceeds a certain small value. The fact that the curves show no change in direction at the iso-electric point makes it probable that the mode of action of an ion is the same, both in reducing and increasing a potential difference.

It was, I believe, Lord Rayleigh who used the phrase "incomplete chemical change" in speaking of contact difference of potential. In the preceding pages I have endeavoured to give some definiteness to the expression. The interaction is conceived as being limited to a thin shell perhaps 10 water molecules deep,[1] which includes portions of each phase and is itself therefore in a sense a separate phase maintained by opposed osmotic and electrostatic forces. The formation of the phase is traced to the relative immiscibility of one ion.

Certain cases of contact potential remain to be dealt with, namely, those in which the solid is insoluble and apparently from its chemical constitution incapable of ionising. Cellulose and water may be taken as an example. Perrin's[2] experiments show that the former is negative to water, the potential difference being of the order of 5×10^{-3} volts.

Orme Masson[3] considers that water "undergoes continuous osmotic diffusion" into cellulose, forming with it a solid solution. When one phase has mechanical rigidity and the other phase is a fluid the law of distribution can be best explained by supposing that, in the equilibrium state, the fluid forms a diffusion column in the solid so that the concentration of the solid solution decreases the further one goes from the surface.[4] If the relative mobility of the ions were the same in the solid as in pure water, there would be a tendency to a partial separation of ions of different velocities, the solid solution, or at any rate its deeper parts, having the charge of the more mobile ion. Action of this kind would, however, cause the solid to be positive to water whereas it is negative. Some other factor besides simple diffusion must therefore come in, namely, the relative specific inductive capacities of the phases.

In the state of equilibrium as many molecules of water will pass into the solid solution and leave it at any instant of time. But if, as is probable, the specific inductive capacity of the solid solution is less than that of water, ionised molecules of water entering must on the whole coalesce to

---

[1] Helmholtz's value is less than this.   [2] *Loc. cit.*
[3] *Proc. Roy. Soc.* (1904), LXXIV, 230.
[4] Travers, *Proc. Roy. Soc.* (1906), LXXVIII, 9.

neutral molecules. On the other hand, a fraction of the neutral molecules entering the water will ionise. These two actions would appear to cause an excess of ions, having a velocity normal to the common surface, to enter the water. Owing to the higher mobility of the hydrogen ion there would be as an average state a negative charge at the immediate surface and an equal positive charge on the skin of the water.

Action of this kind may occur even when the fluid does not penetrate the solid, for the neighbourhood of a mass of lower specific inductive capacity will depress ionisation in the layer of water immediately wetting it and thus lead to the same general result.

One may note in passing the possibility of a chemical interpretation. Cellulose is a carbohydrate and the members of the group, including cane-sugar, when in solution in water migrate in an electric field as though they were acidic. As aldehydes they are acidic in so far that they very readily undergo oxidation to acids containing the carboxyl group.

The last case to be considered is that in which ionic action is excluded altogether. When the fluid is non-ionising Perrin found no endosmose through a diaphragm. This, however, is not proof of the absence of a contact difference of potential between fluid and solid, if, as I take to have been the case, both were practically perfect insulators. There can be no continued flow of fluid if the conditions are such as to prevent any flow of electricity. The effect of an external field in such a case would be to displace any electric double layer on the opposite faces of the diaphragm until, by increase in the distance separating the two electricities, a potential difference between the faces is established equal and opposite to that due to the external field.

This raises the larger question of the electric charges in non-electrolytic solution. In 1903 the author completely failed to find any solute which did not migrate in an electric field when dissolved in water.[1] Purely negative results were, however, obtained with non-ionising solutes and solvents in fields up to 500 volts per cm., e.g. azobenzene in chloroform. But neither this absence of drift nor the absence of any sign of electrolytic conduction finally prove the absence of any potential difference between solute and solvent.

Electrolytic conductivity implies two things: (1) the dissociation of neutral electricity into charges on the ions, and (2) the possibility of the transference of these charges to the electrodes. Let it be supposed that each molecule of a solute is surrounded by an atmosphere of attached molecules of solvent and that each such system is sufficiently distinct from the solvent to form an internal phase such as Quincke assumes to appear in a fluid mass about to solidify.[2] Let it be further supposed that solution is

[1] Journ. Physiol. (1905), XXXIII, 326.   [2] Proc. Roy. Soc. (1906), LXXVIII, 64.

accompanied by a displacement of electricity, such that each molecule of solute forms the nucleus of a system which is at a different potential from the mass of the solvent. When, as in the case of chloroform, the solvent is non-conducting, such a solution would neither conduct nor would the solute migrate in an electric field, since it is a system capable of statical equilibrium which does not fall under the "fixed condition of continuous chemical transformation".[1]

If solution in non-ionising fluids be accompanied by a displacement of electricity, it should be possible to form non-settling suspensions in such fluids, which would bear the same relation to molecular, non-electrolytic solution that the electrolytic colloids bear to simple electrolytic solution.

Apparently such a non-settling suspension can be realised by washing by decantation a heavy powder with sufficient changes of hot pure benzene in a quartz vessel.

[1] Helmholtz, *Wied. Ann.* (1879), VII, 337.

27

PROTEINS OF BLOOD PLASMA
WITH MRS STANLEY GARDINER

[Journ. Physiol. (1910), XL, 1P]

Plasma or serum or both of the following have been examined: ox, horse, sheep, pig, dog, rabbit and bird. From any plasma it is easy to separate three fractions. (1) A remarkable complex which is freely soluble in a very narrow range of salt concentration. It can be purified by repeated solution and precipitation with $MgSO_4$ (not $AmSO_4$, since this salt rapidly makes it entirely insoluble) and then forms a limpid colourless solution of rotation (ox) $-42°$.[1] On throwing this into a large volume of water the complex instantly decomposes into an insoluble stringy protein which appears to be ordinary fibrin, and a globulin which is wholly insoluble in saturated $MgSO_4$ and is closely associated with a large amount of a brown-gold pigment. This fraction we call the "clot fraction". It is remarkable in the fact that, though it yields a pigment when decomposed, when intact it is absolutely pigment free.

(2) Serum globulin insoluble in sat. $MgSO_4$ at 33° and in $\frac{1}{2}$ sat. $AmSO_4$, and (3) serum albumen. To separate completely (2) and (3) so that the former is quantitatively insoluble in the salt concentrations given above, about seventeen solutions and precipitations are needed. One then has a sky-blue fluid absolutely free from serum pigments. The globulin set free by the decomposition of the clot complex differs from serum globulin in having pigment associated with it. It is very doubtful whether it corresponds with any fraction of serum globulin. In serum of horse, globulin forms 60–70 per cent., albumen 30–40 per cent. of proteins.

Each fraction (serum globulin and serum albumen) when quantitatively purified is a complex, the former of globulin and cholesterin esters (± 5 per cent.), the latter of albumen, cholesterin esters (± 10 per cent.) and two pigments, one lemon-yellow freely soluble in alcohol, ether or chloroform; the other gold-brown and most intractably insoluble in all solvents when once separated from the protein. Neither pigment shows bands, both absorb the light of short wave-length.

For a colloidal solution the grade of solution of serum is very high, yet the solution complex decomposes into one substance soluble in water, namely albumen, and six substances insoluble in water, namely globulin, three cholesterin esters and two pigments.

[1] All rotations are for light of wave-length 656·3 $\mu\mu$.

In order to study the proteins it was necessary to eliminate the esters and pigments. This was done by the following procedure. Serum or plasma is cooled to 0°, added to several times its volume of alcohol [or acetone] previously cooled to −8°. The precipitate filtered off at 0°, washed with dry cold ether at 0° until the alcohol or acetone is removed; thoroughly extracted with boiling ether, filtered off and dried over H_2SO_4 *in vacuo.* The product is a white powder, freely soluble in distilled water, which clots normally if from plasma, is physiologically unchanged in, *e.g.,* the fact that any antitoxic value is preserved, and shows only the change in rotation to be expected from removal of the esters. A solution can be fractionated by salt just as can the original plasma or serum, but, in the absence of esters, the serum globulin is wholly soluble in saturated NaCl, otherwise the fractions have normal solubilities. Owing to its insolubility in alcohol, ether, etc., the gold-brown pigment is not removed by this procedure. The product when dissolved shows the normal alkalinity of serum, which therefore is not due simply to alkaline carbonates.

In order to save duplication of figures the remainder of the communication will deal only with horse serum.

The albumen fraction from normal serum was crystallised, as many crops as possible being taken.

Crystals are yellow and contain both pigment and esters. Rotation from different horses −42·5, −42·6, −42·8, −42·8: mean −**42·6**.

Mother liquor rotation varying from −40·9 to −24·8, rotation falling as successive crops of crystals are taken out, finally rising.

This difference in rotation is wholly due to two causes: (1) the excess of esters in the mother liquor, and (2) the reaction. It does not indicate the presence of two proteins, for after removal of the esters we have

$$AmSO_4 \begin{cases} \text{Crystals} & -44·5, \quad -45·5, \quad -44·9, \quad -44·9 \quad \text{mean } \mathbf{44·9} \\ \text{Mother liquor} & -43, \quad\quad -42·5, \quad -42·5 \quad\quad — \quad\quad \text{mean } \mathbf{42·7} \end{cases}$$

The small difference is due almost solely to the difference in reaction of a solution of crystals and the mother liquor. In the process of crystallisation as Inagaki showed, the salt present is decomposed ($AmSO_4$, $MgSO_4$, or Na_2SO_4), the crystals taking the acid and leaving the mother liquor relatively alkaline. Inequalities in the readings for crystals are also due to variation in acidity. In the absence of lipoids 70 per cent. of the albumen crystallises out.

The rotation of the albumen is a function of the reaction of the solution:

	Faintly acid (HA)	Faintly alkaline (Am)	Diff.
Crystals containing esters and pigment	−42·6	−39·4	3·2
Ester free	−44·9	−41·4	3·5

The difference is not due to production of acid or alkali albumen since it occurs in about 91 sec. and is reversed by reversing the reaction. Further, the crystalline form varies, that of the alkaline phase being a blunt oval, that of the acid phase fine needles in clusters.

The globulin fraction presents many difficulties. Whether esters are present or absent it can readily be split by salt fractionation into a fraction of lower solubility which is opalescent in solution and gummy on the filter-paper (1), and a fraction which is limpid in solution and granular on the filter-paper (2).

Usual rotations: (1) very difficult to read $\pm -35 \cdot 8$,
                          (2) varies from -34 to $-40 \cdot 6$.

It was noticed, however, that (1) appeared more readily when great care was not taken to fractionate with carefully neutralised $AmSO_4$. Fractionation conducted throughout at from $0°$ to $-5°$ in neutral medium resulted in a series of fractions, the one which came down with minimal salt being now as limpid as the rest, but it clouded at room temperature in the course of 24 hours. Fractionation at a low temperature with neutral solutions gave products all of identical rotation, namely $-38 \cdot 5$. We are therefore not satisfied that there exists two globulins in intact serum.

As in the case of albumen the rotation of globulin is a function of reaction, being higher in presence of a trace of acid and lower in presence of a trace of alkali. It varies also with the salt present, being higher with $MgSO_4$ [-38] and lower with $AmSO_4$ [-36].

Effect of acid and alkali on rotation. Serum globulin, serum albumen, and crystalline egg albumen agree in the fact that the rotation is immediately (90 sec.) raised by a trace of acid, lowered by a trace of alkali. The rotation is a function of the strength of the acid or alkali and of time.

The effect of time may be illustrated as follows:

Ester-free serum albumen. Rotation $-44 \cdot 5$. Trace of ammonia added until reaction just alkaline to litmus:

Time after adding alkali	90 sec.	15 min.	45 min.	20 hours
Rotation	$-42 \cdot 8$	$42 \cdot 2$	$42 \cdot 0$	$41 \cdot 9$

Ester-free globulin. Rotation $-37 \cdot 0$. Ammonia added as above:

	90 sec.	90 min.	20 hours
Rotation ...	$33 \cdot 4$	$35 \cdot 66$	$36 \cdot 7$

The characteristic secondary rise is probably not due to decomposition of the protein, since the electric conductivity remains constant.

The case of albumen would be satisfied by the hypothesis that the rotation of the free protein was greater than that of the compound with either acid or base. That is

$$A = x\alpha_{protein} - (1 - x)\,\alpha_{protein\ acid\ or\ alkali},$$

where A is observed rotation, α the rotation of the protein itself, and x the fraction hydrolysed.

But the behaviour of globulin is not explained by any such simple dissociation hypothesis. It suggests the presence of two dynamical isomers.

28

NOTE ON THE SURFACE ELECTRIC CHARGES
OF LIVING CELLS

With H. W. HARVEY

[*Proc. Roy. Soc.* (1911), B, LXXXIV, 217]

The movement of free living cells suspended in a fluid through which an electric current is passing towards one or other of the poles has been described by many observers. In almost every case the movement has been observed in thin films of fluid under a cover-glass mounted in the way usual for microscopical examination. The cells do not always all move in the same direction; some migrate towards the anode, others to the cathode, and Thornton[1] found that in mixed suspensions of diatoms and amœbæ, or yeast cells and red blood corpuscles, the animal cells migrated to the anode, the vegetable cells to the cathode. He infers from this that animal and vegetable cells are oppositely electrified, the former being negative, the latter positive, to the fluid.

It is obvious at the outset that there are exceptions to this generalisation, for Becholt[2] describes a movement of bacteria towards the anode, the direction being reversed after agglutination. Dale[3] and Lillie[4] also have described movements of animal cells to the cathode, but Thornton points out with some justice that in these cases the cells were not in their normal habitat.

The objection does not, however, apply to a natural culture of *Gonium*, *Vorticella*, and *Amœba*. In thin films such as Thornton used we found that the first two moved towards the cathode, while the amœbæ moved to the anode.

The movement of living cells, or indeed of any suspended particle, in films of liquid a millimetre or less in depth enclosed between glass plates is not open to the simple interpretation which Thornton places upon it. Arising from a contact difference of potential at the glass-water interfaces the upper and lower surface films of the water are dragged along in the electric field with considerable velocity on account of the ions they contain, and the flow along the boundary so produced is compensated under hydro-

---

[1] *Proc. Roy. Soc.* (1910), B, LXXXII, 638.
[2] *Zeits. Phys. Chem.* (1904), XLVIII, 385.
[3] *Journ. Physiol.* XXVI, 219.
[4] *Amer. Journ. Physiol.* (1893), VIII, 273.

static pressure by a return flow in the middle or round the edges of the stratum of water, if it be thick enough. The velocity of a particle past the observer is the sum of the velocity of the fluid and the velocity of the particle *through* the fluid, and since water is usually positive to glass and to particles suspended in it, the apparent velocity is commonly the sum of two velocities of opposite sign. Both movements, that of the water and that of the particle, are remarkably dead beat in thin films. If the bodily flux of fluid throughout the whole thickness under one field of the microscope were zero, and the stream lines were constant, the average velocity of particles taken through this entire thickness would, of course, give the mean velocity relative to the fluid.

When a suspension of yeast cells and red corpuscles in isotonic sugar is observed in a U-tube wide enough to reduce the flux of fluid practically to zero, *both* migrate to the anode, but the corpuscles travel much the faster. In Thornton's experiment, therefore, the yeast cells move past the observer towards the cathode because they are unable to stem the current of water which is travelling in that direction. Yeast and blood cells under the conditions of the experiment are not oppositely electrified. Both are negative to the fluid, but the yeast cells migrate more slowly.

When the depth of the fluid is increased to 2·5 mm., the yeast and blood cells are seen to move in the same direction in the middle regions and in opposite directions in the film of fluid next the glass floor of the cell, and also in that next the surface.

Troughs of various shapes were used by us to observe these movements. Good results were got with one 2 cm. long, 3 mm. wide, and 2·5 mm. deep, with parallel glass sides, which opened at each end into a wide portion, divided into two compartments by a porous plate. The outermost compartment at each end was filled with a saturated solution of zinc sulphate, into which dipped electrodes of amalgamated zinc. The fluid was not covered in any way.

The current was between 0·0001 and 0·002 ampère, and was not allowed to run in the same direction for more than a few seconds at a time. With a current of more than 0·01 ampère, the vapour density rose to a point at which it deposited on the front of the objective—a remarkable result, which can be attributed only partly to heating. The image was, as a rule, completely "fogged" by dew when the current had run 4 sec., the front glass of the objective being about 4 mm. above the surface of the fluid.

The difference in the apparent movement of red corpuscles and yeast cells in the layer next the glass, and in the middle of a stratum of liquid 2·5 mm. or more deep, is not due to a difference in the interface between the cells and the fluid, for the velocity of one kind of cell with respect to the other was the same in both regions.

Countings of the number of divisions of a micrometer scale which contiguous yeast and blood separated from each other during a run of the current gave, for the layer next the glass, where they moved in opposite directions, 0·79 division per second, and for the middle region, where they moved in the same direction, 0·81 division per second—an agreement within the limits of error of observation.

Electrification of its surface, due to contact with the medium in which it lives, must modify endosmotically the passage of substances into or out of a living cell; one might expect, therefore, that a part of the work of the cell would be expended in controlling this polarisation. It is unfortunately difficult to get reliable information on this point.

·The fact that yeast and blood corpuscles migrate to the anode in isotonic sugar at different rates probably means that the negative charge per unit area on the red corpuscle is greater than that on the yeast cell, for, according to theory, the velocity, due to shedding of the charged fluid layer, is independent of the size or shape of a particle, provided the slip at the interface be small compared with the dimensions of the particle.[1] This last condition is usually held to be fulfilled by solid particles of finite size, but it must be remembered that the interface between the enormous molecules of living matter and a fluid possibly differs widely in its properties from that between inert solid and fluid. Some features in the transport of fluid through living membranes seem to point to a very high coefficient of slip.

Another difficulty is that observation must be on cells in their natural habitat. Yeast and blood corpuscles in isotonic sugar solution are not in an indifferent medium, which leaves their properties unchanged. Isotonic sugar solution washes electrolytes out of muscle fibres, for instance, and so induces paralysis. The diffusion of salts out of yeast and blood corpuscles will polarise the surface to an extent determined by the osmotic properties of the surface and the nature of the salts.

The effect of poisons may be explained in this way. Chloroform, toluene, or traces of mercuric chloride reverse the sign of the charge on living cells, a second reversal, that is a return to the original charge, occurring after 2 or 3 days. The death change, however, is known to be accompanied by the liberation of salts, which previously were not "free",[2] and the change in the polarisation of the surface may be referred to the diffusion of such salt out of the cell. The electrification of the surface certainly does not depend upon the intactness of the cell, for fragments of yeast cells broken up by pounding in a mortar moved in the same way, and at much the same rate, as did intact cells.

[1] Lamb, *Brit. Assoc. Report* (1887), p. 501.
[2] Macdonald, *Proc. Roy. Soc.* (1905), B, LXXVI, 322; Macallum, *ibid.* (1906), B, LXXVII, 165.

In spite of this, we incline to the view that the surface charge does vary with variations in the state of activity of the living cell, for in a natural mixed culture of *Gonium*, *Vorticella*, and *Amœba*, the fact that different cells of *the same species* migrated at different rates was very noticeable. The observations were made in the water in which the cells had been living, exposed to air, so as to leave the respiratory exchange normal. Red blood corpuscles are living cells, with very slight or no intrinsic chemical activity. In correspondence with this, they were found to migrate in blood serum to the anode at a remarkably uniform rate.

Contact potential at the free surface of water. When finely powdered graphite was sprinkled upon distilled water contained in the observation cell already described, and the current, led through non-polarisable electrodes, was not more than 0·002 ampère, the following phenomena were noticed: Of the graphite particles some broke through the surface of the water and sank slowly, others floated unwetted; the latter therefore served as an index of the movements of the actual skin. Except near the upper and lower surfaces the graphite particles migrated to the anode, just *below* the free surface and just above the glass they migrated to the cathode. The unwetted floating particles either did not migrate at all, or performed relatively slow irregular movements, which were not reversed on reversing the direction of the current and were due to heating. The movements of the particles contained within the water were dead beat, and reversed with the current. We may take it (1) that the actual surface skin is not propelled at all, or so slowly that the movement escapes detection in a period of, say, 5 sec., during which submerged particles immediately below have hurried half across the field of view; (2) that the layer of fluid immediately below is driven by the field past this skin in the same direction and with the same order of velocity as the water past the glass. If additional proof of this were wanted, it is to be found in the fact that yeast and red blood corpuscles move in opposite directions in the layer immediately below the free surface, just as they do in the layer next to the glass, and for the same reason, namely, because the more slowly migrating yeast cells are unable to stem the current of water.

The stationary layer is exceedingly thin. With oc. 4, ob. B, focused on the floating graphite, submerged particles showing rapid movement are scarcely out of focus, and the spectacle produces a remarkable impression of the presence of a tenacious skin which has sufficient rigidity to act as a relatively fixed layer past which the subjacent water is being driven.

The flow of water in electric endosmose is due to "relatively enormous electric forces acting on the superficial film, and dragging the fluid (as it were) by the skin through the tube".[1] At the free surface of a fluid, there-

[1] H. Lamb, *Brit. Assoc. Report* (1887), p. 501.

fore, there must be relatively[1] enormous forces dragging the surface skin and the water in opposite directions if the movement of the water be due to a difference of potential between it and a surface film of impurities condensed from the air or neighbouring solids. The only escape from this conclusion is that the movement of the water is due to a circulation produced by the endosmotic movement of the layer touching the glass, but any compensating circulation would be opposed in direction to the flow at the glass face, whereas the surface flow is in the same direction—it is, in fact, precisely what it would be if the air and surface film were replaced by a plate of glass.

It seems difficult to avoid the conclusion that the film acts in the electric endosmose as though it were a rigid solid, and its properties are the same, when all ordinary precautions are taken to avoid contaminating the surface, as when a very thin layer of oil is allowed to spread over the water.

If the surface film really acts, as it would seem to, as a fixed layer past which the water is driven, since the stresses would be purely tangential, it is only necessary to regard it as having tenacity and as being anchored all round to the unwetted glass walls, and the apparent tenacity of the film will be partly true tenacity due to the forces between its component molecules and partly due to the work needed to rupture the film and expose a fresh water-air interface.

When the floating particles move at all, the movements are slight, irregular in direction (that is to say, they may be at an angle to the stream lines), and the direction is not reversed when the current is reversed. When the electrodes are placed directly in the distilled water, so as to cut out the large resistance of the end-plates of porous earthenware, and the current thereby increased to 0·01 ampère or more, these movements are more rapid, and the submerged particles also now move in the same general direction as the floating particles, and their movement ceases to reverse when the electric field is reversed.

These movements, at first sight puzzling, admit of a very simple explanation. In the first place the direction is determined by the trough used and not by the current. That is to say, if the particles move from right to left no matter how the current is running, and the trough is displaced end for end, they now move from left to right. If we regard the gain of heat per unit of time from the current as being symmetrical with respect to the electrodes, the observed effects would be produced by an unequal *loss* of heat at the two ends of the trough, due to the disposition of the materials, to differences in their specific heat, or to asymmetrical conductivity of heat. The result would be an unequal rise of temperature in the two halves of the chamber, and consequent differences in surface tension. If this explanation

[1] Relative, that is, to the surface stresses in ordinary flowing due to differences of hydrostatic pressure.

be correct, though the direction of the movement of submerged particles is independent of the direction of the current, the velocity past the observer should vary. This was found to be the case.

An analysis of the movements of the floating particles based on this hypothesis shows that in stronger fields the surface skin itself is dragged along. The following is an example: Field approximately 35 volts per centimetre. Movement of floating particles always towards the right, but by reversal of current the velocity towards the cathode was 2,* towards anode 10. The drift due to heating therefore was 6, and the migration 4 divisions per second, and the latter was towards the anode. The surface film therefore was negative to the subjacent water.

Appendix
The Electrification of Surface Films

The observations recorded in the preceding paper upon the endosmotic drift of the water in contact with a surface film involving foreign matter throw some light upon the range of molecular attraction. Under the conditions of the experiments, and for the short periods during which the current was on, it may be taken that there was no sensible hydrostatic pressure established due to change of level between the two ends of the trough. Under these conditions, if u be the velocity of the water past the anchored surface film, we have

$$u = \frac{d\phi}{dx} \, \sigma \, \frac{1}{\gamma}, \dagger$$

where σ is the electric density, and γ is a coefficient of sliding friction of water over the film.

The surface film acts as a thin sheet past which the fluid can flow, just as when the thickness of a soap film exceeds the range of molecular forces the interior mass may flow past the surface films which act as fixed boundary walls.

Whatever view be taken of the physical significance of the coefficient γ, it must be related in some simple way to the forces of attraction of the water for the superficial film. So long as the depth of the effective film is greater than the range of the molecular forces, the attractive forces across the interface will be constant for films of the same composition and at the same temperature. When the thickness of the film is less than this range the Laplacian pressure at the interface, and therefore γ, must diminish and the velocity of the water under unit electric field increase.

The most probable assumption is that γ varies directly with the intrinsic pressure at the interface.

---

* Measured in divisions of the micrometer scale.
† H. Lamb, *Brit. Assoc. Report* (1887), p. 495.

Let the attraction of a molecule of water upon a molecule of the film be $mm'\phi(f)$, where f is the distance between them. Then, if z be the depth of the film and dz an infinitely thin plate, the attraction of the whole mass of the water on the film is

$$2\pi m\rho \int_z^\infty \pi(f)f\,df, \quad \text{where} \quad \pi(f)=\int_f^\infty \phi(f)\,df.$$

The density of the water may be taken as uniform. The density of the film will vary rapidly. Call its density ρ' and let

$$\psi(z)=\int_z^\infty \pi(f)f\,df.$$

The pressure at the interface will now be

$$2\pi\rho \int_0^z \rho'\psi(z)\,dz.$$

Leaving out of account for the moment the variation of density $\rho'(dz)$, and putting ρ equal to unity, we have

$$p=2\pi \int_0^z \psi(z)\,dz, \quad \text{which is equal to} \quad 2\pi\left[z\int_z^\infty \pi(f)f\,df+\int_0^z \pi(f)f^2df\right].$$

Putting $\pi(f)=K\beta^{-1}e^{-\beta f}$ as an analytically simple hypothesis, this integral reduces to

$$p=2K\pi\beta^{-4}[2-e^{-\beta z}(z\beta+2)],$$

where p is the pressure at the interface.

Rücker[1] gives as the range of molecular attraction $50\,\mu\mu$. The estimate is based upon measurements of the thickness of soap films made by himself in association with Reinold, and upon a critical analysis of measurements by Quincke and others. If β be put equal to 10^6 the force is approximately $\frac{1}{3}$ at $10\,\mu\mu$, $\frac{1}{150}$ at $50\,\mu\mu$, and vanishingly small at $100\,\mu\mu$. Thus $\beta=10^6$ approximates closely to Rücker's estimate.

With this value, and on the assumption stated above, I find that the pressure at the interface would vary as follows:

Depth of film, in $\mu\mu$	% of maximal pressure
50	95
40	90
30	80
20	66
10	45
5	25

The pressure would, however, not increase so rapidly as this with increase in the thickness of the film, owing to the variation of density in the film itself which is not taken account of in the above calculations. This is obvious

[1] *Journ. Chem. Soc.* (1888), p. 222.

when we remember that the film as it gains in thickness also gains in mean density owing to compression by the increase in the mean Laplacian pressure. The compressibility of the film will be relatively great, since it is a transition layer between gas and fluid.

Turning to the observations themselves, the film was always so thin as to produce very slight effect upon the movements of shreds of camphor. This is what might be expected, since the distilled water was drawn from the bottom of a large glass reservoir, and all the chambers were thoroughly rinsed. From Rayleigh's measurements of such films[1] the thickness may be put with tolerable certainty as less than $2\,\mu\mu$—probably $1\cdot5\,\mu\mu$.

At $2\,\mu\mu$ the interfacial pressure will be considerably less than 10 per cent. of its maximal value, and the coefficient of sliding friction γ should have diminished proportionately.

Therefore, for the same potential gradient, the velocity of the water past the film should be much greater than it would be past a film $100\,\mu\mu$ thick, or past the glass if we assume that the layer of electric density at the glass-water interface does not differ widely from that at the film-water interface.

So much for theory. Observation shows that the velocity of the water past the surface film differs very slightly from that past the glass at the bottom of the trough. Thus a surface film of a thickness far below the accepted estimate of the range of molecular action acts like a mass of solid of, relatively, infinite thickness.

In considering this surprising result the three variables on which the relative velocity at an interface depends have to be remembered. The external electric field being taken as the same in all cases, they are: (1) the electric density at the interface, (2) the coefficient of sliding friction (γ), and (3) variations in the attraction of water for different substances.

Taking these in the order mentioned, so far as I know it the literature of electric endosmose without exception supports the view that the electric density on surfaces in contact with water varies within narrow limits. The velocity of a submerged visible particle is independent of size and shape, and varies directly with the electric density on the particle. It was easy in our experiments to see chance fragments, motes of dust, and living cells, travelling with velocities which agreed to within 1 or 2 per cent. The evidence, therefore, is in favour of the view that the electric density at the film-water interface did not differ much from that at the glass-water face.

By hypothesis γ and $\phi\,(f)$, the coefficient of sliding friction and the intermolecular force, are dependent variables. If the thickness of the matter on each side of the interface exceeds the range of molecular

[1] Proc. Roy. Soc. (1890), XLVII, 364.

attraction, γ varies directly as $\phi(f)$, where $\phi(f)$ refers only to the molecular attraction across the interface.

Here, again, there is evidence that γ does not vary. Putting the external electric field at unity, the velocity of a particle is given by the equation

$$V = -\sigma/\gamma,\text{[1]}$$

that is, in particles of $1\,\mu$ diameter and upwards, the velocity is independent of size and shape. But if $\phi(f)$ and therefore γ were different for different substances, the velocity should depend upon the nature of the particle.

Instead of this being the case we find protein masses, metals, and motes of dust in water, all moving in unit field with velocities of from 10 to 20×10^{-5} cm./sec., and the variations within this range can be traced to the influence of the chemical nature of the particle upon the polarisation of the interface.

We are thus driven to the conclusion that the adhesion of the film to the water practically reaches its maximum when the thickness is still much less than the accepted value for the range of the molecular forces.

In the case of a small sphere at a potential different from the water urged along by an electric field, the hypothesis which has been adopted would make γ sensibly constant until the diameter of the sphere fell to about $300\,\mu\mu$, when the pressure at the interface would be about 90 per cent. of its maximal value.[2] This agrees with the fact that down to a diameter of $500\,\mu\mu$ the velocity still appears to be independent of size and shape.

It may be well, in conclusion, to emphasise the significance of the experiments. They seem to prove either that the coefficient of sliding friction between two phases is independent of the Laplacian pressures at the interface, or that the range of the molecular attraction is much less than Rücker's estimate—$50\,\mu\mu$.

[P.S. During the present hot weather, when the water in the laboratory stands at 28° C., the film was found to have diminished in tenacity to a great extent. In order to give it the same degree of fixity under electrical stresses which it possessed at temperatures between 15° and 20°, it had to be thickened with oil until a blue film was produced, which almost entirely stopped the movements of camphor.]

---

[1] Lamb, *Brit. Assoc. Report* (1887), p. 502.

[2] The integral for the pressure at the surface of a sphere in a vacuum is given by Rayleigh (*Phil. Mag.* (1890), (2), 30, 456), as $2\pi \int_0^{2r} f^2 \pi(f)\,df - \dfrac{\pi}{f}\int_0^{2r} f^3 \pi(f)\,df$. Putting $\pi(f) = \dfrac{K}{\beta}e^{-\beta f}$, this reduces to $\dfrac{\pi K}{\beta^2}\left[e^{-2r\beta}\left(4\dfrac{r}{\beta} + 8\dfrac{1}{\beta^2} + 6\dfrac{1}{r\beta^3}\right) - 6\dfrac{1}{r\beta^3} + 4\dfrac{1}{\beta^2}\right]$.

29

THE GENERAL THEORY OF COLLOIDAL SOLUTIONS

[Proc. Roy. Soc. (1912), A, LXXXVI, 601]*

A large number of observations upon the filterability and the optical properties of colloidal solutions prove that they are coarse grained and consist of particles dispersed throughout a continuous fluid phase. For this reason these solutions are commonly described as multiphase systems in which the dispersed particles form one of the phases. This description is wrong unless it be qualified by the further statement that the particles do not constitute a phase *in esse*, but only a phase *in posse*.

Consider a heterogeneous fluid such as is described above in contact with its own vapour, and let the components be two, *e.g.* gelatine and water. If the fluid really consisted of two phases there would, with the vapour, be three phases. Let us distinguish the vapour phase by v and the exterior and interior fluid (or solid) masses by $''$ and $'$ respectively. Then, since the surface enclosing the interior fluid (or solid) masses is curved, we have

$$p^v = p'' < p'.$$

The variables, therefore, are the two components, temperature, and two pressures, or five in all. Since by hypothesis there are three phases, the system has two degrees of freedom. Therefore, in order completely to define the system, it will be necessary to fix the temperature and one of the pressures, or the composition of the phases. But on this hypothesis the two fluid phases are separated by a curved surface, and this introduces a difficulty; for if the surface be freely permeable by the components the potentials and pressures in the two phases must be the same, and, as neither component is confined to the interior phase or to the interface, that phase is unstable.[1] The phase rule is applicable only to contiguous masses which are thoroughly stable, whereas the stability of the interior masses in colloidal solutions is open to question. From the fact that the state of the solution at any instant is determined not only by temperature, pressure, and composition, but also by previous history,[2] we may conclude that neither the exterior nor the interior fluid masses conform to Gibbs' criterion of stability, namely, that the value of the expression $\epsilon - T\phi + Pv - \Sigma Mm$ when ϵ, ϕ, v, m are respectively intrinsic energy, entropy, volume, and mass, and $T, P,$

[1] Gibbs, *Trans. Conn. Acad.* III, 406.

[2] In order to obtain comparable results for osmotic pressure or viscosity, it is necessary to start the system from a fixed point and allow it to proceed along a particular path.

and M constants, shall be zero for the particular phase and either zero or positive for any other phase of the same composition.

"If equilibrium subsists (in such systems) without passive resistance to change it must be in virtue of properties which are peculiar to small masses surrounded by masses of different nature and which are not indicated by the fundamental equation."[1]

There is no intention at this stage of attempting to elaborate a complete theory, and enough has been said in support of the contention that it is almost as erroneous to speak of colloidal fluids as multiphase systems without qualification as it would be to ignore their heterogeneity altogether.

The nature of the problem and the particular questions which need investigation may be brought out by considering a simplified diagrammatic scheme of a system consisting of a heterogeneous fluid in contact with its own vapour. Let it be supposed that a solution of A in B is homogeneous and forms a simple phase when the concentration of A is sufficiently small, but becomes heterogeneous by the separation of small globules rich in A and poor in B when the concentration rises above a point defined by temperature and pressure.[2] It would obviously be possible to produce the required concentration by compressing the mass in a cylinder with a piston permeable only to the solvent B. Let this be carried out and let the system now consist of vapour of B, pure B, and a heterogeneous fluid consisting of a weak solution of A in B, with globules of solution rich in A dispersed through it, and let us agree to distinguish the vapour phase by v, the pure solvent by s, the globules by ′ and the external solution by ″. The piston which separates solvent from the complex solution is maintained in position by a force which is obviously equal to the osmotic pressure of the solution. Call this pressure P.

Let the volume of the space under the piston be V, and let there be only one globule present. For a small compression which expels dV of solvent and causes the globule to increase by dv' we have

$$dv'' = dV + dv', \tag{1}$$

and, if m_1 be the mass and μ_1 the potential of the solvent, and m_2 and μ_2 the mass and potential of the solute, and t the tension and a the area of the surface of the globule, we have for the thermodynamic potential

$$(d\psi)_\theta = -p^v dV + p'' dv'' - p' dv'$$
$$+ \mu_1{}^s dm_1{}^s + \mu_1' dm_1' - \mu_1'' dm_1'' + \mu_2' dm_2' - \mu_2'' dm_2'' + d\,(ta), \tag{2}$$

an equation which is true only if the specific volumes of the components be reckoned the same throughout the fluid parts of the system.

----

[1] Gibbs, *Trans. Conn. Acad.* III, p. 159.

[2] For the limitation of this supposition in connection with infinitesimal changes of state see Gibbs, pp. 187–93.

But if the piston is completely permeable to the solvent, and the surface of the globule is completely permeable to both solvent and solute,[1] we have

$$\left.\begin{array}{c} \mu_1{}^s = \mu_1{}' = \mu_1{}'' \\ \mu_2{}' = \mu_2{}'' \end{array}\right\} ; \qquad (3)$$

and since $\qquad dm_1{}^s + dm_1{}' = dm_1{}'' \quad$ and $\quad dm_2{}' = dm_2{}'', \qquad (4)$

all the terms including these quantities disappear.

Also, if P be the pressure on the piston—that is the osmotic pressure—and P' be the excess pressure in the interior of the globule due to the tension of the surface, then

$$p'' = P + p^v \quad \text{and} \quad p' = P' + p''. \qquad (5)$$

Equation (2) now reduces to

$$(d\psi)_\theta = P\,dV - P'\,dv' + d\,(ta).$$

Dividing through by dV, and remembering that $(d\psi/dV)_\theta = P$, we get

$$P'\,dv' = d\,(ta),$$

and from (5)

$$P = -t\frac{da}{dv'} - a\frac{dt}{dv'} - p^v + p' = -\frac{2t}{r} - \frac{dt}{dr} - p^v + p'. \qquad (6)$$

In colloidal solutions there is usually a difference of electric potential between the globule and the exterior fluid due to polarisation of the interface. The electric density must be a function of the chemical nature and density of the matter on each side of the interface, since the polarisation is due to chemical and osmotic forces acting across it. Therefore, if the globule be so large as to contain in its interior a phase fully formed (that is if the radius be greater than the effective range of any of the molecules there present) the electric density will be sensibly independent of the size of the globule, being, like the pressures and potentials of the interfacial film, fixed by the phases on each side supposed constant. But when the radius is less than the range of molecular action the composition and physical properties of the globule will vary, and therefore the density of the charge on the surface will also vary.

An expression for the energy of an interfacial layer expressed in terms of tensions may be got by the method of Dupré:[2] "If T_{12} denote the interfacial tension, the energy corresponding to unit of area of the interface is also T_{12}, as we see by considering the introduction (through a fine tube) of one body into the interior of the other. A comparison with another method of

[1] The latter part of this assumption probably is not true of actual colloidal systems, for if any degree of mechanical stability be imparted to the interface, by *e.g.*, an electric charge, the great difference in diffusibility between solute and solvent (leading to great differences in the rate of transference across the interface) will operate in finite time as though the surface were only partially permeable to the solute.

[2] Lord Rayleigh, *Phil. Mag.* (1890), (5), xxx, 461.

generating the interface similar to that previously employed when but one body was in question will now allow us to evaluate T_{12}.

"The work required to cleave asunder the parts of the first fluid which lie on the two sides of an ideal plane passing through the interior is, per unit of area, $2T_1$, and the free surface produced is two units in area. So for the second fluid the corresponding work is $2T_2$. This having been effected, let us now suppose that each of the units of area of free surface of fluid (1) is allowed to approach normally a unit area of (2) until contact is established. In this process work is gained which we may denote by $4T''_{12}$, $2T''_{12}$ for each pair. On the whole, then, the work expended in producing two units of interface is $2T_1 + 2T_2 - 4T''_{12}$, and this, as we have seen, may be equated to $2T_{12}$." Hence

$$T_{12} = T_1 + T_2 - 2T''_{12}.* \tag{7}$$

Experiments with plane surfaces (see paper No. 30) suggest that the term T''_{12} consists mainly of the electric energy per unit area of interface due to electric polarisation. But as this is not established it will be well to proceed in a more general way.

Let T represent the tension which the interface would have if no polarisation occurred, then putting σ as the electric density on the surface of the globule, we have for t the actual tension

$$t = T - \phi(\sigma), \tag{8}$$

a relation which implies that the total energy per unit area of interface is fixed by the mass of the components and the temperature. From (6) and (8) we get

$$P = -\frac{2T}{r} - \frac{dT}{dr} + \frac{d\phi(\sigma)}{dr} + \frac{2\phi(\sigma)}{r} - p^v + p'. \tag{9}$$

As an approximation, the electric displacement at the interface may be represented by two concentric shells, each infinitely thin, and with electric density σ and $-\sigma$. The radius of the inner shell being the radius of the globule r, and the distance between the shells x, the energy due to polarisation is then given by

$$8\pi^2 r^4 \sigma^2 \frac{x}{r(r+x)}.$$

Putting x constant, we have

$$P = -\frac{2T}{r} - \frac{dT}{r} + \frac{8\pi^2 x r^2 \sigma}{r+x} \left[\frac{\sigma(2r+3x)}{r+x} + 2r \frac{d\sigma}{dr} \right] + \frac{16\pi^2 r^2 \sigma x}{r(r+x)} - p^v + p'. \tag{10}$$

If the solute A were an electrolyte, account would have to be taken of the electric work gained or lost in phases ' and " during the compression.†

---

* Equation (7) is rigorously true, for, if account be taken of any work gained or expended in the formation or condensation of vapour, the balance of such work is zero, if Dalton's law be assumed.

† In order to simplify the problem, any kinetic energy which the globule may possess, owing to progressive motion, has been neglected. This quantity has been experimentally investigated by Perrin for certain suspensions.

Equations (9) and (10) show that the osmotic pressure of a schematic colloidal solution such as we are considering depends upon the functions dt/dr and $d\sigma/dr$.

When r exceeds the range of molecular action, the globule contains in its interior a phase (the interior phase) fully formed, but under a pressure greater than the pressure p'' of the external phase by the quantity $P' = 2t/r$. Let us make the probable assumption that, when the globule has grown to this size, so as to constitute a phase in mass, the pressure P' has fallen so low as to cease to affect sensibly the composition or properties of the internal phase.

Since the pressure and potentials of the layer of transition between phases ' and " are fixed when those phases are fixed, dt/dr and $d\sigma/dr$ become equal to zero, and the equation becomes

$$P = -\frac{2t}{r} + \frac{2\phi\,(\sigma)}{r} - p^v + p'. \tag{9}$$

The point where r is equal to the extreme range of mutual action of any of the molecules present is therefore one of great importance. Three things therefore need discussion: they are the forms of the functions dt/dr and $d\sigma/dr$ and the range of molecular action.

The surface tension of the globule. Lord Rayleigh[1] has shown that, according to the Young-Laplace theory of capillarity, the tension of a very thin film of matter should increase as the film thickens, according to the square of the thickness. Experiment with actual films has failed to confirm this generalisation. The experiments of Reinold and Rücker[2] and of Johannot[3] prove that, for soap films in air, the curve connecting tension and thickness shows a series of maxima and minima (Fig. 1).

The tension in the case under consideration is that of a drop of liquid forming in the interior of a mass of liquid. It is therefore that of a fluid-fluid interface. If, however, the function dt/dr has maxima and minima, certain important conclusions follow. For let phase " be supersaturated with respect to phase ': by raising the osmotic pressure P_0 to P_1, the result would be the formation of a globule of tension t_1 and radius r_1. But if the system contains globules of radius $> R$, and the osmotic pressure be reduced by admitting solvent, globules of tension T_2 and radius R_1 will be formed. At pressure P', globules of radius r_1, r_2, r_3, r_4 can co-exist, but those of radius r_2 and r_4 will probably be unstable.

According to the accepted theory, the tension of a film is constant so long as the thickness is greater than twice the range of molecular action. That portion of the curve in the figure which lies between $r = 0$ and R relates to a system whose globules are so small that the radius is less than the

---

[1] *Phil. Mag.* (1892), (5), XXXIII, 468.   [2] *Phil. Trans.* (1886), CLXXVII, 679.
[3] *Phil. Mag.* (1899), (5), XLVII, 501.

range of molecular attraction. It follows, therefore, that so long as the globule substance has not the properties of a phase in mass, *i.e.* so long as $dt/dr > 0$, a state in which the interior mass is present as globules is stable, and the globules may be of different radii, also the osmotic pressure of the solution will not be fixed by fixing the ratio of the masses of the components and the temperature. These are the characteristics of a colloidal solution, and it is therefore obvious that a determination of the form of the function dt/dr is of fundamental importance in the theory of such solutions.

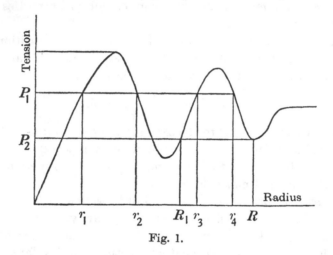

Fig. 1.

The evidence that in certain films tension varies discontinuously with thickness is complete. Reinold and Rücker determined the thickness of the black area in a soap film to be about $12\,\mu\mu$, and found that at the edge of the area the film thickened abruptly to $50\,\mu\mu$ or more. Since the tension of a horizontal film must be everywhere the same, the tension of the film $12\,\mu\mu$ thick is also that of the thicker film. The abruptness of the transition between thin and thick regions can only mean that films of intermediate thickness are unstable,[1] and that between certain thicknesses tension increases as thickness diminishes. Johannot's experiments make it probable that when the tension of the film is considered as a function of the thickness there is more than one maximum.

It is commonly held by physicists that these results obtained by the study of films of fluid immersed in air can be applied to variations of the tension of a globule of fluid immersed in its own vapour.[2] The likelihood of this being true depends upon the view that is taken of the significance of these maxima and minima.

---

[1] Reinold and Rücker, *Phil. Trans.* (1886), CLXXVII, 679.

[2] Thomson, *Conduction of Electricity through Gases*, 2nd edition, p. 183; Thomson and Poynting, *Properties of Matter*, p. 168.

If they are due to something fundamental and fixed in the nature of the field of force about an atom or a molecule, they will appear in all expressions for the energy of a mass or layer of matter whose dimensions are less than the range of the force.

Maxwell[1] showed that if the mutual attraction between the molecules of a fluid changes to a repulsion when the distance separating the molecules is less than a certain quantity, the tension of a film would begin to increase with decreasing thickness when the thickness was equal to this quantity, and Reinold and Rücker[2] suggest that the increase of tension which they found in the films $50\,\mu\mu$ thick is due to this cause. J. J. Thomson[3] finds in the corpuscle theory of atomic structures justification for the view that the extra atomic field of force has zones of attraction and repulsion. It is obvious on this theory that the field must be complex, for consider the simplest model of an atom—a central core of stationary electrons surrounded by a plane ring of rotating electrons. There will be a field due to the stationary electrons in which the electric intensity will vary inversely with the square of the distance from the centre of the atom. And there will also be an electromagnetic field due to the rotating ring in which the intensity varies along that radius which is at right angles to the plane of the ring inversely with the cube of the radius.

On purely theoretical grounds, therefore, Boscovich's conception of a field of force about an atom whose sign changes as the distance from the centre of the field increases has some justification, and, as is well known, Kelvin[4] applied this conception to capillary phenomena. The assumption that maxima and minima occur in the variation of the tension with the radius of a drop of fluid also offers a singularly simple and direct explanation of the influence of electrically charged nuclei upon the condensation of supersaturated vapour,[5] and, as we have seen, it would furnish an equally direct interpretation of the peculiarities of many colloidal systems.

Unfortunately much may be urged against the view that these maxima and minima in the tension of a soap film have any direct relation to the field about an atom or molecule.

It must be borne in mind at the outset that a drop of water immersed in its own vapour is a system composed of one component, and that variations in the tension of the drop with the radius can, therefore, be due only to the nature of the field of force between the molecules of water. That is to say, it depends solely upon the form of the function $\phi\,(f)$ in the Young-Laplace theory of capillarity. But the soap films with

[1] *Encyc. Brit.* art. "Capillarity".   [2] *Phil. Trans.* (1886), CLXXVII, 681.

[3] *The Corpuscular Theory of Matter* (London, 1907), p. 158.

[4] *Mathematical and Physical Papers*, III, 398 and 409.

[5] J. J. Thomson, *Conduction of Electricity through Gases, loc. cit.*

which Reinold and Rücker and Johannot worked contained two components at least, and were composed of colloidal solution, that is, of a peculiarly complex type of fluid. If maxima and minima did exist in the tension of a drop of pure water of radius less than the range of the mutual attraction of the molecules, it should be possible to form films of pure water just as it is possible to form films of soap solution. This is not so. A film of water breaks with very great rapidity; in fact, it behaves as it should if tension decreased very rapidly with the thickness of the film.

The repulsion suggested by Maxwell, and needed to account for the peculiarities of soap films, can perhaps be found in quite another direction. In a previous paper I described the existence of a remarkable potential difference between a film of impurity,[1] such as an oil, floating on the surface of water and the subjacent water. It occurred to me, when conducting the experiments, that if such a surface were lifted up to form a free film, as it can be lifted up by slowly withdrawing a solid frame from the fluid, it

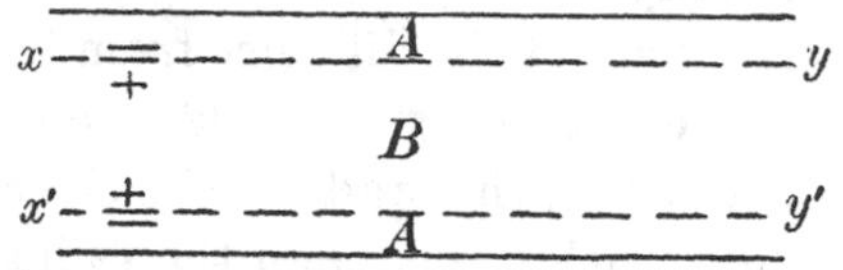

Fig. 2.

would consist of three layers, two outer ones at a potential higher or lower, as the case may be, than the interior layer. The arrangement would be such as is shown in the diagram (Fig. 2).

A, A are films of oil which may be from 0·5 to 10 $\mu\mu$ thick, and at the interfaces $x'y'$, xy, a constant difference of potential is found due to polarisation of the matter on either side.

Let the fluid B drain away. The positively charged portion of the fluid on either side is of finite thickness, owing to molecular movements. When B has drained away so that the extreme oscillations of the positively charged molecules at x are brought into the electric field about the positive molecules at x', there will be a repulsion. If the film thin further it must first entirely reconstruct itself. The first state may be that of soap film showing Newtonian colours; the second, that is after reconstruction, may be that of a black film.

This theory finds justification in a study of the actual films. The experiments and apparatus will be described more fully in a later paper. For the present it will be sufficient to refer briefly to two typical cases.

A film, 1 × 2 cm., of soap, starch, or saponin solution in water, with air

---

[1] *Proc. Roy. Soc.* (1911), B, LXXXIV, 220; compare also Reinold and Rücker, *Phil. Trans.* (1893), A, CLXXXIV, 527.

on each side, was so arranged that it could be inspected under the microscope while a constant current was passed through it. Lycopodium dust, or specially prepared French chalk, was dusted on to the film, and also immersed in it, and it was noticed that the unimmersed particles moved in the electric field in a direction opposite to that of the immersed particles.

Measuring movement in the divisions of a micrometer eyepiece covered in 5 min., I found:

	Direction + −	Distance
Saponin-water film:		
Unwetted particles on the surface	—	0
Middle of the film...	→	12
Starch-water film:		
Unwetted on surface	→	10⎫ 15⎫
Immersed	←	10⎭ 7⎭

That is to say, it was possible to observe particles at different levels in the film moving in opposite directions at the same time, the direction of the movement of each particle being reversed on reversing the direction of the electric current. These observations can, I think, mean only one thing, namely, that such films are composed of layers between which there is a contact difference of potential.

Let us agree to call the various surfaces of such composite films interior and exterior surfaces. There will then be in the case of a film of fluid in air two exterior and two interior surfaces, and in the case of a film of one fluid on the surface of a mass of another fluid two surfaces, one interior and one exterior. In the same way there will be in the former case one interior and two exterior layers, and in the latter an exterior layer and an interior mass.

When all the layers have a thickness exceeding twice the range of action of any molecules present in the layers, the sum of the tension of the layers and the electric density at the interfaces will be constant and independent of a variation in thickness of any or all the layers. When the thickness of all or of any one layer is less than this magnitude the tension and the electric density vary, and may be considered functions of the thickness. It is of importance, therefore, first of all to determine whether the range of molecular action is a fixed value or whether it is dependent upon the chemical nature of the molecules concerned. The question is considered in the next paper.

30

THE TENSION OF COMPOSITE FLUID SURFACES AND THE MECHANICAL STABILITY OF FILMS OF FLUID

[Proc. Roy. Soc. (1912), A, LXXXVI], 610]

The tension of a composite surface composed of a fully separated layer of one fluid (say oil) spread over a mass of another fluid (say water) is given by the equation

$$T = T_A + T_{AB}, \qquad (1)$$

where A and B denote respectively the oil layer (exterior layer) and the water (interior mass).

If z be the depth of the exterior layer, then, for a given temperature θ, the quantity T_A will reach a constant value when z is either $>$ twice the range of action of the oil molecules on each other, or $>$ the sum of two values, namely, the range of action of oil molecules on each other, and of the oil molecules on the water molecules. Let Z be the lowest value of z for which T_θ is constant: $\frac{1}{2}Z$ is equal to the greater of two quantities, the range of action of the molecules of oil, or the mean of this value plus the range of action of these molecules upon the molecules of water.

The attractive force between two molecules in the theory of capillarity is a force which decreases rapidly as the distance between the molecules increases. Young assumed the force to be negative, that is to say, a repulsion, over a very small distance, positive beyond this distance, and constant until it vanished at a distance large in relation to the depth of the zone of repulsion, but very small relatively to distances directly appreciable by our senses. Maxwell assumed the force to vary inversely with the fifth power of the distance. Whatever the form of the function may be, that which is measured when tension is taken as an index of the range is not necessarily the absolute limit of the field of force about a molecule, but the limit of its sensible action.

It can be proved that

$$T_{AB} = T_A + T_B - 2T'_{AB}, \qquad (2)$$

the term $2T'_{AB}$ being the work gained per unit area when a surface of A is allowed to approach normally a surface of B.[1] The full expression for the tension T of a double surface is therefore

$$T = 2T_A + T_B - 2T'_{AB}. \qquad (3)$$

---

[1] See the preceding paper, p. 502; also Lord Rayleigh, *Phil. Mag.* (1890), (5), xxx, 460.

This expression, and the equation $T = T_B$, are the values of T for the limits $z = Z$ and $z = 0$ respectively, and it should be possible to pass from one to the other without a break by gradual increments of the fluid A. For all thicknesses of $A > Z$, T is constant, therefore it would appear that Z could be determined by measuring the minimal thickness of A for which T is constant.

Obviously A must be a pure substance, for if it contained an impurity whose molecular relations to B were such that it reduced the tension of B more effectively than pure A, the tension T_{AB} would be constant and minimal only when the mass of the external layer (Aa) per unit area was great enough to saturate the interface with a.

The conditions which determine the spreading of one fluid over another, or rather between two others, one of which is air, may be stated as follows: Denoting the third fluid, air, by C, then at the edge of a flat drop of A three forces have to be resolved due to the three tensions. "If the three fluids can remain in contact with one another, the sum of any two of the [tensions] must exceed the third, and by Neumann's rule the directions of the interfaces at the common edge must be parallel to the sides of a triangle, taken proportional to T_{AB}, T_{BC}, T_{CA}. If the above-mentioned condition be not satisfied, the triangle is imaginary, and the three fluids cannot rest in contact, the two weaker tensions, even if acting in full concert, being incapable of balancing the strongest. For instance, if $T_{CB} > T_{AB} + T_{AC}$, the second fluid spreads itself indefinitely upon the interface of the first and third fluids."... "The experimenters who have dealt with this question—Marangoni, van der Mensbrugghe, Quincke—have all arrived at results inconsistent with the reality of Neumann's triangle. Three pure bodies (of which one may be air) cannot accordingly remain in contact. If a drop of oil stands in lenticular form upon a surface of water, it is because the water-surface is already contaminated with a greasy film."[1] Further, the Young-Laplace theory of capillarity leads, according to Lord Rayleigh, "to the important conclusion, so far as I am aware hitherto unnoticed, that, according to this hypothesis, Neumann's triangle is necessarily imaginary that one of three fluids will always spread upon the interface of the other two". The preceding pages present what I believe to be the current theory of the tension of composite surfaces; my own experiments prove that it is too narrow for the facts.

When some ricinoleic acid (Kahlbaum) is placed upon a clean surface of water the quantity first added spreads with great rapidity, but when the film is somewhere about $40\,\mu\mu$ thick, any further quantity added refuses to spread, but remains permanently gathered into a lens. When more acid is added to the lens it enlarges until the whole surface is covered with a layer of the oil which has been forced over it by gravity. Clearly, therefore,

[1] Lord Rayleigh, *Phil. Mag.* (1890), (5), xxx, 463.

whereas the first added oil spreads because it lowers the tension, the sign of the effect changes at a certain stage in the thickening of the layer A, further thickening increasing the tension. The spreading of the oil is therefore resisted, and a lens is formed. With some substances (olive oil, cymene, heavy oil), before the stage is reached at which a drop refuses to spread, small lenses, just visible to the naked eye, appear dotted all over the surface. In the case of cymene these tiny lenses are just visible when the film A is of an average thickness of $100\,\mu\mu$, calculated on the assumption that its density is that of cymene in mass. The phenomena can be followed in detail by adding successive drops of "heavy oil"[1] at the same spot. The result is a patch which is bounded by rings. Each drop allows the one last added to contract to a ring. The rings are of the same breadth when the drops are of equal size. The outermost ring of all—very visible when heavy oil is used—is the contracted field of contamination on the surface before the oil is added. The colour of the rings slowly changes. A drop of heavy oil spreads first to a patch of superb blue, then changes to purple with a defined yellow edge, 1 mm. wide, the patch being (say) 5 cm. across. The patch, up to now uniform, begins to become mottled, the red gives way to a beautiful pattern of peacock green, steel blue, and bronze yellow, the various colours being sharply marked off from one another even under considerable magnification. In the centre of each blue patch, which in an undisturbed surface is accurately circular, is a small lens of oil, appearing as a point only to the naked eye.

The mode of formation of these lenses is clear. Each is a droplet of oil condensed about some nucleus, for the most part solid, much more rarely a tiny bubble of gas. They appear on a surface of distilled water greased with the heavy oil when the average thickness of the film is approximately $200\,\mu\mu$. The surface except at first is therefore highly complex—coloured areas and lenses being in tensile equilibrium.

In these cases when the general surface is nearly saturated with oil, that is, when the minimum of surface energy is nearly reached, condensation occurs on any nuclei which may be present. The ultimate relation is the same as that found for ricinoleic acid, or, indeed, for all the substances experimented with (olive oil, castor oil, croton oil, ricinoleic acid, benzyl cyanide, "heavy oil", cymene and benzene): a uniform sheet of a third fluid in mass[2] can only be inserted between the fluids air and water by the operation of an external force, namely gravity.

It is obvious from equation (1) that T must have a minimal value unless the rate of decrease of T_{AB} is equal to the rate of increase of T_A. For, let it

---

[1] Distilled from Price's "Motorine A", see later.

[2] That is a sheet whose depth is $>Z$, so that the substance is present as a fully separated phase.

be supposed that oil could be spread evenly and continuously on water until it reached the thickness Z at which it forms a separate phase, the first oil spread on the surface causes the tension T_B to fall, but the last of the oil added is causing tension T_A to rise, and the rise may more than compensate for any further fall in T_B. An approximate equation for the surface may be got as follows. Consider a large lens of oil, say 10 cm. in diameter and about a millimetre in thickness. Such a lens can readily be formed. Its surface is sensibly flat, but at the edges it curves down to meet the general surface of the fluid, which is also depressed at the edge. Let the tensions of the flat surfaces be denoted by T_1, T_2, T_3. Then

$$T_1 = T_A + T_C - 2T'_{AC}, \qquad T_2 = T_A + T_B - 2T'_{AB}. \tag{4}$$

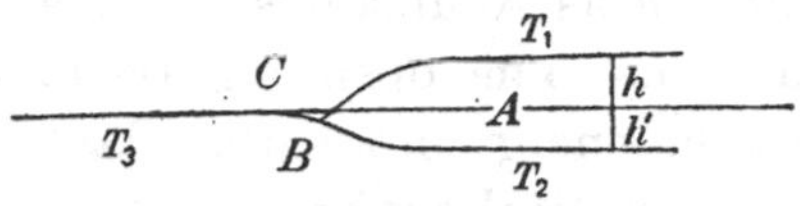

Fig. 1.

Let a be the area of a horizontal section of the lens, h the height of the upper surface above the mean level of the fluid, h' the depth of the lower plane surface below the mean level, and ρ the density of A, ρ' of B. Then for the potential energy of the drop we have, if the effects of curvature at the edge be neglected,

$$\psi = [T_1 + T_2 - T_3 - \tfrac{1}{2}g \, (\rho h^2 - \rho h'^2 + \rho' h'^2)] \, a. \tag{5}$$

All the terms in this equation are constant, except h, h', and a, which are related variables.[1] We may therefore put

$$\frac{d\psi}{da} = 0, \tag{6}$$

and

$$\frac{d\psi}{da} = T_1 + T_2 - T_3 - \tfrac{1}{2}g \, (\rho h^2 - \rho h'^2 + \rho' h'^2) - g \left(\rho h \frac{dh}{da} - \rho h' \frac{dh'}{da} + \rho' h' \frac{dh'}{da} \right). \tag{7}$$

dh'/da and dh/da are very small quantities, so that the whole of the last term may be neglected.

From (4), (6), and (7) we get

$$T_3 = T_A + T_C - 2T'_{AC} + T_A + T_B - 2T'_{AB} - Kg,$$

in which the term depending upon gravity is written Kg for brevity. The tension of the gas T_C may certainly be neglected, and probably T'_{AC}, which represents the energy per unit area of the gas or vapour condensed on to the surface of the lens.

There remains $\qquad T_3 = T_A + (T_A + T_B - 2T'_{AB}) - Kg, \tag{8}$

and the term in brackets is equal to T_{AB}.

---

[1] ρ appears only in the term relating to gravity, and may be taken as constant over that range in which this term has a sensible value.

If T_3 were equal to T_{AB} then $T_A = Kg$, that is to say the lens is maintained against the action of gravity solely by the tension of its air surface, that is by the tension of the oil itself.

But there is no necessary equality between T_3 and T_{AB}. They refer to essentially different things. T_3 is the potential per unit area of a "film of discontinuity", to use Gibbs' phrase, between water and air whose tension is reduced by a concentration Γ per unit area of a component A. T_{AB} on the other hand refers to a surface of discontinuity between A and B both in mass.

In the application of capillary theory to the solution of the equilibrium of three fluids it is usual to consider the tensions as forces which meet at a point. The theorem known as Neumann's triangle may be quoted as an instance. The method seems to me open to grave objections. It is true, as Gibbs has shown, that the energy of a surface layer may be equated to a strain limited to a mathematical surface, the "surface of tension", and, therefore, at the meeting-place of three fluids these surfaces will intersect in a line, but since the range of action of molecular forces is finite the tensions of the surfaces cannot be the same for a finite distance from the line of intersection, the distance being the greatest distance at which any of the molecules present act upon one another. The alternative assumption is absurd, namely, that there is no mutual attraction between the material composing the three films right up to the mathematical intersection of the "surfaces of tension".

At the edge of the lens there is equilibrium, so that

$$t_3 = t_1 \cos \theta_1 + t_2 \cos \theta_2; \qquad (9)$$

but t_1, t_2, and t_3 must not be identified with T_1, T_2, and T_3.

All interfaces which I happen to have observed have been highly polarised, the work T'_{AB}, therefore, in those cases must include at least two terms, one for the work per unit area needed to produce electrification of density σ and $-\sigma$, and another for the work due to simple molecular attraction. If the sum of all the molecular forces which contribute to the self-attraction of the matter be included in the phrase "molecular attraction", the electric energy ceases to be a separate term. But if any attempt be made to distinguish between the kind of attractions, e.g. chemical or physical, two terms must be employed.

The actual phenomena are undoubtedly better represented by two terms. Consider the tension T_2, which is equal to T_{AB}. We have, as before,

$$T_{AB} = T_A + T_B - 2T'_{AB}.$$

In forming this surface, by allowing unit areas of surfaces A and B to approach normally, let the work $2T'_{AB}$ be gained in two steps, that is to say, when the surfaces come into contact, let work be first gained by the operation of the mutual attraction of unchanged molecules of A and B; call this work per unit area $2T'_{ab}$. The molecules then proceed to alter one another so that polarisation occurs and work per unit area is done equal to $f(\sigma)$. Since the necessary condition of equilibrium of a film of discontinuity is that the tension shall be "less than that of any other film which can exist between the same homogeneous masses",[1] we may conclude that the effect of polarisation is to make the tension less. We have, therefore, the relation

$$2T'_{AB} = 2T'_{ab} + f(\sigma) \qquad (10)$$

and

$$T_{AB} = T_A + T_B - 2T'_{ab} - f(\sigma). \qquad (11)$$

Owing to molecular movements, the polarised molecules occupy layers of finite thickness, but we may with Helmholtz consider the two electricities as distributed on two parallel surfaces as in a condenser, the plates of which are separated by a very small distance x and have an area a. Polarisation will be complete, that is, the electric density will be maximal, only when the film of A shall have reached a certain thickness, which may or may not be the same as the extreme range of attraction of the molecules of A and B. In other words, σ and x and, therefore, the electric energy of the surface, are functions of the mass of A per unit area.

$$f(\sigma) = 2\pi x \sigma^2$$

and with (8) and (11) we have

$$T_3 = 2T_A + T_B - 2\left[T'_{ab} + \pi x \sigma^2\right] - Kg. \qquad (12)$$

It is important to realise distinctly the meaning of the terms T'_{ab} and $f(\sigma)$. The former is that fraction of the work gained when unit area of a surface of pure A approaches normally unit area of a surface of pure B due to the Laplacian attraction of molecules of pure A, say oil, for pure B, say water. On the corpuscular theory of matter, it is the term which expresses the work gained from the mutual influence of the external or stray fields of the molecules. The latter term, $f(\sigma)$, is that portion of the work, assumed to be expended entirely in producing polarisation, which is due to chemical action between the molecules. It is therefore the term which expresses the sum of the change in the internal fields of force of the molecules. If A does not react chemically with B at the interface, $f(\sigma) = 0$, and T'_{AB} is equal to the work gained by simple Laplacian attraction. This distinction may seem artificial, but it will be seen that it is needed in order to explain the actual phenomena. It must also be made in any complete specification of the intrinsic energy of some fluids. Consider, for instance, benzene and water. In the former, self-attraction, so far as we know, is entirely due to the stray field of the stable ring of atoms which forms the molecule. In the latter, a fraction of the total self-attraction, a very small fraction it is true, is due to dissociated molecules, that is, to the opening out of the internal molecular fields.

[1] Gibbs, *Trans. Conn. Acad.* III, 403.

H

It must be borne in mind in considering the polarisation of an interface between such bodies as oil and water that the phenomena may be partly due to electrolytes present as "impurities". Certainly they cannot be without influence, and equally certainly they cannot be wholly excluded (*e.g.* the carbonic acid of the air). But in colloidal solutions the charge on the colloid masses is always greatest when the concentration of electrolytes is least, and I have found that a slight leakage of zinc sulphate from the non-polarised electrodes into the observation chamber completely arrests any migration at an interface. According to our present knowledge such interfaces are discharged, not charged, by soluble electrolytes, but further experiment is both needed and difficult.

Experimental methods. The tension was determined in two ways. In the first, usually called Wilhelmj's method, the weight needed to balance the pull exerted by the fluid on a thin blade hung vertically is measured by a balance; in the second, the weight needed to detach a flat plate from the surface is measured.

Five glass blades, each 0·15 mm. thick, were mounted in a light frame so that they were parallel to each other and 1 cm. apart. The frame and blades were suspended from one arm of a balance and the level adjusted by means of a fine screw so that the lower edge of the blades was as nearly as possible at the mean level of the fluid when the pointer of the balance stood at zero. If l be the total length of the blades, m the weight needed to balance the pull, the tension is given by $T = mg/2l$. This formula assumes that the angle of contact of fluid and glass is zero. The assumption is sensibly true for clean water and nearly true for a contaminated surface. The factor to correct for the angle is its cosine, and as this decreases in value very slowly for small angles, the correction is small. But, though small, it must not be lost sight of. The effect of ignoring this correction is to make the calculated values of the tension too low when the water surface is much contaminated. The formula cannot be used to obtain the tension of a double surface A and AB, for the angle of contact of the AB surface with the glass seems in all cases to be considerable.

In prolonged experiments, it is very necessary to inspect the blades from time to time with a hand lens, to see that no "beading" has taken place owing to impurity condensing on to the surface of the glass from the air.

The formula usually given for the plate method also assumes the angle of contact between fluid and plate to be zero at the moment of breaking. The formula is at best an approximation, and the error may be considerable. As the best approximation, I used the following, $T = (m \times 503)^2/g\rho$, and the weight m was always taken as the mean of the weight which the surface would just hold, and the weight which broke the plate away. The plate method has this advantage, that it can be applied when only a small quantity of fluid is available. It was therefore used only to obtain an approximate value for the tension of the substances used to form the film on the water.

The thickness of the film was varied either by Miss Pockel's method,[1] in which the mass of the film is constant and the area varied, or by gradually thickening a film by increments of material, the surface being constant.

The ordinates of the curves in all cases are tension in dynes per centimetre, the abscissæ are the thickness of the film calculated from the weight of material placed on the surface, and the area of the surface, the density of the film being taken as that of the substance in mass. I did not succeed in weighing the minute quantities of croton oil used, and the abscissæ for this curve are therefore on an arbitrary scale, the unit 1 being probably much less than $1 \mu\mu$.

Except when the contrary is stated, freshly distilled water from a silver still was used. The trough, which was of tin plate, measured $71 \times 10 \times 1$ cm. The most satisfactory barriers were made of tin plate 1·5 cm. wide, with a vertical strip of plate slightly less than 10 cm. long, and about 0·75 cm. deep, soldered (with tin) to the middle. Each blade in transverse section was thus T-shaped. Two blades were always used, and kept about half a centimetre apart or less, so as to obviate leakage from a contracted film. Direct leakage was, I think, practically absent—at any rate, lycopodium dusted on to the cleaned surface just near the blade failed to reveal it. But, of course, the contracted and the cleaned surfaces are in communication with the mass of the liquid, and, if the substance used to contaminate the surface is at all miscible with water, solution from the contracted surface and condensation on to the clean surface will go on. The vertical strips on the blades were placed there to "wire-draw" any diffusion stream. The influence of solubility on the equilibrium of a surface needs careful consideration; for the present, it may probably be neglected owing to the extreme insolubility in water of the substances experimented with. The water surface was cleaned by sweeping any contamination to one end with one of the blades. It was then got rid of by a quick jerk, which threw the surface out of the trough. Lord Rayleigh[2] has measured the tension of clean and contaminated surfaces of water by the ripple method and by the blade method. The values obtained are more divergent the greater the degree of contamination:

	Blade method	Ripple method	Difference
Clean water	74	74	0
Camphor point	57·7	53	4·7
Saturated with olive oil	[About 48]*	41	7

* Measured by myself.

[1] *Nature* (1891), XLIII, 437; also Lord Rayleigh, *Phil. Mag.* (1899), (5), XLVIII, 331.
[2] *Phil. Mag.* (1890), (5), XXX, 398.

Owing, however, to the hysteresis of the surface, the ripple method itself cannot be relied upon to give correct results.

The sources of error in the blade method are of contrary sign, and should, therefore, to a certain extent neutralise one another. As contamination increases three changes have to be considered, (1) an increase in the angle of contact of the surface with the blades, (2) a thickening of the edge of the film of fluid on the blades, which may result in "beading", and (3) any change in the tension of the fluid where it wets the blades owing to an attraction of the solid for the oil. (1) would give the relation $T_{\text{calc.}} < T_{\text{real}}$, from (2) $T_{\text{calc.}} > T_{\text{real}}$, and from (3) $T_{\text{calc.}} > T_{\text{real}}$. So long as there is no sign of beading or irregularity when the blades are inspected by a good lens, and the angle of contact appears to be nearly zero, $T_{\text{calc.}}$ is probably not far removed from T_{real}. In order to centre attention on the main point, namely, the relation of tension to chemical constitution, the hysteresis of composite surfaces is accepted throughout this paper without criticism. There is no doubt in my mind that it is a real phenomenon of the surface and not merely an effect of adjustments at the blades, since it appears when the layer of contamination is excessively thin: also Reinold and Rücker found the properties of a film to depend upon its age.

It will be well here briefly to recapitulate Lord Rayleigh's results. The theory of the movements of camphor on the surface of water, due to van der Mensbrugghe, implies that they will take place so long as the tension of the surface is greater than that of a saturated solution of camphor. At a certain tension, therefore, which Lord Rayleigh found by the blade method to be 0·78 that of a pure water surface, the camphor movements cease.[1] To this particular tension he gives the name "camphor point". On the assumption that the density of a film of olive oil is the same as that of the oil in mass, Lord Rayleigh found the thickness of the film at the camphor point to be $2\,\mu\mu$.[2] Using castor oil, not olive oil, to form the film, Lord Rayleigh measured the tension when the thickness of the film was varied by Miss Pockel's method. His curves show that up to a thickness of about $1\,\mu\mu$ the oil scarcely alters the tension, then a rapid fall sets in until the film becomes about $2\,\mu\mu$ thick. Further addition of oil has only a slight effect on the tension. The most important and the most quoted conclusion which Lord Rayleigh draws from these curves is that "if we begin by supposing the number of molecules of oil upon a water surface to be small enough not only will every molecule be able to approach the water as closely as it desires, but any repulsion between molecules will have exhausted itself. Under these conditions there is nothing to oppose the contraction of the surface—the tension is that of pure water."

[1] *Phil. Mag.* (1899), (5), XLVIII, 334.
[2] *Proc. Roy. Soc.* (1890), XLVII, 365.

"The next question for consideration is—at what point will an opposition to contraction[1] arise? The answer must depend upon the forces supposed to operate between the molecules of oil. If they behave like the smooth rigid spheres of gaseous theory, no forces will be called into play until they are closely locked. According to this view the tension would remain constant up to the point where a double layer commences to form. It would then suddenly change, to remain constant at the new value until a second layer is complete.... If we accept this view as substantially true, we conclude that the first drop in tension corresponds to a complete layer one molecule thick, and that the diameter of a molecule of oil is about $1\,\mu\mu$."[2]

The first comment to be made on these conclusions is that the oils with which Lord Rayleigh experimented are quite exceptional in their power of reducing the tension of water, and a similar train of reasoning applied to other substances seems to lead to impossible conclusions. The molecules of heavy oil and of cymene would have a diameter of 20–$40\,\mu\mu$. Or, taking the camphor points given in the following table as the measure of a layer two molecules thick, we should have to accord diameters of from 200 to $500\,\mu\mu$ to their molecules.

Substance	Tension (approximate)	Density	Viscosity	Average thickness of film on water at camphor point in $\mu\mu$
Castor oil	28·6 at 22°	0·9351	Thick oil	1·59
Croton oil	27·8	0·9047	„	<1·59
Olive oil	—	—	—	2
Cymene	22·7	0·8700	Labile spirit	300– 600
Heavy oil	26·4	0·9086	Will only just flow	300–1000
Benzene	28·36 at 15°	0·86	Light spirit	Cannot stop the movements of camphor

Benzene spreads on a clean water surface, but a flat lens 1/10 mm. in average thickness has no observable effect upon the movements of camphor. To give this result, however, the benzene must be carefully purified by distillation and crystallisation. Impure benzene behaves very differently. Camphor is quite "dead" upon pure benzene in mass, but still active upon a water surface carefully saturated with this substance. A surface of

[1] In one sense of the word contraction it may be said that the clean surface of water does not contract at all. When the blades are moved so as to diminish the area of a surface, motes floating on it do not move at all. This is the best test I know of the cleanliness of a surface. There is therefore no tangential contraction of the surface such as occurs and is readily seen when any "skin" of impurity is present. The diminution of area is effected solely by movement of the surface layer normal to itself. So long as this kind of adjustment can take place and only so long is there no hysteresis. All the elements of the surface (cf. Gibbs, *Trans. Conn. Acad.* III, 468) are in complete equilibrium with one another throughout any changes of area.

[2] *Phil. Mag.* (1899), (5), XLVIII, 335.

Cambridge tap water is very slightly contaminated. The surface must be contracted to one-tenth its area in order to produce a readable difference of tension. On such a surface of 700 sq. cm. area, 4 c.c. of benzene poured on to it forms lenses 3–4 cm. in diameter. These lenses show regular expansions and contractions, and smaller lenses, that is those of less diameter, move across the surface and fuse with larger ones. The expansions and contractions of the lenses are due to the tension alternately falling below and rising above that of the plane surface in the process of adjustment of the tension by horizontal spreading of a sheet of benzene from a lens and its removal by evaporation. The explanation of the pulsations therefore is the same as that which accounts for the "tears" of wine.

A touch of ether (redistilled) caused the benzene-water film to contract, owing to its tension being greater than that of the ether-water film. This is the more remarkable when it is remembered that ether is fairly soluble in water (1 in 12), and that, therefore, a large part of the ether added must be removed from the surface to pass into solution. After the addition of ether the small lenses of benzene behave differently. They move as before towards the large lenses but union seems to be impossible. When only a short distance separates small from large the former burst, the nearest point of the large lens being at the same time deeply indented. The bursting of the small lens is, therefore, accompanied by a local and temporary lowering of tension.

If an "active" substance be one which lowers the tension of water when a film less than $2\,\mu\mu$ thick is spread on the surface, they fall into two groups which, in order of activity, are

Active	Relatively inactive
Croton oil	Benzene
Castor oil	Cymene
Olive oil	Heavy oil

The first group contains bodies of characteristic chemical instability, namely, the esters. They are salts of unsaturated fatty acids and glycerine. Castor oil, for instance, consists mainly of the glyceride of ricinoleic acid. The "inactive" group are as characteristically stable, the heavy oil being composed of paraffin, while cymene is a benzene derivative. Benzene, the least active, has a very stable ring structure. The simple paraffins are at least as stable as benzene and the relatively large activity of the heavy oil may, therefore, be due to the presence of the "active" impurity. The "heavy oil" was obtained by distilling Price's "air-cooled oil A" in a vacuum. The fraction which boils at about 400° was used and denoted by the term "heavy oil". Since the entire Price's oil A actively reduces tension this fraction may well contain a trace of active substance. In cymene the benzene ring is no longer simple, as the formula shows, and all such modifications of the benzene ring decrease its stability:

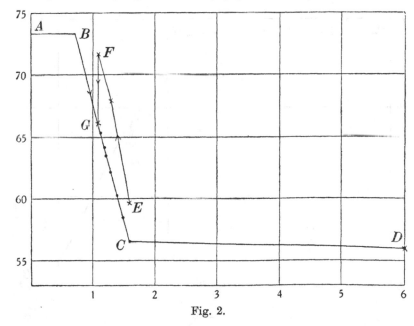

Benzene   Cymene

The chemical stability of benzene is associated with a remarkably slight influence on the tension of water. The molecules of the "active" substances, being organic salts, might be expected to polarise readily under an axial stress such as must be exerted at the AB surface.

CASTOR OIL AND CROTON OIL

The curves (Figs. 2, castor oil, and 3, croton oil) differ from those obtained by Lord Rayleigh in the sharp points of inflection. In order to

Fig. 2.

settle whether the inflection is really as abrupt as it appears to be, the first part of the curve for croton oil was mapped with great care (Fig. 4). In the figure the dots represent one series of measurements, the crosses another series, obtained in both cases by Miss Pockel's method of varying the extent of surface. Inflection at B appears to be quite abrupt, that is to say, the crosses which lie in the first part of the descending limb lie on a straight line, which would meet the part AB at an angle. The portion AB, which is

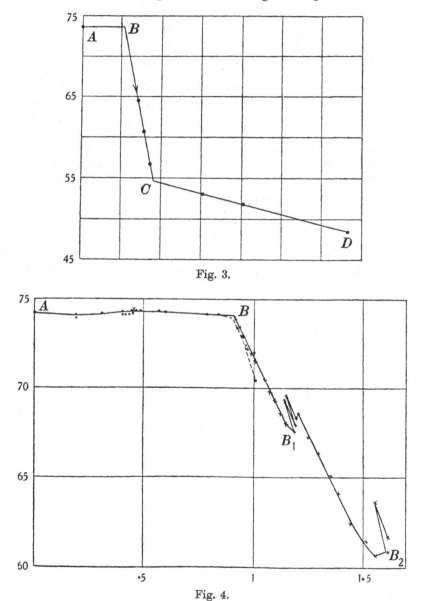

Fig. 3.

Fig. 4.

seen much magnified in curve 4, undulates, from which one may infer that the first traces of oil throw the surface into an unstable condition—obviously, it cannot be uniform, since there is not enough oil to make a layer one molecule deep, and adjustment by horizontal diffusion or spreading will be slow and difficult. In this region contraction of the surface caused a rise of tension, for instance:

Croton Oil, a trace

Barrier	Tension
68	74·1
41	74·17
31	74·27
68	74·1
31	74·17
28	74·2
22	74·2

Successive positions of the barriers are given in centimetres from that end of the trough at which the tension was being measured, and it will be noticed that, with slow contraction, the tension at 31 rose to 74·27, but, with rapid contraction, the barriers being moved quickly from 68 to 31, it rose only to 74·17.

The part BC appears to be accurately a straight line. In curve 4 it is seen to be made up of two straight lines, which are not continuous, but are joined by a cluster of points. The explanation brings out the most interesting property of these surfaces. Owing to the hysteresis, it is obvious that any accidental expansion occurring while one is mapping a curve of contraction will displace the curve—the subsequent values belong, in point of fact, to quite another curve (compare, for instance, Figs. 5 and 6). The nebulæ of points therefore represent attempts which I made to return to the original curve when I had accidentally left it. The part of the curve CD slopes very slightly downward, but, as the formula for computing the tension is probably far from exact in the region CD, the particular form of the curve is here not of great importance.

Cymene. I am indebted to my friend, Dr Ruhemann, for a sample of this hydrocarbon. By ordinary tests it is insoluble in water. The sample was of ordinary chemical purity, probably sufficiently so to give the true value for a water-cymene-air surface. Fig. 5 needs some explanation. The curves are plotted to two distinct scales of thickness of film. The upper curves are plotted to the upper scale of $0–200\,\mu\mu$, the lower to the scale of 0–2000. Three sets of measurements are recorded: the simple dots were obtained by contracting or expanding a surface with blades, the ringed dots and crosses by adding known quantities of cymene to a surface of constant area. The agreement between the results proves that solubility of cymene in water does not produce a sensible effect, at any rate, over the first part of the curve, for the trough used contained $1\frac{1}{2}$ litres of water, while for the measurements with constant area of surface only 70 cm. of water was used. The slight divergence in the later parts of the curves may be due to the difference in the mass of water employed or to the different type of surface produced by moving a barrier so as to contract the area to what is formed when successive additions of cymene are made to a surface of constant area. It has been pointed out already that a surface in contracting follows a curve different to that of an extending surface—the figure of hysteresis for cymene illustrates the point. Now, in Miss Pockel's method, a surface starts from an arbitrary point, and thenceforward is a pure contraction surface. But, when a second drop of cymene or any other substance is added, it is rapidly extended by the already existing surface, and then, after a few oscillations, comes to rest. The surface, as formed, is thenceforward composite, being partly a surface of expansion, partly a surface of contraction.

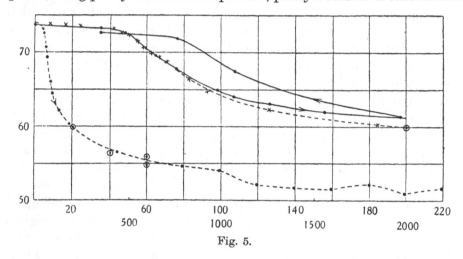

Fig. 5.

Cymene is a relatively inactive substance. With the diminished activity, the sharp points of inflection on the curve disappear, and a curve of gradually changing slope is obtained. The hysteresis is seen in the curve of expansion which follows on and is continuous with the curve of contraction.

Heavy oil. From its source may be regarded as being mainly composed of paraffins. The curves for the heavy oil are again those characteristic of an inactive substance. The slope is gradual throughout. The initial rise of tension is well marked.

A surface of clean distilled water was greased by four drops of the oil applied at intervals of 10 min. The surface was then allowed to remain quiet for 70 min. in order, if possible, to attain equilibrium. In all, 4·4 c.mm. was spread on a surface of 68,000 sq. mm. area. The result was to lower the tension to the first point indicated by a ringed dot (Fig. 6). The crossed points refer to an entirely different experiment, in which the disc method was used. The surface was contracted slowly, and the curve *AC* represents the results. At *C* it was found that the tension rose when the surface was allowed to remain quiet, and reached a stable point at D; it was then extended rapidly, the points measured being indicated by the ringed dots. It was again contracted, this time rapidly, and now followed curve EF, rising again to D, and again extended slowly, the curve being DE. The whole time occupied in plotting curve DE was 100 min., and no sign of variation of tension with time was noticed at any point. The surface was then again expanded along EF. It will be noticed that changes of direction appear at exactly corresponding points in the curve of slow contraction and slow expansion. The four drops of oil spread on the surface do not fuse.

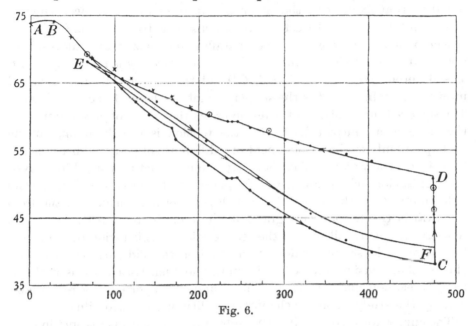

Fig. 6.

The last added drop, after 70 min., could be seen by reflected light to form a large patch with sharply defined edges. The surface is therefore composite, and the breaks in the curves may provisionally be referred to this.

The question arises whether the curve of contraction AC would not be identical with the curve ED if time were allowed for the surface to adjust itself after each phase of contraction—expansion of the surface is, so to speak, dead-beat, as is shown by the ringed dots lying on the curve DE of relatively slow expansion. The evidence available does not suffice to settle the point. This, like the many other questions raised, must be made the subject of separate experiment. It may be noticed, however, that the curve of rapid extension of the heavy oil surface coincides with that of slow extension. Also the curve of very rapid contraction EF, which was taken firstly in four rapid stages, and secondly in one jump, i.e. in about four seconds, lies above and not below curve EC, which was taken quite slowly, the time occupied in the contraction being about 1 hour.

A comparison of these results shows that the influence of a substance upon the tension of a water surface is remarkably independent of its own tension, or viscosity. Cymene, a light spirit, has much the same type of curve as that given by an oil so viscous that it will scarcely flow. Benzene, with a surface tension $= 28$, has relatively little influence on the tension of water. Croton oil, with about the same tension, is incredibly potent. All physical factors have an influence which is insignificant compared to that exerted by the chemical nature of the substance. Organic salts, that is to say the esters, are intensely active, and have sharply inflected curves. Stable substances, paraffins and benzene derivatives, have given curves only slightly inflected, and their activity is relatively slight. The different portions—AB, BC, CD—of the sharply inflected curves I interpret as follows: AB is the stage in which the layer A is discontinuous. The surface is unstable, and the tension rises above that of pure water when the surface is contracted. In BC the tension is rapidly falling, owing perhaps mainly to the development of a difference of potential between the film A and the water B. That is to say, we may picture partial hydrolysis and ionisation of the ester taking place at the interface. The second inflection at C marks the point at which polarisation is maximal, and from C onwards the tension falls gradually to a minimum. Measurements of tension, however, fail to give the real relation in this region owing to the formation of lenses by condensation of oil on to solid particles. The conclusion that tension falls to a minimum and then rises again is probable from considerations already advanced, and is supported in a remarkable way by the study of the mechanical stability of composite films.

The curves for relatively inactive substances are sharply distinct in type. There is an absence of sharp inflections, and the second inflection may be almost absent. The curve for the heavy oil suggests that the whole effect in reducing tension may be due to an impurity (see p. 529), the bulk of the oil being not relatively but absolutely inactive. The slow decrease in tension with increase in concentration of the substance on the surface of the water I ascribe to the feebleness or absence of chemical reaction at the interface.

The mechanical stability of composite surfaces. When a bubble of air is allowed to rise in a fluid the surface is lifted and a film is formed which proceeds to thin owing to the fluid being drained away by its own weight. After a certain time the film becomes so thin as to be ruptured by the tension. The time of persistence of the film obviously will depend upon the viscous resistance to the flow of fluid which thins it, and the speed with which adjustment of tension can be effected by tangential diffusion of the components.[1] These facts also determine the durability of a flat film. Without attempting precisely to analyse what it means, let us agree to speak of the mechanical stability of a film and to regard the time which elapses between formation and rupture as a direct measure of this property.

Bubbles of the same size were formed as nearly as possible at the same distance (1·5 cm.) below the surface in a trough, and their persistence on the surface measured by a stop-watch. The thickness of the film A was calculated, and the tension measured by the blade method. For each point the duration of at least ten bubbles was recorded and the mean taken. As the bursting of bubbles on a surface almost invariably raises the tension,

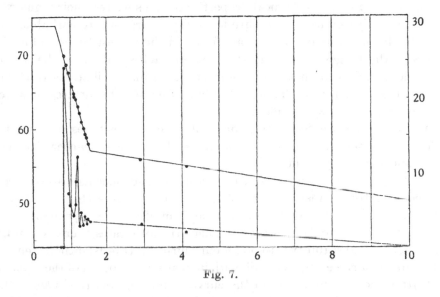

Fig. 7.

this was measured before and after each set of ten, and the mean taken as the tension corresponding to the mean duration. The rise of tension due to the bursting of ten bubbles corresponds on the average to about 0·05 gr. in the balance pan.

Castor oil-water. The upper curve (Fig. 7) is the curve of tension, and the scale of the ordinates appears to the left. The lower curve shows the changes in mechanical stability with changes in thickness of the film. The ordinates for the curve denote seconds and the scale appears to the right. The abscissæ for both curves are to the same scale of $\mu\mu$. The zero of the time scale is the time taken to burst by a bubble formed on a clean surface of water. It is as nearly as can be measured 0·3 of a second. At the second minimum in the curve the duration of the bubbles is even less than this; I was unable even to estimate it and entered it in my notes simply as 0.

[1] Gibbs, *Trans. Conn. Acad.* III, 475.

The lower curve shows a remarkable series of oscillations over a region which corresponds to a composite surface whose tension is varying very rapidly. The regular decrease in the amplitude of the oscillations is remarkable. It is obvious that oscillations must occur in the means of any finite series of numbers, for not only is the tension of the surface varying rapidly with variations in the thickness of the film, but also it is subject to large hysteresis effects. Therefore, the durability of a film formed from the same surface will vary with variations in, *e.g.*, the velocity of ascent of the bubbles, the distribution of solid nuclei in the film, etc. The oscillations may, therefore, be a function of the means rather than an indication of a real variation in the mechanical properties of the surface. Their regularity and the orderly decrease in amplitude, however, are most remarkable, and as precisely the same oscillations are found in the longest times and in the means of the longest and shortest times for each set it is, on the whole, probable that they are not merely numerical. Similar oscillations appear in measurements of the mechanical stability of composite surfaces of water with cymene or benzyl cyanide.

In the region of the first oscillation bubbles have sometimes persisted for as long as 70 sec. It is possible, therefore, that an apex earlier than the one found may be present.

Cymene-water. The duration of the bubbles in seconds as ordinates are plotted directly against tension (abscissæ) in the curve (Fig. 8). Fewer points were taken, and therefore the true form of the curve probably does not appear. The oscillations are obvious. In both cases there is a stage when mechanical stability is that of a pure surface. It lasts until the tension of the mixed surface begins to fall, with increase in the depth of the layer A. Just after the first inflection of the curve of tension and thickness a film suddenly acquires very great mechanical stability because when it thins, owing to draining away of the fluid, the thinning of layer A, which must also occur, at once raises the tension. When a surface is brought as nearly as possible to the point where mechanical stability appears the record of persistence of bubbles is very remarkable.

Tension	Duration of a series of bubbles in seconds														
70·1	0·3	0·3	0·3	0·3	0·3	0·3	0·3	21	0·3	0·3	0·3	0·3	0·3	0·3	0·3
	0·3	0·3	0·3	0·3	0·3	0·3	0·3	0·3	1	0·3	0·3	15	0·3	0·3	0·4
	0·3	0·3													
69·7	10	50	55	20	23	22	11	20	22	11					

In the first series a few bubbles, so to speak, make films with an order of stability which rightly belongs to the next series. If those bubbles of exceptional structure be disregarded, the curve may be said to rise abruptly from the base line to the maximum of mechanical stability.

When the surface of croton or castor oil and water is near this region, bubbles which burst as rapidly as possible give way when the barrier is moved 1 mm. to bubbles which persist for 20 sec. or more and show a full play of Newtonian colours.

It will be noticed that there are two points at which the mechanical stability of the surface is minimal, one when the tension is sensibly that of clean water, and another when it is heavily contaminated. The second minimum occurs when the *mean* thickness of the layer of castor oil is about $13\,\mu\mu$, and of cymene is of the order of $2000\,\mu\mu$. The line AA in Fig. 8 marks a thickness of $1980\,\mu\mu$. Beyond this I have no measurements.

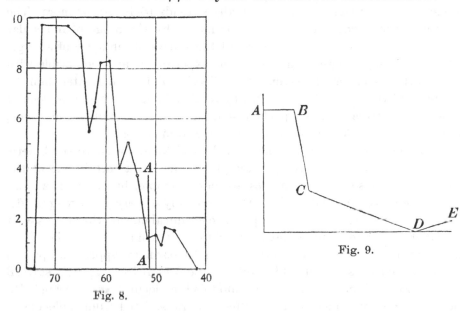

Fig. 8.

Fig. 9.

Bubbles of the composite surface at the second minimum seem to burst even more rapidly than do those on a clean water surface. In my notes I have simply entered the time as 0. At both minima, and there only, the bubbles burst with a sharp crackle strongly suggesting an electric spark. Some active liberation of energy is needed to account for the rapidity of the rupture at the second minimum. It is not the rate of bursting of bubbles in, *e.g.*, pure castor oil, for that oil is a very viscous fluid. It is a true minimum and beyond it the curve rises again. For instance, a surface was oiled with a large excess of castor oil and the mean duration of the bubbles rose to 3·1 sec.

The existence of the second minimum of mechanical stability confirms the view already arrived at that between the limiting values $T = T_B$ of a clean surface and $T = T_A + T_{AB}$ of a double surface there is at least one minimal value of T less than either. For let the curves of tension have the form shown in Fig. 9. From A to B the tension of a free film decreases as the thickness diminishes, and therefore it is unstable. Also the tension of a free film formed from the composite surface at D would increase with both increase and decrease in thickness; that is to say rupture would occur even more rapidly than in the region AB. But, from considerations already advanced, the extreme range of action of the molecular forces at the surface must be equal to at least half the thickness of the film when T is minimal, and the method adopted for calculating this thickness, which itself gives minimal values, would fix it as $9\,\mu\mu$ for castor oil, and some hundreds of millimicrons for cymene or heavy oil. Are we to conclude from this that the range of the Laplacian attraction really extends to $100\,\mu\mu$ or more? The answer depends entirely upon what is meant by the range of molecular action. If by it is meant the radius of the sphere of influence of a single molecule of, say, a gas, then I think the experiments are compatible with the view that this has an extremely small value, not much greater than the diameter of the molecule itself. Quite another kind of range of molecular action is possible in close packed structures such as fluids and solids, namely a strain transmitted from molecule to molecule.

At the interface AB, which has sensible thickness owing to the kinetic energy of the molecules, the matter present has to a certain extent the properties of a solid. The stresses are not isotropic, but there is a component along the normal to the surface probably of very great magnitude. This stress may, as we have seen, produce chemical effects, that is to say, it may distort the intramolecular fields of force, it will also strain the extra-molecular fields. The strains on the intramolecular fields appear as a partial ionisation of esters, or partial hydrolysis. Looked at from a purely statical standpoint, we may regard these modified molecules as reacting with molecules on *both* sides, and the effect so transmitted from molecule to molecule on each side of the mathematical interface. From the kinetic standpoint we may consider the give and take between these strained molecules and those further away from the interface which must occur in the course of their vibrations. Molecules diffusing out of the interfacial zone are changing to normal molecules, those diffusing in are becoming abnormal, just as at an interface between water and water vapour, vapour molecules entering the zone on the average become polymerised to water molecules, while molecules leaving it to enter the vapour must on the average be in process of depolymerisation. It is this irradiation of strain in a close packed structure which may extend so deeply as to modify the molecular state of a skin some hundreds of microns in depth.

Appendix

After the foregoing paper was completed, certain new facts came to light which make it possible to extend and confirm the conclusions arrived at. As will be remembered, a suspicion was expressed that the effect of the oil called "heavy oil" upon the tension of water might be due to the presence in it of some active impurity. I wrote to Messrs Price Patent Candle Co. asking them for information as to the chemical constitution of the oil "Motorine A", and especially whether the distillate I employed under the name "heavy oil" might be taken to be composed entirely of paraffins. I did so because, as the simple paraffins are more stable than benzene, the chemical theory of the tension of composite surfaces demanded that the effect of paraffins upon the tension of water should be even less than that of pure benzene.

The courteous letter which I received from Messrs Price informed me that the Motorine A contains glycerides, which would account for its activity; further, that my distillate also might contain glycerides, that is esters, or products of the decomposition of glycerides. Messrs Price sent to me a sample of a paraffin oil about as heavy as Motorine A, and about the same time I came, by accident, across another sample of an oil containing chiefly paraffins. These oils I will call B and C respectively. Both of them refuse completely to spread upon pure water. The statement is founded upon the most rigorous tests I could devise. One experiment was as follows: The trough and blades were scrubbed with strong caustic potash, and left in running tap water for 1 hour. The trough was then rinsed out with water freshly distilled in a silver still, filled with water, and the surface scraped at intervals for an hour. It was now not possible to detect any contraction of surface by the movement of motes. Carefully cleaned blades were inserted, and the tension found to be constant when the surface was contracted as much as possible. The barriers were placed at the extreme end of the trough. A very few grains of lycopodium were then dusted on to one spot—very few in order to avoid risk of contamination. While the position of a particular cluster of grains was being observed along a fixed line of sight a small drop of oil B was placed on the surface 1 cm. distant from the cluster. The cluster did not move at all, and the tension did not vary. A tiny drop of oil C was then placed 2 mm. away from the cluster, again with no result. Each drop floated on the surface as a tiny lens. The barriers were now moved, and the surface contracted as much as possible, with no effect upon the tension. In contracting the surface it was noticed that the lenses of oil did not move until they were impelled forward by the upward slope of the surface to the barriers. That is to say, the surface was so pure as to be "non-contractile". They were

H   34

driven in this way right up to the blades. Both barriers were now lifted out, without any sign of expansion of a "skin", and placed touching one another in the middle of the length of the trough. They were then rapidly moved apart to each end and on to the fresh surface so formed a few grains of lycopodium were placed, and oil drops beside them, all as rapidly as possible. The oil did not spread at all.

The great chemical stability of the paraffins makes chemical interaction with water impossible, and with the absence of chemical action at the interface the term $f(\sigma)$ in equation (11) vanishes, and the term in brackets in equation (12) reduces to $2T'_{ab}$. Some degree of chemical action, therefore, would seem to be necessary to make one fluid spread as a film between two others (air and water). This leads to a purely chemical theory of the miscibility of fluids, for fluids mix when T_{AB} is negative, and given the relation

$$T_{AB} = T_A + T_B - 2T'_{AB},$$

this occurs when $2T'_{AB} > T_A + T_B$, that is, when the energy of chemical action per unit area of interface is sufficient to satisfy the condition

$$2\,[T'_{ab} + f(\sigma)] > T_A + T_B.$$

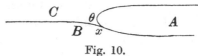

Fig. 10.

Oil C is completely colourless and transparent. The contour of the edge of the lens and the nature of the oil-water surface can therefore be followed with ease. Careful examination of the lens shows that the water-air surface is drawn under its edge. The contour of a large flat-topped lens, 2 or 3 cm. in diameter, is somewhat as shown in Fig. 10. The significance of this is obvious, namely, that $t_{AB} > t_{BC}$, for, if the relation be one of equality, the angle θ must be 102° in order to satisfy the relation

$$t_{AC} = 2\,(t_{BC} \cos \tfrac{1}{2}\theta),$$

when $\dfrac{t_{AC}}{t_{BC}}$ is put $= \dfrac{T_{AC}}{T_{BC}} = \dfrac{30}{74}$,[1]

a relation which is certainly sufficiently exact to prove the proposition. From this point two possibilities confront us: either that at the edge x three fluids do actually exist in contact, in which case Marangoni, Quincke, and Lord Rayleigh are wrong in concluding that Neumann's triangle is necessarily illusory; or that $T_{AB} > T_{AC} + T_{BC}$, in which case the lens of A cannot touch B but is separated by a pellicle of air. The choice between these two alternatives must be based upon a study of the optical properties of the

[1] Quincke's value for the tension of paraffin is 30. My own measurements of the tension of oil C give 30·5 at 12°.

surface. The general appearance of the AB surface does not suggest the presence of a pellicle of air at all.

It is interesting to note that the second alternative may imply a repulsion of A for B. This obviously follows from the relation $T_{AB} = T_A + T_B - 2T'_{AB}$ if the quantities due to air be ignored and T_A, T_B be put respectively equal to T_{AC}, T_{BC}. The quantity T'_{AB} must then necessarily be negative.

The corpuscular theory of matter traces all material forces to the attraction or repulsion of foci of strain of two opposite types. All systems of these foci which have been considered would possess an unsymmetrical stray field—equipotential surfaces would not be disposed about the system as concentric shells. If the stray field of a molecule, that is of a complex of these atomic systems, be unsymmetrical, the surface layer of fluids and solids, which are close packed states of matter, must differ from the interior mass in the orientation of the axes of the fields with respect to the normal to the surface, and so form a skin on the surface of a pure substance having all the molecules oriented in the same way instead of purely in random ways. The result would be the polarisation of the surface, and the surfaces of two different fluids would attract or repel one another according to the sign of their surfaces.

The statement, so commonly made, that a drop of one fluid refuses to spread on the surface of another only when the latter is contaminated by a film of impurity is remarkably far from the truth unless the contamination in question be the air—an interpretation not usually included in the statement. I have found lenses of oil C on pure water unchanged after 24 hours, the edge still being tucked in. The lens, however, at once flattens if the clean water surface be very slightly oiled with an active substance such as castor oil, and the edge takes the form shown in Fig. 1. Castor oil reduces the quantity T_{BC}, and therefore the result is at first sight paradoxical. The explanation is simple—I owe it entirely to an observation made by Mr Stevens. When a coloured body such as impure dibromricinoleic acid is used to contaminate the water surface it may be seen to be drawn in as a film under the lens. The quantity T_{AB} is therefore also reduced in value.

An instance of the fact that spreading is made possible, not prevented, by a film of impurity, which must not be too thick, is furnished by oil B, which spreads slowly on tap water but refuses to spread if the surface has been first thoroughly scraped. When a lens of fluid A stands on a surface of water which is coated with a film of A, it is possible that the layer of discontinuity AB of the general surface is continued under the lens. In that case the composite surface between lens and water would be composed of two surfaces of discontinuity of different molecular structure, namely, the surface of mixed oil and water, which is an extension of the general surface, and the surface of the oil.

31

ON THE FORMATION OF A HEAT-REVERSIBLE GEL

[*Proc. Roy. Soc.* (1912), A, LXXXVII, 29]

In the course of his study of the cyclic ketonic compounds, Dr Ruhemann has discovered a substance which is, I think, destined to play an important part in the development of the theory of gelation. The substance is 5-dimethylaminoanilo-3 : 4 diphenylcyclopentene-1 : 2 dione,[1] the formula being

$$\text{C}_6\text{H}_5\text{C}\text{----}\text{C : NC}_6\text{H}_4\text{N(CH}_3)_2$$
$$\text{C}_6\text{H}_5\text{C} \qquad \text{CO}$$
$$\text{CO}$$

It is a solid at ordinary temperature, and occurs either as orange-coloured needles or as small dark red plates; the former separate from dilute solutions, the latter from concentrated solutions in absolute alcohol.

"If a little water is added to the hot alcoholic solution, no crystals separate on cooling, but the whole sets to a transparent jelly, which, according to concentration, is yellow or yellowish red. This gel, when kept at the ordinary temperature, gradually liquefies, and in the course of one or two days completely disappears, with separation from the resulting solution of the azomethine in orange needles." Dr Ruhemann has therefore furnished us with a substance which offers special facilities for studying the relations between the crystalline and the colloidal state.

The colloidal characters do not appear in any of the substitution products of the azomethine which have been prepared. They completely disappear when a single hydrogen atom is displaced by, *e.g.*, bromine. Why so small a change in the molecule should cause the complete disappearance of such striking physical characters is an important problem, which I understand Dr Ruhemann is investigating. I have to thank Dr Ruhemann for his kindness in providing me with a small quantity of the azomethine, and for allowing me the opportunity of examining the remarkable changes of state. In the following pages is given an account of a purely qualitative study.

Influence of the solvent. The action of water in promoting the change of solutions in alcohol to the colloidal state (and I may also add solutions in aldehyde and glacial acetic acid) led the authors to suggest that the phenomenon is accompanied by the formation of a hydrate, which spontaneously loses water, and so yields the original compound in the crystalline

---

[1] Ruhemann and Naunton, *Trans. Chem. Soc.* (1912), CI, 42.

form. My own experiments prove that, if water be necessary for the change, the most minute traces are sufficient.

I found it easy to obtain a gel at 10° with a good sample of absolute alcohol, provided the solutions, made at the boiling-point of the alcohol, were sufficiently concentrated. Water lowers the concentration of solute needed for gelation. It was immaterial whether red or orange crystals were used for making the solution. The substance is completely insoluble in water, which, therefore, may be said to act by lowering the degree of super-saturation needed for gelation, which occurs only in solutions super-saturated with respect to crystals. In unit volume the added water displaces so much alcohol, and combines with some of the alcohol actually present, and so in both ways increases the mass ratio of solute to alcohol. Water also delays crystallisation, and thereby alters the time relation in such a way as to make it easier to reach the limit of overcooling, at which gelation occurs without deposition of crystals. This is especially seen when ether is the solvent.

More rigorous experiments to eliminate water are the following:

Absolute alcohol, prepared by standing on calcium chloride for a week, and prolonged boiling with calcium chloride under a vertical condenser, and subsequent distillation, was again kept over a large quantity of freshly fused chloride for a week. It was then distilled slowly, the first quantity to come over being rejected, and the middle portion received directly, with exclusion of air as far as possible, into a quartz test-tube, which had been strongly heated, and allowed to cool in a desiccator. Crystals of azomethine which had been exposed to an air current for 4 hours at 120° were at once added, and a solution made by heat, but without boiling. In 3 min. the solution was poured off the remaining crystals into another dry quartz tube, and placed in a freezing mixture of ice and salt. It at once set to a firm gel.

Ethylic ether made anhydrous by distillation from sodium and exposure to a large surface of sodium for a week was then tried in a similar way, but as the substance is only slightly soluble in ether it was necessary to concentrate by rapid boiling. When the solution saturated at the boiling-point of ether had been concentrated to one-third its volume it was placed in the freezing mixture and gelation occurred.

Carbon tetrachloride is a very good solvent for the substance and deep orange-red solutions are readily obtained. This solvent is immiscible with water. A solution made in a dry quartz tube of red crystals with solvent which had stood for 48 hours over a large excess of freshly fused calcium chloride gave a firm gel in a freezing mixture. The most probable conclusion from these and similar experiments with other solvents is that water is not necessary. With the help of a freezing mixture a gel has been obtained from solutions in ethylic alcohol, ethylic ether, aldehyde, acetone, carbon tetra-chloride, carbon bisulphide and glacial acetic acid, and, doubtfully, in chloroform and benzene. The substance is insoluble in the paraffins (petro-leum ether) and in water. Some of the solvents which give gels are associating liquids, that is to say the liquid molecule is a complex of simple molecules,[1] others are not. Gelation occurs at least as readily in carbon tetrachloride, which does not associate, as it does in alcohol, which is an associating liquid. An important negative conclusion follows, namely, that gelation is not founded upon an intrinsic tendency of the solvent molecules to form complexes amongst themselves. Nor, if the view of the mechanism of solution put forward by Walden and Abegg[2] be accepted can gelation be traced to the association of solute molecules with molecules of solvent, for, according to this view, such associations occur only in solvents which themselves associate and do not occur in non-associating solvents. From whatever point of view it be regarded the slight influence exerted by the molecular state of the solvent is remarkable and unexpected. The process of gelation and the curious structural features of the gels are the same in solvents of both types.

All the gels examined liquefy on standing with deposition of crystals. The change occurs at room temperature (15°) in a few minutes from an ether gel, a few hours from gels of absolute alcohol or aldehyde, and in some days from a carbon tetrachloride gel. The presence of water delays the lique-faction of the gel, that is it stabilises the gel state so that gels of aldehyde water and alcohol water may persist for days. Sooner or later at about 15° the gel liquefies; it is, therefore, at this temperature labile to the system saturated solution-crystals. At temperatures below say + 5° the gels change very slowly. At higher temperatures the gel melts and will set again on cooling. If it remains melted crystallisation rapidly occurs. Above a certain temperature gelation will not occur at all, but crystallisation occurs direct. The limiting temperature rises as concentration of solute increases. For instance, alcohol 97·5 per cent. (by specific gravity) as solvent, a strong solution supposed to be saturated at the boiling-point set to a gel at 32°, but deposited crystals only at 36°. On again testing after further boiling with crystals it now formed a gel at 35·17° which completely liquefied with deposition of crystals in 3 min., the temperature rising to 35·55° in this process. At this temperature the borderland between gel and crystals is as narrow as it is with ether at temperatures about the freezing-point. The temperature range of solutions in alcohol may, therefore, be divided into an upper region above about 35° in which gelation does not take place, a middle region in which the gel is labile to crystals, and a lower region below

---

[1] Ramsay and Shields, *Phil. Trans.* (1893), p. 647.
[2] *Zeits. f. anorg. Chem.* (1904), XXIX, 330.

about $+5°$ in which the gel persists for so long a time as to be sensibly stable.

The azomethine is, so far as I know, unique in the wide range of solvents which it will convert into gels. It is true that Graham described gels of silica and alcohol or silica and sulphuric acid, etc., but these were obtained by displacing the water of an already formed hydrogel by diffusion—a process probably akin to the replacement of one fluid in the interstices of a sponge of solid by another. Ruhemann's substance seems capable of forming gels directly with any solvent.

The structure of the gel. Whatever the molecular process underlying gelation may be, it in many most remarkable features resembles crystallisation. To this I now turn. When a strong solution is cooled, it becomes supersaturated with respect to crystals and to gel. One characteristic of over-cooling is that, if the fall of temperature below the point of change be sufficiently great, the appearance of the new phase is prevented.[1] In this sense a solution may be over-cooled with respect to gelation. When a solution of 97·5 per cent. alcohol was rapidly cooled in ice, it remained fluid, but set to a gel when allowed to warm to room temperature, or at once on rapid shaking. I have observed the same feature of over-cooling in solutions of gelatine in water rapidly brought to the freezing-point; and Garrett, working under Quincke, found that the changes in a cooled solution of gelatine in water were hastened by sowing with already formed gel—an observation which clearly points to a similar state of affairs.[2] An over-cooled and therefore supersaturated solution contains a quantity of heat H_1 which must be dissipated before gelation can occur, and a further quantity, H_2, which is dissipated when the gel liquefies with separation of crystals, and, from what has been already said, the ratio H_1/H_2 must be regarded as varying with temperature, and becoming equal to zero at some temperature about 35° for solutions in strong alcohol. Since the gels of this type, that is heat-reversible gels, are liquefied by a rise of temperature, gelation may be taken to be an exothermic change. Attempts were made to measure the heat given off or absorbed, but the experimental difficulties proved to be considerable, and the results were not satisfactory. They will be renewed with a new form of calorimeter. Gelation of the opposite type, namely, that which is not reversed by warming but produces a gel which shrinks on warming, occurs with absorption of heat. For instance, when allowance is made for the heat of mixing of the solvent with chloroform, it is easily possible to prove experimentally that the gelation of a solution of celloidin in absolute alcohol and ether, which is caused by chloroform, is accompanied by a small absorption of heat.

[1] Cf. H. A. Wilson, *Phil. Mag.*
[2] Dissertation, Heidelberg, 1903.

Nuclei, as is well known, favour crystallisation, and nuclei seem necessary for gelation. The gel starts always at distinct points on the wall of the vessel which holds the solution. I failed to discover the nature of the nuclei, but their influence is unmistakable and profound. If a test-tube of glass or quartz containing a solution about to gel be tilted, the gel may be watched rapidly forming at a number of points on the wall. Each mass grows rapidly until it meets neighbouring masses, on which it seems to press. The result is that a fully formed gel, though it appear transparent and homogeneous, has in reality a remarkable structure determined by the number and nature of the nuclei. It is composed of masses more or less imperfectly joined together and of very various sizes in alcohol gels, and very uniform in size in carbon tetrachloride gels. The formation of these masses may be readily followed, and they are easily obtained isolated, either by pouring away the fluid from a partially formed gel or by disintegrating a gel with careful shaking.

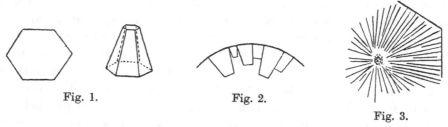

Fig. 1.   Fig. 2.

Fig. 3.

For example: 130 c.c. of a 0·7 per cent. solution was made in 97·5 per cent. alcohol, and allowed to cool slowly by standing on the laboratory bench. It set to a continuous, firm, transparent gel the colour of amber. At the end of an hour about 0·5 c.c. of fluid separated and floated on top of the gel. In the fluid were fine crystals. Seen by transmitted light the gel was seen to be composed of facetted masses, the base of each where it abutted on the glass being a regular hexagon. A sketch, the actual size, is given in Fig. 1, and the arrangement is shown diagrammatically in Fig. 2, which is an optical and horizontal section of part of the gel. Sections cut with the free hand of a gel show, under the microscope, polygonal areas, and in the centre of some of these a confused mass appears, from which darker lines radiate regularly to the periphery (Fig. 3). This figure was drawn from a gel entirely free from true crystals. These masses have all the appearance of crystals. They are not crystals, however. They are singly refractive, and when they form in free fluid the surface is rounded and not at all facetted. In fact, each nucleus is the centre of a sphere of gelation which continues as a sphere until it meets neighbouring masses, when mutual pressure produces a polygon. The true crystals of the azomethine are doubly refractive. The appearance of radiating lines reproduced in Fig. 3 puzzled me greatly until I came to study gels of carbon tetrachloride and carbon bisulphide. In these gels the gel masses are easily studied, for they are always very distinct, giving to the gel a curious cloudy opaque appearance. The following description applies to gels of tetrachloride. The gel is seen by a hand lens of good power to be composed of spheres, very uniformly 0·14 mm. in diameter. These spheres hang together to form a framework, and there is fluid between them. I described gels with similar structures many years ago.[1] Each sphere is opaque and cloudy, and the cause of this is found when a sphere is examined under a higher magnification (Oc. 4, Ob. D). Each sphere is then seen to be built of close packed smaller masses, which are polygonal owing to the close packing, and are very uniform in size, an axis being $10\,\mu$. The smaller masses are arranged in a pattern which radiates outward from the centre of the larger sphere. They are singly refractive, and at the free surface the facetted type seems to give place to a spherical type—to globuliten, in fact.

The gels of azomethine are thus built up of at least two orders of structure, and the air surface of a transparent and apparently homogeneous alcohol gel exhibits by reflected light a pattern of large hexagons. The further history of a gel can be followed under the microscope. The gel never changes directly into crystals. The true gel is completely stable with respect to crystals, but the component large masses melt at their surfaces only, and the fluid so formed deposits crystals.

These facts concerning the gels of azomethine, together with what is known of gelation in aqueous solutions, lead to the following account of the process in heat reversible gels.

The over-cooled fluid solution changes to a solid solution by a process which, starting about some nuclei, spreads equally through the fluid, so that spherical masses of solid are formed. Each solid mass is isotropic, and it may be compared to a glass. The foundation of gelation, therefore, is a formation of masses of glass, which start from independent centres. The process has many points of resemblance to crystallisation, but the masses are not crystals unless the sphere be admitted as a crystalline form. To the masses the old term globuliten fitly applies, and the change may be called vitrification. The particular glass of azomethine gel is a solid solution. But the globuliten which appear in a cooling colloidal solution may be at first fluid, and continue fluid over a determinate range of temperature, just as ordinary glass will melt to a viscous fluid. This appears very conclusively from Garrett's investigation of the viscosity of solutions of gelatine at various temperatures. The temperature range divides itself into three parts: (1) from 100° to about 25° the viscosity of a 2 per cent. solution of

[1] *Journ. Physiol.* (1899), xxiv, 167; *Proc. Roy. Soc.* (1900).

gelatine does not change with time; (2) a middle region, between about 25°
and 21°, in which viscosity rises with time to a maximum—in this region,
although the viscosity is high, gelation does not occur; and (3) below 21°,
in which viscosity continually rises with time, until it becomes infinite when
the state of a solid gel is reached. The existence of the middle region may,
without doubt, be traced to the separation of a colloid rich phase, which is
fluid, as Garrett himself points out. It is the region in which vitrification is
incomplete, a fluid glass separating. Vitrification may involve all the
solution, or a portion of fluid may persist, in which case two phases result.
Both conditions are readily obtained and studied in carbon tetrachloride
gels of azomethine. The phase which separates on cooling as spheres is not
always the colloid rich phase. Thus, when gelatine is dissolved in 40 per
cent. alcohol, if the percentage by weight is less than about 25, spheres of a
solid solution rich in gelatine separate, and adhere together in anastomosing
lines, so as to form a brittle framework of solid. But when the concentration
of gelatine lies between about 25 and 36, spheres of fluid poor in gelatine,
and uniformly $10\,\mu$ in diameter, may be seen to separate at 20°, and the
remainder vitrifies as a continuous solid mass. I succeeded in pushing the
concentration of gelatine to 40 per cent.; the mixture now melted at 70°
and set at 35° to a milky gel with the remarkable feature that increasing
the concentration of the colloid increases the volume of the spheres of
fluid—some being $25\,\mu$ in diameter.[1]

Agar water containing 1–2 per cent. of agar unquestionably vitrifies
incompletely. The colloid poor phase can be expressed from the gel with the
utmost readiness either by centrifuging, by squeezing the gel in one's hands,
or by gravity, as, *e.g.*, if a hole is made in a mass of such gel it rapidly fills
with fluid. Why do not the masses of the colloid rich phase fuse completely?
I suggested in an earlier paper[2] that the "passive resistance to change—
introduced by the formation of a solid phase" would account for it. The
influence of this resistance is undeniable; once the solid state is reached,
progress to complete equilibrium between the phases will take place mainly
by slow interchange due to diffusion—mixing by convection becomes
impossible. Difficulties in the way of dissipation of surface energy will also
play their part—electrification of the surface of the spheres would, for
instance, delay or prevent the diminution of surface which follows fusion.
All these features have been already insisted upon,[3] but the azomethine
series of gels afford convincing evidence that the distinct vitreous masses
which start from the nuclei, although they may and do adhere to one
another, still do not fuse in the sense of losing their individuality. Each
continues to persist as a quasi-independent system of molecules grouped

[1] Hardy, *Journ. Physiol.* (1899), xxiv, 167.
[2] *Proc. Roy. Soc.* (1900).   [3] *Loc. cit.*

about and related to its own nucleus, and each, when the mass changes, changes independently of the others at its own rate. Whether the mass would be annealed to a truly continuous glass by very slow cooling is a matter for speculation, but it may be some such process of slow annealing, that is the adjustment of the individual vitreous masses of different size and nature to a common level, which accounts for the slow change in vapour tension and elasticity which reversible gels are known to exhibit.

If gels of the azomethine are typical of all reversible gels, then these are the glass state of solutions supersaturated with respect to crystals. Why, therefore, do not gels of gelatine or agar slowly liquefy and deposit crystals? It is not because of the large size of the chemical molecules, for many proteins can be obtained in the crystalline form. The answer lies, I think, in the fact that, in the case of solutions of azomethine, between the region of concentration and temperature in which the gel is permanent, and the region of simple saturated solutions, there is interposed a region in which the glass melts to a true fluid which is supersaturated with respect to crystals. In the case of solutions of gelatine or agar, this intervening region does not exist; instead, we have the glass softening to a viscous fluid, which passes insensibly to colloidal solution.

That solutions of azomethine are not singular in the fact that gelation starts from centres, and only slowly involves the whole mass, is proved by an observation made by me many years ago. A solution of silicic acid was dialysed for some days against running tap water until it formed a firm gel in the dialysing tubes. The gel was then removed from the tubes, broken up, and centrifuged for 3 hours, with the result that a small quantity of fluid separated at the top of the gel. This was pipetted off and, on adding a trace of magnesium sulphate, it at once set to a firm gel. Vitrification proceeding from nuclei in the solution of silicic acid had therefore not yet had time to involve the whole mass of the solution.

A stage in which a melted viscous "glass" separated was directly observed by me in a solution of 55 parts by weight of gelatine in equal parts by volume of absolute alcohol and water. When sufficiently heated, this forms a fluid which appears homogeneous under a magnification of 450 diameters. On cooling, it may be seen to separate into two fluids, one of which, as the temperature continues to fall, becomes solid.

[*Postscript*. The gels described in the preceding pages behave like ordinary watery gels in the fact that dyes such as eosine and methylene blue diffuse into them from alcoholic solution and "stain" the substance of the gel.]

32

THE INFLUENCE OF CHEMICAL CONSTITUTION
UPON INTERFACIAL TENSION

[*Proc. Roy. Soc.* (1913), A, LXXXVIII, 303]

If T_B be the pull per linear centimetre exerted by the surface tension of a pure fluid B, then, if the surface be increased by 1 sq. cm., the increase in surface energy is also T_B. Let a thin film of a second fluid A immiscible with B spread over the surface of B; the tension of the composite surface so formed is less than that of pure B, otherwise spreading would not occur. When the film of A has become sufficiently thick two independent surfaces are formed, one of pure A, the other the interface between A and B. Denoting these respectively by T_A and T_{AB} the energy of the two for each unit of the original surface of pure B is

$$T = T_A + T_{AB}, \qquad (1)$$

and this quantity T is not affected by further addition of A. Therefore in the spreading of A upon B we have, as the limits of the change of surface energy, $T = B$ and $T = T_A + T_{AB}$.

In an earlier paper[1] I gave reasons for believing that the changes in the value of T between these limits depend mainly upon the chemical constitution of A.

Between the limits $T = T_B$ and $T = T_A + T_{AB}$ lies a series of values of T when A is not present in mass. The phenomena over this region are complex, therefore, in order to obtain a simpler presentation of the influence of chemical constitution it is necessary to have A as well as B present in mass. The equation then reduces to the simple form $T = T_{AB}$. It is with measurements of the quantity T_{AB} that we are concerned in this paper.

Following Dupré[2] we may write

$$T_{AB} = T_A + T_B - 2T'_{AB}, \qquad (2)$$

in which $2T'_{AB}$ is the work done per unit of area of interface by the attraction of A for B when a surface of A is allowed to approach normally and touch a surface of B. Since the quantity $2T'_{AB}$ is the total work expended in forming unit area of interface by the molecular forces which operate between A and B, it follows that evidence of the influence of chemical constitution upon surface energy must be looked for in a com-

---

[1] *Proc. Roy. Soc.* (1912), A, LXXXVI, 610.

[2] *Théorie Mécanique d. l. Chaleur* (Paris, 1869), p. 369. See also Lord Rayleigh, *Phil. Mag.* (1890), (5), XXX, 461.

parison of this quantity when different fluids A form an interface with a constant fluid B—in this case water. The tension T_{AB} will not serve our purpose so well since it depends directly upon the tension of A itself.

The first person to introduce the critical constants into considerations of surface energy was Eötvös,[1] and he was followed by Ramsay and Shields.[2] The argument may be stated as follows:

In the gas equation $pv = R\theta$, if v be kept constant p decreases with fall of θ until, where $p = 0$, $\theta = -273$ or absolute zero; the origin of the scale of temperature is thus also the origin of the scale of pressure. The absolute zero of surface energy is the critical temperature, and if an equation analogous to the gas equation be used, namely $Ta = K\tau$, in which a is the area of surface, the temperature τ must be measured in centigrade degrees downwards, the critical temperature being taken as zero. Corresponding to this equation the relation between tension and temperature should be a linear one—a relation proved experimentally by Ramsay and Shields. Near the critical point, however, the linear relation is departed from, with the result that the line which represents the relation between Ta and τ cuts the axis of temperature at a point some degrees below the critical point. The equation for the linear portion of the curve, that is for temperatures other than those near to the critical temperature, is therefore

$$Ta = K (\tau - d).$$

In comparing different substances that surface must be taken which is formed by the same number of molecules. Such a surface, as Eötvös pointed out, is the side of the cube which contains 1 gram-molecule. If v be the volume of a gram-molecule in cubic centimetres, that is the molecular volume, the area of the side of the cube will be $v^{2/3}$. Call this a molecular surface. The equation now becomes

$$Tv^{2/3} = K (\tau - d). \tag{3}$$

This equation enables us to calculate the constant for the pure fluid A from measurements of T_A and T_{AB}.

Putting $(\tau - d) = 1$ in equation (3) we see that K is the gain in molecular surface energy per degree of temperature. An equation precisely similar to (3) for an interface would have as its zero of temperature, not the critical points of the pure fluids A and B, but the temperature at which A and B become completely miscible. This temperature is unknown for the pairs of substances used; an equation for the interface may, however, be derived from equations (2) and (3), namely

$$K_A = \frac{(T_{AB} + 2T'_{AB}) v_A^{2/3} v_B^{2/3} - K_B (\tau_B - d) v_A^{2/3}}{(\tau_A - d) v_B^{2/3}}. \tag{4}$$

[1] Wied. Ann. XXVII, 452.
[2] Phil. Trans. (1893), A, p. 647.

This equation enables us to calculate the constant for the pure fluid A from measurements of T_A and T_{AB}.

Experimental methods. These need somewhat detailed discussion. The surface tensions were measured by determining the weight of the drops of liquid delivered by certain capillary tubes, of which No. 1 was straight and No. 2 bent at right angles. The procedure may best be described by a random selection of an experiment.

Pure oleic acid was poured over some distilled water in a cylindrical vessel until it formed a layer about 3 cm. deep. The straight tube (No. 1), previously cleansed with sulphuric and chromic acids and washed out with distilled water, was dipped into the lower layer (distilled water) and enough fluid drawn up. The point of the tube being raised so as to be in the upper layer, the contents were allowed slowly to form drops which were counted while 3·56 c.c. of fluid left the tube. The operation was repeated with this result:

Temperature 20°. Tube 1. Capacity 3·56 c.c.; number of drops 44, 44·5, 43·8, 43·8.

Tube 2, bent at right angles, was next filled with the upper layer (oleic acid) and drops allowed to form and float up when the point was immersed in the lower layer (water).

Tube 2. Capacity 1·517 c.c.; number of drops 37·1, 37·2.

Tube 1 was then cleaned, dried, filled with oleic acid at 20° and a series of 10 drops allowed to fall into a weighed stoppered weighing bottle.

Weight of 10 drops 0·1625, 0·1647, 0·1617, 0·1617. Mean 0·1626.

Loss of weight by evaporation, determined by control experiment, 0·0000.

From these data we now have to calculate the quantities T_{AB} and T_A. The formula frequently used to compute the surface tension from the weight of a drop delivered by a tube is $T = mg/2\pi r$, when m is the weight of the drop and r the radius of the tube; but this formula gives little more than half the true surface tension. The weight of the drop cannot be calculated from statical consideration: the detachment of the drop is a dynamical effect. A complete solution of the dynamical problem is impracticable. It was attempted by Dupré[1] and the argument has been restated by Lord Rayleigh,[2] who found the formula $T = mg/3·8r$ to be a close approximation for thin-walled tubes.

The tubes used by myself had the following dimensions at the orifice:

No. 1. External radius 0·1165 cm. Internal radius 0·0615 cm.

No. 2. ,, ,, 0·0917 ,, ,, ,, 0·0629 ,,

For these tubes the factor 3·8 was found too low. It applies apparently only to tubes of larger radius. This is at once seen when the tension of water

[1] *Théorie Mécanique d. l. Chaleur* (Paris, 1869), p. 327.
[2] *Phil. Mag.* (1899), (5), XLVIII, 321.

is calculated from the weight of the drop delivered by tubes used by Lord Rayleigh.[1] The first five on his table give the following values:

	15°	
r (cm.)	m (gr.)	T (factor 3·8)
0·11175	0·0375	86·7
0·17017	0·0526	79·72
0·24256	0·0712	75·7
0·254	0·0755	76·7
0·32512	0·0923	73·2

For the smaller tubes therefore the tension comes out too high, the value for water at 15° being 74·45. I therefore determined experimentally the factor for water for tube 1 and found it 4·413 at 16° and 4·391 at 18·5°, and these values were used in computing my results. With Lord Rayleigh's values I now get for his smallest tube:

$$15°: r \ 0·11175; \ m \ 0·0375; \ \text{factor} \ 4·413; \ T \ 74·64;$$

which agrees closely with the standard value 74·45.

For the quantity T_{AB} we have for the weight of the drop, if v be the volume of fluid delivered, n the number of drops and D_A the density of the upper layer, D_B that of the water: $m = v \left(D_B - D_A \right) g/n$ and the formula becomes

$$T_{AB} = \frac{v \left(D_B - D_A \right) g}{nrF}.$$

Returning to the actual experiment, when the lower layer forms drops in the upper layer since water wets the glass of the capillary tube they detach themselves from the whole face of the end of the tube, therefore r is here the external radius or $r = 0·1165$ cm. But when oleic acid is delivered into water by tube 2 it does not wet the walls and the drops are detached from the lumen, and r is the internal radius or 0·0629 cm. When oleic acid is delivered from tube 1 in air, however, it wets the walls and the external radius must be used.

We therefore have:

Tube 1. Mean number of drops 44, r 0·1165; T_{AB} 15·66⎫
Tube 2. ,, ,, 37·15, r 0·0629; T_{AB} 14·64⎭ Mean 15·15.

Tube 1. Weight of one drop of oleic acid in air 0·1626; T_A 31·15.

T_B, the tension of water at 20°, being 73·70, we get from equation (2)

$$T'_{AB} = 44·85.$$

Inspection of Table III shows that the values of T_{AB} obtained for different substances with water sometimes agree closely for the two tubes,

[1] *Phil. Mag.* (1899), (5), XLVIII, 324.

sometimes differ widely. How far may the means be trusted to give the true value? The best evidence that they do so is to be found in the consistent way in which the quantity T'_{AB} follows chemical constitution. The chief error arises from viscosity, for with very viscous fluids the drop takes an appreciable time to break away, therefore if the column of fluid in the capillary tube be moving with any velocity *at the time of breaking* the drop is over filled and the tension comes out too high. As an extreme case take the very viscid ricinoleic acid at 16°, just above the temperature of solidification. When each drop occupied about 40 sec. in forming and breaking away the numbers were—Tube 1: 20·1, 20, 19·8. But when the time was reduced from 40 to 10 sec. the number fell to 8. Now with respect to this source of error the conditions are entirely dissimilar in the two tubes, in tube 1 a drop attached over the whole face is being filled by a stream of relatively small section, while in tube 2 the whole column of fluid which supports the drop is in movement. The concordance of the means leads one to conclude that in the first case the kinetic energy leads to over-inflation of a drop, while in the second case it causes the drop to break away too soon and therefore to be too small. Lord Rayleigh says that "sufficient time must be allowed for the normal formation of the drop", but the most important condition to observe is that the movement of the column of fluid in the tube should be as near as possible zero at the moment when the drop breaks away.

Where the critical constants are known the accuracy of the results may be tested by equations (3) and (4). Ramsay and Shields find that $K = 2 \cdot 21$. I get the following, putting $d = 6$ as they do.

TABLE I

	T_A	$v^{2/3}$	$t°$	$(\tau - d)$	K_A
Carbon bisulphide	30·88	15·396	17	252	1·80
Carbon tetrachloride	24·51	20·999	17	260	1·98
Benzene	27·59	19·624	20·5	261·5	2·07
Toluene	26·82	22·154	22	292·5	2·03
Octylene	21·07	28·87	17·2	275·6	2·20
Cyclohexane	24·21	22·73	16·8	257	2·14
Ethylbenzene	27·83	24·56	17·5	322·4	2·12
Octane	20·63	29·70	18·5	272	2·25

Rejecting the value for carbon bisulphide as being clearly too low, the order of agreement is the same as that found for various substances by Ramsay and Shields, and the mean is 2·11. Similarly from equation (4) we may calculate the constant K_A, using the interfacial tension T_{AB} as one of the data. The constant K_B, since the molecules of water associate in the fluid state to some multiple of the chemical molecule, is less than 2·11. The standard on which all measurements in this paper are based is $T_B = 74$ at

18°. Taking 365° as the critical temperature of water, equation (3) gives $K_B = 1.49$. Equation (4) now gives the following values for K_A:

TABLE II

Carbon bisulphide	...	1·88	Octylene	...	...	2·17
Carbon tetrachloride	...	1·98	Cyclohexane		...	2·16
Benzene	... ...	2·10	Ethylbenzene		...	2·12
Toluene	... ...	2·04	Octane	...	...	2·25

Here again the agreement justifies the conclusion that the mean value of T_{AB} is not far removed from the actual value. The most serious source of error, especially in the case of substances which have little power of lowering the tension of water, lies in the presence of more active impurities. Rigorous circumspection is necessary both in the handling of the water, and of the other fluids, and in their preparation. The effect of a trace of impurity is hardly to be credited. Cyclohexane, a saturated hydrocarbon, does not spread upon water. In some of it 0·00015 per cent. by weight of oleic acid was dissolved, making a $0.0004 N$ solution. The fluid now flashed over a water surface and the tension of the interface with water was found to have fallen from 55·15, that of pure cyclohexane-water, to 32·37. It cannot be without interest to notice that the concentration of oleic acid employed is of the order of concentration needed in the case of such physiologically active substances as adrenalin reckoned as percentage of the body weight of the animal; also, if FitzGerald's view be true that the force of contraction of a muscle is derived from changes in the tension of internal surfaces, such a concentration of a similarly active body would reduce the absolute force by about 50 per cent. While dealing with the question of chemical purity certain special cases must be mentioned.

A sample of octane purchased from Kahlbaum gave the following values: T_{AB} 29·3, T'_{AB} 32·6. As these differed widely from the values obtained for other saturated bodies the sample was freed from unsaturated substances, by shaking for 4 weeks with many changes of concentrated sulphuric acid. It was then washed with water and distilled from metallic sodium, and the fraction which came over at 123–126° now gave the values T_{AB} 53·48 and T'_{AB} 20·54, which accord with the values found for other saturated bodies.

Octylene, as obtained from Kahlbaum, gave T_{AB} 12·9, T'_{AB} 41. It was then distilled and a fraction of B.P. 122–124° collected, giving the values T_{AB} 22·48, T'_{AB} 36·24. Like octane it was found to be impossible to get a fraction giving a perfectly constant boiling-point. A series of fractionations was therefore undertaken and two fractions collected, (A) 120–122°, (B) 122–124°, Bar. 758. The B.P. of octylene is given in Beilstein as 122–123°, and as 124·6° at 796·6 mm. Fraction A gave T_{AB} 22·61, T'_{AB} 36·18. Fraction B, T_{AB} 23·06, T'_{AB} 35·95.

The mean of these values is the best that can be got for octylene by fractional distillation.

The substances employed were for the most part purchased from Kahlbaum and, when possible, distilled to a constant boiling-point just before use. Some specimens were lent to me by Dr Ruhemann, to whom I owe also a great debt for his kindness in directly superintending and helping in the purification of each substance. Without the aid which his profound technical knowledge afforded I could not have succeeded in making the measurements.

The effect of any slight degree of mutual solubility of the fluids A and B upon the quantity T_{AB} needs consideration. Let T_a and T_b be respectively the tension of the saturated solution of B in A and of A in B which can form contiguous phases. We then have

$$T_{ab} = T_a + T_b - 2T'_{ab}. \tag{5}$$

Since $T_B > T_A$ in all the cases dealt with, $T_a > T_A$, and $T_b < T_B$, the relation of T'_{ab} to T'_{AB} so far as magnitude is concerned is uncertain, terms of opposite sign being introduced by the presence of A and B on both sides of the interface. When the two solutions are brought into contact the work done is derived from (1) the attraction of A for A and of B for B across the interface, and (2) the attraction of A for B. When pure A and B come into contact (1) is zero, and any degree of miscibility increases the work due to (1) and decreases that due to (2). Therefore, since mutual miscibility introduces effects of opposite signs into the right-hand members of equation (2), the relation of T_{ab} to T_{AB} cannot be predicted.[1] Whatever the theoretical conclusion practically the value of T_{AB} under the conditions of measurement was found to be sensibly independent of the state of the phases.

The alcohol cyclohexanol is soluble in 28 parts of water. Dry cyclohexanol on water gave the following values:

16·2°. Density alcohol 0·948, water = 1.

T_{AB}. Tube 1 3·951; Tube 2 3·665. Mean 3·808.

The layers were then thoroughly shaken together in a stoppered vessel and left 24 hours. For the layers mutually saturated I got

16°. Density alcohol layer 0·9538; density water layer 0·9986.

T_{AB}. Tube 1 3·912; Tube 2 3·693. Mean 3·802.

Benzene is slightly soluble in water. For two samples I found the following values. Benzene and water not shaken together T_{AB} (17°) 38·01, (20·5°) 36·84; mean 37·42; and after shaking together and leaving for 24 hours, T_{AB} (17°) 37·56, (20·5°) 37·58; mean 37·57.

---

[1] T_{ab} obviously does not include the loss of potential due to the mixing of A and B except at the interface, for, if it did, T_{ab} would be a function of the ratio surface/(mass of A + mass of B).

The experimental results are gathered together in Table III. It will be noticed that whereas the interfacial tension (T_{AB}) only roughly follows the chemical constitution of the fluid A, the quantity T'_{AB} follows it very closely. That is to say, when unit area of a free surface of a fluid is brought into contact with unit area of the free surface of water, the loss of potential, or the work done by molecular forces, is determined by the chemical constitution of the former.

To illustrate this we have the first group of saturated substances, in which I have included oil "C", composed mainly of high boiling-point paraffins. T'_{AB} lies between 20·8 and 24·5. The introduction of an unsaturated linkage into the paraffin octane raises the value of 36 (octylene). But in the case of a ring compound unsaturated linkages produce a smaller effect, the number rising only to 32 (benzene).

The introduction of the OH group into a ring compound increases the quantity by 30 (cyclohexane and cyclohexanol, $51·43 - 21·63 = 29·8$); but the effect in the case of the paraffin chain is only about one-half. The presence of the carboxyl group produces the same effect as the OH group, the quantity being increased in the case of capryllic acid by 10·4.

For the esters we have 41 for ethyl hydrocinnamate. The introduction of one more unsaturated linkage increases this only slightly, namely, to the 43 of ethyl cinnamate, and the double ester ethyl phthalate again is only slightly higher, namely, 46. Other relations might be pointed out, but the figures speak for themselves. I prefer to pass on to the more important negative relations.

In the saturated bodies there is a very small but I believe real difference between the paraffin chain and the ring formation, and again between these and the compounds CS_2 and CCl_4. The two cyclic alcohols have practically the same value, and the three fatty acids agree closely. Similarly benzene and the allied substances give values close together. But the remarkable feature of these correspondences is that they are for unit area of interface and not for areas on which equal numbers of molecules impinge.

In the gas equation $Pv = R\theta$ the constant R refers to a volume which contains 1 gram-molecule. Similarly, in the comparison of surface energy, as Eötvös showed (loc. cit.), surfaces of the same molecular value must be compared. Such a molecular surface is the face of a cube which holds a gram-molecule of the fluid—or $v^{2/3}$. Tables I and II prove that the values found for T_A and T_{AB} are consistent with this conception of equi-molecular surfaces. But the values for T'_{AB}, though derived from T_A and T_{AB}, are in respect to their relation to chemical constitution independent of the area of the molecular surface. To put the matter in another way—agreement in the values for a particular type of substance, say, saturated compounds, or acids, would by analogy be between the product of the quantities T'_{AB},

TABLE III

Substance	Boiling-point	Temp. (°)	Density (D)	T_A	T_B	T_{AB} I	T_{AB} II	T_{AB} Mean	T'_{AB}	v	$v^{2/3}$	$T'_{AB} \times v^{2/3}$
Octane	123–126°	18·5	0·7034	20·63	73·93	53·74	52·18	52·96	20·80	161·9	29·7	615
Cyclohexane	Abs. const.	16·8	0·779	24·21	74·2	56·00	54·31	55·15	21·63	108·4	22·73	492
Oil "C"	—	17·0	0·876	30·12	74·15	64·72	57·99	61·35	21·46	—	—	—
Carbon bisulphide	Abs. const.	16·0	1·253	30·88	74·3	58·15	54·15	56·15	24·51	60·42	15·396	377
Carbon tetrachloride	—	17·0	1·63	24·51	74·15	53·2	47·16	50·18	24·24	96·24	20·999	509
Octylene	122–124°	17·0	0·7255	20·82	74·15	22·91	22·07	22·48	36·24	155·18	28·87	1046
Benzene	Abs. const.	20·5	0·879	27·59	73·77	37·68	36·75	37·49	31·93	86·95	19·624	627
Toluene	Abs. const.	22·0	0·8646	26·5	73·40	37·35	34·70	36·02	31·94	104·3	22·154	708
p-Cymene	Abs. const.	13·5	0·8633	27·0	74·57	43·15	41·36	42·25	29·66	155·2	28·87	856
Styrene	144–148°	19·0	0·9028	30·54	73·85	39·70	38·20	38·95	32·72	114·5	23·58	771
Ethyl benzene	Abs. const.	17·5	0·879	27·83	74·07	33·93	32·90	33·41	34·24	121·56	24·56	841
Benzene iodide	Abs. const.	16·8	1·831	36·6	74·2	44·88	40·65	42·76	34·00	111·3	23·14	787
Ethyl iodide	Abs. const.	16·0	1·943	26·31	74·3	38·40	37·90	38·15	31·23	80·2	18·59	580
Capryllic acid (n)	Abs. const.	18·1	0·9116	27·3	74·0	8·43	8·35	8·39	46·45	157·9	29·21	1357
Oleic acid	Crystallises (Kahlbaum)	20·0	0·898	30·99	73·70	15·73	14·71	15·22	44·73	314·0	46·19	2066
Ricinoleic acid		16·0	0·9538	34·31	74·3	15·68	14·50	15·09	46·76	312·0	46·73	2185
Cyclohexanol	Abs. const.	16·2	0·9480	32·36	74·3	3·951	3·665	3·808	51·43	105·5	23·77	1223
Octyl alcohol	Abs. const.	17·5	0·8293	26·19	74·07	8·053	7·882	7·967	46·14	156·6	29·04	1340
Benzyl alcohol	Abs. const.	22·5	1·0442	37·93	73·33	4·898	4·891	4·894	53·18	101·6	21·77	1158
Ethyl phthalate	Abs. const.	20·5	1·1175	35·26	73·48	17·10	17·06	17·08	45·83	167·7	30·41	—
Ethyl cinnamate	Abs. const.	19·5	1·0497	36·59	73·63	24·01	24·48	24·24	43·00	175·2	31·31	1308
Ethyl hydrocinnamate	Abs. const.	21·5	1·014	33·23	73·48	24·30	24·73	24·5	41·02	—	—	1284
Castor oil	—	17·0	0·966	35·27	74·15	26·2	19·7	23·0	43·26	—	—	—

v is the volume of 1 gram-molecule, $v^{2/3}$ the molecular surface.

which refers to unit area, and the molecular surface ($v^{2/3}$). The last column in Table III shows that this product follows no regular system.

This remarkable fact may be interpreted in various ways, but from whatever aspect it is considered the result is likely to be barren unless molecules are taken account of. If the figures in Table III are interpreted to mean that the energy peculiar to an interface between two fluids is determined by the chemical reaction between the fluids, we may conclude that such action will be complete and independent of the molecular volumes, since the mass of the fluids on either side of the interface is practically infinite. Against this view, however, it is to be noted that chemical action between water and such stable substances as paraffin or cyclohexane, both in mass, is unknown. If it did occur at an interface it could only be as the result of great stresses.

Table III shows most clearly that the quantity T'_{AB} increases with what may be called the chemical reactivity of the fluid B, and especially that it is greatest when the molecules are of the salt type—acids, alcohols, or esters. Such molecules by their constitution readily exhibit electric polarisation, and we have here additional evidence for the fact mentioned in the earlier papers that the chief modifying factor in all interfaces is the development of a contact difference of potential, due to polarisation of the molecules by stresses normal to the interface. If the quantity T'_{AB} represents mainly the specific electric polarisation of the interface, then its value for paraffin, cyclohexane, etc. may represent polarisation due to the water molecules, which are of this salt type. For a surface between two saturated substances not of salt type the quantity might well be zero, in which case we should have

$$T_{AB} = T_A + T_B.$$

From the fact that T'_{AB} is equal for molecules of the same chemical type two conclusions follow. Let each molecule of A attract a molecule of B by a force which is a function only of the distance which separates them, the density term in the Laplacian theory then becomes the reciprocal of the molecular volume and, following the usual notation,

$$T'_{AB} = \frac{\pi}{v_A v_B} \int_0^\infty z \psi_{AB}(z)\, dz,$$

v_A, v_B being respectively the molecular volumes, and z the axis normal to the interface. Then on Young's hypothesis, that molecular attraction is a force which is of constant value over the range a, we have for two similar chemical substances, for which therefore T'_{AB} is the same, $a_1/a_2 = v_A'/v_A''$. That is to say, the range would be proportional to the molecular volume.

The alternative assumption, that the attractive force of a molecule of A for one of B falls off according to some power of the distance which separates them, yields the result that for similar chemical substances, since T'_{AB} is equal, the molecular volumes vary inversely with this power. Thus, if the attractive force $\phi(f)$ be put $= e^{-\beta f}$ then $v_A'/v_A'' = \beta_{\shortparallel}^5/\beta_{\shortmid}^5$.

33

THE TENSION OF COMPOSITE FLUID SURFACES
NO. II

[*Proc. Roy. Soc.* (1913), A, LXXXVIII, 313]

With the figures in Table III of the preceding paper (p. 548) as a guide the problem of the spreading of one fluid over the surface of another may be approached with some sense of security. In an earlier paper[1] the equation of a composite surface was found to be $T_S = T_A + T_{AB} - Kg$, where T_S is the tension of the composite surface, and Kg a term depending upon gravity. Putting $Kg = 0$, it is seen that spreading will occur only when $T_S > T_A + T_{AB}$, and, at the limit, $T_S = T_B$, that is to the tension of pure water in the experiments under consideration.

Taking $T_B = 74$, the tension of pure water, we have from the last paper:

TABLE I

	T_A	T_{AB}	
Cyclohexane	24·21	+ 55·15	= 79·36
Octane	20·63	52·96	73·59
Oil "C"	30·12	61·35	91·47
Carbon disulphide	30·88	56·15	87·03
Carbon tetrachloride	24·51	50·18	74·69

That is to say a drop of any of these saturated substances should not spread upon a surface of pure water except octane, and in actual fact none of them do spread, except octane, a drop of which slowly expands on water to form a very thin plate. In the case of all the other substances examined the quantity $(T_A + T_{AB}) < T_B$, and they all flash over a water surface in characteristic fashion.

The word "spreading" here refers strictly to the expansion of a lens of A over a clean surface of B due to the tension of B being greater than the sum of the tensions of the upper and lower surfaces of the lens. The vapour of A will condense on to the surface of B since $T_{AB} < T_B$. At the edge of a lens there is therefore a condensation of vapour of A on to the water, and the sheet of lowered tension so produced is pulled outwards away from the lens by the higher tension of the water surface. The sheet will also be pulled inwards to the lens, a quantity of the vapour of A so finding its way back into the lens. When the saturated substance has a high vapour pressure, such as, *e.g.*, cyclohexane, the presence of unstable sheets of condensed vapour about each lens is easily detected and readily explains the

---

[1] *Proc. Roy. Soc.* (1912), A, LXXXVI, 610.

interesting attractions and repulsions of lenses for each other. When undisturbed a pair of lenses will often continue alternately to attract and repel one another, producing a quite regular pulsation.[1] The movements cease when the space above the surface is saturated with the vapour of A.

When the vapour tension of A is practically zero a lens shows no trace of spreading of any kind. This is the case with oil "C", which boils somewhere about 400°. For the sum of the tensions of the upper and lower surfaces of a lens of this oil we have $T_A + T_{AB} = 30\cdot12 + 61\cdot35 = 91\cdot47$. It is obviously impossible for the tension of water (74) to pull such a lens out into a sheet.

When the fluid A is a pure chemical substance, or when it is composed of substances having identical influence upon the tension, the phenomena of spreading are always of the simplest. The first added fluid forms a continuous sheet on the surface of the water (B), which may be thickened until spreading ceases, when the excess remains as a single lens. I have met no exception to the rule that, when pure A spreads on water, the equilibrium state is a single lens in tensile balance with a uniform composite sheet of A spread evenly on B. This follows from experiment and also from a consideration of vapour pressures; for let the space above be enclosed, and let a number of lenses of different horizontal diameters be formed. Lenses of greater curvature will lose vapour, and lenses of less curvature will condense it until a single lens is formed. A number of lenses of precisely the same curvature could co-exist, but the equilibrium would probably always be unstable. The condition of equilibrium therefore is twofold, (1) that the lens and the plane surface are in tensile equilibrium according to the equation $T = T_A + T_{AB} - Kg$, and (2) that the vapour tension of the lens and of the plane surface shall be the same.

When a large lens of a fluid whose vapour pressure is negligible is placed on water if it is capable of spreading the lens is at once extended to form an irregular sheet, which then proceeds to contract to one or more lenses which are in tensile equilibrium with a composite surface. Equilibrium is reached quickly when salts are present in the water and slowly when the water is of low conductivity. Taking, for instance, ricinoleic acid as an example, equilibrium is reached with tap water in 10 sec., and with distilled water only after perhaps 20 min. When distilled water is used there appears to be a large tangential viscosity which impedes contraction of the excess oil. As I have already pointed out[2] the film is at a different electrical potential to the underlying water, and the tangential viscosity may be due to the low

---

[1] Cantor (*Wied. Ann.* (1895), LVI, 492) is wrong in his conclusion that the vapours of fluids which do not "wet" the surface of water will not condense on to the surface. The saturated vapours of the substances mentioned in Table I will condense on to tap water at the same temperature, as a dew of fine lenses, though in the case of carbon bisulphide owing to the high value of $T_A + T_{AB}$ the dew forms only with difficulty.

[2] *Proc. Roy. Soc.* (1911), B, LXXXIV, 220; also (1912). A, LXXXVI, 608.

conductivity of distilled water delaying the dissipation of electrical energy. Whatever be the true explanation the contrast is remarkable. When distilled water is used the tangential viscosity is so great as to allow the sheet of acid to develop wrinkles and folds as though it were a solid.

What we may picture as happening in all cases is that since $T_B > T_A + T_{AB}$ the lens is at once pulled out to form a sheet. From the visible edge of the sheet fluid is spreading on to the surface of B, at first rapidly, then more slowly, the flow being impeded by the viscosity of the film and possibly by other causes. As a consequence of this streaming from the edge the tension of the plane composite surface falls, and as it does so the extended lens contracts. The sheet formed by the first rapid extension of the lens of A frequently is unstable and ruptures. Circular spaces appear which are occupied by a composite surface similar to the composite surface outside the lens. The chief features on which I would insist here are (1) the relatively gradual formation of the composite surface by increase in the mass of A per unit area until it is in tensile equilibrium with a convex lens of A, and (2) the fact that when a lens of A is extended to form a sheet, the sheet is unstable if its thickness fall below a certain quantity.

When the fluid A has a sensible vapour tension—such as benzene or toluene—other features appear. Let us start with excess benzene placed on water, and exposed to air, the benzene is pulled out into an irregular sheet. When the air space is covered in so that it may become saturated with benzene vapour the sheet contracts to one or more lenses which are highly convex and quite immobile. These lenses are in tensile equilibrium with a composite surface of benzene and water. If the cover to the space be removed benzene evaporates from the composite surface, the tension rises, and the convex lens (or lenses) is at once pulled out to a sheet. It is very easy again to notice that an over extended sheet of benzene is unstable and ruptures. It will be noticed that whatever the vapour tension of A, whether it be high or insensible, the system settles down to the same state, namely, a composite sheet in tensile equilibrium with a convex lens of A. Now the tension of one face of the lens is T_{AB}, and of the other is T_A, or if B be soluble in A some quantity slightly greater than T_A. Therefore, since the lenses are convex we have for the equilibrium state the important relation $T_S < T_A + T_{AB}$.

In the case of the chemically saturated substances in Table I the lens is not pulled out to form a sheet, and fluid cannot be drawn from its edge on to the surface of the water since $T_A + T_{AB} > T_B$. But the vapour condenses on to the water since $T_{AB} < T_B$. This may readily be proved by pouring the vapour of octane, or carbon bisulphide, etc., on to a clean surface of water and noticing the movement of lycopodium grains dusted on to the surface.

When A is not a pure substance but a mixture of substances which individually produce effects of different magnitude upon the tension of B, the phenomena of spreading are more complex. Mixtures of, *e.g.*, cyclohexane or oil "C" and oleic acid or stearic acid at first flash over the surface as a continuous sheet which, viewed at certain angles, is uniformly and brilliantly coloured. From the particular colour we may put this sheet as from 300 to $1000\,\mu\mu$ thick. The sheet is unstable and ruptures, circular spaces appearing. As these spaces extend a horizontal net is formed, and finally the bars of the net rupture and the isolated masses contract to lenses. The total time occupied in forming the lenses, and the relative duration of particular stages depend upon the ratio of the components of A, upon their nature, and upon the concentration of electrolytes in the water. Increasing the number of components as a rule lengthens the total time and the complexity of the phenomena, and the vivid play of Newtonian colours is then very beautiful. By increasing the number of components a composite sheet may be formed of films of different thicknesses which pass abruptly into each other. The whole surface shows sharply determined coloured areas, red, blue, green, bronze, etc., which endure for days.

The phenomena ultimately depend upon the accumulation by diffusion of the most active constituents at the interface, and their complexity and durability are due to the slowness with which tangential diffusion can occur owing to the small depth of the film. The phenomena, occurring as they do on the surface of water, may be said to take place in two dimensions. If they occupied the mass of the fluid, that is to say, if they were in three dimensions, the unstable sponges, films, and lenses would yield the phenomena of the colloidal state, and especially of gelation. The horizontal networks which appear, and which are in an unstable state so far as surface energy is concerned, must represent in a crude way the mechanism of a muscle reduced to two dimensions, for there can be little doubt now that the force of muscular contraction is derived from changes in the energy of surfaces in the interior of each muscle fibre.

Curves of the change of tension produced by the spreading of A upon water, obtained by Wilhelmj's method of measuring surface tension, were given in an earlier paper (*loc. cit.*), and it was noticed that, when disturbance due to hysteresis of the surface is avoided, the curve for certain substances consists of a series of straight lines. I add here the curve for pure oleic acid (Fig. 1). The vapour pressure is so low as to render the error due to vaporisation from the surface negligible, but, unfortunately, the acid leaks past the barriers used to contract the film by diffusion through the body of the water, so that the observed tension tends always to be too high. In time, the oleic acid completely escapes from the control of the barriers, thus a contracted surface of tension 46·2 was left overnight, two barriers close together confining it. Twenty hours later the tension had risen to 60·9. By again sweeping as much of the acid as possible on to the same restricted area the tension fell to 47·6 at the point marked with a cross in Fig. 2. The

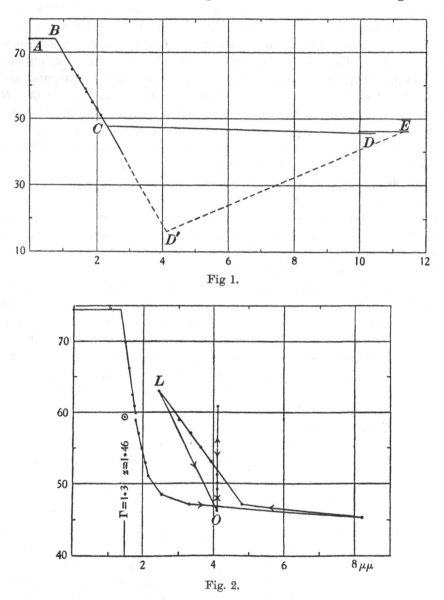

Fig 1.

Fig. 2.

curve for oleic acid is again the same type of curve—straight lines and sharp inflections. In the previous paper other curves are given in which there is only one sharp inflection and no part beyond this forms a straight line. With the fuller knowledge now in my possession I regard such curves as characteristic of films composed of two substances at least, one having an effect on the tension of the surface less than that of the other.[1]

Measurements of the tension T_{AB} throw some light upon the singular features of these curves. Let it first be supposed possible by avoiding the formation of a lens (as, *e.g.*, by allowing vapour to condense on to the surface in the absence of nuclei) continuously to increase the mass of A per unit of surface area, a point will be reached when an independent surface of pure A is formed. The curve therefore begins at $T_S = T_B$ and ends at $T_S = T_A + T_{AB}$, both points being determinable by experiment. Of the intervening curve the first portion AB is horizontal or slightly undulating.[2] Lord Rayleigh interpreted it to be the region in which the quantity of A placed on the surface is insufficient to form a continuous sheet. The objection to this interpretation is the dimensions of the molecule which follow from experiment. Taking, for instance, pure oleic acid, the depth of the film of acid at the point B is of the order 13×10^{-8} cm. Putting 10^{-8} as the diameter of the hydrogen atom, a molecule of the acid regarded as a sphere would have by the Barlow-Pope theory a diameter of only $4 \cdot 8 \times 10^{-8}$ cm. Perhaps a way out of the difficulty may be found ultimately by treating the point B as a critical point in the electric polarisation of the surface, also the calculated thickness of the film of A may be largely in excess of the real thickness, for complete immiscibility of A and B cannot be postulated, and some loss must occur owing to diffusion into the mass of B.

Lord Rayleigh remarks that "an essentially different result would seem to require a repulsive force between the molecules (of oil), resisting" contraction of the film. There is, as I have already pointed out, evidence to show that the oil spread on the surface is at a different electrical potential to the water. A repulsion due to the charge on each molecule of oil must therefore exist. But this only increases the difficulty, since such a tangential repulsion would, if it operated alone, bring about a fall in the tension before a continuous layer of oil was formed on the surface. It is open to us to suppose that one of the first effects of the oil is to undo a state of affairs at the surface of the water, namely, an average orientation of the water molecules themselves under the influence of the inwardly directed force of attraction, and thereby to increase the tension of the water, but a hypothesis which goes so far beyond ascertained fact cannot be so satisfactory as the direct explanation offered by Lord Rayleigh.

The significance of the inflection at C (Fig. 1) is, I think, clear. It is the point where the continuous uniform sheet of A spread on B ceases to be stable and any further added quantity gathers into a lens. This occurs when

---

[1] The sample of cymene used previously was found to contain a small percentage of an impurity with a high boiling-point.

[2] *Proc. Roy. Soc.* (1912), A, LXXXVI, 623.

the tension of the composite plane surface is equal to $T_A + T_{AB}$. The following table confirms this conclusion. T_C is the tension at the point C, as measured by Wilhelmj's method:

Castor oil	$T_C = 58\cdot0$	$T_A + T_{AB} = 58\cdot27$
Oleic acid	$= 47\cdot5$	$= 46\cdot30$
Ricinoleic acid	$= 48\cdot8$	$= 48\cdot34$
,, ,,	$= 45\cdot0$	
Ethyl hydrocinnamate	Between 55·64 and 58·16*	$= 57\cdot73$
Benzene	$= +64\cdot0$†	$= 65\cdot4$

* Camphor is still active on a surface on which a lens of this ester is standing, the space above being saturated with the vapour. Rayleigh fixes the camphor point at 57·7.

† In order to maintain the surface against loss by evaporation it is necessary to have a flat lens of benzene, therefore Kg has a sensible value, and the tension observed is slightly lower than T_C.

The slope of the line CD is determined partly by the form of the lens or lenses on the surface, that is to say, by the quantities T_A, T_{AB}, and the density of the fluid A or D_A; and partly by the circumstance that the abscissæ are calculated from the total mass of A placed on the surface divided by the area. They do not, therefore, any longer represent the true mass of A per unit area of the composite surface, but an average of that quantity and the mass absorbed into the lens. Since this is always greater than the true value, the line CD should slope more steeply than it does in the figure. The point C, therefore, is not the lowest point of the curve. This conclusion follows from other considerations. In the case of chemically saturated substances C ($= T_A + T_{AB}$) lies on or above the level of T_B. In spite of this, vapour of any of these substances condenses on to a surface of pure water and in so doing lowers the tension. Also, as we have already seen, the relation $T_S < T_A + T_{AB}$ holds for all surfaces in which excess of A forms a convex lens on a composite surface of A and B, and such lenses are formed by the condensation of the vapour of carbon tetrachloride, benzene, etc., on to the surface of water. Lastly, the mechanical stability of a film formed from the composite surface is not minimal at C but at a point of lower tension produced by still further concentration of A on to the surface.[1] The true curve of the variation of tension of the composite surface of A and B, therefore, must fall below the point C to a minimum at D' and from C to D' the surface when realised is unstable with respect to lens formation. This explains an otherwise anomalous result which I have thrice obtained with exceedingly pure oleic acid spread on to a very clean surface of tap water (temperatures 14·5° and 11·5°), namely, that at a certain point further thickening of the film of acid by contraction of the surface caused a sharp rise of tension which thenceforward remained constant at the higher level even when the film of acid was thickened as much as possible by extreme

[1] *Proc. Roy. Soc.* (1912), A, LXXXVI, 630.

contraction of the surface. Each step in the contraction now gave rise to a sharp rise of tension, which at once gave way, the tension falling to the constant figure. Inspection of the surface with a hand lens left no room for doubt as to the explanation of the phenomenon, for it was seen to be covered by a dew of very tiny lenses. On extending the surface the dew disappeared just as though the acid had evaporated and the tension now rose and fell normally with movement of the barrier. On again contracting the surface the dew reappeared and the tension at once became practically independent of the area. It is probably not owing merely to accident that this striking phenomenon of supersaturation has been observed only when a substance of high chemical purity was used.

At D' some new physical factor must come into operation. One supposition open to us is that at D' the attraction of B for A is fully satisfied. This would make $\Gamma_{D'}/D_A$ equal to the range of the attractive force between the fluids B and A if the density of A on the composite surface be put equal to that of A in mass. Γ is the mass of A per unit area of surface.

The part of the curve $D'E$, of unknown slope, would then relate entirely to the work expended in forming a layer of a new phase, namely pure A, and the process is complete at E when the tension of the upper face becomes T_A, and that of both faces $T_A + T_{AB}$. The two tensions are now equal to the tension at C, which, however, is that of a single true composite surface.

The diagram (Fig. 3) may serve to make the explanation clearer. On it are shown curves for four substances, the point D' being in each case

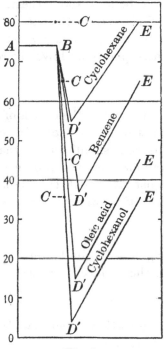

Fig. 3.

put equal to T_{AB}. The scale of the abscissæ and the slope of the lines is entirely arbitrary. At the point C on each curve the tension of the composite surface has fallen to the equality $T_S = T_A + T_{AB}$, save in the case of the chemically saturated substance, when T_S is always $> T_A + T_{AB}$. From C to D' the composite surface is stable with respect to infinitesimal variations of the mass of A per unit area, but unstable with respect to finite variations. That is to say, a lens cannot form spontaneously, because any tendency to a local accumulation of the fluid A will be resisted by the tension of the surface. But if suitable nuclei are present, or if a lens of A be placed on the surface, condensation must occur. If the curve from

B to D' be a straight line we have, as the equations of this part,

$$\frac{dT_s}{d\Gamma} = -c \quad (1) \qquad \text{and} \qquad T_s = C - c\Gamma, \qquad (2)$$

Γ being the mass of A per unit area of surface.

The part $D'E$ is a region of complete instability, hence, as has already been pointed out, when a lens of A is extended to form a sheet, the sheet ruptures when its depth is diminished below a certain minimum. The form and slope of this part are entirely unknown.

The changes of the vapour pressure of A in equilibrium with the composite surface cannot be followed with any certainty, though something may be said about them. At A on the curve the vapour pressure of the fluid A in a closed space above the surface would of course be zero, and at E it is that of the vapour saturated in contact with pure A in mass. Between these points we have the relation established by Gibbs,[1] namely

$$dT_s = -\frac{\Gamma}{\gamma}\, dp, \qquad (3)$$

in which Γ is the excess of A concentrated on unit area of the surface over what would be there if no concentration occurred. Γ therefore has sensibly the same significance as that given to it in this paper; γ is the density of the vapour. As an approximation we may put $\gamma = p/R\theta$, in which R is the gas constant, and θ the absolute temperature on the thermodynamic scale. Equation (3) now becomes

$$dT_s = -\Gamma R\theta\, \frac{dp}{p},$$

which for limits Γ_0 and Γ_1, when the vapour is compressed isothermally on to the surface, becomes

$$(T_1 - T_0) = -R\theta \int_{\Gamma_0}^{\Gamma_1} \frac{\Gamma}{p}\, \frac{dp}{d\Gamma}\, d\Gamma,$$

an equation which can be integrated only if we know the form of the function $dp/d\Gamma$.

Equations connecting the vapour pressure with the tension of the composite surface are given by Warburg[2] and Cantor,[3] but they cannot be trusted owing to faulty reasoning. In both cases they are derived by equating the balance of work gained to the change in the potential of the surface when fluid is evaporated from A and condensed on to the surface of B, the series of operations performed upon A being essentially those used by Helmholtz[4] in his calculation of the change of free energy when a quantity dm of water is evaporated from the plane surface of a mass of

[1] *Trans. Conn. Acad.* (1878), III, 398.
[2] *Ann. Phys. u. Chem.* (1886), XXVIII, 394.
[3] *Ibid.* (1895), LVI, 492.   [4] *Mem. Phys. Soc.* I, pt. II.

pure water, and condensed into a solution of salt in water. In the final process, when the water vapour is condensed into the salt solution, Helmholtz puts the pressure constant, and the work therefore simply equal to pv. This is correct only if the mass of the salt solution be infinite. Warburg and Cantor follow the procedure of Helmholtz in that they put the pressure of the vapour of A constant while it is being condensed on to the surface of B. This is equivalent to putting T_S constant throughout the operation, and is obviously wrong.

Something may be said in answer to two questions, namely, (1) at what point on the curve is the vapour pressure of A maximal? and (2) at what point is it equal to that of the vapour tension of pure A?

Equation (3) may be written

$$\frac{dT_S}{d\Gamma} = -\frac{\Gamma}{\gamma}\frac{dp}{d\Gamma}.$$

Since Γ/γ is necessarily positive the curve would begin to ascend when $dp/d\Gamma$ becomes negative—that is, when for any further addition of A to the surface its vapour pressure falls. This would happen along $D'E$; and the fall in vapour pressure would be strictly analogous to that which occurs when spheres of fluid in equilibrium with vapour about them fuse to form a larger sphere or a plane sheet. At D', therefore, the vapour pressure of A would be maximal. It would also be supersaturated with reference to a plane surface of A, since along $D'E$ it is falling to the pressure of saturated vapour at E.

Passing backwards from D' towards C the vapour tension of A at the composite surface is at first greater than the vapour pressure of pure A, and the surface would be in both tensile and vapour equilibrium with a convex lens of A. At some point the tension of the vapour at the composite surface falls to an equality with that of a plane surface of pure A. The point at which this happens will depend upon the pressure of the saturated vapour and, since T_{AB} changes only slowly with temperature, the point would fall lower and lower on the line CE' as the temperature rises. That this is so appears from measurements made by Clark[1] of the tension at the interface of ethyl ether and glycerine, and of a surface of glycerine in contact with a vapour space in which the vapour was maintained in equilibrium with the pure ether. The quantity T_S is always $> T_{AB}$ except near the critical point of ether, when it is $< T_{AB}$. It is interesting that in both cases the curves which connect temperature and T_S and T_{AB} respectively are sensibly straight lines. The curve for T_S is at ordinary temperature high above that of T_{AB} but, falling more rapidly, it cuts the latter at about 170°—the critical temperature of ethyl ether being 194°.

[1] *Proc. Amer. Acad. of Arts and Sciences* (1906), XLI, 361.

The way in which vapour is as it were flung off a surface during spreading points very decisively to a large rise in the tension of the vapour of A at some stage in the process. Ethyl hydrocinnamate boils at 247–249° according to Beilstein, and as it is immiscible with water I hoped this ester would serve for measurements of the change of tension produced by thin films spread on water. It proved, however, to be quite impossible to obtain any measurements owing to the fact that quite large quantities of the ester evaporate practically completely from the clean surface of water in less than a minute. The evaporation is accompanied by violent currents and eddies on the surface. Measurements of the tension prove that evaporation is not from the fully extended sheet of ester, for, though the excess of each drop placed on to the surface evaporates, a residue remains as a film which causes a permanent fall of tension. Also, when the sheet of ester has been sufficiently thickened by a succession of drops (the larger part of each being lost by violent evaporation) and in that way the tension lowered permanently to $\pm$ 56 dynes, the rapid loss of vapour ceases and a thin extended sheet of ester will lie quietly upon the surface. It would occupy too much space to describe the complicated changes which occur when a lens of this ester is trying to spread over a surface of high tension, but, taken together, they show that the expulsion of vapour occurs only near the edge of an extended lens. Consider what happens when a lens is extended by a higher tension. In the processes of expansion and thinning of the sheet it must be dragged through the whole region of instability and high vapour pressure from E through D' to C. The only way of escape is for the unstable sheet to gather into small lenses, and this is what continually happens, very small lenses being formed by the cast off from the edge of the main sheet. The result in the main is that the potential energy of a lens of ester standing on a surface of high tension is partly expended in boiling the ester off the surface.

Any explanation of the variation of tension with the varying depth of the film of A to be adequate must include an explanation of the remarkable movements of, e.g., a pair of lenses of carbon bisulphide on a clean water surface. To exhibit the movements to perfection the lenses must be small and highly convex. If they come within about 1 or 2 cm. of each other they are violently attracted and move directly towards each other until the edges are nearly in contact, when equally violent repulsion occurs. In this way rapid alternate attraction and repulsion occurs, always, if undisturbed, accurately along a line joining the centres.

The explanation is I think as follows: Consider each lens when out of the sphere of influence of the other. Vapour is being condensed on to the water face to form a film which spreads as a film centrifugally until it is destroyed by evaporation. The lens is thus the centre of an area of lowered tension, which therefore forms a depression. When two lenses come sufficiently close together these depressed areas fuse and a trough is formed between the centres along which they move—the trough deepening as they approach.

With the near approach of the lenses the vapour sheet is thickened until the tension falls to D', beyond this there would be an abrupt rise of tension and a repulsion. The obstacle to the fusion of the lenses which is so obvious is the portion $D'E$ of the curve and a pair of lenses must acquire a critical quantity of kinetic energy before they can break through it.

A relation found by Antonoff from measurements of interfacial tensions would, when it holds, fix the vapour pressure of the fluid A at the point C on the curve as very near to that of a saturated solution of A in B. Using partially miscible substances, Antonoff[1] finds that the interfacial tension is equal to the difference between the tensions of the two phases. That is, in our notation,

$$T_{ab} = T_b - T_a, \tag{4}$$

and at the limit when the fluids are immiscible

$$T_{AB} = T_B - T_A.$$

Clearly the relation is not universal, nor even a very common one, for, taking fluids sensibly immiscible with water, we have

TABLE II

	$T_B - T_A$	T_{AB}		$T_B - T_A$	T_{AB}
Cyclohexane	50·0	55	Carbon tetrachloride	50·0	50
Octane	53·4	53	Castor oil	38·7	23
Oil "C"	44·0	61	Ethyl cinnamate	37·0	24
Carbon bisulphide	43·0	56	Ethyl hydrocinnamate	40·5	24

It will be noted that cyclohexane, oil "C", and carbon bisulphide are exceptions to Quincke's rule that T_{AB} is always less than $T_B - T_A$.

I am at a loss how to criticise Antonoff's values, since they purport to be calculated by the erroneous equation $T_S = m/2\pi r$. This would give results 40 to 50 per cent. wrong, but his figures agree with mine in the very low value for the quantity T_{AB} for alcohols, and in the value found for benzene. The values for T'_{AB} calculated from his figures also agree with those found by me for alcohols.

	T_{AB}	T'_{AB}
Isoamylic alcohol	4·43	44·59
Isobutyric alcohol	1·76	46·75
Benzene	32·6	34·2

It is possible that though the faulty equation is quoted with approval, the experimental results were obtained with tubes standardised by measuring the tension of some pure fluid whose tension is known.

[1] *Journ. de Chim. Phys.* (1907), v, 372.

The explanation of Antonoff's relation is simple, and its theoretical significance not great. If water is saturated with a fluid which causes the tension of water (T_B) to fall, the surface will also become saturated with this fluid. When this happens T_B is reduced to the tension at C on the curve, which, as we have already seen, is that of a surface saturated with A and is equal to $T_A + T_{AB}$. We therefore have $T_b = T_A + T_{AB}$, and equation (4) now becomes

$$T_{ab} = T_A + T_{AB} - T_a,$$

which, since T_a is always sensibly equal to T_A, means that T_{ab} is sensibly equal to T_{AB}, and this has already been found to be the case.

The portion of the curve BD' in the diagrams is drawn as a straight line for reasons which are given in an earlier paper. The curves which Lord Rayleigh gives of the changes of tension with variation in the thickness of a film are gradually bending lines,[1] and from theoretical considerations he concludes that the tension of a composite surface of A spread on B should vary as the square of the thickness of the film of A.[2] The question, in the long run, must be decided by experiment, and I therefore add a table of constants (Table III) calculated from measurements of tension by Wilhelmj's method, the thickness of the film of A being varied by contracting the surface with movable barriers.

The method of thickening a film by sweeping it up to a contracted area by blades is open to many objections. Though leakage of the surface past the barriers may be avoided, loss by diffusion must occur to some extent, since perfect immiscibility cannot be postulated. A contracted film of pure oleic acid in 20 hours had again distributed itself evenly over the entire surface, largely, so far as I could determine, by diffusion through the body of the water. There is thus always a decrement which tends to make the curve bend away from the axis of tension. The result of many measurements is to show that the more accurately and rapidly the series of measurements are made the more closely does the curve approximate to a straight line.

The most successful series I was able to obtain with oleic acid is plotted in Fig. 4, and the constants are given in the third column of Table III. Three barriers were used to contain the film and, as far as possible, kept touching each other throughout. The water used was carefully cleaned tap water. The readings were taken rapidly and were very steady until the first inflection of the curve at M; here, owing to some unknown cause, leakage occurred, the tension rising while the weight was being recorded. At N the tension rose abruptly to O owing to the formation of exceedingly fine lenses, each of which appeared under a hand lens as little more than a point. A further slow rise from O to P occurred in 9 min., and during

[1] *Phil. Mag.* (1899), (5), XLVIII, 331.
[2] *Ibid.* (1892), XXXIII, 468.

this 9 min. some lycopodium dust drifted 2·5 cm. away from the barriers on the side remote from the contracted film of oleic acid. The rise OP therefore was due to a very slow leakage past the barriers.

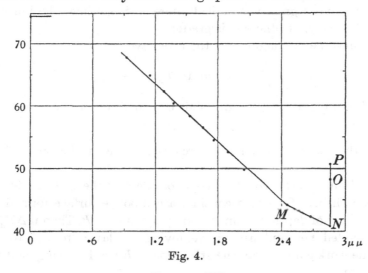

Fig. 4.

TABLE III

T_s	C_i	C_{ii}	C_{iii}	C_{iv}
72·4	23,528			
71·8	26,188			
70·85	25,612			
69·9	27,892			
68·9	28,652			
67·8	—	—	15·658	
67·0	27,450			
66·0	—	—	—	15,238
65·4	—	—		15,036
64·9	—	—	15·127	
63·9	—	—	—	15,070
63·5	—	—	—	15,316
63·0	26,503			
62·5	—		15·562	
62·0	—	17,390	—	15,180
60·4	—	—	15·830	15,198
59·0	26,831	18,067	—	15,252
58·5	—	—	15·523	
58·0				
57·0	20,045			
56·5	—	18,000	15·443	14,960
55·0				
54·6	—	—	15·694	
53·0	—	17,415		
52·6	—	—	15·587	
51·0	—	17,405		
49·9	—	—	15·796	
44·25	—	—	15·486	
42·5	—	—	14·831	
40·9	—	—	14·457	

C_i, C_{ii}, C_{iii} refer to three separate measurements made with pure oleic acid. C_{iv} to castor oil. C_i, C_{ii}, C_{iv}, the constants are in arbitrary units. C_{iii}, the units are dynes per linear centimetre and 10^{-7} cm.

The particular form of the curve near the second inflection (at point C) depends upon the conditions which favour or hinder lens formation such as the number and nature of solid particles or other nuclei in the surface film, and is not of very great theoretical interest. Sometimes the inflection is remarkably sharp, at others it is gradual.

If the curve be a straight line, then we have

$$\frac{dT_S}{d\Gamma} = -c \quad \text{and} \quad T_S = C - c\Gamma,$$

C being the integration constant. This last equation may be written

$$T_S = T_B - c\,(\Gamma_S - \Gamma_0), \tag{5}$$

Γ_0 being the mass of A per unit area of surface at the point C in the curve.

The relation $dT_S/d\Gamma = $ constant is of considerable theoretical interest if it can be established, for let us consider a composite surface formed by removing a layer of depth Δ from A and applying it to B. Then if ΔT_A is the work expended per unit area in removing the layer from A to ∞ and $2\Delta T'_{AB}$ the work gained from the attraction of B for A in bringing it to rest on B we have

$$T_S = \Delta T_A + T_B - 2\Delta T'_{AB}.$$

Let ρ_A, ρ_B be respectively the density of the fluids A and B, then if $m\rho_A\psi_A\,(z)$ is the attraction between a mass m of A distant z from an infinite mass of A and the latter, the work needed to remove to infinity a sheet of thickness dz is, per unit area, $\rho_A{}^2 \int_z^\infty \psi_A\,(z)\,dz$.

Call this integral $\theta_A\,(z)$, then ΔT_A becomes equal to $\rho_A{}^2 \int_0^\Delta \theta_A\,(z)\,dz$.

Similarly, $\qquad 2\Delta T'_{AB} = \rho_A\rho_B \int_0^\Delta \theta_{AB}\,(z)\,dz,$

and we have $\quad T'_S = T_B + \rho_A{}^2 \int_0^\Delta \theta_A\,(z)\,dz - \rho_A\rho_B \int_0^\Delta \theta_{AB}\,(z)\,dz$

and $\qquad\qquad \frac{dT_S}{dz} = \rho_A{}^2\theta_A\,(z) - \rho_A\rho_B\theta_{AB}\,(z).$

But $\qquad\qquad \Gamma = \Delta\rho_A \quad \text{and} \quad d\Gamma = dz\rho_A, \tag{6}$

therefore $\qquad\qquad \frac{dT_S}{d\Gamma} = \rho_A\theta_A\,(z) - \rho_B\theta_{AB}\,(z); \tag{7}$

and, since $\dfrac{dT_S}{d\Gamma}$ is by experiment equal to a constant,

$$\theta_A\,(z) = \frac{\rho_B}{\rho_A}\,\theta_{AB}\,(z) + \beta, \tag{8}$$

where β is another constant.

$\theta_A(z)$ is the function of the potential of A at the surface of A, and $\theta_{AB}(z)$ of A at the surface of B, hence we may conclude that if $dT_S/d\Gamma$ is equal to a constant the form of these functions must be the same.

We may perhaps proceed a stage further. Laplace assumed that the function of the attractive force between matter at minute distances is the same in all cases, "the attraction differing merely by coefficients analogous to densities in the theory of gravitation." On this assumption we may write equation (7)

$$\frac{dT}{d\Gamma} = \alpha\theta_A(z), \tag{9}$$

and $\theta_A(z)$ is then seen to be a constant, a result which is in agreement with the suggestion of Young that molecular attraction is a force which is constant in magnitude over the very minute range through which it acts. But the range in question must be less even than the thickness of the films of oil which when spread upon water reduce the tension, for, if $\theta(z)$ be constant, then since $d\theta(z) = -\psi(z)$ this last function is equal to 0. The physical significance of this last relation would be that each layer of molecules of A spread on the surface is attracted only by the layer of molecules previously there with which it comes into contact, a result not inconsistent with many aspects of this difficult question but altogether inconsistent with the view that the attraction of B for A ceases only at the point D' in the curve.

Though such conclusions, based as they are upon an assumption of uniform density throughout an interface, cannot have great value, they are interesting as pointing to an unexpected simplicity in molecular forces. The simple spatial relations which are the essence of the Barlow-Pope theory of the molecular structure of close-packed forms of matter also seems to demand some simple law. If there were, for instance, alternating zones of attraction and repulsion about each molecule, more than one arrangement in space would satisfy the condition of minimal potential, and it would be possible by adequate pressure to compress a fluid to a volume which it would continue to occupy when the pressure was lowered.

The unexpectedly simple relation $dT/d\Gamma = -$ a constant would appear to hold for an interface between solid and fluid. A large amount of work has shown that when a solute such as iodine is condensed on to the surface of, for instance, animal charcoal, the equilibrium reached is given by the empirical equation $m = kc^{1/n}$, where m is the mass of the solute condensed on to the surface, c the final concentration of the solution, and k and n are parameters. Putting the area of the surface of the solid as unity, this may be written

$$\gamma = \alpha\Gamma^n,$$

which on differentiating gives

$$d\gamma/d\Gamma = \alpha n\Gamma^{n-1} = n\Gamma/\gamma.$$

By comparison with Gibbs' equation $\Gamma = -\dfrac{\gamma}{R\theta}\dfrac{dT}{d\gamma}$, we get

$$\frac{dT}{d\Gamma} = -nR\theta.$$

Hence for an isothermal change $dT/d\Gamma$ is a negative constant.

Another significance of the constant c in (5) must not be lost sight of. Comparing the relation $dT_s = -cd\Gamma$, with the similar equation for volume energy $(dP) = Kd\gamma$, in which γ is the mass in unit volume, we see that c corresponds to K which in the gas equation is the gas constant R. But, whereas R implicitly refers to a zero of temperature, namely, the absolute zero, the constant c is related to the temperature of complete miscibility of A and B when T_s is zero.

If the fields of force about a molecule are not symmetrical, that is to say, if the equipotential surfaces do not form spheres about the centre of mass, the arrangement of the molecules of a pure fluid must be different at the surface from the purely random disposition which obtains on the average in the interior. The inwardly directed attractive force along the normal to the surface will orientate the molecules there. The surface film must therefore have a characteristic molecular architecture, and the condition of minimal potential involves two terms, one relating to the variation in density, the other to the orientation of the fields of force.[1] When the composite surface is formed its architecture is determined by the interaction of the fields of force of molecules of A and B under the influence of the attraction of B for A. When the structure is complete the tension T_{AB} is reached and any further addition of molecules of A to the surface does not disturb the architecture. But, just as there are in many cases two arrangements of the same molecules in the solid state, that of the glass, and that of the crystal, the former containing the greater quantity of energy per unit of mass, so in the formation of these films of matter the architecture actually reached may not always be that of least potential.

Description of figures. Save for the diagram Fig. 3 the scale of the ordinates is dynes per linear centimetre, that of the abscissæ 1×10^{-7} cm. ($\mu\mu$), calculated from the weight of acid spread on the surface on the assumption that the density of the film is that of oleic acid in mass. Fig. 1 is plotted from a particular experiment with pure oleic acid—that is to say with the purest acid which Kahlbaum prepares, further purified by fractional crystallisation. The line BC has been followed by Miss Pockell's

---

[1] Lord Rayleigh (*Phil. Mag.* (1892), XXXVIII, 309) has shown that the deviations from Fresnel's formula for the reflection of light at a liquid surface may be traced to the presence of a film of impurity on the surface. The residual deviation which persists when all such films are swept away may, perhaps, be attributed to the real surface film of the pure fluid.

method into the unstable region CD', until the end of the continuous line. The horizontal line through E marks the tension which is equal to the sum of the tension of the pure acid and that of the interface between acid and water or $T_A + T_{AB}$. Fig. 2 illustrates a prolonged experiment with oleic acid. The curve is interrupted at one point where there was an accident to the balance, the result being to alter the slope of the curve. The curve was followed without pause, save for the accident just alluded to, to the point O, when the barriers were left in position. The tension rose in 10 min. to the first dot, and 20 hours later it was found to be at the top of the vertical line. By again concentrating as much as possible of the oleic acid on to the same area the tension fell to the point marked with a thick cross. The contraction of the surface from L to O was made very rapidly. The ringed dot marks the tension when 0·1 mg. of oleic acid was spread on to a surface of clean distilled water 685·7 cm.² in extent. For this tension therefore $\Gamma = 1\cdot3 \times 10^{-7}$ gr. and $z = 1\cdot46\,\mu\mu$. The scale of the abscissæ is based upon this measurement. The remaining figures are adequately explained in the text.

APPENDIX I

In a previous paper[1] I pointed out that the mechanical stability of a composite surface is maximal just beyond the point B in the curve when the tension begins to fall. The measure of mechanical stability was the time which elapsed between the formation and bursting of bubbles of a particular size on the surface. The rise in mechanical stability from zero to a maximum was found to occur at a point some distance lower on the curve than B. The reason for this displacement is simple. When a film is formed at the surface by allowing a bubble of air to ascend from below, the surface is stretched and the quantity Γ, that is the concentration of the fluid A spread upon the water, thereby diminished. The effect is to put the state of the surface backwards along the curve towards the inflection at B. If the film is sufficiently stretched, that is if the bubble is large enough to make $\Gamma < \Gamma_\beta$, the bubble at once bursts. The amount of displacement of the rise of mechanical stability forwards along the curve is thus a function of the radius of the bubble of air. With infinitely small bubbles it would coincide with the inflection at B, where theory would place it.

APPENDIX II

An attempt to measure the tension of a composite surface in equilibrium with the saturated vapour of cyclohexane failed and the failure is instructive. Let a large lens be formed on the surface of water in a closed space. When the vapour pressures are in equilibrium since the lens of cyclohexane

---

[1] *Proc. Roy. Soc.* (1912), A, LXXXVI, 627.

is sensibly flat the vapour tension must be that of saturated vapour. In order to calculate the tension of the composite surface of water and condensed vapour of cyclohexane we have the relation

$$T_S = T_A + T_{AB} - \tfrac{1}{2}g \, (D_A h^2 - D_A h'^2 + D_B h'^2)$$

and the hydrostatic equation

$$D_A h = (D_B - D_A) \, h'.$$

h and h' refer respectively to the height of the upper surface of the lens above the surface of the water and the depth of the lower surface below it; h and h' very approximately can be got by measuring the area of the lens and its volume, and taking the thickness as uniform.

I found that when placed on tap water of tension 73·6, the vapour space being small, 0·5 c.c. cyclohexane formed a lens 6·4 cm. in horizontal diameter, 2 c.c. a lens 13·5 cm. diameter. The last term is negligible for lenses so thin (0·015 cm. and 0·014 cm. respectively), and T_S calculated is 79·34 and 79·35. Clearly, therefore, T_{AB} must have been reduced by the accumulation at the interface of some trace of impurity. The changes observed agree with this, for the cyclohexane when first put on forms a convex lens which slowly flattens out. The flattening cannot be due to an increase in T_S above T_B since the condensed vapour lowers the tension, as is readily seen by admitting a little air so as to relieve the concentration of the vapour of A, when the flattened lens at once contracts.

Appendix III

The arguments employed in this paper throw light upon a suggestion put forward by Laplace, and incidentally on that vexed point the physical significance to be attributed to the term "density" as used by Laplace. By the Young-Laplace theory we have for the intrinsic pressure and the surface tension of a fluid respectively

$$\kappa = 2\pi\rho_2 \int_0^\infty \psi \, (z) \, dz \tag{10}$$

and

$$T = 2\pi\rho_2 \int_0^\infty z\psi \, (z) \, dz. \tag{11}$$

Laplace assumes that $\psi \, (z)$, like gravity, is a function depending only on the density of the substance, and we may therefore write

$$T = \rho^2 T_0 \tag{12}$$

and

$$T'_{AB} = \rho_A \rho_B T_0. \tag{13}$$

From (10) and (12) follows

$$T_{AB} = (\rho_A - \rho_B)^2 \, T_0.$$

It has been objected that the facts do not accord with this relation. This is true, but the cause may lie either in the fact that T_0 is constant only for similar fluids—that is for fluids of the same chemical type, or in the difficulty in identifying the density of Laplace's theory with a particular physical property.

Comparing equations (3) of the paper preceding and (12) we get

$$\rho_0{}^2 T_0 = \frac{(\tau - d)}{v^{2/3}} \kappa.$$

If the density be taken as a molecular quantity then $\rho = 1/v$ and

$$T_0 = (\tau - d)\, \kappa V^{-\frac{2}{3}}.$$

From this we can derive equation (4) of the preceding paper as the expression for the interfacial tension.

The expression $T_{AB} = (\rho_A - \rho_B)^2\, T_0$ is now seen to be wrongly derived, the false assumptions being the identity of κ_A with κ_B, and of $(\tau_A - d)$ with $(\tau_B - d)$. And for a similar reason the expression $T'_{AB} = \rho_A \rho_B T_0$ is inadmissible.

34

NOTE ON DIFFERENCES IN ELECTRICAL POTENTIAL WITHIN THE LIVING CELL

[*Journ. Physiol.* (1913), XLVII, 108]

The appearance of a paper of much interest by G. L. Kite[1] on the physical properties of living matter has led me to write a brief account of some experiments which were carried out in 1903–4. The results were interesting and suggestive, but they raised difficulties which at the time seemed insuperable. The growing root tips of onions were used, and the first object aimed at was the exploration of the cell for interior differences of electrical potential. This led me to examine the effect of fixatives acting at low temperatures.

Many experiments were made to determine the effect of fixatives at low temperatures. Flemming's fluid, osmic acid, and formalin vapour were employed at temperatures ranging from $-2 \cdot 5$ to $25°$ C. In a typical case some root tips were fixed at room temperature as controls, others were slowly cooled to as low as $-2°$ C., the process occupying some hours, and then placed in fixatives previously cooled to the same temperature. They remained in the fixative in the ice-chest 20 hours, were then washed with ice-cold water, and transferred by slow dialysis in the cold to 95 per cent. alcohol, for cutting in paraffin, or at once removed from ice-cold water to gum for cutting frozen. Saffranin or iron-hæmatoxylin was used to stain. It will be seen that abrupt changes of temperature or of concentration of fluids was scrupulously avoided. Shrinkage, however, commonly occurred during the process of embedding in paraffin, the nuclear substances, for instance, pulling away from the nucleolus.

Nuclei fixed at about $0°$ C. which were not exposed to subsequent heating presented the following appearances. The nucleus consists of a homogeneous nuclear substance in which are embedded a dense nucleolus, and numerous granules. The granules vary in size in different nuclei but are fairly uniform in size in the same nucleus. They may be so small as to be almost a dust, or so large as to rival the nucleolus. The increase in the size of the granules is the only optical evidence of the ripening changes which precede mitosis. When the granulation has become very coarse the nucleus divides. So long as the granules are small there is a sharply defined nuclear membrane which, however, is absent from nuclei with very coarse granulation. Obviously such appearances cannot be insisted upon as representing

[1] *Amer. Journ. Physiol.* (1913), XXXII, 146.

normal nuclear structure owing to the large number of uncontrolled, and apparently uncontrollable, variables. But the observations are of interest because they agree more closely than does the figure seen in nuclei fixed at ordinary temperatures with the description of living nuclei given by Kite, who by the aid of a peculiar technique has succeeded in dissecting out the nucleus of various animal and vegetable cells. Kite, however, does not seem to have succeeded in recognising a nucleolus in any of the nuclei he has examined.

In order to detect interior differences of electrical potential root tips were laid horizontally between non-polarisable electrodes, the final lead to the tissue being through some of the fluid in which the roots had been growing. A field having a mean value of from 5 to 20 volts per centimetre was established for from 2 to 10 min. when the root was instantly fixed with acetic absolute. Longitudinal sections were cut and stained with iron-hæmatoxylin or saffranin. The strength of the field to which the living matter was actually exposed cannot be calculated. It seemed hardly worth while to measure the resistance of the root tip and of the electrodes in view of the uncertainty which must obtain as to the variations in the field within the tissues of the root.

The effects produced were decisive and consistent. They varied only in degree with variations in the strength of the field and the time of exposure.

The nucleus was usually altered slightly in shape. From a sphere it was pulled out so as to be ellipsoidal with the long axis parallel to the stream lines. It maintained its position in the middle of the cell—that is to say, taking the external boundary of a cell as the surface of reference, there was no movement of the nucleus as a whole. The solids of the cell substance were collected usually at the negative end of the cell, though not infrequently they were condensed into an equatorial plate. But prolonged exposure to a strong field condensed them at the positive end (Fig. 3). Within the nucleus the most characteristic and marked migration occurred, the bulk of the solids being collected at the positive end; single or double loops, spiral threads, or an open net projecting thence into the cleared space at the negative end. At the negative end, owing to the clearing away of the solids, the nuclear membrane was left very sharply defined. The nucleolus when visible sometimes maintained a central position, but mostly it was swept with the other solids to the positive end of the nucleus. The most remarkable feature, however, was in the absence of any effect where one would be most looked for, namely, in cells during nuclear division. There was no sign of any orientation or indeed of any displacement whatever of spindles or chromosomes. Even when currents were pushed to the point of general disruption of the tissue it could be said only that it was difficult or impossible to distinguish the mitotic figures.

In seeking for an interpretation of these appearances we have for guides the fact, first noticed by Faraday, that when a current traverses a gel the watery and the solid constituents move in opposite directions so that a rod of gel is swollen near one pole and constricted near the other. In most sols there is, as is well known, free migration of the colloid particles in an electric field. If the solution be of proteins, save in rare cases (globulins dissolved in a dilute solution of a neutral salt), the first effect is an aggregation and precipitation of the protein at one of the poles, the second a reversal of the

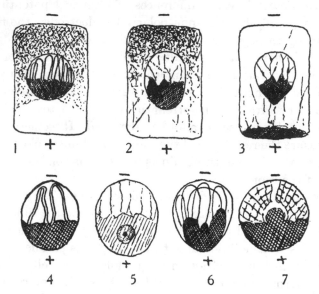

Figs. 1, 2, 3, 4, 6 and 7 are from preparations stained with iron-hæmatoxylin. Fig. 5 from one stained by saffranin. Figs. 4–7 are of nuclei only. The double loops in Fig. 4 are remarkable.

electric charge on the protein, its consequent repulsion and secondary migration to the other pole. Lastly Kite finds that nuclear substance and cell substance are in most cases a mixture of sols and gels.

The appearances in the cell substances are such as would be due to simple migration of colloid particles in a sol, that is to say of electrically charged particles distributed throughout a fluid. The sign of the particles being at first positive, they move towards the negative end of the cell (Figs. 1, 2), where reversal of sign of the charge occurs accompanied probably by alteration in the chemical nature of the proteins. The reversal of movement of the protein may end in the formation of a dense plate at the positive end as in the remarkable case shown in Fig. 3.

Within the nucleus convection also occurs, but the charge on the colloid particles is negative. The loops and threads which are, as it were, left behind represent the masses of gel described by Kite. The appearances throw no light upon the existence or non-existence of potential differences within such masses, but I think they are inconsistent with the existence of any potential difference between nuclear sol and nuclear gel.

A puzzling feature is the absence of any movement of the nucleus as a whole—the distortion in shape of the nucleus is such as might be produced by mechanical stresses due to movement of the contents. The freedom of movement displayed by the solids of the cell substance precludes the idea that the nucleus is anchored in its place by strands of gel. We seem to be driven to the unexpected conclusion that, although the colloid particles inside the nucleus are negative, while those outside are positive, the general balance of colloid and salts is such that there is no electric charge upon the surface of the nucleus. Similarly the appearances point to the absence of a potential difference between cell substance and cell wall.

During actual mitotic division the effect upon the nuclear structures is, as I have said, nil. An explanation is afforded by the work of Kite. His dissections of living cells and nuclei show that in dividing cells all the surrounding cell substance can be cut away, and the spindle itself cut to pieces in Ringer's fluid without the pieces undergoing any change in shape. Thus the rigidity of the contents of a dividing cell is such that the electrical stresses are unable to produce obvious mechanical effects.

35

AN APPLICATION OF THE PRINCIPLE OF DYNAMICAL SIMILITUDE TO MOLECULAR PHYSICS[1]

[Proc. Roy. Soc. (1915), A, XCII, 82]

1. Consider a molecular system subject to the condition of stationary motion, and let the forces acting between or within the molecules be conservative. A statistical specification of the energy content of such a system can be obtained either by considering a mass containing a large number of molecules, and expressing the energy as the sum of the energies of individual molecules at an instant of time, or by averaging the energy of individual molecules over a sufficiently long time, and expressing the whole energy as the sum of these time averages.

Let us choose the former method. At any instant of time some of the molecules will be in material contact, some free, and, if the molecules are not all alike and some are formed by the more or less complete fusion of others there will be molecules in process of combination to form a distinct type, and molecules of this type in process of breaking up into constituent molecules.

Material contact between molecules is to mean any encounter in which forces of mutual repulsion come into operation. This definition is exact so long as radiation between the molecules is excluded (as it is by the imposition of the condition of conservatism). For the term "repulsive forces" is here to refer to those forces which confer upon the individual molecules the property of extension in space. It is not necessary for the moment to enquire whether these forces arise wholly or in part from the variation of the kinetic energy of internal degrees of freedom.

As the molecules are the moving parts, it is necessary to specify the energy in terms of their number and kind. The molecules may be classified in two ways, which yield totally different results, namely, according to their chemical structure or according to their dynamical characters, *e.g.*, the number of degrees of freedom. The latter is more general; it includes the former, and it is necessary to follow it in this paper.

---

[1] A reference to a paper by Kamerlingh Onnes (*Verh. d. K. Akad. d. Wet.* 3, Amsterdam (1881), vol. XXI) first drew my attention to this principle. On turning to the paper I found it to be in Dutch, and therefore sealed from my comprehension. This communication gives the results of an entirely independent research, which will, I hope, be found to cover fresh ground.

We can recognise three conditions: free molecules, molecules in contact, and transition stages. The free molecules will include any multiple from unity upwards of the chemical molecule or molecules which form the system. Let the free molecules be divided into n classes in any convenient manner: the transition stages then are stages of combination or breaking apart of molecules of some of these classes to form molecules of other classes. Stages in the combination of single chemical molecules to form complexes would come within this category. Obviously, in order to avoid statistical ambiguity, the n classes must be so defined as to exclude transition stages, which then form a class apart. A sufficient definition is that molecules of the n classes are structures in stable equilibrium; transition states, on the contrary, are in process of internal change, and are therefore labile states.

Let the word "encounter" refer to molecules which are merely in material contact. Such molecules can be enumerated as members of the n classes, for during an encounter they do not lose their individual dynamical characters.

If E is the whole energy of the mass, and E_a, E_b, etc. that of the molecular species, a, b, etc.,

$$\bar{E} = \bar{E}_a + \bar{E}_b + \ldots + \Delta,$$

where Δ is the energy of transition states. Since in a steady state the distribution of the energy amongst the different molecular species is constant, we can write

$$\Delta / \bar{E} = \text{constant}, \tag{1}$$

the constancy being supposed to refer to averages taken over different masses of the same chemical fluid each containing a very large number of molecules and each in the same state as defined by temperature, pressure and specific volume.

The kinetic energy of any one molecule of a species, say species a, is $\frac{1}{2}\sum_{0}^{s} c\eta^2$, where s is the number of degrees of freedom. The potential energy similarly may be regarded as the sum of a number of terms representing respectively the potential energy due to the attraction of the other molecules and the elastic forces within the molecule, including repulsive forces, that is to say those forces which confer extension in space upon the molecule. Since the forces are conservative the potential energy is a pure function of the co-ordinates which specify position, therefore each term in the potential energy will have the form

$$\phi = \int (X\,dx + Y\,dy + Z\,dz), \tag{2}$$

where X, Y, Z are the resolved forces.

Averaging over all the molecules of class a, we have as the total energy:

$$\bar{E}_a = \sum \left\{ \frac{1}{2} \sum_{0}^{s} \overline{c_a \eta_\alpha}^2 + \sum \bar{\phi}_a \right\}, \tag{3}$$

$\frac{1}{2}\overline{c\eta}^2$ is the average kinetic energy of a degree of freedom, and $\overline{\phi}$ the average potential energy of any one system of forces. If n denote the number of distinct species of molecules, the energy of the mass is

$$\bar{E} = \overset{n}{\underset{0}{\Sigma}} \left[\tfrac{1}{2} \overset{s}{\underset{0}{\Sigma\Sigma}} \overline{c\eta}^2 + \Sigma\Sigma\overline{\phi} \right] + \Delta. \qquad (4)$$

The quantity Δ includes both kinetic and potential energy. It is the energy of molecular states which are not in stable equilibrium—that is of labile states. It is a familiar fact that molecular reactions involving the formation of new species often can take place only when molecules which do not form part of the new species assist. An extreme instance of this is found in ferment actions. In general a reaction between two molecules leading to structural changes may be supposed to be possible only in certain special local states of the general internal field of force as regards intensity and, especially if the forces are electrical, orientation. Thus from whatever aspect the energy Δ is regarded it is seen to be energy which cannot be considered as belonging to the chemical molecule or to molecules formed by association of chemical molecules, for these are molecules which are in stable equilibrium. It is obvious that Δ vanishes when the system is expanded to the state of an ideal gas, and increases on the whole as it is condensed, until the crystalline state is reached, when the gas molecule as such ceases to exist.

Let there be two systems, distinguished by the suffixes 1 and 2, which differ only in their linear dimensions. There will be two equations similar to equation (4):

$$\bar{E}_1 = \overset{n}{\underset{0}{\Sigma}} \left[\tfrac{1}{2} \overset{s}{\underset{0}{\Sigma\Sigma}} \overline{c_1\eta_1}^2 + \Sigma\Sigma\overline{\phi_1} \right] + \Delta_1, \qquad (5)$$

$$\bar{E}_2 = \overset{n}{\underset{0}{\Sigma}} \left[\tfrac{1}{2} \overset{s}{\underset{0}{\Sigma\Sigma}} \overline{c_2\eta_2}^2 + \Sigma\Sigma\overline{\phi_2} \right] + \Delta_2. \qquad (6)$$

The limits of summation will be the same in both, and

$$\bar{E}_1/\bar{E}_2 = \Delta_1/\Delta_2. \qquad (7)$$

The one equation will be changed into the other if

$$\left. \begin{array}{l} \bar{c}_1{}' = \mu\bar{c}_2{}' \\ \bar{c}_1{}'' = \mu\bar{c}_2{}'' \end{array} \right\} \quad (8) \qquad\qquad \left. \begin{array}{l} \bar{\eta}_1{}' = \beta\bar{\eta}_2{}' \\ \bar{\eta}_1{}'' = \beta\bar{\eta}_2{}'' \end{array} \right\} \quad (9)$$
$$\text{etc.} \qquad\qquad\qquad\qquad \text{etc.}$$

$$\left. \begin{array}{l} X_1{}' = FX_2{}' \\ X_1{}'' = FX_2{}'' \end{array} \right\} \quad (10) \qquad\qquad \left. \begin{array}{l} x_1{}' = lx_2{}' \\ x_1{}'' = lx_2{}'' \end{array} \right\} \quad (11)$$
$$\text{etc.} \qquad\qquad\qquad\qquad \text{etc.}$$

$$\Delta_1 = Fl\Delta_2 \quad (12) \qquad \text{and} \qquad \mu\beta^2 = Fl. \quad (13)$$

On comparing equations (7)–(13) with equations (5) and (6) it is seen that

$$\bar{E}_1 = \mu\beta^2\bar{E}_2. \qquad (14)$$

2. Equation (8) practically limits the application of the principle of dynamic similitude to molecular systems composed of one single chemical substance, and to chemical series such as the paraffins, alcohols, etc. For such a series the parameter μ may be identified with the ratio of the molecular weights, or if M is the gram-molecule

$$\mu = M_1/M_2. \tag{15}$$

The parameter β offers greater difficulties. β is the ratio of corresponding linear velocities in the two systems. The theorem of equipartition of kinetic energy states that

$$\overline{c'\eta'^2} = \overline{c''\eta''^2} = \ldots = R\theta, \tag{16}$$

or the mean energy absorbed by each degree of freedom is the same. R is a constant having the dimensions of energy, and θ may be taken to be a pure number. We now have for two systems

$$R_1\theta_1/R_2\theta_2 = \overline{m_1 u_1^2}/\overline{m_2 u_2^2}, \tag{17}$$

in which m is the mass of the chemical molecules, and u a component of the velocity of the centre of gravity of the molecule.

If the units of energy are the same in both numerator and denominator, $R_1 = R_2$ and we get from (9) and (17)

$$\beta^2 = \tau_1 m_2/\tau_2 m_1 = \tau_1 M_2/\tau_2 M_1, \tag{18}$$

where τ_1 and τ_2 are the absolute temperatures of the two systems when dynamically similar, that is to say when they are in corresponding states.

The propriety of extending the law of equipartition of energy to condensed systems is, however, still open to debate.[1] But the statement that the temperature of a substance is proportional to the mean kinetic energy of translation of the molecule, taken in conjunction with equations (8) and (9), gives relation (18) independently of the distribution of the energy amongst the degrees of freedom. I do not think it can be maintained that there is any proof of this statement. It is founded upon two facts of experience: (1) that if a number of systems are placed in material contact, that is to say in a relation such that mutual repulsive forces come into play, they reach equality of temperature in the sense that if the physical state of any other system, say a column of mercury, be fixed upon as an indicator, and placed in contact with each of the systems in turn, it will indicate the same temperature for each; and (2) that potential energy of one system cannot be directly converted into the potential energy of another system. The fact that one of the contiguous systems may be a permanent gas enables us to identify the temperature scale with the kinetic scale.

The parameter l offers no difficulties, for corresponding volumes will be the volumes occupied by corresponding masses. If v is the specific volume,

---

[1] Cf. for instance Burberry, *Kinetic Theory of Gases*.

Mv will be a corresponding volume, namely, the volume occupied by a gram-molecule. We then have

$$l = (M_1 v_1)^{1/3}/(M_2 v_2)^{1/3}. \tag{19}$$

3. The real difficulty in the application of the principle of dynamical similitude to actual substances lies in the identification of corresponding states. Equation (4) may be written

$$\bar{E} = (a' + a'' + \ldots)\,\bar{E} + (b' + b'' + \ldots)\,\bar{E} + c\bar{E}, \tag{20}$$

the various coefficients being less than unity. In a steady state the coefficients are constant and corresponding states are such that $a_1' = a_2'$, $a_1'' = a_2''$, etc. Theoretically, therefore, any physical property such as temperature, pressure, specific volume, or magnetic susceptibility which can be used to measure the ratio of any one of the terms on the left (say $a'\,\bar{E}_1/a'\,\bar{E}_2$) can be used to identify corresponding states, since $\bar{E}_1$ and $\bar{E}_2$ will bear the same ratio, and these are quantities which in certain cases are subject to actual measurement.

Comparing equations (8), (9) and (18), we get

$$\bar{E}_1/\bar{E}_2 = \overline{\mathbf{M}_1 V_1{}^2}/\overline{\mathbf{M}_2 V_2{}^2} = \mu\beta^2 = \tau_1/\tau_2. \tag{21}$$

$\mathbf{M}$ here is the mass of a molecule of any one of the molecular species and $\bar{V}$ the average velocity of progressive motion.

There is no difficulty in fixing upon certain corresponding states, namely, that at the absolute zero, that of the ideal gas, and, with less certainty, the critical state. Putting θ_c for the critical temperature,

$$\tau = n\theta_c, \tag{22}$$

and for two systems in corresponding states

$$n_1/n_2 = (\mu\beta^2)_\tau/(\mu\beta^2)_c. \tag{23}$$

We may also write

$$\frac{n_1}{n_2} = \left(\frac{\overline{\mathbf{M}_\tau V_\tau{}^2}}{\overline{\mathbf{M}_c V_c{}^2}}\right)_1 \div \left(\frac{\overline{\mathbf{M}_\tau V_\tau{}^2}}{\overline{\mathbf{M}_c V_c{}^2}}\right)_2 = \frac{(\mu_1\beta_1{}^2)_{\tau.c.}}{(\mu_2\beta_2{}^2)_{\tau.c.}}. \tag{24}$$

Therefore, when $n_1 = n_2$, the systems 1 and 2 are not only dynamically similar at the corresponding temperature τ, but there is also mutual dynamical similarity along the temperature scale at temperatures defined by constant differences in the value of n.

It is important to notice that the theory affords no justification for the current view that when $n_1 = n_2$ there is no molecular association. It shows only that whatever degree of association there is in the one system will be reproduced exactly in the other system when there is correspondence.

In comparing different systems it is, of course, necessary to compare masses composed of the same number of molecules. The gram-molecule may be taken, for if two systems are dynamically similar the number of moving parts, that is the number of molecules including transition states, will then be the same in both, irrespective of the degree of association of the chemical molecules to form complex molecules.

4. The limitation of conservatism does not make the theory inapplicable to actual substances. A mass of fluid is in a steady state when its temperature, pressure, and volume are constant. It is then receiving energy from without by conduction or radiation on the average as fast as it is losing energy. In this sense the system is conservative, and when two steady states are compared as to their energy content that portion of the energy belonging to forces whose potential energy is not a single valued function of position may be dealt with as the quantity Δ was dealt with. It may be put constant for each steady state, and, therefore, as bearing a constant ratio to the whole energy content.

5. *Internal latent heats.* At corresponding temperatures such that $\tau < \theta_c$ we have for each substance two conjugate states, that of the fluid and that of its vapour. The work done in changing from the one state to the other against the forces of cohesion may be identified with the product of the internal latent heat and the total mass. For this heat is the total energy required to convert unit mass of fluid into vapour *less* energy absorbed by surface forces such as an external pressure.

Let $\bar{E}'$ be the energy of a gram-molecule of the substance as fluid, and $\bar{E}''$ as vapour, both at the same temperature. Then, if L be the internal latent heat of unit mass, we have

$$LM = \bar{E}' - \bar{E}'' \quad \text{and} \quad L_1 M_1/L_2 M_2 = (\bar{E}_1' - \bar{E}_1'')/(\bar{E}_2' - \bar{E}_2'').$$

From equations (14), (15) and (18) we get, since $\theta' = \theta''$,

$$(\bar{E}_1' - \bar{E}_1'')/(\bar{E}_2' - \bar{E}_2'') = \tau_1/\tau_2 \quad \text{or} \quad L_1 M_1/L_2 M_2 = \tau_1/\tau_2. \tag{25}$$

Equation (25) follows directly from the law of equipartition of kinetic energy if this be assumed. Let S be the total number of degrees of freedom of the system, then, if all the equations apply to corresponding states, we have

$$\bar{E} = SR\tau + \phi,$$

and

$$\phi = \alpha SR\tau,$$

$$\therefore \quad \bar{E}_1/\bar{E}_2 = R\tau (S + \alpha S),$$

and for two states of the same substance

$$\bar{E}' = R\theta' (S' + \alpha' S'),$$
$$\bar{E}'' = R\theta'' (S'' + \alpha'' S'').$$

Let one state be fluid and the other vapour in equilibrium with it, then $\theta' = \theta''$ and

$$LM = \bar{E}' - \bar{E}'' = R\theta [S' (1 + \alpha') - S'' (1 + \alpha'')],$$

but for systems in corresponding states the term in square brackets is equal.

$$\therefore \quad L_1 M_1/L_2 M_2 = \tau_1/\tau_2.$$

6. The equation for corresponding states, therefore, is

$$LM = a\tau, \tag{26}$$

where a is a function only of n in equation (22). Since dynamical similarity between molecules of different chemical types is not to be expected, this relation cannot be a general one. A chain compound is as little likely to correspond dynamically to a ring compound as is a reciprocating engine to a turbine. If dynamical correspondence is to be found it must be looked for in members of the same chemical series.

The difficulty mentioned in paragraph 3 now confronts us. Two courses are open, we may either make some assumption as to corresponding temperatures and develop the consequences, or identify corresponding temperatures by equations (25) and (26). I propose in this paper to confine myself to the first alternative, and to examine the consequences of the assumption that corresponding temperatures are identified by the relation

$$n_1 = n_2 = n_3, \text{ etc.}$$

That is to say corresponding states are those whose temperatures are equal fractions of the critical temperature. Since van der Waals' time such states so defined have been called corresponding states, though, so far as I know, the property of dynamical similitude has been attributed to them only by Kamerlingh Onnes.[1] Ample excuse for the examination of this assumption in detail is to be found in the immense superstructure of theory as to molecular association, etc., which has been based upon it.

7. Mills[2] has calculated the internal latent heats of a number of compounds. His figures furnish data for four chemical series: (1) the benzene series, namely, benzene, cyclohexane, and the four halogen substitution products of benzene; (2) the paraffins, including from pentane to octane of the normal series and the isomers isopentane, diisopropyl, and diisobutyl; (3) the esters; and (4) the alcohols.

The internal latent heats were calculated in order to test the truth of the formula $L/(\sqrt[3]{\rho_f} - \sqrt[3]{\rho_v}) = \text{constant}$, where L is as before internal latent heat and ρ_f and ρ_v are the densities of the fluid and vapour respectively, and published in 1904 and 1909.[3] Since then Mills has recalculated the values and improved them. The altered values are not available, but in all cases they are said to make the approximation to the formula closer. ·Certainly Mills' formula is very closely followed by the benzene series and paraffins, and where his figures lie off my curves the value given by the curve fits his equation more closely than do his own values. Thus as an example the values for brombenzene at $\theta = 433$ to $\theta = 463$ are below the curve and give values for Mills' constant which are too low. For this reason the value for pentane at $\theta = 273$ is too high, and those from $\theta = 303$ to 313 are too low. Therefore the point for pentane on the curve for $n = 0·5$ is fixed where it is. Mills thoroughly criticises the accuracy of his calculations. The great care he took is revealed in his papers and his relation given above is so close an approximation as to have led to corrections being made in Young's data.[4]

[1] Cf. van der Waals, *Zeits. phys. Chem.* XIII, 657, Appendix 2.

[2] *Journ. Phys. Chem.* (1904), VIII, 383.

[3] *Journ. Phys. Chem.* (1909), XIII, 512.

[4] Mill's figures for diisobutyl I found intractable. They would not fall into any regular curve, therefore I do not give them. The specimen Young used seems to have been impure.

It is assumed that when $n_1 = n_2 = n_3$ in equation $\tau = n\theta_c$, the substances are in corresponding states.

8. *Benzene series.* Below are given the values of a in equation (26). According to theory they should be constant for corresponding temperatures.

n	Cyclohexane	Benzene	Fluorbenzene	Chlorbenzene	Brombenzene	Iodobenzene
0·488	—	—	28·77	29·99	29·65	28·34
0·5	27·1	27·43	27·88	28·34	28·87	28·71
0·6	20·3	20·53	21·59	21·33	21·37	21·38
0·7	15·4	15·49	16·22	16·00	15·93	16·05
0·8	11·3	11·46	11·57	11·85	11·87	12·05
0·9	7·5	7·66	7·81	7·81	7·86	7·84

The agreement with theory is more exact than that usually accepted as being satisfactory in papers dealing with molecular physics, but in spite of this it would be wrong to conclude that equation (26) is complied with.

In Fig. 1 the molecular latent heats are plotted against absolute temperature. On each curve certain points surrounded by a ring are at corresponding temperatures, namely, $0\cdot8\theta_c$, $0\cdot7\theta_c$, etc. The oblique lines which pass through certain of these points, therefore, by assumption, pass through corresponding points, and they are the curves of equation (26). They satisfy the equation in that they are straight lines, but they do not satisfy it in that when produced they do not pass through the origin at absolute zero. The curve for $n = 0\cdot6$ does within the limits of error pass through the origin. The concordance between theory and fact is close, but the equation of corresponding states, instead of $LM = a\theta$, is $LM + b = a\theta$. It will be noticed that the equation applies only to the halogen derivatives, that is to say, these substances form a dynamically similar series to which neither benzene nor cyclohexane belong.

The parameters a and b are functions of n in the equation $\tau = n\theta_c$.

$n =$	0·5	0·6	0·7	0·8
a	33·1	21·4	15·6	13·3
b	+1096	−138	−182	+744

9. *The paraffin series* (Fig. 2). The normal paraffins form a series whose equation for corresponding temperatures again is $LM + b = a\theta$, but b is numerically larger, and is always positive.

$n =$	0·5	0·6	0·7	0·8	0·9	0·95	0·99
a	[66·2]	43·85	29·58	22·62	15·97	12·62	6·09
b	[+9380]	+6538	+4505	+4164	+3519	+3262	+1558

Of the isomers isopentane is indistinguishable from the series, but diisopropyl clearly falls out of it.

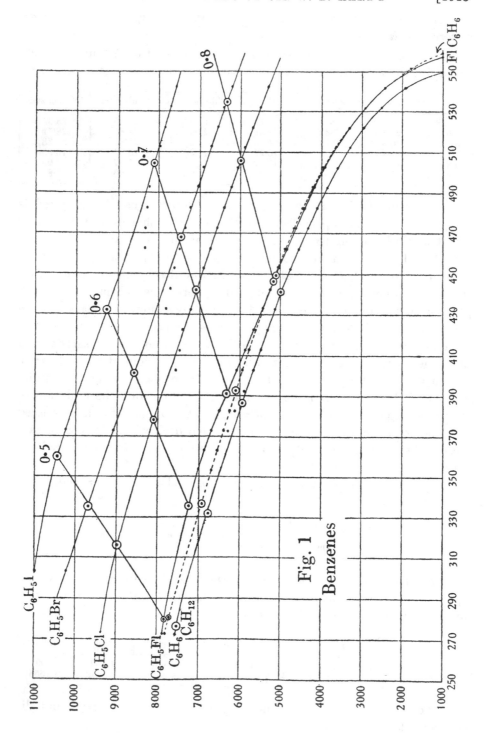

Fig. 1

Benzenes

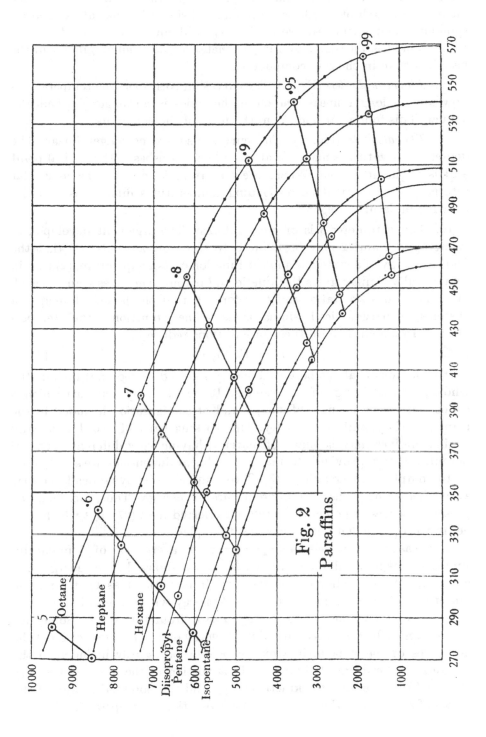

Fig. 2
Paraffins

The benzene derivatives and normal paraffins thus form two series, the members of each of which are in dynamical similarity, but at each corresponding temperature the series is displaced uniformly. The degree of displacement is measured by the parameter b, and is much greater in the paraffins than in the ring compounds.

10. *The methyl esters* (Fig. 3). The curves for corresponding temperatures now are no longer linear, therefore the series is no longer dynamically similar. This follows at once from the form of the equations.

11. *The alcohols* (Fig. 4). The curves also are no longer linear. The degree of curvature is maximal at $n = \pm 0.95$, vanishes at the critical point and is slight at $n = 0.5$. Where the curvature is greatest the molecular internal latent heat at corresponding temperatures diminishes with increasing molecular weight.

12. The parameter b is of great interest. The argument developed in paragraphs 1 and 2 may be reduced to three propositions: (1) that the kinetic energy of a system is a linear function of its temperature, (2) that in dynamically similar states the kinetic and potential energies stand to each other in a constant ratio, and therefore (3) that the potential energy of systems when dynamically similar is also a linear function of their temperatures. In algebraic form these propositions reduce to

$$\mu\beta^2 = Fl = \tau_1/\tau_2. \tag{27}$$

It is obvious that the potential energy referred to need not include the whole potential energy of the systems. It includes only that part which is concerned in the changes of energy and of configuration included in the principle of dynamical similarity. The working parts of two dynamically similar engines may be made of material having very different chemical potential energy provided the masses and linear dimensions are so adjusted as to comply with equation (13). Two guns may be dynamically similar even when firing shells with different bursting charges. Thus, there may be present reserves of potential energy not referred to in (27), but which may be drawn upon when certain changes take place.

If the assumption that $n_1 = n_2 = n_3 =$ etc. is a criterion of dynamically similar states be valid, the relation $LM + b = a\tau$ shows that the energy $a\tau$ is made up of the quantity LM received from without as heat, and the quantity b drawn from some reserve of energy within the fluid. b is not necessarily the whole of the energy drawn from this store. Some, for instance, may be converted into kinetic energy of progressive motion of the molecules of the vapour, in which case it is necessarily included in the temperature term $a\tau$. It is merely that portion of the reserve of energy absorbed in increasing the kinetic energy and the potential energy of the forces of attraction which is not a linear function of temperature.

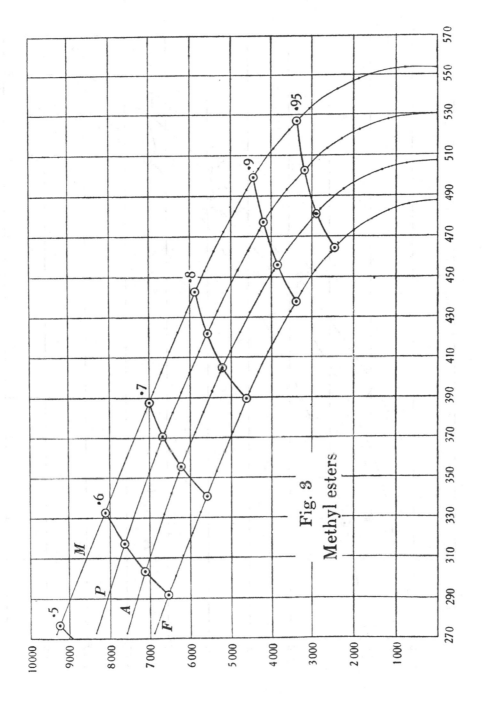

Fig. 3
Methyl esters

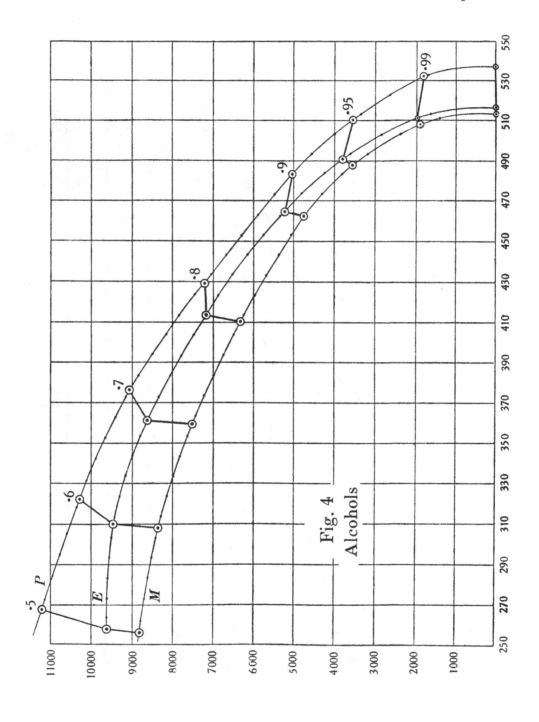

Fig. 4
Alcohols

The reserve in question must be intramolecular energy. Calling, as before, the forces which confer on the molecules the property of extension in space repulsive forces, the reserve is potential energy of repulsive forces. It is not suggested by anyone that the mutual attraction of molecules vanishes at the absolute zero; therefore, except in the single case of rigid molecules, the

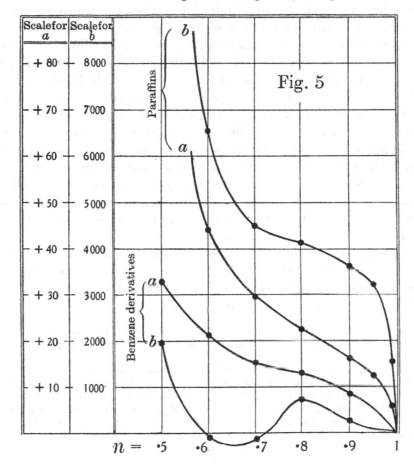

potential energy of repulsive forces vanishes only in the state of an ideal gas, increases on the whole as the system becomes more condensed, and is probably maximal at the absolute zero.

The quantity b in general decreases as the absolute temperature rises, and vanishes at the critical point. The variation is shown in Fig. 5, where both a and b are plotted against n. It is easy to see that this must be so. If it were possible to convert fluid into vapour at a temperature just above the absolute zero, we should pass at once from the state of maximal condensation to maximal expansion, and the whole potential energy of repulsive forces would be converted into kinetic energy and the potential energy of

attractive forces. Passing up the temperature scale the fluid is more expanded and the vapour less expanded, with the result that the reserve of
energy is less in the former and greater in the latter, until they finally
become equal at the critical point. Since b is the whole or part of the difference, it will, on the whole, decrease as temperature rises to the critical point.

In order to include the small negative values of b, which the halogen
derivatives of benzene seem to show, it is necessary to drop the narrow
conception of simple repulsive forces, or, rather, to enquire what is meant
by repulsive forces. We may conclude from what is known of the structure of
matter that the property of extension in space of individual molecules, even
at the absolute zero, is due to the possession of kinetic energy by internal
degrees of freedom. It is known that the store of energy within a molecule
is enormous judged by any ordinary standard, and a mere inspection of
specific heats at low temperatures shows that this store is not exhausted
when $\theta = 0$. The elastic resistance to any change of shape or volume of
the molecules which may occur when the system, as a whole, expands will
depend upon the whole variation of their internal energy, both potential
and kinetic. The small positive values of b sometimes found show that the
balance sometimes represents a gain of intramolecular energy.

The change in the internal energy of the molecules, which would thus
seem to occur when fluid expands to vapour, may fitly be put beside the
observed discontinuity between the magnetic susceptibility of a crystal and
of the same substance when fused.[1]

In the present state of the controversy as to the structure of matter and energy it is
of interest to note that the equation $LM = a\theta - b$ can be derived directly from the
theory of quanta.

Let it be supposed that the kinetic energy of a fluid is distributed between (1) the
movements of progressive motion of the molecules, (2) the rotation of the molecules,
and (3) the vibrations of an indefinite number of vibrators within the molecules. If
(1) and (2) between them have n degrees of freedom, we may, without violating the
quantum theory, put their energy as sensibly equal to $\frac{1}{2}nR\theta$. The energy of each
vibrator is

$$\tfrac{1}{2}R\theta \, \frac{h\nu/R\theta}{e^{h\nu/R\theta} - 1},$$

where h is Planck's constant and ν the frequency.

As a first approximation when ν is not too large—that is in fluids at ordinary
temperatures—this is equal to

$$\tfrac{1}{2}R\theta - \tfrac{1}{4}h\nu.$$

Therefore, in the absorption of an amount of energy E from without by a fluid, the
distribution of the energy amongst the degrees of freedom would be given by an
equation of the form

$$E = \tfrac{1}{2} \, (n - n') \, R\theta - \tfrac{1}{4} \overset{n'}{\underset{0}{\Sigma}} h\nu.$$

The frequencies of the vibrators will depend upon configuration since they are
determined by the damping effect of the fields of neighbouring molecules. The number

[1] Oxley, *Phil. Trans.* (1914), A, ccxiv, 109.

of vibrators n' will depend upon the chemical structure of the molecule, that is to say, upon the number and nature of the atoms and their spatial arrangement. It will also depend upon the configuration of the system as a whole, as Weiss's work on the magneton shows. We may, therefore, suppose that for corresponding states of systems whose molecules belong to the same chemical series, the last term in this equation either is constant or varies from substance to substance in the series by some multiple of a constant term.

13. Obviously, it is important to settle whether b really is constant. In compiling the following table, the method of interpolation adopted was not sufficiently accurate to give the units in the values of the internal latent heat. Within the degree of accuracy aimed at, b is seen to be a true constant for the paraffins. It is only approximately constant for the benzene derivatives.

$n =$	0·6	0·7	0·8	0·9	0·99
Isopentane	6455	4574	4171	3475	1559
Pentane	6571	4528	4140	3488	1554
Hexane	6562	4537	4192	3598	1602
Heptane	6424	4401	4171	3460	1495
Octane	6594	4557	4154	3532	1582

$n =$	0·5	0·6	0·7	0·8
Fluorbenzene	+ 1078	− 180	− 202	+ 767
Chlorbenzene	+ 1137	− 101	− 178	+ 716
Brombenzene	+ 1027	− 124	− 134	+ 748
Iodobenzene	+ 1142	− 149	− 215	

Knowing the immense pains taken by Mr Mills to secure accuracy in his calculation of the internal latent heats, it is almost ungracious to suggest that the consistent negative values found for $n = 0·6$ and $0·9$ are within the limits of error. Yet a glance at the curves shows that in this region the values refuse to plot to a curve which can be made to include the values between $n = 0·5$ and $0·6$. It is therefore possible that b really never is negative.

14. The dynamical similarity of the benzene derivatives is not to be expected from *a priori* considerations. When an atom of hydrogen is replaced by a halogen atom the symmetry of the benzene ring is destroyed. The loading would introduce an additional degree of freedom, and therefore, since the loaded molecules are corresponding masses, the loading must be such as to keep the centre of mass of the whole molecule in geometrically corresponding points. It is clear therefore that the halogen atom cannot simply replace a hydrogen atom so that the centre of mass of the former occupies the same position as the latter. This follows from the fact that the mass $C_6H_5 = 77$ is constant whilst the load varies from $F = 19$ to $I = 126·6$. The load must either be at the centre, or the ring must be distorted proportionately to it. Neither benzene (C_6H_6) nor cyclohexane (C_6H_{12}) belong

to the series, since they are symmetrical unloaded molecules, probably with fewer degrees of freedom.

15. The esters and alcohols do not exhibit dynamical similitude because they do not conform to the narrow assumption that corresponding states are to be identified by the relation $n_1 = n_2 = n_3 = \ldots.$

When this holds we have (equation (24))

$$(\mu \beta^2_{\tau.\theta_c})_1 = (\mu \beta^2_{\tau.\theta_c})_2 = (\mu \beta^2_{\tau.\theta_c})_3, \text{ etc.} \tag{28}$$

Choosing any two corresponding temperatures τ' and τ''

$$(\mu_1 \beta_1{}^2)_{\tau'\tau''} = (\mu_2 \beta_2{}^2)_{\tau'\tau''}, \tag{29}$$

and therefore $$(F_1 l_1)_{\tau'\tau''} = (F_2 l_2)_{\tau'\tau''}. \tag{30}$$

It is with the last relation that esters and alcohols probably fail to comply.

J. J. Thomson has given reasons for concluding that the molecules of alcohols are polarised electrically,[1] and the study of the tension of interfaces gives reason to believe that the molecules of esters if not actually polarised very readily acquire that property.[2]

Let us assume that the molecules in both cases contain electric doublets. There will then be attractions and repulsions between the molecules due to their electrification, and there will also be mutual induction leading to clustering of the molecules.[3] The distribution of the energy amongst the molecules must be very complex, little likely to have the simple relation to temperature denoted by equations (26) and (30).

Sutherland, who was the first to suggest that cohesion might be of electric origin, has calculated the forces between molecules holding electric doublets. His argument briefly is as follows:[4]

The sum of the attractions and repulsions averaged over a space containing a large number of doublets equally spaced tends to zero. Therefore such a system would have no cohesion. But when the doublets are in motion there is a preponderance of attraction over repulsion because when two doublets are approaching they tend to pull one another's axes into the straight line joining their centres. This follows from the fact that the force which one doublet exerts on another forms a couple, and while the attractions and repulsions vary inversely as the fourth power of the distance separating the centres, the couples vary inversely as the third power. When a pair are near together the couple is thus much more potent than either attraction or repulsion, therefore orientation of the axes of approaching molecules in the stable end-on position tends to occur before the latter forces are very great. Now, if two doublets are approaching with the axes

[1] *Phil. Mag.*   [2] *Proc. Roy. Soc.* (1913), A, LXXXVIII, 303.
[3] Cf. Thomson, *loc. cit.*
[4] *Phil. Mag.* (1902), (6), IV, 625.

similarly directed, the attractive force increases to a maximum at collision, while repulsive forces arrest approach before a maximum is obtained. In general there is the same tendency for attractive forces to increase as for repulsive forces to decrease.[1]

Sutherland draws attention to a property of such a system of doublets which has apparently been entirely overlooked, namely, that the range of the force is a function of the violence of the molecular agitation. For though the range of the force between two doublets has no upper limit the movement of neighbouring molecules tends to fix a limit, since, as the distance from the centre of any doublet increases, the end-on orientation, on which the excess of attraction over repulsion depends, will be diminished by the influence of other doublets until at a distance which can be only some small multiple of the average distance between the molecules, the disposition of the axis of any particular doublet considered with respect to the axes of surrounding doublets becomes purely random and the sum of the attractive and repulsive forces on it vanishes. In this way the form of this force becomes a function of temperature, of the angular moment of the molecule, and of the density of the fluid. The linear relation of equation (28) is then not likely to hold.

Summary

The principle of dynamical similitude is developed and applied to the case of the internal latent heat of evaporation. It is found that if temperature be proportional to the mean energy of progressive motion of the molecules, the internal latent heats of dynamically corresponding states should be given by the equation $LM = a\tau$, where L is the latent heat, M the gram-molecular weight, a a constant, and τ the temperature. This equation may be used either to identify corresponding temperatures or to test some assumption as to corresponding temperatures. In this paper the assumption made is that the critical states are dynamically corresponding states and that corresponding temperatures are equal fractions of the critical temperature. The equation for internal latent heats is then found to have the form $LM + b = a\tau$ for normal paraffins and for certain halogen derivatives of benzene. The conclusion is drawn that, subject to the assumption being correct, the potential energy of repulsive forces acting between the molecules contributes to the process of evaporation. An examination of the alternative view that corresponding temperatures may be identified by the relation $LM = a\tau$ is deferred for the present.

---

[1] There is no abrupt change at the fusion point of glasses in any physical property. Hence one may conclude that a random disposition of molecules persists in glasses. Therefore electric cohesion of glasses should vanish at the absolute zero.

36

SOME PROBLEMS OF LIVING MATTER
THE GUTHRIE LECTURE

[*Proc. Physical Soc.* (1916), XXVIII, (ii), 99]

I am, I understand, to discuss to-day physical problems raised by the study of living matter. There is no lack of such problems. I am embarrassed by their number and variety.

Let me begin by some general account of living matter. A simple single-celled animal of the type of amœba has a volume roughly equal to a sphere of about 30μ radius. It is composed of a translucent turbid material showing no structure. Recently such cells, in spite of their minute size, have been dissected while still living. The operation was carried out under the microscope with fine needles of hard glass, operated by a mechanical device. The result has been to remove some classical misconceptions. One is that living matter has a foam structure. Dissection shows that the appearance of spaces filled with fluid is due to the presence of fat drops, which can be dissected out of the living cell. Such drops are probably merely food reserve, and are external to the living substance. The latter, freed from such masses, is in some places a rigid gel, in others a fluid of high viscosity—that is to say, a slime. Somewhere in the cell is a mass having distinctive properties called the nucleus. It may be either a colloidal solution enclosed in a very tough membrane of transparent gel, or a tough gel containing many granules from $0·5$ to $1·5\mu$ in diameter, which is rigid in the sense that minute pieces broken off do not round up under the influence of molecular attraction. The living substance is sometimes a puzzling combination of fluidity and elasticity. This is especially the case with the substance of a muscle fibre. A glass needle may be moved about in it in all directions freely, but if a portion be broken off it can be pulled out into a filament of extraordinary length, which when released will almost regain its previous shape.

The ordinary microscope, therefore, teaches us only that the actual living substance when freed from food matters and stored chemical products is a translucent gel or slime. The ultra-microscope carries us a stage further. It shows the translucent material to be composed of an optically homogeneous substance, in which are embedded a multitude of minute particles. Usually the particles are in Brownian movement; the mass, therefore, as a whole is more fluid than solid. But the Brownian movements sometimes cease for a time, to be renewed later. The degree of fluidity, therefore, is variable, the mass, or portions of it, may alter in the direction of a rigid gel, and the change can be reversed. Such a change in the direction of an increase in rigidity seems to entail the expenditure of energy, since it is described as occurring in response to an external stimulus. It is also characteristic of the death change when the free energy of the cell runs down with rise of temperature.

The appearance of minute particles shown by an ultra-microscope is common to colloidal systems, so we may in general regard the living cell as a mixture of colloidal slimes of varying degrees of fluidity. This seems a curiously insufficient basis for the prodigious potentialities of, e.g., an ovum or spermatozoon!

One feature may suggest more to my audience than it does to myself—namely, that though the substance of the living cell is uniformly translucent to light of ordinary wave-length, some parts are remarkably opaque to ultra-violet light. This is true of the nucleus. By taking advantage of this difference, quite sharp photographs have been obtained of the nucleus in the living cell.

The activities of these simplest organisms are at first sight as simple as their structure. Consider, for instance, their mode of progression. It is no more than a flowing of the protoplasm which causes the whole cell to roll along a surface. A few grains of charcoal placed on the upper surface of an amœba are carried forward—i.e. in the direction of motion of the whole animal over the forward edge and backwards along the face in contact with the surface over which it is crawling. It reminds one of Helmholtz's account of the displacement of electricity when a dielectric is rubbed over a surface. The simple movement of an amœba is what would occur if the contact potential difference between the animal and the solid were greater at the front than at the back end.

At intervals the animal takes into its interior particles of food by flowing over them. It dissolves out the portion profitable to itself and extrudes the remnant. But this seemingly purposeful act is very closely copied when drops of chloroform in water are fed with particles of glass coated with shellac. The particles are pulled into the interior of the drops, the shellac dissolved and the indigestible masses of glass thrust out.

We have merely to observe the marvellous variety of animal and plant life to realise that this appearance of simplicity must cover a real complexity. The appearance of simplicity is, indeed, but an example of the characteristic of living matter, the adaptation of the means to the end. The phenomena are simple only so long as the object to be attained is simple. Take movement as an example. I have watched one of these simple cells endeavouring to bend and break a chain of bacteria so as to get short lengths capable of being ingested. During the operation a particular cell became extended between the ends of a chain until it had the form of a cylinder about 30 μ long. The movements now took on an entirely different character. The cylinder swelled out at one end, and the swelling progressed as a slow wave forward and backward over the length of the cylinder. Simple flowing movements were replaced by an orderly rhythm of waves of shortening.

Let us take another illustration. Some animals of the simple type we are considering get their living as the spider does, by spreading a net. The net has not the geometrical regularity of those which are hung in the hedgerows where they condense the atmospheric water to such delightful patterns, but not the less are they nets spread to catch prey. The threads are actual processes of the body of the animal—the pseudopodia of the zoologist— and each single thread is a cylinder only a few μ in diameter. Such a cylinder considered as a figure of tension must be highly unstable. The tension of the surface probably is small compared with that of water. Certain evidence would fix it about 45 dynes. The maintenance of this thread of semi-fluid material, therefore, either is due to a solidification of its surface or needs continuous expenditure of energy. The matter does not end here however, for the thread is not a simple cylinder of protoplasm. The substance is in movement; the motion of the minute particles floating in it show that there is an outward and an inward stream or flow, and at the face where the two streams meet the frictional stresses produce double refraction.

Consider also the puzzling discriminations exercised by the surface of these simplest cells. When two parts of the same animal come into contact the surfaces soften and the parts fuse, but this does not occur when different individuals touch.

The living cell is, in fact, a machine, capable of carrying out movements, and of complex growth and differentiation of substance. Oxidative processes are the source of its energy, but the oxidation is not direct, as in the burning of a candle. There is a store of energy which is set free in response to a stimulus, and the intake of oxygen occurs only *afterwards*. The oxygen *restores* the potential. A large variety of ferments are present either actually or potentially. They are set free, and can be identified when the cell is broken down and its material dissolved.

Now, living matter contains from 80 to 90 per cent. of water. In any ordinary dead gel or slime with so much water, diffusion would obliterate differences of state so far as simple chemical substances such as salts are concerned. Initial differences in the state of aggregation of the colloid would not so readily disappear owing to the slow rate of diffusion, but if electrolytes were present they would, by their diffusibility and profound influence upon colloid structure, tend to produce a uniformity. Living matter is remarkable in the fact that we have freedom of interchange of matter, otherwise nutrition would fail, combined with the preservation of sources and sinks of energy in a material which is essentially a fluid.

The most fundamental, and to my mind the most puzzling, problem of living matter lies in this contradiction between the functional and chemical complexity of living matter and the apparent simplicity of its structure. Both microscope and ultra-microscope tell us simply that it is a colloidal sol of high but varying viscosity. The material basis of life is apparently much less structural than, for instance, a simple gel such as, *e.g.*, that of azomethine. This is at first sight merely a translucent yellow glass, but the microscope shows it to be a weld of masses, each of independent origin; each of these masses is a weld of smaller masses arranged in an orderly pattern, and these again are built of ultra-microscopic particles.

The questions which I specially wish to discuss are raised most directly by the phenomena of growth. The growth of living matter has been compared to many things—to the growth of a crystal, among others. In the growth of a crystal, molecules of the same nature as those which constitute the crystal are aggregated to a definite formal pattern. In the growth of living matter, extraneous molecules also are ordered to a particular pattern, but the molecules are not the same as those which constitute the growing mass. Growth is part of a cycle of chemical and physical changes. The case of the crystal is helpful as furnishing an example of the directive influence of a pre-existent molecular pattern. Perhaps it is not generally known that similar directive action is manifested by dead colloids. Many colloids are systems following a path of change whose form is determined by their previous history. The course of the change with time and at constant temperature of the viscosity of, for instance, a solution of gelatine is determined by the temperature at which the solution was made and by the rate of cooling. If, now, this solution is sown with small portions of another system, say with bits of gel, the path of change is altered completely.

We thus have analogies in the world of dead matter for individuality in the characteristic path of change of colloidal systems, and for the directive influence which underlies growth in both crystals and colloids; but a remarkable feature of living matter is the distance over which such directive influence appears to be exerted. The regeneration of cut nerves perhaps best illustrates this.

Let me first premise that the growth of any part of a cell is controlled in some unknown way by the nucleus. When individuals of the single celled type are cut into pieces most of the fragments grow and regenerate the entire animal; but this power of regenerative growth is limited strictly to those fragments which contain a portion of the nucleus.

38-2

Consider now a nerve cell and its related nerve fibre. The latter is a process which grows out from the former, and is covered by certain insulating sheaths. The cell body containing the nucleus varies much in size, but we may put it as roughly equal in volume to a sphere of 50–100 μ in diameter. The cell process or axon divested of its insulating sheaths is a delicate rod of protoplasm, 5–10 μ in diameter, which, in a large animal such as a whale, may be some metres in length. The integrity and the power of growth of the whole of the axon is dependent upon the nucleus. If the nucleus is destroyed the whole structure perishes, and if the nerve fibre is cut that portion of the axon which is still in connection with the nucleus grows, whilst the part cut off disintegrates and dissolves. The nerve fibre, therefore, furnishes us with an example of a system capable of growth only so long as it is united with a portion of matter, which may be some metres distant and to which it is connected only by a cylinder a few μ in diameter. I do not think that there can be any transport of actual nutritive material by such a tenuous connecting link. The relation may be a static one, the molecular pattern of the nerve fibre being unstable when isolated from the nerve cell. But such scanty evidence as is available tends to show that it is a dynamic one in the sense that the integrity of the molecular structure of the entire axon is maintained by expenditure of energy. The nucleus and cell body, in upholding the molecular pattern, may be said to exert a directive influence along the whole length of the nerve fibre.

The immediate environment of the growing face of a cut nerve fibre also exerts a remarkable directive influence upon its growth. It must be known to all of you that the severed ends of a nerve will heal so as to give complete restoration of function. The nerve in question may be a bundle of some thousand fibres, along which the nervous impulses which produce the exactly controlled movements of skilled action are transmitted, and the restoration of complete function after section is like the joining together of a bundle of cut telephone wires in such a way as to restore all the connections.

For complete restoration of function it is not necessary to fit the cut ends of the nerves closely together. They may be left in the wound separated by a space of some centimetres. Imagine them so separated. From the central end—that is to say the one whose fibres are still connected with the nerve cells—the axons grow out. The axons of the peripheral end break up and dissolve. Now the growth of the axons is at first aimless in direction, the filaments making their way along lines of least resistance in the spaces of the tissues. After a time, however, they come under some influence emanating from the other cut end of the nerve. They turn and grow as directly as possible towards it, and finally invade the empty sheaths and grow down them. What is the nature of the directive influence? It is without doubt chemical. Some substance diffusing out determines the direction of growth, for when two tiny tubes of celloidin are placed near the cut end, one filled with an emulsion of liver and the other with an emulsion of nerve, the axon filaments grow towards the latter. They are, as it were, attracted by it.

The direction of growth must be determined by the density of a diffusion column. If we take the end of a single axon as the end of a cylinder $5\,\mu$ in diameter, and if it is inclined at a small angle to the axis of diffusion, the difference in the rate of growth of different parts of the cylinder needed to orient it must be produced by differences in the concentration of the diffusing substances which occur in distances less than $1\,\mu$, measured along the axis of diffusion. I need not emphasise how minute these differences must be.

It looks as though we might have to claim for the axon the perfect sense of direction ascribed by a student to amœba. Asked to describe the response of this animal to an electric current, he said: "When subjected to electricity an amœba withdraws all its pseudopodia except one, and then directs itself to the North Pole!"

Let us consider the directive influence which underlies growth from another point of view. The complex organisation of the higher animals exhibits well-marked periods of growth, decay and death, and the duration of these periods is characteristic of each species—the three score years and ten of man, for instance. But these periods are not obvious in the history of single-cell types, in which the whole cell simply divides to form a new generation. Indeed, the question arises whether decay and death are intrinsic properties of these animals. Individuals of the species *Paramœcium* have been isolated under normal and healthy conditions. Each individual was the starting point of a series of generations, there being on the average two generations in three days, and the rate of division was recorded, the records furnishing the basis for a curve of vitality. Such a curve shows fluctuations of a fairly regular character—"rhythms," they have been called. The curve alternately rises and falls, and each complete rhythm, a rise and a fall that is, lasts about a month. The curve as a whole, however, was found steadily to fall, until at about the hundred-and-seventieth generation the race dies out. This was held to establish the appearance of senile decay in strains isolated from mixture with other strains but otherwise in completely healthy surroundings. The discovery that unfertilised eggs of sea-urchins could be made to develop by immersing them for a few hours in more concentrated sea water suggested the possibility that senility could be cured in Protozoa. Therefore, when the period of decay had arrived, and individuals were dying off rapidly, the effect of placing the individuals in various infusions was tried. The infusions of animal tissues were found to give the required result. After exposure to them for a short time the rate of growth and of reproduction reached the normal level, death ceased, and a strain was maintained in full vigour for 860 generations. The living matter of these cells seems to be potentially immortal.

The point of immediate interest in these experiments is that, in the periods of depressed vitality, the power of heredity in determining form becomes imperfect, and many "monsters" are produced. We have here, then, a wearing out of the directive influence which underlies growth.

In the process of artificial rejuvenescence it cannot be a chemical insufficiency which is made good; it must be a physical state which is restored. This is obvious when we consider a special case. Thirty minutes' immersion of an individual paramecium in 0·1 per cent. potassium phosphate was found to restore vitality, and the effect persisted for 282 generations. The effect cannot be due to the presence in individuals of a trace of the salt, for each generation would halve the amount, so that as early as the twentieth generation less than a millionth part would be left for each individual.

It will be remembered that when a solid surface has been washed by a salt solution the contact potential of the surface with pure water is altered, and, according to Perrin, simple washing with water fails to restore the original state. It is, I expect, the wearing out of this kind of state which leads to senility.

Recent work on nutrition suggests reconsideration of these experiments on Protozoa. A diet must, of course, contain water and a supply of fuel in the shape of proteins, fats and carbohydrates. It must also contain substances which do not contribute energy, such as simple salts and a class of substances whose presence has only been recently detected called vitamins. Nothing is known of the chemical nature of these substances nor how they act, but incredibly minute quantities are sufficient.

The existence of and need for vitamins first appeared, I believe, in connection with the disease beri-beri. This is a disease of the nerves, and it was noticed that birds fed on polished rice—that is to say, upon rice free from its pericarp—developed beri-beri, but recovered if the polishings themselves, or an alcoholic extract of the polishings, was added to the white rice.

Once attention was drawn to these accessory foodstuffs instances of their occurrence multiplied quickly. They have been found necessary for mammals, birds and minute plants. Scurvy seems to be due to their absence from preserved foods. Their influence would appear to be limited to the processes underlying growth. The following two cases illustrate this and also indicate the minute quantities necessary. When young rats are fed upon artificial milk made by mixing together in the right proportions the component substances of ordinary milk, previously separated and purified, they lose weight and die. Now, to convert artificial milk into a perfectly adequate foodstuff all that is needed is the addition of 2 per cent. of ordinary milk. Certain Continental workers challenged this result, but later they had to confess their error, which arose in an interesting way. The commercial pure lactose which they had used to make the artificial milk was found to contain vitamins. Even after four recrystallisations two grams a day carried enough vitamin for a 50-gr. rat. Only by many recrystallisations is vitamin free lactose produced.

Another case. The common brown diatom of the sea litoral will not grow at all in artificial sea water, but will grow very freely in such water to which as little as 0·2 per cent. of natural sea water has been added.

The intimate connection between vitamin and growth is illustrated by the case of cancerous tumours. Mice can be kept at constant weight on a diet of purified wheat protein, starch, lard, lactose, salts and water. Some mice, previously inoculated with a rapidly growing sarcoma, were put on the artificial diet, others similarly inoculated upon a normal diet of bread and corn. The cancer was found to grow much more slowly in those on the artificial diet. To give an example, a mouse after 52 days of artificial diet showed a tumour only 4 mm. in diameter. It was then put upon normal diet, with the result that in 30 days the tumour was nearly as large as the mouse itself. Too much must not be based upon this result, interesting as it undoubtedly is, until it is fully confirmed.

It has been suggested that vitamins intervene in growth by contributing to the fixation of the molecular pattern of the newly formed substance. Their action has been compared to that, say, of a dextrorotary crystal which by its presence causes a mass of fluid indefinitely larger than itself to deposit right-handed crystals. The analogy can at the best be but rough. The influence of a crystal is limited to directing the system along alternative paths of change, all of which increase the entropy. In the growth of living matter, however, the local change, at any rate, is from simple to complex, from a lower to a higher content of free energy.

The nerve fibre of an adult animal grows only when it is cut. The faculty is latent in the completely functional structures; therefore, though growth and function affect each other, they must be distinct processes in the sense that the whole cycle of change of energy and matter which constitutes function may occur without growth. If vitamins intervene in growth by a stereochemical effect upon the construction of new matter, the other class of accessory foodstuffs—namely, the simple salts—stand in interesting contrast, since they exert a directive influence chiefly upon the changes of energy which constitute function. The amplitude, period and form of the contraction wave of the heart are, for instance, determined by the nature of the electrolytes present. The rhythm of the heart muscle appears to be based upon the antagonistic action of univalent and bivalent cations. The voluntary muscles have no intrinsic power of rhythmic movement. They respond only to the nervous impulse or to some external stimulus. But one of these muscles immersed in a bath of the right kind and concentration of electrolytes, will beat as regularly as the heart itself. The regulation of respiration—that is to say, the adjustment of the rate and depth of breathing to the needs of the body—also is based upon the extreme sensitiveness of a portion of the brain to variations in the hydrogen-ions concentration in the blood. Normally this has the value $10^{-7.3}$. Scrambling up a mountain 1000 ft. ascent in 30 min. alters this only to $10^{-7.1}$.

The physical basis of this hold which electrolytes have on the activities of living matter is clear. The configuration of a colloid is determined as much by the electrolytes present as by its own chemical nature. The inertia of a colloid—that is to say, the degree of resistance it offers to change—is determined by electrolytes, since they provide what Helmholtz calls the "first condition of electrical distribution", as between the colloid particles. It is possible by eliminating electrolytes to raise the sensitiveness of some colloidal solutions to the point at which they are stable in quartz vessels, but are at once precipitated when transferred to clean glass vessels.

Exposure to distilled water also increases the sensitiveness of the surface of some living cells without destroying it. They will live and thrive in pure distilled water, but a trace of metallic impurity (estimated at 1 in 70,000,000) will quickly destroy them.

Here, I think, a certain broad conclusion is forced upon us. Since electrolytes control the configuration of colloidal systems in respect to the size, number and distribution of the colloid particles, and also shape the path of change of energy in living matter, we may infer that the functional processes of a cell are conditioned by the configuration of the colloid in the sense mentioned above.

The material which manifests the phenomena of life is, as I have said before, nothing more than an optically homogeneous fluid medium, in which are suspended particles. When the store of potential energy is being drawn upon—that is to say, in the functional state of activity—the fluidity is decreased and a transitory stage of gelation may occur. Thus, when the cell substance contracts under stimulation, and in the death change when the potential energy is dissipated as heat fluidity decreases to the point at which Brownian movement of the particles is suppressed.

Also the establishment of the field of force which brings about division of cells in the process of multiplication is accompanied by a local gelation.

What physical significance can be attached to this decrease in fluidity? According to the theories of Einstein and Hatschek a decrease in the fluidity is due to an increase in the volume of the colloid particles following upon an intake of water from the continuous medium. I do not think so simple a theory covers the facts. Their formula fails except in the case of the simplest hydrosols and suspensions. Actual observation of the process of gelation under the microscope reveals only an inconsiderable change in volume. The main events seem to be, firstly, a damping down of the Brownian movements; and, secondly, an agglutination of the particles into rows which form the fibres of a quasi-rigid framework.

It is practically certain that the process does not go so far as this in normal functional states. The colloid particles maintain their independence, but their relation to each other and to the continuous medium are changed in such a way as to make their potential energy a function of their position. The molecular mechanism of these changes is quite obscure. We may picture to ourselves the colloid particles as strain centres—the microscope justifies so much—and the continuous medium as having the mechanical properties of an unannealed glass. Mechanically a living cell has many points of similarity to a Rupert's drop; indeed, many forms violently disrupt if the surface "skin" is cut through. Unfortunately we do not know the basis of the simple rigidity of glasses. The absence of any discontinuity in the change from fluid to glass similar to the discontinuity of energy and property which marks the change from fluid to crystal shows that a glass is a fluid which has lost its characteristic property of fluidity, but to the question why the molecules lose their freedom of movement there is at present absolutely no answer.

One fact may be seized upon—namely, that a change from fluid to a solid of the gel or glass type may be due to quite a small local change in a complex molecule.

Consider, for instance, the substance azomethine, which has the formula:

$$C_6H_5C \underline{\hspace{3cm}} C:NC_6H_4N(CH_3)_2$$
$$C_6H_5C \qquad\qquad CO$$
$$CO$$

This substance has the property of forming rigid gels with *any* solvent; with ether, alcohol, benzene, chloroform, etc. But these very exceptional colloidal properties totally disappear if one hydrogen atom of the side chain is replaced by bromine. I think we are here in a region of molecular physics which is practically unexplored.

The damping of Brownian movements in the course of the functional activity of living matter is, I think, a clue of the first importance to the mechanics of the cell. It suggests a much more extended "grip" of the colloid particle upon the continuous medium than would be indicated by current conceptions of the range of molecular action.

There are properties of simple hydrosols and of suspensions of living cells which point in the same direction. The simplest hydrosols, for instance, seem to have saturation points; that is to say, the number of colloid particles in a cubic millimetre cannot be increased beyond a certain point without occasioning precipitation, and the saturation point seems to be reached when the particles are still widely separated. A similar relation appears in a much more complex system—namely, a suspension of bacteria in a nutritive medium. As the bacteria grow and divide, the rate of growth slackens and finally ceases, but this is not due to exhaustion of the medium, for if the bacteria are filtered or centrifuged off, and the material freshly inoculated, a fresh and copious growth results.

Very little is known about these curious saturation points, but I think you will agree that what is known suggests a range of influence of one particle upon another much greater than the range ascribed to molecular action. Something of the same kind is seen in thin sheets of fluid. A layer of carbon tetrachloride on water exhibits curious mechanical instabilities even when it is of the order of a millimetre in thickness. Such a sheet of liquid is like an unannealed colloid. The tension of the upper face, the air face, is 24 dynes, whilst that of the water face is 50 dynes, and I think there is much to be said in favour of the view that these unequal stresses make themselves felt throughout the entire thickness of the sheet.

I am inclined to suggest the phrase "range of molecular action" must be held to refer to two distinct things—the one is the range of an isolated molecule, say, a molecule of a gas in its free path; the other is the orienting influence of molecules upon one another in the close packed states of fluid, glass, or gel.

The true range of molecular action is that of the gas molecule. It is probably very small, of the order of 10^{-8} cm. But in the close packed states it is probable that the asymmetrical field of force at an interface produces distortion of the external fields of the molecules and orientation extending on either side until it is upset by the heat motion. On this view the transition layer at an interface is a region of more or less fixed molecular pattern, akin, perhaps, to that of a glass. Before the war I had begun to observe

---

[1] The evidence, however, is not wholly one-sided. In Einstein and Hatschek's formula for the viscosity of two-phase systems the viscosity is given as a simple fraction of the ratio between the volume of the dispersed phase and the total volume. When the degree of dispersion is high so that the particles of the one phase are ultramicroscopic the viscosity is greater than would be given by the formula. If, now, it is assumed that the discrepancy is due to the volume of the dispersed phase being increased by what might be called adsorption envelopes, it is found that these latter need be only $0·87\,\mu$ in thickness. But the formula also fails in other respects.

particles of interfaces under the ultramicroscope with the object of detecting signs of such structure. It is certain that the Brownian movements are sometimes damped in a remarkable way in this region. The depth of these transition layers might be expected to be increased when the molecules involved are large, or, if small, when the external field of each molecule is markedly asymmetric about the centre of mass. It is significant that the influence of electrolytes upon living matter and upon complex organic colloids with large chemical molecules such as proteins, is determined by the volume of the ion as well as by the charge carried; whereas the effect upon simple interfaces between fluid and solid and upon simple suspensions and hydrosols is determined solely by the charge.

Mere increase in size of the molecule must of itself introduce new considerations into molecular physics. The molecular weight of the red pigment of blood, itself a complex protein, is about 16,000; its volume would be equal to some hundred water molecules. The energy related to it when in solution will be its own kinetic energy, and the surface energy of the water molecules surrounding it. Also if it is to be regarded as an elastic structure there will be potential energy due to bulk compression produced by the intrinsic pressure of the fluid.

We are compelled, I think, to attribute surface energy to these large molecules, not, of course, strictly to the molecule itself, but to the displaced water, by the fact that they are aggregated by minute traces of electrolytes in a way resembling the effect upon coarse colloid systems and suspensions. A molecular solution of hæmoglobin is possible only in very pure water.

It is curious that large molecules sometimes exhibit simple physical relations. Most proteins obstinately refuse to form states simpler than complex solutions or complex slimes or gels. Some, and hæmoglobin is one of these, readily crystallise. Now hæmoglobin when it crystallises preserves its peculiar optical characters unaltered. It has the same colours and the same absorption bands. It also continues to manifest the same special relations to gases as it has when in solution. Essentially it is unchanged. This continuity of physical properties exhibited by so large a molecule is, I think, startling when considered in relation to Prof. Bragg's account of the structure of crystals.

I have no time to deal with the changes of energy in living matter. The subject is too complex for brief treatment, but I should like to offer a special problem. In general the energy changes of the living cell manifest the ordinary features; they are, for instance, accompanied by the dissipation of a fraction of the energy on heat. One case, however, presents puzzling features—that of the nervous impulse. This is a short wave of molecular change which progresses along the axon filament at about 30 metres a second. It can be started by a local chemical, electrical, or mechanical change. The passage of the wave over any region is followed by a brief period of complete inexcitability, followed by rapid recovery of function. The molecular change involved in the wave is, therefore, one which needs a finite, though very short, time to be recovered from. One would conclude that the wave is a discharge of energy which has to be made good before another discharge can take place.

This conclusion is borne out by the fact that recovery of function can occur only in presence of oxygen, and also by the fact that if the wave passes through a region in which its amplitude is diminished by cold or narcotics it is restored to its full extent so soon as it emerges into a normal region.

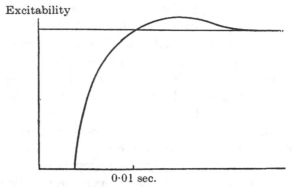

Excitability

0·01 sec.

Curve showing recovery of excitability in a nerve after activity.

The wave would, therefore, appear to be one of chemical change, but if this be so it is one of exceptional character. The temperature coefficient is very low for a chemical change, being only 1·78 per 10°, and there appears to be absolutely no concomitant liberation of heat. The absence of any heat change is, however, I suppose, no more remarkable than the limitation of the rays emitted by phosphorescent animals to visual rays, and the energy of these rays is certainly due to oxidation.

The profound difference between living matter and a simple gel is seen in the selective permeability of the former. Selective absorption of dyes is a common property of gels. Thus, gelatine gels condense basic dyes more than acid dyes. Gels of agar condense acid dyes more than basic dyes. But the permeability of living matter is both curiously selective and variable. The act of fertilisation of the ovum—the entry of the spermatozoon, that is—is followed by a passing rise in the permeability of the surface. In some cases permeability is said to be unidirectional. When the living skin of the frog is used as an osmometer membrane it is said to permit the passage of carbonic acid in only one direction—namely, from its internal to its external surface.

Permeability has been most completely studied in the kidney, and it appears that selective permeability is controlled by or due to an expenditure of energy by the living membranes of the kidney tubules.

The business of the kidney is to maintain the blood of a certain normal composition as regards certain constituents. For our purpose we may regard the blood as a solution of certain colloids—mainly proteins—and of other solutes of simpler molecular character such as urea and salts. The kidney also appears to be chiefly responsible for maintaining the proper reaction of the blood—that is to say, the normal ratio of hydrogen and hydroxyl ions.

Consider first the removal of excess water. This is governed by simple considerations of osmotic and hydrostatic pressure. Incidentally we get clear evidence, if physicists were in need of it, that large molecules, such as those of proteins, exert osmotic pressure. In an osmometer whose membrane is freely permeable by salts, sugar, urea, etc., the blood shows a permanent osmotic pressure of about 35 mm. of mercury. This is the osmotic pressure of what we will call for short the colloids of the blood. It is a critical pressure as regards the elimination of water, for when the blood pressure in the renal arteries falls to about this level the flow of urine ceases, because the hydrostatic pressure of the blood is then equal to the opposing osmotic pressure.

Now there is a certain solution of salts called Ringer's fluid, much in use in biological investigations, which represents very closely the salts of the normal fluids of the body. If some of this solution be injected into the blood there is an immediate flow of urine which is just the Ringer's fluid, or its osmotic equivalent, which has been added. The cause of the flow is obvious. The added Ringer dilutes the blood colloids and lowers the osmotic pressure. The result is that the balance between osmotic pressure and the hydrostatic pressure of the blood is upset and fluid flows out.

That the kidney plays a passive part is proved by the fact that in such a case a copious flow of urine occurs without any increase in the rate of oxidation in the organ. There is thus no expenditure of energy in the process.

That the colloids of the blood plasma exert their influence simply in virtue of their osmotic pressure, and not because of their chemical nature, is proved by the fact that if gelatine be added to the Ringer's fluid in quantity sufficient to give the osmotic pressure of the blood colloids and the fluid injected there is no diuresis. But if soluble starch be added until the viscosity of the Ringer's fluid is, say, twice that of blood (3 per cent. starch), the diuresis is much the same as with simple Ringer's fluid. This is because starch does not sensibly increase the osmotic pressure of Ringer's fluid.

But a flow of urine may be provoked by the injection of alkaloids or of urea or of a simple salt such as sodium sulphate. Consider the case of sodium sulphate. The injection of this salt essentially alters the chemical character of the blood. A diuresis is at once set up and a urine rich in sodium sulphate eliminated. In this case the living membrane discriminates in a remarkable way between sodium sulphate and the normal salts of the blood. That is the usual way in which the physiologist puts it, but it is as well to remember that the discrimination may be due to the remarkable colloid complex which is present in blood plasma. At any rate, be that as it may, a differential treatment is accorded to the sodium sulphate or to, for instance, urea. Such selective diuresis is accomplished only at the expenditure of energy, as appears from the increased intake of oxygen by the organ. The physical processes underlying the transference of solutes or water in the body are probably very complex. For simple transference of fluid through a membrane one's mind turns at once to electric endosmose. When a current is passed through a rod of gel there is an endosmotic movement of water, due to the fact that the colloid particles are usually at a different potential to the fluid bathing them. We may conjecture, therefore, that an electrostatic field mantained across a living membrane would cause transference of fluid. When an expenditure of energy takes place in living matter its potential rises above that of the neighbouring inactive medium. For instance, the contracting portion of a muscle fibre, or the part of a nerve fibre actually conveying a nervous impulse, shows a raised potential. Also, when a flow of saliva is excited, the electrical potential of the gland as a whole is altered. There can, therefore, be no question of the ability of a living membrane to produce a difference of potential between its two faces. But in simple electric endosmose the transport of fluid is due to an external field acting upon that portion of the fluid which is charged, owing to contact with the containing walls. In the living gland there is nothing to take the place of this external field. If a difference of potential is established between the two sides of the living membrane, it is due to a change in the colloid walls themselves.

Suppose a gradient of contact potential were maintained along the length of a tube holding fluid, would a flow of fluid result? I am inclined to think it would, the necessary flux of electricity taking place through the fluid or through the walls.

The transference of fluid in a living cell, instead of being a simple endosmotic phenomenon, is most probably based upon variations in the water-holding power of colloid systems in different states. The imbibition of fluid by a body such as gelatine is a very obscure phenomenon partaking of the character of solution and of endosmotic movement. Whatever its origin, the water-holding power is easily altered. Exposure to the vapour of formaline will completely change the character of the gel. Its structure changes, the gelatine is altered so that it ceases to be soluble in water, and, though there is no immediate change in volume, the gel which has been "coagulated" by the formaline has its vapour pressure raised. It holds water much less firmly. The partial coagulation which occurs in some processes of life, the change from a slime to a rigid gel, for instance, are unquestionably accompanied by variations in the potential, in Gibb's sense, of the water. If the surface film of a colloid membrane separating two masses of fluid were to change in such a way as to lower the potential of the water in it, water would enter the region from both sides simply. But if the change of state were to be propagated as a wave of change starting at one face and dying out at the other face, water would be carried along from the one side of the membrane to the other. A succession of such waves would maintain a flow of fluid.

I should like, in conclusion, to say a word as to the physical knowledge which the biologist is now in need of. The biologist has to deal with the molecular mechanics of gels and slimes. When the formation of a gel is followed by the microscope or the ultra-microscope, the change of state is seen to be due to an increase in size of the colloid particles, followed by their coalescence to form the fibres of a sponge. The fibres of the sponge, in many cases at any rate, are amorphous solids, like a glass. Now we know that there is no break between the state of fluid and that of glass or gel such as there is between fluid and crystal. Even viscosity shows no discontinuity. Generally speaking, all physical properties pass over without interruption from the one state to the other. A gel or a glass, therefore, is in this sense a fluid which has lost the characteristic property of fluidity. What the biologist first of all wants to know, then, is the molecular mechanism of this change. When a fluid becomes a glass the fields of force about the molecules, as I picture it, become interlocked, so that, although the crystal state is not reached, the potential energy of a molecule in the interior of the mass ceases to be independent of its position. Osborne Reynolds' work upon dilatation is helpful in this connection, but I do not think it carries us far.

To understand the gel or slime it is first necessary to understand the fluid state, and, unfortunately, though physics has much to say of the molecular structure of gases and crystals, that of fluids is still very obscure.

Colloidal systems frequently are systems undergoing spontaneous change. As Graham said, they contain "inergia". But in a dead colloid the "inergia" is due to differences of concentration which right themselves slowly owing to the slow rate of diffusion of the solute and to delays caused by internal surfaces. The remarkable hysteresis of surface films similarly springs from the slowness of tangential diffusion, as Gibbs pointed out.

The available energy of a dead colloid or of an interface is in the main physical in character. That of living matter is essentially a chemical potential, and the chemical potential gives way in the response to a stimulus, and is restored by an intake of oxygen. The activities of a living cell appear to be due to a cunning combination of the chemistry of the dead space with the hysteresis of the colloid. The well-known drop of mercury in a very dilute acid solution of a chromate might perhaps be taken as a model. The drop is thrown into regular pulsation by intermittent contact with a wire. But the drop of mercury must be reduced in size until the chemistry of its surface becomes a rapidly varying function of its radius, such that if a displacement occurs a train of pulsations results.

ABSTRACT

The lecture dealt with the physical aspect of some of the phenomena of living matter. At the outset some of the snares and pitfalls which beset the path of the physiologist in search of physical facts were indicated. One of these was the effect of instinct or intelligence in anticipating altered circumstances, and in setting the appropriate physiological process in action even before the actual change in circumstance had taken place. For example, in some experiments to determine the time lag of increased respiration, it was found that the subject—a cyclist—always anticipated the application of the load and the acceleration of his breathing commenced *before* the actual increase in effort of which it should be the consequence. Then, again, many of the long-accepted theories of the structure of living organisms were ultimately found to be due wholly to the method of killing the organism and preparing it for examination. Exactly similar structures could be obtained in pieces of gelatine if treated in the same way.

Many phenomena were quoted to illustrate the close analogy of many of the functions of living matter to processes familiar to the physicist and physico-chemist. In some of these—for example, the digestive processes of certain organisms—the analogy was so complete that the whole process could be reproduced exactly with non-living material.

The principles underlying growth and development were outlined, and it was shown that growth depended on the presence in the food of *vitamins*. Thus, in the case of white mice fed on artificial milk made by mixing together the purified constituents of ordinary milk, but containing no vitamins, growth ceased altogether, while if 2 per cent. of ordinary milk were added to the artificial milk the growth rapidly became normal. Another example of the influence of these vitamins was that, in pure distilled water, it was impossible to obtain a development of living organisms, but if the slightest trace of tap water were added—putting a stopper in the bottle which had previously been in a bottle of tap water was sufficient—the organisms would thrive quite normally. The fact that such small proportions of nutriment containing vitamins was sufficient suggested a parallel between their action and that of a few crystals thrown into a supersaturated solution of a salt in initiating a crystallisation of the whole.

A further example of the subtle influence of physical laws in physiological phenomena was contained in the explanation of the joining up of severed nerve processes. The directive action involved in this was stated to depend on extremely minute concentration gradients.

Many other interesting phenomena were described and explained.

37

NOTE ON STATIC FRICTION AND ON THE LUBRICATING PROPERTIES OF CERTAIN CHEMICAL SUBSTANCES

With Flight-Lieutenant J. K. HARDY, R.A.F.

[Phil. Mag. (1919), XXXVIII, 32]

In the *Philosophical Magazine* for February 1918, Lord Rayleigh describes experiments which were undertaken to examine more particularly the well-known fact that a few drops of water wetting the parts in contact will prevent a cup of tea from slipping about in a saucer. A glass carriage with three legs terminating in three feet of hemispherical form was made to slide over a plate of glass or copper. The horizontal pull needed to cause movement was found to be 42 gr. where the surface was covered by a film of oil estimated as being of the order of $1\,\mu$ in thickness. The superposition of a layer of water on the film of oil decreased the lubrication, the threshold value of the force rising to 126 gr., that is a threefold increase, and the effect was the same when the water layer was a film deposited by the breath as it was when the plate was completely flooded.

Paraffin (lamp) oil gave a similar result, the force needed to bring about slipping being least when the layer of oil was of insensible thickness. Therefore, in Lord Rayleigh's words, the "friction is greater with a large dose than with a minute quantity of the *same* oil, and this is what is hard to explain".

We started with the object of clearing up if possible this paradoxical phenomenon, and in a certain sense this limited aim has been attained. The two cases, namely, water on a grease film, and lamp oil used alone, are similar in that the phenomena in both cases are due to chemical heterogeneity. Lamp oil appears to consist of substances of high lubricating power dissolved in a more volatile fluid with little or no lubricating power. In the process of forming a thin film the former are concentrated on the surface by evaporation of the latter.

It was felt at the outset that no progress would be made unless individual chemical substances were used. Ethyl ether, ethyl alcohol, and benzene were tested, the specimens in each case being supposed to be pure. Each fluid was found to act to some degree as a lubricant, but no consistent figures could be obtained. This suggested the presence of impurities. By following up this clue it was found that none of the three substances mentioned above had when chemically pure any power of lessening the pull needed to cause the one glass face to slip over the other. Chemical substances in fact fall into two classes according to whether they are active or inactive in this respect under the conditions of the experiments. Water is an inactive substance, and this is the basis of the tea-cup experiment.

In our observations watch-glasses weighted with lead run into them whilst hot were used. A small arm projecting from the lead was attached to a silk thread which passed over a pulley to a pan which held the weights. We were at some pains to procure a glass plate for the watch-glasses to slide upon having an "optical" face. But it was found that ordinary plate glass gave the same value as the plate with an accurately plane surface. Measurements were carried out on a levelling table.

Plates and glasses were cleaned by washing with soap powder and then rubbing under a vigorous flow of tap water with the finger tips, previously washed, until the peculiar clinging stage was reached which sets a finger bowl vibrating when the wet finger tip is rubbed round the edge. Both plates and glasses were drained and dried in air. Contamination will creep over the clean surfaces from solids touching them, therefore contact during the drying process must be reduced to a minimum.[1]

A solid surface dried in air will retain a layer of water in equilibrium with the aqueous vapour, and some of the fluids used, such as acetic and sulphuric acids, absorb water. For these reasons it was necessary to carry out the measurements in air free from aqueous vapour. This was done by using a chamber through which a rapid stream of dry and dust-free air could be passed. The results recorded below will not be obtained unless this precaution be taken. In such a chamber a cleaned surface retains its purity for hours, the substances used to dry the air seeming completely to remove the lubricating matter which condenses to form the "grease" film on surfaces exposed in the ordinary way.

According to the accepted view a "clean" surface as defined above would differ from the "raw" surface produced by splitting a solid in a vacuum by the presence of a layer of condensed gas. There does not, however, appear to be conclusive proof of the existence on all solid faces of such a layer. It would seem, therefore, that we can assert nothing as to the presence of a layer of foreign matter on a "clean" surface of glass in contact with dry air.

The tangential force was applied gradually about 1 cm. above the centre of gravity of the watch-glass by slowly withdrawing support from the pan. The effect was to cause the watch-glass to rock forward until equilibrium was reached, when the static friction balanced the tractive force. If the

---

[1] In using a Pockell's trough to measure the surface tension of water under various conditions, in order to obtain really steady readings, one of us found it necessary to reduce contact between the trough and the table by interposing a couple of glass rods.

latter were increased beyond a certain quantity determined by the nature and state of the surfaces static equilibrium was not reached, but the cohesion gave and the watch-glass moved bodily forward. The limit of static friction is the force which just fails to bring about this movement forwards. It is called the "threshold value".

The watch-glasses were of the same pattern and make, but differently weighted. The threshold values in grammes were:

Watch-glass	Weight gr.	Pull gr.	Pull per gr. wt.
1	54·25	46	0·85
2	58·5	50	0·85
3	57·79	51	0·85
4	170·2	143	0·84

For plates of a rather greenish glass:

Watch-glass	Weight gr.	Pull gr.	Pull per gr. wt.
5	14·58	13·5	0·93
	„	16·2	1·1
6	58·25	54·2	0·93
7	46·95	45·0	0·96

Over a considerable range therefore and for clean surfaces the threshold value is, as a first approximation, equal to the total weight multiplied by a constant which is different for different kinds of glass.

The effect of temperature was tested by altering the temperature of the stream of dry dust-free air passing through the chamber, the actual temperature of the surfaces being taken to be that of a thermometer whose bulb was inside the chamber. Over the narrow range available no effect of importance was detected.

Glass 2

Temp. ° C.	Pull gr.
15·5	49·5
19·0	49·5
23·0	50·0
23·7	51·0

In order to be certain that the watch-glass and plate were actually at the temperature of the air in the chamber, each stage occupied about an hour. The experiment therefore affords striking testimony to the steadiness of the state of clean surfaces for over 4 hours in clean dry air.

These observations were undertaken merely to satisfy ourselves that small fluctuations of temperature did not introduce sensible error.

The forward movement is always accompanied by an actual tearing of *both* glass faces. The length of the tear on the plate is of course determined by the distance forward the watch-glass travels. The length of the tear on the watch-glass is much the same, and is caused by the glass rocking backwards when the seized faces give.

The appearance to the naked eye or hand-lens is that of a fine scratch pointed at one end. A high magnification (1000–2000 diameters), however, shows that there is no regular continuous cutting of the faces. On both watch-glass and plate the scratched line is composed of an irregular collection of very shallow pits and very thin plates which have been torn from the opposite face. The track begins at the pointed end, it rapidly widens, and the sides, which are more or less clearly defined, become parallel. It ends abruptly with a square or bluntly rounded end. At the pointed end, where the movement starts, it consists of delicate flakes or pits of the order $1\,\mu$ across.

These features prove that cleaned faces cannot slide over one another. The forces of cohesion come into play and they seize. Once seized, the substance of the glass fractures before the seized faces will give. Very slight normal pressure is needed to cause seizing. An unweighted watch-glass cannot be moved on a plate even quite slowly without producing these characteristic torn tracks, if both faces are really clean.

The breadth of the track may be as much as $50\,\mu$. Why does it broaden? Why does it not continue as it begins at a breadth of about $1\,\mu$? The widening cannot be due to the abrasion of the curved surface, since the contact is continually changing owing to the rocking backwards of the watch-glass. The appearances suggest that it is owing to minute and very rapid vibrations or oscillations of the watch-glass at right angles to the direction of pull. When the glass tears the "give" is not directly in the line of the pull, the watch-glass is thus thrown out of its balance to one side or the other with the result that it chatters forward vibrating rapidly from side to side. The breadth of the track is determined by the amplitude of these vibrations. When a watch-glass is moved by hand, owing to the unsteadiness of the motion the track commonly begins in more than one (five or six) fine lines which keep separate for say $40\,\mu$ and then blend into one wide track.

The appearance under the microscope is consistent with the view that the faces are seized before movement starts, and the thesis developed in these pages is that static friction both of clean and lubricated faces is due to cohesion which, in the case of clean faces, causes the glass itself to seize and in the case of lubricated faces causes the film of lubricant on the one face to seize to that on the other face.

We can find no justification for the view commonly held that static friction is due to inequalities of the surfaces. Actual measurements proved that the threshold value for glass with an "optical" face is the same as that for ordinary plate glass, whilst the value for ground glass is lower probably because it is impossible to clean it. If static friction be due to inequalities these must be of insensible magnitude, and "inequalities" of this order are indistinguishable from the attractions of individual molecules[1] across the interface.

The weight of the watch-glass will form a depression in the plate, and it might be supposed that resistance to slip is due to this fact, the surface of the plate being heaped up into a crest which ruptures and so starts the tear when the stress exceeds a certain limit. This amounts to supposing that the watch-glass acts as a cutting tool. Deformation of both surfaces must occur, but the action cannot be of the kind just described since *both* faces are torn, and always in such a way as to detach very thin flakes. The microscope reveals no sign of a burst through.

The resistance offered to slip is large. If it be due to cohesion, why should there be no sensible resistance to displacement of the surfaces along the normal?

Two answers may tentatively be given to this very difficult question. The first is that, in movement along the normal, the curved is as it were peeled off the flat surface, so that the force required for an infinitely small displacement is infinitely small.

This answer is more general than would appear at first sight. The range of the force of cohesion is so small that all surfaces, no matter how carefully trued, may be considered to touch only at the summits of elevation. The play of Newtonian colours between glass plates affords pretty testimony to this.

The second answer is, that at the surface of any solid or liquid the molecules are orientated by the normal component of the forces of attraction, and that this contributes specially to resistance to slip.

It is not easy to see why cohesion should resist slip since the potential energy of the forces of attraction will be the same wherever contact is made.

If both faces were formed of continuous solids there would be no initial force to resist slipping, though there would be dissipation of energy in compression waves *during movement* if the bodies were elastic.

Resistance to slip must be due to the discontinuity of matter. When the applied faces are at rest the molecules on either side of the interface take up the position in which the potential of the attractive forces is minimal. Consider a single molecule: it will "seize" in a position of least potential and a tangential force applied to it produces a displacement until the internal and external forces on it balance one another. If the molecules of a solid were able to change partners freely, as they can do in a fluid, slipping would occur, as it of course does occur in a fluid. The fact that solid faces will not slip past one another when external force is applied, means that the

---

[1] This word is used merely to denote the units out of which the matter is built.

uncompensated force on the molecules at the surface increases rapidly for displacements from the equilibrium position which are small in comparison with the distances between their centres. Why, then, is there not resistance to slip, that is to say, initial resistance as distinct from the dissipation of kinetic energy during relative motion, in a fluid? The answer is to be found in the fact, which van der Waals emphasised, that the problem is dynamical and not statical. Increase in the heat energy decreases the damping of the heat vibrations of a solid until a point is reached at which the molecules are able to change partners freely.

Resistance to slip is, however, not due merely to the short range of the forces acting on the molecules, but to their orientation. In 1913 one of us showed that the work done by the forces of cohesion in the formation of an interface is determined by the chemical nature of the substances concerned, and that it is greatest when the molecules of these substances are of a salt type such as esters, acids, or alcohols. It was then pointed out that such molecules are readily polarisable, and that they would be oriented by the forces acting across the interface[1] and that this orientation is the cause of contact difference of potential. From this it was inferred that "The surface film (of fluid or solid) must therefore have a characteristic molecular architecture and the condition of minimal potential involves two terms, one relating to the variation of density, the other to the orientation of the fields of force" of the molecules.[2]

The theory of surface forces has since been developed along these lines in a very striking and beautiful way by Harkins[3] and Langmuir.[4]

Any polarisation of the molecules at the surface must introduce a factor in the resistance to slip which is absent from any resistance there may be to displacement along the normal, for the former includes the resistance which the molecules may offer to disturbance of their orientation, and this might be as great as or even greater than the resistance to displacement of the molecule as a whole along the axis of the normal to the interface.

If this view be correct, the effect of a tangential force would be to produce a fresh orientation of the molecules at the interface, and this would, amongst other things, alter the contact electrical potential between the faces. In the case of fluids, the new orientation would disappear when the external force was removed, but in the case of solids the new orientation would be irreversible if it exceeded a certain small amount,[5] and the work

[1] *Proc. Roy. Soc.* (1912), A, LXXXVI, 610; *ibid.* (1913), A, LXXXVIII, 312.

[2] *Ibid.* (1913), A, LXXXVIII, 330.

[3] *Journ. Amer. Chem. Soc.* (1917), XXXIX, 354 and 541.

[4] *Ibid.* (1917), XXXIX, 1848.

[5] Since the above was written we have come across two lectures by Sir Alfred Ewing, in which he suggests that friction may be due to the attraction between molecules (*The Molecular Process in Magnetic Induction*, Royal Institution of Great

done in causing slip would be in part consumed in producing this new setting of the molecules.

There is fortunately direct evidence available to prove that irreversible changes in molecular configuration are produced when clean glass faces are forced over one another. The original interface is preserved at the junction of the thin flakes, torn off one solid, with the other solid, and these flakes are found to be doubly refractive.

LUBRICATED FACES

The effect of the depth of the layer of lubricant must be taken account of. No attempt was made to measure this, but three stages were distinguished —the film, the smear, and complete flooding. A film is a layer completely invisible, of depth insufficient to give Newtonian colours and almost certainly of the order of $1 \mu\mu$. A smear is a visible but thin layer. In complete flooding the watch-glass moves in a pool.

The difference between the film and the two other states of the surface is more than one of mere depth of the layer. We know from experiments with fluids that when a layer of one fluid spreads upon another, the surface energy is at first a function of the depth of the layer. Thus, when olive oil spreads on clean water, the surface tension is a function of the quantity of oil per unit area until this exceeds a certain limit at which two independent interfaces are formed, that of oil-air and that of oil-water. Over the range of varying tension the properties of the surface, such as its chemical potentialities, its electric charge, and its mechanical tension, are a function of the thickness of the film on the surface, the chemical composition of the material composing the film and the fluid on which it lies, and the temperature. Surfaces of this kind, whose properties depend upon the interaction of two kinds of states of matter, have been called by one of us "composite surfaces".[1]

The experiments described in this and the following paper prove that in the "film" stage of lubrication we are dealing with a composite surface.

The feature of a composite surface which is of most importance in lubrication, is that the energy per unit of mass of the film is a function of position on the axis of the normal. Two consequences follow—the film resists tangential displacement, and therefore has tenacity, and, since the

Britain, May 1891: "The inner structure of simple metals," *Journ. Inst. Metals* (1912), VIII). When one face is made to slide past the other "the polar forces continue to act across the plane of sliding, causing first a quasi-elastic turning (of the molecules); but when a certain very limited range of movement is exceeded there is dissipation of energy through the original bonds being broken and new bonds established with oscillation of the particles."

[1] A soap-bubble is composed of two composite faces placed back to back, cf. *Proc. Roy. Soc.* (1912), A, LXXXVI, 609.

potential energy of the molecules composing it is a function of their position, the film is not completely fluid even though it be formed from material which is a fluid when in mass at the same temperature and pressure. This is true of composite surfaces of fluids as well as of solids.

Owing to this defect in fluidity composite surfaces are capable of seizing, and the static friction of such surfaces is, in our opinion, due to this fact.

By seizing is meant the capacity for offering resistance to slip when both faces are at rest. Though it is due to the operation of the same forces of attraction as bring about cohesion, resistance to slip and cohesion are not identical. Any internal surface of a fluid may be considered to be formed by bringing together two fluid faces which cohere, but there is no initial resistance to slip gained thereby. Composite faces resist slip because of their defect in fluidity.

If the matter stopped here, if, that is to say, the defect in fluidity were due solely to the surface energy being a function of the thickness of the film, static friction between composite plane faces would be purely an edge phenomenon and would vanish if the area of the faces was infinite. The way in which the energy of an interface depends upon chemical constitution, the fact, already noted, that it is closely related to the polarisability of the molecules, proves that the surface energy is a function of the orientation of the molecules as well as of their position on the axis of the normal, and it is to this that we may look for the source of the static friction of composite surfaces as well as of clean faces. There is, indeed, no final distinction between composite faces as defined and clean faces, for there is an orientation of the molecules at the surfaces of any fluid or solid, and this skin, which has its own peculiar configuration, constitutes, as Gibbs pointed out, a separate phase.

The distinction between the film and the smear or flooded states of a surface is now seen to be not merely one of degree but one of kind. This follows not only from what has been just said, but also from the way in which the surface energy varies with the thickness of the film. The well-known phenomenon of the grey and black area of soap-bubbles, and the fact that a layer of fluid of small but sensible depth when spread on another fluid or on a solid is not in equilibrium, but breaks under the influence of surface forces into a film and thick sheets or lenses which are in tensile equilibrium with the film, proves that the surface energy is a discontinuous function of the thickness. When the layer is about $1\,\mu$ thick the function changes its sign. There is thus a region of instability, or regions of instability, which definitely mark off the composite surface from the true double surface.

In static lubrication a layer of fluid is interposed between two solid faces. Now, though we know that the energy of a composite surface is a discontinuous function of the depth of the layer of fluid when the surface is bounded on the one side by air or vapour, we know nothing directly of the form of the function when it is bounded by another solid. Let us assume that it is discontinuous also in this case, and consider what must happen when the two solid faces are forced together by a normal pressure. The fluid will be squeezed out the less rapidly the greater its viscosity until a certain critical thickness is reached at which the layer becomes unstable. It will then collapse until the thickness is reached at which the surface energy begins to increase as the thickness decreases. The film now has gained tenacity and lost fluidity. For a certain normal pressure therefore a certain critical thickness of film will persist, and therefore a certain definite resistance to slip. The critical thickness may be expected to vary very little over a wide range of pressure since the energy of thin films is a rapidly varying function of their depth.

The layer of fluid must have some tenacity to be able to maintain itself, and as increase in tenacity and loss of fluidity go together, a limit will be set to the lubrication. The resistance to slip will never wholly vanish as it should do if the surfaces were separated by true fluid. Thus, what Osborne Reynolds calls "boundary conditions" must always operate in static friction, whereas they may be absent in kinetic friction. The existence of a discontinuity in the variation of energy with the thickness of the layer of lubricant would tend to confine this maximum of lubrication within narrow limits, and observation shows this to be the case.

As the facility for slipping increases the character of the movement changes. When the surfaces are clean the pull is 0·85 or 0·93 gr. per gram normal pressure, and steady slow sliding motion cannot occur owing to the violent seizing. As the facility for slipping increases seizing decreases, and the motion assumes more and more a gliding character. When the threshold value has fallen to 0·1 gr. per gram weight true gliding is completely established. Any small increase in the tangential force beyond the threshold value now causes a slow glide without noticeable acceleration. In other words, the state of the surfaces is such as to dissipate any small increments of energy as heat as fast as they are gained.

No fluid was found which would increase the facility for slipping beyond this point. It appears to be a true maximum for the kind of surfaces employed. Between this maximum and the minimum of clean surfaces there is a greater or less degree of seizing which must be broken away. The force necessary to effect the break away is too great to be absorbed as heat, and the movement therefore, instead of being a steady glide, is characterised by more or less marked acceleration.

Films of lubricant can be deposited on the surfaces in many ways, such as by bubbling the air which comes to the chamber through the fluid (ethyl alcohol, ethyl ether, benzene, water, ammonia fortiss., acetic acid); by flooding the surface and evaporating off the excess with dry pure air until nothing visible remains (acetic acid, tripropylamine, and triethylamine); by flooding the surface, washing off excess with a vigorous stream of tap water, and draining and drying the plate and watch-glass in the way described earlier (acetic acid, oleic acid, sulphuric acid, castor oil, and paraffin); or by flooding the surface with a very dilute solution in pure dry ether and evaporating off the ether with dry air (oleic acid, castor oil).

Technically, the most beautiful method of forming a film is by taking advantage of the fact, dealt with more fully in the next paper, that a totally invisible film is formed about a drop of some fluids when placed upon a clean glass plate, provided water vapour is completely excluded. With a substance such as tripropylamine, in order completely to alter the state of the whole surface of the plate it is sufficient to place near one corner a small drop of the fluid.

It is not permissible to form a film by polishing off excess with "clean" linen since such linen for these purposes is not clean. Simply polishing the surface with clean linen effectively lubricates it by leaving behind an invisible film.

If the static friction of lubricated surfaces is determined by the variation of the surface energy, we should expect to find it closely dependent upon the chemical constitution of the lubricant. This is the case, and the most unexpected fact is that certain fluids have no power of lowering the friction.

The *inactive fluids* discovered were: ethyl alcohol, ethyl ether, benzene, water, and ammonia fortiss. sp. gr. 0·880. Films of insensible thickness and visible layers were without any influence upon the friction, as also was light and heavy flooding. Finally, these fluids when dried off left the surfaces uncontaminated and the cohesion unchanged.

It is of course obvious that when a layer of any fluid in mass is maintained between solid faces, the friction will be fluid friction and the facility for slipping infinite. Thus, when two glass plates are flooded with water or ammonia and placed face to face, the one plate can be readily slipped past the other since a layer of fluid is maintained between them by capillarity. But a very slight normal pressure suffices to displace such a layer, and the faces then seize. Inactive fluids therefore are lubricants only in this limited sense. *They are inactive in the sense that they have little or no power of so altering the solid surfaces as to facilitate slipping.*

This statement may be put in another way. The available energy of a composite surface is a function of the thickness of the layer of fluid. If the energy increases as the thickness is reduced the layer will resist displacement. In the case of inactive fluids the variation of the available energy must be such as to confer little or no capacity for resisting displacement. The behaviour of ordinary commercial glycerine is interesting in this connection.

Glycerine is remarkable in that it can increase the facility for slipping nearly if not quite to the maximum, but only when the surfaces are heavily flooded, when, owing to its viscosity, and in consequence of the rocking forward of the watch-glass, a thick layer of fluid is maintained between the solid faces. A film of glycerine had no effect whatever, thus:

Glass 2. Clean: Threshold value 51 gr.

	Films*	Smear	Plate flooded and drained	Heavily flooded
Threshold value	51	43	30	6
Per gr. pressure	0·87	0·73	0·51	0·10

* Deposited from a 0·002 per cent. solution in alcohol, the plate being flooded with the solution and dried off without draining four successive times without change in the threshold value.

Glycerine therefore belongs to the class of inactive fluids as defined above.

By looking for the Newtonian colours one can form some conception of the depth of an inactive fluid which must be maintained between the plates to permit freedom of movement. With strong ammonia (0·880) most violent seizing occurred so soon as these colours began to appear. With water, seizing was perhaps less violent and Newtonian colours well seen before it took place.

Active fluids are those which facilitate slipping when present on the surfaces in layers of any thickness. They operate when the layer is of insensible thickness and of the order of 1 μ in depth. It may be claimed therefore that it is not *qua* fluid that they act, but because they react with the solid face to form a composite surface having a lower available energy and therefore a lower capacity for seizing.

The active substances differ amongst themselves, as might be expected, in the extent to which they reduce friction; some only are capable of reducing it to the limit, noticed in a previous section, at which a slow steady glide is possible. Threshold value in grammes per gramme normal pressure:

	Film	Smear	Flooded
Acetic acid	0·62	0·73	0·73
Butyric acid	0·5	0·5	—
Oleic acid	0·1	0·1	0·17
Sulphuric acid	0·58	—	0·75
Strong hydrochloric acid	—	—	0·63
Trimethylamine, strong solution	—	—	0·6
Triethylamine	0·39	—	0·32
Tripropylamine	0·26	—	0·27
Pyridine (impure sample)	—	0·46	—
Castor oil	0·1	0·1	0·1
"Paraffin"	0·29	0·15	0·22

The influence of chemical constitution is seen in the sharply contrasted action of acetic acid and ammonia, the former is active the latter inactive; in the alkaline series, ammonia, trimethylamine, triethylamine, and tripropylamine. It will be noticed, too, that friction slightly increases as the quantity of an acid present on the plate increases, and decreases slightly as the quantity of an alkali present increases. In both acids and alkalis the activity of the fluid increases as the molecular weight increases.

If these several relations are confirmed by more extended investigations, the case "hard to explain" of lubrication diminishing as the quantity of lubricant increases will arise, but, when the bulk of the measurements given above were made, the extreme sensitiveness of the faces to minute traces of water was not appreciated by us (see the following paper). The figures given in the table were obtained under comparable conditions, but they are not free from the suspicion of the influence of water. The sample of butyric acid almost certainly contained a trace of water. Acetic acid in the complete absence of water vapour gave the values:

Weight of watch-glass gr.	Film	Flooded
46·95	0·47	0·36
170·2	0·64	0·58

The film, however, is subject to intense evaporation, due to the rapid current of dry air. The maximum of lubrication was therefore probably not reached. Acetic acid frozen on the plates gave the values:

Weight of watch-glass gr.	Film	Flooded (*i.e.* sensible layer of ice)
65·25	0·23	0·23

The sample of tripropylamine was specially purified, and the values given in the table were obtained in the complete absence of water. The vapour pressure of this substance is low, and the values therefore are probably close to its maximum of lubrication.

The evidence, so far as it goes at present, is against the conclusion that lubrication is a function of the quantity of lubricant on the plate when the lubricant is a single pure chemical substance, and when its viscosity is not very great.

The fluid called "paraffin" is the rectified "paraffin" of the British Pharmacopœia. It was somewhat uncertain in its effects, behaving as though it were a mixture of active and inactive substances.

The basis of the original observation on the tea-cup is revealed by these results. The sticking of the cup is not due to the fact that a thick layer of homogeneous fluid lubricates less than a thin layer, but to the fact that an inactive fluid, in this case water, diminishes the effect of an active fluid. There is much of scientific interest in the way the water acts. It does not remove or even temporarily detach the film of lubricant, for the full influence of the latter is restored when the water is dried off. On the other hand, it would seem to lessen the grip of the film on the glass, for the latter can be detached by lightly rubbing the surface *under the water*.

Glass 2. *Paraffin. Film polished with linen: Threshold value 10·5 gr. 17° C.*

Flooded with water	18 gr.
Rubbed lightly under water with the finger tip, which had been freed from grease by washing	36 ,,
Rubbed more vigorously	41 ,,

Glass 3. *Castor Oil. Polished film: Threshold value 11 gr. 19° C.*

Flooded with water	Pull 18 gr.
Rubbed under water with cleaned finger tips	,, 43 ,,

That the water should act by reducing the tenacity of the film is quite in accord with the classical theory of capillarity. When the film is formed of a *solid*, flooding the surfaces with water has no effect, the solidity of the film apparently enabling it to resist the tangential stress. A solid film was formed by coating a clean surface of glass with melted solid paraffin. The excess was then polished off with linen until nothing visible remained. The threshold value was found to be 0·1 gr. per gram weight. Flooding the surface with water did not alter this value.

Attention has already been drawn to the fact that ordinary "pure" ether and benzene contain a lubricant in solution and leave behind a lubricating film on evaporation. Rectified spirit also contains an impurity which causes it to behave in an interesting way. Friction was lowered when the plate was flooded with the spirit, but rose to the " clean " value when the fluid was completely dried off. The point of interest is that the full value was not restored until the layer of fluid had been reduced by evaporation to well past the Newtonian colour stage.

Actual values in grammes are:

Clean 49; flooded 43; Newtonian colours seen about 45; colours vanished 46; and 1 hour later 49.

The question now arises whether the difference between active and inactive substances is one of degree or one of kind. The final answer must be left to further investigation with more refined apparatus. All that can be said now is that the inactive fluids were inactive for the lightest as well as for the heaviest watch-glass used. On the other hand, an insensible film of an active fluid lubricates the surface for the heaviest watch-glass used, and in the case of tripropylamine, oleic acid, castor oil, and paraffin just as effectively as for the lightest glass. The insensible film formed by spreading from a drop of tripropylamine under the influence of surface forces gave for instance the following values:

	Weight of watch-glass gr.	Threshold value per gr. weight
Flooded Film	46·95	0·24
	46·95	0·25
	170·21	0·26

Temperature 10° C.

SUMMARY

One of us has shown that the variation of the surface energy at an interface between two fluids and of a composite surface is closely related to the chemical constitution of the substances concerned. The inference to be drawn is that the work done in forming the interface or the composite surface is done by chemical forces. In the theory of capillarity as developed by Young and Laplace cohesive forces are taken to be, like gravity, incapable of saturation, and unlike gravity only in being of very short range. If cohesive forces are chemical they are capable of saturation, or, as a chemist would put it, of neutralisation, and the film of matter on a composite surface may be said to reduce the surface energy by neutralising to a greater or less extent the forces at the surface of a solid or fluid. The function of a lubricant is to reduce the energy of the surface, and thereby to reduce the capacity for cohesion and the resistance to slip when two composite surfaces are applied the one to the other. The function of a lubricant therefore is the opposite to that of a flux. A good deal is to be gained, in our opinion, by recognising lubrication as a special case of that incomplete chemical reaction characteristic of surfaces in which the law of multiple and definite proportions does not hold.

Evidence for the orientation of the molecules at an interface or on a composite surface which we take to be the source of static friction·is to be found in the fact that chemical substances whose molecules are by their nature readily polarisable such as those of acids, bases, and esters, produce the greatest changes in surface energy,[1] and that there is contact difference of potential between the film of a composite surface and the matter on which it lies.[2] This holds even when no part of the matter concerned can with certainty be said to be present in mass as in soap-bubbles. In these "free films" the films on each face are at an electrical potential different from that of the middle portion.[3]

[1] *Proc. Roy. Soc.* (1913), A, LXXXVIII, 303.
[2] *Ibid.* (1911), B, LXXXIV, 220.   [3] *Ibid.* (1912), A, LXXXVI, 608.

The theory of static friction which seems best to accord with the facts is that it is due to cohesion between the faces. When a lubricant is present we may consider the friction as operating at an imaginary surface situated in the lubricant parallel to and midway between the solid faces. This surface is an interface between two composite surfaces, and may be considered as being formed by bringing these two composite surfaces together. Work is done by the forces of cohesion when the film of lubricant is applied to the solid face to form each composite surface, and the surface energy is decreased by this quantity. A further quantity of work is done when the two composite surfaces are applied to one another with a further change in the surface energy. The static friction is an unknown function of the total change in surface energy.

38

THE SPREADING OF FLUIDS ON GLASS

[*Phil. Mag.* (1919), xxxviii, 49]

The phenomena which accompany the spreading of an immiscible fluid over the surface of water under the influence of surface forces have been known for a long time and studied in great detail. The salient facts are these—when a small drop of some, not all, fluids is placed on a clean surface of water on which a few dust particles rest, a film of insensible thickness rapidly spreads from it in all directions, driving the dust particles before it. Sometimes, indeed usually, the drop itself also expands to a plate of sensible thickness which is in tensile equilibrium with the invisible film. This plate usually is not in equilibrium. It thins in places and, after certain changes, settles down to an irregular or regular pattern, the spaces of which are occupied by an invisible film, the counterpart of that which surrounds the expanded drop.

There are thus two distinct processes, the spreading to form the invisible film, and the spreading of the drop to form a layer. Some confusion has crept into the literature of the subject by the use of the same word to describe these two dissimilar processes. I propose therefore to distinguish the first as primary spreading and the surface formed by it as a "primary composite surface", and the second as secondary spreading which forms a "secondary composite surface". A primary surface is covered by a film of the order of 1μ in thickness, whilst a secondary surface is covered by a layer from 50–500μ in depth.

By employing pure chemical substances I was able to verify Lord Rayleigh's suggestion, that the complicated secondary changes in a secondary composite surface are due to the chemical heterogeneity of the surface layer, and to prove further that when a drop of a rigorously pure chemical substance is placed on a clean surface of water, a primary composite surface is formed by the spreading from it of an invisible film, and there the matter ends, since the equilibrium state for a single pure chemical substance spread on water is such a surface in tensile equilibrium with a lens-shaped drop into which the excess of the substance is gathered.[1]

The existence of the invisible film is made manifest by the fall in surface tension occasioned by its presence, by the sweeping away of motes of dust on the surface as it spreads, and by the fact that it can be scraped off the

---

[1] *Proc. Roy. Soc.* (1913), A, lxxxviii, 314.

surface, the surface tension being thereby raised to the full value of clean water. The movement of motes during the process of scraping shows that the invisible film confers on the surface the property of tangential elasticity or contractibility. This is an exceeding delicate test for its presence. The film is at an electrical potential different from that of the water; there is a contact difference of potential between it and the water.[1]

In the paper already referred to I described how drops of certain fluids of low chemical reactivity, *and whose vapour pressure was negligible*, such as a paraffin of high boiling-point, could be placed on a clean surface of water without either displacing motes on the surface or lowering its tension, or making the surface "contractile."

Some fluids therefore will not spread on water *at all*, and that despite the fact that they are capable of lowering the tension of water. Saturated paraffins of low boiling-point will form a film, but that only by condensation of the vapour, and not by direct spreading of the drop, for reasons given in the earlier paper.

The point of interest and, so far as I know, of novelty, is that all these relations hold quite simply for a clean solid face. From a drop of certain fluids primary spreading occurs, if the fluid in question is not rigorously pure secondary spreading follows, and certain fluids will not spread at all.

A drop of acetic acid which has been purified by recrystallisation when placed upon a surface of glass which has become contaminated by exposure to the atmosphere does not spread. On clean glass the drop flattens out rapidly, dust particles or drops of water being swept away. It was found, however, that the flattening was due to the water vapour present in the air. The remarkable character of the action as an instance of the operation of small quantities of matter upon surface energy cannot be exaggerated.

Attention was drawn to it by the following observation. A clean plate of glass was placed in a chamber through which a rapid current of dry air was flowing. After the plate had been in the chamber for 30 min., the cover was lifted slightly for a moment to allow of a drop of acid being placed on the plate from a fine capillary tube. The drop at once spread to an even layer, thin but distinctly visible. After being for some minutes in the dry air the layer altered. It thinned in places, and the thicker parts contracted into a number of bead-like drops which then slowly moved towards one another and fused until finally one or a few drops remained, the rest of the plate being apparently but not really free from acid. On lifting the cover even for a moment the drops at once expanded again, and the layer again proceeded to contract when the cover was replaced.

The question therefore arose whether acetic acid would spread if water

---

[1] *Proc. Roy. Soc.* (1911), B, LXXXIV, 220; (1912), A, LXXXVI, 608.

vapour were rigorously excluded. By a simple device drops were placed on the glass plate without lifting the cover, and no flattening occurred. But though the drop remained where it was placed without visible change, there had in fact spread from it over the plate an invisible film. In other words, the single pure chemical substance acetic acid behaves on glass as does a single chemical substance on water; from it a film spreads to form a primary composite surface, and if any secondary spreading of the drop itself occurs it is due to an impurity—in this case water.

It is not possible to measure the surface tension of a solid in the simple way applicable to fluids, but the existence of an invisible film of acid covering the surface can be proved by the fall in static friction which it produces.

The static friction of a clean glass plate measured in the way described in the preceding paper was found to be 0·92 gr. per gram weight of the watch-glass. A small drop of acetic acid was placed on the plate near one corner. It remained apparently unchanged, but the friction of the general surface was found to have fallen to 0·47 gr. per gram weight. The surface was therefore covered by an invisible film of acid. In the complete absence of aqueous vapour, pure acetic acid forms a primary composite surface on glass with the excess acid gathered into a lens just as, for instance, oleic acid or benzene will form a primary surface in tensile equilibrium with a lens on clean water.

Tripropylamine, just distilled to a constant boiling-point, exhibits precisely similar phenomena. A drop placed near a corner of a plate lowered the static friction over the whole surface to 0·25 gr. per gram weight of the watch-glass.

When water is present in the acetic acid itself, or as vapour, the acid behaves on glass just as does an impure chemical substance on water. A drop spreads to form a secondary composite surface. The concentration of water needed to upset the equilibrium of a drop of pure acid and induce secondary spreading must be incredibly small. A drop of acid standing quietly on the surface at once spreads if the cover to the chamber be slightly lifted for a moment *whilst the current of dry air is pouring in.*

For reasons to be given presently, it is important to note that these observations on the behaviour of acetic acid were made at 18° C.

Not all fluids will spread to form a primary composite surface on water, and this is true for glass. At 18° C. neither castor oil nor B.P. "paraffin" formed such a surface when drops were placed on a plate.

When more than one drop of acetic acid is present on a plate at 18° C., the existence of an invisible film about them is manifested in a curious way. Drops which are not more than one or two centimetres apart attract one another. They become oval in outline, the long axes being directed towards one another, and slowly move across the plate until they meet and fuse. The edge of the plate acts like another drop, a single drop tends to move to the edge and cling there, but the edge does not attract it so strongly as does another drop. The experiment is a simple one, but very striking. The invisible film is nowhere greater than $1\,\mu$ in thickness, yet it is capable of pulling large drops of fluid along. The stress per unit area of transverse section must be enormous.

The reason for the movement of the drops is probably to be found in the evaporation due to the current of dry air. The film everywhere is thinned by this, and the tension of the composite surface therefore greater than what it would be if the space above were saturated with the vapour of acetic acid. The evaporation is continually made good by the drops which are as continually feeding the film. The surface tension in a line joining two drops will therefore always be *lower* than elsewhere. Why are the drops not pulled away instead of towards each other? The answer can only be that on the general surface of the plate the layer is so thin and so closely applied to the solid face as to have lost fluidity. It is practically a non-contractile solid. Between the drops it is thick enough to preserve some degree of fluidity and it contracts, pulling the drops together. This explains the fact that drops more than 1–2 cm. apart do not attract each other at 18° C.

There is much of interest in considering how the water vapour acts when it causes the drop of acetic acid to spread over the primary composite surface. The most obvious suggestion is that it is condensed on to the composite surface where, by diluting the film of acetic acid, it raises the tension and the drop of acid is pulled out to a plate. Of course, the same result would be produced if it diluted the drop of acid and lowered the tension of one or both faces, but the quantity of water operating would seem to be too small to produce a sensible change in a large drop of acid.

The suggestion that the water vapour condenses on to the primary composite involves a paradox, namely that a vapour can condense on to a surface whilst raising its surface tension. At first sight this looks like a spontaneous increase in free energy. A parallel case can, however, be found. When the area of a primary composite surface of pure oleic acid spread on water is contracted by narrowing the boundaries, it is possible sometimes to force the surface tension down to a very low level. The surface is then supersaturated with the acid, and it not infrequently changes, apparently spontaneously, the excess acid gathering itself into lenses, and this change is accompanied by a *rise* in the tension.[1]

Returning to the case of acetic acid, the fact that a crop which has spread rapidly contracts when in dry air points to the aqueous vapour being condensed on to the composite surface and not on to the lens; for in

[1] *Proc. Roy. Soc.* (1913), A, LXXXVIII, 326.

the latter case the acid is present in mass, and water would not readily distil off from a very concentrated solution of acetic acid. If the water distils off the primary composite surface more readily than the primary acid, we may conclude that it lowers the surface energy of glass less than does the acid; and this is in agreement with the fact that water does not, and acetic acid does, lower the static friction of glass.

When a layer of sensible thickness of a pure fluid such as benzene is formed on water and allowed to thin by evaporation, it becomes unstable when the thickness falls to a certain point. The layer then thins in places so as to form primary composite surfaces, and the excess benzene is gathered into a lens-shaped drop or drops. The surface energy is a function of the depth of the layer of benzene, but it is a discontinuous function. At the edge of a drop of a pure fluid which is on water and in tensile equilibrium with a composite surface, a peculiar state must therefore obtain, for at the junction of the drop with the film the instable thicknesses must be some-how bridged over, just as are the instable thicknesses at the junction between the black or grey areas with the rest of a soap-bubble. There is then this difference between the formation of a primary and a secondary composite surface by spreading from a drop: in the latter the edge state moves bodily over the surface, and in the former the substance of the drop has to be, as it were, dragged through the instable thickness. It is this latter fact which I think accounts for the remarkable phenomena exhibited by a drop of ethyl hydrocinnamate on water.[1]

The importance of the vapour in jumping the barrier at the edge of a drop must not be overlooked. Experience of the formation of primary composite surfaces on water by many pure chemical substances leads me to conclude that the surface is formed by actual spreading of the substance of the drop itself only when the sum of the tensions of the upper and lower surfaces of the lens-shaped drop is considerably less than that of water, or, using Gibbs's notation, in which the vapour phase is indicated by the numeral 1, the water by 3, and the substance which is to form the composite surface by 2, when $T_{12}+T_{23}$ is considerably greater than T_{13}.

Taking oleic acid and ethyl hydrocinnamate as examples:

	$T_{12}+T_{23}$	T_{13}
Oleic acid	$31 + 15 = 46$	< 74
Ethyl hydrocinnamate ...	$33 + 24 = 57$	< 74

When the sum of the tensions is only slightly less than that of water, the drop does not flash over the surface, it remains apparently unaltered, but it is the vapour phase which condenses to form the primary composite surface, *e.g.*:

Octane ... ... $20 \cdot 6 + 53 = 73 \cdot 6 < 74$

[1] *Proc. Roy. Soc.* (1913), A, LXXXVIII, 324.

These relations are clearly indicated by Gibbs,[1] but for the purposes of some part of his argument the surface 13 is the primary composite surface, and not the water-air surface.

No fluid amongst those which I examined was found to flash over glass. Judged by the fall in static friction, a film takes some seconds to spread a few centimetres from a drop of acetic acid.

Whether primary or secondary spreading does or does not occur on a fluid face depends mainly upon the relative value of the surface tensions, but on a clean solid face it must depend wholly upon the vapour tension. For consider a drop of any fluid placed upon a clean solid surface. The tension of the latter (T_{13}) may be much greater than $T_{12} + T_{23}$: but, since the 13 surface is not contractile owing to its solidity, the disparity of force acting at the edge of the drop is inoperative. If vapour is given off from the drop it will condense to form a primary composite surface, and this *being contractile* may pull the drop itself out to form a secondary surface. The fact that B.P. "paraffin" and castor oil do not spread at all on clean glass is due to their low vapour pressure, and is no evidence as to the tension of the glass-oil interface.

I confess I should not have expected that a film of fluid of the order of $1\,\mu$ in thickness on a solid face would be spontaneously contractile, but the migration of the drops of acetic acid appears to afford conclusive evidence on this point. Any primary composite surface will contract only if a further thickening of the film lowers the surface tension. A fully formed primary surface, that is to say one in which the film has thickened to the point at which the function dT/dx (where T is the tension, and x the thickness of the film) changes its sign, would have no tendency to contract; indeed it would resist further contraction. The fact that primary surfaces saturated with fluid actually do contract spontaneously when exposed to air is due frequently to the thinning of the surface film by evaporation. This is readily demonstrable with pure benzene on water. Substances such as benzene on water and acetic acid on glass behave in fact as though the film on a primary composite surface varied in thickness with variations in the vapour pressure of the fluid of which it is composed. If this be so the film on a composite surface cannot be always and of necessity one molecule thick as some writers suppose.

[1] *Trans. Conn. Acad.* III, 422.

39

STATIC FRICTION. II

[*Phil. Mag.* (1920), XL, 201]

A. CHEMICAL CONSTITUTION AND THE LUBRICATION OF BISMUTH

These experiments are a continuation of earlier work on the static friction of glass faces (*Phil. Mag.* (1919), ser. 6, XXXVIII, 32). A slider having a curved surface was applied to a plane surface, both slider and plate being of bismuth.[1] This metal was chosen because it is highly crystalline and at the same time takes a high degree of polish. It therefore offered unusual facilities for comparing the friction of the amorphous state of the metal found on a burnished face[2] with that of the crystalline state.

One of the faces being curved, contact was over an area defined by the weight and the elasticity of the material. The extent of this area is unknown, and the normal pressure over it not uniform. There are therefore unknown quantities which make it impossible to express the static friction in terms of a normal pressure. The observations of Burgess[3] show that nothing would be gained in this respect by employing two plane surfaces. The best obtainable surfaces touch only at points, and if of glass, show Newtonian colours.

The tractive force was applied slowly, and in such a way as to make the slider rock forwards. The static friction therefore is that between freshly applied faces. The method is apt to give undue importance to viscosity, and care is needed to distinguish between transitory effects due to viscosity and true lubrication (see the earlier paper).

Passing from glass to bismuth faces, one enters another world. Meticulous care perfected by practice is needed to secure clean glass faces, and, once secured, they readily contaminate by the spreading of a film of matter from solids with which they are in contact or by condensation from the air. The edges of a plate oppose the spreading of most if not all fluids, and it is this interesting property of edges which renders exact work with glass possible.

Bismuth, on the other hand, can be cleaned by simply rubbing the face with wash-leather, and the clean plate may be handled freely if actual contact with the burnished face is avoided. To clean the faces the plate and

---

[1] I am indebted to my friend Mr Heycock for a specimen.
[2] Beilby, *Proc. Roy. Soc.* (1903), LXXII, 218.
[3] *Proc. Roy. Soc.* (1911), A, LXXXVI, 25.

slider were washed in ethyl alcohol 98 per cent., drained, and burnished with wash-leather. Wash-leather clings to a clean face of bismuth. When the surface is rubbed with it, there is a characteristic harsh feel and characteristic notes of high pitch are given out *if the surface be clean*. Rarely a lubricant is found which will not be displaced from the surfaces in this simple way: it is then necessary to clean them with rouge.

What is the test of cleanliness? There is no criterion other than static friction itself. When static friction is maximal it is assumed that the surfaces are clean, or, to put it more exactly, "clean" faces are defined as those whose static friction reaches a certain level such that a force of 34,300 dynes just fails to produce movement in the slider used throughout, whose weight was 70·5 gr. and radius of curvature 25·5 mm. This assumption is justified by the fact that every substance tried was found to reduce this datum value when applied to the surface. In this bismuth differs from glass. Many substances are neutral to glass in that they do not alter the static friction. No substance neutral to bismuth was found.

This section, with the limitations noticed in paragraph 2, is confined to a study of one variable—namely, the chemical constitution of the lubricant. One point should, however, be mentioned and reserved for future discussion. A few lubricants appear to abolish the static friction of bismuth altogether (*e.g.* ricinoleic acid). In these cases the value given is that at which the tractive force produced sliding so slow as to be just detectable by unaided vision.

In testing a fluid the surfaces were flooded so that the slider moved in a pool. The thickness of the film of lubricant then was that determined naturally by capillarity acting in opposition to the normal pressure. Solid lubricants were deposited in a layer, thick enough to dull the burnished surfaces, from dilute solutions in ethyl alcohol, benzene, or ether. There is a danger lest the friction of bismuth against the solid in mass be mistaken for what is sought for, namely, the friction of bismuth against bismuth lubricated by the solid. The former obviously is an important limiting value when the lubricant is a solid. Whether it is the only value and identical with the latter will be considered on some future occasion. In the meantime it may be noted here that the value of the friction of bismuth lubricated with cholesterol was the same for an obvious smear of the solid and for the invisible film left when the smear had been wiped away by cotton-wool moistened with alcohol. On the other hand, the value of the coefficient for a visible layer of stearic acid deposited from benzene or ether was 0·20 and for an invisible film 0·15. I incline to the view that the difference is due to the difficulty of getting rid of the last traces of the solvent from some substances.

The force which just fails to move the slider *immediately it is applied* is taken as the measure of the friction. Precision in this matter is needed, because static friction is sometimes a function of the time during which the external force has acted. The significance of this time factor is not yet clear.

Solid bismuth. Surface burnished: some of the fluids used, however, etched the surface so as to expose the crystals: such are the acids formic, acetic, propionic, and valeric, and the sulphur compounds thiophenol and benzylhydrosulphide. Temperature 11–14° C. Measurements made in a current of dry air. Weight of slider 70·5 gr.; radius of curvature 25·5 mm.

The results given in the column headed Static friction are the values of the ratio applied force in grams divided by the weight in grams of the slider.

Static friction is a function of the molecular weight of the lubricant; and in a simple chemical series of chain compounds such as fatty acids and alcohols or paraffins a good lubricant will be found if one goes high enough in the series. But it is not a simple function, as inspection of the charts and curves shows. The friction, for instance, rises sharply in moving from $CHCl_3$ to CCl_4 and from phenol to catechol and quinol. The influence of molecular weight is overshadowed by the influence of chemical constitution.

In some simple chemical series the relation appears to be a linear one. Examples are paraffins; the series benzene, naphthalene, anthracene; and, making allowance for the fact that the ammonia was a solution in water, the series ending with propylamine.

In the aliphatic alcohols and acids the chain is weighted at one end with the ˙OH or ˙COOH group, and the simple linear relation to molecular weight is disturbed thereby.

The relation of lubricating qualities to viscosity broadly resembles that to molecular weight. In a simple chemical series lubrication and viscosity change in much the same way with molecular weight, but that there is no fundamental relation between viscosity and lubrication is shown by the following figures:

	Viscosity at 20°	Static friction
Carbon tetrachloride	0·0096	0·43
Chloroform	0·0056	0·30
Acetic acid	0·0122	0·40
Octyllic acid	0·0575	0·19
Benzene	0·0065	0·39
Toluene	0·0058	0·28
Benzyl alcohol	0·0558	0·31

Fluidity of the lubricant has no constant significance, as indeed might be expected on the surface-energy theory of lubrication. The curves for acids, alcohols and paraffins show no break where with increasing molecular weight the lubricant becomes a solid at the temperature of observation.

Compare also benzene, naphthalene and anthracene; menthone and menthol; thymol and carvacrol.

The upward trend of the first part of the curve for the aliphatic alcohols is in agreement with the fact that methyl alcohol is abnormal in some of its physical properties such as specific gravity.

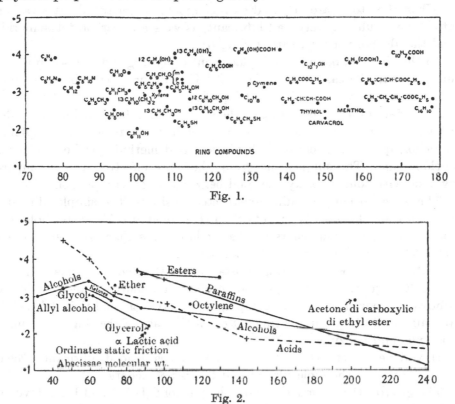

Fig. 1.

Fig. 2.

In their qualities as lubricants of bismuth, ring compounds are the converse of chain compounds: thus the effect of a double-bonded atom is to decrease the lubricating value of the former and to increase that of the latter. As examples, compare naphthoic acid with double-bonded oxygen, with naphthalene, menthone with menthol, cyclohexanone with cyclohexane, benzoic acid with benzene. As examples of double-bonded carbon compare cinnamic ester with hydrocinnamic ester, dipentene, having two unsaturated carbon atoms, with menthol and cyclohexane. On the other hand, the presence of unsaturated atoms increases the lubricating qualities of chain compounds, whether it be the double-bonded oxygen of ketones or acids, or carbon of olefines and alcohols; but this rule is departed from (in the case of acids) when the chain becomes much elongated.

Whatever view be taken of the structure of the benzene ring, it must be admitted to be less saturated than cyclohexane, and we find consistently that the more saturated cyclohexane and its derivative are the better lubricants.

When ring and chain are joined as in butylxylene, the result is a better lubricant than either.

When the atoms are disposed with complete symmetry about a carbon atom the result is a very bad lubricant, as we see in carbon tetrachloride and the alcohol penterythritol $C(CH_2OH)_4$.

In the ring compounds the replacement of hydrogen decreases lubricating power in the case of N, :O, or 'COOH, and increases it in the case of other groups in the order $C_2H_5 < CH_3 < OH$.

The effect of a second group of the same or of a different kind is to decrease the effect of the first. Compare for instance toluene with xylene; catechol, quinol and cresol with phenol; and methyl cyclohexanol with cyclohexanol. The simpler the group the more effective it is. Compare cymene with toluene or xylene; and benzyl alcohol with phenol.

The esters occupy a quite unexpected position. The simple aliphatic esters are much worse lubricants than their related acids or alcohols. On the contrary, the ring esters are better lubricants than are their related acids (*e.g.* ethyl benzoate and benzoic acid).

Perhaps the most interesting substances are the hydroxy acids with OH and COOH groups. This conjunction produces a remarkable increase in the lubricating power of a chain compound (α lactic acid and ricinoleic acid), and almost destroys lubricating action in the case of the ring compounds (salicylic and benzylic acids).

It will be noticed that no ring compound is a good lubricant. Even cholesterol with the molecular weight 366 is no exception.

The group SH acts much as OH, thiophenol C_6H_5SH and benzylhydrosulphide $C_6H_5.CH_2SH$ resembling phenol and benzyl alcohol respectively.

It need hardly be stated that these conclusions are presented not as generalizations, but as a summary of the relations which actually obtain amongst the limited number of compounds studied. The fact that the influence of chemical constitution differs widely in degree if not in kind when glass is lubricated would alone enforce caution.

Mixtures of lubricants were not specially studied, but it is certain that they will reveal complex relations of great practical importance. In the experiments solids were deposited in thin layers on the plate from very weak solutions in ether, ethyl alcohol, or benzene. In most cases the change in friction was followed whilst the solutions were drying and nothing abnormal noted, the friction changing rapidly to the value for the solid in the last stage of drying, and in such a way as to suggest that the

Faces clean. Static friction 0·5

	Static friction		Static friction

CHAIN COMPOUNDS

Alcohols

Methyl	0·29	Isopropyl	0·32
Ethyl	0·32	Isobutyl	0·30
Propyl	0·34	Allyl	0·29
Butyl	0·30	Glycol	0·30
Amyl	0·27	Glycerol	0·22
Octyl	0·25	Penterythritol	0·40
Cetyl	0·17		

Acids

Formic	0·45	Stearic	0·15
Acetic	0·40	Oleic	0·10
Propionic	0·31	Ricinoleic	0·02
Valeric	0·28	α Lactic	0·20
Capryllic, fluid	0·19	Glyceric	0·22
Capryllic, frozen on plate	0·05		
Acetone	0·32	Ethyl ether	0·33
Methyl ethyl ketone	0·29		
Ethyl acetate	0·36	B.P. "paraffin"	0·20
Ethyl valerianate	0·35	Solid paraffin, m.p. 30·5	0·09
Tristearin	0·24	Solid paraffin, m.p. 46	0·07
Triolein	0·14	Carbon tetrachloride	0·43
Acetone di-carboxylic di-ethyl ester	0·29	Chloroform	0·30
		Amylene	0·26
n Hexane	0·37	Octylene	0·28
n Heptane	0·346		
n Octane	0·32	Butyl xylene	0·27

RING COMPOUNDS

Benzene	0·34		
Ethyl benzene	0·32		
Iodo benzene	0·30	Thiophenol	0·22
Toluene	0·28	Benzylhydrosulphide	0·23
Xylene	0·30		
p Cymene	0·31	Pyridine	0·33
		Piperidine	0·32
Phenol	0·25		
Catechol	0·39	Naphthalene	0·29
Quinol	0·40	Anthracene	0·26
m-Cresol	0·26	β Naphthol	0·38
Benzyl alcohol	0·31	Naphthoic acid	0·39
Benzoic acid	0·38	Carvacrol	0·23
Phthalic acid	0·37	Thymol	0·24
Cinnamic acid	0·27	Menthol	0·26
Benzilic acid	0·45	di Pentene	0·31
Salicylic acid	0·41	Camphor	0·24
Ethyl benzoate	0·33	Active ethyl ester of Camphor oxime	0·33
o-Phthalic ester	0·27	Iso-Cholesterol	0·27
Ethyl hydrocinnamate	0·28		
Ethyl cinnamate	0·32		

CYCLIC COMPOUNDS

Cyclohexane	0·31		
Methyl cyclohexane	0·30		
1.3 di-methyl cyclohexane	0·29	Cyclohexanone	0·35
		1.2 methyl cyclohexanone	0·32
Cyclohexanol	0·20	1.3 methyl cyclohexanone	0·35
1.2 methyl cyclohexanol	0·28	1.4 methyl cyclohexanone	0·33
1.3 methyl cyclohexanol	0·25		
Ammonia fortiss.	0·34	Castor oil	0·03
Triethylamine	0·30		
Tripropylamine	0·26	Water	0·33

combined effect of the two substances was merely additive. In some cases, however, *e.g.* phenol and cetyl alcohol, abnormal values were noted. Just before drying was complete the friction fell to a comparatively low level (0·07 cetyl alcohol, 0·15 phenol).

The values for paraffins can be considered only as approximate. Octane was the purest specimen employed. The solid paraffins were identified by their melting-points and were not wholly free from unsaturated substances, but the displacement of the observed frictions due to this may be considered as being small.

The lower members of the fatty acid series etched the surfaces so as to expose the crystals, the action being slow and slight in the case of valeric acid. Capryllic acid did not etch the surface. This action no doubt contributed to the bending downwards of the first part of the curve.

I am indebted to Prof. Lapworth for examples of cyclic compounds, to Sir George Beilby, Mr Dootson, and Dr Ida Maclean for various specimens, and especially to Sir William Pope for permission to raid his large collection of substances.

B. Influence of the Beilby Film

In 1903 Beilby described how, in the process of burnishing, or polishing, the substance of the solid actually flows so as to cover the surface with a film of amorphous material. The formation of the Beilby film can be readily followed on bismuth, and in order to test its influence some measurements of friction were made after it had been etched off by dilute acid so as thoroughly to expose the coarse crystalline structure of the metal. The following values were obtained:

	Burnished	Etched	Ratio
Benzene	0·34	0·39	0·87
Pyridene	0·33	0·4	0·83
Ethyl alcohol	0·32	0·39	0·82
Butyl xylene	0·27	0·37	0·72
Octyl alcohol	0·25	0·36	0·7
Cyclohexanol	0·20	0·33	0·6

The substances are arranged in the order of the value for static friction of burnished faces. The values for etched faces do not follow this order, whilst the ratios do. This may be merely coincidence, but it raises questions which must be reserved for discussion until more facts are available.

C. Adsorption of Lubricants

The theory outlined in the first paper (*Phil. Mag.* (1919), ser. 6, XXXVIII, 32) embodies two propositions: the first, that resistance to slipping is due to cohesion even when a lubricant is present, and that a lubricant decreases friction by partially or wholly masking the cohesive forces of the solids; the second, that a lubricant maintains its position against the normal pressure because its surface energy is a function of the thickness of the layer.

The capacity for decreasing friction, then, is a function of the potential of the attractive forces between lubricant and solid integrated through the depth of the layer, whilst the stability of a layer of a given thickness is a function of the differential coefficients of the interfacial energy taken with reference to the thickness.

The integral is the work done by the cohesive forces acting between lubricant and solid when the layer of the former is interposed between the faces of the latter. This may be expressed in terms of a tension, and thereby become measurable, if one solid face alone is considered. Let t be the work in ergs done per unit area in removing a layer from the surface of a mass of fluid, the layer being so thin that the fluid composing it is not in mass; t will then also be the tension of the free layer. Now apply the layer to the solid face. The forces of attraction between the two will do work. Let this be t' per unit area. The tension T of the composite surface so formed will then be

$$T = t + T_s - t',$$

where T_s is the tension of the solid.

For the difference between the tension of two composite surfaces formed on the same solid but with different fluids a and b we have

$$T_a - T_b = (t_a - t_b) + (t_b' - t_a').$$

If we assume that t_a bears the same ratio to t_b as do the tensions of the fluids in mass and choose for the purpose of experiment two fluids whose surface tensions are equal, the term $(t_a - t_b)$ will vanish, and the left-hand side of the equation be positive or negative according as the term $(t_a' - t_b')$ is positive or negative.

In an earlier paper (*Phil. Mag.* (1919), ser. 6, XXXVIII, 49) I described how films of insensible thickness form on a plate of glass about drops of certain fluids, and how the drops are moved over the surface of the plate by the contractility of the films. This property may be utilized to measure the sign of the term $(t_a' - t_b')$. Let one drop of each of two fluids a and b be placed on the plate: there will form about each a composite surface of tensions T_a and T_b respectively. If T_a is greater than T_b the drop a will move away from drop b and the latter will pursue it and, if the surface tensions of the fluids are equal, such movement will show that the adhesion of the film of b to the surface is greater than that of a.

Benzene and propionic acid are a pair of fluids whose surface tensions are practically equal, while the former does not and the latter does lubricate the surface of glass. A drop of each was placed on a clean glass plate a few millimetres apart with striking results. The drop of acid chased the drop of benzene to the edge of the plate where, owing to the characteristic edge repulsion, the latter was split into two. We have, I think, in this observation direct evidence that the forces of attraction operate more strongly between a solid face and a good lubricant than between it and a bad lubricant.

The better lubricant is more strongly adsorbed by the solid face. Olefines are better lubricants than paraffins, and one of the methods in use for freeing the latter from the former is by taking advantage of the fact that olefines are more strongly adsorbed by a solid. The impure paraffin in commercial practice is filtered through a dry powder to clear it of unsaturated substances.

Of the pair of fluids acetic acid–water, a drop of the latter pursues strongly a drop of the former on a plate. The surface tension of water is much higher than that of acetic acid, and, since water does not and acetic acid does lubricate glass, the decrement due to interaction with the solid face is, according to theory, less for water. Therefore, in respect of both terms of the right-hand member of the equation the advantage lies with the insensible film of water.

In both of these pairs the result was the same whichever one of a pair was placed first on the plate. Another pair, benzene and acetic acid, gave uncertain results. According to theory benzene should move away and the acid follow, and this usually happened. Sometimes, however, when a drop of acid was placed on a plate on which a drop of benzene already stood, the latter darted away, as though the vapour of acetic acid had lowered the tension of the insensible film of benzene between the drops; sometimes the drops simply moved away from each other. These complications are to be expected, for we have the tensions of three films to consider—those of the relatively pure films outside each drop, and that of the film of mixed origin between them. If the vapours condense in the last in such proportions as to produce a film of tension less than that of either pure film, the drops will be pulled apart.

Reasons are given in the earlier paper for believing that an insensible film about a drop is formed always by condensation of vapour and not by direct spreading.

40

PROBLEMS OF LUBRICATION

[Royal Institution of Great Britain, 1920]

In lubrication a fluid or other body is used to decrease the friction between opposed solid faces. The lubricant may act in one of two ways. It may separate the faces by a layer thick enough to substitute its own internal friction, modified by the mechanical conditions in which it finds itself, for that of the solid faces; or it may be present as a film, too thin to develop its properties when in mass, which reacts with the substance of the solid faces to confer upon them new physical properties. In the latter case the solid faces continue to influence each other, not directly, but through the intermediation of the film of lubricant. There are indications that these two types of lubrication—the one in which the solid faces intervene only owing to their form, rate of movement, etc., and not by their chemical constitution; the other in which the chemical constitution is directly involved—are discontinuous states, in that the one cannot be changed gradually into the other by simply thinning the layer of lubricant. The change from the one to the other is probably abrupt.

It may by no means be asserted that resistance to relative motion is always least when the solid faces are floated completely apart; it would indeed probably be truer to say of the best lubricants, that friction is least when the "boundary conditions", to use Osborne Reynolds's phrase, are fully operative.

This address is concerned wholly with "boundary conditions", and we get directly to the heart of the problem by certain simple experiments. If a glass vessel, such as a bottle, be placed upon an inclined pane of glass at a certain angle, it slips smoothly down. The glass plate is an ordinary plate cleaned with a cloth. In the usual sense of the word the surface of the plate is not lubricated, the surface is "dry". The lower half of the plate is then wetted with water, and the bottle is now found to slip on the unwetted part and to be sharply pulled up by friction when it reaches the wetted part. It is not sufficient, therefore, to interpose a liquid film between solid faces to get lubrication. Indeed, as the experiment proves, water increases the friction; it is an anti-lubricant for ordinary faces of glass.

Is then the quality of lubricant a property of a fluid? Does water fail to act merely because it does not possess that property to which the name "oiliness" is sometimes given? Another simple experiment supplies the answer. Instead of a glass plate let us use a plate of ebonite. The glass plate does not readily slip on this. The angle at which slipping occurs is steeper than when a glass plate is used. Now, when the lower half of the ebonite plate is wetted it is found that a glass bottle encounters relatively high friction on the unwetted part, but slips quite freely on the wetted part. Water, in short, is an admirable lubricant for glass on ebonite.

Here is another plate, picked up at random in the laboratory of the Royal Institution. Its composition is unknown. Tested in the same way, water has no detectable influence on the friction between glass and the surface of this plate.

It will be well to confess at once that these simple experiments raise questions which are as yet without an answer, and that much of what follows concerning them is merely tentative. They seem to establish two things, the first being the curious paradox that a film of fluid introduced between two surfaces does not always decrease friction; it may, indeed, very much increase it. The second, that the quality of "oiliness", the quality, that is, which enables a substance to act as a lubricant, seems to be not the property of a given fluid, but only of that fluid considered in reference to a particular surface.

It is necessary at this stage to clear away a possible explanation of the paradox. When two solid faces are separated by a thin film of fluid, capillary forces operate and in certain cases, at any rate, they resist slipping. They will so act, for instance, when the movement of the one face past the other increases the area of the free surface of the film. Water has a high surface tension; the capillary forces to which it gives rise are unusually large. Therefore it is pertinent to ask whether, when a layer of water diminishes the facility for the slipping of glass on glass, it is owing to capillary action? A qualitative answer is to be found in the fact that water does in some cases, as when glass is applied to ebonite, increase the facility for slipping; and Lord Rayleigh furnished the quantitative answer. He calculated the magnitude of the capillary effect and found it negligible compared to the actual friction of glass on glass wetted with water. An appeal to capillary forces of this type will not solve the paradox.

Some light is thrown upon it when we enquire into the state of the surface of glass whose friction is increased by water. Surfaces of glass "cleaned" in the ordinary way by rubbing with a glass-cloth, or glass faces which have been simply exposed to the air, are in point of fact not clean; they are highly lubricated with a film of matter derived from the cloth or condensed from the atmosphere. This "grease" film is of invisible thinness. It is probably of the order of $1\,\mu\mu$ in thickness, that is to say, one-millionth of a millimetre. It can be removed by soap and water, which in turn must be removed by a stream of water, and the plates dried in clean air out of contact with solids. The film reforms quickly, very quickly in London air, and less quickly in the country. A "grease" film also creeps over a cleaned glass face from ordinary solids with which it may be in contact. Still, when due precautions are taken, and they are many, it is possible to get a glass face which seems to be really clean.

The first property of such faces is that their friction, one for the other, is very high; indeed it is impossible to make them slip past one another. One glass plate may be forced past another, but true slipping does not take place, they tear at the point or points of contact. It is easier in short to disrupt the actual substance of glass itself than to get the surfaces to slip over one another. Clean glass faces "seize" when they touch.

When chemical substances are tested as lubricants on clean glass faces a remarkable fact emerges—namely, that some are quite neutral in that they do not alter the resistance to slip in the least, such are water, alcohol, benzene, strong ammonia. Other substances have some lubricating action great or small. That is to say, they decrease the force needed to produce slipping; such are the alkalis, trimethylamine and tripropylamine, the fatty acids, e.g. acetic acid, and the paraffins. Those fluids which act as lubricants are not necessarily fluids of any considerable viscosity, indeed a high viscosity is compatible with the absence of any true lubricating action. Thus glycerine facilitates the slipping of clean glass on clean glass only when it is present in quantity sufficient to float the surfaces apart. On the other hand, acetic acid and tripropylamine, substances of low viscosity, are admirable lubricants.

No fluid amongst those tested has been found to raise the friction of clean glass faces. A fluid either is neutral or decreases friction to a greater or less extent. The property of increasing the friction of glass faces which neutral fluids, such as water, possess, is due not to their action on the glass, but to the fact that they interfere with the effect of the invisible grease film. Water on an ordinary glass face acts as an anti-lubricant; on really clean glass it is "neutral".

All solid faces, however, do not distinguish chemical substances into those which are "neutral" and those which possess lubricating properties. Nearly one hundred substances have been tested on burnished faces of bismuth, and in every case some decrease of friction was observed.

A comparison of the lubricating action of simple chemical substances on clean faces of glass, and of bismuth, would seem to show that the quality of oiliness is due to some reaction between the substance and the solid face. Much is still obscure, but certain facts seem to be capable of interpretation in no other way. Thus water and ethyl alcohol have no detectable lubricating action on clean glass, whilst both are moderate lubricants for clean bismuth.

The thickness of the layer needed to lubricate is astonishingly small. It is quite invisible, and probably only one or a very few molecules thick. To

H

41

discuss this adequately would take too long, but the fact may be instanced by an experiment of great beauty. A tiny drop of, say, acetic acid or tripropylamine is placed near one corner of a plate of clean glass 6 cms. square, nothing detectable by the senses happens, the drop is there and that seems to be all. But the whole surface of the plate has in fact been changed fundamentally. It is now fully lubricated by an invisible film which has spread rapidly over it from the drop. The presence of this film may be detected by measuring the friction or by following the migration of two drops of fluid over the face of the plate. It will be found that the drops attract one another under conditions which point to the cause being the contractility of the invisible film.

This brings me to the second part of my subject—namely, the relation of lubricating power to chemical constitution.

In particular experiments with bismuth a slider having a curved surface was applied to a plain surface of metal, both surfaces being highly polished, and the force required to initiate movement was measured. This force measures what is usually called static friction as opposed to the kinetic friction when the surfaces are in relative motion. The static friction was found to be a function of the weight of the slider. Therefore, as a relative measure, the ratio of the weight of the slider to the weight needed to move the slider was used. The results appear in the table on page 635.

It will be seen that static friction is a function of the molecular weight of the lubricant; and in a simple chemical series of chain compounds, such as fatty acids and alcohols or paraffins, a good lubricant will be found if one goes high enough in the series. But it is not a simple function. The friction, for instance, rises sharply in moving from $CHCl_3$ to CCl_4 and from phenol to catechol and quinol. The influence of molecular weight is overshadowed by the influence of chemical constitution.

In some simple chemical series the relation appears to be a linear one. Examples are paraffins; the series benzene, naphthalene, anthracene.

The relation of lubricating qualities to viscosity broadly resembles that to molecular weight. In a simple chemical series lubrication and viscosity change in much the same way with molecular weight; but that there is no fundamental relation between viscosity and lubrication is shown by the following figures:

	Viscosity at 20°	Static friction
Carbon tetrachloride	0·0096	0·43
Chloroform	0·0056	0·30
Acetic acid	0·0122	0·40
Octyllic acid	0·0575	0·19
Benzene	0·0065	0·39
Toluene	0·0058	0·28
Benzyl alcohol	0·0558	0·31

Fluidity of the lubricant has no constant significance. The curves for acids, alcohols and paraffins show no break where, with increasing molecular weight, the lubricant becomes a solid at the temperature of observation. Compare also benzene, naphthalene and anthracene, menthone and menthol, thymol and carvacrol.

Perhaps the most unexpected result is the distinction between ring and chain compounds. The simple ring compounds benzene, naphthalene and anthracene show the linear relation to molecular weight, and the values are much the same as those for paraffins of the same molecular weight. The similarities, however, end here, for any change in the molecular structure produces opposite effects according as it takes place in a chain or ring. Thus a double bond decreases the lubricating action of a ring compound, but increases that of a chain compound. As examples, compare naphthoic acid with double-bonded oxygen, with naphthalene, menthone with menthol, cyclohexanone with cyclohexane, benzoic acid with benzene. As examples of double-bonded carbon, compare cinnamic ester with hydrocinnamic ester, di-pentene, having two unsaturated carbon atoms, with menthol and cyclohexane. Also the more saturated cylic compounds are better lubricants than the less saturated ring compounds.

When a ring or chain are joined, as in butylxylene, the result is a better lubricant than either.

The esters occupy a quite unexpected position. The simple aliphatic esters are much worse lubricants than their related acids and alcohols. The ring esters, on the contrary, are better lubricants than are their related acids (e.g. ethyl benzoate and benzoic acid.)

Perhaps the most interesting substances are the hydroxy acids with OH and COOH groups. This conjunction produces a remarkable increase in the lubricating power of a chain compound (lactic acid and ricinoleic acid), and almost destroys lubricating action in the case of the ring compounds (salicyclic and benzylic acids).

In the ring compounds the replacement of hydrogen decreases lubricating power in the case of N, :O, or .COOH, and increases it in the case of other groups in the order $C_2H_5 < CH_3 < OH$.

The effect of a second group of the same or of a different kind is to decrease the effect of the first. Compare, for instance, toluene with xylene; catechol, quinol, and cresol with phenol; and methyl cyclohexanol with cyclohexanol. The simpler the group the more effective it is. Compare cymene with toluene or xylene, and benzyl alcohol with phenol.

When the atoms are disposed with complete symmetry about a carbon atom, the result is a very bad lubricant, as we see in carbon tetrachloride and the alcohol penterythritol $C(CH_2OH)_4$.

It will be noticed that no ring compound is a good lubricant. Even cholesterol, with the molecular weight 366, is no exception.

The group SH acts much as OH, thiophenol C_6H_5SH and benzylhydro-sulphide $C_6H_5.CH_2SH$ resembling phenol and benzyl alcohol respectively.

Concerning one matter, and that the most fundamental, some conclusion must be come to, even though it be upset later. What is friction due to? The *Encyclopædia Britannica* is in no doubt as to this. Friction, it says, is due to inequalities of the surface. This conclusion cannot, I think, be accepted. Why, if it be true, should clean burnished faces of glass or bismuth refuse to slide over one another? It does not even accord with such simple facts as we now know. For instance, the friction of an optical face of glass was found to be the same as that of ordinary plate glass within the limit of accuracy aimed at. And both the optical face and ordinary plate were found to give higher values than ground glass.

The subject cannot be fully discussed here, but I think we may conclude with some confidence that the friction both of lubricated and of clean faces is due to true cohesion—to the force, that is, which binds together the molecules of a solid or of a fluid. If there were no seizing, there would be no friction. The function of the lubricant is to diminish the capacity for seizing by saturating more or less completely the surface forces of the solid. In some cases it seems to abolish it completely so that static friction vanishes.

The subject of lubrication is of interest to the engineer, but it is of perhaps more interest to the physicist, for it offers a means of exploring the most difficult region of the physics of boundary zones—namely, the surface energy of solids. It will, for instance, I believe, enable us to prove that the simplest chemical change at the surface of a metal takes place only when the surface energy is decreased thereby. The film of oxide or sulphide which forms on copper acts as a very effective lubricant, and it acts also like a grease film in preventing water from wetting the surface; and from both of these facts we may conclude that the presence of the film lowers the surface energy of the metal.

41

BOUNDARY LUBRICATION. THE PARAFFIN SERIES

With I. DOUBLEDAY

[*Proc. Roy. Soc.* (1922), A, c, 550]

In what is often called complete lubrication, the kind of lubrication investigated by Towers and Osborne Reynolds, the solid surfaces are completely floated apart by the lubricant. There is, however, another kind of lubrication in which the solid faces are near enough together to influence directly the physical properties of the lubricant. This is the condition found with "dry" or "greasy" surfaces. What Osborne Reynolds calls "boundary conditions" then operate, and the friction depends not only on the lubricant, but also on the chemical nature of the solid boundaries. Boundary lubrication differs so greatly from complete lubrication as to suggest that there is a discontinuity between the two states. In the former the surfaces have the property of static friction, and the resistance is some inverse function of the viscosity of the lubricant. In complete lubrication static friction is absent and the resistance varies directly with the viscosity of the lubricant. Boundary lubrication is alone considered in this paper.

The enquiry is limited to the lubricating qualities of normal paraffins and their related acids and alcohols. The molecules of the substances employed, therefore, consist either of a simple chain of carbon atoms to which are attached atoms of hydrogen, or of such a chain loaded at one end with the hydroxyl group—OH, or the carboxyl group—COOH.

Attention was concentrated chiefly on three variables—the quantity of lubricant present, the composition of the solid faces, and the chemical constitution of the lubricant. The relations disclosed by the experiments are of surprising simplicity, whilst their interpretation is difficult. For this reason the paper is divided into two parts, the first devoted to the results of experiment and the second to theory.

The tables include values of the coefficient of friction for polished surfaces of bismuth. They are taken from an earlier paper, but it must be observed that a much higher order of accuracy was arrived at in the measurements with glass and steel than in those made previously with bismuth. The sensitiveness of the apparatus has been improved and very much greater care taken to secure purity in the substances used.

Of the chemicals used, some have been made by one of us (I. D.) and all have been specially purified until chemical tests failed to detect any impurity. Impurities present in too small amount to be detected by chemical means will produce a measurable effect on the very sensitive surfaces employed. It will be noticed that the value for undecane falls slightly below the curve for paraffins. The specimen had a marked odour, and, according to our experience with, *e.g.* octane, the smell of normal paraffins decreases as purification proceeds. We believe, therefore, that there was a residual impurity of vanishingly small amount in the sample of undecane.

We are specially indebted to Prof. R. Robinson, F.R.S., for some of the higher paraffins and their related acids and alcohols.

Part I. Experiment

The experimental method employed to measure friction was that described in the earlier papers. The sliding piece or slider had a spherical surface, which was applied to a plane surface, the plate, both surfaces being highly polished. From the middle of the slider a short stem projected, from which a fine thread passed over a light pulley to a light pan for holding weights. The thread was adjusted so that the pull on the slider was parallel to the plane surface.

The use of a curved surface has the advantage that some measure of definiteness is given to the area of true contact. Two plane surfaces, trued as carefully as possible, touch only at a few points, as Budgett's experiments prove.

Both slider and plate were enclosed in a small chamber through which a current of air was passed, which had been purified by exposure (in order) to sulphuric acid, solid potash, calcium chloride, and phosphorus pentoxide, and finally passed through a column of glass-wool. In the earlier experiments the air was purified simply by bubbling through sulphuric acid and then passing it through a column of glass-wool, to trap any possible spray, but the vapour of sulphuric acid was found very seriously to interfere with the results by acting itself as a lubricant, and it was necessary to introduce solid potash to remove it. When the surfaces were covered only by an invisible film of lubricant the vapour of sulphuric acid was found to lower the friction by no less than 17 per cent. Sulphuric acid is needed as a cleansing agent to remove organic impurities from the air.

Three solids were used, glass, steel and occasionally bismuth. The test of cleanliness, as was pointed out in an earlier paper, is the development of a high and constant value for friction. Each solid demands its own peculiar treatment to secure cleanliness, and when one embarks on a new solid, such as steel, it may be a month or more before a suitable method is found.

The "clean" state is one from which the grosser of the impurities which contaminate all naturally occurring surfaces have been removed. What remains, whether, for instance, there is left a film of condensed gas, is unknown. The state is, however, a perfectly definite one, in that it can be reproduced time after time with complete certainty. A plate of the glass used in these experiments is "clean" when its coefficient of friction is 0·94; the steel plates when clean gave the value 0·74. Other kinds of glass give values differing from 0·94. A negative may perhaps be ventured on. It is that "clean" surfaces are not the same as surfaces freshly produced by fracture. It is said that when steel is fractured under mercury the new surfaces amalgamate. "Clean" steel plates in clean air will not amalgamate.

Glass plates were cleaned by first washing with soap and water, rinsing under the tap, and allowing to dry by draining. They were then heated for about half an hour in a solution of chromic oxide in sulphuric acid, rinsed thoroughly under the tap (the plate being held with clean tongs) and allowed to drain. During draining, the lower edge of the plate rested upon two clean glass rods, the upper edge leaning against a clean glass bottle.

The glass sliders, being weighted with lead, could not be cleaned with chromic acid. They were, therefore, washed with soap and water, rubbed vigorously with the finger tips under a rapid flow of water until the "clinging" state was reached, and suspended in air until dry.

Steel plates and sliders were cleaned by washing with soap and water, rubbing vigorously with the finger tips in a stream of tap water until water wetted the entire surface, rinsing with perfectly dry pure alcohol, and allowing to drain in air. During the rinsing and draining processes the steel was not touched with the fingers at all, but held in tongs which had previously been cleaned by strong heating.

Variables. The variables which have to be considered are the weight of the slider, the curvature of its surface, temperature, the thickness of the layer of lubricant, and the chemical nature of the lubricant and of the solids respectively. Of these one alone was not studied, namely, temperature, because the walls of the chambers which have been tried have given off vapours at moderate temperatures in quantity sufficient completely to vitiate the measurements. A chamber of special design is being constructed which will, it is hoped, relieve us of this difficulty.

The fundamental law of sliding friction is usually known as Coulomb's[1] law. As a matter of fact it was formulated 86 years earlier by Amontons,[2] we, therefore, propose to call it Amontons' law. According to it the friction is proportional to the weight of the slider and is independent of the area of contact. Coulomb regarded this law merely as an approximation and his view has been accepted by subsequent writers. We have made many measurements to verify the law, and they prove definitely that it is not an approximation, but an exact law, which holds so long as the surfaces, or

---

[1] Coulomb, *Mem. de l'Acad. Roy. des Sci.* (1785), x, 161.
[2] Amontons, *ibid.* (1699), p. 206.

rather the whole system, solid surfaces and lubricant, remain unchanged. Many solids suffer viscous flow from the pressure. This is the case when wood is used, and then the law ceases to hold. It is a rigid law for hard solids such as glass and hard steel.

According to Hertz's equation the area of contact varies as the cube root of the weight of the slider divided by its curvature. Careful measurements showed that the coefficient of friction (μ = friction divided by weight) is independent of the weight and of the curvature. It is, therefore, independent of the area. The values of μ are given in Tables I and II.

TABLE I

Glass					
Weight in gr.	21·63	31·64	41·63	51·63	61·63
Propyl alcohol	0·6321	0·6354	0·6301	0·6298	0·6352
Butyl alcohol	0·6061	0·6042	0·6059	0·6076	0·6012
Amyl alcohol	0·5821	0·5800	0·5833	0·5861	0·5847
Octyl alcohol	0·5112	0·5126	0·5095	0·5093	0·5113
Heptylic alcohol	0·4001	0·4018	0·4076	0·3982	0·3873
Caprylic alcohol	0·3412	0·3392	0·3428	0·3413	0·3371
Heptane	0·6751	0·6703	0·6728	0·6741	0·6743
Octane	0·6551	0·6542	0·6508	0·6581	0·6548
Steel					
Weight in gr.	15·46	25·46	35·47	45·47	55·49
Amyl alcohol	0·3658	0·3679	0·3708	0·3649	0·3651
Octyl alcohol	0·2991	0·3032	0·2987	0·2999	0·3061

TABLE II. *Steel*

Radius of curvature (cm.)	14·70	3·58	1·78
Butyl alcohol	0·4018	0·4067	0·3917
Octyl alcohol	0·3006	0·2991	0·2939
"Clean"	0·7408	0·7421	0·7427

Influence of the Quantity of Lubricant

Fluid lubricants. Three methods of lubricating the surfaces were adopted:

(1) A drop of lubricant was placed on the plate so that the slider stood in a pool of fluid. This is the *flooded* state.

(2) When a drop of a lubricant which has a sensible vapour pressure is placed anywhere on a clean plate nothing detectable by the senses occurs. The drop to all appearances remains where it is placed without change. An invisible film, however, spreads from the neighbourhood of the drop so as to cover the whole plate. The presence of the film can be detected by the fall in friction. For reasons which will appear later (Appendix II) we call the invisible film the *primary film*, and the process by which it is formed *primary spreading*.

Obviously the invisible film would be subject to intense evaporation if the ordinary procedure were followed, in which a stream of clean air is passed through the chamber. This procedure was, therefore, modified as follows: The lower part of the chamber was covered with a layer of fresh dry calcium chloride, and thoroughly washed out with a stream of dry air. The air stream was then cut off and it was found that clean plates remained completely uncontaminated in the chamber for 10 hours.

(3) When the lubricant has a sufficiently high vapour pressure it can be deposited upon the surfaces by passing its vapour into the chamber. To accomplish this the stream of dry clean air was divided into two, one of which was bubbled through the lubricant whilst the other was led through a bypass. The two streams were then led into a glass bulb provided with a device for thoroughly mixing them, and thence to the chamber. The proportion of the stream passing through the lubricant was varied by taps suitably placed and in this way the concentration of vapour in the chamber could be varied. In the case of one substance, ethyl alcohol, the concentration was determined directly by collecting samples of the stream as it entered the chamber and analysing them.

Primary spreading does not take place rapidly on a solid face. The friction falls slowly and finally reaches a steady value, it may be in an hour or so, which is uniform over the whole surface. It is this value which is recorded. The mechanism of spreading is discussed in Appendix II.

TABLE III

Lubricant	Flooded	Primary film	Saturated vapour
Glass			
Ethyl alcohol	0·6512	0·648	0·6491
n-propyl alcohol	0·6301	0·6329	0·6283
n-butyl alcohol	0·6061	0·6107	0·597
Amyl alcohol	0·5853	0·5921	0·577
Caprylic alcohol	0·5371	0·5356	—
Undecyl alcohol	0·4450	0·4452	
Butyric acid	0·5721	0·5793	0·5812
Valerianic acid	0·5259	0·5218	0·5309
Hexoic acid	0·4653	0·4582	0·4648
Heptoic acid	0·4051	0·4008	—
Caprylic acid	0·3418	0·3412	—
Pentane	0·7102	—	0·7158
Hexane	0·6908	—	0·6913
Heptane	0·6753	—	0·6709
Octane	0·6557	—	0·6591
Steel			
Butyl alcohol	0·3922	0·3906	0·3901
Amyl alcohol	0·3751	0·3687	0·3716
Caprylic acid	0·2008	0·2073	—

In Table III the values of the coefficient of friction (μ = friction divided by the weight) are given for flooding, for primary films, and for saturated vapour. The figures show that the friction is independent of the quantity of lubricant present, provided that there is enough to cover the surfaces with the invisible primary film.

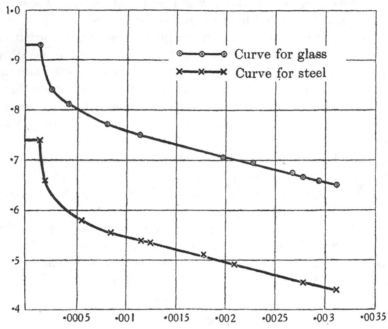

Fig. 1. Abscissæ, number of gram-molecules of C_2H_5OH per litre of air. Ordinates, coefficient of static friction.

It is obvious, since the primary film is in equilibrium with a drop of the parent fluid in a closed space, that its vapour pressure is the same as that of the free surface of the drop. Neglecting the very slight effect of the curvature of the drop in raising its vapour pressure we may, therefore, say that the vapour pressure of the primary film is that of a vapour saturated at the particular temperature. The critical value of friction, the value, that is to say, which is independent of the quantity of lubricant present is, therefore, that of faces so lubricated as to be in equilibrium with saturated vapour. The results obtained by using the lubricant as vapour confirm this view. The critical friction is also the lowest value which can be obtained in boundary friction.

The curves relating the coefficient of friction to the concentration of the vapour of ethyl alcohol are given in Fig. 1, and the figures in Table IV. The concentration of the vapour is in gram-molecules per litre.

The curves are straight lines save where the vapour pressure is very low, when the results are liable to be upset by that degree of contamination from chance vapour which no amount of care in experimental procedure is likely to get rid of.

TABLE IV. *Ethyl Alcohol Vapour-Pressure Values*

Glass		Steel	
Gr.-mol. C_2H_5OH per litre	μ	Gr.-mol. C_2H_2OH per litre	μ
0·00014	0·94 (clean)	0·00013	0·74 (clean)
0·00025	0·841	0·00018	0·661
0·00042	0·816	0·00053	0·580
0·00081	0·773	0·00085	0·555
0·001127	0·752	0·001146	0·543
0·001973	0·705	0·001225	0·538
0·00228	0·696	0·001783	0·512
0·002675	0·675	0·002102	0·489
0·00278	0·668	0·002781	0·454
0·002936	0·660	0·00311	0·441 (flooded value)
0·003111	0·651 (flooded value)		

The fall in friction, therefore, is proportional to the concentration of chemical molecules of the lubricant in the gas phase, that is to say, each such molecule exerts the same influence as every other. The changes of any one molecule sticking to the solid surface will be proportional to the concentration in the gas phase, and therefore we may conclude that each chemical molecule in the film on the solid faces contributes an equal share to the fall in friction. This is not unlikely, since the fall in surface tension due to the presence of oleic acid on the surface of water is also proportional to the number of molecules of the acid per unit area.[1]

The most important features of these curves connecting vapour pressure and the coefficient of friction are: (1) that they do not if prolonged meet the μ axis at the clean value, though the measurement made with the lowest vapour pressure coincides with this value; and (2) that the curves are parallel to each other over the linear portion.

The clean value obviously must form one end of the curves and in the figure they are continued so as to cut the μ axis at this value. The first part of each curve is, therefore, horizontal. That is to say, a finite concentration of lubricant on the surface is needed to produce any fall in friction. It is well known, since Rayleigh pointed it out, that a finite concentration of a substance spread on the surface of water is needed to cause a fall in surface tension. Rayleigh suggested that this critical concentration occurs when the molecules spread on the surfaces are close enough to be within the range of each other's attractions and repulsions: when, that is to say, a

[1] *Proc. Roy. Soc.* (1913), A, LXXXV, 317.

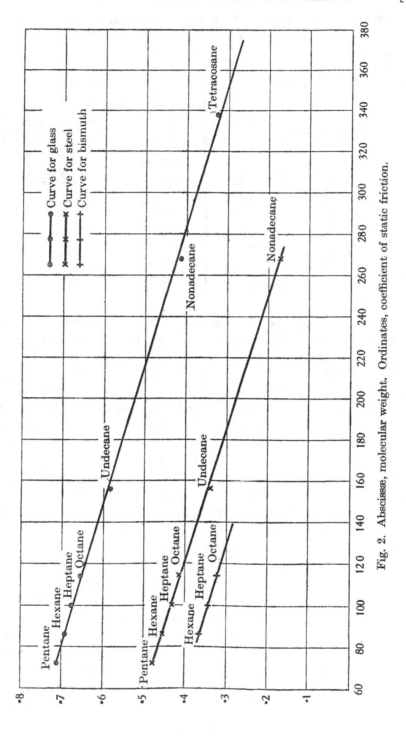

Fig. 2. Abscissæ, molecular weight. Ordinates, coefficient of static friction.

continuous film is produced. Do the inflection points in the curves in Fig. 1 also mark the formation of a continuous sheet?

The parallelism of the curves for glass and steel shows that the effect which each molecule of the lubricant produces is independent of the nature of the solid surface to which it is applied. The fall in friction due to each molecule is, therefore, a pure function of its field of force, or, in other words, of its atomic configuration.

Solid Lubricants

It is more difficult to follow the influence of the quantity of lubricant when it is a solid, because neither primary spreading nor the vapour phase can be made use of.

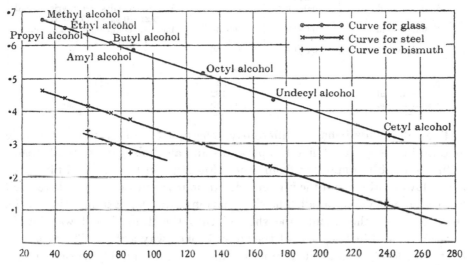

Fig. 3. Abscissæ, molecular weight. Ordinates, coefficient of static friction.

Surfaces were lubricated with solids either by smearing the lubricant over the plates with a glass rod or by depositing a thin film from solution in ether. A film of insensible thickness was obtained by polishing off as much as possible of the lubricant in the way described below.

In the case of paraffins and alcohols, as the ether evaporates the friction falls until a very low value is reached, which coincides with the presence of the last trace of the solvent. The value persists only for a few seconds, the friction then rising to a steady value which is that of the pure substance. The final steady values lie on the curve for the series to which the lubricant belongs (Figs. 1 and 3). The first set of values shows that for certain mixtures of solvent and solute, in this case ether and the particular paraffin or alcohol, the friction is much less than that obtained when either is used by itself. The fact that a mixture of two substances may lubricate better than either when used alone was noted in an earlier paper.[1] The figures are given in Table V.

TABLE V. *Solid Lubricants*

Lubricant	Polished film	Film deposited from ether	
		Initial value	Final value
Glass			
Undecyl alcohol	0·017	0·0781	0·4472
Cetyl alcohol	0·019	0·0203	0·3284
Nonadecane	0·0467	0·0231	0·4119
Tetracosane	0·0321	0·0192	0·3251
Decoic acid	0·0632	0·0738	0·2006
Dodecoic acid	0·0404	0·0391	0·0983
Myristic acid	0·0671	0·0382	Variable results, *e.g.* myristic acid = 0·24 − 0·52 − 0·17 − 0·31
Palmitic acid	0·0715	0·0340	Palmitic acid = 0·26 − 0·23 − 0·37 − 0·8 − etc.
Stearic acid	0·0318	0·0296	
Steel			
Nonadecane	—	0·0213	0·1783
Cetyl alcohol	—	0·0182	0·1126
Palmitic acid	—	0·0113	Value irregular

The solid acids behaved differently. The first two used, namely, decoic acid and dodecoic acid, like the paraffins and alcohols, gave normal steady values for a film deposited from ether (Fig. 4), and the polished film gave much lower values, but the higher acids from myristic acid onwards refused to give steady values at all. The explanation is simple. Examination of the surface under the microscope showed that the film of acid was highly crystalline and brittle, so that the slider broke away a crust. The observed friction, therefore, was due mainly to this crust and could be varied merely by moving the slider about on the surface before taking a reading. For example, values from $\mu = 0·17$ to $\mu = 0·52$ could be obtained with myristic acid. The crystallisation did not, however, seem to involve the primary film since very thin films deposited from ether were found to give values as low as polished films. Thus, with stearic acid the lowest value ($\mu = 0·035$) obtained was got by depositing a film from an exceedingly dilute solution. The increase in the hardness of the crystals with increase in molecular weight probably accounts for the difference in the behaviour of myristic, palmitic and stearic acids, from that of decoic and dodecoic acids.

Attempts were made to polish off both fluid and solid lubricants, with interesting results. For this purpose fragments of fine linen were used which had been exhausted of all "greasy" matter by prolonged extraction with benzene. The linen was regarded as being clean when the friction of a

[1] *Phil. Mag.* (1920), (6), XL, 201.

clean surface was not lowered after having been vigorously rubbed with it. In all the operations the linen was not held in the hands but by clean tongs. When the lubricant was a fluid, polishing completely removed it from the surfaces, the friction rising to the clean value, but it was not found possible to polish off a solid lubricant. Polishing a plate covered with a solid lubricant until nothing was visible under a microscope left a surface which gave a very low value for friction. If the normal value of the particular lubricant be that which lies on the curve for its particular chemical series,

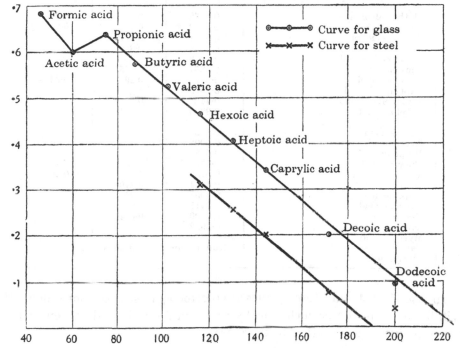

Fig. 4. Abscissæ, molecular weight. Ordinates, coefficient of static friction.

then polishing produced an abnormally low value. An interesting case is that of undecyl alcohol. This substance melts at 11° C., it was easy, therefore, to apply it either in the fluid or solid states. In the former the alcohol could be completely polished off a surface, whilst in the latter polishing merely served to produce an invisible film having an abnormally low friction. Abnormally low values are also obtained when fluid lubricants are polished, but they are transitory. The figures for polished films appear in Table V.

Influence of Chemical Constitution

Table VI gives the coefficient of friction of each substance for glass on glass and steel on steel. The values for bismuth are taken from an earlier paper.

TABLE VI

Lubricant	Glass	Steel	Bismuth
n-pentane	0·7102	0·4763	—
n-hexane	0·6908	0·4528	0·37
n-heptane	0·6751	0·4307	0·346
Iso-heptane	0·6683	—	—
n-octane	0·6552	0·4112	0·32
Undecane	0·5903	0·3421	—
Nonadecane (deposited from ether)	0·4119	0·1785	—
Tetracosane (deposited from ether)	0·3251	—	—
Methyl alcohol	0·6772	0·4610	0·29
Ethyl alcohol	0·6512	0·4408	0·32
Propyl alcohol	0·6301	0·4173	0·34
Butyl alcohol	0·6061	0·3924	0·30
Iso-butyl alcohol	0·6273	—	—
Amyl alcohol	0·5854	0·3752	0·27
Octyl alcohol	0·5176	0·2981	0·25
Caprylic alcohol	0·5373	—	—
Undecylic alcohol	0·4455	0·2298	—
Cetyl alcohol	0·3253	0·1143	0·17
Formic acid	0·6823	—	0·45
Acetic acid	0·6003	—	0·40
Propionic acid	0·6387	—	0·31
Butyric acid	0·5721	—	—
Valeranic acid	0·5259	—	0·28
Hexoic acid	0·4654	0·3108	—
Heptoic acid	0·4051	0·2556	—
Caprylic acid	0·3417	0·2003	0·19
Decoic acid	0·2006	0·0741	—
Dodecoic acid	0·0983	—	—
Myristic acid ⎫ Palmitic acid ⎬ Stearic acid ⎭	Readings for smeared faces variable		

In Figs. 2, 3, 4 and 5 these values are plotted against molecular weight. It will be seen that for each chemical series, and for each solid, the curve is a straight line. The equation is, therefore,

$$\mu = b - aM,$$

where M is molecular weight and a and b are parameters. The effect of the nature of the solid face is unexpectedly simple. In changing from glass to steel the curve for a series is merely moved parallel to itself, and in moving from steel to bismuth there is a further shifting. Therefore, in the equation the parameter a is independent of the nature of the solid face and dependent only on chemical type, varying from one chemical series to another. The parameter b, on the other hand, is dependent upon the nature of the solid face as well as upon the chemical series.

When other variables are not involved the value of the coefficient of friction is always in the order glass > steel > bismuth. For the two metals this is in the order of their respective hardness. But it does not depend only on hardness, since the steel used was harder than the glass.

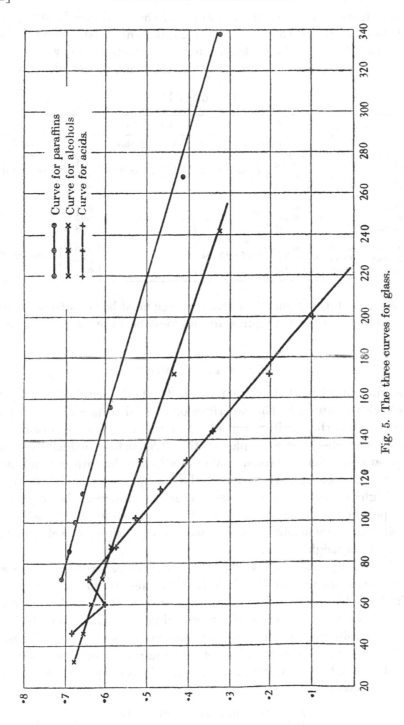

Fig. 5. The three curves for glass.

The influence of the chemical nature of the solid face is unexpectedly simple in other ways. Thus, for the three lubricants tried, when the slider is of glass and the plate of steel, the value of μ is exactly midway between that for glass on glass and steel on steel. The following figures illustrate this:

TABLE VII

Slider	Plate	Lubricant		
		Butyl alcohol	Amyl alcohol	Octyl alcohol
Glass	Glass	0·606 ⎫	0·585 ⎫	0·5176 ⎫
Steel	Steel	0·3924 ⎬ mean 0·4992	0·375 ⎬ mean 0·48	0·2981 ⎬ mean 0·4078
Glass	Steel	0·493 ⎭	0·48 ⎭	0·41 ⎭
Glass	Glass	0·606 ⎫	—	0·5176 ⎫
Bismuth	Bismuth	0·30 ⎬ mean 0·453	—	0·25 ⎬ mean 0·3838
Bismuth	Glass	0·451 ⎭	—	0·38 ⎭
Steel	Steel	0·3924 ⎫	—	0·2981 ⎫
Bismuth	Bismuth	0·30 ⎬ mean 0·3464	—	0·25 ⎬ mean 0·274
Bismuth	Steel	0·348 ⎭	—	0·27 ⎭

To get this simple result the curved surface must be of the softer material. If it be of the harder, the figures are different because the softer surface is cut.

PART II. THEORY

The current view that friction is due to the interlocking of asperities dates back to Coulomb. The asperities of a polished face are no doubt insensible, but earlier writers conceived of them as real elevations of the surface. Herschel, for example, writes: "A surface artificially polished must bear somewhat the same kind of relation to the surface of a liquid, or a crystal, that a ploughed field does to the most delicately polished mirror."[1]

Rayleigh[2] came to the conclusion that the difference between a fluid and a polished surface was not great, the elevations of the latter being of molecular dimensions. Beilby's work on polishing is well known. It confirmed Rayleigh's view.

Coulomb's hypothesis, that friction is due to asperities acting like inclined planes, implies at the limit that the actual surface of the solid is frictionless. This does not seem to be possible, since we know that the surface of a solid is the locus of a powerful field of attraction which is the cause of capillary phenomena, and this attraction, though it may be modified by the film of contamination on all naturally occurring solid faces, is not destroyed by it, otherwise, for example, a drop of water would not cling to a glass rod.

[1] *Enc. Met. Art. Light.* (1830), p. 447.
[2] *Polish*, Royal Institution, March, 1901.

The view taken in earlier papers[1] is that friction is due to molecular attraction operating across an interface, in short, to cohesion. If matter were continuous, and if the equipotential surfaces of its superficial attraction field were parallel to a plane surface, it would be frictionless. But matter is discontinuous, and it is to this that friction is due. The asperities required by Coulomb are in fact the atoms and molecules, for, owing to the short range of molecular and atomic attractions, the equipotential surfaces are by no means simply parallel to the general material surface. The action of a tangential force does not merely move the atoms and molecules in their own tangent planes, but also rotates them and produces movements of compression along the normal, and the friction—the tangential reaction as French writers call it—is the sum of the resistances to displacements of the centres of mass and to rotation.

When two solid faces are separated by a layer of lubricant the molecules of the latter must be supposed to be highly oriented. The fact that there is a finite resistance to slip shows that the substance of the layer, though it may be formed from material which is fluid at the temperature of observation, has lost its fluidity. This can be due only to the influence of the attraction fields of the solids, which destroys the random arrangement of the molecules which permits of fluidity, and substitutes for it a configuration in which the potential of a molecule is a function of its position and orientation.

The system, therefore, is one which varies rapidly in state along a normal to the solid faces, supposed to be plane, but is homogeneous along surfaces which are parallel to the solid faces. It may be regarded as consisting of a series of imaginary surfaces each having different mechanical properties. When the two solid faces are identical in nature and parallel to each other there may be supposed to be symmetry about a plane surface midway between them, such that surfaces equidistant from this median plane will have identical properties.

When a tangential force is applied there will be "give" at each plane, which will increase with the growth of the applied force until a yield point is reached, when slipping occurs, and, owing to the heterogeneity of state along the normal, the slip is confined to one, or at most two, planes.

The fact that the friction of surfaces flooded with lubricant is the same as that of the same surfaces when covered only with a primary film shows that the layer of lubricant actually present between the opposed faces is identical in thickness and structure in both cases. When the lubricant is in excess this layer will be that which can support the weight of the slider owing to the operation of capillary forces. Let a surface be drawn in the interior of a mass of lubricant resting on a plane solid face at a distance

[1] *Phil. Mag.* (1919), (6), XXXVIII, 32; *ibid.* p. 49; (1920), (6), XL, 201.

from the solid face equal to the thickness of a primary film. Such a surface
must have peculiar significance in lubrication. It is easy to see why this is
so. When a fluid such as, *e.g.*, benzene is poured over a clean surface of water
in a vessel, covered so that the space above may become saturated with
benzene vapour, it will be noted that, if the sheet of benzene is not too
thick, say about 1 mm., it ruptures in places, and contracts to form lenses
which are in equilibrium with a composite surface covered with an invisible
film of benzene. The continuous sheet of benzene is, therefore, unstable, and
it ruptures in the plane which separates the primary film from the mass of
the fluid; the benzene below this plane remains adherent to the water,
while the benzene above contracts on itself to form lenses or a single lens.
The surface mentioned above is, therefore, at that critical distance from the
water face at which the attraction of benzene for itself is greater than its
attraction for water. It marks what might be called a natural flaw in the
structure of the benzene. For this reason it may be expected to be a natural
plane of slip when tangential stress is employed, and therefore to be of
fundamental importance in lubrication.

Consider two solid faces each covered with a primary film, and let them
be applied the one to the other, and let the area of one face be much greater
than that of the other. The layer of lubricant may either consist of the two
primary films or be greater or less than this, the thickness being determined
by the pressure and capillary forces. If it be greater, then the excess must
be drawn in between the solid faces by capillary forces from the primary
film elsewhere on the surfaces. The balance of probability is that the layer
is not thicker than two primary films, for capillary attraction might be
expected to draw in more from a flooded face than from a face covered only
by a primary film, and the equality in the coefficient of friction proves that
this does not happen. Measurements made with sliders loaded excentrically
point strongly to the same conclusion (Appendix I). It is not likely to be
less than two primary films in thickness, because the coefficient of friction
in that case would not equal the flooded value, as the study of the relation
to vapour pressures shows.

The parameter a measures the contribution of the solid faces to the
friction. When two different solids, 1 and 2, are employed the contribution
of each, as experiment shows, is $\frac{1}{2}a_1$ and $\frac{1}{2}a_2$, respectively. Therefore each
solid face contributes its effect upon the friction independently of the other.
The question, however, to which an answer is not readily forthcoming is,
Why does the nature of the solid influence the friction of lubricated surfaces
at all?

The slipping which occurs when the stress reaches a certain intensity may
be due either to a general breakdown of structure in the lubricant or the
structure may persist and the slip be confined to a plane, which is probably

the plane midway between the solid faces *when these are composed of the same material*. The fact that there is no break in the curve connecting friction with molecular weight when the lubricant becomes a solid is against there being a general breakdown of structure, for it is unlikely that tangential stress would confer upon a film of a solid lubricant the property of fluidity. A further objection is found in the fact, noted by many writers, that kinetic friction is equal to static friction and is independent of the velocity. This can only mean that static friction measures the maximum tangential stress which a certain molecular structure is capable of bearing. If the structure broke down during relative motion the stress must, we conceive, be changed in amount.

Let us assume, then, that slip is confined to one plane and that the friction measures the tangential reaction at this plane. The reaction depends, as we have seen, not only upon the chemical nature of the lubricant, but also on that of the solids. The field of attraction of the latter must therefore make itself felt in a decisive way at the plane of slip. In the language of the classical theory of capillarity, the range of the force of attraction must be at least equal to half the thickness of the layer of lubricant.

The point at issue will be made clearer by considering this fact in terms of a particular theory of the structure of primary films. When a substance such as, *e.g.*, oleic acid is allowed to spread over water no fall of surface tension is produced until the concentration of the acid on the surface has reached a certain very small value. Rayleigh suggested that this concentration marks the point when the acid forms a continuous layer one molecule thick, when, that is to say, the molecules of the acid are close enough together to be within the range of each other's attraction and repulsion. Increase in concentration of the acid causes the surface tension to fall until a second point is reached, at which further additions cease practically to produce any further effect. The surface is now saturated with the acid and any excess merely collects into lenses of acid in mass. The second point of inflection, therefore, is the point at which the primary film is completely formed. Rayleigh[1] suggested that at the second point the layer of acid is two molecules thick.

Rayleigh's first suggestion has been accepted by all who have since worked at the subject. His second suggestion, namely, that the primary film consists of two layers of molecules has not found favour. The view which has replaced it we owe to Devaux,[2] who considers that the primary film is always of a single layer of molecules, and that the second point mentioned above occurs when the molecules are packed as closely as possible.

[1] *Phil. Mag.* (1899), (5), XLVIII, 334.
[2] *Rev. Gén. des Sciences* (February 4, 1913).

The film, in short, is regarded as an elastic structure, which is at its maximal extension when it first begins to affect the surface tension, and at its maximal tangential compression when the surface is saturated. Further compression then results in some of the molecules mounting on top of the layer. This is the mono-molecular theory of the structure of these surface films. One of us pointed out that in such films and at all interfaces between immiscible fluids the molecules must be highly oriented with respect to the surface by the asymmetrical field of attraction.

The theory of surface films has been developed on these lines in a striking way by Langmuir[1] and Harkins.[2] Consider, for example, an acid like palmitic acid. Each molecule consists of a chain of carbon atoms with attached atoms of hydrogen, with a carboxyl group at one end. The carboxyl group is soluble in water, for whereas the paraffins are typically insoluble, the related acids of the lower members of the series are soluble. The carboxyl group may, therefore, be regarded as being much more strongly attracted by water than is the chain of carbon atoms. Langmuir, therefore, pictures a primary film of palmitic acid on water as being composed of molecules attracted to the water by the carboxyl group only, the carbon chain being disposed at right angles to the water surface. The forces of attraction he considers to be of such short range as to extend only from one atom to its immediate neighbours. The influence of the attraction field of the water would, therefore, not extend beyond the atoms of the carboxyl group, the rest of the molecules, consisting of the carbon chain, would be wholly beyond its range. This represents the monomolecular theory pushed to its limits. Imagine the attraction of a solid limited in the same way to the carboxyl group, friction would depend only upon the lubricant and would be independent of the nature of the solid, for cohesion at the plane of slip would then be due only to the terminal group of the chain, namely, the CH_3 group. We know, on the contrary, that the friction is dependent upon the nature of the solids, and that, therefore, the field of attraction of the solids does not merely serve to anchor the molecules of the lubricant, but must also, in some way, make itself felt at the plane of slip.

One of us has urged that the range of attraction in gases or vapours is quite different from what it is in condensed states of matter. In gases or vapours the range is no doubt very short, so short as to be comparable to the distance between the atoms of a molecule. In fluids, however, there must be an irradiation of effect. If the field of one atom or molecule tends to orientate the axes of neighbouring atoms or molecules, these in turn will tend to orientate other molecules, and so the effect will spread until, in the case of a fluid, it is finally lost in the heat motions. Such an irradiation of effect must be specially great at interfaces where the general attraction

[1] *J. Amer. Chem. Soc.* (1917), XXXIX, 1848.   [2] *Ibid.*

field is itself oriented, and it is this extended range, and not the absolute range of attraction in a vacuum, which is operative in the theory of capillarity. The solids may be supposed to modify cohesion right through a layer of lubricant by spreading of strain in this way, even though the *absolute* range of their attraction may be as small as Langmuir supposes it to be.

The most puzzling questions are, Why should friction fall with increase in the molecular weight of the lubricant, and why should it vary directly with the normal pressure? Coulomb took the fact that friction is independent of the area of the surfaces to prove that it could not be due to cohesion because this obviously varies with the area. Cohesion, he admitted, must play some part, but a small part compared to that due to the "engaging" of asperities. Amontons' law, in that case, would be only an approximation, as Coulomb indeed took it to be. It is, on the contrary, a rigid law, save in those cases in which the traction alters the texture of the solids.

There seems to us to be an unanswerable objection to Coulomb's theory. Let us suppose that the molecules of the lubricant are anchored by the attraction fields of the solid, and that they reduce friction by reducing the inequalities of the solid surfaces. The capacity for reducing inequalities, or filling up depressions, might well be supposed to increase as the molecules got larger, but there are no serious differences in the size of molecules of paraffins and the related acids or alcohols which have the same number of carbon atoms. In other words, the theory fails wholly to explain the connection between chemical constitution and lubrication.

A picture which, however, is purely qualitative, of how molecular weight may influence lubrication is afforded by the assumption that the molecules of the lubricant are oriented by the attraction fields of the solids so that their long axes are at right angles to the solid faces. To give precision to the picture let us assume that the layer of lubricant is composed of two primary films the plane of slip being between them. Each CH_2 group which is added to the molecule lengthens the carbon chain and, therefore, increases the distance of the plane of slip from the solid faces. But the external tangential force is applied at the solid faces and as the distance is increased the moment of the force at the plane of slip will be increased. Lengthening the carbon chain will, therefore, have an effect similar to the lengthening of a rod when a couple is applied at the ends.

We are, however, no nearer explaining why each carbon atom added to the chain should produce the same effect. If the chain were straight the moment of the tangential force would be proportional to the number of carbon atoms, but it is certain that the chain is either a zigzag or a flat spiral. We are almost driven to the conclusion that in the primary films the atomic pattern of the molecules suffers rearrangement, somewhat like that which occurs in crystal formation. The primary film in a sense may be said to be crystallised, but the pattern is determined by the attraction fields of the solids. The view that a sub-crystalline structure obtains even at interfaces between immiscible fluids was put forward by one of us many years ago. It helps us to understand how the presence of hydroxyl and carboxyl groups may influence lubrication by modifying the attraction field of the solid.

It must be admitted that no explanation of Amontons' law is forthcoming. Like Hooke's law, and the simple linear relation between the concentration of oleic acid on the surface of water and the fall in surface tension, it is probably one of those laws of average values which mask the innate complexity of matter. The most promising way of stating the law is as follows: For the same solids and the same lubricant the tangential reaction (the friction) per unit area is dependent only upon the pressure. The area now disappears from the equation since it occurs both as numerator and denominator.

An explanation of this relation will be forthcoming only when the theory of boundary friction is developed by some one fully cognisant of the theory of elastic solids. We content ourselves with a suggestion. An equipotential surface drawn above the general plane surface of a solid is not likely to be a plane, because the solid is particulate and each particle is surrounded by a complex field of force. Any equipotential surface is a frictionless surface in the sense that a particle of matter small enough not to distort the surface could be moved over it without effort. In measurements of static friction we can neglect inertia; therefore, if there were no normal pressure due to an external force on the particle, any force having a tangential component would cause it to move over the equipotential surface on which the particle is maintained by attractive and repulsive forces. If there were a normal pressure due to an external force a tangential force would experience resistance to the motion of the particle and the resistance would be proportional to the pressure.

The work done by the external force in the condition of static friction is absorbed by the elastic forces between the atoms and molecules of both lubricant and solid. If the force be applied with sufficient slowness the action is probably thermodynamically reversible, the local cooling or heating effects being reversible. The strain produced by the external force will consist in displacements of the centres of mass and in rotation of atoms and molecules. When slipping occurs, that is to say in the condition of kinetic friction, we picture the slip as being confined to one plane because the yield point of this plane has been reached. The yield point fixes a limit and therefore (as many observers, from Coulomb onwards, have noted) the tangential reaction of kinetic friction is equal to that of static friction. It follows that the strain during the application of the external force is the same in both states, and the dissipation of energy in kinetic friction is due to the recovery from the strained position of the atoms and molecules which occurs immediately *behind* the slider.

The curves connecting friction and molecular weight, if prolonged, will meet the base line; therefore, when the molecular weight is increased sufficiently static friction vanishes. This point is difficult to observe, for the following reasons. When lubrication is bad the motion which occurs when the slider breaks away is rapid, as though the tangential force necessary to cause the breaking away were in excess of that needed to maintain movement.[1] As lubrication is improved the motion takes on more and more the nature of a slow glide, which may be so slight as to be barely detectable. Ricinoleic acid, which is a fluid at ordinary temperatures and a very good lubricant (cf. *Phil. Mag.* (1920), XL, 201), does abolish static friction in the sense that the smallest tangential force which could be applied with our apparatus, namely, that due to the light aluminium pan (= 0·6 gr.), caused the heaviest slider to glide very slowly.

Solid lubricants which should lie on the curve below the base line seem to give finite, but very small, values for static friction. There may, therefore, be a discontinuity between the fluid and solid states in this region. The apparatus, however, was not sufficiently sensitive to settle the point. The values obtained are given below, for what they may be worth:

Tetracosane on steel ...	...	...	...	$\mu = 0\cdot085$
Dodecoic acid on steel	...	...	...	0·031
Dodecoic acid on steel (polished film)		...		0·026
Palmitic acid on steel	...	...	...	0·029
Stearic acid on steel ...	...	...	...	0·023

APPENDIX 1

Treatises on mechanics deal exclusively with frictionless systems. Painlevé, in 1895, took up the question whether it is possible to develop a general theory for systems with friction analogous, for example, to the application of the equation of Lagrange. His analysis led him to the conclusion that Amontons' law contradicts the equation of motion of a solid, at least in certain realisable cases. This conclusion led to a controversy maintained by French and German writers for about 15 years, which turned on the question whether the condition of absolute rigidity postulated by Painlevé was not responsible for his paradoxical conclusion.

Amongst the many papers is one by Chaumat,[2] in which he states that the friction depends not only on the pressure, as Amontons' law states, but

---

[1] This at first sight appears to contradict the equality between kinetic and static friction so often observed, but this equality has been noted in experiments in which both surfaces have had considerable extension. When the surface of the slider is curved and that of the plate plane, motion of the slider tends to cause it to mount on a thickened pad of lubricant.   [2] *Comptes Rendus* (1903), CXXXVI, 1634.

also on the degree of asymmetry of the loading. An oval slider, for example, which has a weight placed near one extremity will not conform to the law. In our own experiments the loading was always symmetrical in the case of the steel sliders, which were themselves accurately turned, and any extra weight was added in the shape of circular discs. Loading was not, however, always exactly symmetrical in the case of the glass sliders, since the melted lead with which they were weighted tended to run into an irregular pattern. Measurements were, therefore, made with sliders loaded as unsymmetrically as possible, the weight of lead being either wholly in front or wholly behind the point of application of the external force. In the first case the slider was tilted forward, and in the second case backward by the weights.

The following table gives flooded values of μ for forward-loading, backward-looking and symmetrical loading, in which the external force was applied so as not to cause the slider to rock sensibly:

Loading	Lubricant	
	Octyl alcohol	Caprylic acid
Forward	0·5192	0·3423
Backward	0·5028	0·3356
Symmetrical	0·5176	0·3412

The value for the backward loading is slightly but definitely lower. This, however, does not in our opinion confirm Chaumat's conclusion. Consider a curved slider standing in a pool of lubricant. Chance oscillations tend to bring excess of lubricant between slider and plate in the region of "contact". General earth vibrations, due for example to the slamming of a door in the building, were found sometimes completely to upset the readings. A backward-loaded slider was found to be mechanically unstable to vibration. Readings were obtained only by exercising the utmost patience. A backward tilted slider is like a backward tilted plane surface, in that any movement forward due to small vibrations will tend to thicken the layer of lubricant. We conclude, therefore, that the slightly lower value of friction given by backward loading was due to secondary causes.

This is confirmed by the fact that when lubrication was by a primary film, there being no excess of lubricant on the surfaces, the friction was independent of the nature of the loading, as the following values for heptylic alcohol show:

Forward   ...   ...   0·4090
Backward   ...   ...   0·4088
Symmetrical...   ...   0·4050

This result would seem to prove that in lubrication by primary film the slider is not seated on a thickened pad of lubricant drawn in by capillary forces. The layer is indeed composed of a double primary film.

APPENDIX II

When a drop of a single pure chemical substance is placed upon a clean surface of water one or both of two things may follow, namely, the spreading from the drop of an invisible film, and the spreading of the drop itself by actual flattening to form a layer which, at any rate at first, is of sensible thickness. These processes are distinct, because the first may occur without the second. Spreading of the drop itself occurs only when the surface tension of water is greater than the sum of the tensions of the upper and lower surfaces of the drop. It is this kind of spreading which is contemplated in the classical theory of capillarity. Normal paraffins with more than eight carbon atoms, and stable saturated substances like carbon tetrachloride, do not manifest this kind of spreading; but, if they have a sensible vapour pressure, an invisible film is formed, not directly from the drop, but through the mediation of the vapour phase. The formation of the insensible film is called primary spreading, whilst the flattening out of the drop is called secondary spreading.[1]

There can be no manner of doubt but that primary spreading on solid faces occurs through the intervention of the vapour. Only those lower members of a series which have a sensible vapour pressure manifest it, and, as the molecular weight increases, and the vapour pressure decreases, the film forms more and more slowly. The time taken, after the drop was in position, for the friction to become uniform all over the plate at the critical or "flooded" value was, in minutes:

Lubricant	Glass	Steel
Methyl alcohol	5	—
Ethyl alcohol	5	—
Propyl alcohol	10	—
Butyl alcohol	20	20
Octyl alcohol	35	40
Acetic acid	5	—
Butyric acid	10	—
Heptylic acid	40	—
Caprylic acid	50	45
Undecane	5	5

The gradual formation of the film is shown by the following diagram, in which the position of the drop of butyl alcohol on the steel plate is indicated by a circle, and the figures give the value of the coefficient of friction at the place where they are printed, 10 min. after the drop was placed on the plate.

```
0·44
0·42
0·41
0·40
0·39
0·38
0·38
 ○
```

[1] *Proc. Roy. Soc.* (1913), A, LXXXVIII, 313; *Phil. Mag.* (1914), XXXVIII, 49.

42

BOUNDARY LUBRICATION.
THE TEMPERATURE COEFFICIENT
With IDA DOUBLEDAY

[Proc. Roy. Soc. (1922), A, CI, 487]

The experimental method employed was that described in earlier papers.[1] A slider having a spherical face is made to slide over a plate in an atmosphere of rigorously clean and dry air. The friction measured is static friction and the object of the experiments the determination of the effect of temperature. This has now been studied over a range of 15–110° C., and it may be said at once that the relations discovered are of a totally unexpected character.

More than one attempt to study the effect of temperature was defeated by the fact that lubricating vapours were given off from the walls of the chamber in which the plate and slider were enclosed. This difficulty was completely removed by using a chamber with double walls, the inner wall being a continuous sheet of nickel. Between the walls were placed the electric grids for heating the chamber. The stream of dry air with which the chamber was flooded was also heated by being passed through a tube of silica, which was maintained at the required temperature by a coil of wire through which a current was passing. The temperature of the stream of air and the temperature of the chamber were recorded electrically.

Temperature might influence friction by altering the state of the solid faces or of the lubricant; it was, therefore, necessary to study clean faces first.

Clean Faces

Glass was found to be totally unsuitable for use because the physical state of its surface was profoundly altered by a relatively small rise of temperature. The figures in Table I illustrate this.

Table I. *Glass*

Time elapsed min.	Temperature (° C.)	Pull gr.	μ
5	15·5	25·5	0·94
30	15·5	25·5	0·94
70	73	13·6	0·51
100	74	12·1	0·46
130	74	12·1	0·46
180	15·5	12·0	0·46
240	15·5	12·0	0·46

[1] *Proc. Roy. Soc.* (1922), A, C, 550.

It will be noticed that the coefficient of friction ($\mu = $ pull $\div$ weight of slider) falls as the temperature rises, and that the changed state of the surface which this indicates is not reversed when the temperature falls.

The fall in friction was greater the higher the temperature to which the glass was exposed. Thus glass at 74° reached a steady state in about 30 min. with $\mu = 0\cdot46$, but at 110° in 45 min. with $\mu = 0\cdot36$. The effect is due to the formation of a film which prevents water from wetting the surface, and which cannot be removed by soap and water; it is, however, destroyed by hot chromic acid and the friction then returns to the full "clean" value characteristic of the particular sample of glass.

Quartz, when cleaned in the way ordinarily used for glass, namely, by heating in a strong solution of chromic acid and rinsing in tap water, behaved like glass as the values given in Table II show:

TABLE II. *Quartz*

Time min.	Temperature (° C.)	μ
0	17	0·73
30	17	0·75
65	72	0·13 (slow glide)
120	110	0·09
175	20·5	0·08

By this time we were aware of the fact that the friction of steel and bismuth is not affected by temperature. This suggested that the behaviour of quartz might be due to a film obstinately adhering to the surface, and not to the substance itself. The suggestion proved to be correct, since the fall in friction observed with rise of temperature did not appear when the quartz faces were vigorously rubbed with the fingers under a stream of water, after having been heated in the chromic acid bath. It is not unlikely that this film on quartz, which resists hot chromic acid and needs actual traction for its removal, is itself a glass due to surface reaction of the silica with alkali from the air or from polishing or cleansing substances.

The values in Table III are for fully cleaned quartz. It will be noticed that the coefficient of friction of clean quartz is less than that of glass.

TABLE III. *Quartz*

Time min.	Temperature (° C.)	μ
0	16·5	0·770
45	16·5	0·770
120	110	0·765
145	110	0·765
210	18	0·763

Steel and bismuth, when clean, show no change whatever in the friction over the range of temperature which was tried.

The result of these experiments with clean faces then is that the friction of a face of quartz, steel or bismuth, when thoroughly freed from adherent films, is sensibly independent of temperature over the range 15–110° C.

LUBRICATED FACES

Steel and quartz alone were used in these experiments; there is, of course, no reason to suppose that bismuth would have behaved differently.

Surfaces flooded with lubricant. When the lubricant was fluid throughout the whole temperature range the friction was independent of temperature.

The following values were given, without sensible change, over the range 15–110° C.:

TABLE IV

Lubricant	Steel μ	Quartz μ
Undecane	0·34	0·49
Caprylic acid	0·20	0·306
Pelargonic acid	0·14	0·238
Butyl alcohol	0·39	0·545
Octyl alcohol	0·29	0·457
Undecyl alcohol	0·23	0·39

When the lubricant was solid over part of the range the effect of temperature was different below and above the melting-point of the solid. The friction was always found to fall with rise of temperature up to the melting point. At this point there was a remarkable discontinuity, the friction falling to zero. When the lubricant was fully melted the friction suddenly reappeared at a higher level than before the discontinuity, and was constant over the remainder of the temperature range. When the temperature was allowed to fall from the highest point (110° C.) there was no change in friction until the melting-point was again reached, when there was a sudden fall, but not to zero. Below this point it was not possible to go, because the slider became set fast in the solid mass of lubricant. These relations are illustrated in Table V, and by the curve in the diagram.

The lubricant was applied by placing a few crystals of the solid on the plate in the chamber which had already been heated above the melting point. The slider was placed on the plate now covered by a layer of fluid lubricant and the chamber allowed to cool slowly. When at room temperature, the slider was lifted from the plate with the whole visible mass of lubricant attached to it, replaced, and observations of the friction were then taken.

Nonadecane, docosane, tetracosane, decoic acid, undecoic acid, palmitic acid, stearic acid, and cetyl alcohol were tried. It will be sufficient to give the figures for one of these:

TABLE V. *Palmitic acid on quartz. Melting-point, 62° C.*

Time min.	Temperature (° C.)	μ
45	19·0	0·163
65	28·4	0·140
90	34·2	0·121
110	39·8	0·098
145	48·5	0·063
160	54·5	0·038
195	59·3	0·021
210	62·0 (melting)	0·0
212	62·0 (liquid)	
245	110	less than 0·03
290	63	
295	61·8 (solidifying)	

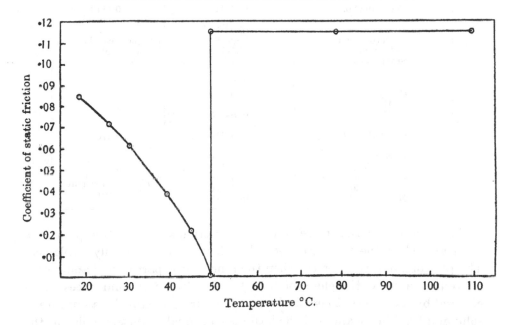

The discontinuity at the melting-point would appear to be due to the fact that, when the lubricant is still solid, the friction that is measured is that of the lubricant on itself. This might well be expected to fall with rise of temperature so near the melting-point. This conclusion derives a certain amount of support from the fact that the coefficient of friction of a glass slider on the solid paraffin $C_{24}H_{50}$ is, at 16°, $\mu = 0·07$, whilst the coefficient of glass on glass lubricated by a film of this paraffin is $\mu = 0·325$. Above the melting-point the friction, on the contrary, is that of solid faces, as modified by a primary film of the particular lubricant. In order to eliminate the friction of the lubricant on itself, films of insensible thickness were deposited on the faces from solutions in ether. The discontinuity now disappeared, and the friction became constant over the whole range, both when the temperature was rising and when it was falling, and we may therefore conclude that, over the range of temperature 15°–110° C., the friction of quartz and of steel, lubricated with normal paraffins and the related normal acids and alcohols, is sensibly independent of temperature.

TABLE VI

Lubricant	Temperature (° C.)	μ
Steel		
$C_{19}H_{40}$	15·6–110	0·179
$C_{22}H_{46}$	17·3–110	0·110
$C_{24}H_{50}$	17·0–110	0·068
Cetyl alcohol	16·5–110	0·1143
Decoic acid	15·0– 75	0·075
Undecoic acid	15·0– 80	Negligible—glides
Palmitic acid	16·5–110	with weight of
Stearic acid	17·0–110	scale pan
Quartz		
$C_{19}H_{40}$	15·0–110	0·324
$C_{22}H_{46}$	15·8–110	0·26
$C_{24}H_{90}$	16·5–110	0·23
Cetyl alcohol	19·0–110	0·276
Decoic acid	17·2–110	0·18
Undecoic acid	16·5–110	0·114
Palmitic acid	16·5–110	
Stearic acid	18·3–110	Glides with weight
Myristic acid	18·0–110	of scale pan

The curious fact that there is no static friction at the melting-point of a lubricant when the faces are flooded with it is not so readily explained. The phenomenon is observed when the lubricant is partly solid and partly melted. We may therefore picture to ourselves the solid faces, each covered by its adherent film, and separated during melting by a mixture of solid and fluid lubricant, and the latter is continually being supplied at the expense of the former. Under these conditions the friction would be fluid friction, and therefore static friction would be absent; and the friction of the solid faces, as lubricated by the particular lubricant, would not reappear until melting was complete and all the fluid squeezed out by the weight of the slider, leaving the solid faces separated by two primary films of lubricant, as described in the earlier paper. The fact that the static friction reappears instantly and at full value confirms the view there taken that solid friction does not change gradually to fluid friction, but that the one changes to the other abruptly at a certain critical thickness of the film. The state of the material in these lubricating films is determined by the attraction field of the solid as well as by the mutual attraction of the atoms and molecules of which it is composed. Since there is no change in the curve friction/temperature at the melting-point of the lubricant, we may suppose that the former is more important than the latter. The fact, known since Leslie's time, that heat is liberated when a solid is wetted by a fluid, shows that the degree of freedom of a molecule in these films is less than that of one in the interior of a fluid mass of the same lubricant.

The curves connecting static friction with molecular weight, which were given in the preceding paper, cut the base line; that is to say, above a molecular weight which is different for each chemical series, static friction should vanish. More careful measurements made since the preceding paper was written show that this actually is so when the lubricant is present only as a film of insensible thickness. Then the slider moved with the smallest weight which could be applied, namely, 0·7 gr. It moved slowly, as might be expected, since the accelerating force was so small—the observed rate being of the order 0·4 mm. per hour.

When the lubricant is present in excess there frequently is resistance to movement due to surface tension, for consider a small drop symmetrically disposed about the area of contact between slider and plate. Movement of the slider will increase the area of the interfaces between the drop and air and the drop and the plate. In certain cases this absorbs work. For example, a steel slider in a small drop of melted palmitic acid on a steel plate needed a weight of 3·6 gr. to move it, when the drop was smeared about the plate a weight of 0·9 gr. was sufficient, but when the acid was present on the surfaces only as a film of insensible thickness the weight of the scale pan (0·7 gr.) alone sufficed to make the slider glide slowly.

43

BOUNDARY LUBRICATION. THE LATENT PERIOD AND MIXTURES OF TWO LUBRICANTS

With IDA DOUBLEDAY

[Proc. Roy. Soc. (1923), A, CIV, 25]

Part I. The Latent Period

When a lubricant is applied to a solid surface the friction, with certain definite exceptions, does not at once reach its steady value, but starting from a relatively high level, falls in longer or shorter time to a constant value. The existence of this latent period has been known to us for a long time and put aside for further enquiry, the values of μ considered in the earlier papers having always been the final or "steady" values.

The latent period seems to be merely the time occupied in the orientation of the molecules of the lubricant at the solid face.

The carbon chains of fatty acids or alcohols are loaded at one end with the —COOH or —OH groups respectively. They may, therefore, be likened to rods loaded at one end. Placing a pool of fluid upon a surface is like flinging at it a handful of such rods grasped at random. Some will hit and stick by the loaded ends, others by the unloaded ends. The rods may be supposed to be attracted to the solid more strongly by their loaded ends, and the process of evaporation into the overlying fluid and condensation from it will increase the proportion of rods sticking to the solid face by their loaded ends until a steady state is reached. The steady state will not be one in which all the rods of the primary film have their loaded ends directed inwards, since, in the give-and-take with the overlying fluid, some of those condensing are likely to arrive unloaded end foremost. The steady state will, in short, be an average state depending upon the rate of evaporation and condensation, and upon the difference between the attraction of the solid for the loaded and unloaded ends respectively.

Similar considerations hold when the primary film is formed by condensation from vapour, but then the rods are flung singly.

Let us suppose that a steady state has been reached for both slider and plate, for example by exposing both to the saturated vapour of the lubricant, and let the slider then be placed on the plate. There now are two primary films enclosed between two solid faces and, therefore, shielded from evaporation or condensation. If the molecules of these films still possessed sufficient freedom of movement to effect end for end changes, a further latent period would be observed with further fall in friction until the final steady state was reached in which all the molecules were oriented with loaded ends attached to the solid faces. This seems to happen, and it explains the fact, which was at first so puzzling, that the lowest steady value of friction can be got with fatty acids and alcohols only by leaving the slider for some time in position on the plate.

The proof that this view of the latent period is correct lies in one significant and striking fact, which seems to us to be open to no other interpretation. It is that molecules composed of a chain of carbon atoms, both ends of which are alike, do not manifest a latent period. Normal paraffins are such substances. In their case the rods are unloaded, and no matter how applied to the surfaces the first recorded value of the friction is the same as the last.

The latent period must be of importance in practical lubrication, for it measures the rate of repair of the lubricating film. It is true that the object of the engineer is to keep lubrication within the limits of complete fluid lubrication, in which the solid surfaces are separated by a layer thick enough to exclude boundary conditions. But he rarely succeeds, since every particle of metal worn away is evidence of failure. Disaster happens in the region of boundary lubrication, and there the rate at which a ruptured or partly ruptured primary film can reform itself must be no unimportant factor.

The phenomena of the latent period, as they first presented themselves to us, were discouragingly complex. This could not be otherwise, because there are three possible steady states, namely, equilibrium with fluid, equilibrium with vapour, and the equilibrium of primary films enclosed between two solid faces, and the attempt to measure the value of the friction in either of the two first states causes it at once to begin to change to the third, because it necessitates placing the slider in position on the plate. All that can be done, therefore, is to take a reading of friction as quickly as possible, and regard it as an approximation to the value required.

There are three possible states and there are four variables which affect the duration of the latent period, namely, the chemical nature of the lubricant and of the solid faces, temperature, and mechanical agitation such as is produced by moving the slider about upon the plate.

Material, glass throughout. Figures are for the coefficient of friction μ, and the latent period is given in minutes.

Case I. A large pool of lubricant placed on the plate and the slider put into it at once.

	First reading	Steady value	Latent period min.
Caprylic acid	0·57	0·34	60
Heptoic acid	0·50	0·40	45
Octyl alcohol	0·59	0·52	15

Case II. Surfaces in equilibrium with fluid lubricant before being placed in contact. The slider and plate were covered with lubricant in excess and left in the chamber for three hours. The slider was then put on the plate, of course without opening the chamber.

	First reading	Steady value	Latent period min.
Caprylic acid	0·26 rising rapidly to 0·44	0·34	10
Heptoic acid	0·31 rising rapidly to 0·46	0·40	10
Octyl alcohol	0·47 rising rapidly to 0·54	0·52	10

The first recorded value taken as soon as possible after the slider has been placed on the plate is low, and there is a rapid initial rise in friction: this is characteristic of surfaces which have been allowed first to come into equilibrium with the *fluid* lubricant.

These two cases taken together show that the condition of the lubricant is profoundly changed by contact with the solids, and we picture the sequence of events as follows:

When a pool of one of these lubricants is placed on a solid, the molecules throughout have at first that random arrangement of the fluid in mass which confers upon it its characteristic viscosity. Orientation begins at once at the solid face and, as the primary film is formed, extends inwards into the overlying fluid until finally a steady state is reached of orientation maximal at the solid face and decreasing thence as it is more and more upset by the heat motions.

The orientation in the layer over the primary film will alter the viscosity, tending to increase it, and, therefore, the slider will take a sensible but brief time to squeeze out the excess fluid, until only the two primary films remain. Therefore, the first recorded value will not be for the primary film but for a layer of lubricant thicker than this, and the friction is lower than that given by completely oriented primary films. When the slider is placed in position at once, that is before orientation has had time to become established, the excess fluid is squeezed out so rapidly that the first reading records the high friction of the imperfectly oriented primary films.

Experiments with glycerol confirm this view. This substance has a high viscosity in mass and, when used as a lubricant, the first readings of friction are vanishingly small. They rise, however, with lapse of time so that in an hour or so the friction has risen to the value for the clean unlubricated surfaces. Glycerol indeed, like water, is a neutral substance for glass and steel in that it does not lower static friction, but, owing to its high viscosity, considerable patience is needed to detect the fact.

If the initial low value of friction be the reading taken whilst the slider is sinking through fluid whose molecules are partly oriented, and whose viscosity is thereby greatly increased, then it should not be observed when the surfaces before being placed in contact are in equilibrium not with fluid, but with vapour. This is Case III.

Case III. Surfaces in equilibrium with the vapour of the lubricant before being placed in contact. Saturated vapour was passed for three hours into the chamber, in which were both plate and slider but not in contact. Slider then placed on plate. The initial low value is not now observed.

	First reading	Steady value	Latent period min.
Caprylic acid	0·45	0·34	5
Heptoic acid	0·48	0·40	5
Octyl alcohol	0·54	0·52	5

It will be noticed that the first reading for surfaces in equilibrium with fluid and with vapour respectively is, within the limits of unavoidable error (the state being in process of change after the slider is put on the plate), the same. As the friction is, *ex hypothesi*, a measure of the degree of orientation, we conclude that primary films in equilibrium with fluid are in the same state as those in equilibrium with vapour.

In what follows the latent period given is the time which elapses between putting the slider on the plate as soon as possible after lubrication and the attainment of the final steady state. Agitation, either thermal or mechanical, shortens the latent period. The effect of temperature is illustrated by the following figures which give the time required to reach the steady state after the slider has been placed in a pool on the plate.

Temperature	Latent period min.	First recorded value of μ
Heptoic acid on glass ($\mu = 0 \cdot 405$)		
15° C.	45	0·50
48° C.	15	0·46
72° C.	5	0·44
Octyl alcohol on glass ($\mu = 0 \cdot 517$)		
13° C.	30	0·59
50° C.	10	0·55
60° C.	2 to 3	0·53

The value of μ *in the final steady state* is independent of temperature, within the limits of accuracy of the experimental method (*Proc. Roy. Soc.* (1922), A, CI, 487), as perhaps might be expected since the primary films are then sheltered from evaporation and condensation. The variation in the first recorded value of μ, that is, in the friction taken as soon as possible after the slider is in position, indicates, however, that the degree of orientation at surfaces *still in contact with fluid or vapour* does vary with temperature.

Mechanical agitation, such as is produced by moving the slider backwards and forwards on the plate, has a remarkable effect. For example:

Slider in pool of caprylic acid

Slider at rest	Latent period 60 min.
Slider moved up and down the plate ...	Latent period 10 min.

Equilibrium is hastened and the latent period thereby shortened by movement of the slider, even when the solid surfaces are covered only by an insensible film, as the following example shows:

The slider was resting on the plate in the chamber, and both were clean. A small drop of caprylic acid was then placed near one corner of the plate. In one hour the acid had spread over the surfaces, to cover them with an invisible but complete primary film, and the coefficient of friction had fallen to the steady value $\mu = 0\cdot34$. The slider was then lifted quickly to another part of the plate when a reading taken as rapidly as possible gave $\mu = 0\cdot41$. The slider was then moved about this part of the plate, when the value fell almost at once to $0\cdot34$ and remained constant—that is to say, movement of the slider did not lead to any further change. The slider was then lifted to another part of the plate and left quietly in position, when μ fell to its steady value $0\cdot34$ in 10 min.

It is to be observed that the sliders used had a spherical surface of radius $14\cdot7$ cm. The area of contact, therefore, was so small as to make it practically certain that, in spite of precautions which made the rocking insensible, the application of the external force did cause the slider to rock insensibly to a new contact. The effect of the slider in sheltering the primary films from evaporation into vapour or liquid is, therefore, probably not confined to the area of contact.

The properties of mixtures of lubricants belong to the next section. It will, however, be convenient to notice here the effect on the latent period of the presence of indifferent molecules—that is, of molecules which by themselves do not manifest the phenomenon. Water is an indifferent substance—indeed, it is a neutral substance in that it does not alter the friction of clean plates of glass on steel. Normal paraffins also are indifferent substances. In

Lactic acid and water on glass

Clean value $\mu = 0\cdot92$. Initial value of μ in *all* cases about $0\cdot8$.

Percentage of water	Latent period min.	Final value of μ
100	0	0·92
99·59	5	0·67
74·4	20	0·62
21·4	30	0·57
1·7	45	0·56
0·0	60	0·55

Caprylic acid and Octane on glass

Initial value of μ between 0·5 and 0·6 in each case.

Percentage of acid	Latent period min.	Final value of μ
0	0	0·65
0·0007	0	0·65
2·05	2	0·34
15·59	3	0·34
39·89	5	0·34
99·08	40	0·34
100·0	45	0·34

both cases the greater the relative number of indifferent molecules the shorter is the latent period. This is an entirely unexpected relation, because the indifferent molecules might be expected to interfere with the rate of evaporation and condensation of the active molecules.

When an alcohol and an acid are mixed the latent period is longer the greater the percentage of alcohol.

For example, in a mixture of octyl alcohol and caprylic acid.

Percentage of alcohol	Latent period min.
98·24	40
0·58	75

The practical interest of the latent period has been noted.

The theoretical interest lies in the proof it furnishes of the view put forward by one of us in 1913 (*Proc. Roy. Soc.* A, LXXXVIII, 330) that orientation extends mathematically from any interface on both sides to infinity, but that its sensible extension is fixed by the heat motions which tend to upset it, and, we must now add, the righting forces on atoms in a crystal. This sensible extension is the "range of attraction" in the classical theory of capillarity and not the range of atomic or molecular attractions in a space otherwise empty of matter.

PART II. MIXTURES OF TWO LUBRICANTS

Save for casual reference, boundary lubrication with a single pure chemical substance has alone been considered so far in these papers. We now turn to a preliminary study of lubrication with two substances. The examples fall into two classes, that in which the lubricants were miscible and mixed before being applied, and that in which the lubricants whether miscible or not were applied successively. The object of the second series was in the main the study of the effect of the presence of two primary films upon each solid surface. It may be well to note here that when a single substance is used as lubricant the solid faces are separated by two primary films; whereas in the case just mentioned where two primary films dissimilar in character have been deposited upon both surfaces, these latter when opposed are at least potentially separated by four primary films.

The following facts, set out in detail later, prove, in our opinion, that when two immiscible substances such as a paraffin and water are applied in succession to each face and the faces then placed in contact, there actually are four films between the solid faces. When water vapour is admitted to a surface which is already covered by a film of a lubricant *miscible with water*, this film is disorganised and the friction thereby raised: but when the lubricant is *immiscible with water*, the admission of water vapour always lowers friction. The disposition in this second case, namely, of superposed films, is however, unstable and readily broken up, with rise of friction, by mechanical agitation.

I. Lubricants Mixed before being Applied

Examples of all the forms of curves discovered are gathered together in the figure. There are three types.

Type I. For all mixtures of aliphatic acids or alcohols with a paraffin. The curve for the pair undecane and caprylic acid is chosen as the example.

Obviously, the acid or alcohol is condensed from the solution on to the surface so that, save when less than about 0·7 per cent. is present, the friction is the same as that for the pure acid or alcohol. The mixture cannot, however, be considered, *qua* lubricant, as the equivalent of the pure acid or alcohol. It may be better or worse according to circumstances. The latent period of the mixture is, for example, shorter, and, therefore, the rate of repair of the primary film greater.

The form of the curves suggests that there is a critical point in the dilution of the better lubricant, below which it can maintain a primary film formed completely of its own molecules, and above which the film includes molecules of both species.

These cases conform to the rule that the better lubricant is the one more strongly adsorbed by a surface, and, therefore, the one which most reduces the potential of the forces of attraction. If they were the only cases, the whole theory of boundary lubrication might rest securely on that basis.

Type II. For mixtures of aliphatic acids and alcohols whose coefficients of friction when pure differ greatly (palmitic acid and octyl alcohol), and for lactic acid with water. The curve shows a critical point at which its direction suddenly changes, but there is no horizontal portion. Throughout the whole range of concentration the worse interferes with the action of the better lubricant. Some factor intervenes to restrain adsorption of the more active component.

Type III. For mixtures of aliphatic acids and alcohols whose coefficients of friction do not greatly differ. There is no change of direction and no sign of preferential adsorption, the curve being a straight line.

Comparison of these three types shows that the form of the curve is not a simple function of the difference of level at the ends as it would be if boundary lubrication were determined wholly by adsorption. It is possible that the degree of miscibility of the components in the primary film is a factor, just as two components may be miscible as fluids but immiscible in the same crystal.

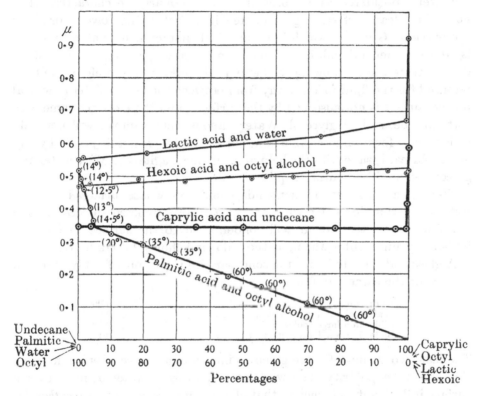

It should be mentioned that these curves are isotherms, except that for the mixture of palmitic acid with octyl alcohol. Owing to the limited miscibility of these substances it was necessary to raise the temperature as the proportion of acid was increased. The figure in brackets gives the temperature at which a measurement was made. It will be noted that there are no breaks in the curve corresponding to the changes of temperature.

It must be observed that the three types of curves given are not the only ones. In the course of this work many instances have been met when a lubricant has been applied to the faces in solution in a volatile solvent in which, as the last traces of solvent have been in process of evaporation, momentary values have been recorded much lower than the friction of the better lubricant when present alone. This is not shown in any of the curves possibly because the effect is confined to so narrow a range of concentration as to have escaped notice.

II. Lubricants applied in succession

Pairs only were examined of which either water or benzene formed one.

Water, considered as a lubricant, is of a remarkable character. It is neutral to clean surfaces of glass or steel in that it neither lowers nor raises the friction. Glycerol, which in its physical characters is so akin to water, is another neutral substance, but considerable patience is needed to demonstrate the fact, owing to the high viscosity and, therefore, the time required for the fluid to flow away from between the faces. Water is neutral not because it is not adsorbed by the surfaces, and therefore the same may certainly be said of glycerol. Water vapour, for example, will not only condense to form an insensible film, but such a film is not displaced by any good lubricant, save with the help of mechanical agitation. Why water and glycerol should be neutral is quite obscure.

Water is a lubricant to other solids, even those whose surfaces it will not wet. For example, water will not wet a freshly exposed surface of solid paraffin, yet water vapour will condense on to the surface to form an invisible primary film which, as the following shows, lowers friction.

A glass slider coated with the solid paraffin $C_{24}H_{50}$ on a freshly prepared surface of the same paraffin:

In dry air	...	...	...	...	...	...	$\mu = 0.060$
Water vapour admitted	...	...	...	...	$\mu = 0.025$
Slider in a large pool of water	...	...	...	$\mu = 0.025$
Slider in a small drop of water	...	...	...	$\mu = 0.11$

The last two values of μ are given to illustrate a source of error, namely, that due to capillarity. The work needed to force water in mass over a surface it does not wet, such as that of a paraffin, *in spite of the fact that this is covered by an insensible film of water*, is great enough to be readily detected when a small drop of water is used, for the contour of such a drop is varied by the least motion of the slider.

Water vapour also condenses on to surfaces already covered by an insensible primary film of other lubricants, and the effect is to lower friction if water is immiscible with the lubricant, and to raise friction if the contrary is the case. The disposition in the former case is that of two primary films on each face, the upper being of water, for the arrangement is unstable and readily broken up by the mechanical agitation due to moving the slider about.

Let us suppose the slider to be on the plate and the faces separated by four primary films, and let those films next the solids be called the outer films. Then, no matter whether water form the inner or outer pairs, the effect of the inner pair is always to lower friction below what it would be if the outer pair came together. In other words, a film of water deposited on a film of a lubricant immiscible with it lowers friction, and a film of an immiscible lubricant deposited on one of water also lowers friction. This may be of the nature of a general law. Many pairs will have to be studied before the effect of mechanical agitation on these doubled films is understood. All that can be done here is to point out the paradoxical results.

When water forms the inner pair of films, moving the slider about raises friction to a steady value which, however, is never the "clean" value; the value, that is, of water alone. The new steady value is reached much more quickly if the lubricant conjoined with water is fluid. In other words, if the outer pair are formed from a solid lubricant the system is more stable than when they are of a fluid lubricant.

When water forms the outer pair the effect of mechanical agitation might be expected to lower friction by permitting access of the lubricant to the solid surfaces. In fact it raises friction.

When the lubricant is miscible with water and present on the surfaces only as an insensible film, the admission of water vapour to the chamber always raises friction, and mechanical agitation by moving the slider about does not contribute to the effect.

Glass

Pure lubricant			Water vapour admitted
Lactic acid	...	...	$\mu = 0.49$	$\mu = 0.56$
Ethylene glycol	...	$\mu = 0.506$	$\mu = 0.53$
Triacetin	...	...	$\mu = 0.35$	$\mu = 0.72$

Flooding the surfaces with water still further raises the friction, *e.g.* in the case of ethylene glycol to $\mu = 0.83$. This again, however, is not the value for water alone, which is that of clean glass, namely $\mu = 0.92$.

When the lubricant is immiscible with water, and present as a film only, water vapour always lowers friction. Mechanical agitation now raises friction above the value for the pure lubricant.

Glass

Pure lubricant		Water vapour admitted
Oleic acid	... ...	$\mu = 0.31$	$\mu = 0.17$
B.P. paraffin ...	...	$\mu = 0.34$	$\mu = 0.24$
Undecyl alcohol	...	$\mu = 0.44$	$\mu = 0.33$
Undecoic acid	...	$\mu = 0.15$	$\mu = 0.13$
Dimethyl decyl carbinol	$\mu = 0.41$	$\mu = 0.37$

The effect of moving the slider vigorously up and down the plate raised the value for undecyl alcohol + water from 0·33 to 0·48.

The following are solid lubricants:

Tetracosane	...	$\mu = 0.32$	$\mu = 0.11$
Cetyl alcohol	...	$\mu = 0.32$	$\mu = 0.09$
Palmitic acid	...	$\mu = 0.0$	$\mu = 0.06$

In all these cases water formed the inner pair of films. The reverse arrangement was obtained by placing the slider in a pool of a lubricant which, owing to its low vapour pressure, was known not to spread as a primary film over the rest of the surfaces (see *Proc. Roy. Soc.* (1922), A, c, 573). Water vapour was then admitted and the friction read first *within* the pool, and then the slider carrying part of the pool of lubricant with it was moved gently to another part of the plate which was covered with an insensible film of water.

Lubricant alone			On surface already covered with film of water vapour
Oleic acid (imp.)	...	$\mu = 0.31*$	$\mu = 0.62$
B.P. paraffin ...	...	$\mu = 0.34$	$\mu = 0.73$

* This is higher than the value for pure oleic acid. The impurity present seemed to consist in an acid of lower molecular weight.

In both these cases the full rise of friction was obtained only by moving the slider about, and, indeed, in the second case there was no rise of friction until this was done. The only possible interpretation of this is that B.P. paraffin lubricates the composite surface of glass covered by a film of water as well as it does a clean glass face, the only difference being that the succession of films is unstable to mechanical agitation, and that, when broken up, the new arrangement is less effective as a lubricant.

The question naturally arose in the course of these experiments whether the clean face of glass, seeing that water is used to cleanse it, is not already covered by a film of water. All that can be said in reply is that no procedure which suggested itself to us altered the frictional properties of the face. The most drastic treatments were boiling in absolute alcohol and drying in dry air; or, after boiling in absolute alcohol, heating in an oven to 150° C. and cooling in dry air.

When there is only one pair of primary films between the solids the plane of slip may reasonably be assumed to be between them, that is, midway between the solid faces. When there are four primary films the matter is more dubious, since there are now five surfaces of discontinuity. The initial slip which defines the value of static friction is probably confined to the median surface or one of the other pairs of surfaces, because the yield point is not likely to be the same for surfaces which are differently placed; for example, the yield point of a surface between the two primary films of water, supposing water to form the inner pair, is not likely to be the same as that between water and the lubricant, or between the lubricant and the solid. The probability is still in favour of the median plane being the plane of slip, because the presence of an inner pair of primary films lowers friction. If this probability be granted then the figures prove that the attraction field of the solid is transmitted through *two* primary films to this plane.

Benzene, unlike water, lowers slightly the friction of substances even when it is miscible with them. This is not because benzene itself is a lubricant; indeed it is so nearly a neutral substance for glass as to lead one to attribute the very slight lubricating effect to an impurity, especially as very few samples of benzene, and those only after most tedious purification, will give the low lubricating value recorded below, namely:

Clean glass   ...   . $\mu = 0.92$
Lubricated with benzene   ...   ... $\mu = 0.89$

The admission of the vapour of this benzene gave the following results:

	Alone	Saturated benzene vapour admitted
Dimethyl decyl carbinol	0.41	0.39
Lactic acid	0.49	0.48
Cetyl alcohol	0.32	0.30
Tetracosane ($C_{24}H_{50}$)	0.325	0.316
Palmitic acid	Not measurable	Not measurable

SUMMARY

1. When the lubricant is composed of molecules whose structure is a chain of carbon atoms loaded at one end by a carboxyl or hydroxyl group, some time elapses after its application to the solid surfaces before the friction reaches a steady state. This interval is called the latent period.

2. The latent period seems to be the time required to attain that degree of orientation of the molecules in the primary film possible under the particular circumstances. When, therefore, the lubricant is a normal paraffin in which both ends of the chain of carbon atoms are alike, there is no latent period.

3. The latent period is shortened by rise of temperature, by mechanical agitation, and by dilution with a substance such as a normal paraffin which itself does not manifest a latent period.

4. The lubricant consisting of two pure substances mixed together:

(*a*) for aliphatic alcohol or acid mixed with a paraffin, the friction is identically that of the alcohol or acid until its percentage has fallen to about 0.7:

(*b*) for a mixture of an aliphatic alcohol with an acid, the difference in lubricating properties being considerable the friction rose slowly with increase in the percentage of the worse lubricant until a critical point was reached, at about 5 per cent., when the curve bent sharply upwards.

Lactic acid and water conformed to this type:

(*c*) when the difference in lubricating properties was slight, the curve was a straight line joining the values of the coefficient of friction for the pure substances.

5. When two substances were applied to the faces in succession then:

If the substances were immiscible the application of a second lubricant in the form of vapour always lowered friction.

If the substances were miscible, when water was the second to be applied, it always raised the friction; when the second lubricant applied was benzene it always slightly lowered friction.

Mechanical agitation such as is produced by moving the slider on the plate raised the friction when two immiscible lubricants had been applied in succession to the surfaces.

REFERENCES

Rayleigh, *Phil. Mag.* (February 1918).
Hardy, W. B. and J. K., *Phil. Mag.* (1919), xxxviii, 32.
Hardy, W. B. *Phil. Mag.* (1919), xxxviii, 49 and (1920), xl, 201.
Hardy and Doubleday, *Proc. Roy. Soc.* (1922), A, c, 549 and (1922), A, ci, 489.
Doubleday, *Trans. Chem. Soc.* (1922), cxxi, 2875.

44

BOUNDARY LUBRICATION. PLANE SURFACES AND THE LIMITATIONS OF AMONTONS' LAW
BAKERIAN LECTURE

With IDA BIRCUMSHAW

[*Proc. Roy. Soc.* (1925), A, cviii, 1]

INTRODUCTION

The earlier papers of this series deal with experiments in which the slider had a spherical face, the plate being plane. This paper is concerned with experiments in which the face of the slider was plane. The relations are strikingly different—for example, when the plate is flooded with excess lubricant and the slider is curved, after the slider has been placed on the plate the friction *falls* to its steady value. When the slider is plane, however, the friction *rises* during this latent period and the final steady values are not the same in the two cases.

The differences are unusually confusing, but they seem to be capable of explanation by an important extension of the theory of boundary lubrication to the effect that there are two distinct stages or states of lubricated surfaces which come under this designation, and that the passage from the one to the other is determined by the normal pressure—that is, the load divided by the area of the bearing surface.

In Fig. 1 the curves ABC give the steady value of the coefficient of friction μ plotted against the load, the slider being plane. It will be noticed that in the first part of the curve μ falls as the load increases, whilst in the second part BC it is sensibly independent of the load.

In the first place, therefore, Amontons' law, according to which the friction is equal to the load multiplied by a parameter, does not hold, whilst in the latter it does, either rigidly or to a very high degree of approximation.

When the slider has a spherical face the area of contact between it and the plate is enormously reduced and the pressure in consequence very high. The values of μ, therefore, then come wholly into the flat part of the curve, but a long way on in the scale of pressure.

This is shown in Fig. 2 (which is not to scale). The scale of pressure is supposed to be broken between C and D, and the part DE belongs wholly, so far as our experiments go, to sliders with a spherical face. The portion DE is parallel to the axis of pressure—that is to say, any deviations from

Amontons' law are so slight as to escape detection over a fiftyfold increase of the load. *DE*, however, is definitely, though very slightly, lower than *BC*.

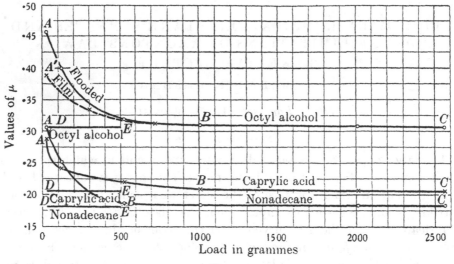

Fig. 1.

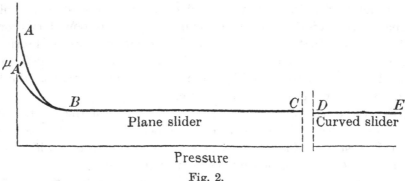

Fig. 2.

It seems certain that along *AB* the lubricant forms a layer between the faces which is several molecules in thickness, and is held in place by those ordinary capillary forces which cause fluids to run between two plates. The adjustment of the friction to the pressure is by variation in the thickness of this layer; as it thins, friction rises, but not so fast as the pressure, and therefore the coefficient μ ($=$ friction/pressure) falls. From *B* onwards this mechanism is almost wholly replaced by another; the thickness of the layer is no longer altered by a gross squeezing out of the lubricant, it becomes sensibly constant, being altered only by elastic compression. From *B* onwards, therefore, the adjustment of the friction to the pressure is due to the operation of the elastic forces between the atoms, and not to capillarity in the ordinary sense.

That there is a real difference in the mechanism involved appears from the following: In the region AB any addition to the load raises friction, but the change takes time, because squeezing out the lubricant is a slow process, and it may be hours before a new steady value is reached. There is, therefore, a latent period of rising friction following an increase of load, and of falling friction following a decrease of load. In the region $BCDE$ there is no latent period following a variation of load, provided the lubricant has been allowed previously to come to a steady state. There is, it is true, a latent period in this region under certain circumstances, but it is always a period of falling friction due to increase in degree of orientation of the molecules of the lubricant.

Other differences are the existence of a coefficient of temperature over AB, whilst the friction is independent of temperature over $BCDE$ within the limits explored. Finally, the curve connecting the coefficient of friction μ, with the molecular weight of the lubricant, is linear for loads in the region $BCDE$ and curved in the region AB.

When a spherical slider is used, the friction is the same when the slider stands in a pool of lubricant as it is when the solid surfaces are covered only by an invisible film, called by us a primary film, deposited from the saturated vapour of the lubricant. This is because the spherical slider cuts through to a limiting layer of great mechanical stability. Is this layer to be identified with a layer composed of two primary films, one on the face of the slider, the other on the plate? Apparently not, because the portion AB of the curve is not entirely eliminated even when lubrication is only by primary films. The curve for μ/pressure then starts at a lower level, but there is a region $A'B$ (Figs. 1 and 2) over which Amontons' law does not hold, and where the latent period and the relations to temperature are those characteristic of the region AB. We are thus forced to conclude that a lubricating layer composed only of two primary films is capable of being reduced in thickness by actual squeezing out of material. A primary film, in equilibrium with saturated vapour, therefore, is not monomolecular.

A primary film is of insensible thickness. That a layer of lubricant composed only of two such films should thin under pressure by some portion being actually squeezed out from between the faces is not easy to believe, yet the facts lend themselves most simply to that hypothesis. The alternative hypothesis that the molecular structure of the layer of lubricant is a function of pressure over the range from A to B, and becomes independent of pressure from B onwards, is not in our opinion so well supported by experiment.

The reason for the slight difference in level between BC and DE is now apparent. No load that we were able to place upon a plane slider was able to force it down to the limiting layer reached at once by a spherical slider.

H   44

It may be noted in passing that even if there be a limiting layer or limiting thickness beyond which simple pressure will not thin the layer of lubricant, and even if a spherical slider penetrates to this layer, the friction may not be due wholly to it. In the case of a spherical slider, contact over this layer will be limited to a small circle, about which is a perimeter in which there is a thicker layer, certainly more than two molecules thick, but possibly capable of contributing to the total friction.

The most remarkable feature of boundary lubrication remains to be noted. The coefficient of friction is a measure of the efficiency of a lubricant with respect to one variable, the pressure between the faces. The inevitable result of our experiments is that this efficiency increases as the layer of lubricant thins, until it becomes constant when some limiting layer is reached. The following suggestion may be put forward to explain this unexpected fact.

The layer several molecules in depth, through which a plane slider falls in the region AB, cannot throughout any part have the same random disposition of the molecules which obtains in the interior of the fluid. It has indeed lost its fluidity in so far that it can bear a finite stress without slip. The change of state can be due only to the spreading of the orientation of the molecules from the solid faces. In general, the molecules may be supposed to be more firmly held and the orientation to be most complete at the interface, and to diminish along the normal, until, in the interior of the fluid, it is destroyed by the heat movements. These studies in friction show that the efficiency of a given lubricant increases as the degree of orientation increases, and, therefore, this gradient of orientation may explain the curious relationship between efficiency and the depth of the layer.

Another important relation follows, namely, that over the region AB, where orientation is likely to be greatly affected by the heat motions of the molecules, friction may be expected to be a function of the temperature, whereas in the region $BCDE$, where interatomic forces alone are concerned, the effect of temperature is likely to be small. This expectation is realised. Over the range of temperatures available, namely from $10°$ to $120°$, there was a temperature coefficient for loads lying in the region AB, and no detectable temperature coefficient in the region $BCDE$.

To prevent confusion, it may be well to note here that the orientation spoken of in the last paragraph occurs with normal paraffins as well as with their related acids and alcohols, in which the carbon chains are loaded at one end.

METHODS

The friction measured was always static friction, and measurements were made in dry air, rigorously clean save for the vapour of the lubricant. The slider used was a triangular piece of steel with three hardened steel studs, of circular cross-section, passing through the corners. The ends of the studs were ground and polished, and each had a diameter of 3 mm. The weight of the slider was 20·4 gr. It was drawn along by means of a thread attached to a hook, which was fixed at one of the angles of the triangular piece of steel.

The obvious possible mechanical sources of error were explored. The friction was found to be independent of any shift in the point of attachment of the thread in the vertical plane. A shift in the direction of pull in the horizontal plane produced a fall in the observed friction. Errors due to setting the slider in the horizontal plane would, however, lead to random variations in the values—the consistency of the measurements shows that the error was negligible.

It was thought that the friction of a plane slider might be increased by what may be called edge effects, that is by some increase of potential at, or purely mechanical action of, the sharp edges of the sliding faces. The edges therefore were slightly rounded by careful grinding, but without sensible effect upon the friction.

Error due to absorption of work in varying the area of the surfaces of a pool of lubricant was eliminated by using a large pool, in which all three feet of the slider stood. The difference in the values of μ for large and small pools was considerable only when the molecular weight of the lubricant was high, as the following figures show:

	Small pool	Large pool
Stearic acid	$\mu = 0 \cdot 0132$	$\mu = 0 \cdot 0052$
Palmitic acid	$0 \cdot 0124$	$0 \cdot 0075$
Dodecoic acid	$0 \cdot 0276$	$0 \cdot 0183$
Docoic acid	$0 \cdot 1295$	$0 \cdot 1293$
Cetyl alcohol	$0 \cdot 2632$	$0 \cdot 2391$
Tetracosane	$0 \cdot 2010$	$0 \cdot 1652$
Nonadecane	$0 \cdot 3122$	$0 \cdot 3050$

For acids, alcohols, or paraffins of lower molecular weight the differences were negligible.

Gross capillary effects of this kind were of course absent when the surfaces of slider and plate were covered only by a film of insensible thickness.

A. FLUID LUBRICANTS

Section 1. *Influence of the Load*

In this section only steady values of friction are considered. The variations during the latent period are described in the next section.

Plane Slider

Surface flooded with lubricant. The values of μ for varying pressures (load ÷ area of contact) are given in the following tables, and are plotted against the loads in Fig. 1 (curves *ABC*). The values for a spherical slider are shown also over the range of loads tried (curves *DE*). It will be noticed that at about the point *B* the curves for the plane slider become sensibly horizontal, and that the curves *DE* for the spherical slider are always very slightly below the curves *BC* for the same lubricant. (See Table I.)

TABLE I. *Plane slider—steel*

Pressure in gr. per cm.^{2}	Nonadecane	Caprylic acid	Octyl alcohol
96	$\mu = 0.305$	$\mu = 0.288$	$\mu = 0.457$
568	0.252	0.241	0.400
2,455	0.187	0.220	0.318
4,755	0.184	0.209	0.310
9,482	0.184	0.207	0.308
12,092	0.184	0.206	0.307
	Spherical slider—steel. Radius of curvature—3.58 cm.		
120,000* 2,000,000	$\mu = 0.181$	$\mu = 0.204$	$\mu = 0.306$

* Assuming the area of contact to be a circle of diameter 0.01 cm.

Surfaces lubricated by a primary film of insensible thickness deposited from saturated vapour. From the conditions of the experiments the layer of lubricant between the faces must be composed of two primary films, one belonging to each face, unless some of the vapour is drawn in and condensed between the bearing faces.

That vapour can condense to form a layer between the faces is proved by the experiment described in section 2; the process, however, does not produce a layer exceeding two primary films in thickness, because, if it did, there would be evidence of the fact in the existence of a latent period of falling friction after the slider had been placed on the plate, when both had previously been covered with a primary film. When lag due to orientation is excluded by using a normal paraffin as lubricant, there is no such latent period (cf. p. 698).

A curve for lubrication by primary film, the slider being plane, is shown in Fig. 1, *A'BC*, and the values of the coefficient are given in Table II.

Over the region *A'B* the relations to temperature, and the latent period, are the same as they are over *AB*; we are, therefore, forced to conclude that a layer of lubricant composed only of two primary films can be reduced in thickness by squeezing out lubricant. It follows that the primary film in equilibrium with saturated vapour cannot be monomolecular.

The fact that the curves for primary films are lower than those for flooded surfaces when loads are light, proves that boundary conditions extend more deeply into the body of the lubricant than the depth of a primary film.

TABLE II. *Surfaces lubricated only by a primary film.*
Slider plane: steel: lubricant, octyl alcohol

Load gr.	μ
20·4	0·389
300·4	0·335
720·4	0·317

Latent period

Initial load, 20·4 gr. Times in min.	700 gr. added μ
1	0·284
10	0·297
20	0·303
30	0·312
60	0·312
90	0·312
120	0·312

Spherical Slider

μ is now critically independent of the quantity of lubricant on the plate, as the following figures show:

TABLE III. *Mean values of μ for four alcohols, three acids, and four paraffins on glass (see Table III,* Proc. Roy. Soc. *A,* c, 555)

Slider spherical

	Flooded	Primary film formed by	
		Spreading	Saturated vapour
Alcohols	0·618	0·621	0·613
Acids	0·521	0·520	0·521
Paraffins	0·683	—	0·684

The spherical slider must, therefore, owing to its shape and the consequent great pressure between the bearing surfaces, cut through the lubricant until a limiting layer is reached of great mechanical stability. The lubricating properties of this layer are defined by the equation

$$\mu = f/p$$

where f is the friction per unit area of bearing surface, and p the pressure or load ÷ area. In other words, for this limiting layer and for it only, Amontons' law holds rigidly, as the following figures show (Table IV).

The value of μ is, however, slightly lower than the lowest values which were obtained with a plane slider loaded as heavily as was possible (Table I).

TABLE IV. *Steel: slider, spherical*

Load gr.	Caprylic acid	Octyl alcohol	Nonadecane
34·5	$\mu = 0.2047$	$\mu^2 = 0.3062$	$\mu = 0.1813$
84·5	0·2047	0·3060	0·1798
145·5	0·2039	0·3049	0·1817
184·5	0·2046	0·3058	0·1801
534·5	0·2041	0·3067	0·1808
	Mean 0·2044	0·3059	0·1807

Section 2. *The Latent Period*

Some time necessarily elapses after the slider has been placed in position before a reading of friction can be taken—the phrase "no latent period" therefore needs qualifying. It means that the friction measured as soon as possible—namely, in 25 sec.—after the slider touches the plate is unchanged by further lapse of time.

Lag is due to two entirely distinct things—the delay in squeezing out or drawing in lubricant beneath the plate, and the time taken to reach the equilibrium degree of orientation. The latent period due to the first of these will obviously be a function of variations in the load.

Consider a light plane slider in a pool of lubricant so large that effects due to variations in the shape of the pool may be neglected. The slider will fall until its weight is supported by the increase of pressure in the layer of fluid next each solid face, due to the attraction of the solid for the fluid. This pressure might well be called Leslie's pressure; it was he who first pointed it out as the cause of the rise of fluid in a capillary tube, and Clerk-Maxwell accepted the explanation as adequate.[1]

The slider, in the steady state, is floating in the lubricant, so that if the load be reduced it rises, and if the load be increased it falls. Lubricant is drawn in or forced out, as the case may be, from between slider and plate, and there will be a latent period whilst this adjustment of the thickness of the layer is taking place.

The following is an example:

Steel. Lubricant, octyl alcohol. Load, 420·4 gr. $\mu = 0.33$.
Load increased by 50 gr. μ falls to 0·32 in 10 min.
Load decreased by 50 gr. μ returns to 0·33 in 5 min.

It will be noticed that though the variation in the load, and therefore in the pressure between the faces, is the same in the two cases, the duration of the latent period is not. This difference is due perhaps to a difference in the viscosity of the fluid, that squeezed out from between the faces being more

[1] Leslie, *Phil. Mag.* (1802), XIV; Maxwell, *Encyc. Brit.* 9th edit., Art. "Capillary Action".

viscous, because more oriented, than that drawn in from the general body of fluid.

The pressure required to float the slider may be looked at from another point of view, which, however, is not fundamentally different from that of Leslie, namely, from Gibbs's conception of an interfacial phase. These studies of friction, and the analysis of the structure of films of grease spread upon solids which has been made by Bragg and his colleagues, show that such a phase has a real objective existence. It is formed usually with evolution of heat, as Leslie first pointed out; it is the seat of prodigious electric stresses, which manifest themselves as the so-called contact potential differences; and it possesses a molecular structure peculiar to itself. In boundary friction we are concerned mainly with the properties of one-half of such a phase, and these experiments show that this half is several molecules deep.

This half will have a certain depth, determined, in the absence of any external constraint, by the field of force at the actual interface, the temperature, and the atomic configuration of the molecules. Following Gibbs, let a surface be drawn enclosing this depth, then the space between this imaginary surface and the actual interface will have certain properties. Within it the "state" of the liquid in mass will be unstable. There is a pressure urging the liquid molecules into the interfacial pattern, just as there is a pressure in a thin layer of supercooled fluid covering a crystal which urges the molecules into the crystalline pattern. Both the degree of instability and the pressure are not uniform throughout the space, but vary along the normal to the interface. If the space be incompletely filled with fluid, as it is when the solid face is only partially covered, there will be a pressure tending to fill it.

It is this pressure which detaches solid faces which have seized together by inserting a film of lubricant between them, but this property is, as perhaps might be expected, limited to substances with relatively small molecules (see later, p. 700).

In Table V are given examples of the change of friction, recorded as the pull in grammes, which occur after the slider is put into the pool of lubricant, on the plate. The load was 20·4 gr., and pressure 96 gr. per cm.², corresponding to the point A in the curves.

Prof. G. I. Taylor has worked out the equations for the fall of bodies through a viscous fluid on to a plane surface. We quote them here with his permission.

(1) Case of a spherical surface of radius of curvature R, mass m gr., falling through a fluid of viscosity η. The time T is given by

$$T = \frac{6\pi\eta R^2}{mg} \log_e \frac{h_1}{h_2},$$

where h_1 is the initial height of the lowest point of the slider above the plane and h_2 is the final height.

(2) Case of a flat disc of radius a

$$T = \tfrac{3}{4} \frac{\pi \eta a^4}{mg} \left(\frac{1}{h_2{}^2} - \frac{1}{h_1{}^2} \right).$$

TABLE V[1]

	Pull in gr.	μ
Bayonne Oil:		
1 sec.	0·0	0
5 ,,	0·400	0·02 (approx.)
1 min.	2·86	0·11
5 ,,	4·08	0·20
10 ,,	5·22	0·256
15 ,,	5·75	0·282
30 ,,	5·94	0·291
45 ,,	6·04	0·296
60 ,,	6·20	0·304
90 ,,	6·47	0·317
120 ,,	6·59	0·323
150 ,,	6·69	0·328
180 ,,	6·75	0·331
210 ,,	6·83	0·335
240 ,,	6·83	0·335
270 ,,	6·85	0·336
8 hours	6·90	0·338
Octyl Alcohol:		
1 sec.	0	0
3 ,,	2·65	0·13 (approx.)
10 ,,	4·30	0·21 ,,
1 min.	7·14	0·35 ,,
5 ,,	7·75	0·38
10 ,,	8·51	0·417
15 ,,	8·96	0·439
25 ,,	9·22	0·452
30 ,,	9·34	0·458
45 ,,	9·44	0·463
60 ,,	9·57	0·469
90 ,,	9·73	0·477
120 ,,	9·79	0·480
150 ,,	9·83	0·482
180 ,,	9·89	0·484
240 ,,	9·91	0·485
6 hours	9·91	0·485
Undecane:		
1 sec.	3·06	0·15 (approx.)
5 ,,	6·53	0·32 ,,
1 min.	8·77	0·43 ,,
5 ,,	9·85	0·483
10 ,,	10·57	0·517
15 ,,	10·83	0·531
30 ,,	10·98	0·538
45 ,,	11·14	0·546
60 ,,	11·14	0·546
90 ,,	11·20	0·549
120 ,,	11·20	0·549
180 ,,	11·20	0·549

[1] The steady values differ from those given in Tables I and II because different steel was used.

Concerning these equations Prof. Taylor writes as follows:

"It will be observed that in case 1, though the time for $h_2 = 0$ is infinite, it is a very low order of infinity; therefore, though the formula gives infinite values for actual contact, it gives small values for all distances right down to molecular dimensions.

"A numerical example will illustrate this. Putting $R = 5$ cm., $\mu = 0.011$, $m = 50$ gr., $h_1 = 1$ mm., $h_2 = 10^{-7}$ cm., then $T = 1.5 \times 10^{-3}$ sec.

"The formula for case 2, on the other hand, also gives infinite values, but of a totally different order of infinity. I should expect a flat surface to take a long time to get down to molecular distances from another flat surface. It is clear that slight bending of the surface will make a very great difference.

"Taking the case when the slider has a load of 100 gr. on each of its three legs, and assuming that the diameter of each leg is 3 mm., the formula shows that when water is used as a lubricant its thickness decreases to 10^{-7} cm. in 3.3 hours and to 10^{-6} cm. in 2 min."

During the latent period, when the flat disc is falling through the lubricant, the friction is rising. This is what might be expected, the capacity for bearing tangential stress rising as the layer thins.

Over the region AB the layer of lubricant between the bearing faces must be more than two molecules in thickness, and since a finite traction can be applied without producing detectable slip, the fluid lubricant must have been solidified by the attraction of the solids. The state of the lubricant is not, however, capable of easy definition, since it is solid in the sense mentioned above, and fluid in that when the load is varied it can flow in or out of the space between the solid faces.

If the above argument is valid, a primary film in equilibrium with saturated vapour must also be more than one molecule deep, because, when the solid faces are lubricated only by a primary film, the changes following a variation in the load are the same in kind as they are when the plate is flooded with fluid. The figures are given in Table II.

The latent period is a remarkable function of the load.

Consider a plane slider in a pool of lubricant and let the load be increased by a fixed amount at different points along AB, that is, with different initial loads.

TABLE VI. *Steel: slider plane, lubricated with large excess of liquid nonadecane, latent period measured for increase in weight of slider*

Initial load gr.	Original (steady value)	Weight added gr.	After adding weight (steady value)	Latent period min.
120.4	$\mu = 0.2523$	50	$\mu = 0.2441$	12
420.4	0.2105	50	0.2055	5
821.2	0.1871	50	0.1857	± 1

The duration of the latent period decreases as the initial load increases so that, in this restricted sense, the latent period may be said to decrease in length along AB and finally to vanish at B.

Let two different loads representing two points on AB be indicated respectively by $'$ and $''$, the single accent representing the lower load, and let D be the distance between the solid faces, η the viscosity of the lubricant between the faces, and p the excess of pressure in the layer of lubricant. Then for the same increase to the load we have for the latent period:

$$\frac{\delta T'}{\delta T''} = \frac{\eta' . D'' . \delta p'' . \delta D'}{\eta'' . D' . \delta p' . \delta D''}.$$

We may write $\qquad \eta' < \eta'', \quad D'' < D', \quad \delta p'' = \delta p'.$

Therefore, the relation found experimentally shows that the decrease in the depth of the lubricating layer for a given increase in load must diminish rapidly as the total load increases.

In the region $BCDE$ (Figs. 1 and 2) there is no latent period following a variation in the load. There is a latent period of the type just described when a plane slider is first placed on the plate, because the region AB or $A'B$ has first to be traversed. A spherical slider, however, cuts through the lubricant so rapidly that this latent period is not detectable, save in one case, to be described below.

Spherical slider. The changes in the latent period were fully discussed in an earlier paper.[1] They may be summarised as follows:

(1) Lubricants whose molecules are loaded at one end, such as the aliphatic alcohols and acids, have a latent period of *falling* friction during which orientation of the molecules at the solid face is increasing until it reaches a maximum.

(2) When, as in the normal paraffins, the molecules are symmetrical end for end there is no latent period.

(3) In one case only does the friction *rise* during the latent period, namely, when the faces of both slider and plate are in equilibrium with fluid lubricant *before* being placed in contact. Even a curved slider then takes a sensible time to fall through a layer of lubricant rendered highly viscous by the spreading of orientation. There is then a rapid initial rise in friction whilst the slider is falling through the lubricant, followed by a much slower fall due to increasing orientation.

The argument, which has been stated in the earlier paper, may be briefly recapitulated:

In the acids and alcohols the terminal groups of the molecule are different; at the one end is the —CH_3 group, at the other the —$COOH$ or —CH_2OH group. Let us call the latter ends the loaded ends of the elongated

[1] *Proc. Roy. Soc.* (1923), A, CIV, 25.

molecule and assume that the attraction of the solid face for the loaded ends is greater than that for the unloaded ends. When an acid or an alcohol is applied either as fluid or vapour to the solid face, some molecules will stick by the loaded, others by the unloaded ends, and the percentage of the former will be increased by evaporation and condensation into and from the fluid or vapour until a steady state is reached. The steady state is not likely to be one with all the loaded ends attached to the solid, for evaporation and condensation continue, and some of the molecules which enter the layer will do so unloaded end foremost.

When the lubricant is a normal paraffin there is at both ends of the chain of carbon atoms the —CH_3 group and neither end is loaded. No time will then be occupied in changing the molecules end for end. The assumption in this argument is so obvious that it need not be stated.

Plane slider. It is not now so easy to detect the lag due to orientation, because the plane slider takes so long a time to fall through the lubricant. One has, therefore, ordinarily only the latent period of the type already described, which is shown by paraffins as well as by acids and alcohols.

This latent period can, however, be cut out completely, so as to leave only effects due to orientation, by lubricating with an insensible primary film deposited on the solid surface from saturated vapour, and allowing it to get into a steady state before the slider is placed on the plate. The orientation of the molecules of the films raises the viscosity to the point at which the weight of the slider is unable to squeeze any of the lubricant out. Friction now falls after the slider is put on the plate, and the fact that this fall occurs only with acids and alcohols, and not with paraffins, proves that it is due to end for end readjustment of the molecules. The case is similar to that mentioned earlier on p. 698, and discussed on p. 27, vol. CIV, of the *Proceedings* of the Society, with the significant difference that the high pressure under a spherical slider is able to do what the low pressure under a plane slider is unable to do, namely, cut through or squeeze aside some of the lubricant despite the raised viscosity due to orientation.

Over the region *BCDE* it was found for both types of slider that, once the molecules of the lubricant had been allowed to reach the greatest attainable degree of orientation, variations of the load were not followed by a latent period, provided the variations were not large enough to bring the load into the region *AB* or *A'B*. The adjustment of the friction to the pressure took place too quickly for the process to be detected. It must be remembered that orientation reaches its maximum only after the slider has been placed on the plate, and the lubricant between the faces thereby shielded from evaporation into, or condensation from excess fluid or vapour (cf. *Proc. Roy. Soc.* A, CIV, 26).

The region *BCDE* is therefore characterised by two things—Amontons' law holds and the adjustment of the friction to variations of pressure is sensibly instantaneous if the lubricant is in the condition of maximal orientation in the case of acids and alcohols, and always in the case of paraffins. And these characters can be due only to the fact that friction is adjusted to the pressure solely by the operation of the elastic forces between the atoms.

Temperature coefficient. There is yet another significant difference between the two states of boundary lubrication. Over the region *BCDE* friction is independent of temperature over the range at our command, namely, from 15 to 120° C. (cf. *Proc. Roy. Soc.* (1922), A, CI, 487). In the region *AB*, however, friction falls as the temperature rises.

TABLE VII. *Slider plane: steel. Load:* 20·4 gr.

Lubricant		
	15·5° C.	100° C.
Caprylic acid	0·288	0·267
Undecyl alcohol	0·361	0·348
Undecane	0·571	0·509
	63° C.	110° C.
Palmitic acid	0·0075	0·0034

This is what might be expected, since the friction in this region of small pressures is always a function of the degree of orientation, which, again, is a function of temperature.

Seizing of clean faces. Clean faces seize so that, when the slider is moved tangentially, both faces are torn. When the slider is moved along the normal—that is, lifted from the plate—there is no sensible resistance other than that due to gravity if the slider is spherical, and considerable resistance if the slider is plane. The first case, that of the spherical slider, was discussed in an earlier paper.[1]

Seized faces can be set free by applying lubricant, but, when the slider is plane, only those with a sensible vapour pressure are effective. This may mean either that the vapour phase alone can then penetrate between the faces, or that it is a question of the size of the molecules irrespective of the phase.

Examples

Curved slider. Faces clean. Weight in scale pan needed to move 22 gr.

18 gr. were placed in the scale pan, which was left hanging freely. A drop of lubricant was placed about 3 cm. distant from the contact between slider and plate.

Methyl alcohol. Slider shoots forward in about 2 min.

Octyl alcohol. Slider shoots forward in about 30 min.

[1] *Phil. Mag.* (1919), XXXVIII, 32.

But if a pool of a lubricant with no sensible vapour pressure, such as oleic acid or medicinal "paraffin", is made to touch the contact, the slider also shoots forward.

Plane slider. Faces clean. Weight in scale pan needed to move about 140 gr.; 80 gr. placed in scale pan, which was left hanging freely.

If the saturated vapour of a lubricant be admitted to the chamber, or if a drop of a lubricant be placed anywhere on the plate, the slider shoots forward after a time which is longer the lower the vapour pressure of the lubricant.

Lubricants with no sensible vapour pressure, however, will not set a plane slider free, even though a pool be made to touch the edges of the contacts. The following examples of such lubricants were tried; oleic acid, linoleic acid, medicinal "paraffin", tetracosane and cetyl alcohol, the last two at temperatures just above their melting-points.

Since vapour can condense between plane faces, the question whether the fall of friction observed when such faces, *previously in equilibrium with saturated vapour*, are brought together, is due to further condensation of vapour between them, or to variation in the degree of orientation, is a real one.

The following experiment proves that it is due to increased orientation. A plane slider of hard steel, and weight 20·4 gr., and a plate of the same steel were left in the saturated vapour of the lubricant for some hours before the former was placed on the latter.

Lubricant	First reading	Steady value
Octyl alcohol	$\mu = 0.413$	$\mu = 0.389$
Caprylic acid	$\mu = 0.240$	$\mu = 0.226$

In both cases the steady value was reached in about 15 min.

Lubricant	First reading	Steady value
Undecane	$\mu = 0.409$	$\mu = 0.409$
Octane	$\mu = 0.496$	$\mu = 0.498$

The lubricating layer was, by the conditions, composed initially of two primary films. If the layer had been merely deepened by condensation of vapour between slides and plate, there would have been a latent period while the process was going on for the paraffins as well as for the acids and alcohols. The increase in the degree of orientation of alcohols and acids, beyond what obtains when they are in equilibrium with vapour or fluid phases, was noted earlier in experiments with spherical sliders[1] and attributed to the elimination of the disturbing influence of those phases.

[1] *Loc. cit.*

The fact that vapour alone can penetrate between plane faces when seized probably explains why a fitter, wishing to start a seized joint, uses, not a good lubricating oil, but ordinary lamp paraffin, which contains bodies with a sensible vapour pressure. And the fact that lubricants with no sensible vapour pressure can set free seized faces when one of them is spherical, may not be due to the ability of the fluid phase to insert itself between the faces, but to slight rocking of the slider owing to earth tremors.

Section 3

The experimental evidence for the view that the adjustment of the friction to the load is effected along AB by the Leslie pressure, and along $BCDE$ by elastic forces between the atoms, has now been given. It remains to discuss why the coefficient of friction should be slightly higher along BC than along DE.

The value of the friction is determined by the physical state as well as by the chemical nature of the solid faces, the same plate or slider, for example, will sometimes give slightly different values after being refaced; therefore, though for purposes of comparison plane slider and curved slider are made of the same steel, the physical state of these surfaces produced by the polishing may be slightly different. This would be ample to account for the magnitude of the difference, but it would not explain the fact that it is always the plane slider which gives the lower value; and there is the further fact that the value for the plane and curved sliders used in these comparisons actually did coincide when the lubricating film was reduced to its physical limit by polishing (cf. p. 706). Other possibilities, therefore, need to be explored.

There can be no doubt but that the great pressure under a curved slider cuts through the lubricant down to a layer of great mechanical stability. What is this limiting layer?

Let us consider first what a primary film is. When a fluid composed of a single chemical substance is poured over a clean solid face, or, if it be immiscible, over a clean water face, to form a layer not more than a few millimetres deep the layer ruptures and the excess gathers into flat-topped lenses, leaving a film of insensible thickness over the rest of the surface. If the space above be enclosed, both the lenses and the primary film get into equilibrium with saturated vapour; the complete primary film, therefore, is enclosed between an imaginary surface drawn in the fluid parallel to the solid or water faces, where the attraction of the fluid for itself becomes greater than its attraction for the solid or the water as the case may be, and where the vapour pressure becomes equal to that of the saturated vapour of the fluid. This plane is a plane of natural cleavage in the fluid.

When the slider has a spherical face the friction is the same, when it stands in a pool of lubricant, as it is when the surfaces are covered only by a primary film deposited from saturated vapour. We may, therefore, suppose that the effective lubricating layer is the same in the two cases. The lubricating layer is, therefore, as was pointed out in an earlier paper, defined to a certain extent by this plane of natural cleavage. The primary film is, however, not the limiting layer spoken of above, because even a plane slider, by increasing the load, can be forced through it to a certain extent, but apparently not so far as is reached by a spherical slider.

The friction observed with a spherical slider must to a certain extent be of mixed origin. We may conclude from what has gone before that the capacity for resisting traction without slipping exists throughout a layer of lubricant of unknown depth. Let the total depth be h, measured along the normal to the plate, then the friction as measured by a spherical slider is the reaction to the traction integrated over a circle whose circumference lies where the normal h cuts the surface of the slider. Within this circle a smaller circle may be supposed to lie concentrically, within which the limiting layer, if such indeed be reached, alone persists.

The total observed friction may, therefore, be regarded as composed of two parts—that due to the limiting layer of the inner circle, and that due to the perimeter between the latter and the outer circle. The contribution of the perimeter may, however, be very small, since the friction at each point is, at any rate approximately, equal to the pressure at that point multiplied by a parameter, and the pressure may be almost vanishingly small over the perimeter. It is tempting to suppose that the small difference in the value of the plane and curved sliders is due to the former failing to reach the limiting layer at all. This much, at any rate, is certain from experiments (see later), that polishing off the lubricant gives the only surface in which the coefficient of friction is the same for both plane and curved sliders, and polishing may be supposed to obliterate entirely the perimeter mentioned above.

There is another possibility which deserves mention. According to Ramsay and Shields,[1] the angle of contact between liquid and solid (glass) is zero when only the vapour phase of the liquid is present. If air is present the angle has a positive value. The primary film therefore would be slightly thicker when air was excluded. A plane slider will exclude air more effectually than a spherical slider, and this may be the cause of the lower value of μ for primary films.

B. Solid Lubricants

The problems presented by solid lubricants are two—the first is to account for the fact that, when a spherical slider is used, there is no break in the curve for the coefficient of friction and molecular weight at the point where the change from fluid to solid occurs. The curves for a spherical

[1] *Phil. Trans.* (1893), A, clxxxiv, 647.

slider are straight lines and the values (Fig. 3; see also *Proc. Roy. Soc.* c) for solid members of a series take their proper place on these lines.

Solid lubricants in our experiments were deposited on the plate and slider from weak solutions in ether; therefore, the film left, though thin, was of no determinate thickness. Why then should the value of μ be "normal"? The answer lies in the fact that the value is not normal, but too low when the spherical slider is first applied to the plate, and becomes normal only after the slider has been moved quietly backwards and forwards on the plate, so as mechanically to disturb the layer of lubricant.

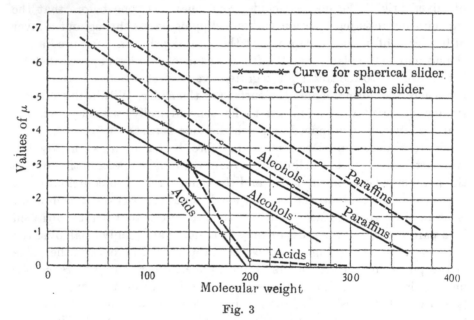

Fig. 3

During the movement the friction rises at first, but soon settles to a steady value which further movement of the slider does not disturb. The period of rising friction is not a true latent period, for time alone will not alter the value; movement is necessary.

There can be only one explanation of these facts, namely, that a natural plane of cleavage exists in a solid lubricant, just as in a fluid lubricant. Whether this plane is a surface of discontinuity between those molecules, oriented by the attraction field of the solid, and the crystal lattice of the mass, or whether it is defined merely by a critical capacity for withstanding traction, must be left uncertain.

That a real limiting layer is reached with a solid lubricant when the slider is curved is shown by the fact that, once it has been reached, the state is independent of the load, so that if the load be increased, further movement of the slider fails to alter the value of μ.

TABLE VIII. *Steel, lubricant solid nonadecane, deposited from solution in ether. In the last column is the time occupied in quietly moving the slider on the plate and which elapsed before the friction became steady*

Load in gr.	Traction in gr.	μ	Min.
	Slider spherical		
34·5	6·25	0·181	10
84·5	15·20	0·180	0
134·5	24·44	0·182	0
184·5	33·23	0·180	0
534·5	96·63	0·181	0
Slider plane, of weight 20·4 gr. $\mu = 0.230$. Load increased by 500 gr.			
—	79·22	0·152	1
—	83·32	0·160	2
—	92·31	0·177	5
—	96·04	0·184	10
—	96·32	0·185	20
	Slider plane		
20·4	4·70	0·230	20
70·4	15·39	0·219	5
120·4	25·20	0·209	5
520·4	96·32	0·185	5
1008·4	186·38	0·185	5
2563·6	473·10	0·185	5

A plane slider on the other hand fails to reach this limiting layer. Movement raises the friction, it is true, to a steady value, but, if the load be increased, more movement causes μ again to fall. In other words, whereas a plane slider will fall through a fluid lubricant until the load is balanced by the capillary pressure, it will not simply fall through a solid lubricant, but if moved about will thin the layer of lubricant, until a steady value is reached, incapable of further change by the quiet movement mentioned above. The layer now left between the faces is, however, not a limiting layer, because, if the load be again increased, movement again causes the friction to rise, until a new steady state is reached in which μ has a value less than it had with the smaller load. The behaviour of a plane slider is thus fundamentally the same with both solid and fluid lubricants, for in both cases, as the load is increased, it may be made to descend through the lubricant, and as it does so the value of μ decreases.

The second problem presented by solid lubricants is to account for the abnormally low value of friction when the lubricated surfaces are polished with fine linen which has been extracted by pure benzene until it is incapable of depositing anything which alters the friction of "clean" faces (*Proc. Roy. Soc.* A, c, 561). Fluid lubricants can be polished off a surface completely, solid lubricants cannot, but an invisible film of abnormally high lubricating power is left on the solid face (see the case of undecyl alcohol, *loc. cit.*).

H

45

Polished faces have three unique properties: (1) the friction is the lowest which can be got from the particular lubricant; (2) they are the only faces whose friction is the same for plane as it is for curved sliders; and (3) there is no latent period of any kind associated with them.

Polishing, therefore, produces a lubricating film of great mechanical stability, which finally rids us of all effects due to capillary pressure, to variations in the depth of the lubricating layer, or to varying degrees of orientation. It is impossible to decide whether the properties of polished faces are due to thinning of the layer of lubricant or to change of structure, but the evidence leans perhaps towards the former alternative, because the values for polished faces are the meeting points of the curves of load/μ for both types of sliders, and similar abnormally low values of friction are obtained when a fluid lubricant is in process of being polished off the surfaces.

One is tempted to suppose that polishing, and polishing alone, can reduce a layer of lubricant to the final limiting layer spoken of earlier, that this layer is monomolecular, so that the lubricating layer between slider and plate is bimolecular, and that Amontons' law $\mu = kp$ holds rigidly only for this limiting layer.

The distinctive differences between solid and fluid lubricants of the same chemical series then really reduce to two:

(1) If the load be decreased, there is a latent period of falling friction, if the lubricant is a fluid, due, according to the theory developed in this paper, to fluid being drawn in between the faces. If the lubricant be solid there is no such latent period.

(2) Fluid lubricant can be completely polished off a solid face, solid lubricant cannot.

[*Note added, April* 3, 1925. In looking through my own note-books I came across the following:

When a film has been polished on to both slider and plate and the former is placed on the latter, the lubricating layer is composed of two polished films. When only one surface, either that of slider or plate, is covered with a film, there will be only half the thickness of lubricant between the faces. I found on two separate occasions the friction of the single just one-half that of the double film.

This curious result seemed at the time beyond credibility, but, as my notes were positive, I asked my colleague to repeat the experiments. She found friction much less with the half film, but failed to get the value exactly halved. It must therefore be accepted that, with these long-chain lubricants at any rate, friction diminishes literally to a vanishing thickness of the layer. W. B. H.]

C. The Coefficient of Friction as a Function
of the Molecular Weight

In Table IX the steady values of the coefficient of friction are given for a plane slider of weight 20·4 gr.; and in Fig. 3 they have been plotted against molecular weight. The curves for a spherical slider of the same steel are given for comparison.

It will be noticed that the coefficient of friction is consistently higher for the plane slider, and though the curves converge as the molecular weight rises, they do not meet, and in the case in which it was followed far enough, the curve for a plane slider does not cut the base line.

It has already been pointed out that with a plane slider μ decreases as the load increases for a given lubricant; each curve for a plane slider is therefore merely one of a family of curves which flatten as the pressure between the bearing surfaces is increased, and the limiting curve is the straight line given by a spherical slider.

TABLE IX. *Material, hard steel. Plate flooded with the lubricant. Each solid lubricant was tested at a temperature above its melting-point*

Lubricant	Steady value of μ
Pentane	0·6891
Hexane	0·6509
Heptane	0·6265
Octane	0·5983
Undecane	0·5175
Nonadecane (in liquid state)	0·3050
Tetracosane (in liquid state)	0·1652
Ethyl alcohol	0·6491
Propyl alcohol	0·6160
Butyl alcohol	0·5832
Octyl alcohol	0·4575
Undecyl alcohol	0·3637
Cetyl alcohol (in liquid state)	0·2391
Hexoic acid	0·4456
Heptoic acid	0·3672
Caprylic acid	0·2884
Decoic acid (in liquid state)	0·1293
Dodecoic acid (in liquid state)	0·0183
Palmitic acid (in liquid state)	0·0075
Stearic acid (in liquid state)	0·0052

D. The Fundamental Equation

The equation for the limiting curve, that is, the curve which belongs wholly to the region $BCDE$ of the μ/pressure curve, is given by a spherical slider and has the form

$$\mu = b - aM,$$

where b is dependent both upon the nature of the solid and the chemical series to which the lubricant belongs and a is dependent only on the chemical series. When the slider and plate are of different solids, (1) and (2), experiments show that this equation becomes[1]

$$\mu = \tfrac{1}{2}\,(b_1 + b_2) - aM.$$

Since a depends only on the chemical series of the lubricant and the members of the series differ from each other only in the number of carbon atoms in the chain, aM may be written $c\,(N-2)$, where N is the number of carbon atoms.

The parameter, b, depends upon the nature of the solids and the chemical series. Let us write $b = b_0 - d$, when b_0 depends only on the nature of the solids and d only upon the chemical series of the lubricant. We now have for a pair of solids, $\mu_1 - \mu_2 = (b_0)_1 - (b_0)_2 =$ a quantity dependent only upon the nature of the solids.

And for a pair of lubricants of series $'$ and $''$,

$$\mu' - \mu'' = (d'' - d') - N\,(c' - c'') - 2\,(c'' - c')$$
$$= \text{a quantity dependent only upon the chemical}$$
$$\text{series of the lubricants.}$$

These relations have been found to hold over a series of solids including different kinds of steel, a bronze, and glass, and for the normal paraffins and their related acids and alcohols.[2]

The full equation for a curved slider now is

$$\mu = b_0 - d - c\,(N-2),$$

and a physical significance can be given to each term, namely:

b_0 is the friction of the totally unlubricated "clean" surfaces. It is a pure function of the chemical nature and physical state of the solid faces.

c is the decrement in friction due to any single atom in the carbon chain other than those in the end groups, and this leaves

d as the decrement due to the end group or groups of the molecule.

This last quantity can be calculated from the equation $b = b_0 - d$. The following table (Table X) gives the value for b and for d calculated from the values of μ given in Table I (*loc. cit.*). The solids were:

A 50-ton steel, B medium carbon steel, C nickel chrome steel, D mild carbon steel, E phosphor-bronze.

The quantity d clearly is independent of the nature of the solid in these cases.

The calculation of these values led to the detection of an error in Table I in the paper cited above, which may be corrected here. The "clean" value for the mild steel should be 0·79 instead of 0·74. The value changed from the earlier figure 0·74 after the plate had been refaced to remove scratches; the "clean" value therefore depends upon physical state as well as upon chemical composition.

[1] *Proc. Roy. Soc.* (1922), A, c, 563.
[2] *Proc. Roy. Soc.* (1924), A, cvi, 341.

Table X

Lubricant	a		A	B	C	D	E
		b_0	0·88	0·83	0·93	0·79	0·94
Heptane		b	0·67	0·63	0·73	0·59	0·74
Octane	0·00156						
Undecane		d	0·21	0·20	0·20	0·20	0·20
Propyl alcohol		b	0·60	0·56	0·65	0·52	0·66
Octyl alcohol	0·00169						
Undecyl alcohol		d	0·28	0·27	0·28	0·27	0·28
Heptoic acid		b	0·65	0·62	0·69	0·59	0·69
Caprylic acid	0·00259						
Decoic acid		d	0·23	0·21	0·24	0·20	0·25

In an earlier paper in this series, the field of attraction of the solid was spoken of as "making itself felt" at the plane of slip (*Proc. Roy. Soc.* A, c, 567). We now see that the field of each solid face is a fixed quantity, from which each atom of the molecule of the lubricant subtracts a quantity determined only by the configuration of the molecule—that is, by the chemical series. The capacity of an individual carbon atom in the chain, other than the atoms at each end, is the same, no matter where it may be placed in the chain, and is determined solely by the nature of the end groups.

It is obvious that these relations can be conceived of more simply if the limiting lubricating film be composed of two monomolecular layers, one on each face, with the plane of slip between them.

When the lubricating layer is more than two molecules thick, as it appears to be for light loads, the field of attraction of the solid still makes itself felt at the plane of slip. If the function of the solid face were simply to fix the first layer of molecules in position, and these in their turn fixed other molecules, this would not be so—the conditions at the plane of slip would then be determined solely by the nature of the lubricant. This is never the case even with the lightest loads, as Table XI shows.

Table XI. *Slider steel: face plane. Load 20·4 gr. corresponding to point A on pressure/μ curve. Lubricant octyl alcohol*

Plate	μ
Medium carbon steel	0·463
Mild carbon steel	0·457
Nickel chrome steel	0·469
Phosphor-bronze	0·510
Bismuth	0·438
Glass	0·523
Quartz	0·472

SUMMARY

1. When the slider has a plane face the coefficient of friction is a function of the load, decreasing as the load increases, until a point is reached beyond which the coefficient is independent of the load.

2. When the slider has a spherical face the coefficient is always independent of the load.

3. The coefficient is a measure of the efficiency of the lubricating layer with respect to one variable—the load. Recollecting that the pressure between the bearing surfaces must be very great when the slider has a spherical face, the above results show that with low pressure the efficiency of the lubricant increases as the pressure increases, until a limit is reached, beyond which Amontons' law holds.

4. It is probable that, during the first period when Amontons' law does not hold, the slider is floating on a layer of lubricant whose thickness is a function of the pressure, whilst in the second period where Amontons' law holds, all lubricant which can be squeezed out has been squeezed out, and a layer of constant molecular composition has been reached.

5. In the first period friction is adjusted to the load by variations in the thickness of the layer of lubricant, and in the second period by the elastic forces between the atoms.

ON THE SPREADING OF FLUIDS ON WATER AND SOLIDS: AND THE THICKNESS OF A PRIMARY FILM

[*Institut International de Chimie Solvay*, Brussels, 1925]

1. When a volatile substance such as benzene is poured over clean water to form a layer of the order of a millimetre in depth the layer is unstable. It breaks at points, and, by a series of interesting changes, is finally resolved into a number of lenses which are surrounded by a composite surface of benzene and water.

The composite surface is so called because its physical properties are neither those of a surface of benzene nor a surface of water. It is covered by a film of benzene of insensible thickness which I will call the primary film.

If the space above be opened to the air, the lenses of benzene dance about over the surface in a remarkable way, the small ones moving to the large and sometimes fusing with them. To fuse, a pair of lenses must have kinetic energy of approach sufficient to overcome a striking repulsive effect which is obvious when the distance separating the edges of the lenses is of the order of a millimetre.

Let the air space now be covered. The movements diminish as the vapour pressure of benzene rises and finally cease when it becomes saturated. The lenses then slowly move together and fuse, so that the final state is a single lens in tensile equilibrium with a composite surface. The nature of the equilibrium is seen at once when the cover is lifted so that the vapour of benzene can escape. Benzene evaporates off the composite surface and the lens is at once expanded into a thin sheet which shows the characteristic instability mentioned earlier.

The theory of these movements is not clear, and I do not propose to go into it at any length. It will be sufficient merely to point out the alternative explanations of, for example, the expansion of the lens when the cover is lifted. We may regard the equilibrium between the lens and the composite surface as being either a thrust at the margin of the former, due to the tendency of both lens and composite surface to expand, or a tension that is a pull due to the tendency of both to contract.

The latter explanation is in accord with the classical theory of surface tension and my own feeling is that the classical theory is correct. I believe that critical and conclusive experiments are forthcoming in support of this view, but the subject would need a paper to itself and can only be mentioned here in passing.

2. The phenomena of spreading of a fluid on the clean face of a solid are essentially similar to those on water. If the liquid be composed of a single chemical substance a sheet of sensible thickness is unstable and breaks up into lenses, and the details of the process are precisely like those seen on water. The final equilibrium, in presence of saturated vapour, is similar, namely, a single lens in equilibrium with a composite surface covered by a primary film.

In both cases this sharp equilibrium point is not reached if more than one component be at the interface, if, for example, the benzene is impure; but the effect of impurities is more striking when the face is solid. Almost any degree of impurity will then give an indefinite capacity for local adjustment, as Gibbs pointed out, and sharp equilibria are not reached.

3. Though the resemblances are obvious there exist certain fundamental differences between spreading of a fluid on water and on a clean solid— such as glass or steel. Before describing these it is necessary to distinguish sharply between two kinds of spreading.

A drop of carbon tetrachloride, or carbon bisulphide, placed on clean water is not drawn out to form a sheet, yet it cannot be said not to spread at all. The lens of fluid stands apparently unchanged on the water, yet there spreads from it a primary film of insensible thickness to form a composite surface. It is almost certain that this primary film is formed through the vapour phase because fluids like the higher paraffins with no sensible vapour pressure do not spread over water even in this way. There are, therefore, two kinds of spreading:

I. To form a primary film, the lens remaining without any flattening;

II. The spreading of the lens itself which flattens and flows over the surface.

It is convenient to call these primary and secondary spreading.

4. Coming now to phenomena on solid surfaces: The first peculiar feature is that, so far as my experience goes, secondary spreading never occurs if the substance is rigorously pure and, of course, if the solid face be clean. Therefore a drop of any pure chemical substance stands on such a surface almost like a drop of quicksilver. Primary spreading, on the other hand, does occur but only through the vapour phase. Taking, for example, a chemical series such as the normal paraffins, normal acids, or normal alcohols, members of low molecular weight form primary films, that is to say, a drop placed anywhere on a plate say 10 cm. in area, suffers no change discernible by the senses, but in a longer or shorter time the whole surface of the plate is found to be covered by a film of insensible thickness which reduces friction to the lowest point. Moving up the series, as the molecular weight increases, and the vapour pressure falls, the rate of spread of this film falls until the rate becomes insensible.

5. The primary film formed in this way spreads with great force. If a clean slider be placed on a clean slate the surfaces seize and considerable force is needed to separate them. But if a drop of fluid with a vapour pressure is put on the plate a few centimetres away the primary film condensed from the vapour will insert itself between the seized faces so as to liberate them. It is a remarkable fact that fluids with no vapour pressure are incapable of penetrating between seized faces even when a drop of any one of them is placed in contact with the edges. For example, a steel slider had seized to a steel plate and neither oleic acid, linoleic acid, medicinal paraffin, tetracosane, nor acetyl alcohol poured all round the contact, were capable of liberating the seized faces in a period amounting to hours, whereas the vapour of methyl, or octyl alcohol, or similar lubricant with a good vapour pressure, freed the faces in a few minutes. This explains why a fitter when he wishes to free seized metals uses ordinary burning paraffin oil instead of a good lubricating oil because the former contains a plentiful supply of substances with a high vapour pressure.

6. As I said before, a second component at an interface causes spreading over a solid and an incredibly minute amount is sufficient. The most striking example is furnished by acetic acid on glass. A drop of anhydrous acetic acid remains drawn up in a lens on a clean glass plate provided the air is completely dry. If the cover to the chamber be lifted for a moment so as to admit air not dry, the drop of acetic acid at once flashes over the whole surface.

7. The whole subject of spreading on solids needs to be worked at systematically. The above observations are based on some years experience with a variety of solids and fluids, but the observations were only made episodically in the course of researches directed to the elucidation of quite other problems.

8. It may probably be said with safety that according to the views most widely held a primary film is monomolecular. Studies of friction throw doubt upon this view. They make it probable that a complete primary film—that is a primary film in equilibrium with saturated vapour—is very rarely composed of only a single layer of molecules. Consider for a moment the question from the theoretical side, and let fluid in mass with a plane surface be in equilibrium with a plane composite surface covered by a primary film of the fluid. The only condition to be satisfied by the composite surface is that the primary film should, on the average, receive in unit time as many molecules from the saturated vapour as it loses. Is it likely that a monomolecular film alone is able to fulfil this condition? It is in my opinion more likely that it will only do so very rarely.

9. When the friction of a slider standing in a large pool of lubricant on a plate is measured and the coefficient of friction μ ($=$ friction divided by

the load) plotted against the pressure between the bearing faces we get the curve $ABCDE$, Figs. 1 and 2 from *Proc. Roy. Soc.* (1925). In the horizontal part $BCDE$, the film of lubricant has been reduced by the pressure to a certain basal film of great mechanical stability. In the region AB the slider has not been forced down to this basal layer, but is floating on a pad of lubricant, the thickness of which is a function of the load. When the

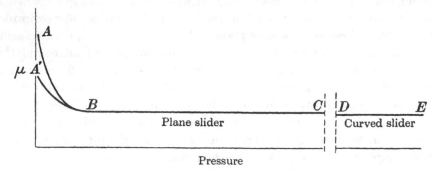

Fig. 1. The scale of pressure is broken at C. Pressure is the load divided by the area of the bearing surface, or surface of contact between slider and plate.

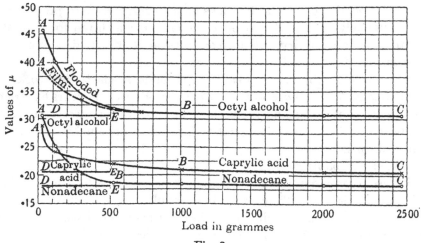

Fig. 2.

slider is placed in the pool of lubricant it takes some time to fall through the fluid until it reaches the point at which the pressure in the layer of lubricant between the faces is sufficient to carry the load. The rate of fall of a slider with a plane surface through a viscous fluid is given by the following equation which was worked out by Prof. G. I. Taylor:

$$T = \frac{3}{4}\frac{\pi\eta a^4}{mg}\left(\frac{1}{h_2^{2}} - \frac{1}{h_1^{2}}\right),$$

T is the time occupied in falling from the height h_1 to the height h_2, η is the viscosity of the fluid, a the radius of the surface of the slider, and m the weight of the slider. Using this equation for a slider with three legs each with a diameter of three millimetres and carrying a load of 100 gr. on each leg, then if water were the lubricant the thickness of the layer between the phases would decrease to 10^{-7} cm., that is to molecular distances, in 3·3 hours. There would, therefore, be a lag or latent period after the slider had been placed on the plate during which it was falling through the lubricant, and during which the friction would rise as the layer of lubricant thinned. This is actually what happens as the following table shows:

TABLE I. *Bayonne oil*

	Pull in gr.	μ
1 sec.	0·0	0
5 ,,	0·400	0·02 (approx.)
1 min.	2·86	0·11
5 ,,	4·08	0·20
10 ,,	5·22	0·256
15 ,,	5·75	0·282
30 ,,	5·94	0·291
45 ,,	6·04	0·296
60 ,,	6·20	0·304
90 ,,	6·47	0·317
120 ,,	6·59	0·323
150 ,,	6·69	0·328
180 ,,	6·75	0·331
210 ,,	6·83	0·335
240 ,,	6·83	0·335
270 ,,	6·85	0·336
8 hours	6·90	0·338

TABLE II. *Octyl alcohol*

	Pull in gr.	μ
1 sec.	0	0
3 ,,	2·65	0·13 (approx.)
10 ,,	4·30	0·21 ,,
1 min.	7·14	0·35 ,,
5 ,,	7·75	0·38
10 ,,	8·51	0·417
15 ,,	8·96	0·439
25 ,,	9·22	0·452
30 ,,	9·34	0·458
45 ,,	9·44	0·463
60 ,,	9·57	0·469
90 ,,	9·73	0·477
120 ,,	9·79	0·480
150 ,,	9·83	0·482
180 ,,	9·89	0·484
240 ,,	9·91	0·485
6 hours	9·91	0·485

If the slider is floating in the lubricant, then when the load is decreased it will rise and there should be a latent period of falling friction. The following is an example of the lag for both increased and decreased loads:

TABLE III. *Steel: slider plane, lubricated with large excess of liquid nonadecane, latent period measured for increase in weight of slider*

Initial load gr.	Original (steady value)	Weight added gr.	After adding weight (steady value)	Latent period min.
120·4	$\mu = 0.2523$	50	$\mu = 0.2441$	12
420·4	0·2105	50	0·2055	5
821·2	0·1871	50	0·1857	± 1

The pressure which carries the load is that to which Leslie drew attention in 1801[1] as being the consequence of the attraction of the solid for the fluid.

Along $BCDE$ when the load has forced the slider down to the basal film of lubricant, there is no latent period such as that mentioned above. There is no temperature coefficient over the range 15–120°, whereas there is a temperature coefficient along AB, and the coefficient of friction is critically independent of the load. The conditions are, therefore, very much simpler than they are along AB.

The simplest explanation of these relations is that, with a relatively light load in the region AB of the curve, the layer of lubricant is several molecules deep and its depth is varied as the load varies by the actual squeezing out or drawing in of fluid.

It is possible to lubricate not by a pool, but by covering the solid faces with a primary film deposited from the saturated vapour of the lubricant. When this is done and when the slider and plate are brought together it is obvious that the lubricating film consists of no more than two primary films each of insensible thickness. The curve for μ/load or μ/pressure now follows a path $A'B$, and it is to be noted that for each load up to the point B the lubrication is more efficient the thinner the layer. Beyond this, when the basal layer is reached, the curves for lubrication by primary film and by flooding the plate coincide.

If the explanation given above for the region AB is true, namely, that the slider rises or falls in a layer of lubricant several molecules thick, it follows that primary films of the substances experimented with cannot be monomolecular when they are in equilibrium with saturated vapour.

When the coefficient of friction is considered as a function of molecular weight the curves for the region $BCDE$, that is for the limiting layer of lubricant, are linear and the equation is $\mu = b - aM$, where μ is the coefficient of friction and M is molecular weight. The parameter b is a function of the chemical nature of the solid and of the chemical series of the lubricant.

[1] *Phil. Mag.*

Now experiment has shown that the complete equation has the form $\mu = b_0 - d - c\,(N-2)$, where N is the number of carbon atoms in the chain. The parameters b_0, d, c are of unusual interest:

b_0 is merely the friction of the clean faces, steel on steel, glass on glass, etc. It is, therefore, a pure function of the chemical nature of the solid.

c is the decrement due to each carbon atom of the chain other than those in the end groups. It is independent of the nature of the solid and a pure function of the structure of the end group, that is to say of the chemical series.

d remains, therefore, as the decrement due to the end groups.

Obviously if N is large enough μ vanishes, and this actually is realised. With the primary acids when N is greater than 15, the smallest traction that can be applied produces slip.

The remarkable features of this equation are two: The fact that the effect of a carbon atom is independent of its position in the chain and is determined only by the configuration of the end groups, and that the influence of the molecule of lubricant is wholly negative. It contributes nothing positive to the result, and one may well ask what happens to its own field of attraction.

46

CHEMISTRY AT INTERFACES

[*Journ. Chem. Soc.* (1925), CXXVII, 1207]

When your Council invited me to deliver this lecture I was conscious more of the honour than of the responsibility. The sense of the former has not faded, but that of the latter has grown to terrifying dimensions. The fact of the matter is that I cannot hope to inform; the best I can do is to take this unique opportunity for propounding certain conundrums which have been forced on my notice during thirty years' work on interfaces and films.

The material I first worked with was proteins—or proteids as they were then called—and my studies led to a frankly chemical explanation of the properties of this singular group. In a physiological laboratory proteins, the physical basis of life, were still regarded as something outside chemistry and incapable of reactions of the ordinary type.

The protein class which attracted me was the globulins. They are singular in that, whilst themselves insoluble in water, they are dissolved as salts by acids, or alkalis, and, in the absence of acids and alkalis, by neutral salts such as sodium chloride. When dissolved by alkali, that is, when they react as acids, they form systems like soaps. Now the remarkable feature is this—I directed attention to it in the Croonian lecture for 1905—that the two types or states of solution, by acid or alkali, or by neutral salt, are discontinuous in the sense that to pass from one to the other the protein must first be precipitated, and that the protein masses are electrically charged in the first type of solution and are not electrically charged in the second type.

In that lecture I directed attention to the fact that while the electrically uncharged system in its phase relations can be homologated with such a straightforward system as that of water, sodium chloride, and succinonitrile, the electrically charged system, with contact potential differences at internal surfaces, manifests the abnormal viscosity, high degree of inertia to change, and other features characteristic of the colloidal state. The electrical aspect of this state has to my mind been too much neglected of late. I believed then, and I believe now, that colloid chemistry has much to gain from someone who will start from the rich literature of the Lippmann period. He, if anyone, may succeed in rescuing that much-abused word "colloid", and in bringing it within the four corners of a satisfactory definition.

I tried myself—indeed I have tried many times—but it needed a better physicist than I am. The fundamental difficulty is easily stated. Take, for example, Dupré's equation for the free energy between two phases A and B

$$T_{AB} = T_A + T_B - 2T'_{AB},$$

the question is whether the final term, which gives the gain of work in forming the interface, can or cannot be divided into two which can be written

$$2T'_{AB} = 2T'_{ab} + f(\sigma),$$

when σ and $-\sigma$ are the densities of the electricity on either side of the interface.

$2T'_{ab}$ then is the work done by simple forces of cohesion, and $f(\sigma)$ the work done in forming the electrical layers. You may object at once that such a question is beside the mark, since all cohesive forces are now recognised as being electrical. I submit, however, that there is a real question. Contact potential implies a certain orderly disposition of the electric polarities and, on the chemical side, a corresponding orderly configuration of atoms and molecules with respect to the interface. Is all the free energy peculiar to the interface gathered in this orderly way, or is there a portion which does not directly contribute to the definite distribution of electricity which constitutes a contact potential difference? This question, to my mind, includes the question of the different valencies —co-valency, ionic valency, etc.

The first obvious step towards answering this question was to measure the potential difference between organic fluids and water and see what relation it had to chemical constitution. I made more than one attempt and obtained always an impossible result—namely, that the contact potential difference was almost independent of chemical constitution. The cause of failure is of interest. The only experimental method which suggested itself to me was to measure the movement of a small drop suspended in water in an electric field of known strength, but it seemed to be impossible to preserve so small an area of interface from contamination amidst such a relatively large mass of water.

Consideration of electrical internal surfaces does, at any rate, lead us directly to what I still cling to as the distinctive feature of colloidal solutions, namely, their inertia whereby a state is not defined simply by the temperature, pressure, and components, so far as they can be recognised, but also by the previous history. The path of approach has to be reckoned with. But although internal electrified surfaces give a high viscosity and a slow rate of change, they are not, I think, the whole story.

The theory of these salt globulins is simple. The insoluble protein is dissolved as a molecular compound with the neutral salt, in the same way that silver chloride is dissolved by sodium chloride. But it is rare to find a component in a solution in water which is so uncharged as to show no movement in an electric field. After much search, I found only one clear case—azomethine dissolved in chloroform—and here the fact is not evidence of the absence of a charge on the molecules of azomethine, because the electric flux necessary for electrophoresis was probably not possible.

Possibly there is no great physical interest at the moment in elucidating the molecular mechanism of contact potential—physicists seem satisfied with a simple electronic theory and chemists with a simple ionic theory—that of Nernst, but there is more to learn. What part do polar groups play? Where in the molecules oriented at an interface is the contact potential situated? And, above all, what are the conditions which permit of an electric flux?

The conditions of discharge of such contact fields are of peculiar interest to the biologist, and of basal importance in the study of that difficult subject the corrosion of metals.

Let me give an example which shows the need the biologist has for an intimate analysis of surface energy in terms of contact potential, of tension, and of orientation. The blood of lobsters and crabs flows, not in capillaries, but in large spaces called sinuses, spaces so large that, if one were freely opened, the animal would quickly bleed to death. These sinuses are torn open not infrequently in the misadventures of their lives. They fight, for example, and a great claw is often torn away in the conflict. Loss of blood on these occasions is reduced to a minimum by that natural first aid which serves us all, namely, the faculty of the blood when disturbed to clot. Clotting in this case must be swift, and it is surprisingly swift—the sinuses are plugged almost at once.

It is to the first stage in this process of clotting that I want to direct your attention. There are in that blood certain cells or blood corpuscles which have been called "explosive" cells, because on the slightest disturbance to the chemical or physical balance of the blood they actually do explode. The upset is due simply to contact of the blood with a "foreign body" which may be sea-water, or may be a glass slide. Using suitable precautions, one can get a drop of blood on to the under side of a cover glass and watch the process. By the time the microscope is adjusted the corpuscles nearest the glass have already exploded, but further down in the drop there are still intact cells and one can watch the progress of the disruption moving away from the glass face. A wave of change must start at the interface between the glass and the blood, and progress through the latter, involving these sensitive corpuscles in its path. What is it at the interface which starts this wave of change? It has something to do with what might be called interfacial reactivity—that reactivity which lowers surface tension and increases the contact potential difference. So much we know because, if paraffin wax be substituted for glass, the change does not occur. I have long wanted to trace the effect of chemical constitution upon the phenomenon and have never yet had time. The change will start from a quartz face, but it must be remembered that it is unusually difficult to free a quartz face from a film of natural contamination.

The two impressive features of this phenomenon are (1) that a chemical change of catastrophic character can be started at an interface, and (2) that the change can be propagated apparently indefinitely through one of the phases.

You have all heard of nerves, nerve fibres, and nervous impulses, which last are waves, partly electrical, which are carried by nerve fibres. A nerve fibre consists essentially of a fine rod of protoplasm from 2 to 8μ in diameter, and it is an outgrowth from a nerve cell. These fine rods may attain a prodigious length—in a large animal like a whale, the rods which connect the cells in the spinal cord with some of the more remote parts of skin or muscle must be many yards in length, and each starts from a small object, the nerve-cell, which is only, say, 100μ in diameter. Now the structural integrity and capacity for growth—note especially the latter—of this tenuous filament depend throughout its length entirely upon its connection with the nerve cell of which it is an outgrowth, and the interface from which this control irradiates is without doubt that inside the cell between nucleus and cell body. There can be no question of mere gross manufacture and transference of nutriment—the process would require geological time. There must be some kind of static or dynamic balance just as a row of bricks up-ended on an inclined plane would fall if the support at the end were withdrawn.

So far as my limited knowledge goes, there is nothing in the inorganic world to parallel this dependence of structure, relatively or actually far distant, upon the integrity of a particular interface.

Let us form an interface say by bringing a paraffin, or one of the derived acids or alcohols, in contact with a solid face of glass or steel. There will be adhesion and we can now, from evidence derived from more than one source, be certain that the long molecules will be oriented at the solid face with their long axes at right angles to the interface. Does the effect of the interface end there? The forces involved are of very short range, less than any dimension of the molecule. Their range is indeed of the order of the distances which separate the atoms in a molecule. But, if there is a mono-molecular layer of oriented atoms formed at the interface, there will be a secondary interface formed between it and the mass of fluid where a random disposition of the molecules holds, and this will have its effect in orienting a second layer of molecules and so on. In fact, I picture the process going on until it is upset by the heat motions. The study of friction furnishes, I believe, proof that this limit set by the heat motions is not reached until several—one cannot say how many—layers of molecules are involved.

One has, therefore, at the interface a new phase—the interfacial phase, we might call it—which is formed with evolution of heat as Leslie pointed out as far back as 1802, in the formation of which, therefore, degrees of freedom are suppressed, and the phase has a configuration or structure which intrinsically is unstable at the given temperature and pressure, and acquires stability only by the intervention of the forces at the interface. Moreover, the phase is the seat of an electric field of prodigious intensity.

I accept the view that the forces at the actual interface—the adhesional forces of the solid—are, in fact, of exceedingly short range, because all those best qualified to judge are agreed as to this—but in some way they make themselves directly felt right through the interfacial phase so that the capacity for withstanding traction—the friction, in short—is never determined by the chemical nature of the lubricant alone, but by the chemical nature of the solid face as well. If it were merely a matter of setting a number of molecules of a given kind on end, layer on layer, the properties of any plane between the layers would depend merely on the particular kind of molecules bounding it. This never is the case—the physical properties of such a plane depend just as much upon the nature of the solids as upon the nature of the molecules.

Gibbs, it will be remembered, introduced with tremendous effect the conception of an interfacial phase, and what we are particularly concerned with is the depth of this phase. Leslie, more than a hundred years ago, pointed out that the intrinsic pressure—not the quantity p in Gibbs' equations, but the van der Waals' pressure—is different in this phase from what it is in the body of the fluid contiguous to it.

The first evidence I will bring forward to support the statement that the interfacial phase may be many molecules in depth is final and convincing. Bragg and his fellow-workers have been analysing with the aid of X-rays the structure of thin layers of grease spread upon a solid face, and they find that the long paraffin molecules are oriented end-on to the interface, and that they are disposed in rows one above another. There is, in fact, as I claimed many years ago, a kind of crystallisation at an interface with the crystal planes disposed tangentially.

There is one observation made by Bragg and his fellow-workers of peculiar interest to me. It will lead to a digression, but I hope an interesting one. They find that the grease when simply placed on the solid does not show this architecture, but develops it rapidly when rubbed or smeared upon the surface. This is, perhaps, what might be expected; the study of friction shows that orientation comes to its maximum slowly even when mobile fluids such as octyl alcohol, or caprylic acid, are used. With the latter, for example, friction falls, owing to gradually increasing orientation at the interface, for about 60 min. at ordinary room temperature (11°). Such a period might be expected to be indefinitely long in a highly viscous grease.

It is really surprising how much mechanical agitation will shorten the period even when the lubricant has high fluidity. With caprylic acid on glass, orientation reaches a maximum, and friction reaches a steady value, when there is no mechanical disturbance, other than earth tremors, in about 60 min.; but if the surface be disturbed by vigorously moving the slider about, it is completed in about 10 min.

It looks as though the first effect of the attraction field of the solid was to lock the molecules in the random disposition characteristic of the body of the fluid. Put in another way, the effect of the attraction field is to increase the viscosity of the fluid in the neighbourhood of the solid so that rearrangement under the influence of their polar or other attractions is delayed. There is direct evidence of such a change.

Mechanical agitation materially quickens the process of orientation even when the surfaces are covered only by an insensible film deposited from vapour. Consider this case for a moment. Caprylic acid consists of a carbon chain loaded at one end by the carboxyl group. The molecules might therefore roughly be likened to long rods the attraction of which for the solid face is greater at one end than at the other. These rods are, at any instant, disposed at random in the vapour, and therefore, when they move to the solid face, although an excess will strike with the more attracted end, say the carboxyl end, many will strike with the other end. The film as first formed will be far from completely oriented and there is a latent period or lag during which orientation becomes more orderly owing to the evaporation and condensation of molecules into and out of the vapour.

Left to itself, the film reaches a steady state, that is to say, friction ceases to diminish, in about 10 min., but mechanical agitation, produced by motion of the slider about the plate, causes the steady state to be reached in a matter of seconds. Obviously in the second case something more than evaporation and condensation is at work—the molecules in the film must actually rearrange themselves on the spot, and the process of changing ends is hastened by mechanical agitation.

I could multiply instances, especially of the varying degree of sensitiveness to mechanical disturbance. I will take an extreme case, and the case reveals a paradox.

It is possible to deposit films in succession upon a solid face, and friction values reveal that they actually are in succession. The stability of the arrangement obviously depends upon the order in which the films lie. For example, a clean surface of glass or steel can be covered with an insensible film of solid paraffin, or oleic acid, or undecyl alcohol, or undecoic acid, etc., and a second film of water deposited on this film by admitting water vapour. With great care to avoid mechanical agitation, the friction of the doubled film can be observed, but the disposition is highly unstable if the first deposited film is fluid, and almost any degree of movement upsets it with consequent change of friction. Here are the figures for undecyl alcohol:

Alcohol alone	Water vapour admitted	Mechanical agitation at once rose to
$\mu = 0.44$	$\mu = 0.33$	$\mu = 0.48$

I see in this an interesting possibility. Let us suppose there is a righting mechanism which, by expenditure of energy, would, given the opportunity, restore the original disposition of the films. I am, of course, thinking of living matter. Such layered films might reach any degree of sensitiveness to mechanical shock, and you obviously have here the germ of the mechanism of a sense organ capable of reacting to any kind of mechanical disturbance. More than that, by having a succession of films rightly arranged one could get an effect magnified, and it is at any rate noteworthy that in what are believed to be organs of the pressure sense—of traction at any rate—one does get a succession of surfaces about a nerve ending, like the coat of an onion.

Here we see the importance of not losing sight of the contact difference of potential. These films, no matter how thin, are at a different potential from that of the substance on which they lie. I was able to observe this as far back as 1911; and the fact is, of course, the basis of frictional electricity as Helmholtz recognised, although he did not specify or, I think, visualise an actual film of foreign matter. Any disturbance of the films, whether single or doubled, is certain, therefore, to produce a re-distribution of electricity; so the way for an effect on the end of a nerve fibre is clear.

To complete the story—the possibility of producing a succession of films depends upon the immiscibility of the first-used substance with water. When the substance is miscible with water, unstable states are not produced and water vapour always and at once increases friction.

Benzene vapour, on the contrary, when used in the same way always lowers friction.

Let me pick up the paradox before I forget it. Water is, it will have been noticed, a lubricant in these cases. This is obvious when a plate of solid paraffin is used. Here is a table of values of great interest:

Glass lubricated with tetracosane	$\mu = 0.325$
Water film formed on the composite surface	$\mu = 0.110$
Tetracosane on tetracosane	$\mu = 0.060$
Water film deposited on solid tetracosane from vapour	$\mu = 0.025$

But water is entirely without lubricating action on clean glass, steel, bronze, quartz, or bismuth. I do not want to be mistaken here. Water, of course, can act as a lubricant in the hydrodynamical sense investigated by Osborne Reynolds. That is to say, if the solids are kept apart by a pad of water held in place by external forces, it has the lubricating properties of a fluid with its physical characters. But in the boundary region, a film of water, or a pool of water, has no power of lowering the friction between these various solids. Yet it can alter the state of the surface of a substance so, in a sense, inimical to itself as a normal paraffin, and reduce the friction by more than 40 per cent. This peculiar feature of water, the limitation of its powers as a lubricant, is at present wholly without explanation.

Water in a sense is an abrasive. It decreases to a startling extent the lubricating powers of all substances with which it is miscible in the boundary zone. A trace of water will destroy the lubricating power of one of the lower alcohols. In this sense, water is a true abrasive, although it has no influence one way or the other on the friction of clean faces.

Glycerol is another of these neutral substances, although owing to its high viscosity it is not easy to get rid of the flotation action. However, if surfaces are forced together through the pad of glycerol, disaster quickly follows. A case was brought to me a few days since of ball bearings, lubricated only with glycerol, in which the balls were quickly torn into fragments. A chemist would, I believe, expect glycerol to behave like water in most respects.

A short while back I postulated a righting moment which should restore a film which had been destroyed. Here is an example which is remarkable, because the righting moment is exerted at a distance and depends for its operation upon the integrity of another interface.

Eggs of certain starfish, because they are large, have been much studied by the methods of microdissection. A thin film can be detected under the microscope covering the surface of one of these eggs. When this film is pierced and the contents of the egg, the living substance, are injured over a limited area, repair is effected by the reformation of this surface membrane between the injured and uninjured parts. This is the process of healing. The injured parts are beyond repair and they are simply cut off by this membrane formation. There is here that righting power which I postulated.

But the righting power depends, as in the example given earlier, upon the intactness of the interface between nucleus and cell body, for if the fine needle is simply pushed into the cell until it punctures the nucleus of the egg, the superficial film, instead of being reformed, perishes over the whole body of the cell and in a curious fashion, for it disappears first about the point of entry of the needle, and then the solution or disappearance progresses slowly over the whole surface. Following on this, the whole cell becomes invaded by water and breaks up.

One must suppose that the influence of the nucleus or the nuclear interface must diminish with distance—the righting power will diminish—and this I cannot but suppose is one of the factors which sets a limit to the size of the living cell.

I will now return from this somewhat long digression to the question of the depth of an interfacial phase and consider it in the light of measurements of friction.

It will be necessary first briefly to summarise the evidence. Consider a flat plate covered with fluid and a heavy body falling under its own weight through the fluid. As it falls it must displace the fluid lying between its lower surface and the plate. The time occupied in falling will, therefore, be a function of the viscosity of the fluid, the area of the plate, the distance through which it falls and, to a remarkable extent, the shape of the surface of the falling body. The equations for the time occupied in falling, which have been worked out by Prof. G. I. Taylor, show that whilst the time needed to fall completely through the fluid to the plate is always infinite, the infinity is of a low order if the under surface of the falling body is curved, and of a high order if it is plane.

Calculated with the help of these equations, the time needed to fall through water to a molecular distance from the plate is a matter of hours for a plane slider—as I will call the falling body—and of seconds for a curved slider.

As the layer of lubricant between the faces is thinned by the fall of the slider, one may expect the friction to increase; therefore, by following the changes in friction after the slider has been put in a pool of lubricant, one can follow its fall. There is, in short, a lag or latent period following the placing of the slider in the pool whilst the former is falling through the fluid.

Now when the slider is plane this lag actually lasts for an hour or more, but when it is spherical it is so short as to be indetectable. It must be remembered, however, that it takes 20 sec. to make a measurement of friction.

Before giving actual figures, I must, to avoid misunderstanding, point out that a lag or latent period before a steady state is reached can be due to one or both of two things—to the time occupied in falling through the lubricant, or to the time occupied in the orientation of the molecules of the lubricant with respect to the normal to the surface of the plate. Time does not permit me to give the evidence in detail. It must suffice to say that when paraffins or their derived acids or alcohols are used, lag due to orientation is exhibited only by the latter two classes, in which the two ends of the elongated molecules differ. No lag due to orientation has been detected with certainty when the lubricant is a normal paraffin.

There is another way to discriminate between the two causes. When the lag is due to orientation, friction diminishes during the latent period, but when it is due to thinning of the layer of lubricant friction increases during the latent period.

Here is an example of a latent period of increasing friction due to fall of the slider through the lubricant.

Undecane.

	Pull in gr.	μ		Pull in gr.	μ
1 sec.	3·06	0·15 (approx.)	30 min.	10·98	0·538
5 ,,	6·53	0·32 ,,	45 ,,	11·14	0·546
1 min.	8·77	0·43 ,,	60 ,,	11·14	0·546
5 ,,	9·85	0·483	90 ,,	11·20	0·549
10 ,,	10·57	0·517	120 ,,	11·20	0·549
15 ,,	10·83	0·531	180 ,,	11·20	0·549

What is the steady state? Is it a state in which the layer of lubricant has reached a constant value? In other words, has the slider ceased to fall because a layer of definite thickness and of great mechanical stability has been reached? Apparently not, because if additional weight is placed on the slider there is a further rise in friction and another latent period of rising friction which ushers in a new steady state. Similarly, if the weight is reduced there is a latent period of falling friction. In short, the slider appears to move up or down the fluid according as the load is decreased or increased.

Plane slider of steel, lubricated with a large excess of liquid nonadecane.
Latent period measured for increase in weight of slider

Initial load (gr.)	Original μ (steady value)	Weight added (gr.)	μ after adding weight (steady value)	Latent period (min.)
120·4	0·2523	50	0·2441	12
420·4	0·2105	50	0·2055	5
821·2	0·1871	50	0·1857	± 1

This process, however, has a limit, for when the load is high enough variations of load are not followed by a latent period.

These various features are best exhibited by a curve (Fig. 1) in which the coefficient of friction, $\mu = $ friction/load, is plotted against the pressure between the bearing faces or load/area.

You will notice over the first part of the curve μ is falling. Now this is itself remarkable, for μ is a measure of the efficiency of a lubricant with respect to a single variable, the load. Over this part, AB, there is the latent period characteristic of variations in the depth of the layer of lubricant owing to the slider rising or falling in the fluid. We have, therefore, the curious fact that the efficiency of a lubricant increases as the layer thins until it reaches a maximum at B.

At B over the horizontal part, there is no latent period of the kind mentioned. If any point in $BCDE$ is taken and the load varied, but not enough to trespass on the region AB, the adjustment of the load to the friction is sensibly instantaneous.

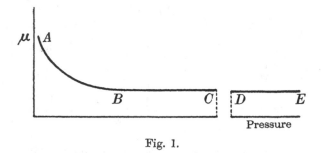

Fig. 1.

There is a fundamental and surprising law of friction, which was formulated by Amontons in 1699, according to which the friction is equal to the load multiplied by a parameter which of course then is the coefficient of friction μ. This law holds rigidly over $BCDE$ and does not hold over AB.

The part DE refers to a spherical slider. It is obvious that from its shape the pressure under a spherical slider must be immense, and to indicate this I have broken the scale of pressure. A spherical slider owing to its shape at once cuts right through the lubricant until the limiting layer characteristic of the whole region $BCDE$ is reached, and this alone would explain the fact that one never gets any indication of rising or falling of a spherical slider in the lubricant.

It is not my business here to discuss this curve in detail, that will be done elsewhere. It will suffice if I say that over the region AB the layer of lubricant would appear to be many molecules thick, and that there the adjustment of friction to the load is effected by an actual flowing in and out of lubricant from between the surfaces of slider and plate, whilst over the region $BCDE$ this process ceases and the adjustment of friction to variation in the load is by the elastic forces between the atoms, and for this reason it occurs with a rapidity too great to be detected. It is only when interatomic forces alone are concerned that Amontons' law holds.

I can now take up the question of spreading of effect from the interface.

At A the layer of lubricant is several molecules in thickness, yet the state is far removed from that of the interior of the fluid. The layer can exhibit static friction, that is to say, it can carry a stress without continuous deformation. Somehow, thoughout its entire depth, the molecules are to this extent locked in place by the attraction fields of the solids. It is structured throughout and the maintenance of the structure is due to the fields of the solid.

It is legitimate to assume that the plane of slip is midway between the solid faces. If the state of the lubricant were unchanged there, the friction would be independent of the nature of the solid and be a function only of the constant which defined the state of the fluid. This is not so—the same lubricant at the point A in the curve gives entirely different values of friction for different solids. The following table illustrates this:

Steel slider. Face plane. Load 20·4 gr., corresponding to point A on pressure/μ curve. Lubricant octyl alcohol

Plate	μ
Medium carbon steel	0·463
Mild carbon steel	0·457
Nickel chrome steel	0·469
Phosphor-bronze	0·510
Bismuth	0·438
Glass	0·523
Quartz	0·472

I do not want to press this evidence too far, because the experimental conditions were not rigidly controlled, but, under conditions in which there would appear to have been a layer of water several molecules in thickness between the faces, we find water:

> Antilubricant for glass on wood.
> Lubricant for glass on ebonite.
> Neutral for glass on sulphur.
> for glass on glass.
> for glass on steel.
> for steel on steel.

The physical properties of water, which depend upon its molecular structure, are changed by the force-field of the solid and the change penetrates some distance into the fluid.

Further experiments—measurements in the region AB with different solids—will, I hope, enable us to formulate what might be called the law of transmission, that is, the variation in the effect of the field of the solid with

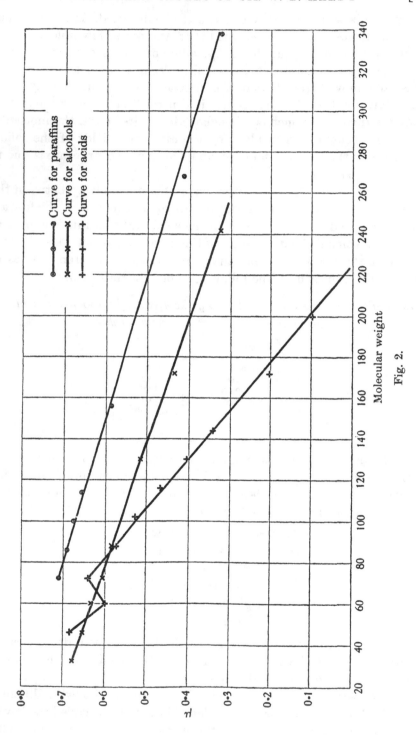

Fig. 2.

distance along the normal. That has been done only in the region *DE* and to the results I will now turn.

In the region *DE* the increased pressure has reduced the layer of lubricant to a layer of limiting thickness and of great mechanical stability. I am going to assume that this layer consists of two monomolecular layers, one for each face, that the long molecules in each of these layers are placed on end on the surfaces, and that the plane of slip lies between the two layers midway, that is, between the solid faces. This is pure assumption, justified

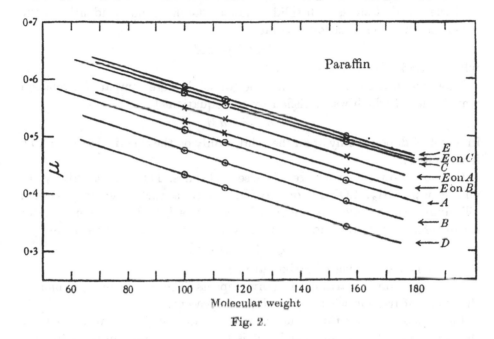

Fig. 2.

perhaps by the analogy of certain films on the surface of water, and by the fact that it alone makes it possible to visualise the relations I am about to describe.

Many solids and many lubricants have been investigated in the region *DE*. I will confine my observations to normal paraffins and their related acids and alcohols. When the coefficient of friction is plotted in relation to molecular weight, we get the curves shown in the diagram (Fig. 2). The curves are straight lines.

I may note, in passing, that in the region *AB* (Fig. 1) the curves are not straight lines, but concave with the convexity turned to the horizontal axis. There is in fact an infinite number, a family of curves, each corresponding to a particular load and the straight-line curves are merely the limit of this family. We know almost nothing at present of this family.

Going back to the straight-line curves. When a change is made from one solid to another, say from a steel to glass, or one steel to another, the effect is merely to shift all the curves parallel to themselves.

The equation for the curves therefore is

$$\mu = b - aM,$$

where b is a parameter depending upon the chemical nature of the solids and a only upon the chemical series, paraffin, acid or alcohol, as the case may be.

If two different solids are used, if the slider, for example, is of bronze and the plate of steel, each solid produces its effect independently of the other, so that the equation becomes

$$\mu = \tfrac{1}{2}(b_1 + b_2) - aM,$$

where 1 and 2 refer to the two solids.

Now, when the curves for a number of solids were compared a relation was disclosed which was satisfied by the equation

$$\mu = b_0 - d - aM,$$

in which b_0 depends only upon the solid, d only on the chemical series, and a is as before.

Consider now the structure of these molecules. They consist always of end groups CH_3, $CO{\cdot}OH$, or $CH_2{\cdot}OH$, with a carbon chain between. Eliminating aM, which obviously refers to the carbon chain, the parameter d can refer only to the end groups, so we may write as the final form

$$\mu = b_0 - d - c(N - 2),$$

where N is the number of carbon atoms.

This equation is startling in many respects. The parameter b_0 is merely the value of friction when no lubricant is present.

Let us assume that this value is some function of the attraction field at the surface of the solid—indeed this is the only assumption open to us—then the equation means that each atom of the lubricant is responsible for a decrement in this field which is a pure function of its nature and of the configuration. Each atom is, as it were, capable of saturating or neutralising a fixed fraction of the field.

I want you to notice one thing which will mean more to you as chemists than it does to me. The parameter c is the decrement due to each carbon atom. It is a quantity completely independent of the nature of the solid, being the same for solids so different as glass, steel, and bismuth. It is not, however, a fixed property of the carbon in the paraffin chain, but is determined by the nature of the end group. The curves slope differently for acids, alcohols and paraffins.

Obviously, if the chain is long enough, if N is large enough, the field will vanish and there will be no friction. This we have actually observed to this extent, that the smallest traction we could apply produced slip.

Let this happen—let N be large enough to reduce friction to zero—what happens to the field of attraction of the molecules itself? That is the remarkable feature of the equation, the purely negative part played by the molecule of the lubricant.

I raised the question quite early in this lecture, whether all the work done by the cohesive forces other than that dissipated as heat was expended in setting up the contact potential difference. Here is an indication, no more, that it is not.

Friction when the limiting layer alone is involved is, as we have just seen, a linear function of the molecular weight; but the capacity for reducing the contact potential between water and mercury emphatically is not, as Guyot's measurements show. Like the total work of adhesion to mercury, the capacity for reducing the contact potential increases as the carbon chain lengthens, but not in proportion to the lengthening.

We may suppose that b_0, the friction of clean unlubricated faces, is some function of the cohesion field on the face of the solid, and the equation given above then tells us that this field is gradually extinguished as the carbon chains lengthen.

The cohesion between these paraffin substances and the solid faces should therefore increase as the chain lengthens. The cohesion—that is, the work of forming an interface—between these substances and water has been measured and found to depend almost wholly upon the end groups. It is independent of the length of the chain in the alcohols and acids and increases only slightly in the case of the paraffins (e.g. C_6, 19·99; C_{10}, 24·12 dynes per sq. cm.).

But, according to Harkins, with the metal mercury the work of cohesion at the interface increases greatly with the length of the chain (e.g. paraffins, ±5 dynes per carbon; alcohols, ±10 dynes per carbon).

There is a simple way of showing the effect of a lubricating film upon cohesion. A plate of glass is cleaned, lubricated with a film of insensible thickness, and roughly cleaned as to one-half by passing the flame of a Bunsen burner quickly across it. Clean mercury is now finely sprayed on to the surface, and the plate is then turned upside down; the drops of mercury fall off the lubricated part and remain cohering to the clean part.

Something may be gleaned from the properties of free films as to the distance through which the state at any interface may extend.

It was, I believe, Gibbs who first suggested that a soap film consisted of two walls or skins with fluid between. In Reinold and Rücker's papers each wall is taken to be merely the superficial parts wherein surface effects are resident—the depth of each "wall", therefore, was equal to the "range" of the cohesive force.

It is, however, quite easy to construct a film in which the surface wall is without doubt a chemically distinct structure. If a film of, say, oleic acid be formed upon clean water, the density of the film being less than that required to produce the full fall of tension, and a ring of, say, copper wire— that is what I used in my experiments—be gently withdrawn from the body of the water, a film is formed over the ring which is of astonishing stability if the right concentration of the oleic acid on the surface has been hit off. I have passed such a film in lecture round a big class without its breaking.

There can be no doubt here as to the structure of the film. The com posite surface has merely been folded together so that there are two skins of oleic acid, AA' in the diagram (Fig. 3); one on each surface with a plate, B, of water between them. By observing the motion of particles under the microscope when a weak electric field is established in the film, it is possible to convince oneself that the skins AA' are at a different potential from the water between them. The skins of oleic acid are always negative to the water.

Many years ago—in 1885—Reinold and Rücker described how the passage of a small current through a soap film would prevent it thinning under the action of gravity. The difference of potential between surface skins and middle plate noted above accounts for this. When one watches such a process under the microscope one sees that the superficial skins show no movement at all, they seem quite rigid, but the internal plate of fluid flows under the influence of the electric field.

Now, if the internal plate of fluid were merely fluid as in mass and un- affected by the interfaces on either side of it, the tension of the film and its specific electric conductivity would be independent of the thickness. This, indeed, is true of a soap film until it thins to about $50\,\mu\mu$, when, as you all know, tension becomes a rapidly varying function of the thickness. If x is the thickness, the sign of the quantity dT/dx changes from negative to positive, and remains so until the thickness falls to that of the black film, which is about $11\,\mu\mu$.

Reinold and Rücker, following a suggestion made by Clerk Maxwell, attributed this change to instability—because the film obviously is unstable when the coefficient is positive—to a change in the force of cohesion from an attraction to a repulsion. They supposed that the force was a repulsion from 50 to $11\,\mu\mu$ and then again became an attraction.

No one I suppose holds that view now. The accepted theory of the stability of free films is based upon the variation of superficial tension of water on whose surface is spread some "active" substance, that is, some substance which lowers the tension. If a free film thins under its own weight for example, the concentration of active substance on its surface diminishes and the tension rises. So long as this happens, the film is stable.

This theory, however, fails to account for the curious instability over certain variations of the total thickness, because in my opinion it concentrates attention on the surface films to the neglect of the middle plate: yet the microscope shows that it is in the middle plate where adjustment is made to imposed stress.

McBain and Jordan Lloyd, using the ultra-microscope, have shown that the colloidal particles in a solution of soap or gelatin become attached to one another to form threads. In certain solutions of protein, the growth of such threads can be followed with the ordinary microscope, and they are seen to become attached to one another so as to form a framework or sponge.

In a stretched structure such as a soap film the direction of such threads with respect to the tangent will be determined by the lines of stress. I was able to observe this in protein solutions. Let them be disposed in the middle layer, B (Fig. 3), at right angles to the surface with their ends

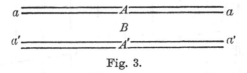

Fig. 3.

anchored in the interfaces aa, $a'a'$. When the middle layer thins so that the threads can stretch completely across from aa to $a'a'$, a critical point in the history of the film will be reached, for it is known that such threads tend to shorten—the property of shrinkage, or "synæresis" as Graham called it, of jellies depends upon this property—and drainage of free fluid from the film will be hastened. If this occurred more rapidly at any point than elsewhere, the tension at that point would fall, and rupture would occur were it not for the stiffening due to the framework.

A soap film is symmetrical in the sense that the films A, A' on each face are the same. When they are not similar, other phenomena appear.

It is easy to make an unsymmetrical film by liberating a bubble of air under water which is covered by a skin of some substance such as oleic acid which is immiscible with it. If the water be chemically pure and freed as far as possible from active substances by scraping its surface, the amount of contamination available for forming a skin on the inner face of the bubble will be very small, and the time available for condensation at the face very short, namely, the time occupied by the bubble in rising through a centimetre or so of water. The density of the skin on the upper surface of the water is under the control of the experimenter and may be varied as desired.

When the bubble has risen to float on the surface, the free film therefore has an exceeding small amount of matter condensed on its lower surface and can have any density of a known component the operator may desire on its upper surface.

The mechanical stability can be gauged from the time the bubble lasts on the surface. During this period the film will be thinning by drainage of the water in the middle plate, B, and the apex obviously will thin most rapidly. The bubble bursts when, owing to the thinning, the tension begins to fall at the apex.

Now one finds that the duration of the bubbles, that is, the amount of drainage they can stand before they become unstable and burst, is a remarkable function of the tension of the composite surface oleic acid-water, that is, of the density of the acid on the water face.

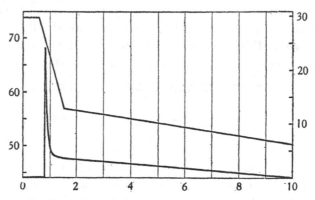

Fig. 4. The scale of tension of the general surface of the water is given on the left, and the upper curve is the curve of tension. The time scale in seconds is given on the right, and the lower curve gives the duration of the bubbles. The abscissæ give the calculated thickness of the layer of oleic acid on the surface of the water in $\mu\mu$.

Here is the curve, with the curve of tension also (Fig. 4). You will notice that there are two points at which the bubbles have no stability, that is to say, when they burst at once and with a peculiar, sharp click. The first is the region in which the tension is that of pure water, and the second is that in which the surface of the water is saturated or rather oversaturated with oleic acid. (Other substances give the same result.) But the most striking thing is that one passes from extreme instability to maximum stability at once. There is, in fact, a discontinuity of state.

Why should stability decline as tension falls? The variation of tension with the concentration of oleic acid cannot be the cause, because the curve relating them appears to be a straight line. In other words, $dT/dF = \text{constant}$.

I suggest that the mechanical stability of the film falls as it becomes less symmetrical owing to the increasing difference in the density of A and A' (Fig. 3). The amount of contamination on the inner face (A') must be exceedingly small and therefore the two surfaces of the film are most nearly alike in state when the contamination, say oleic acid, on the upper surface, A, is also small in amount.

The water in B is, in short, modified throughout by the interfaces aa, $a'a'$, and when those interfaces are not similar it is like an unannealed plate.

Another instance of the influence of the forces at an interface extending through a layer of fluid many molecules thick is found in the way liquid will cause a ball to stick in a cylindrical hole when it has a clearance of 0·0001 inch. This case is discussed by A. A. Griffith (*Phil. Trans.* (1920), A, CCXXI, 196), who supposes that the ball is bound to the inner surface of the cylinder by chains of molecules stretching from interface to interface.

A last word as to why a biologist, in spite of the handicap of his imperfect training for the task, is justified in sticking so obstinately to this problem of the distance to which the matter on either side is modified by the field of force at the interface. It is because it offers a slight and precarious but definite foothold in the most obscure region of biology. Growth, and its fellow repair, considered as physical processes are almost wholly beyond existing knowledge. The fidelity with which structure repeats itself is obvious in heredity.

Recent work, if anything, deepens the mystery of living matter, of how it preserves its space pattern, how it can be the *milieu* of chemical processes of peculiar and special kind, how it can maintain within itself sinks and sources of energy. It contains it is true a multitude of suspended particles, but these seem to be merely enclosures. When they are driven to one side, as they can be by centrifuging, the material is an optical vacuum. Take the nucleus—what is it? apparently no more than a pellicle or skin, a mere bladder containing fluid!

47

FILMS

[*Royal Institution of Great Britain*, 1926]

The title of this discourse is perhaps misleading. It has, I find, suggested to my friends a popular form of entertainment. Let me hasten to set this right. I shall have nothing to say about the cinema! My subject is something quite different—namely, those thin films of matter, familiar to all in the form of soap bubbles or lubricating films of oil.

The subject is of great scientific interest. You all of you know of the three states in which matter can exist—solid, fluid and gas or vapour. These films are no less than a fourth state, because, choose what physical constant we may, it will have a different value for any particular kind of matter in this state to what it has in any of those other states which are more easily apprehended by the senses.

Life itself depends upon this fourth state of matter. There is a film spread over the surface of each living cell which seems to control the passage of substances into or out of the cell. This film is actively maintained by the expenditure of energy on the part of the cell. The new technique of micro-dissection, by which living cells so small as to be almost or quite invisible can be dissected, has increased our knowledge of this surface film. If it be punctured at one place, the living matter in the neighbourhood of the puncture becomes curdled in appearance and dies, but the membrane grows in at the back of this dead substance cutting it off from the rest of the cell. This is the fundamental surgery of life.

I cannot hope to do more in one hour than deal in haphazard fashion with this vast subject. I propose to begin with an experiment which, in spite of its simplicity, shows how ubiquitous films are, and how our most elementary impressions of the external world depend upon them.

Take, for example, smoothness. It is not a property of solid matter in mass, but of this fourth state of matter. A tea-cup has the delicate velvety feel of a polished surface; but neither porcelain nor ware is really smooth in that sense. Their surface, like that of all other naturally occurring surfaces, is covered by a film of greasy matter, which can come from the atmosphere or from the "clean" cloth with which the object has been dried. If that film be removed the surface feels harsh and rough because, to use the engineer's phrase, one's finger-tips, if they are freshly washed, seize to it.

It is not possible here and now to remove the film. The necessary procedure would take too long, and in any case the film would quickly reform in this atmosphere. I can, however, destroy its effectiveness by taking advantage of a curious property of water. That substance is not only not a lubricant for vitreous surfaces, but it is also an anti-lubricant in that it destroys the effect of the natural lubricating film. All I have to do, therefore, is thoroughly to wet the surfaces of the tea-cup and saucer, and the tea-cup ceases to slide in the saucer.

A tea-cup suggests a storm, and that suggests the curious power which oil has of smoothing the sea. The oil spreads over the surface of the water until the layer is only about the five-millionth of a millimetre in thickness. A figure of that kind is apt to mean little; I will therefore try to give you an impression of the minute quantity of oil needed in another way. In 1919 an oil ship was wrecked inside the Lizard. The oil tanks were burst open and the oil rapidly escaped. There has been no sensible quantity of oil in the wreck for the last six years, yet sufficient still escapes to the surface of the sea to produce an obvious "smooth" for a mile or more to leeward. The effect of a film of oil of quite invisible thickness upon the sea is very real. A vessel labouring in a sea-way or running before a gale can, and does, find some measure of safety by streaming bags filled with oil to windward, and Pliny records how the oyster fishers used oil to calm the surface of the sea so that they were more easily able to work.

It is obvious that the presence of this oil film cannot seriously modify the energy of great seas, say, a quarter of a mile from crest to crest; but when seas enter a "smooth" they change their character with dramatic suddenness. They lose their viciousness, and the moment they are in the "smooth" take the character of those relatively harmless undulations which do not break on to a vessel, but merely make her roll and pitch. The question how the oil film, so tenuous as to be of invisible thickness, curbs the seas is an interesting one, and the attempt to answer it will inevitably introduce us to the chief properties of films on water.

In the late 'nineties a most ingenious method of demonstrating the existence of films on water, and of controlling them for experimental purposes, was devised by a German lady, Fräulein Pockels. I think I may say without exaggeration that the immense advances in the knowledge of the structure and properties of this fourth state of matter which have been made during this century are based upon the simple experimental principles introduced by Miss Pockels. I propose now to make use of her method. Here is an oblong trough of metal filled with water. On the surface of the water, quite invisible to you because it is even thinner than the invisible dead black portion of a soap film, there is a layer of greasy contamination. If I lay upon the trough a strip of glass or metal so that it touches and is wetted by the water, and move it along, I can compress the superficial film in front and expand it behind. Both processes are easily rendered visible by scattering lycopodium dust on the surface.

The capacity which these films have of expansion is easily shown by sweeping the natural film to one end, thus leaving a tolerably clean surface of water behind. Some lycopodium dust is now placed at one end and the surface touched with a platinum wire, the extreme tip of which has just been dipped into an oil. The dust particles are swept swiftly away in front of the advancing film of oil, although the film itself is absolutely invisible.

The film tends to spread, but the surface of the water in virtue of its surface tension tends to contract. It is this same surface tension which rounds up drops of fluid to spheres, or as near an approach to the spherical

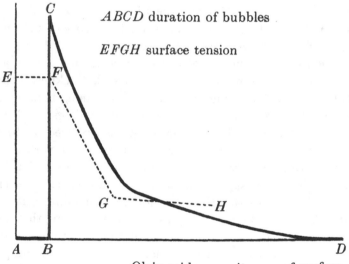

Oleic acid per unit area of surface

shape as other forces which may be operating, such as gravity, permit. There are therefore opposing influences: the tendency of the water to contract, opposed by the tendency of the film to expand, with the result that composite surfaces of oil and water have a surface tension less than that of pure water. Composite fluid surfaces have also an enhanced mechanical stability. When a ring of wire a few centimetres in diameter is withdrawn from clean water no film is formed across it, but when the surface of the water is coated with oil it acquires the property of forming free films, which may have an endurance comparable even with that of a soap bubble. In the figure the curve *EFGH* gives the surface tension plotted against the quantity of oleic acid per unit area of surface, and the curve *ABCD* gives the duration of bubbles which have been formed on the surface by allowing air to escape slowly and regularly from an orifice within the trough. It will be noticed that the bubbles are most stable when the film of contamination is just dense enough to begin to alter the surface tension, and that the bubbles have no stability when contamination is either vanishingly small or very great.

Do these properties of composite surfaces, namely, the lowered surface tension and the increased mechanical stability, explain the calming of the sea? In my opinion the answer is "No", but current doctrine would perhaps say "Yes". It has been pointed out that the special capacity of composite surfaces to resist extension and their mechanical stability, which is only another special aspect of the same thing, tends to prevent the inevitable expansion of the surface which occurs when a wave is formed. This has been held to be a sufficient explanation. I do not think it is, and for two reasons. The first is that the surface of the sea always is contaminated by something which lowers its surface tension and gives to it a remarkable measure of mechanical stability. One of the most striking aspects of a heavy gale is the "windrows", which are due to the foam formed when a sea breaks being blown by the wind in long lines over the surface. Foam after all is no more than a collection of bubbles; obviously therefore these naturally formed bubbles have great stability.

The surface of the sea is already covered by a film of matter the nature of which must remain uncertain. Sometimes it is composed of substances, like saponin of vegetable origin, derived from the masses of seaweed flung upon the coast. Foam of this kind is remarkably stable. I have seen it on the day following an on-shore gale knee-deep in the hollows above Flamborough Head. The bruising and shattering of seaweed is however a coastal happening and "windrows" are deep-sea phenomena.

The true explanation was, I think, furnished by Benjamin Franklin in 1773. His discussion is worth reading. It has the spacious dignity and charm which the hurry and specialisation of to-day have of necessity banished from scientific papers. He tells how he was at sea in 1757 with a convoy of 96 sail, the wind being very fresh, and how he noticed a "smooth" in the wake of two of the vessels. He enquired the cause of one of the officers and was told with some degree of contempt, it being a thing which every fool should know, that the "smooth" was due to the fact that the cook had just thrown greasy water over the side. In those days tallow was used to coat the bottoms of vessels to keep them clear of growth, and Franklin also notes the "smoothness" in the wake of vessels which had been freshly tallowed.

Franklin's explanation is based entirely upon friction. The oil makes the sea so very smooth that the wind cannot "catch upon it". I confess Franklin's explanation did not appeal to me at first, but I believe he is right. The comparative safety of a "smooth" is due not to the fact that the seas in it are sensibly smaller than those outside of it, but to the fact that they have been deprived of their viciousness. Now the viciousness of a sea, the degree of danger it carries to the mariner, is measured by its instability. It is when the head of the sea topples over and becomes a mass of water moving with a high velocity that it is dangerous. Within the limits of a "smooth" produced by oil the seas cease to break, or to "crack" as Cornish fishermen say. The wind not only drives a sea forward by its horizontal pressure, but also draws the crest upwards by friction against the surface of the wave.

If the friction between the air and the water be greatly reduced, the wind fails to lift the crest of a wave to the point at which it is blown bodily over by the horizontal pressure. The wave then sinks down to a relatively harmless "swell".

The "catch" of the wind upon the waves is not however confined simply to a direct frictional pull, and here it is that the surface tension phenomena perhaps come in. It is easy to convince oneself that an oil film prevents the formation of ripples—that is, of the very smallest kind of wave. When there is no oil film a great wave carries countless ripples and wavelets each of which gives the wind a direct thrust on the surface. It is to the suppression of ripples and wavelets that the characteristic smooth appearance is due, and when they cease to be formed the chief "catch" of the wind upon the sea is lost.

48

LIVING MATTER

[Colloid Symposium Monograph (1928), VI, 7]

My object when I began this address was to sketch the present position of what might be called the mechanistic theory of life, but I soon found that, if a theory be an inference from positive knowledge, there is no such theory. There is, it is true, a hypothesis, or in more direct phrase, a guess, and about it in the literature of the subject an accretion of analogies and scraps of physics and chemistry which, as biological tests are rarely applied to them, are probably irrelevant to the real issues.

The fact is, a biologist to-day is pretty much where an engineer would be if he knew even in detail the cycle of chemical changes which took place within an internal combustion engine but was wholly ignorant of the disposition of the moving parts.

I must not seem to belittle attainment. This century has seen immense advances in biology, but when one is trying to think in terms of mechanism they lie off the picture, and for two reasons, either because they are inexplicable, or because they are advances in our knowledge of the results of the activity of living matter and not of the working of the machine itself.

In my student days some forty years ago, there was a great body of knowledge of form and function, but living matter itself was accepted in a curious way as something inevitable and mysterious. The great third edition of Foster's *Text Book of Physiology*, which may fairly be said to have made both the British and American schools, contains only three pages of generalities concerning living matter. Strictly speaking it is silent as to the inner nature of protoplasm.

The mystery of life is as great now as it was then, and it were wise to recognise the fact.

My thesis soon narrowed itself down to the ungrateful task of displaying the difficulties of the problem, and as a first step, let me state a belief which I have held for thirty years or more. It is that nothing is to be gained by claiming living matter as colloidal. There is as much reason to call it crystalloidal as colloidal and that much is just no reason at all. When a sufficient definition of colloid is forthcoming, it will be time seriously to consider whether living matter conforms to it. At present the colloidal kingdom seems to be an Alsatia wherein difficult states of matter find refuge from a too exacting enquiry.

A definition of the colloidal state which certainly had a vogue was in terms of the size of the units. An emulsion which turned up fortuitously in the course of some work on lubricants must be held to have disposed finally of a definition of that kind. It was a stable emulsion whose droplets were uniformly 3 mm. in diameter. By shaking it violently the droplets could be broken up but, on standing, they reverted slowly to the standard size—3 mm. in diameter! By no test was it other than a typical solution save in its Gargantuan dimensions.

The most curiously life-like substance to my knowledge in the non-living world was a wax whose behaviour was described by Kohlrausch more than fifty years ago. Obviously it was a plastic solid but there is no justification save an unwarranted extension of the term for calling it colloidal. Kohlrausch states that a cylinder of this wax hung from one end, then bent in succession say north, east, north, south and so on, would when left to itself, repeat the movements but in reverse order. Now the life-like quality lies not in the running down of a store of energy but in the existence of a timing device.

Looking back to my student days I seem to see an advance towards a knowledge of the mechanism of life in three directions. Of these, the first and most important is the accumulation of evidence which points to the existence of two processes distinguishable from one another and corresponding to the structural division into nucleus and cell substance. The chemical cycle of the former involves nitrogen and it is concerned with growth and repair. It is doubtful whether this process can cease without involving the living cell in that irreversible change called death. The chemical cycle does not include the production of simple acids such as lactic acid.

The other process is characteristic of the cell substance, that is to say, of the portion of the living cell which envelops the nucleus and which is immediately responsive to those variations in the environment which the physiologist calls stimuli.

This process—I use the singular though it may vary in character from mass movement to the specialised molecular wave known as the nervous impulse, or the movement of water and complex substances in the act of secretion—this process, I say, is characterised by the oxidation of carbohydrate to produce the necessary energy, and by the production of acid. There is also a change of electrical potential, the material in the active state being galvanometrically negative to that in the resting state.

The best known example, the one that has been most fully worked out, is the contraction of skeletal muscle. The normal cycle includes two phases, an active phase which is anaerobic and in which lactic acid appears, and an aerobic phase where glucose is oxidised and the energy set free utilised in part to restore the "loaded" state. This normal cycle, however, is conformed to only when the fuel glucose and oxygen are present in adequate quantities. With the elusiveness characteristic of living matter, wide departure from the normal cycle occurs when either fuel or oxygen is deficient.

I have ventured to include the act of secretion of a gland cell, such as a cell of the kidney or salivary gland, in this class though I believe there is still doubt as to whether this cell involves either a variation of electrical potential or the appearance of acid. With respect to the former, I find it difficult to believe that water-holding electrolytes can be drawn through even living matter without there being a potential gradient. It is certain that the normal act of secretion involves the expenditure of energy for reasons familiar to every physiologist, the chief being the fact that it is accompanied by an uptake of oxygen.

The measurable phenomena of an organ are the sum of a very large number of changes. They are statistical phenomena and difficulty in detecting an electromotive change may be due merely to lack of orientation of the units.

The proof of the existence of two distinguishable fundamental processes in the living cell can hardly be summarised, it is so abundant and diverse. There is, first of all, the many experiments ranging from cutting operations on unicellular animals, such as *Stentor*, or on the nerve cell, to the analysis of the properties of the ovum.

A most profound observation was made by Chambers. When the nucleus is injured, repair is impossible, the entire cell perishes, and there is no production of acid. But when the cell substance is injured, repair and recovery of state commonly follow and at the seat of injury acid is formed.

Set this last fact alongside the other well-known fact that the death change in skeletal muscle consists in, amongst other things, a catastrophic destruction of the carbohydrate reserve and the appearance of lactic acid.

If the detection in a general way in the working of the machine of two fundamental processes is the first great advance, the second assuredly is the attainment of some degree of certainty as to the nature of the moving parts.

It will be well here to admit complete ignorance of the nucleus machine. We do not even know that it is a machine in the proper sense of the word. Are there cyclic changes of energy in which the free energy is directed to particular purposes? If anyone has the answer to this question, I must confess ignorance of it.

But, despite this ignorance, there are two orders of facts which finally fix upon proteins as the moving parts, no matter what the processes may be. The first is the certainty that the diversity in the machine which determines the differences between the strains of living matter on the earth—between the hundreds of thousands of plants and animals—is based not upon their chemically active parts, such as their engines, but upon differences in the chemical constitution in the proteins out of which the living substance is mainly built up. The processes of digestion and such phenomena as anaphylaxis and the precipitin reactions prove this.

The second line of evidence is furnished by what we know of muscular contraction. The moving parts of any machine are those parts which direct the available energy and are not themselves implicated directly in the chemical cycle. To take the second character first—it is certain that proteins, despite the fact that they form some 80 per cent. of the solids, are not directly involved in the chemical cycle of contraction of muscle. A muscle is an internal combustion engine which normally burns carbohydrate.

This assurance as to the significance of proteins is a great advance on the position in my student days when their chemistry was only guessed at, and when they partook of the mystery of life itself. I recollect receiving reproof from one of my elders when I attempted a straightforward explanation of the behaviour of certain proteins in solution.

The third great advance is in the recognition of ferments as the active agents in the machine. In my early days ferments were still classified as organised and unorganised. The former were living cells, such as yeast cells, the latter free ferments produced by living cells. I cannot recall anyone bold enough to suggest that inside the cell were ferments whose business it was to control chemical change.

Buchner made the first great necessary step when he conducted alcoholic fermentation in the absence of living cells but by ferments extracted from living cells. Since then endoferments have been demonstrated in plenty. Meyerhoff, for example, has prepared from muscle an extract practically free from protein which contains a catalyst, or catalysts, which reproduces *in vitro* the changes in carbohydrate characteristic of intact muscle.

Let us look a little more closely at this question of catalysts, for it is not without its difficulties. Clearly their activities are controlled during life because at death they become a disorderly mob who pull the very fabric to pieces, protein and all becoming involved in the breakdown. They are during life rigidly restrained, subject to a coordinating timing device in the intact machine.

It is a plausible suggestion, due to Jordan Lloyd, that the proteins exert the control. There are plenty of instances in the chemistry of living matter. Hæmoglobin, the red pigment of blood, carries oxygen from the air in the lungs to the tissues, and the dissociation constants of the hæmoglobin of different animals are such as fit it for the peculiar needs of that animal.

Now hæmoglobin is composed of an active molecule hæmatin which, though capable of oxidation and reduction, is totally unsuited to carry oxygen in the blood. It is only when combined with protein that it is fitted for its task. And there is more. Hæmatin seems to be the same in all animals in which it is found, but the protein ally is different. The inference is so obvious that it need not be stressed.

A few lines earlier I said that at death the catalysts escape from control. Why? Because the structure of the moving parts gets disorganised, but structure here is not gross but molecular structure. Let us look into this for it leads to a curious problem.

At the death of an animal, its muscles die sooner or later, apparently because in the absence of oxygen it cannot work and to maintain itself it needs must expend energy. It is true that the firing mechanism can be put out of gear without general destruction. Foster and Fearon, for example, found that frog's muscles at 0° and in abundant oxygen became non-irritable, but they were not dead for recovery took place when the temperature was raised. There was not even carbohydrate destruction and production of lactic acid *provided the tension of oxygen was high enough*.

When oxygen is withheld and death occurs, the proteins are hydrolysed and the change can be followed quantitatively by estimation of the soluble nitrogen. In about four days a steady state is reached, but further breakdown can be started by freezing the muscle. And now the catalysts have finally escaped from control—hydrolysis shows no signs of stopping even in sixteen days.

This effect of freezing and thawing has been ascribed to mechanical injury produced by the ice crystals. This is not likely since a typical degradation or coagulation of the proteins takes place when the expressed and filtered juice from the muscles is frozen and thawed.

This kind of change is common in colloidal solutions. Milk, a solution of chlorophyll-like muscle juice, can be coagulated by freezing and thawing, but the molecular changes need time for their accomplishment and therefore, as is well known, they can be avoided if both cooling and thawing are sufficiently rapid. I do not recall a colloidal solution which does not give this evidence of a time lag in the process of coagulation.

Now here are two facts which seem to me to have immense significance. By no procedure has it been found possible to freeze and thaw living muscle without causing instant death. Whatever the key structure may be, the presence of what means life, there is no evidence of a time lag in its destruction.

The second fact is that it is not cold which destroys, but dehydration. The freezing-point of frog's muscle is $-0.42°$, and muscle can be frozen and thawed without causing death unless a critical temperature is over-passed, namely $-2 \cdot 0°$. The quantity of water withdrawn in the form of ice at $-2°$ can be calculated. It is $77 \cdot 5$ per cent. of the total water and Moran, to whom these facts are due, finds that $77 \cdot 5$ per cent. of the water can be removed from living frog's muscle by simply drying over calcium chloride without killing it. The muscle when the water is restored is again irritable and has the alkaline reaction and other qualities of living muscle. Muscle is not killed by cold. It can be overcooled to $-4°$ for days without impairment of its properties.

That the living machine is in part, at any rate, a surface energy machine is an hypothesis due, I believe, in the first instance to the physicist FitzGerald. The only case in which it has been possible vigorously to examine the hypothesis has broken down. A. V. Hill finds that the surface tension needed within a muscle fibre is 4800 dynes—an impossible figure. Mines' careful and indeed brilliant work furnishes the best evidence I know of, that the living machine is a surface energy machine, but after all what does the immense literature on the influence of electrolytes amount to? No more than this—that electrolytes like the electric current and other agents, mechanical and chemical, can act as stimuli to alter the rate of working. It is not even certain that the presence of electrolytes is necessary for the maintenance of the machine though, if present, a due balance of concentration is necessary. I hasten to say that the firing mechanism seems to depend for its working upon the presence of electrolytes.

For the maintenance of the machine, four cations are needed, those of calcium, magnesium, potassium, and sodium, and in certain relative concentrations. We do not at all know how they act, though the work of Chambers seems to suggest that they control the internal viscosity and therefore probably electrical potential differences. The chapter on electrolytes may almost be said at present to begin and end with Macallum's profound observation that the necessary electrolytes are those of sea water where living matter probably had its origin, and the necessary relative concentrations also are those of sea water.

The muscle fibre is, so far as its proper function movement is concerned, an internal combustion engine working within singularly narrow limits of temperature. The thermodynamic problems raised by this, I have not space to deal with. Since it works only within narrow limits of alkalinity and is subservient mainly to cations it must, I suppose, be accorded the general characters of an electro negative colloid.

The only exception which, so far as I know, can be taken to the statement that the structure of the machine can be maintained only by continued expenditure of energy is the case of drying mentioned above. It is not possible to conceive of normal chemical change continuing in a muscle dried to the consistency of a tough glass.

The movement of certain forms of marine amœbæ has been analysed by Pantin. Each individual is a minute cylinder and it progresses straightforward in the direction of the major axis. The more external part or wall of the cylinder is viscid, the internal part more fluid, and movement is due to the internal fluid flowing forwards and bursting through to the exterior when it at once assumes the consistency of the wall. In the meantime, the wall at the hind end becomes more liquid and flows in.

This simple example of movement is worth mentioning because it is an instance of the adaptation of a widespread and possibly universal property of living matter to a simple end. The property is the increase in viscosity which accompanies the active state.

It is well to remember that this simple animal does much more than move. It is sensitive to changes in its surroundings, it selects its food, it grows and multiplies, preserving its type.

Growth and reproduction are perhaps the most abiding characteristics of living matter. Neither is to be confused with the growth of a crystal for it is a dynamic state which increases in volume and reproduces itself.

Keeping in mind the unit of life, the cell, we see at once that the capacity is limited. A whale and a flea are complexes of cells and the individual cell is of much the same size in both.

Broadly, it may be said that when a unit of living matter grows to a certain size it becomes unstable and divides, producing two daughter units. But the limiting size varies from that of the submicroscopic, filter-passing virus of many diseases to those curious crawling plants, the *Myxomycetes*, in which the volume of an individual may amount to a cubic centimetre.

There does, however, appear to be a fundamental relation, for these larger cells always have more than one nucleus. It looks as though the volume ratio of cell substance to nuclear substance cannot exceed a certain value.

When we bear in mind Chambers' observation on the influence of the nucleus upon repair, it seems safe to conclude that the limiting factor is the distance at which the nucleus ceases to be able to exert its control. The range of influence of the nucleus estimated in this way is of the order of the range of the capillary attraction of a solid surface in orienting the molecules and so controlling the structure of a fluid with which it is in contact.

Here apparently is a secure conservative conclusion, yet, with characteristic elusiveness, living matter furnishes an exception so stupendous as to be derisory.

A nerve fibre consists essentially of a delicate cylinder about 3 mm. in diameter. This cylinder, called the axon, is the very tenuous outgrowth of a nerve cell. It may, in a large animal, be metres in length, yet the capacity for growth and repair of every part of this process is wholly dependent on the nerve cell from which it has its origin—that is, upon a microscopic speck of matter only 50–100 μ in its major axis. Let the axon be cut at no matter what the distance from the nerve cell and the part cut off dies and disintegrates, whilst the part which retains connection with the nerve cell not only persists but, if opportunity offers, it will ultimately grow down to remake contact and restore function.

This fact is a commonplace of the text-book of physiology. To me it has always been almost the most intriguing mystery of living matter. How shall the capacity for gathering unformed and unlike matter to itself persist in so specialised a structure as the axon of a nerve fibre? And how shall the capacity be controlled by a portion of matter of negligible size situated a yard away?

Any theory of the mechanism of life must take account of its amazing elusiveness. Considered from the standpoint of purpose, we may say that it can become as labile as it needs or as obstinately persistent. Of its fidelity to type, heredity offers countless examples. Of its labile qualities, let this instance suffice:

In the course of his remarkable experiments on newt embryos, Sheeman has found that most tissues transplanted at an early age from one situation to another adopt the character of the new environment. Potential skin, for example, transplanted to brain becomes brain, whilst potential brain transplanted to skin becomes skin. When we contrast the qualities of skin, a mere integument, and of brain, the seat of the most mysterious aspects of life, this mutual convertibility becomes simply fantastic. It is even possible to overpass the limits of species. A portion of potential skin of species A planted amid the potential brain of species B becomes brain, but it remains a portion of the brain of A habited within the brain of B!

Let us in conclusion consider the ovum as a physical system. Its potentialities are prodigious and one's first impulse is to expect that such vast potentialities would find expression in complexity of structure. What do we find? The substance is clouded with particles but these can be centrifuged away, leaving it optically structureless but still capable of development.

On the surface of the egg there is a fine membrane, below it fluid of high viscosity, next fluid of relatively low viscosity, and within this, the nucleus which is, in the resting stage, merely a bag of fluid enclosed in a delicate membrane. How shall sources and sinks of energy be maintained in a fluid composed, as to over 0 per cent. of water? They undoubtedly are there for the egg is a going concern, taking in oxygen and maintaining itself by expenditure of energy.

Clearly the ovum is possible only as a paradox. It is no pangenetic structure—a mosaic of all the parts to which it will give origin. Tristram Shandy's theory is false. To play its part it can only be the simplest form of living matter, but its simplicity is neither that of a machine nor of a crystal but of a nebula. Gathered into it are units relatively simple but capable by their combinations of forming a vast number of dynamical systems into which they fall as the distribution of energy varies. After all, a nebula holds within itself the beginnings of a history more complex even than that of an ovum and yet, so far as structure is concerned, it is but a simple affair!

The more there is known about living matter the more there is revealed a curious simplicity. Sheeman finds skin transmuted to brain or brain to skin, but the agent which effects the change appears to be a chemical substance probably of quite ordinary character. You may lead living matter as you may a donkey with a carrot—but you have to choose the carrot with some care.

Biology halts on the mechanical side because it needs the services of men who are at once real physicists and real biologists—both faculties being within the same brain. Biochemistry has made its great advances because it has been served of late by men who are both real chemists and real biologists.

Our difficulties need restatement by a physicist with ingrained knowledge and wide experience of the properties of matter. I can perhaps illustrate my meaning by a phenomenon of great biological importance, namely chemotaxis. When microscopic free swimming organisms are present in water of uniform quality, they distribute themselves at random; but if a gradient of chemical reaction be established by the introduction of a trace of weak acid into one part of the water, they gather in the more acid or less acid region according to their nature and previous history. The biologist has been accustomed to express facts such as these in terms of purpose, as an attraction towards or repulsion from the acid.

The facts are, however, capable of expression in much more concrete terms. Let us assume that the acid slows the rate of movement of the organisms; what will be their final distribution? Kinetic theory tells us that they will gather where their velocity is least.

If, as I firmly believe, biology stands now in urgent need of real physicists, how shall such men gifted with biological insight be obtained? Most readily by making a new form or crystallisation centre of knowledge.

All of you will recollect that curious intellectual ferment which led in the seventeenth century to the foundation in Italy, France and Great Britain of so many academies devoted to learning. It was preceded and in part accompanied by the promulgation of schemes, often fantastic but always interesting, for "Colleges", all of which bore the stigmata of their derivation from the monastic ideal. Each scheme included a set of rules which should govern conduct.

Let me give you my ideal of a Biological College. It should have three floors—a ground floor for molecular physics, a first floor for biophysics, and a top floor for cell mechanics. And of the staff the professor of molecular physics should have no responsibility for biology, the professor of biophysics should be a Mr Facing-both-ways, responsible to physics and to biology, whilst the professor of cell mechanics should be a biologist pure and simple. That college I should expect to provide the new synthesis of knowledge of which biology stands in need.

Of rules there should be none and of precepts only one—To publish is good but not to publish is better. I would that precept could have been followed with respect to this address!

49

A MICROSCOPIC STUDY OF THE FREEZING OF GEL

[*Proc. Roy. Soc.* (1926), A, cxii, 47]

[PLATE XIII]

The curious spheres described by Moran, consisting as they do of a succession of shells, afford unmistakable proof that the formation of the ice phase inside a gel may not only vary in rate but actually intermit. This study was undertaken in the hope of throwing some light upon this phenomenon. It has revealed two unexpected facts, namely, that, save in very dilute gels, the course of internal freezing is usually intermittent, and that, instead of pure ice, a solid solution of gelatin and ice separates. Pure ice can and does sometimes form in the shape of rounded crystals scattered throughout the gel, but in the common type of freezing, by spheres or rays spreading from centres of crystallisation, it is always a solid solution which separates.

The current conception that the spongy structure found in gels after being frozen and thawed is due to crystals of ice is wrong. It is due to the desolution on rise of temperature and fall of pressure of the solid solution mentioned above. Actually, so far as my observations go, when crystals of pure ice melt, the water is reabsorbed at once by the surrounding gel, leaving only a tiny cleft.

Neither the optical properties nor the behaviour on thawing of the ice phase support Moran's suggestion that it is at any stage a mixture of ice crystals and particles of dehydrated gel.

PART I. MICROSCOPICAL OBSERVATIONS

Freezing was watched under the microscope in cold chambers at $-7°$, $-11°$ and $-12·7°$ respectively. All appliances and reagents were at the temperature of the chamber. With the exception of numbers 1, 2, 8 and 9, the figures are from free-hand sketches made as carefully as the rigorous conditions permitted of.

The process was followed in plates of gel, roughly 0·5 mm. thick, prepared by placing a drop of melted gel on a slip of glass, covering it with a very thin sheet of glass, and allowing it to set at room temperature. Ordinary medicinal paraffin was run round the edge to prevent evaporation, and the preparations were stored at 0° for a few days before use.

The following types of freezing were found:

         (1) Circles.
         (2) Rays.
         (3) Disseminated.

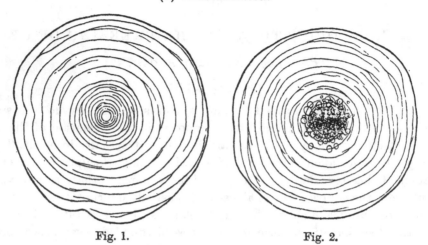

Fig. 1.   Fig. 2.

40 per cent. gel, frozen at −11° (× 100 diameters, from photographs).

Fig. 1.   A circle.

Fig. 2.   Secondary areas forming in central part of a circle.

Circles were undoubtedly the equivalent of the spheres observed by Moran. Each was about 0·5 mm. in diameter, and consisted of a central circular area surrounded by rings (Fig. 1). With high magnification the rings were seen to be separated from one another by membranes (Fig. 3 (*a*)),

(*a*)   (*b*)

Fig. 3. 38 per cent. gel, frozen at −11°. Part of a circle highly magnified: (*a*) before, (*b*) after, rapid thawing. To save time, the secondary areas, which now completely occupy zones 1, 2 and 3, were sketched in only in places.

about 0·5 μ thick, of dense gel, which were curved in a vertical plane. The structure, therefore, was that which would be produced by compressing one of Moran's spheres between two planes. The membranes separated zones of optically homogeneous material which appeared to be pure ice (Fig. 3 (*a*)), but which, on thawing, was found to be a solid solution of ice and gelatin (Fig. 3 (*b*)). Freezing obviously had been intermittent.

Rays, like circles, were always products of intermittent freezing, being divided by curved membranes into compartments filled by homogeneous solid solution (Fig. 5). Circles formed at first rapidly and then slowly. In the first period, growth was too rapid, and in the second too slow, to be followed. Rays, however, advanced at a rate which allowed the process to be followed under a high power with ease.

Rays and circles are merely minor variants of the same type of freezing. Sometimes a circle would stop growing when the diameter had become about 0·5 mm. Others at or near this limit would continue growing by rays, which often advanced with the same velocity, so as to preserve the circular contour. The membranes of the circle could be seen bulging into the base of the rays (Fig. 6).

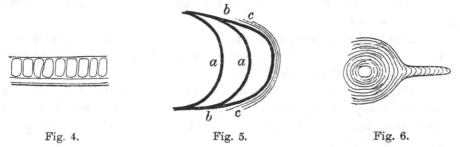

Fig. 4.   Fig. 5.   Fig. 6.

Fig. 4. 40 per cent. gel, frozen at −11°. Transferred to 3° to allow secondary areas to develop fully.

Fig. 5. Tip of a ray showing membranes (*a*), luminous zone (*b*), and fine etched lines (*c*).

Fig. 6. Sketch of a circle with a single ray growing from it.

Disseminated freezing was of two types: (1) The gel was everywhere closely studded by crystals with rounded edges very regular in size, each being about 20 μ in diameter. Each crystal was of pure ice, and each began as a minute sphere whose growth was rapid in very dilute gels, and too slow to follow in gels from, say, 20 per cent. upwards. Freezing did not appear to be intermittent.

Sometimes circles, with or without rays, would form at a few centres, say, 6–10, and after a while cease growth. Eight or more days later the remainder of the gel would be found to be occupied by crystals.

(2) The gel was occupied by minute spheres, all of the same size, namely, about 3 μ in diameter (see Moran's Fig. 2). This kind of freezing was found by Moran only at temperatures below −19°, when cooling was very rapid. Probably in this second form of disseminated freezing ice began to form during the fall of temperature, but ceased when the viscosity became too great, in which case the two types are merely different stages in the same process.

Nothing is known as to the nature of the centres of crystallisation, but it is clear that they were in two classes, a few (not more than, say, a dozen in a square centimetre) which became active with moderate cooling and gave rise to circles and rays, and others, many hundreds in number, which in strong gels became active with moderate cooling only after a long latent period lasting for days and at once with great cooling ($-19°$ or more). The less the degree of cold, the fewer of this second class became active.

Prolonged study would be needed to evaluate all the variables, especially as some unknown factor operates. All the preparations from the same mass of gel, for example, do not give the same results. Thus, in ten preparations of 15 per cent. gel exposed to $-11°$ for one day, there was no freezing at all in one preparation, and circles, with or without attached rays, from two to five in number, in all the others. The following conclusions, however, appear to be certain:

Freezing is intermittent, with separation of a solid solution in gels from 2 to 40 per cent. (the latter the strongest used) when exposed to temperatures from $-6°$ to $-13·5°$.

Freezing is always disseminated and very rapid, with separation of a multitude of crystals of pure ice, in gels of less than 2 per cent. This same type of disseminated freezing may occur in stronger gels, but it takes a week or more to appear.

Exposed to $-19°$ C. in gels between 15 and 40 per cent., freezing was always disseminated, and only minute spheres of ice were formed. Weaker gels were not tried.

At $-2·6°$ to $-3°$, no spontaneous freezing occurred.

Callow found that when he removed gels in which spheres had begun to form at $-11°$ to $-3°$, pure ice was deposited about them; therefore a solid solution separates only at temperatures below $-3°$. It will appear later that the solid solution is unstable at $-3°$.

It might be supposed that the separation of a solid solution depended upon the rate of cooling. Slow cooling was tried, ten days being occupied in the change from $0°$ to $-10·4°$. Freezing was intermittent, and a solid solution separated. Owing to the capacity for overcooling, however, slow cooling does not imply slow freezing.

Membranes. The end of a ray has a smooth rounded contour suggesting a surface moulded by surface tension. The gel ends in a luminous zone (b, Fig. 5), which may be a diffraction halo or may be a zone of denser gel. It is about $0·5\,\mu$ wide. Beyond it in the gel are finely etched lines (c in the figure). Sometimes lines appear also to be within the luminous zone, but this appearance is probably due to the curvature of the surface in the vertical plane. Fine etched lines, similar to those described above, are also found in the gel at the outer edge of circles.

As growth proceeds membranes may be seen to become detached, so as to divide the ray into compartments (Figs. 5 and 6), each of which is filled completely by optically homogeneous transparent material.

It might be supposed that each compartment represented a single block of solid solution and that the appearance of membranes is really due to diffraction halos at the surfaces separating them. This is negatived by the fact that the membranes are singly refractive, while the contents are uniformly doubly refractive. With crossed Nicols the former are dark against a luminous background. The membranes also persist on thawing.

Sufficient cause is found in Part II why freezing should be intermittent, the pauses being due either to an increase in the internal friction, or to the concentration of the gel at the ice face increasing until it is in equilibrium with the ice phase. The microscope shows, however, that when a pause occurs freezing starts again, not at the original face but at a new face within the gel, thus leaving the characteristic membrane of dehydrated gel behind.

The explanation probably is simple. During intermittence the temperature at the face will fall and the hydration of the gel rise. The fine lines described above show that cleavage occurs in the gel owing to uncompensated stresses, and each cleft will be the locus of a thin layer of dilute solution. Such a solution will have a higher freezing-point than that of the concentrated gel on either side, and therefore freezing will start in it. The layer of dilute solution will, like the layer of fluid of insensible thickness on the surface of overcooled gel referred to by Moran, be the locus of centres of crystallisation of higher potential than any in the interior of the gel. The clefts at the end of a ray are seen as they appear after thawing in the photograph (Fig. 9).

Thawing. By a simple device it was possible to raise the temperature of a preparation on the stage of the microscope either quickly or slowly, and thus watch the changes.

Two distinct events happen—distinct because the first can happen without the second. They are a separation of the solid solution into ice or water and concentrated gel at about $-6°$, and a violent transference of water from the sponge of concentrated gel so formed to the surrounding gel, which occurs at about $0°$ C. The sponge commonly described in the interior of gels which have been frozen and thawed is due to the first of these changes of state, and the final volume occupied by the sponge when thawing is complete is very much less than the volume of the ice phase, owing to the second process. Neither process is arrested by the use of strong fixatives such as formaldehyde. Some years ago[1] I drew attention to the artifacts caused by fixatives. No more striking instance could be

[1] *Journ. Physiol.* (1899), XXIV, 158.

furnished than the false picture given by strong fixatives of the process of freezing in gels, even when used as Moran used them, in the most favourable fashion.

As temperature slowly rises, small spheres appear in the solid solution, which slowly increase in size. In circles they appear first in the central area. Fig. 1 shows a circle at $-11°$, and Fig. 2 a circle in which small droplets are beginning to appear at the centre owing to a slow rise of temperature.

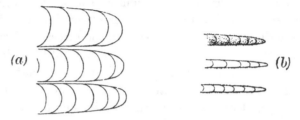

Fig. 7. 15 per cent. gel, frozen at $-11°$. Rays before thawing (a), and after rapid thawing (b). The fine grained structure produced by rapid thawing is indicated in only one of the latter.

As the word "sphere" has been used to denote an entire system, it will help to avoid confusion if these spherical droplets are called secondary areas. The secondary areas are smaller and more numerous the more rapid the rise of temperature. One has therefore in the solid solution, as in entire mass of gel, a great number of centres or nuclei, of which the number which become active is determined by the rate of change of temperature. Secondary areas can be developed to their limit of size by taking a gel from, say, $-11°$ and keeping it for a day at $-3°$. They are then found to occupy the whole distance between the membranes (Fig. 4).

Figs. 3 (a) and (b) illustrate the effect of more rapid warming. Very rapid warming or flooding with 40 per cent. formaldehyde or absolute alcohol—of course, after removal of the cover glass—produces a fine-grained structure composed of a multitude of secondary areas not more than $0\cdot1\,\mu$ in diameter. The variation of the size of the areas with the rate of warming proves that they are not preformed.

Secondary areas do not appear to communicate with one another—a fact which makes it difficult to understand the expulsion of water which takes place on further rise of temperature. It is impossible to follow this process; it takes place with such startling rapidity. The quantity of water lost is considerable—rays shrink by 30–50 per cent. of their volume, and the water so lost is at once taken up by the surrounding gel. Figs. 7 (a) and (b) show the appearance of rays before and after rapid thawing. The serrated edges indicate where the membranes have folded together.

Expulsion of water from the solid solution occurs also in circles, but a large part is trapped between the concentric layers. The result is that the membranes are split. This is clearly shown in Figs. 3 (*a*) and (*b*), which show part of a circle highly magnified before and after thawing.

The capacity of the gel for reabsorbing water is shown also in the case of disseminated freezing. Each ice crystal disappears, and in its place is left a small cleft. The space occupied by the crystal therefore vanishes almost entirely, but, as might be expected, the collapsed walls do not join together. It is to be observed, however, that such extensive reabsorption of water occurs only when the gel is fairly concentrated (20 per cent. and upwards); when it is very dilute (under 2 per cent.), the spaces occupied by the crystals round up on thawing, with but little decrease of volume. The result is an open sponge, the spaces in which, however, do not appear, at any rate at first, to communicate with one another.

Polarised light. Before any freezing occurs, the gel is singly refractive, even though it has been overcooled for some days. When freezing has taken place, that part of the plate unoccupied by circles, rays or crystals is doubly refractive; the solid solution is also doubly refractive; the membranes are singly refractive. After the formation of secondary areas their walls of concentrated gel are singly refractive.

Even when thawing is complete the plate of gel remains doubly refractive. The molecular structure imparted by freezing therefore persists. Whether it ultimately vanishes was not determined.

These observations suggest an explanation of a striking fact discovered by Callow.[1] He seeded cylinders of gel overcooled to $-3°$ at one end and observed the rate of advance of the ice. Up to a concentration of 2 per cent. the velocity was of the order found in pure water. At 2 per cent. it suddenly dropped from 960 cm./hour to 40 cm./hour. My own observations were made entirely upon spontaneous freezing, but they showed that at some low concentration of the gel intermittent freezing appeared and disseminated freezing with separation of pure ice became very infrequent. The drop in velocity noted by Callow may therefore have been due to intermittent freezing replacing a continuous process.

These observations throw some light upon Moran's dilatometer curve (Fig. 9)[2] which shows that as temperature rises there is a sudden contraction at between $-6°$ and $-7°$. This is the temperature at which secondary areas appear; it is, therefore, a decrease of volume due to desolution of the solid solution. When thawing is complete, the volume does not always return to its original value—there is persistent slight increase. This is probably due to the persistence of that molecular pattern into which the gel is thrown by the stresses set up about the places where actual

[1] *Proc. Roy. Soc.* (1925), A, cviii, 307.   [2] *Proc. Roy. Soc.* (1926), A, cxii, 30.

separation of an ice phase takes place, and which is manifested by the persistence into the thawed stage of double refraction.

Attempts were made to determine whether the separation of a solid solution was due to the rate of freezing by exposing gels to $-3°$. They were not successful owing to the overcooling. Gels from 2 to 40 per cent. failed entirely to freeze at $-3°$. Callow obtained considerable nodules of ice in the interior of a large mass of gel in a test tube, but each was deposited about a minute sphere which had previously formed at $-11°$.

My acknowledgments are gladly given to Mr Hale for microphotographs taken under most trying conditions, and for making all the preparations of gel needed.

PART II. THEORETICAL

Moran distinguishes between internal and external centres of crystallisation and points out that ice formation is confined to the latter when the overcooling is not too great.[1] He also finds that when ice forms wholly on the surface of the gel a true phase equilibrium between ice and gel is reached in what is, for the colloidal state, a short time.

Phase equilibrium of the kind described is so rare in the case of colloidal systems (I cannot recall another instance) as to deserve some thought. It is no doubt conditional equilibrium and not the absolute equilibrium which simple solutions exhibit, because it will certainly depend upon the previous history of the gel upon, for example, the temperature at which the gelatin was dissolved and the rate at which gelation took place, since such things influence the structure of gels. The point is that when the structure had been finally established the phase relations with ice became a pure function of temperature and, as will appear later, of pressure.

Moran took care to assure himself that the phase relations observed by him were between pure ice and gel. The interface was effectively plane, for, in the thin discs employed, the curvature of the rim was small and its area only a small fraction of the whole surface.

Consider the formation of ice on the surface of one of his thin discs. The ice face would advance inwards at a rate equal to the mass of ice (m) deposited in unit time multiplied by its specific volume (S_i). The gel face would retreat at a rate equal to the same mass multiplied by the specific volume of water (S_w) if the minute contraction which occurs when gels of medium strength absorb water be neglected. Since S_i is greater than S_w a pressure would be set up which would crack the shell of ice at the edges of the thin discs, and Dr Moran tells me that the shell of ice always was found

[1] *loc. cit*

so cracked. We may therefore take it that the equilibrium obtained was not only at a plane face but also sensibly under constant pressure.

When ice is being formed we may, following H. A. Wilson, suppose that the water is being driven from the gel to the ice by a pressure A which is equal to the difference in the internal pressure W_g of the water in the gel and of the ice W_i. That is, $A = (W_g - W_i)$. This is the pressure which is equal in magnitude to that which would have to be applied to a piston impermeable to ice to stop freezing. For small values of A it may be put equal to the difference in the vapour pressures of gel and ice at the interface multiplied by a constant.[1]

The pressure A may be supposed to drive water on to the ice face through a layer of gel of depth a proportional to the range of molecular forces, against a frictional resistance η. The velocity of ice formation V will then be

$$V = \frac{A}{a\eta} \qquad (1),$$

which, for small overcooling, may be put $= C\,(\theta_0 - \theta)$.

The effect of an external pressure upon the internal and vapour pressures of a gel is not, so far as I am aware, known, but, save perhaps for very concentrated gels of gelatin in which all the water is absorbed with considerable evolution of heat and contraction of volume, increase of pressure is certain to have much the same effect as it has on water. Therefore, if both ice and gel are subjected to an external pressure P instead of a pressure applied only to the ice, W_g will be increased, but not to the same extent as W_i. Therefore, though P can stop ice formation, it will have to be much greater than A to do so.

The pressures A and P are the only ones which have to be considered if either ice or gel, or both, are free from external constraint, as they would be, for example, if they were contained in a cylinder with frictionless walls open at one end. This condition is practically realised when a cylinder of gel enclosed in a test tube is seeded at one end if the gel be dilute, because then its adhesion to the glass will be slight. This is the condition which obtained in Callow's measurements of the velocity of ice formation.[2]

When both ice and gel are under external constraint, as they would be if they were enclosed in a rigid envelope, there is a third pressure π normal to the ice face and due to the expansion of water on freezing. This pressure will diminish the effective overcooling by lowering the freezing-point until freezing ceases, when the ice phase will be in equilibrium with gel at the temperature θ and the pressures P and π.

When freezing occurs in the interior of the gel these conditions are realised, except that the interface is no longer plane and the rigid walls are

[1] H. A. Wilson, *Proc. Camb. Phil. Soc.* (1898), x, 25.
[2] *Proc. Roy. Soc.* (1925), A, cviii, 307.

replaced by the elastic mass of gel. Freezing must be stopped at some point by the elastic compression of the ice by the surrounding gel unless the latter is fractured, which it never seems to have been. As a matter of observation, with moderate overcooling freezing did start at only a few centres in the interior, and after a relatively short time ceased.

When ice forms from water the direct influence of the ice face may be supposed to end at the distance a measured along the normal. When it forms from gel, however, owing to the fact that the interface is impermeable or only slightly permeable to gelatin, a diffusion column is formed beyond the limit a.

Let us call the layer of depth a next to the ice M and the diffusion column N. The movement of water in the latter is due to a gradient in the internal pressure so that the velocity through any elementary layer is given by

$$V_w = \frac{dW_g}{dx}\frac{1}{\eta} \tag{2}$$

when η is the frictional resistance reckoned at the layer.

The expression $V = \dfrac{A}{a\eta}$ differs from the similar expression for the freezing of water in two particulars; the quantity A is a function not only of the degree of overcooling and the pressures P and π, but also of the concentration of gelatin in the layer M. This follows from the phase relations found by Moran.

The resistance η also is a function of concentration as well as of the degree of overcooling and pressure.

The internal pressure of water in the gel W_g is also a function of temperature, pressure and concentration, but the coefficient dW_g/dx in equation (2) is either independent of temperature and pressures or is not the same function as the other quantities.

This must not be taken to mean that if the magnitude of A is changed by, for example, a fall of temperature, the diffusion column N will not change. It means simply that if the gradient of concentration dc/dx be everywhere kept constant throughout N and the temperature or pressure alone changed, there is no evidence to show that the gradient of internal pressure will change. We may therefore assume that dW_g/dx is a pure function of dc/dx.

Let $\theta_0 - \theta$ be the overcooling, c' the concentration of gelatin in the layer M. We have then:

$$A = \frac{\phi'\,(\theta_0-\theta)}{\phi''\,(c',\,P,\,\pi)} \tag{3},$$

$$\eta = F\,[(\theta_0-\theta),\,c',\,P,\,\pi] \tag{4},$$

$$\frac{dW_g}{dx} = f\left(\frac{dc}{dx}\right) \tag{5}.$$

Let the efficiency of the diffusion column have its obvious meaning, namely, the rate at which water is brought to the layer M. It is easy to see that the velocity of freezing will depend not only upon expression (1) but also upon the efficiency of the diffusion column. It is also obvious that if one of the variables, temperature or pressure, be altered, the result will depend upon the rate of change of different processes such as the rate of addition of water to and of its removal from the layer M; it is therefore the second differentials taken with respect to time which are of importance.

The intermittent character of the freezing is expressed algebraically by saying that dV/dt is not always positive. Moran's study of the phase relations shows that at these low temperatures it may even have a negative value. The value will be zero when the concentration of gel in layer M is high enough to be in equilibrium with the ice phase at the local temperature and pressure, but expressions (1) and (2) show also that it may become sensibly zero if the quantity η becomes large enough.

Consider, for example, the effect of a sudden increase in the degree of overcooling. The change would, by increasing η, decrease the efficiency of the diffusion column. It would also, if not too great, increase the rate of removal of water from the layer M and the result of the two processes would be a rapid rise in the concentration of the gel in M.

The chief cause of accumulation of gelatin in the layer M with consequent rise of concentration is, however, the impermeability, or relatively slight permeability, of the interface to this substance. It is easy to see that by reason of this impermeability a plane face of ice would not advance along a cylinder of gel at a constant rate. Callow (loc. cit.), it is true, found the velocity of crystallisation to be remarkably constant, but what he observed was the rate of advance of the ends of rays of ice along a cylinder of gel seeded at one end and not the total ice formation. The gels he used were of low concentration and the overcooling slight. Probably owing to the low concentration the growing points of the rays pushed aside the accumulated gelatin. The ice face also was curved and the efficiency of a diffusion column is greater over a curved than over a plane surface.

The quantity η is an important one in the theory of freezing. The form of the curve connecting the velocity of crystallisation with the degree of overcooling is determined mainly by it. H. A. Wilson[1] points out that in one-component systems the pressure term A increases more rapidly than η for small overcooling, but when overcooling is great η becomes so large as to stop freezing.

In a single component system η is the pressure needed to drive unit mass of the fluid at unit velocity through itself. For a gel it is the pressure needed to drive unit mass of water through the gel in layer M at unit

[1] Proc. Camb. Phil. Soc. (1898), x, 25.

velocity. In reckoning the quantity, however, regard must be had to the movement of the framework of the gel which the movement of the water brings about. Owing to the impermeability of the interface to gelatin, the internal pressure A is called upon actually to compress the framework. η obviously is a more complex term than it is in single component systems, and in any complete analysis it would probably be necessary to express the frictional resistance by two terms.

Since η is taken to include the resistance of a solid framework built of enormous hydrated molecules of gelatin, it is likely to increase rapidly with fall of temperature; it is therefore not a matter of surprise that, save in dilute gels containing much free water, freezing ceases at a very early stage when the overcooling is only as much as 19°.

The framework of gels is not a purely passive structure. In some gels, such as those of silica or fibrin, the framework spontaneously shrinks and water is expelled. Graham gave to this process the name synæresis. Let us call the synæresis of such gels positive. In other gels synæresis is negative, up to a point—that is to say, in contact with water such gels imbibe water and increase in volume. A gel of gelatin has negative synæresis.

The sign of its synæresis must be an important factor in the freezing of a gel, as is obvious if dehydration and hydration are taken in two stages. Let the gel first lose water to the ice. If synæresis is positive, rehydration by absorption from neighbouring gel will be resisted. Synæresis acts like an internal friction which may be very great.

Over the range of concentrations we are considering the synæresis of gelatin gel may be taken to be negative. Nothing is known of the effect of temperature upon it, but the extraordinary rapidity with which water is reabsorbed on thawing shows that it can by no means be neglected.

In any case the framework of a gel with no synæresis is not one which offers no resistance to change of form, because synæresis is a measure only of the intrinsic capacity for spontaneous change.

The internal pressure of water W_g in the layer M next the ice face is a function of the concentration c' of the gel in that layer, but for dilute gels in which some of the water is "free" it will be independent of concentration, and become dependent only for more concentrated gels. The limit between dilute and concentrated, however, is unknown, but Moran's observations appear to fix it at about 50 per cent. All the gels used in this enquiry were below that concentration, but we cannot conclude that A was independent of concentration because the value referred to is the local concentration in the layer M at the face. All we can say is that it would be difficult or perhaps impossible for local accumulation of gelatin seriously to decrease the quantity W_g when the concentration of the general mass of gel is very low, and this, no doubt, is one of the reasons why intermittent freezing was not found in very dilute gels.

Two of the observed relations seem susceptible of simple explanation. The prepotency of the centres of crystallisation on the surface of a gel can be accounted for by the presence there of an insensible layer of very dilute solution, and to the absence of the normal pressure π; and the separation of a solid solution in the interior within certain limits of concentration and temperature is due to the pressure π and the degree of overcooling, since Moran found pure ice deposited at $-3°$ on the surface only where $\pi = 0$, but at lower temperatures pure ice was deposited on the surface and at the same time, as his figures show, solid solution in the interior, and the only difference between surface and interior was that π was zero at the surface and had a positive value in the interior. If, however, the specimen was removed from, say, $-11°$ to $-3°$, pure ice was deposited in the interior about the spheres of solid solution.

Freezing at a spherical surface. Let a sphere of the ice phase form inside a mass of gel large enough for the distribution to be symmetrical about its centre. The pressures at the surface of the sphere are A and $(P + \pi)$. P is the atmospheric pressure. The pressure π gives rise in the gel to a radial pressure and a circumferential tension, both of which vary inversely with the cube of the radius.

Since the internal pressure of water in the gel is increased by pressure, the effect of this distribution of radial pressure will be to decrease the steepness of the gradient of internal pressure of water in the diffusion column about the sphere of ice, so that, if the gradient of concentration remained unchanged, the effect of introducing the radial pressure would be to decrease the rate at which water moved to the ice face. On the other hand, the velocity of the diffusing water through each shell required to keep the rate at which it arrives at the ice face constant varies inversely with the square of the radius. We therefore have as a consequence of the form of the surfaces two effects of opposite sign, that with the negative sign being some function of the inverse cube, and that with the positive sign varying with the inverse square of the radius.

Whilst the ice phase is forming heat will be liberated at the surface of the sphere. If the quantity formed in unit time were constant, and loss of heat by radiation be neglected, the sphere would be at a constant temperature. If the rate of formation of ice varied about a mean value, the sphere would act as a reservoir of heat, so that oscillations of temperature due to variations of the rate would decrease as its radius increased.

The gradient of falling temperature about the sphere will have an important effect upon the efficiency of the diffusion column. Let a given diffusion column at uniform temperature deliver water on to the ice face at a certain rate. Now let the external temperature be varied so that, whilst the temperature at the ice face remains constant, a gradient of temperature falling from the ice face outward is set up. The result will be to increase the frictional resistance η everywhere except at the ice face. From geometry it is obvious that the gradient $d\eta/dr$ so produced about a spherical face will be greater than $d\eta/dx$ at a plane surface if the diffusivity of the gel is the same.

If the diffusivity of the gel were very low, enough heat might accumulate in the sphere to stop freezing until some of it was dissipated. This possibility was explored. Mr Adair was good enough to measure for me the effect of concentration upon the diffusivity of the gel, and found it to be the same as that of still water over the range of concentration examined.

Probably the most important effect of curvature of the ice face lies in its effect upon the redistribution of gelatin. The framework of the gel is being pushed back by the ice face and is also retreating owing to the transference of water to the surface of the sphere. Therefore, when any shell in the gel expands from r_1 to r_2, there will be motion of the molecules of gelatin both radially and tangentially. Since the greater part of the gelatin is contained in a solid framework, the rate of redistribution will be rather that of a solid than of a fluid; therefore, unless the rate of ice formation is very low, low enough to permit of redistribution of the stresses, actual fracture of the structure is likely to occur. The microscope shows that fracture does occur. Clefts in the gel appear about the ice face (Figs. 8 and 9) and, much more rarely, radial clefts appear as fine radial lines.

Moran found that gels in which the concentration was greater than 65·5 per cent. could not be made to freeze. His phase curve becomes horizontal at this concentration. From this he infers that at this concentration none of the water in the gel is available for freezing because it is bound chemically to the gelatin. There is an alternative explanation. It is stated in textbooks on colloids that the freezing-point of water absorbed in swelling may be lowered as much as 100°. The statement has no particular significance unless it means that at, say, $-100°$ ice has been found to separate. It is certain, however, that, in the strict sense of the word, the freezing-point of water in gels of high concentration is lowered considerably. At the same time the internal friction η increases as concentration increases and as temperature falls; it is possible, therefore, that freezing ceases at high concentration because the forces tending to form ice are not able to overcome the internal friction and that all the points on Moran's curve are determined by this equation. If this were the case, however, since η is of the nature of a friction, one would not expect the complete reversibility which he found.

Plate XIII

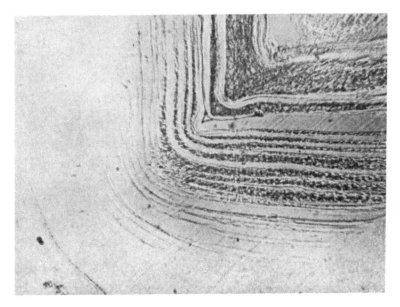

Fig. 9. 23 per cent. gel. Top of ray after complete thawing. Note cleavage lines in the gel. Mag. about 150 diams.

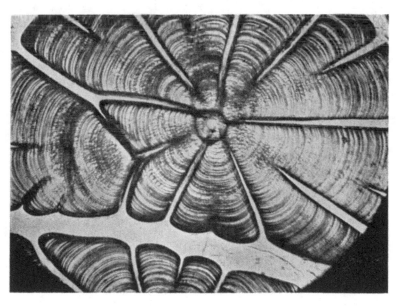

Fig. 8. 40 per cent. gel. Frozen at − 9°C. A circle which has formed rays early. Note secondary areas appearing at about − 6°.

50

STUDIES IN ADHESION. I
With MILLICENT NOTTAGE

[Proc. Roy. Soc. (1926), A, cxii, 62]

Friction measures the tangential reaction at an interface to external forces, and certain relations to time, temperature, pressure and chemical constitution have been described in earlier papers. It seemed worth while to examine the relations of the normal reaction, but nothing has been attempted beyond a preliminary survey of what has proved to be an interesting field.

For the purposes of this paper the word adhesion means simply the normal force needed to detach completely a cylinder from a plate. Measurements of this force are described in an interesting paper by Budgett,[1] which will be referred to later.

The difficulty in discovering the laws of adhesion lies in the fact that, when the lubricant is fluid, anything between zero and a high value can be obtained by varying the time relation and the method of placing the cylinder and lubricant on the plate. To get comparable values one has to seek out mechanically "corresponding" states, to borrow the convenient notation of chemists. One of these states is dealt with in this paper.

Static friction, strictly speaking, is the tangential force per unit area which just fails to cause slipping. It cannot be observed because, owing to the fallibility of our senses, a certain rate of slip enters into all observations. What actually is observed in experiments upon static friction is the force which produces a certain tangential acceleration and it is noticeable that the acceleration varies widely for different lubricants. As a broad rule, it is high when the molecular weight of the lubricant is low, and low (merely a gentle slide) when the molecular weight is high.

Let R_0 be the reaction to the traction just before slipping occurs, then the observed reaction is

$$R = R_0 + \int_0^v \frac{dR}{dv}\, \delta v.$$

v might be called the velocity of release.

As nothing exact is known of boundary conditions in kinetic friction we cannot say whether R_0 really does carry over from the static to the kinetic state. It may, however, be worth while enquiring what happens if it does.

---

[1] *Proc. Roy. Soc.* (1911), A, lxxxvi, 25.

The second term on the right then becomes, with the sign changed, the force producing acceleration.

R_0 therefore (= minus the true static friction) would, on this assumption, be less than R (observed) for lubricants of low molecular weight, and the two, R and R_0, tend to equality as the molecular weight rises. The enquiry cannot be carried further without more knowledge derived from experiment.

The fact that the observed static friction is independent of the quantity of lubricant on the plate, of whether the lubricant is solid or fluid, and of temperature within the limits explored, is perhaps assurance that the observed value is very close to the limiting value. No such assurance, however, is forthcoming for studies of adhesion. It is difficult to settle what exactly is being measured, save in one group of cases, namely, when a solid lubricant is employed, when the force needed to break the cylinder away from the plate without doubt measures the tensile strength of the joint, so that, though acceleration comes in as it does in static friction, the theoretical reaction is clear.

Any value can be obtained for the adhesion produced by a fluid lubricant, for any normal force given time enough will, if it be sufficient to overcome the relatively slight resistance offered by the surface tension of the lubricant, lift the cylinder. When the normal force reaches a certain value, however, the break away occurs instantaneously, and this is a true limiting value because any addition to the force fails to alter the result. As it is a limiting value, it is identifiable. It is not the only identifiable value, that given by solid lubricants, for example, is another. Under certain circumstances it becomes a corresponding value which we will call Value A.

Value A

Methods. Both cylinders and plates were ground to "optical" faces. Each cylinder, no matter of what material it was made, weighed 5·6 gr., and had a diameter of 1 cm. A normal force was applied by a cord attached to the cylinder in such a way that the force was central, and led over a light pulley to a pan for carrying weights. The normal force is the weight in the pan less the weight of the cylinder. The pressure between the faces was varied by placing weights on the top of the cylinder; these weights, with that of the cylinder itself, are called the load. Unfortunately it was necessary to remove the added weights before a measurement could be taken and, as adhesion decreases when the load is reduced, the recorded value is somewhat less than the true one. The measurements were carried out in a chamber filled with clean dry air.

The lubricant was added in one of two ways: a large pool was made on the plate and the cylinder then put into it; or the cylinder was first put on the plate, a little fluid then placed touching its edge and, when fluid had ceased to be drawn underneath by capillary forces, and when therefore fluid was visible all round the edge, more was added to form a large pool.

G. I. Taylor has calculated the rate of fall of a flat disc through fluid on to a flat plate and his equation shows that it would take infinite time to get within molecular distance of the plate.[1] This equation is abundantly verified by these experiments. If sufficient time were allowed the cylinder placed in a pool would fall until its weight was borne by the Leslie pressure[2] due to the attraction of the solid faces for the fluid.

This equilibrium position can, however, be reached quickly by starting from the other end, that is to say, by placing the cylinder on the plate, and allowing the fluid to run underneath. The capillary forces then are enormous and equilibrium is reached in a few seconds. *Value A was taken always from this equilibrium position—it is therefore the force needed to break the cylinder away instantaneously when the thickness of the layer of lubricant is such that the Leslie pressure carries the load.* For all the loads employed, this thickness includes hundreds, if not thousands, of molecules, as is proved by the fact that if the temperature be allowed to fall sufficiently to freeze the lubricant, and the cylinder be then broken away, the layer is found to be of sensible thickness to be measured in fractions of a millimetre rather than in μ even for the heaviest loads employed. This alone is proof, if further proof be needed, that the attraction field of the solids modifies the state of the lubricant throughout a layer many hundreds or thousands of molecules in thickness.

The latent period is the interval which elapses between placing the cylinder in the pool, or forming the pool about it, and the time when adhesion attains a steady value. It may be the time taken by the cylinder in falling or rising in the pool; or the time occupied in the orientation of the molecules of the lubricant in the attraction fields of the solids. When the cylinder is falling the value increases, and the opposite when it rises. The latent period of orientation is always a period of increasing values.[3]

The latent period of orientation can be obtained by following the normal procedure, namely, placing the cylinder in position and allowing the fluid to run under. The values given in Table I were obtained.

It might be supposed that some part of this latent period was occupied by the flowing of fluid between the surfaces, but it must be remembered that the pool was not formed and the measurement was not taken until fluid had ceased to be drawn in. A similar latent period was also found in the study of friction.

[1] *Proc. Roy. Soc.* (1925), A, cviii, 12.   [2] *Loc. cit.*
[3] The latent period in friction is discussed in *Proc. Roy. Soc.* (1924), A, civ, 25, and (1925), A, cviii, 9.

TABLE I. *Load* 5·6 *gr.* *Latent period of orientation*

	Min.	Viscosity
Octane	0	0·0054 at 20°
Cyclohexane	0	0·0089
p-Cymene	0	
Methyl ethyl ketone	20	0·0042 at 20°
Acetophenone	20	
Cyclohexanone	20	0·0280 at
1-3 methyl cyclohexanone	20	
1-4 methyl cyclohexanone	20	
Ethyl alcohol	20	0·0108 at 25°
Butyl alcohol	20	0·0260 at 25°
Octyl alcohol	20	0·07215 at 25°
Benzyl alcohol	25	0·0528 at 25°
1-2 methyl cyclohexanol	30	
1-3 cresol	40	0·1878 at 20°
Carvacrol	40	
Heptylic acid	60	0·0435 at 20°
Caprylic acid	60	0·0575 at 20°

Octane, in which both ends of the carbon chain are alike, and the saturated ring compound cyclohexane, gave no latent period of orientation, that is to say, the first value obtained was always the same as that found after an hour, no matter what the load might be. Why paracymene should show no measurable polarity must be left to chemists to discuss.

The latent periods of orientation for static friction of the three 8-carbon compounds are reproduced here for comparison: octane, none; octyl alcohol, 15 min.; caprylic acid, 60 min.

The latent period seemed to increase slightly with increase in load, but this might be due to the defective experimental procedure which involved removal of added weights before a measurement could be taken.

Paracymene was chosen for the study of the latent period due to rise or fall in the pool, because it gave no measurable latent period of orientation. A pool was first made on the glass plate, a steel cylinder placed in it, and after a known interval the force needed to detach it instantaneously was measured. Cylinder and plate were then cleaned and another measurement made after a longer interval. In this way the curves were obtained for loads 5·6, 115·1, and 259·6 gr. The values of *A* for these loads—that is, the value which would have been reached had the cylinder had time to fall to its equilibrium position—are plotted at the end of the dotted lines (Fig. 1). It is obvious from the form of the curves that it would take a very long time to reach these steady values; these curves, therefore, are completely in accord with G. I. Taylor's equation.

A relation of great theoretical importance which confirmed a similar relation found in the study of friction was got by starting from the equilibrium condition and varying the load.

Steel on glass: p-*cymene*

Load 259·6 gr. Loaded cylinder placed on the plate; fluid then run under, and a pool formed: 254 gr. then *removed* and reading taken after the interval shown in the first column. Force needed to detach in right-hand column.

As quickly as possible			15·2 gr.	= A for a load of 259·6 gr.
30 sec.	...	...	7·5 ,,	
60 ,,	...	...	6·0 ,,	
90 ,,	...	...	5·5 ,,	
2 min.	...	...	5·0 ,,	
3 ,,	...	...	4·7 ,,	
10 ,,	...	...	4·7 ,,	
30 ,,	...	...	4·7 ,,	= A for a load of 5·6 gr.

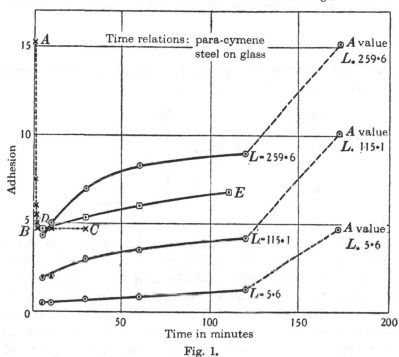

Fig. 1.

The figures are plotted in curve *ABC*, Fig. 1. The cylinder moved from one equilibrium position to the other, rising in the pool, in 3 min.

Cylinder in place, fluid run under as before and pool formed. Load then *increased* from 5·6 to 259·6 gr. and readings taken (curve *DE*, Fig. 1):

As quickly as possible			4·7 gr.	= A value for a load of 5·6 gr.
30 min.	...	...	5·3 ,,	
60 ,,	...	...	6·0 ,,	
120 ,,	...	...	6·8 ,,	

Value of A for load 259·6 gr. is 15·2 gr. If π is the Leslie pressure and P the normal pressure (load divided by area), the condition at the beginning of the first case was $P - \pi = -\dfrac{254}{0\cdot8} = -317\cdot5$ gr., and of the second

$P - \pi = +317 \cdot 5$ gr., why, then, should equilibrium be reached so rapidly when the cylinder rose and so exceedingly slowly when it fell? The answer offered is the same as that given in the papers upon friction[1]—that when the cylinder rises fluid of low viscosity is drawn in, when it falls it presses out lubricant whose molecules are locked in place by the attraction fields of the solids. It is the difference between drawing in a light spirit and expressing a jelly. If this view be correct, the viscosity η in G. I. Taylor's equation must be treated as a variable which is a function of time and the distance between the solid faces.

The time taken for the cylinder to rise to the top of the pool and break away depends, as might be expected, upon the normal force. For example:

OCTYL ALCOHOL

Steel cylinder placed on glass plate, the fluid then run under and a pool formed. Load 5·6 gr.

Normal pull	Time
0·4 gr.	Rose slowly to top of pool in 30 sec.
2·4 ,,	Rose slowly and broke away in 23 sec.
4·4 ,,	Broke away in 9 sec.
6·4 ,,	,, ,, 5 ,,
9·4 ,,	,, ,, 2 ,,
14·4 ,,	,, ,, . 1 ,,
15·9 ,,	,, ,, 1 ,,
[21·2 ,,	,, ,, instantaneously.]

The last of these is the A value.

With the help of a telescope magnifying 10 diameters, the rise of the cylinder was followed under small normal pulls. It moves at first very slowly, but rapidly accelerates until the final break away occurs. The impression is that of a pause followed by rapid movement. At the limiting value there is no apparent pause, and this gives the curious impression that there is no resistance on the part of the cylinder. This apparent disappearance of resistance is characteristic of the A value and of great value in experiments by marking a sharp end-point.

The A value probably is not a measure of the tensile strength of the lubricant. Worthington found the tensile strength of alcohol to be 8165 gr. per sq. cm. The A value for ethyl alcohol for a load of 5·6 gr. was only 9 gr. per sq. cm. The elastic give of the lubricant appears to be sufficient to allow of a tangential flow being established before rupture takes place. The A value in that case is a measure of the viscosity of the lubricant, the time value being arbitrarily fixed by the condition "instantaneous". This question can be more profitably pursued, however, when the adhesion produced by solid lubricants has been described.

[1] *Proc. Roy. Soc.* (1923), A, CIV, 27.

STEADY VALUES

The A values for a number of substances with different solids and loads are given in grams in Table II. The viscosity of the lubricant in mass at a temperature of 20° C. is given in the second column, the figures being taken from various sources. In the third column are the loads in grams.

The value > 110 means that the adhesion was higher than the apparatus would measure.

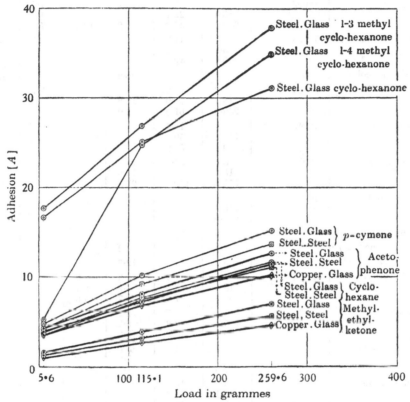

Fig. 2. Load and adhesion (*vide* Table II).

Effect of the pressure. Static friction when both faces are plane increases with the pressure (that is load ÷ area) but not so fast as the pressure. When the latter reaches a certain value the friction varies with the pressure so that the coefficient μ (= tangential force ÷ load) becomes constant. This was attributed to the thinning of the layer of lubricant until a layer of great mechanical stability alone remained.[1]

In these experiments on adhesion it was not possible to reach the pressures found necessary to make μ independent of the load; all values, therefore, lie in the region of varying μ.

[1] *Proc. Roy. Soc.* (1925), A, cviii, 1.

TABLE II. *Value A*

Temperature 18° except where otherwise mentioned			Steel on glass	Steel on steel	Copper on glass	Copper on steel	Copper on copper
	η	L	A	A	A	A	A
Octane	0·005	5·6	1·2	1·1			
		115·1	3·0	2·8			
		259·6	3·8	3·3			
Ethyl alcohol	0·009	5·6	7·2	5·2	2·7	1·7	
		115·1	22·1	14·2	7·7	4·7	
		259·6	33·9	23·2	12·2	7·2	
Butyl alcohol	0·024	5·6	11·7	9·7		6·5	3·7
		115·1	30·2				
		259·6	42·9				
Undecyl alcohol	5·6			26·2		23·2	20·2
Octyl alcohol	0·064	5·6	21·2	18·7		15·7	13·0
		115·1	44·9	36·2			
		259·6	61·9	50·9			
Heptylic acid	0·043	5·6	12·7				
		115·1	50·9				
		259·6	54·9				
Caprylic acid	0·057	5·6	25·9	18·9			
		115·1	99·9	73·9			
		259·6	>110	97·9			
Cyclohexane	0·009	5·6	3·7	3·5			
		115·1	7·7	7·4			
		259·6	11·7	11·2			
Undecane (51°)		5·6		1·5			
Nonadecane (51°)		5·6		8·2			
Tetracosane (51°)		5·6		12·7			
Methyl ethyl ketone	0·004	5·6	1·5	1·3	1·0		
		115·1	3·9	3·2	2·7		
		259·6	7·0	5·7	4·6		
p-Cymene		5·6	4·7	4·2			
		115·1	10·2	9·2			
		259·6	15·2	13·7			
Acetophenone		5·6	4·2	3·9	3·5		
		115·1	8·0	7·2	6·8		
		259·6	12·7	11·5	10·2		
Cyclohexanone		5·6	16·7				
		115·1	25·2				
		259·6	31·2				
1-3 methyl cyclo-hexanone		5·6	17·7				
		115·1	26·9				
		259·6	37·9				
1-4 methyl cyclo-hexanone		5·6	5·2				
		115·1	24·9				
		259·6	34·9				
Benzyl alcohol	0·046	5·6	14·7	12·2	9·7		
		115·1	34·9	27·2	19·9		
		259·6	48·9	38·2	27·9		
1-2 methyl cyclo-hexanol		5·6	68·9	66·9			
		115·1	>110	101·9			
		259·6		>110			
1-3 Cresol	0·187	5·6	74·9	67·9	59·9		
		115·1	>110	94·9	83·9		
		259·6		>110	94·9		
Carvacrol		5·6	105·9	98·9	90·9		
Cyclohexanol	0·500	5·6	>110	>110			
Glycol	0·173 (25°)	5·6	>110	>110			
Glycerol	5·413 (25·6°)	5·6	>110	>110			

The curves in Fig. 3 show that the coefficient α ($= A \div$ load) decreases as the load increases. Whether it would become independent of the load at higher pressures must be left uncertain.

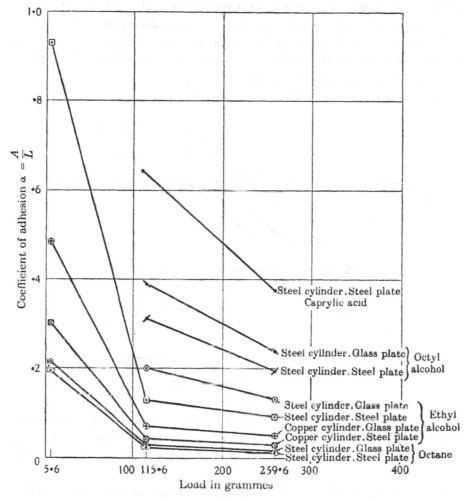

Fig. 3.

Effect of the nature of the solid. The value of A depends not only upon the chemical constitution of the lubricant but also upon that of the solids. The following order was always found glass > steel > copper.

In the case of friction the effect of a change in the nature of the solids was merely to shift the curve for μ and molecular weight parallel to itself, and the curve for two different solids was half-way between the curves for each solid by itself.[1] These same relations seem to hold for the A value, as

[1] *Proc. Roy. Soc.* (1921–2), A, c, 563.

the curves in Fig. 4 show. The equation for the coefficient α $(A \div \text{load})$ is

$$\alpha = \frac{(r_1 + r_2)}{2} - sM,$$

where r is a function of the nature of the solid, the load, and the temperature, s a function of the chemical series, and M the molecular weight.

The influence of the nature of the solid wall is so striking as to make it a matter of surprise that it is not taken into account in certain of the standard methods of measuring the viscosity of fluids.

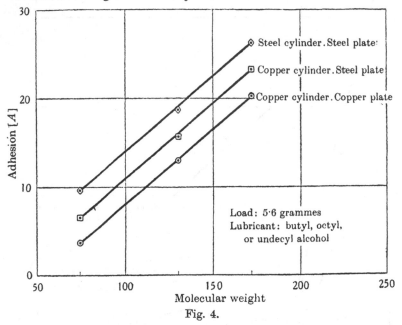

Fig. 4.

Effect of molecular weight. The values for the normal alcohols and normal paraffins are plotted against their molecular weight in Figs. 5 and 6. The relation, like the similar one in static friction, is linear for the same chemical series.

The value for cetyl alcohol was calculated from measurements made at 51° and 75° by using the linear relation to temperature. Since these measurements were made with steel on steel it was necessary to apply a correction to bring the value to that for steel on glass. The value plotted for cetyl alcohol is therefore for the *fluid* state.

Effect of temperature. The value of A for octyl and cetyl alcohols at different temperatures are given in Table III and plotted in the curves, Fig. 7.

It will be seen that the curves for different solids are parallel, but the position of the curve for the same solid varies with the load. The slope of the curves depends upon temperature and the load. The curves are linear.

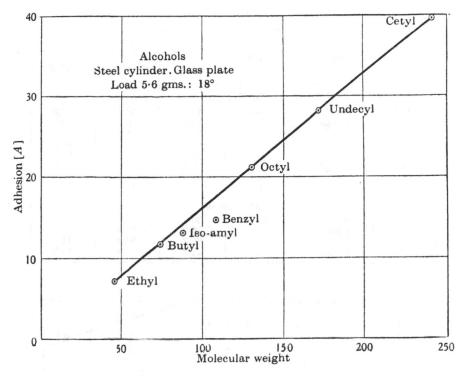

Fig. 5.

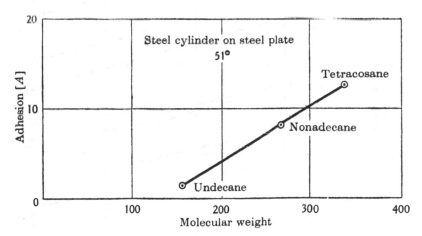

Fig. 6.

TABLE III

Temp.	Cylinder	Plate	Load		
			5·6 gr.	115·1 gr.	259·6 gr.
		Octyl alcohol			
18°	Steel	Steel	18·7	36·2	50·9
18°	Copper	,,	15·7	26·3	35·9
35°	Steel	,,	15·9	31·9	45·9
35°	Copper	,,	12·2	22·2	30·9
51°	Steel	,,	12·2	27·9	41·9
51°	Copper	,,	9·2	18·2	26·2
75°	Steel	,,	8·2	22·7	34·9
75°	Copper	,,	4·2	13·2	19·7
		Cetyl alcohol			
51°	Steel	Steel	30·7	36·9	78·9
51°	Copper	,,	26·7	46·9	63·9
75°	Steel	,,	25·7	50·9	71·9
75°	Copper	,,	22·2	40·9	55·9

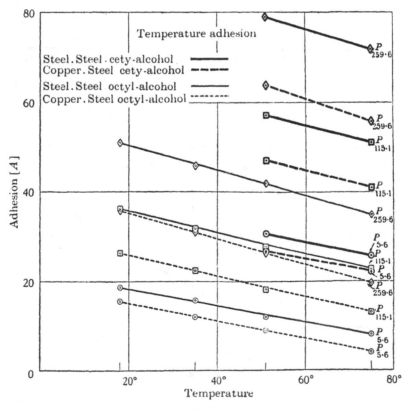

Fig. 7.

The equation therefore is

$$A = \alpha - \beta\theta,$$

where α is a function of the nature of the solid and the load and where β is a function of temperature and load.

These relations are similar to those found for friction in the comparable region where the pressure was not great enough to make the coefficient of friction independent of the load. When that coefficient is independent of the load it is also independent of temperature.[1]

Summary

1. Any force, if it be great enough to overcome the slight resistance offered by the surface tension of the lubricant, can lift a cylinder standing in a pool of lubricant upon a plate; certain values of adhesion are, however, identifiable, and one of these is the normal pull required to break the cylinder away instantaneously. Adhesion here means simply this normal pull.

2. In order that the value shall be comparable, cylinder, plate and lubricant must be in a mechanically corresponding relation. One such is when the load is in equilibrium with the Leslie pressure. To the identifiable value for this corresponding state the name A value is given.

3. The latent period which elapses before this state is reached is due either to orientation of the molecules of the lubricant or to the cylinder rising or falling in the pool.

4. The A value is a function of the chemical constitution of the lubricant and of the solids, of the load, and of temperature.

[1] *Proc. Roy. Soc.* (1925), A, cviii, 16, and (1922), A, ci, 487.

51

STUDIES IN ADHESION. II

With MILLICENT NOTTAGE

[*Proc. Roy. Soc.* (1928), A, cxviii, 209]

Part I. Experimental

[In this paper the results of measurements of the tensile strength of a joint between a cylinder and a plate are described. Such measurements are made with difficulty owing to the sensitiveness of the joint to vibrations. It is therefore the more important to state that the values recorded in the following pages are entirely due to my colleague's patience and skill.—W. B. H.]

An admirable summary of previous work on adhesion is given by McBain and Lee[1] in a paper which breaks new ground, and should be read in conjunction with what follows.

To obtain the measurements a clean metal cylinder was placed upon a clean metal plate in a chamber filled with clean air and warmed above the melting-point of the lubricant,[2] some of which was also placed in the chamber. When the lubricant was melted and the temperature steady, some of the now fluid lubricant was allowed to be drawn under the cylinder by capillary attraction until the space between cylinder and plate was completely filled. Enough lubricant was then added to form a large pool. After a time had elapsed sufficient to permit orientation of the molecules by the attraction fields of the solids to become complete, the lubricant was frozen by allowing the temperature to fall to 18° C. The excess lubricant was then trimmed away from the base of the cylinder, and the plate with the attached cylinder removed from the chamber to a special clamp which could be levelled. A pan was then attached to the end of the cylinder and weights added until the joint broke. This gave the adhesion value B. In other cases the pan was connected with the cylinder as close as possible to the joint and in such a way as to apply the external force tangentially. The tangential pull required to break the joint is the value S.

The cylinder had always the same diameter and weight as those used in

---

[1] *Proc. Roy. Soc.* (1927), A, cxiii, 606.

[2] Throughout this series of papers upon friction and adhesion the word "lubricant" has been used to indicate the substance present between the solid faces. It is retained in this study of the tensile strength of a solid joint.

the previous paper,[1] namely, 1 cm. and 5·6 gr., and both cylinder and plate were polished to "optical" surfaces.

If it was desired to vary the load, weights were placed on the top of the cylinder before the fluid lubricant was run underneath.

Up to this stage the procedure exactly resembled that followed in obtaining value A of the preceding paper—value B and value A refer therefore to mechanically corresponding states.

The temperature fell from 56° to 18° in about 20 min., from 84° to 18° in about 40 min., and adhesion was measured at any time from 3 to 20 hours after the joint had been made. The effect of errors in the form of the surfaces was tested by comparison between the values taken after prolonged use and immediately after refacing.

The pressure-adhesion curve obtained after refacing was more regular and the values for each pressure more concordant, but the difference was not great.

Steel on steel: Myristic acid.
Pressure varying from 7·1 to 3160 gr.

	Mean values	Mean difference between highest and lowest values for each load
Before repolishing	19120	91 = 0·6 per cent.
After repolishing	20260	34 = 0·2 „

The effect of untrueness of the surface was, as might be expected, greater the heavier the load—that is to say, the thinner the layer of lubricant—this appears from the curves, Fig. 1.

The temperature at which lubrication took place had little or no influence. The following values illustrate this.

Cetyl Alcohol. Adhesion (B) in grams. per sq. cm.

Temperature of lubrication	Cylinder ...	Steel	Copper	Steel	Copper
	Plate ...	Steel	Steel	Copper	Copper
54°		—	—	14550	13480
56		15900	14550	14580	13360
84		15590	14470	14545	12380

The time occupied in applying the breaking strain varied from 11 to 20 min. The variation had no detectable influence on the result.

The latent period is the time taken to reach a steady state. It may be due either to the cylinder rising or falling in the pool of lubricant to a position of equilibrium, or to the alteration in the structure of the lubricant due to the orientation of the molecules by the attraction of the solids. It has been discussed fully in earlier papers.[2] In these experiments the age of

---

[1] *Proc. Roy. Soc.* (1926), A, cxii, 62.
[2] See especially *Proc. Roy. Soc.* (1923), A, civ, 25.

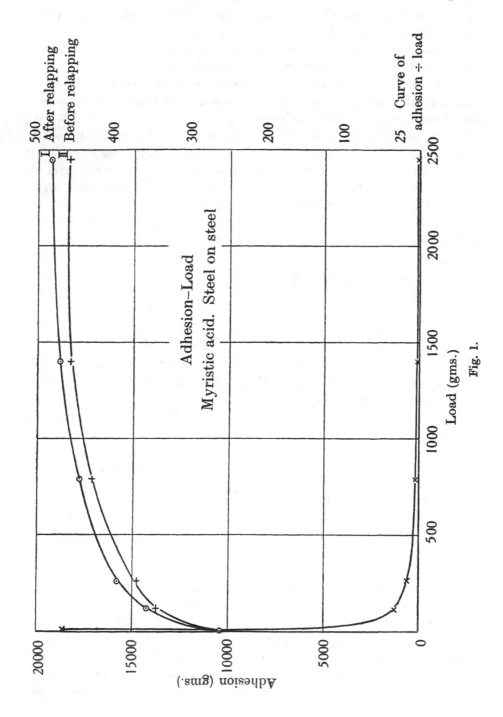

Adhesion–Load
Myristic acid. Steel on steel

Load (gms.)

Fig. 1.

Adhesion (gms.)

I After relapping
II Before relapping

Curve of
adhesion ÷ load

the *solid* joint had no influence upon adhesion; changes took place only whilst the lubricant was fluid, and the latent period appeared to be due wholly to orientation. This follows from the fact that, though each of the acids gave a latent period of about 60 min. and the one alcohol available a period of about 30 min., the symmetrical paraffins gave no detectable latent period. The following is an example.

Palmitic acid M.P., 62°. *Temperature of lubrication,* 68°

Liquid state maintained for	Adhesion	
	Steel on steel	Copper on steel
Min.		
40	13950	11550
50	14940	13720
60	15880	14980
70	15860	15030

Eicosane M.P., 36·7°. *Temperature of lubrication,* 45°

Liquid state maintained for	Steel on steel
Min.	
5*	8652
20	8553
70	8590

The values underlined are steady values independent of time.

The time value "starred" was the least possible, namely, that taken up by the solidification of the lubricant when the heat was cut off at once.

During the latent period the structure of the lubricant is becoming more orderly in the sense that its molecules are moving into positions of equilibrium with respect to the attraction fields of the solids. Therefore, since adhesion invariably increased when lubricants with polar molecules were maintained in the fluid state long enough to permit the molecules to orient themselves, we may conclude that orderliness as defined above increases adhesion just as it decreases friction.

This conclusion is in conflict with the finding of McBain and Lee (*loc. cit.*), namely, that disorderly structure increases adhesion. The study of the microscopical structure of the plate of lubricant leaves no doubt as to the orderliness of the disposition reached when a critically pure lubricant was allowed to come into equilibrium with the forces acting upon it. All that need be said at this stage is that "order" and "disorder" need careful definition and that no generalisation is yet possible. No exception has been found to the rule that adhesion rises during the latent period when the lubricant is composed of a single pure chemical substance unless this be a paraffin, in which case there is no latent period. When the lubricant is composed of two chemical substances sometimes adhesion falls sometimes rises during the latent period when the structure is becoming more orderly.

Effect of chance impurity. On a few occasions the supply of pure air to the chamber failed after the lubricant was in place, but whilst it was still fluid, and on some of these occasions the procedure was carried through in the ordinary way and the adhesion measured. It was always found to be from 60 to 70 per cent. below its value for the pure substance.

The purity of the lubricants. Normal paraffins, their related normal acids and one alcohol were tested. In studies of adhesion as in those of friction when the lubricants are critically pure, and the tests made under mechanically corresponding states, orderly results of remarkable simplicity are obtained; the criterion of purity adopted is therefore of great importance in the logical scheme. It was neither more nor less than the attainment of this same orderliness. This may seem to be reasoning in a circle, but the assumption that orderliness is evidence that the number of variables is small and under control is implicit in all tests of chemical purity.

The final stages of purification were carried out by recrystallisation. This was continued until adhesion became constant and the character of the fracture regular.

A few figures are quoted below to illustrate two things: that the adhesion given by impure lubricants was characteristically variable, and that purification *raised* the value for all the long chain acids used and for the single

Substance	State	Cylinder ... Steel — Plate ... Steel	Copper — Steel	Steel — Copper	Copper — Copper
Cetyl alcohol	Impure	14270	12840	12420	11100
	Intermediate state	14510	12600	11800	11730
	Steady values	15860	14540	14540	13380
Lauric acid	Impure	8736	6533	6279	5414
	Intermediate state	9563	6606	8085	7977
	Steady values	10690	9872	9924	9095
Palmitic acid	Impure	Figures too erratic to quote			
	Several recrystallisations	8954	—	8013	5740
	Later stages	12050	10820	7941	6606
		13120	11620	9636	9311
	Steady values	15860	15030	14980	14190
Eicosane	Impure	11610	10740		
		12290	8948		
	Nearly pure	8736	7473		
		8553	7362		

alcohol, and *lowered* it for all paraffins. The irregularity in the values is without doubt related to the irregular character of the surface of break of impure lubricants.

The adhesion of acids and of the alcohol is much greater than that of paraffins, and therefore the effect of impurities is perhaps what might have been expected. The want of concordance in the values obtained at the same stage of purification is owing to the fact that two time values are involved in the changes of state of an impure lubricant, namely, the time occupied by the orientation of the molecules of the lubricant itself, and the time occupied by the positive or negative adsorption of the impurity on to the face of the solids and at interfaces within the layer of lubricant. The layer of lubricant on fracture was always found to be crystalline, and as purification proceeded the crystals became larger. An impurity would condense at the interfaces between crystals, and the small size of the latter was no doubt due to barriers of impurity being set up between centres of crystallisation. The rate of cooling would therefore be likely to have a great influence upon the structure of impure lubricants. The irregularity in the values, the irregularity in the fracture and the small size of the crystals are different aspects of the same phenomenon, namely, the redistribution of two or more components during fluidity and crystallisation.

Structure and thickness of the layer of lubricant. On breaking the joint, a crystalline cake of lubricant was left on either the cylinder or the plate or partly on one and partly on the other. The cake was easily visible to the naked eye and obviously became thinner as the load was increased. As it thinned it showed Newtonian colours.

Lubricant, myristic acid

Pressure	Colour	Texture
gr. 7·1	White	Long narrow plates laid flat and in fan-shaped masses
147	Very faint red	Crystals smaller and narrower
331	Decided red and green	—
1009 1783 3122	Bright blue and yellow	—
8756	Uniformly a brownish yellow	Crystals very much smaller and narrower

The character of the break was studied in detail under the microscope. When the lubricant was impure the break occurred anywhere and the crystals were small. The surface of break therefore was irregular. As purification proceeded it became more and more regular until the break occurred, so far as the microscope or the unaided eye could detect, at the interface between solid and lubricant (see Pl. XIV, figs. 2B, 3A). The break actually occurred, however, not at the metal-lubricant interface, but within the lubricant at a distance from the interface so small as to leave behind on the metal a film of totally insensible thickness. That the film was there was proved by comparing the friction of the surface of apparently clean steel or copper with that of the clean surface of the plate on which lubricant had not been placed.

> Cylinder, steel; plate, copper.
> Lubricant, eicosane.
> Friction measured with a glass slider.
> Where the lubricant had been broken away $\mu = 0.176$.
> Clean part of the same plate $\mu = 0.8$.

A plate of chemically pure lubricant which has been frozen *in situ* is therefore composed of a central plate of flat crystals enclosed by two primary layers, one on each of the enclosing solids. Each of these primary layers is possibly one and certainly not more than a few molecules in thickness, and the break under either normal, and, as we shall see later, under tangential stress appears to occur at one or both of the surfaces between the primary film and the plate of crystals.

These surfaces mark the limit beyond which the attraction fields of the solids are unable to prevent crystallisation. They do not, of course, mark the limit to which orientation spreads from the metal face into a *fluid* lubricant. This is fixed by the kinetic energy of the molecules in the fluid state. By X-ray examination Trillat[1] found stratification and orientation of long chain molecules to persist up to at least 5μ from a solid face.

They might mark the limit of the adsorbed layer of an impurity in the lubricant which was more strongly adsorbed by the solids than the latter. The known effect of added impurity, however, does not favour this interpretation (see later).

The position of the surface of break, that is to say, whether it was at one or at both of the surfaces mentioned in the last paragraph but one, was found to depend upon whether cylinder and plate were or were not of the same material.

A diagrammatic section through the layer of lubricant when both solids are the same is given in Fig. 2. The surface of rupture is a, b, c, d. It is shown not at the interface but at an insensible distance within the lubricant for reasons just given. A break of this kind can be seen in Pl. XIV, figs. 1 and 4.

The process of purification was by its nature asymptotic, and is legitimate to assume, recollecting the extreme sensitiveness to impurity, that, at the theoretical limit when both solid faces are identical physically and

---

[1] *Ann. de physique* (1926), x, 5.

chemically, the lubricant absolutely pure, and the pull perfectly normal, the break would occur simultaneously near both solid faces.

When the solids were different the surface of rupture was near the metal with weaker attraction field, as in Fig. 3. Pl. XlV, figs. 2 and 3 show that fragments and sometimes a whole crystal were left attached to the copper when the other surface was of steel, but again, at the unattainable limit of experimental precision, it is certain that the surface of break would be close to the copper.

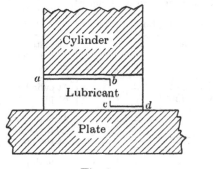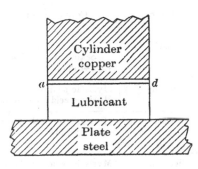

Fig. 2.　　　　　　　　　　　　　　　Fig. 3.

The crystalline structure was apparent in the thinnest films obtained by the heaviest loads but, as the load increased, the crystals became smaller (Pl. XIV, fig. 4). The law defining the position of the break stated above held for the thinnest films and for all the lubricants tried. Discussion of the significance of this law must be postponed until after the effect of the nature of the solid has been further described.

The structure of the plate of lubricant gives objective reality to the conception which has been reached inductively of the presence in colloids of free and bound water, the latter being water so firmly attached to the solute as to be incapable of being frozen.[1] The free and bound portion of the lubricant are sharply divided as the diagrams show, but the freezable portion could not be expected to have the same relations to temperature as the same substance in mass, but would be best described as "capillary". If the solids were moved further apart we should finally have really "free" lubricant between the faces.

Effect of added impurity. The position of the surface of break could be altered at will by impurity localised at one or other interface. To effect this, silk, freed from contaminating substances by prolonged extraction with purest benzene, was lightly rubbed on the contaminant and then lightly rubbed on one or both of the metal faces previously heated to above the melting point of both contaminant and lubricant when the latter was

[1] *Proc. Roy. Soc.* (1926), A, cxii, 30 and 47.

myristic acid (M.P. 54°), and to a temperature below that of the contaminant and above that of the lubricant when the latter was eicosane (M.P. 36·7°). The quantity of contaminant conveyed in this way to the surface must have been exceedingly small. A pool was then formed of the pure lubricant and adhesion measured in the ordinary way. (See table below.)

The figures give the adhesion in grams per sq. cm.

Load 5·6 gr.

Lubricant	Contaminant	Cylinder ...	Steel	Copper	Steel	Copper	Position of break
		Plate ...	Steel	Steel	Copper	Copper	
Myristic acid	Eicosane on plate		10360	8845	—	—	At plate
	Eicosane on cylinder		10820	7511	—	—	At cylinder
	Eicosane on plate		—	—	7291	5702	At plate
Eicosane	Myristic acid on plate		9636	8085	—	—	At cylinder
	Myristic acid on cylinder		9854	8736	—	—	At plate
Myristic acid	Eicosane on both plate and cylinder		9249	—	—	—	At both plate and cylinder
	—		—	7578	—	—	At the plate
Eicosane	Myristic acid on both plate and cylinder		10540	—	—	—	At both plate and cylinder
	—		—	9017	—	—	At the cylinder
Myristic acid	None		13260	—	12440	11580	—
Eicosane	None		8644	—	7416	—	—

This table merits careful consideration. It illuminates vividly the whole subject and has much practical significance. The adhesion value of myristic acid is greater than that of eicosane, and of steel greater than that of copper. The table shows that an insensible film of contaminant at one interface leads to the following:

(1) When one surface only is contaminated the break occurs at the contaminated surface when the adhesion of the contaminant is weaker than that of the lubricant; and at the uncontaminated surface when the lubricant is the weaker adhesive.

(2) When both surfaces are contaminated the ordinary relation between the position of the break and the adhesion values of the solids is not interfered with.

(3) A weak contaminant lowers the adhesion value and to its own value in one case. A strong contaminant raises the adhesion though the surface of break may not be near the contaminated but near the uncontaminated surface.

Therefore, a film of contaminant of the order of one molecule in thickness introduces a surface of weakness *or of strength*, the latter being the more significant theoretically, and imposes its own value on the adhesion of the entire joint, though as relation (2) shows not to the extent of obscuring the characteristic effect of the nature of the solid. The position of the surface of break proves that the contaminant did not diffuse throughout the lubricant.

PART II. STEADY VALUES

The steady value which we call value B is the adhesion per square centimetre. Its relations to the variables are described in this section.

Temperature. A small chamber containing a coil of tubing was built about the clamp and cylinder, and the temperature at which the solid joint was broken controlled by fluid run through the tubing. Manipulation was difficult and to this may be ascribed the irregularity in the values. In the following table the temperature is that at which the joint was broken.

Load 5·6 gr.; Cylinder and plate of steel

Lubricant	Temperature (°)	Adhesion value B
Myristic acid (M.P. 54°)	4	12970
	8	13190
	18	13260
	38	13710
Palmitic acid (M.P. 62°)	4	15810
	6	15700
	8	15430
	14	15500
	18	15860
	42	15820
	56	15474

Within the limits explored adhesion seems to be independent of temperature. It is remarkable that the value should be the same so near the melting-point of the lubricant. This relation proves that the break was uncomplicated by a viscous flow of the lubricant, and the most careful microscopical examination of the lubricant after the joint had been broken failed to reveal any signs of such flow.

The nature of the enclosing solids. Throughout these studies both friction and adhesion have been found to obey what might for convenience be called the mean value rule, which may be stated as follows:

When the enclosing solids are different, being composed of substances M and N, the value of friction or of adhesion is the mean of what it is when both solids are composed of M or both of N. The general equation therefore is

$$X_{MN} = \tfrac{1}{2}(X_{MM} + X_{NN}).$$

As it is important from the theoretical point of view to know whether this equation holds exactly or only as an approximation a series of critical determinations of value B were made with the following results.

Load 5·6 gr., Temperature 18° C.

Lubricant	Molecular weight	Steel plate		Copper plate	
		Steel cylinder	Copper cylinder	Steel cylinder	Copper cylinder
Lauric acid	200	10610 10750 10690 } 10683	9891 9854 } 9872	9961 9891 } 9926	9133 9061 } 9097
			9899		
Myristic acid	228	13280 13240 } 13260	12420 12480 } 12450	12420	11550 11620 } 11585
			12437		
Palmitic acid	256	15840 15990 15730 } 15853	15020 15050 } 15035	14980	14250 14110 } 14180
			15007		
Stearic acid	284	18380 18330 18480 } 18397	17690 17620 } 17655	17580 17580 } 17580	16820 16710 } 16765
			17617		
Mean values		14546	13753	13727	12917
Mean for steel on steel and copper on copper					13731
Mean for steel on copper and copper on steel					13740

The mean value rule seems to hold exactly and the figures for the paraffins confirm this.

	Molecular weight	Time (min.)	Steel plate		Time (min.)	Copper plate	
			Steel cylinder	Copper cylinder		Steel cylinder	Copper cylinder
Eicosane $C_{20}H_{42}$	282	10 15	8553 8664 } 8608	7362 7291 } 7326	10 10	7147 —	5919 5991 } 5955
$C_{22}H_{46}$	310	10 15	9495 9387 } 9441	8158 8085 } 8121	5 10	8015 8121 } 8086	6606 6676 } 6641
$C_{30}H_{62}$	422	40 18 9	12810 12710 12710 } 12743	11370 11290 } 11330	10 30	11270 11330 } 11300	9819 9924 } 9871
Mean values			10264			8882	7489
				8876			

It was pointed out in an earlier paper on friction that the mean value rule seemed to prove that the influence of the attraction field of each of the enclosing solids extends to the surface of slip. The direct knowledge of the position of the surface of break shows that it extends right through the disc of lubricant.

Consider for example the acids. When both surfaces were of copper the mean adhesion was 12917, and when both were of steel it was 14546. When

one surface was of copper and the other of steel the break took place near
the copper at a surface removed from the steel by a layer of lubricant at
least $4\,\mu$ in thickness. If the influence of the steel had been only of molecular range the adhesion would have been that of copper. Instead it was
the mean between that of steel and that of copper.

The effect of contaminants is equally remarkable. When myristic acid
was applied to one solid face, both solids being alike and the lubricant being
the weaker adhesive eicosane, the break took place in pure eicosane at a
surface near the uncontaminated face and removed, say, $3\,\mu$ from the

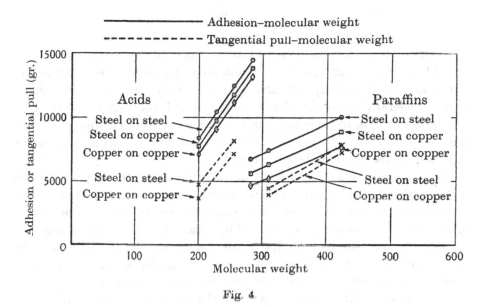

Fig. 4

myristic acid. Yet the adhesion was not that of pure eicosane but always
something larger than this and between the values for pure eicosane and
pure myristic acid.

Molecular weight. The adhesions in the tables on p. 790 are plotted against
molecular weight in Fig. 4. By an oversight the observed values and not
the values per square centimetre were plotted. The relation is linear for
the acids and paraffins. Owing to the depletion of our stock of pure substances it was not possible to confirm this for the alcohols, but the great
body of evidence accumulated in these papers justifies the statement that
for the same chemical series the equation is linear.

Pressure. The pressure is the load in grams divided by the area in square
centimetres.

The values are plotted in Fig. 1, curve I. Curve II gives the values for
the same cylinder and plate but before refacing.

Myristic acid. Steel plate and cylinder. Temperature 15°.

Load	Pressure	Adhesion		$B \div$ pressure
5·6	7·15	13280 13240	13260	1855
115·1	147	18120 18200	18160	123·5
259·6	331	20180 20070	20125	60·8
792	1008	22740		22·5
1400	1783	24040 23890	23965	13·4
2451	3122	24620 24430	24525	7·86

The fact that adhesion increases as the layer of lubricant decreases s part of the general experience of mankind and is indeed recorded in the directions how to use any commercial cement; it may therefore be claimed to be a general property of films enclosed by two continuous solids. It is not, however, easy to explain, for in the case under consideration the break occurs close to one or other or both faces, and pressure merely varies the thickness of the median plate of crystals.

An attempt was made to measure directly the distance between the surfaces of the cylinder and the plate, when the one was resting on the other either loaded or unloaded and with various lubricants. For this purpose a spherometer was specially constructed by Messrs Hilger so that the measurements of the distance between the upper surfaces of the cylinder and of the plate could be made in clean air, and the firm also made two pairs of cylinders and plates. The first pair, owing to the nature of the steel, were the less perfect. Call this pair "A" and the more perfect pair "B".

The first set of measurements revealed the totally unexpected fact that the distance between cylinder and plate was of the same order when the lubricant was clean air as it was when some liquid had been allowed to displace the air, indeed, within the limits of accuracy the distance h was found to be independent of the nature of the lubricant whether the pair "A" or the pair "B" was used. This left us with measurements of the variation in value of h, but without the absolute value because we did not know the length of the cylinders.

The Metrology Department of the National Physical Laboratory were good enough to measure these lengths, which they did by wringing the cylinders down on to a plane surface, a standard gauge being used for reference. They discovered on cylinder "A" two minute bosses which were removed before it was measured. They also subjected the spherometer to critical examination and found the screw to be free from periodic error, but to have a total error over the length used of $-0 \cdot 009$ mm.

Each value in the following tables is the mean of five measurements taken at different points on the top of the cylinder. To show the variation in these individual readings three diagrams are given. The circle represents the top of the cylinder, and the figures are placed where the measurements

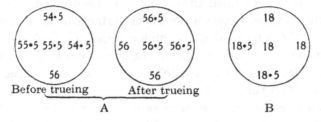

	Before trueing	After trueing	
	A		B

	Lubricant	h_0 mm.	Lubricant	h mm.	Difference
Series I. Cylinder A (before treatment at the N.P.L.). Steel plate. Load 5·6 gr.	Air	0·0070 0·0070 0·0070 0·0070 0·0062 0·0065 0·0067 0·0070 0·0067 0·0070 0·0065	Palmitic acid Cetyl alcohol	0·0070 0·0070	0 0
Mean value		0·0068		0·0070	
Series II. Cylinder A (after treatment at the N.P.L.). Steel plate. Load 5·6 gr.	Air	0·0045 0·0038 0·0037 0·0040 0·0043 0·0045 0·0037 0·0040 0·0042	Octane Caprylic acid Octyl alcohol	0·0042 0·0042 0·0040	Negligible Negligible Negligible
Mean value		0·0041		0·0041	
Series III. Cylinder B (before measurement at N.P.L.). Steel plate	Air	0·0049 0·0044 0·0039 0·0036 0·0049 0·0036 0·0031 0·0041 0·0036 0·0031	Octyl alcohol Caprylic acid Octane Octyl alcohol	0·0049 0·0044 0·0036 0·0036	0 0 Negligible 0
Mean value		0·0039		0·0041	
Series IV. Cylinder B (after measurement at N.P.L.). Steel plate	Air	0·0041 0·0041 0·0036 0·0044 0·0041	Octyl alcohol Cetyl alcohol Octane Caprylic acid	0·0039 0·0041 0·0044 0·0044	Negligible 0 0 Negligible
Mean value		0·0041		0·0042	

were taken and are the readings on the circular scale of the spherometer. Each division of the scale is $2 \cdot 5 \mu$. The examples were taken at random from the notes.

Save where the contrary is stated once the cylinder was placed on the plate it was not moved until all readings had been taken.

If H be the distance between the upper surfaces of the cylinder and the plate, and L the length of the cylinder, then $H - L$ is equal to h, the distance between the opposed faces. In the tables h_0 is the value when cylinder and plate were clean and in clean air, and h the value after liquid lubricant had been run in. The procedure was as follows:

H was measured at 18°, the temperature was raised above the melting-point of the lubricant, and the melted lubricant then run underneath. The temperature was lowered to 18°, and when it was steady, the lubricant being now solid, H was again measured. In those cases in which the lubricant was fluid at 18° it was simply run underneath and H measured at once.

The first series were made before the cylinders and plates had been sent to the National Physical Laboratory and therefore before the two bosses had been removed from the lower face of cylinder "A".

If more accurate methods should confirm the figures, the distance h between the cylinder and plate would be the same for a gas as for a liquid and independent of the chemical constitution of the lubricant and of the enclosing solids.

The decrease in the value of h due presumably to the removal of the bosses on cylinder A is remarkable.

It was inferred from earlier measurements of friction and adhesion[1] that when a cylinder or slider was forced down in a pool of lubricant it rose rapidly when the external pressure was removed. This inference was confirmed. The figures give the readings on the circular scale of the spherometer. Each division is equal to $2 \cdot 5 \mu$.

Cylinder "A". Weight 5·6 gr.

Lubricant	Unloaded	After being loaded with 1400 gr.
Air	55·2	55·2
,,	55·2	55·2
Octyl alcohol	55·5	55·5

The load in the case of air was left on for 24 hours. Measurements completed in from 10 to 15 min. after the load had been removed.

The load (1400 gr.) in the second case was left on for 30 min. which would allow of only a slight descent in the pool of alcohol.

[1] *Proc. Roy. Soc.* (1923), A, CIV, 27; (1926), A, CXII, 67.

The effect of a normal pressure upon the value of h was determined as follows. Readings were taken with the clean unloaded cylinder standing on a clean plate in clean air at 18°. The cylinder was then loaded to the required amount and the temperature was then raised to slightly above the melting-point of the lubricant. Whilst the load was still on melted lubricant was run underneath and a pool formed in the usual way. The temperature was then lowered to 18° when the lubricant froze. The load was then removed and a second set of readings taken.

In the following table the readings on the circular scale are given, those in the third column being for the unloaded cylinder in air at 18°, and those in the fourth column being for the cylinder at 18° but fixed to the plate by the solid joint which had been formed and frozen under the pressure given in the second column. The difference is the distance the cylinder was forced down by the load and h is this value subtracted from the value of h when the pressure was 7·1 gr., namely, 0·007 mm.

Cylinder and plate "A"
(Readings taken before the bosses were removed)

Lubricant	Pressure	Readings		Difference	h
				mm.	μ
Palmitic acid	331	55·5	56·1	0·0015	5·5
	331	55·4	56·1	0·0017	5·3
	1783	55·3	56·7	0·0035	3·5
	3160	55·2	57·0	0·0045	2·5
	5560	55·3	57·5	0·0055	1·7
	7185	55·2	57·5	0·0057	1·3
	8758	55·4	57·8	0·0060	1·0

In Fig. 5 h is plotted against the pressure. The curve for adhesion and pressure (Fig. 1) is of the same form. Both curves tend to become hori-

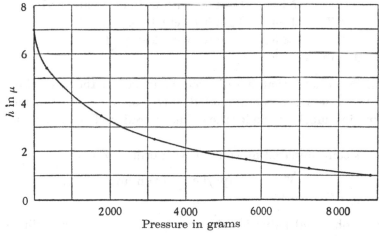

Fig. 5.

zontal as pressure increases. When the pressure was 8758 the cake of lubricant was still clearly visible to the naked eye and still showed Newtonian colours.

The surfaces of cylinder and plate "B" were of proof plane order of accuracy, and they were used to determine whether the distance h was different for clean and contaminated surfaces. If an adsorbed substance neutralises some fraction of the attraction fields of the metals the Leslie pressure which supports the load should be less for contaminated than for clean faces. The cylinder had to be lifted between the measurements to allow of the application of the film of palmitic acid, but octyl alcohol was applied as vapour, the cylinder being left in position.

Cylinder "B" of steel

Plate	Clean in clean air	After contamination by	
Steel	18·3	Palmitic acid	19·2
,,	18·5	,,	19·2
,,	18·7	Octyl alcohol	19·0
,,	—	"Grease" adsorbed from air of room	19·7
Glass	18·5	Palmitic acid	19·3
,,	18·7	Octyl alcohol	19·2
,,	18·7	"Grease" adsorbed from air of room	19·4
Mean	18·6	Difference $0·7 = 1·7\,\mu$	19·3

Therefore $h_{\text{clean}} - h_{\text{contaminated}} = 1·7\,\mu$.

These measurements were repeated with cylinder and plate "A" after the former had been trued.

	Clean	After contamination by
"A"	56·5	Palmitic acid 57·5.
	56·4	Contact with ordinary leather 57·5.
"B"	18·4	Palmitic acid 19·6.
	18·3	Contact with leather 19·7.

Mean difference   1·2

It will be noticed that the differences are all in the same direction, h always being less for contaminated surfaces. When the surfaces were dirty, running lubricant under the cylinder produced no detectable change in h.

Surfaces contaminated by contact with ordinary leather

"A"	57·5	Caprylic acid run under 57·7.
"B"	19·7	,, ,, 19·6.

Octyl alcohol and palmitic acid were also run under the cylinder when the surfaces of both cylinder and plate were contaminated by contact with the ordinary air of the laboratory and the level was not changed. We therefore seem to have:

(1) h independent of the state or chemical nature of the fluid run underneath whether the surfaces are clean or contaminated; and

(2) h is less for contaminated surfaces in pure air than for clean surfaces in pure air.

The method of measurement was only a rough one but it seems to have established the following facts. The thickness of the air film when the surfaces were plane and the normal pressure was 7·1 gr. was of the order of 0·004 mm., and this value was not greatly changed by replacing the air by a liquid. The value was, however, decreased when the surfaces were contaminated. A long chain molecule with 20 carbon atoms would be about $2·8 \times 10^{-7}$ cm. in length, therefore 4μ is equal to about 1500 such molecules arranged end to end.

PART III. TANGENTIAL PULL

A few measurements intended merely to determine the relation to molecular weight and the order of magnitude were made of the tangential pull S required to break a joint. The external force was applied in the plane of the plate and as low down on the cylinder as was possible.

Steel plate. (S value in grams per square centimetre)

Lubricant	Steel cylinder		Copper cylinder	
$C_{22}H_{46}$	5743 5670	5705	5128 4983	5055
$C_{30}H_{62}$	9965 9860	9911	9208 9244	9225
Lauric acid	6030 5958	5994	4731 4585	4658
Palmitic acid	10400 10330	10375	9135 8990	9062

The values in the table are per square centimetre, that is the observed values divided by 0·785. The observed values are plotted against molecular weight in the curve in Fig. 4.

Under the microscope the fracture was seen to take place at the same surface as it did with a normal force, the surface of break therefore was the same in the two cases. S was in each instance much less than B. The break away was sudden, the resistance falling apparently instantly to that of simple external friction.

SUMMARY

1. The adhesion of and tangential pull required to slide a cylinder attached to a plate by a solid joint are described. The joint was formed by running fluid lubricant between the cylinder and plate and freezing it *in situ*.

2. The disc of lubricant was found to be composed of a plate of crystals with an adsorbed layer on each side.

3. The relation of the value observed to the temperature, pressure, chemical composition of the lubricant and composition of cylinder and plate is discussed.

DESCRIPTION OF PLATE XIV

Fig. 1. Lubricant, myristic acid. Cylinder and plate of copper. (A) Cylinder. (B) Plate.

Fig. 2. Cylinder of copper, plate of steel. (A) Plate. (B) Cylinder.

Fig. 3. Plate of copper, cylinder of steel. (A) Plate. (B) Cylinder. Load 5·6 gr. (pressure 7·1 gr.) for Figs. 1, 2, and 3.

Fig. 4. Cylinder and plate of steel. (A) Plate. (B) Cylinder. Load 259·6 gr. (pressure 331 gr.).

Plate XIV

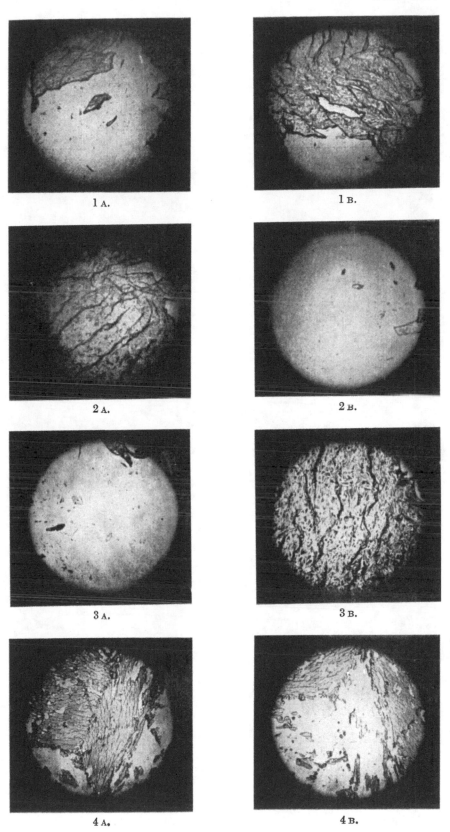

1 A.

1 B.

2 A.

2 B.

3 A.

3 B.

4 A.

4 B.

52

MOLECULAR ORIENTATION IN LIVING MATTER

[*Journ. Gen. Physiol.* (1927), VIII, 641]

Living matter offers many examples of what appears to be action exerted at a distance. I am not attempting to renew the discussion of the classical problem of physics because the action at a distance which I have in mind is not exerted through vacuous space but through matter, and its peculiarity lies in the fact that the intervening matter seems to act in a purely passive way.

The example which has been present in my mind for more decades than I feel happy in contemplating is furnished by the neuro-muscular system. The muscle fibres are controlled by the nervous system with respect to three things: (1) those contractions which move the several parts of the body, (2) their elastic properties which maintain the several parts of the body in their normal posture and (3) the production of heat by the muscles of warm blooded animals in which the skeletal muscles are the chief thermogenic tissues. Of these the second is commonly called by physiologists muscular tone.

The agent, the middle man that is, between a particular cell in the central nervous system and the group of muscle fibres which it controls, is a delicate fibril no more than say 4μ in diameter which consists mainly of protein, and which is enclosed in an insulating sheath of fatty matter. The way this agent acts in controlling the contraction of the muscle fibres is well known. Waves of short length, called nervous impulses, pass down the fibril to discharge some of the stored energy of the muscle fibres, the muscle fibre being an internal combustion engine. Each nervous impulse viewed in one aspect is a wave of molecular displacement involving chemical change, and in another aspect a wave of electrical displacement. The mode of action of the nerve fibre, however, in the other two cases is, so far as I am aware, completely unknown. It is of course possible that muscular tone and heat production also are controlled by nervous impulses which have not hitherto been detected by reason, for example, of their short period. It is also possible, however, that the nerve fibrils act in some purely static way and, should they do so, it would be an example of what I mean by action at a distance in living matter. A simple physical analogy would be a row of bricks on end on an inclined plane supported at the bottom and supporting some weight at the top of the row. There a certain structure transmits stresses which are in equilibrium so long as the structure remains intact.

Let us consider other instances. The nerve fibril itself furnishes one. It is an outgrowth from a nerve cell which has in its interior a little bladder of fluid called the nucleus. The nerve cell may be perhaps $50\,\mu$ across. The filament may be a metre, or metres in length. If the filament be cut that part which is separated from the parent cell dies and breaks up. That part remaining in contact with the parent cell not only persists but grows. In similar cases which have been analysed experimentally the influence which upholds structure and also that capacity for reproducing a particular molecular architecture in space which is called growth, is found to reside in the nucleus. This kind of action at a distance can also be detected in those plates of cells called epithelia, in embryos, in colonial polyps, and elsewhere. I used to wonder whether there was anything similar to this in the world of non-living matter. In a ridiculously humble way I think there is, though its existence has tended to be obscured by the amount of attention which has been rightly paid to monomolecular films.

An interface between two phases is the seat of forces of great magnitude directed along the normal and tangent to the face. These force fields orientate the molecules on either side of the interface, the effect on any individual molecule being greater or less according to the extent to which the equipotential surfaces of the molecule deviate in shape from spheres drawn about the centre of mass. That fact is, of course, familiar to everybody nowadays but to the question how far and to what extent is the orientation transmitted from molecule to molecule into the substances of the phases on either side of the interface no simple answer is forthcoming. Mathematically I believe the orienting influence of an interface to be transmitted to infinity; actually, however, a limit will be fixed by the heat motions, since the righting action of the interface must be supposed to diminish along the normal until it is so far upset by the heat motions as to be sensibly non-existent. I have in mind, of course, fluid states of matter in the interior of which the disposal of the molecules is random.

The greater the eccentricity of the force fields about the molecules, that is the greater their polarity, the thicker will be the orientated layer; also the higher the temperature the thinner will be the layer.

Is there any direct experimental evidence as to whether these orientated layers are a few or possibly very many molecules in depth? I think there is direct and unequivocal evidence that the layer may be at least of the order of a thousand molecules in thickness. The simplest experiment is the following. A weighted cylinder with a plane face is placed on a plate, both faces being clean, and fluid is allowed to be drawn in by capillary attraction with the result that the weighted cylinder is lifted, the work being done against considerable pressure. Every molecule which is drawn in must be under the influence of the attraction field of the solids. The temperature is then lowered until the lubricant is frozen. If the cylinder be now broken away the layer between the faces is found to be of sensible thickness easily visible to the naked eye. One has, then, a layer whose depth must be measured by the thousand molecules and of which no part is beyond the influence of the solid faces. Two physical properties of the layer are open to measurement, its tensile strength when the cylinder is pulled or broken away, and its friction when the cylinder is moved tangentially. In both cases the numerical values depend as much upon the chemical nature of the solids as upon the chemical nature of the lubricant. In both cases too the numerical results are extraordinary in one particular. So far as the measurements have gone at present the following is found both for adhesion and for friction. Let both cylinder and plate be of the same material, say material A, and let a be the value of either friction or adhesion; and let the value be b when cylinder and plate are made of material B. Now let cylinder A be placed on plate B or *vice versa*. The value now always is

$$\frac{a+b}{2}.$$

This is a humble example of transmission to place beside the colossal examples furnished by living matter, but when for years one has contemplated a scientific problem towards the explanation of which nothing could be advanced even the slightest clue is welcome.

53

FREIE UND GEBUNDENE FLÜSSIGKEIT IN GELEN

[Kolloid Zeit. (1928), XLVI, 268]

Zu dem Begriff der gebundenen Flüssigkeit gelangt man auf die allgemeinste Weise, wenn man eine sehr dünne Schicht aus einer der höheren Fettsäuren, wie z. B. Palmitinsäure, betrachtet, welche auf einer Wasseroberfläche ausgebreitet ist. Man nimmt jetzt allgemein an, dass die Moleküle dieses Films in ihrer Lage durch die Anziehungskraft zwischen dem Wasser und der Karboxylgruppe festgehalten werden. Die Paraffinkette ist nicht mit Wasser verbunden und umgeben, nicht weil überhaupt keine Anziehung zwischen dem Wasser und der Kette bestände, sondern weil die Ketten sich gegenseitig stärker anziehen als das Wasser.

Je kürzer die Kohlenstoffkette wird, desto leichter löslich werden die Säuren, weil die kürzeren Ketten sich gegenseitig weniger stark anziehen als die längeren. Schliesslich, wenn die Kette sehr kurz ist, sind Säure und Wasser in allen Verhältnissen miteinander mischbar.

In einem Extrem besteht also völlige Unmischbarkeit, und der Film befindet sich im Gleichgewicht mit Wasser und gesättigter Atmosphäre. Im anderen Fall ist die Fettsäure mit der Flüssigkeitsphase unbegrenzt mischbar, und die Atmosphäre ist mit Wasser und Säure gesättigt.

Zwischen diesen Grenzfällen enthält die Flüssigkeitsphase eine bestimmte Säurekonzentration, welche durch die mittlere Verweilzeit der Säuremoleküle in der Grenzfläche gegeben ist.

Wir werden finden, dass in keinem Falle die Anziehung zwischen allen Teilen der gelösten Moleküle und der Moleküle des Lösungsmittels in Zweifel gezogen werden kann; diese Anziehung unterscheidet sich in nichts von der allgemeinen Kohäsionskraft, welche zusammen mit dem von der translatorischen Bewegung der Moleküle herrührenden Wärmedruck den Zustand der Stoffe bestimmt.[1]

Die spezielle Anziehung, welche die Ionen infolge ihrer elektrischen Ladung auf das Lösungsmittel ausüben, ist für unsere Betrachtungen nichts weiter als eine besondere Art von Kohäsionskraft. Zum Glück brauchen wir uns hier nicht mit dem aussichtslosen Versuch abzugeben,

---

[1] Die Voraussetzung, welche dieser Behauptung zugrunde liegt, ist der Wissenschaft schon seit zwei Jahrhunderten bekannt. Es ist einfach die Feststellung, dass zwischen allen Arten der Materie gegenseitige Anziehung besteht, wenn sie sich in geringem Abstand voneinander befinden, dagegen Abstossung bei noch kleineren Abständen. Ausnahmen von der Regel, welche aber nur die Reichweite der Kräfte betreffen, treten dann auf, wenn die Materieteile elektrische Ladungen tragen.

zwischen Kräften chemischen und physikalischen Ursprungs zu unterscheiden.

Die Anziehung eines gelösten Moleküls wird sich von der eines Lösungsmittelmoleküls sowohl in der Intensität als auch in der Form der äquipotentiellen Oberflächen unterscheiden; jedes der gelösten Moleküle wird deshalb den "Zustand" des umgebenden Lösungsmittels irgendwie verändern. In diesem allgemeinen Sinne trägt es also eine Atmosphäre von Lösungsmittel.

Der Einfluss des gelösten Stoffes auf das Lösungsmittel kann in verschiedener Weise festgestellt und gemessen werden, z. B. durch die Veränderung des Gefrierpunktes oder durch die Ionenbeweglichkeiten. Die letzteren Werte sind dazu verwendet worden, den mittleren Radius der Atmosphäre eines Ions zu berechnen, unter der Annahme, dass das Ion eine ganz bestimmte Menge Lösungsmittel mit sich führt, welche von einer scharfen Oberfläche umgrenzt ist.

Existiert diese Oberfläche in Wirklichkeit? Wenn auch zugegeben werden muss, dass Moleküle des Lösungsmittels von bewegten Ionen mit fortgezogen werden, so kann ich mich doch nicht davon überzeugen, dass ein gelöstes Molekül irgendwelcher Art, geladen oder ungeladen, Lösungsmittel mit sich führt, welches von einer bestimmten Oberfläche umschlossen ist, jenseits welcher die Anziehung des gelösten Moleküls mit einem Male vollständig aufhören soll. Ich bemühe mich deshalb immer besonders, darauf hinzuweisen, wie wenig wünschenswert es ist, diesen Atmosphären bestimmte Radien zuzuschreiben; denn diese Werte, glaube ich, sind weiter nichts als Zahlen, welche lediglich den genauen Ausdruck gewisser experimenteller Tatsachen darstellen, während die Grenzen der Atmosphäre eines gelösten Moleküls unbestimmt sein müssen.

Bei der Diskussion der Beziehungen zwischen Lösungsmittel und gelöstem Stoff wird oft zwischen Kräften von grosser und kleiner Reichweite unterschieden. Elektrische Anziehungskräfte der Ionen werden als solche von grosser Reichweite angesehen im Gegensatz zu der kleinen Reichweite der Kohäsionskraft.

Aber hat diese denn wirklich nur eine kleine Reichweite? Neuere Beobachtungen, auf welche ich jetzt etwas näher eingehen will, über die Eigenschaften von dünnen Schichten, wie Schmiermittelfilmen zwischen zwei festen Körpern, scheinen zu ergeben, dass die Kohäsionskraft eine Reichweite von mindestens $5\,\mu$ besitzt. Wie sollten wir sonst die Tatsache erklären, dass ein Stahlzylinder in reiner Luft unter seiner eigenen Schwere nicht ganz bis zur Berührung mit einer darunter liegenden Stahlplatte herabsinkt? Bei einem von dem Gewicht des Zylinders herrührenden Drucke von 7 gr./cm. war der Luftspalt zwischen Zylinder und Platte $4\,\mu$,[1]

[1] Hardy u. Nottage, *Proc. Roy. Soc.* (1928), A, cxviii.

und dieser Zwischenraum wird nur wenig kleiner, wenn der Luftdruck auf 2 mm. Quecksilber erniedrigt wird.[1] Der Zylinder wird nicht durch elektrische Abstossung in der Höhe gehalten; denn wenn er auch bis zu metallischem Kontakt auf die Platte herabgedrückt wird, so dass beide Oberflächen geschrammt werden, so geht er doch wieder auf seine ursprüngliche Höhe zurück, sobald die hinunterdrückende Last entfernt wird.

Ich glaube, es muss nach diesen Versuchen zugegeben werden, dass die Reichweite der Kohäsion an der Oberfläche von Stahl, Kupfer, Glas oder Quarz überall unerwartet gross ist und dass die Anziehung einfach gleich der Summe der Anziehungskräfte der einzelnen Moleküle ist.

Selbst wenn die Kohäsion nur eine kleine Reichweite hätte, etwa von der Grössenordnung 10^{-8} cm., wie gewöhnlich angenommen wird, so muss noch mit einer anderen Möglichkeit gerechnet werden, mit einem Effekt, auf den zuerst Sutherland hingewiesen hat. Abgesehen von nur ganz wenigen Ausnahmen wird ein gelöstes Molekül die Lösungsmittelmoleküle in seiner unmittelbaren Nachbarschaft orientieren, diese wieder orientieren die hinter ihnen liegenden Moleküle, so dass schliesslich, obgleich die Anziehungskraft des Moleküls nur eine kleine Reichweite hat, ihre Influenzwirkung sich doch über eine ziemlich weite Entfernung auf unserer Längenskala erstreckt. Sowohl die "Influenz" als auch die elektrostatische Anziehung, falls die Moleküle Ladungen tragen, reichen mathematisch unendlich weit, nehmen aber mit wachsender Entfernung vom anziehenden Körper allmählich ab, in Flüssigkeiten wegen der Wärmebewegung der Moleküle und in festen Körpern wegen des "Kristallisationsdruckes".[2]

Die Wirkungsweite eines gelösten Moleküls wird deshalb eine Funktion der Temperatur sein. Ferner wird in einer Flüssigkeit die vollkommen regellose Anordnung der Lösungsmittelmoleküle, die nicht unter dem orientierenden Einfluss des gelösten Moleküls stehen, nicht in einem bestimmten Abstand von diesem beginnen, sondern nur ganz allmählich und ohne scharfe Grenzen die Oberhand gewinnen. In einem festen Körper dagegen existiert möglicherweise eine bestimmte Abgrenzung, weil es in diesem Falle gewöhnlich eine Art Quanteneffekt gibt, dadurch, dass sich die Atome in das Kristallgitter entweder einordnen oder nicht einordnen.

Aber auch in diesem Falle erstreckt sich, wie wir noch sehen werden, der Einfluss des anziehenden Körpers weit in den Kristall hinein jenseits der Grenzfläche, welche Kristall und Adsorptionsschicht voneinander trennt.

An diesem Punkt können wir eine ganz angebrachte Frage stellen. Verdanken Gele ihre Existenz der Abwesenheit dieser Quantenbeziehung? Wenn ja, so wäre Nägeli's Vorstellung, welche er in seiner Mizellartheorie bereits vor 70 Jahren formuliert hat, richtig.

[1] Watson, Unveröffentlichte Ergebnisse.
[2] H. A. Wilson, *Proc. Camb. Phil. Soc.* (1898), x.

Betrachten wir einen in einer Lösung wachsenden Kristall. In der Flüssigkeitsschicht, welche an eine der Kristallflächen angrenzt, besteht ein Druck, der die Moleküle aus der Flüssigkeit in das Kristallgitter treibt. Das ist der schon vorher erwähnte Kristallisationsdruck. Falls die Flüssigkeit eine Lösung bei irgendeiner Temperatur oberhalb des kritischen Punktes ist, so bleibt ein Teil flüssig und das Lösungsmittel in diesem Rest besteht aus denjenigen Anteilen der die gelösten Moleküle umgebenden Atmosphären, welche durch den Kristallisationsdruck nicht in das Kristallgitter getrieben werden können.[1]

Mit fallender Temperatur nimmt der Druck zu, und immer mehr Lösungsmittel geht aus dem Anziehungsbereich des gelösten Stoffes in den des Kristalles über.

Aehnlich bei einem Gel. Das Mikroskop zeigt, dass beim Gefrieren ein Teil des Wassers in eine Eisphase übergeht; dabei wird aber bald eine Grenze erreicht, indem der Rest des Wassers sogar in flüssiger Luft nicht gefriert.

Bei einem Gelatinegel ist das ein ziemlich beträchtlicher Bruchteil, nach Moran[2] bindet ein Gramm trockene Gelatine nicht weniger als 0·53 gr. ungefrierbares Wasser. Das ist eine enorme Menge. Es bedeutet, falls ungefrierbares Wasser ungebundenes Wasser ist, dass jedes Molekül Gelatine etwa 1500 Moleküle Wasser so fest bindet, dass selbst eine Temperatur von $-190°$ sie nicht loslösen kann.

Die Experimente verdienen nähere Betrachtung. Es wurden etwa 2 mm. dicke Gelatinescheiben verwendet. Bei $-3°$ C. bildete sich Eis ausschliesslich auf der Oberfläche, zwischen $-3°$ C. und $-19°$ entstand auch innerhalb des Gels eine Eisphase. Unterhalb $-19°$ schied sich an der Oberfläche der Scheiben kein Eis mehr in merkbarer Menge aus.

Die äussere Eisphase war reines Eis; die innere Eisphase dagegen war eine feste Lösung von Gelatine in Eis, welche infolge der Ausdehnung beim Gefrieren des Wassers wahrscheinlich einem sehr hohen Druck ausgesetzt war.[3] Wenn Moran das Gel zuerst bei $-3°$ mit dem äusseren Eis in Gleichgewicht brachte und erst dann weiter abkühlte, so konnte er das Gefrieren im Innern vollständig vermeiden; unter diesen Umständen fand er, dass reversible Phasenbeziehungen zwischen Gel und Eis bestehen. Bei Erhöhung der Temperatur, natürlich immer unterhalb $-3°$ C., ging Wasser aus dem Eis in das Gel über, bei Erniedrigung der Temperatur wanderte es

---

[1] Wilson's Gleichung für die Kristallisationsgeschwindigkeit V ist $V = \dfrac{A}{\alpha \cdot \eta}$, worin A ein Druck ist, der die Moleküle aus der Flüssigkeit gegen die Kristallfläche treibt durch eine Schicht von der Dicke κ gegen einen Widerstand η, der die Form einer Reibungsgrösse hat.

[2] Moran, *Proc. Roy. Soc.* (1926), A, CXII.

[3] Hardy, *Proc. Roy. Soc.* (1926), A, CXII.

umgekehrt aus dem Gel in das Eis. Bei jeder Temperatur stand das Eis im Gleichgewicht mit einer bestimmten Konzentration des Gels, welche allein von der Temperatur und nicht von dem Wege, auf dem der Endzustand erreicht wurde, abhängig war. Das Phasengleichgewicht wird durch die Kurve in Fig. 1 dargestellt, deren Koordinaten Temperatur und Gelkonzentration sind. Die Kurve ist eine Gefrierpunktskurve; links davon ist kein Eis, rechts ist Eis und unterkühltes Gel. Diese einfachen reversiblen Beziehungen bestanden aber nur dann, wenn das Gefrieren im Innern des Gels vollständig vermieden wurde. Warum? Weil durch den starken

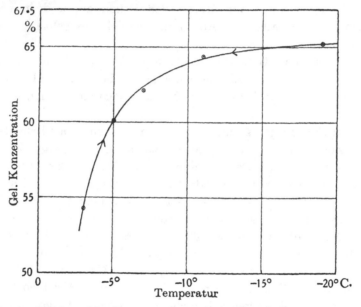

Fig. 1. Gleichgewichtskonzentration zwischen Gel (Gelatine) und Eis.

Druck, der beim Gefrieren im Innern des Gels auftrat, das Gerüstwerk des Gels stark beansprucht und verzerrt wurde. Das Gel wurde infolgedessen doppelbrechend und behielt die Eigenschaft sogar, wenn es wieder völlig aufgetaut war.[1]

Aus Moran's Kurve ist ohne weiteres zu ersehen, dass das Gefrieren bei etwa −19° aufhörte, wobei die Konzentration des Gels etwa 65·5 Proz. betrug. Ein Gel von dieser Konzentration zeigte auch in flüssiger Luft keine Spur von Eisbildung.

Können wir nun daraus schliessen, dass von jedem Gramm Gelatine mindestens 0·53 gr. Wasser gebunden werden?

Die Zahl stimmt gut überein mit den Resultaten von Adair über das Hydratwasser der Gelatine bei 0·0° C.[2] Er fand durch direkte Osmometrie,

[1] Hardy, *Proc. Roy. Soc.* (1926), A, CXII.
[2] *Report of the Food Investigation Board for 1925–7.*

dass ein Gramm Gelatine 0·6 gr. Wasser als Hydratwasser bindet. Andere Werte weichen dagegen beträchtlich ab. So würde beispielsweise die Quellungswärme viel weniger gebundenes Wasser ergeben. Die Volumkontraktion bei der Gelbildung der Gelatine ist von Svedberg[1] bei verschiedenen Konzentrationen gemessen worden, und die aus seinen Zahlen konstruierte Kurve[2] zeigt eine scharfe Biegung bei 0·08 gr. Wasser auf 1 gr. Gelatine (= 92 Proz. Gel).

Das Gleichgewicht zwischen Gelatinegelen und trockenem Azeton, sowie die Kurve der Trocknungsgeschwindigkeit solcher Gele führen zu Werten von weniger als 0·1 gr. Wasser, welches an 1 gr. Gelatine gebunden ist. Allerdings kann gegen Trocknungsexperimente eingewendet werden, dass bei einer bestimmten Konzentration auch Aenderungen in der Struktur mitspielen können. So kann beispielsweise das Leerwerden der kapillaren Räume als Ursache der von van Bemmelen beobachteten Hysteresis des Kieselsäuregels angesehen werden, worauf ich früher schon (1896) hingewiesen habe.[3]

Für die Kurve von Moran ist auch noch eine andere Erklärung möglich. Bei übermässig starker Unterkühlung hört die Kristallisation auf. Wilson schrieb dies dem Anwachsen des Widerstandes η zu, und da wir in einem Gel nicht nur die Strecke κ von der Grössenordnung 10^{-8} cm., sondern auch noch eine beträchtliche Diffusionssäule haben, so wird η in einem Gel sehr viel rascher mit fallender Temperatur zunehmen, als in einer Flüssigkeit. Die Punkte auf der Kurve von Moran sind darum vielleicht Punkte, bei denen der Reibungswiderstand den Uebergang des Wassers in die Eisphase verhindert.[4]

Wir wollen jetzt zu beweisen versuchen, dass selbst wenn eine bestimmte Grenzfläche zwischen einer Adsorptionsschicht und dem Rest des Materials existiert, dieselbe doch nicht die äusserste Grenze der Anziehung der adsorbierenden Substanz ist. Den Beweis werden wir in der Untersuchung der Schmiermittelfilme finden; um ihn möglichst objektiv zu erhalten, werden wir das Verhalten der Filme gegen eine normal gerichtete Kraft statt gegen eine tangential gerichtete zu betrachten haben.

Wenn man die zunächst flüssige Schmiermittelschicht zwischen einem Zylinder und einer Platte zum Gefrieren bringt und dann den Zylinder durch eine senkrecht gerichtete Kraft abreisst, so findet man, dass der Bruch an einer oder an zwei bestimmten Flächen im Innern des Schmiermittels eingetreten ist, von denen die eine in der Nähe der Zylinderoberfläche, die andere in der Nähe der Plattenoberfläche liegt. Jede von diesen Flächen geringster Zugfestigkeit trennt eine Adsorptionsschicht,

[1] Svedberg, *Journ. Amer. Soc.* (1924), XLVI.
[2] Moran, *Proc. Roy. Soc.* (1926), A, CXII.
[3] *Proc. Roy. Soc.* A.   [4] Hardy, *Proc. Roy. Soc.* (1926), A, CXII.

wie ich sie kurz nennen will, von einer kristallinischen Mittelschicht. Die
Lage ist in dem Diagramm Fig. 2 dargestellt. Man kann aus mechanischen
Eigenschaften schliessen, dass die Adsorptionsschichten nicht kristallinisch
sind, sondern aus Molekülen bestehen, welche durch die Oberflächen des
Zylinders und der Platte orientiert sind. Die mittlere Schicht ist, wie die
mikroskopische Betrachtung zeigt, aus flachen Kristallen aufgebaut. Die
Gesamtdicke der Schmiermittelschicht ist abhängig von dem Druck, unter
dem sie gebildet wurde; die grösste beobachtete Dicke war 7 μ.

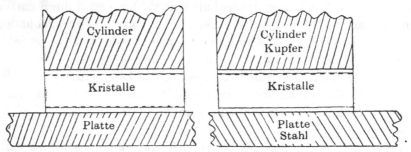

Fig. 2. Strukturbild von zwischen festen Oberflächen gefrorenem Schmiermittel.
1 und 2 sind Flächen, welche eine Adsorptionsschicht von einer Mittelschicht von
flachen Kristallen trennen. Der Bruch tritt bei 1 und 2 ein, wenn Zylinder und
Platte aus demselben Material bestehen, dagegen nur bei 1 oder 2 bei verschiedenem
Material. Die Bruchfläche ist als gestrichelte Linie gezeichnet.

Wenn die festen Körper verschieden waren, beispielsweise der Zylinder
aus Kupfer und die Platte aus Stahl, so trat der Bruch nur in der dem
Kupfer zunächst gelegenen Grenzfläche ein etwa 6 μ vom Stahl entfernt.
Dabei hatte aber die Adhäsionskraft nicht den für Kupfer charakteristischen Wert, sondern den genauen Mittelwert zwischen den Beträgen für
Stahl und Kupfer. Der Einfluss des Stahls ist demnach mindestens 6 μ
weit in das Schmiermittel hineingedrungen.

Es ist jetzt ganz klar, dass die Bruchfläche nicht eine Grenze ist, wo die
Anziehung des Kupfers oder Stahls auf das Schmiermittel plötzlich aufhört,
sondern wo diese Kraft nicht mehr gross genug ist, um die Atome von der
Einordnung in das Kristallgitter abzuhalten. Jenseits dieser Grenze
überwiegt der Kristallisationsdruck den Attraktionsdruck.

Definieren wir nun "gebundene" Flüssigkeit als ungefrierbare Flüssigkeit, so haben wir in diesem Schmiermittelfilm zwei Schichten aus "gebundenem" Schmiermittel, nämlich die Adsorptionsschichten, und zwischen
ihnen sozusagen "kapillares" Schmiermittel. Wenn der Wirkungsbereich
jeder Oberfläche n ist und der Zylinder von der Platte um mehr als 2n
entfernt ist, so muss in dem Film auch noch "freies" Schmiermittel vorhanden sein. Diese Feststellungen haben den Vorzug, nicht auf Speku-
lation, sondern auf ganz objektiven Tatsachen zu beruhen.

In diesem Falle werden gebundenes und kapillares Schmiermittel infolge der schon vorhin erwähnten Quantenbeziehung durch eine bestimmte Fläche voneinander getrennt; ob eine solche Trennungsfläche auch existiert, wenn das Schmiermittel geschmolzen ist, kann nicht gesagt werden.

Sowohl die Reibung als auch die Adhäsion von flüssigen Schmiermitteln hätten uns ebenfalls den Beweis liefern können, dass die Anziehung der festen Oberflächen über die Grenzen des gebundenen Schmiermittels hinaus wirksam ist. Den Fall eines gefrorenen Schmiermittels habe ich nur deshalb gewählt, weil dabei jede Behauptung auf direkte Beobachtung begründet ist.

Wir wollen jetzt annehmen, dass von dem Zylinder und der Platte nur die Oberflächenschichten vorhanden wären und dass eine Kolloidmühle diese Schichten im Schmiermittel verteilen würde. Dann würden wir ein allerdings viel komplexeres, im allgemeinen Typus aber ganz gleichartiges System erhalten, denn auch hier gäbe es zwischen den Teilchen Anziehungszonen, in denen die verschiedenen Kraftfelder übereinandergreifen. Nebenbei sei erwähnt, dass das System in zweierlei Hinsicht sich unterscheiden würde: die Teilchen, zu denen die Metalloberflächen zermahlen würden, würden Translations- und Rotationsbewegung zeigen, ausserdem würde die Bearbeitung durch die Kolloidmühle wahrscheinlich ausserordentlich starke innere Spannungen in ihnen zurückgelassen haben. Aenderungen der Reibung, Adhäsion und der elektrischen Eigenschaften[1] zeigen, dass feste Oberflächen sehr leicht zu dauernden Spannungen neigen, und dass das Kraftfeld durch diese Spannungen verändert wird.[2]

Wir können nun unsere bisherige Definition der gebundenen Flüssigkeit durch eine allgemeinere ersetzen, nämlich, dass gebundene Flüssigkeit solche ist, welche von den Teilchen der dispersen Phase stärker angezogen wird als von allen anderen Materieformen, welche gleichzeitig vorhanden sind. Zwei von diesen anderen Teilen des Systems können reines flüssiges oder kristallisiertes Dispersionsmittel sein. Freies Dispersionsmittel, welches sowohl flüssig als auch kristallisiert sein kann, ist dann solches, welches von den Kraftfeldern des gelösten Stoffes überhaupt nicht beeinflusst wird. Kapillares Dispersionsmittel lässt sich von gebundenem nur dann unterscheiden, wenn die Zustandsformen so verschieden sind, dass sie durch eine bestimmte Grenzfläche voneinander getrennt sind.

Was wird geschehen, wenn wir die Wirkung der Kolloidmühle umkehren und die Teilchen der dispersen Phase dicht zusammen bringen? Das kann durch Gefrieren erreicht werden und das Ergebnis ist bemerkenswert. Wenn man die Temperatur langsam abfallen lässt, werden die Teilchen koaguliert—das heisst, sie haften auch im vollkommen aufgetauten

---

[1] Shaw, *Proc. Roy. Soc.* A.
[2] Es ist übrigens auch keinesfalls anzunehmen, dass die durch den elektrischen Lichtbogen dispergierten Teilchen nichts weiter als ungeladene Bruchstücke sind.

Zustand noch fest zusammen. Bei sehr raschem Gefrieren dagegen findet man, dass die Teilchen nach dem Wiederauftauen noch dispergiert sind.

Die einfache und offenbar ausreichende Erklärung ist, dass bei raschem Gefrieren die Kristallisation an unzähligen Zentren einsetzt. Das trifft sicherlich zu bei einem Gelatinegel, wie die Fig. 3 nach einer Mikrophotographie zeigt. In diesem Falle würden also die Teilchen nicht zusammengehäuft werden.

Fig. 3. Mikrophotographie von 12 Proz. Gelatinegel in flüssiger Luft gefroren.

Eine andere Möglichkeit ergibt sich aus den Versuchen mit Zylinder und Platte. In reiner Luft und bei ganz sauberen Oberflächen bleiben sie nicht zusammen in Berührung, wie ich schon erwähnt habe, auch wenn sie durch kräftigen Druck zu metallischem Kontakt gebracht wurden; wenn die Flächen aber etwas verunreinigt sind, so berühren sie sich auch noch, wenn der äussere Druck beseitigt wird. Langsames Gefrieren wird irgendwelche Verunreinigungen, welche sich in der nicht gefrorenen Phase befinden, konzentrieren. Experimente, bei denen Sole von extremer Reinheit zum Gefrieren gebracht wurden, sind mir nicht bekannt.

Schon vor langer Zeit, im Jahre 1896, konnte ich unter dem Mikroskop die Koagulation eines Sols von denaturiertem Protein beobachten. Ausgehend von einem Sol, welches so hochdispers war, dass es das blaue Licht zerstreute, gab es eine ununterbrochene Serie von Veränderungen, welche zur Koagulation führten und auf sie folgten. Die Synäresis des entstandenen Gels war positiv, das heisst, es zog sich spontan unter Auspressung von Flüssigkeit zusammen.

Die dispersen Teilchen vereinigten sich zunächst zu grösseren. Wenn diese gross genug waren, um sichtbar zu sein, liessen sie Brownsche Bewegung erkennen. Das nächste Stadium wurde augenscheinlich nicht durch weiteres Wachstum der Teilchen erreicht, sondern dadurch, dass sie sich eines an das andere hängte und auf diese Weise schliesslich ein Netzwerk bildete, in dem die flüssige Phase eingeschlossen war. Die kugelförmigen Tröpfchen koagulierten indessen nicht, sie berührten einander nur in kleinen Zonen. Offenbar existierte ein Widerstand gegen die weitere Verschmelzung, welcher auch die Ursache für den quasi festen Widerstand gegen die Deformation der Teilchen war. Die langsame Synäresis des Gels schien davon herzurühren, dass die Teilchen allmählich nachgaben und die Berührungszonen grösser wurden.

Definieren wir die Koagulation als das Zusammenlagern der Teilchen, so besteht zwischen den Zuständen vor und nach der Koagulation ein grosser Unterschied. Im ersteren Falle konnte die Veränderung in jedem Stadium durch passende Einstellung der Elektrolytkonzentrationen aufgehalten werden und das Gleichgewicht wurde in verhältnismässig kurzer Zeit erreicht. Nach der Koagulation dagegen war infolge der geringen Beweglichkeit der kolloiden Teilchen ein grosser Widerstand gegen die Veränderung vorhanden, und die für das Gelstadium charakteristischen Verzögerungserscheinungen traten auf. Die langsame Zusammenziehung des Gels kann, so viel ich weiss, durch Aenderung der Elektrolytkonzentrationen nicht aufgehalten werden, wird aber durch Temperaturerhöhung beschleunigt.

Ein Gel mit positiver Synäresis kann definiert werden als ein Gel, in welchem die Anziehungswirkung der dispersen Phase auf sich selbst viel grösser ist als auf das Dispersionsmittel. Die Zusammenziehung hört auf, wenn die Differenz zwischen den Anziehungswerten gleich irgend einem inneren Widerstand gegen Veränderung ist, der beispielsweise von der festen Beschaffenheit oder von der Aenderung der Oberflächenenergie herrühren kann. Es sei nochmals betont, dass die Definition lediglich die Verschiedenheit zweier Grössen feststellt; sie sagt nicht aus, dass zwischen dispersem Anteil und Dispersionsmittel bei der Synäresis überhaupt keine Anziehungskraft wirksam sei.

Warum hängen sich die Teilchen aneinander und bilden ein Netzwerk, statt sich einfach als disperser Niederschlag abzusetzen? Die Antwort darauf muss sein, dass die Teilchen sehr bereitwillig aneinander festhaften, im Gegensatz zu den Körnchen eines feinen Kristallpulvers, welches nur durch starke Pressung zusammengeschweisst werden kann. Ich bin überzeugt, dass sich für den Kolloidphysiker nichts mehr lohnen würde, als eine so alltägliche Eigenschaft wie den Widerstand fester Körper gegen das Verschweissen beim Zusammenpressen zu untersuchen. Die Sache ist so alltäglich, dass sie bis jetzt unsere wissenschaftliche Neugier noch kaum erregt hat!

Hier will ich eine ebenso alltägliche Erscheinung erwähnen, welche für mich in Beziehung zu diesen Erörterungen von Bedeutung ist. Ein Stück gewöhnliches Waschleder nimmt im feuchten Zustand sehr begierig Wasser auf, wenn es aber trocken ist, wird es nur mit der grössten Schwierigkeit benetzt. Die Ursache ist vermutlich die Steifheit des festen Gerüsts im letzteren Fall. Das Leder nimmt nur dann bereitwillig Wasser auf, wenn seine Fasern genug adsorbiert haben und dadurch geschmeidig geworden sind.

Die Koagulation kann verhindert werden, wenn die Oberfläche der dispersen Phase mit einem Schutzkolloid vergiftet wird, von dem anzunehmen ist, dass seine Wirksamkeit auf der Verminderung des Unterschiedes zwischen den vorhin erwähnten Anziehungswerten beruht.

Es gibt eine Art von nicht kolloiden aber den Emulsionskolloiden verwandten Lösungen, in denen die Fähigkeit des Zusammenschweissens oder Verschmelzens der Faktor ist, der schliesslich die Verteilung bestimmt. Wenn eine gewöhnliche Lösung gefriert, so wird ein Teil des Lösungsmittels dadurch, dass es in das Kristallgitter wandert, dem Einfluss des gelösten Stoffes entzogen. In jenen Fällen schlägt das Lösungsmittel beim Sinken der Temperatur einen anderen und interessanteren Weg ein. Es aggregiert zu einer neuen flüssigen Phase, welche eine kleine Menge des gelösten Stoffs enthält. Ich habe das System Phenol und Wasser im Sinn. Die Ausscheidung geht stufenweise vor sich und es gibt dabei, wie wohl bekannt ist, ein mehr oder weniger ausgedehntes emulsoides Stadium, in dem die ursprüngliche Wasser-in-Phenol-Phase in der Phenol-in-Wasser-Phase dispergiert ist.

Wenden wir die bis jetzt benützte Formel auf diesen Fall an, so müssen wir sagen, dass das in den emulsoiden Teilchen gebundene Wasser stärker vom Phenol angezogen wird als von dem phenolhaltigen Wasser in der äusseren Phase.

Die besondere Eigenschaft, welche dieses System beim Vergleich mit einem Gelatinesol zeigt, besteht darin, dass bei Temperaturerniedrigung die disperse Phase koaguliert, so dass schliesslich im Gleichgewichtszustand zwei durch eine glatte Grenzfläche geschiedene Phasen existieren.

Sicherlich muss die Gruppierung der Wassermoleküle in den beiden Phasen verschieden sein. Das ergibt sich aus der Tatsache, dass, obwohl die Konzentrationen ganz verschieden sind, das Potential doch das gleiche ist, da auch der Dampfdruck der beiden Phasen gleich ist. Die Beschaffenheit der inneren Phase muss so sein, dass bei einer gewissen Temperatur völlige Verschmelzung möglich ist, sie muss also wohl flüssig sein.

Es fragt sich, warum die Teilchen nicht in dem emulsoiden Zwischenstadium verschmelzen. Darauf können viele Antworten gegeben werden, zwei davon will ich auswählen. Eine ganz allgemeine Lösung des Problems wäre es, wenn man den Widerstand gegen das Verschmelzen auf die Form einer Funktion $\dfrac{de}{dr}$ bringen könnte, worin e die Energie und r der Radius eines Emulsoidteilchens ist. Die Beobachtungen von Reinhold und Ruckers an Seifenfilmen sprechen aber dagegen.

Eine andere Antwort, welche eine mechanische Eigenschaft in Betracht zieht, würde sein, dass mit zunehmender Grösse der Emulsoidteilchen auch ihre Fluidität grösser wird, so dass sie sich nach der Berührung nicht wieder voneinander trennen.

Aenderungen in der äusseren Phase dürfen nicht ausser Betracht

gelassen werden. Kehren wir noch einmal zu dem Zylinder und der Platte
zurück, zwischen denen sich Schmiermittel befindet. Wird der Zylinder
der Platte genähert, so erhält das Schmiermittel durch das Attraktionsfeld
eine ausgesprochene Struktur, wodurch es sowohl der normalen als auch
der tangentialen Bewegung, dem Gleiten, gewisse Widerstände entgegen-
setzt, welche ganz und gar auf den Einfluss der angrenzenden festen
Körper zurückzuführen sind. Die Bewegungswiderstände stehen unterein-
ander nicht in einfachem Zusammenhang, wie sich aus den Kurven in
Fig. 4 ergibt. Die erste ist eine gewöhnliche Adhäsionskurve, die andere
steht in einfacher Beziehung zum Schmelzpunkt, der eine Stoffeigenschaft
ist. Der Widerstand gegen senkrechte Bewegung bedeutet Widerstand
gegen eine Kraft, welche den Zylinder gegen die Platte drückt. Betrachten

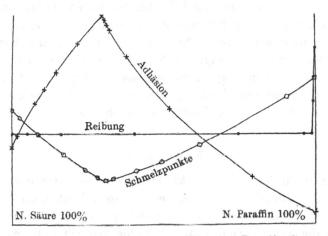

Fig. 4. Reibung und Adhäsion von Fettsäure-Paraffin-Gemischen
zwischen Stahlflächen und die Schmelzpunkte der Gemische.

wir nun zwei kolloide Teilchen, welche sich einander nähern, so wird die
Flüssigkeit zwischen ihnen jeder Relativbewegung in allen Richtungen
einen vermehrten Widerstand entgegensetzen.

Diese Widerstände brauchen Zeit, um zu ihrem Höchstwert anzuwach-
sen, wie aus den Experimenten mit Zylinder und Platte hervorgeht, und
diese Zeitabhängigkeit muss im Falle der Emulsion zu wichtigen Schluss-
folgerungen führen.

Langmuir's Adsorptionstheorie ist allgemein wohl bekannt, und da sie
an dieser Stelle vielleicht Schwierigkeiten zu bieten scheint, so verdient ihre
Begründung kurze Erwähnung. Nach dieser Theorie sind die adsorbierten
Moleküle in Haufen auf aktiven Zonen angeordnet. Die Untersuchung der
Reibung spricht aber nicht für diese Ansicht, denn man kann zeigen, dass
jedes zu einer unvollständigen Schmiermittelschicht hinzugefügte Molekül
die Reibung um einen konstanten Betrag verringert, der von der Natur

des festen Körpers unabhängig ist. Jedes Molekül muss also so angeordnet sein, dass es nicht von anderen gestört wird; das wäre aber dann der Fall, wenn es sich mit einem Molekülhaufen vereinigen würde.[1]

Trotzdem scheinen aber die Tatsachen, welche für die Haufenbildung sprechen, überwältigend. Wenn z. B. geladene Teilchen auf einer Fläche auftreten, so lagern sie sich nicht gleichmässig, sondern in Haufen an.[2] Ferner hat Paneth festgestellt, dass nur 31 Proz. der Oberfläche einer Bleisulfatsuspension von adsorbiertem Farbstoff bedeckt werden.

Waren die zu diesen Beobachtungen verwendeten Oberflächen rein? Bei Paneth's Experimenten war das offenkundig nicht der Fall, denn ausser den Farbstoffmolekülen waren noch eine Reihe von anderen Molekülarten vorhanden, die alle um einen Platz an der Oberfläche mitkonkurrierten. Langmuir verwendete gleichmässige Oberflächen von Glimmer, Glas und Platin, und nach meiner Erfahrung verteilen sich Moleküle keinesfalls gleichförmig auf solchen Oberflächen, wenn diese vorher durch eine andere Art von Molekülen verunreinigt wurden. Sauberkeit und chemische Reinheit helfen wenig, wenn man findet, dass der zufällige Zutritt von Laboratoriumsluft die Werte der Reibung oder der Adhäsion um etwa 60 Proz. ändert, selbst wenn die feste Oberfläche durch eine Schicht von Schmiermittel überdeckt ist.[3]

NACHTRAG

Die Frage, ob kolloide Teilchen kristallinische oder amorphe Beschaffenheit haben und ähnliche Probleme, welche augenblicklich im Vordergrund stehen, sind meines Erachtens viel weniger wichtig als die Frage der Reichweite der Oberflächenkräfte. Bei der kritischen Betrachtung der darüber gewonnenen Resultate zeigt sich zunächst, dass sie vollkommen objektiv sind, auf unmittelbaren Sinneswahrnehmungen beruhen; Berechnungen und Spekulationen über Flächen und Mengen liegen nicht zugrunde. Sie wurden aus Untersuchungen über die Reibung sowie über Zugfestigkeit und Scherfestigkeit von engen, mit Material erfüllten Spalten und Fugen erhalten. Wenn eine dünne Zwischenschicht von festem Schmiermittel zwischen einem Zylinder und einer Platte auseinander gerissen wird, so ist ohne weiteres zu sehen, dass die Schmiermittelschicht aus einer Adsorptionsschicht an jeder Metalloberfläche und einer Schicht von flachen Kristallen dazwischen besteht. Der Bruch tritt immer zwischen der Adsorptionsschicht und der Kristallschicht ein. Ist das System unsymmetrisch, besteht also z. B. der Zylinder aus Kupfer und die Platte aus Stahl, tritt der Bruch in der Nähe des Kupfers ein und kann bis zu

---

[1] *Proc. Roy. Soc.* (1922), A, c, 546.
[2] Henderson, Unveröffentlichte Beobachtung.
[3] *Proc. Roy. Soc.* (1928), A, cxviii, 213.

$6\,\mu$ vom Stahl entfernt sein; dabei ist aber der Adhäsionswert gleich dem Mittel aus den Werten für Kupfer und Stahl. Wie kommt es, dass das Anziehungsfeld des Stahls seinen Einfluss noch in einer so erstaunlichen Entfernung ausübt? Man könnte an eine Polarisation denken, welche sich durch das Schmiermittel ausbreitet, indessen können auch in reiner Luft Wirkungen bis zu denselben Entfernungen beobachtet werden und es ist kaum anzunehmen, dass die Moleküle eines Gases in dieser Art polarisiert werden.

Der Ursprung des Anziehungsfeldes, ob elektrostatisch oder elektrokinetisch, ist im Augenblick unwesentlich, da es vorläufig unter den weiteren Begriff der Kohäsion fällt.

Es wäre möglich, dass die Anziehungsfelder von zwei ausgebreiteten, parallelen Flächen sich gegenseitig verstärken und auf diese Weise Wirkungen über unerwartet weite Entfernungen hervorrufen. Um eine solche "Kondensator"-Wirkung zu untersuchen, wäre eine mathematische Behandlung erforderlich, die über des Verfassers Geschicklichkeit hinausgeht; indessen könnte man höchstwahrscheinlich doch nicht darum herumkommen zu schliessen, dass die Kohäsionskräfte eine viel grössere Reichweite haben als man bisher gewöhnlich angenommen hat. Nach der Aufteilung des Systems durch eine Kolloidmühle wird also die relative Lage der Teilchen ebenso wichtig sein wie ihre Gesamtoberfläche oder die Teilchenzahl in der Volumeneinheit.

Kehren wir für einen Augenblick zu den Versuchen mit Zylinder und Platte zurück. Die Tatsache, dass sich eine direkte Anziehungskraft über die Entfernung von einigen tausend Molekülen ausbreitet, wirft auf gewisse Beziehungen ein unerwartetes Licht.

Aus beobachteten Werten der statischen Reibung und der Adhäsion kann man den Widerstand einer Grenzfläche gegenüber äusseren Kräften in tangentialer und normaler Richtung ableiten. Der Unterschied zwischen beiden Grössen ist sehr erheblich; die normale Kraft, d. h. der Widerstand gegen das Auseinanderziehen senkrecht zur Fläche ist, um tatsächliche Werte anzuführen, für eine Schicht einer normalen Fettsäure zwischen Stahlflächen etwa 20,000 gr. pro q. cm. Der entsprechende Widerstand gegen das Gleiten, d. i. die statische Reibung, ist etwa 2 gr. pro q. cm. Ein Freund machte mich darauf aufmerksam, dass diese Ungleichheit dann zu erwarten ist, wenn die Reichweite der Kohäsion nach μ zu messen ist; denn während der tangentiale Widerstand nur von den Molekülen in unmittelbarer Nähe der Oberfläche herrührt, so wäre dann der normale Widerstand die Summe der gegenseitigen Anziehung von vielen tausenden von Molekülen auf jeder Seite der Zwischenschicht.

So wäre auch zu erklären, warum, wie schon erwähnt, die Reibung in einfacher Beziehung zur Adsorption, die Adhäsion dagegen in ebenso einfacher Beziehung zu einer Stoffkonstante, nämlich dem Schmelzpunkt steht (siehe Fig. 4).

Die statische Reibung darf hier nicht mit der Scherfestigkeit eines Stoffes verwechselt werden, denn es lässt sich zeigen, dass diese mit dem normalen Widerstand in Zusammenhang steht.

Der Aggregatzustand einer Schmiermittelschicht zwischen festen Oberflächen ist ein ganz besonderer. Betrachten wir den Fall eines in Ueberschuss von Flüssigkeit auf der Platte stehenden Zylinders. Die Zwischenschicht ist insofern flüssig, als durch einen auf den Zylinder ausgeübten Druck oder Zug etwas von der Flüssigkeit herausgepresst oder hereingezogen wird. Anderseits ist sie auch fest, da sie offenbar unbeschränkt tangentiale Kräfte unterhalb der statischen Reibung auszuhalten vermag. Sie ist kein plastischer, fester Körper, da das Nachgeben bei tangentialer Beanspruchung nicht langsam, sondern ganz plötzlich wie ein Bruch eintritt; bei geringem Molekulargewicht des Schmiermittels hat sie in der Tat ganz die Eigenschaften eines spröden, zerbrechlichen festen Körpers. Die Schmiermittelschicht besitzt also, wenn sie auch ursprünglich aus Flüssigkeit besteht, eine eigentümliche Anordnung der Moleküle, die durch die Kraftfelder der umgebenden festen Stoffe hervorgerufen wird. Alle mechanischen Eigenschaften einer 4μ dicken Schicht sind in der ganzen Dicke ebenso von der Natur der einschliessenden festen Stoffe, wie von der eigenen chemischen Zusammensetzung abhängig.

So erhebt sich die Frage, ob die Gelbildung etwa auf zwei Vorgängen beruht, welche zusammenwirken oder einzeln eintreten können. Der eine ist das Zusammenhaften der dispersen Teilchen zu einem Netzwerk; es scheint ziemlich sicher zu sein, dass dies eintreten kann und eine der Methoden der Gelbildung ist.

Der andere Vorgang ist das Steifwerden des Dispersionsmittels durch die Kraftfelder der Teilchen, und das ist auch dann möglich, wenn die Teilchen voneinander getrennt bleiben. Ohne Zweifel lassen die Untersuchungen über Reibung und Adhäsion diesen Vorgang nicht nur als eine Möglichkeit, sondern als einen notwendigen und wichtigen Faktor bei der Gelbildung erscheinen.

Die ultramikroskopischen Beobachtungen von Bayliss über die Veränderungen in lebender Materie sprechen dafür, dass das für den aktiven Zustand charakteristische Steifwerden tatsächlich in dieser Weise zustande kommt; in meinen eigenen Aufzeichnungen von 1898 finde ich ferner, dass die Proteinteilchen ihre Brownsche Bewegung verloren, bevor sie sich zusammenlegten, d. h. bevor noch die Koagulation, wie oben definiert, eintrat.

Nimmt man die Existenz eines weitreichenden Anziehungsfeldes um die Kolloidteilchen als gegeben an, so führen Ueberlegungen, wie sie von

Osborne Reynolds in seinem Raumerfüllungstheorem entwickelt wurden, zu der räumlichen Anordnung der Mizellen. Sind beispielsweise die äquipotentiellen Flächen konzentrisch um den Massenmittelpunkt eines Teilchens gelagert, so wäre der Aufbau regulär; dagegen müssten multipolare Einheiten wie Gelatinemoleküle, deren äquipotentielle Flächen ganz unregelmässig sind, im Gleichgewicht in einer viel komplizierteren Anordnung, welche nur schwer zu erreichen ist, gelagert sein. Gele müssten also eine gewisse Struktur haben, deren Bausteine nicht Atome, sondern Kolloidteilchen sind.

Die hier kurz erwähnten Tatsachen sind sehr wichtig für die Bedeutung von Dickenmessungen der Adsorptionsschichten. Selbst wenn man die Methoden der Oberflächenmessung gelten lässt, so geben die erhaltenen Werte doch nur die ungefähre Zahl der Moleküle pro Flächeneinheit, welche die Anziehungskräfte an der Oberfläche gegenüber äusseren wegziehenden Einflüssen festhalten. Mit anderen Worten, was gemessen wird, ist die Stärke der störenden Kräfte, welche die Moleküle und Atome in eine andere Konfiguration als die Adsorptionsschicht bringen wollen, etwa in die Form eines Gases, einer Flüssigkeit oder eines Kristalls.

Es ist deshalb nicht zulässig, aus Messungen der Dicke einer Adsorptionsschicht einen Schluss über die Reichweite von Kohäsionskräften zu ziehen. Versuche mit Zylinder und Platte beweisen, dass diese Kräfte sich weit über die Grenze der Adsorptionsschicht erstrecken, bis hinein in das Kristallgitter, in welches durch den Kristallisationsdruck diejenigen Moleküle hineingezwängt werden, welche nicht in unmittelbarer Nähe der anziehenden Oberfläche liegen.

54

STATISCHE REIBUNG UND ADSORPTION[1]

[*Kolloid Zeit.* (1930), LI, 6]

Eine neuere Arbeit über handelsübliche Oele und besonders über den
Einfluss der Temperatur auf ihre Schmierwirkung ergab einige interessante
Fragen über den Zusammenhang zwischen der statischen Reibung ge-
schmierter Flächen und der Adsorption. Es ist offenbar zu erwarten, dass
jede Theorie über die Beziehungen zwischen Reibung und Adsorption
sowohl die verschiedene Adsorptionsfähigkeit der im Oel vorhandenen
Molekülarten als auch die verschieden leichte Zugänglichkeit der Moleküle
zu den Adsorptionsschichten berücksichtigen muss.

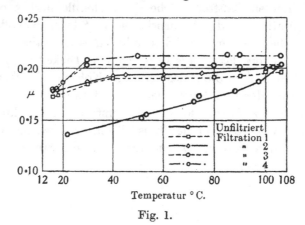

Temperatur ° C.

Fig. 1.

Die Form der Reibungs-Temperaturkurve wird geändert, wenn ein Oel
mit einer festen Oberfläche in Berührung gebracht wird, z. B. wenn man
das Oel durch eine Schicht von Glasperlen filtriert. Die Kurve wird einfacher und lässt sich in verschiedene Abschnitte einteilen, in denen die
Reibung von der Temperatur entweder abhängig oder unabhängig ist
(Fig. 1). Ausserdem wird die Kurve reversibel, insofern als sie stets
dieselbe Form hat, gleichgültig ob die Temperatur ansteigt oder fällt. Die
Vereinfachung geht weiter, bis die ganze Kurve horizontal verläuft und
die Reibung innerhalb des gemessenen Bereiches (von 15° bis 115° C.)
vollständig temperaturunabhängig ist.

Die Veränderungen rühren davon her, dass durch die Adsorption polare
Moleküle entfernt werden. Das ergibt sich daraus, dass die latente Periode

---

[1] Uebersetzt von W. Haller (Leipzig).

(welche für die Orientierung der Moleküle in der Adsorptionsschicht charak-
teristisch ist) nach der Filtration durch Glasperlen verschwindet. Weiter
spricht dafür, dass genau dieselben Aenderungen der Kurvenform durch
Extraktion des Oels mit trockenem Azeton hervorgerufen werden. Est ist
einleuchtend, dass die gleichen Resultate mit jeder beliebigen adsorbierenden Oberfläche erhalten werden können, d. h. die adsorbierende Fläche
braucht nicht aus demselben Material zu bestehen wie der Schlitten und
die Platte, mit denen das Oel geprüft wird.

Die Reinigung ist vollständig, wenn keine Adsorption mehr stattfindet,
d. h. wenn das Oel nur noch als solches adsorbiert wird. In diesem Fall
kann man das Oel ein "Adsorptions-Individuum" nennen. Natürlich ist

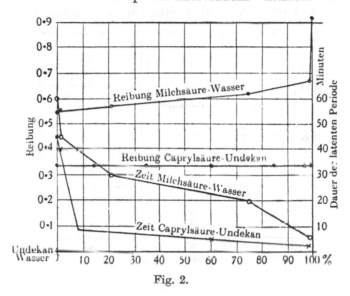

Fig. 2.

es weder sicher noch wahrscheinlich, dass ein Adsorptionsindividuum
auch immer ein chemisches Individuum ist. Die handelsüblichen Schmieröle
(mineralischen Ursprungs) bestehen anscheinend zur Hauptsache stets aus
einem Adsorptionsindividuum mit einer Beimengung von etwa 5 Proz.
an aktiven Stoffen.

Aus einem wagerechten Verlauf der Kurve kann man schliessen, dass die
Zusammensetzung der Adsorptionsschicht von der Temperatur unabhängig ist. Die Berechtigung für diese Annahme ist, dass sie die ein-
fachste ist, welche den Tatsachen entspricht. Bei horizontalem Kurven-
verlauf wird demnach entweder das Oel als Ganzes adsorbiert oder aber es
"dominiert" eine bestimmte Art von Molekülen. Ein einfaches Beispiel
für dieses Dominieren gibt die isotherme Kurve für die Mischung eines
Paraffins mit einer aliphatischen Säure (z. B. Caprylsäure und Undekan,
Fig. 2). Die Reibung ist gleich der der reinen Säure bis herunter zu einer

Konzentration der Säure von 0·7 Proz., bei welcher Verdünnung zum ersten Male Paraffinmoleküle in die Adsorptionsschicht eindringen können.

Wie begrenzt die Gültigkeit einfacher Hypothesen über die Adsorption ist, kann man sehen, wenn man versucht, den Uebergang von dem horizontalen Abschnitt der Kurve zu einem steigenden oder fallenden Abschnitt zu erklären. Betrachten wir z. B. eine Lösung eines festen normalen Paraffins in reinem Paraffinöl ("medicinal" oil). Dieses Oel besteht offenbar aus gesättigten zyklischen Verbindungen. Es wird durch längere Filtrationsreinigung hergestellt und ist ein Adsorptionsindividuum, dessen Koeffizient der statischen Reibung für Stahl auf Stahl $\mu = 0.228$ ist. Dieser Koeffizient ist im ganzen Temperaturgebiet konstant. Wenn nun Paraffinwachs in dem Oel aufgelöst wird, so erhält man die Kurve in Fig. 3. Von

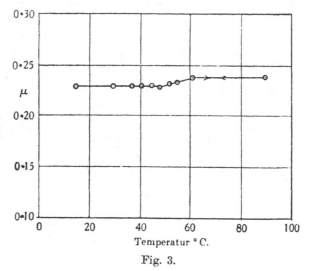

Fig. 3.

18° bis 46° hat das Wachs keinen Einfluss auf die statische Reibung, von 46° bis 62° steigt die Reibung an, um schliesslich von 62° ab wieder konstant zu bleiben. Das Wachs, welches zu diesem in Fig. 3 wiedergegebenen Versuch verwendet wurde, hatte einen Schmelzpunkt von 54° bis 57° und einen Reibungskoeffizienten von $\mu = 0.275$. Dieser Koeffizient war ebenfalls von der Temperatur unabhängig.

Das Paraffin war in dem Oel allem Anschein nach vollkommen gelöst, jedoch waren seine Moleküle im Anfangsgebiet der Kurve offenbar nicht adsorptionsfähig. Die Adsorption begann erst in der Nähe des Schmelzpunktes, wo die Moleküle zum ersten Male in die Adsorptionsschicht gelangen konnten. Der zweite horizontale Abschnitt stellt ein Adsorptionsgleichgewicht zwischen dem Paraffinwachs und dem Oel dar, welches für den ganzen Rest der Temperaturkurve unveränderlich ist. Alle Kurven für Mischungen von Paraffinöl mit Paraffinwachs haben diese typische

Form und die Kurven steigen oder fallen je nachdem, ob der Reibungs-
koeffizient des Paraffins grösser oder kleiner ist als der des Oels. In allen
Fällen schliesst das Temperaturgebiet zwischen den beiden horizontalen
Stücken der Kurve den Schmelzpunkt des festen Paraffins ein.

Die Deutung dieser Erscheinung ist einfach. Oberhalb ihres Schmelzpunktes sind die Wachse in allen Verhältnissen mit dem Paraffinöl mischbar. Wenn wir nun annehmen, dass unterhalb des Schmelzpunktes die
Paraffine nur in assoziiertem Zustande in der Lösung vorhanden sind, und
ferner, dass nur unassoziierte freie Moleküle genügend translatorische
Energie besitzen, um in die Adsorptionsschicht eindringen zu können, so

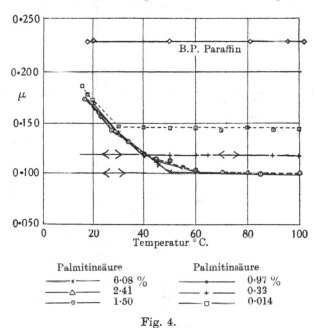

Palmitinsäure   Palmitinsäure
———x———   6·08 %   ———•———   0·97 %
———△———   2·41   ———+———   0·33
———⊙———   1·50   ———□———   0·014

Fig. 4.

haben wir bereits die Bedingungen, welche den Anstieg der Kurven erklären können. Dieses ist ein einfaches Beispiel für gehinderte Adsorptionsfähigkeit.

Es ist schwierig, zwischen der Adsorptionsfähigkeit und der Zugänglich-
keit zum Adsorptionsmittel (availability und accessibility) eine scharfe
Grenze zu ziehen, wenn eine solche überhaupt besteht. Betrachten wir den
folgenden Fall. In Fig. 4 sind Kurven für verschiedene Mischungen von
Palmitinsäure und Paraffinöl wiedergegeben. Man sieht, dass im ersten
Teil jeder Kurve die Zugänglichkeit der Adsorptionsschichten für die
Palmitinsäure unvollständig ist. Sie wird aber grösser und erreicht schliesslich einen maximalen Wert, bei dem die Adsorptionsschicht im Gleichge-
wicht mit der ganzen Mischung steht. Mit zunehmender Konzentration

der Säure steigt die Temperatur, bei welcher dieses Gleichgewicht erreicht wird, bis schliesslich bei 50° die Adsorptionsschicht endgültig mit Säure gesättigt ist, so dass ein weiterer Zusatz von Säure die Kurve nicht mehr ändert. Diese Kurven sind im Gegensatz zu den Kurven mit Paraffinwachs vollständig irreversibel. Nachdem die polaren Moleküle einmal in die Adsorptionsschicht eingedrungen sind, können sie durch Erniedrigung der Temperatur nicht mehr verdrängt werden. Oder anders ausgedrückt, sobald der Knickpunkt der Kurven erreicht wurde und die Zugänglichkeit

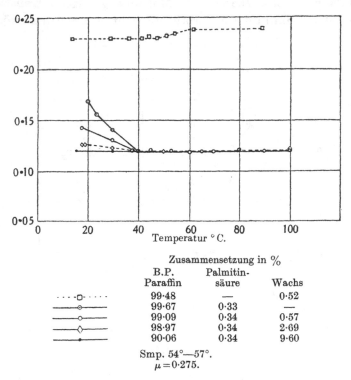

Zusammensetzung in %

	B.P. Paraffin	Palmitin-säure	Wachs
· · · -□- · · · ·	99·48	—	0·52
———⊙———	99·67	0·33	—
———○———	99·09	0·34	0·57
———◇———	98·97	0·34	2·69
———•———	90·06	0·34	9·60

Smp. 54°—57°.
$\mu = 0{\cdot}275.$

Fig. 5.

vollkommen geworden ist, wird *für die spezielle Konzentration der Säure* die Zusammensetzung der Adsorptionsschichten konstant und temperatur-unabhängig, die Kurve verläuft ganz horizontal.

Wenn zu der Mischung aus Oel und Säure noch Paraffinwachs hinzuge-setzt wird, so erhält man ein ganz unerwartetes Ergebnis. Das Endgleich-gewicht wird nicht geändert, aber die Zugänglichkeit der Säure zur Ad-sorptionsschicht bei den niedrigen Temperaturen wird etwas besser, so dass der Anfangsteil der Kurven bei steigendem Wachszusatz immer weniger steil wird, bis schliesslich die ganze Kurve horizontal verläuft. Diese Erscheinung ist in Fig. 5 dargestellt. Die Koeffizienten der reinen Substanzen waren Paraffinöl $\mu = 0 \cdot 228$, Palmitinsäure $\mu = 0 \cdot 00$, Paraffin $\mu = 0 \cdot 275$.

Zusammen mit den drei Prinzipien des "*Dominierens*", der *Adsorptionsfähigkeit* und der *Zugänglichkeit* scheint eine Adsorptionshypothese alle Zusammenhänge zwischen Reibung und Temperatur, welche an handelsüblichen Oelen festgestellt wurden, erklären zu können (siehe *The Analysis of Commercial Lubricating Oils by Physical Methods*, His Majesty's Stationery Office).

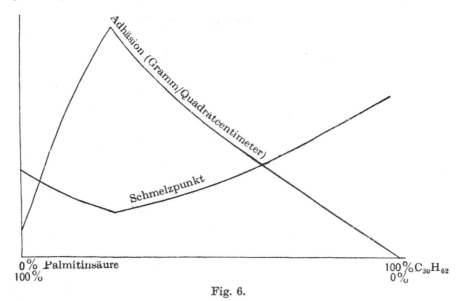

Fig. 6.

Aber einige merkwürdige Paradoxe bleiben doch noch übrig. Das erste ist die latente Periode. Die latente Periode ist nur bei polaren Substanzen zu beobachten, z. B. bei langkettigen Säuren oder Alkoholen und nicht bei unpolaren Substanzen, wie z. B. Paraffinen. Die latente Periode kann deshalb offenbar nichts anderes sein als die Orientierungsdauer der polaren Moleküle in der Adsorptionsschicht (und das kann etwa eine Stunde sein). Wenn das der Fall ist, so ist entschieden die Beziehung paradox, dass die latente Periode um so kürzer ist, je geringer die Zahl der polaren Moleküle ist (Fig. 2). Es hat den Anschein, dass die Orientierung durch die Anhäufung der polaren Moleküle in der Oberfläche etwas verzögert wird.

Zu einem anderen paradoxen Ergebnis gelangt man, wenn man die Beziehung zwischen Adhäsion und Reibung mit der Zusammensetzung einer Mischung von polaren und unpolaren Substanzen vergleicht. Adhäsion ist ein normaler, statische Reibung ein tangentialer Widerstand gegen einen Zug. Bei der Messung der Adhäsion (als Zugfestigkeit) tritt der Bruch immer an einer Grenzfläche zwischen der Adsorptionsschicht und der übrigen Substanz ein, wie man mit dem Mikroskop leicht feststellen kann.[1]

Man könnte deshalb erwarten, dass die Zusammensetzung der Adsorptionsschicht einigen Einfluss auf den Wert des normalen Widerstandes haben werde. Das ist aber nicht der Fall. Die Kurve in Fig. 6 hat keinerlei Aehnlichkeit mit einer Adsorptionskurve und steht offenbar in engem Zusammenhang mit der Schmelzpunktskurve. Die statische Reibung folgt dagegen stets einer typischen Adsorptionskurve (Fig. 2).

[1] *Proc. Roy. Soc.* (1928), A, cxviii, 214.

55

NOTE ON THE CENTRAL NERVOUS SYSTEM
OF THE CRAYFISH

[*Journn. Physiol.* (1929), LXVII, 166]

For the purpose of this note the freshwater crayfish (*Astacus fluviatilis*) is a small edition of the lobster about 3 in. long whose central nervous system consists of a supracœsophageal ganglion connected by circumoral commissures with the post-œsophageal ganglia. These two together constitute the brain.[1]

From the brain there stretches backwards on the ventral aspect of the animal a ganglionated cord with one ganglion in each segment. The cord is enclosed in a tough sheath which forms the wall of a blood sinus within which the cord is suspended. The sheath gives to the cord the appearance of a delicate white thread with swellings at intervals, but when the sheath is opened the cord is seen to consist of two separate cords which are joined by transverse commissures at each ganglion(4).

Operations on the cord are not impossible but difficult owing to its small size, the presence of a longitudinal artery attached below to the sheath, and the relative toughness of the sheath.

The nerve cells of the post-œsophageal cord need special mention. They hang like berries in the blood stream below each ganglion and the single process of each cell forms a T shaped junction with its nerve fibres. The longitudinal commissures connecting these ganglia are entirely free from nerve cells. They correspond wholly to the white matter of the spinal cord of mammals.

Of limbs, the animal has in the thoracic region the maxillipedes which serve the mouth, the great chelæ and the walking legs; and in the abdominal or tail region a pair of swimmerets to each segment save the sixth or last, where the limbs are spread out on each side to form flat steering plates.

SEMISECTION OF THE CORD IN THE THORACIC REGION

This was carried out between the first and second thoracic ganglia and some of the animals made a good recovery. There was loss of tone in the walking limbs, in the abdominal body muscle, and in the sixth abdominal appendage all of the same side as the injury. In about four days the tone of the abdominal muscles returned, the tail ceased to be bent away from

[1] See an interesting paper(2) by the late Professor of Mental Philosophy and Logic in the University of Cambridge—James Ward, *Journ. Physiol.* II.

the injured side and the animals regained the power of swimming, which they do backwards by powerful strokes of the abdomen. It was noticed, however, that they landed invariably on their backs owing to the lack of steering power due to the continued paralysis of the sixth abdominal appendage of the injured side.

After the shock of the operation was over the swimmerets of both sides resumed their regular beat, but the paralysis of the walking limbs persisted. Therefore the last pair of abdominal appendages which take an important share in the orientation of the animal especially during swimming are so completely controlled by the brain as to be permanently paralysed by semisection of the cord, whilst the other abdominal appendages, the swimmerets, are emancipated from cerebral control perhaps as completely as are the lymph hearts of the frog.

A normal animal placed on its back rights itself by swimming over, the steering being done presumably by the sixth pair of abdominal appendages. An animal with the cord semisected is unable to right itself in this fashion but does so slowly by swinging itself from side to side by alternate pushes on the ground with the great chelæ until the pendulum swing is great enough to topple it over.

The ganglionic decussations are complete enough to permit of voluntary movements, namely walking, carried out very imperfectly, and flapping of the abdomen carried out with vigour and so far as the eye could judge with adequate balance between the muscles of the two sides.

Injury to unipolar nerve cells

Attempts were made to cut away the nerve cells from beneath a ganglion with a scalpel made by sharpening a needle to a cutting edge. One case was successful. Subsequent sections showed that almost all the cells had been cut away from the under side of the second abdominal ganglion. The immediate disturbance was remarkably slight since stimulation of the cord in the thoracic region was followed by apparently normal movement of the abdominal appendages and abdominal muscles. The object of the experiment was to decide whether destruction of a unipolar cell affects at once the function of its related nerve fibre. The experiment, so far as it goes, shows what perhaps might have been expected, that there is no immediate effect. The connection between cell and its fibre is solely trophic—if that word has survived into modern physiology.

Relevant papers known to the writer are the following:

(1) Yung. *Archives de Zoologie* (1878), vii, 401.
(2) Ward. *Journ. Physiol.* (1879–80), ii, 214.
(3) Marshall. *Studies Biological Laboratories, Owens College* (1886), i.
(4) Hardy. *Phil. Trans.* (1894), B, clxxxv, 83.

56

PROBLEMS OF THE BOUNDARY STATE

[Phil. Trans. (1931), A, ccxxx, 1]

CONTENTS

References in brackets are to the list of papers at the end, and to the page of the paper referred to. The friction referred to throughout is static friction.

1. INTRODUCTORY

Matter distributed as a thin layer between two continuous phases may fairly be said to be in a fourth state because the energy of every element is shared with those phases. This is not a distinctive character, the energy of any portion of matter in the universe is shared to a greater or less degree with every other portion by the operation of their mutual attractions. The energy of these films is, however, shared so overwhelmingly with the enveloping phases as to merit some distinctive term.

The numerical value of the influence of the enveloping phases depends upon the point of reference (see later, p. 843). One way of estimating it is by a comparison of the tensile strength of the matter in mass with its value when enclosed.

TABLE I

	C_0	C $h=0.007$ mm.	C $h=0.0027$ mm.
Palmitic acid	449	15,853	[31,000]
Stearic acid	636	18,397	[36,000]
Eicosane	454	8,606	[17,000]
Impure wax (M.P. 57–59°)	2907	Not observed	

C_0 is the adhesion in grams per square centimetre across an interface within the solid lubricant[1] in mass. It was obtained in the usual way by breaking a rod. Fracture in each occurred without tangential flow.

[1] This word is retained throughout for the substance between the solid faces.

C the value when enclosed between plane faces of steel, the faces being h apart.

The values in brackets were calculated from measurements made with myristic acid on the assumption that adhesion is always the same function of h.

There is real difficulty in the application of accepted terms to this fourth state of matter. The solid condition, as the microscope shows, is indubitably solid, but the fluid condition is not fluid because it will bear a tangential stress with finite deformation. When a fluid joint is surrounded by liquid it will not, however, bear a *normal* stress(10).[1] The word fluid therefore means no more in this paper than the formation of the joint between slider and plate with material which is fluid in mass.

It does not seem possible to evade this difficulty by calling the "fluid" joint plastic since the resistance to slip is "brittle". My own experiments show that a "fluid" joint will resist for days traction which is less than the static friction.

The experimental basis of this paper is narrow. Though over one hundred substances have been examined as lubricants and the general relations of chemical constitution to friction outlined(2) attention has been confined mainly to long chain compounds, namely, normal paraffins, acids and alcohols and carbinols. The following pages are based on observations made with chain compounds.

A cylinder, 1 cm. in diameter, with a plane face placed on a plate in a large pool of any fluid, air being one, sinks slowly until the capillary pressure between the faces is equal to its weight. If the cylinder be now pressed down or pulled up, it returns to its position of equilibrium when the added force ceases to act((11), p. 225; (9), p. 9).

For each load there is therefore a value of h, the distance between cylinder and plate, at which the capillary pressure is equal to the loading, and this value is constant for all fluids including clean air, and for different solids when the solid faces are clean, but seems to be exceedingly sensitive to errors in the shape of the surface ((11), p. 274). The capillary pressure we have called the Leslie pressure ((9), p. 10).[2]

The most difficult problem of the boundary state is offered by this fact that a solid cylinder will float in air over a plane solid surface at a distance which may be as much as 0·007 mm., is independent within the limits of accuracy of the method of measurement of the solids (steel and glass were tried), and is unchanged when air is replaced by a liquid (see also p. 865).

There must be either an excess of hydrostatic pressure between the faces to support the load or the support is furnished by solid particles adhering to the faces.

[1] See references, p. 866.
[2] Leslie, *Phil. Mag.* (1802), xiv; Maxwell, *Encyc. Brit.* 9th edit., art. "Capillary Action".

That the air gap is due to a true capillary pressure and not to solid particles is inferred from the following facts.

When the cylinder is pressed down by increasing the loading it rises to the same height when the loading is removed. An attempt to destroy or at the least to deform possible particles by moving the cylinder about the plate failed to alter the height. For the same loading the gap is the same. It follows from these that solid particles would have always to be perfectly elastic and of the same size.[1]

The crucial experiment however is the following. When a disc of glass or quartz is placed upon a plate in a high vacuum Newtonian colours appear and the colours persist in the vacuum but disappear when air is admitted.

It is not easy to get this result because the air must be dry. A trace of liquid acts as an adhesive, causing the surfaces to adhere as the following experiment shows. When the disc is pressed down in dry air until colours appear it rises at once when the excess pressure is removed, but the colours persist indefinitely if the surfaces are contaminated with a liquid. A trace will suffice. It is enough to press the disc down with the naked hand, the moisture given off forming the contaminant. But if the hand be covered with a dry cloth the disc rises. Burgess found that steel faces would not adhere when dry (19).

When dry metal surfaces are pressed together until metallic contact is sufficient to scratch them, they separate in dry air when the excess pressure is removed.

Many experiments prove that surfaces held together with liquid will float apart in a liquid miscible with the lubricant. For this reason they will not separate in air.

Solid surfaces which have been wrung together so that they seize separate in a pool of liquid only if that liquid has a sensible vapour pressure ((9), p. 16).

All of the fluid between the faces cannot be forced out by any load which can be applied. There persists a limiting layer of great mechanical stability which resists pressure up to some million grams per square centimetre. There are, therefore, two stages, a first in which the Leslie pressure is adjusted to the load by fluid flowing in or out between the faces, and a second in which the reaction to the load is adjusted by the elastic forces between the atoms and molecules.

In the first stage comparable mechanical states are states of equal Leslie pressure. The proof lies in the fact that comparisons so limited yield surprisingly simple relations between the variables.

[1] Actual measurements will be found on p. 224 of (11). The following were obtained two years later:

	Load ÷ area gr.	Lubricant	Gap mm.
Steel on steel	8·2	Air Cetyl alcohol Naphthalene	0·0039 0·0038 0·0040
			0·0039
Glass on steel	8·2	Air Cetyl alcohol	0·0039 0·0042
			0·0041
Glass on glass	8·2 3·5	Cetyl alcohol Air Cetyl alcohol	0·0042 0·0057 0·0058

On pp. 224 and 225 of (11) the load should have been, cylinder A 6·47 gr., cylinder B 6·48 gr., which give a loading of 8·3 gr./cm.[2]

In the second stage, the Leslie pressure ceases to carry the load and comparable states are simply states of equal loading—that is load ÷ area.

In the first stage the coefficient of friction ($\mu =$ friction ÷ load) varies with the load(9). It is obvious why this is so. The load varies both h and the Leslie pressure and therefore comparison is not between comparable states. In the second stage, however, h is sensibly constant, the Leslie pressure unaffected, and μ is independent of the load. This is Amontons' law.

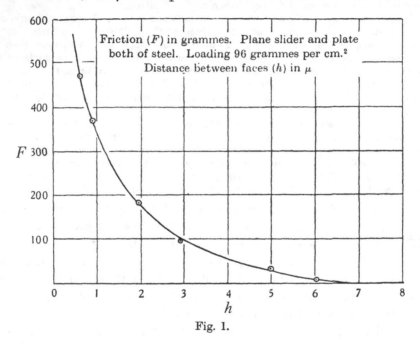

Fig. 1.

The limit of Boundary Lubrication is over-passed when the solid faces are so far apart that a portion of the lubricant is beyond the influence of either. Let a' and a'' be the range of the influence of the two solids, then $h = (a' + a'')$ is the limit of boundary lubrication.

This limit can be fixed with some precision from a curve of any variable and of h. The curve, Fig. 5, for h and the load in the paper of 1928 ((11), p. 226), already referred to, cuts the h ordinate at 0·007 mm., and the curve for h and friction in Fig. 1 of this paper gives the same value. The most perfect plate and cylinder used gave 0·005 mm. for the limit. Either is a surprisingly high value. Watson found the value of h to be constant in air down to a pressure of 2 mm. Hg when it slowly and slightly decreased(15).

It will be convenient to have separate terms for the two ways in which the attraction field at the surface of a solid, or, indeed, of any phase, could affect matter in contact with it. It may do so either by *direct attraction* of relatively long range and of the sort contemplated in Laplacian theory of short range forces or by the spreading from molecule to molecule of a state of strain. Let us suppose that the range of direct cohesive attraction includes only molecules in contact with the face. These molecules will be both oriented and strained and they in their turn orient and strain molecules beyond them until the effect fades out by reason, for example, of the heat motions. Let us call the state of orientation and strain of the molecules "polarisation" and the spreading of polarisation from molecule to molecule *diachysis*. Polarisation and diachysis (διάχυσις) are the basis of the electrical theory of cohesion put forward by Sutherland in 1881 [20].

The limiting layer mentioned earlier may be taken to be composed of two strongly adsorbed layers, the primary films with the surface of slip between them. Bad lubricants, however, seem to shift the surface of slip into the solids [1]. Such lubricants might be distinguished as abrasives, but the same substance may apparently be a lubricant for some solids and an abrasive for others.

A primary film is of insensible thickness. It is completely invisible when deposited on to a clean metal face from vapour of the lubricant or when exposed by breaking a solid joint, yet in both cases the surfaces are as completely lubricated as they would be if the slider were sunk in a pool (N. 554).

The molecules in such a film are highly oriented (see especially [7], [9] and [11]). When the lubricant is a long chain acid alcohol or carbinol, there is a period called the Latent Period, during which friction or adhesion slowly reaches its steady value. The latent period may be due either to the slider rising or sinking in a pool of lubricant until the loading is equal to the Leslie pressure or to the orientation of the molecules of the lubricant by the attraction field of the solids. It is easy to distinguish between the two causes by, *e.g.*, the use of a spherical slider, which, owing to its form, cuts through the lubricant at once until the limiting layer is reached ([9], p. 12). The latent period of orientation is then found to be exhibited only by lubricants, the ends of whose molecules are unlike such as long chain acids which have at one end the carboxyl group —COOH and at the other the group —CH_3 [7].

Normal paraffins whose molecules end for end are symmetrical show no such latent period. This is sufficient evidence that the molecules are oriented. It is also sufficient evidence that the molecules are oriented with the long axis at right angles to the face for, if the molecules were flat on the face, there would be no reason for the long latent period of orientation of molecules of acids or alcohols, 60 min. and 40 min. respectively, and the insensible latent period of paraffins.

Each molecule of lubricant in a primary film exerts its influence independently of the others. This follows from the fact that the friction varies directly with the density of the vapour of the lubricant when the primary films on the surface are deposited from vapour ((4), p. 556).

The properties of primary films are easily studied by the use of a spherical slider, and the fact that the friction is the same when such a slider stands in a large pool of lubricant as it is when the faces of slider and plate are covered only by an insensible primary film deposited from saturated vapour proves that the molecular configuration of such films is the same in both cases (see especially (4), p. 554, and (7), p. 28).

A primary film is there, even when a solid lubricant has been deposited from solution in a volatile solvent, and may be reached by moving a spherical slider about. When this is done the friction rises as the slider cuts through until it reaches a value unchanged by further movement *and identical with the value for the same* lubricant when fluid.[1]

In (9), p. 7, the fact that the coefficient of friction varies with the load and that a latent period of rising friction and falling coefficient followed an increase in the load when the surfaces were lubricated only by a primary film deposited from saturated vapour and the slider was plane was taken to prove that a primary film is more than one molecule in thickness and is capable of being squeezed out from between the faces. That inference was unwarranted. It was drawn before the fact that air is condensed between the faces was discovered ((11), p. 222).

The later knowledge makes it certain that the cylinder was separated from the plate by a layer of condensed vapour and air, about 0·002 mm. in thickness, and the change in friction and latent period following a variation in the load were due to the variation in the depth of this condensed layer.

When the layer of lubricant between the faces is thicker than the limiting layer but yet within the limits of boundary lubrication, the primary films are separated by what we will call the median plate.

If, after the joint has been formed by running fluid between the faces, the temperature be lowered enough to freeze the lubricant in position, the median plate is seen under the microscope on breaking the joint to be crystalline ((11), p. 214). It will be seen from what follows that the crystals throughout are polarised, but nothing is known directly of their structure.

Nothing is known directly of the structure of a fluid joint, but Trillat and Bragg, by X-ray analysis, have found long chain molecules to be disposed on a plane solid face in layers with the long axis at right angles to the face. It is probable, therefore, that such a disposition exists in an enclosed layer of fluid lubricant. There are, therefore, not an infinite number of possible surfaces of slip or break to be considered, but only a relatively small number placed between the layers of oriented molecules because, within each such layer, the limit of resistance to either normal or tangential

---

[1] The reader may be reminded here of the absence of a temperature coefficient.

forces may be taken to be infinite. There are in a fluid joint an infinite number of mathematical tangential surfaces, but only a finite and small number of *"significant"* surfaces.

In a solid joint the number of significant surfaces is reduced to two, namely, those between the primary films and the median crystalline plate (cf. p. 854).

Throughout this paper the condition referred to, save when the contrary is stated, is the steady state of equilibrium between lubricant and solids. The length of the latent period needed to reach the steady state depends upon whether the slider or cylinder is spherical or plane, on how the lubricant is applied, and on the chemical composition of the lubricant. Simple relations between the variables appear only when the state is steady and when "comparable" states are compared.

Measurements were made with solid faces previously cleaned and in an atmosphere cleared from impurities, especially from water.

There is no intention here to consider critically observations made by others. The conditions are defined in order to give point to the remark that they are necessary if a disorderly presentation of values is to be avoided. The simplest most fundamental relations of the values appear only when the lubricant is composed of one pure chemical substance. For example, no indications of true metastable values (Watson and Menon (15), p. 195) were found, save when the lubricant was complex ((7), p. 31). The want of agreement between dielectric strength and friction noticed by these authors might well give way to agreement under more rigid conditions.

SECTION 2

The following values have been measured:

1. The static friction between solid surfaces, either clean or separately coated with solid lubricant (value F_1).

2. The static friction of clean surfaces joined by fluid lubricant (value F_2).

3. The tangential force needed to produce slip in a solid joint between a cylinder and a plate (value S).

4. The normal force needed to break a fluid joint between a cylinder and a plate (value A).

5. The normal force needed to break the same joint when solid (value B).

An attempt to find the relation between these values meets with curious anomalies. Value F_1 obviously is external friction, that is, the friction between readily separable portions of matter. Value F_2 on the other hand, since slip occurs within the lubricant, is internal friction, yet it is completely different from viscosity, which is the internal friction of a fluid in mass, but identical with value F_1. The viscosity of a chemical series, such as the paraffins, increases with molecular weight; values F_1 and F_2, on the other hand, decrease.

$F_1 = F_2$ when the lubricant and solids are the same and the curve connecting friction with molecular weight of a chemical series shows no break where the change from fluid to solid occurs. The orientation of the molecules in a fluid joint must, therefore, be such as to form an internal tangential surface where the discontinuity of structure is so complete as to give to it the properties of an external surface so far as slip is concerned. But the portions of matter are not readily separable at this surface. The position of the surface offers no difficulty. The only measurements available for comparison of the friction with solid and fluid lubricants were made with a spherical slider which in both cases cuts through to the limiting layer. The lubricant therefore consisted in both cases of two primary layers probably monomolecular with the surfaces of slip between them at h_2 (see especially (9), p. 20).

Though value F_2 has all the characters of external friction, it is neither more or less than the strength of a fluid joint in shear and therefore akin to value S. But it differs completely from value S in magnitude and in its relations to chemical constitution. Taking the paraffin $C_{22}H_{46}$ as an example $S/F = 1200$ as a round number.

The real relation of the values one to another appears when their derivation from the fundamental mechanical properties of surfaces, namely, their normal and tangential reactions, is traced. Let the traction on a surface of slip grow from nothing. It is balanced by the tangential reaction until the latter reaches the greatest value of which the surface is capable. This value is a true characteristic of the surface because it is defined by the chemical nature and configuration of the molecules in the capillary layers which enclose the surface.

A second true characteristic is the greatest value of the normal reaction to an external force tending to separate the portions of matter on each side of the surface. Let us enquire into the relation of these two values to the observed values. Here we are helped by two identities. The resistance to a normal force can be identified simply with the normal reaction only when there is no tangential flow of the material under normal stress such, for example, as the flow which causes a rod of metal to narrow before it is broken by a longitudinal pull. Since no sign of tangential flow could be detected, even under the microscope, when a solid joint was broken, value B may be directly identified with the normal reaction, and the tangential reaction T can be safely identified with external friction.

We have, therefore, if N be the normal reaction and T the tangential reaction, the two identities,

$$N = B \quad \text{and} \quad T = F_1 = F_2,$$

and these two identities furnish criteria of the two reactions. Before turning to these, however, it will be well to define the quantities more closely.

Let us consider a tangential surface drawn anywhere within the lubricant. The maximal normal reaction of this surface will depend upon where it is drawn and the adhesion value B is the least of these values. A surface possessing this least value is a surface of break and direct observation shows that there is one, or at most two, of these surfaces of break when the lubricant is solid, the number being determined by the nature of the enclosing solids (11). The quantity B therefore is equal to the maximum normal reaction of a surface of break.

Since solids contract and expand with changes of temperature they are subject to an internal or intrinsic pressure I, which is a function of temperature. When an external longitudinal force of magnitude E is applied tending to break a solid joint, we have at any tangential surface, since there is no viscous flow, the simple condition of equilibrium

$$E = C - I.$$

C is, of course, the cohesive attraction per unit area across the surface, and B is the greatest value of $(C - I)$ which a surface of break is capable of exhibiting. A solid joint breaks when at the surface of break

$$\frac{d\,(C-I)}{dE} = 0,$$

or, if h be the distance separating the surfaces of the enclosing solids, namely, the cylinder and plate in the particular experiments, when

$$\frac{d\,(E-C+I)}{dh} > 0.$$

When friction is external the surface of slip is identifiable if there be no seizing and the value F_1 is the greatest value of the tangential reaction of which this surface is capable. It is also less than the similar value for any surface drawn parallel to the surface of slip. Both B and F_1, therefore, are the least of a series of maxima, and the quantities N and T respectively represent the value of the maximal normal reaction of a surface of break and the maximal tangential reaction of a surface of slip.

The distinctive characters of the normal and tangential reactions now appear when we consider the relation of B and F to the variables, which are the number and chemical constitution of the substances composing the lubricant, the composition of the enclosing solids (the cylinder and plate or the slider and plate as may be), the distance h separating the enclosing solids, the normal pressure (that is, load ÷ area), and temperature.

In dealing with the effect of the composition of the lubricant it is necessary for reasons already given to consider only chain compounds, and, to discover the simplest relation, comparison must be confined to lubricants composed only of one chemical substance. The effect of chemical constitution is then found to include two variables, namely, the chemical series (paraffin, acid, alcohol or carbinol, as the case may be), and for a given series the number of carbon atoms in the chain.

Any of the variables can be used to define features peculiar to B or F, but the one most easily apprehended is the number of carbon atoms in the chain. Call this number n, then if the signs + or − indicate whether the coefficient is positive or negative, we have from experiment

$$\frac{dB}{dn} \text{ is } + \text{ and } \frac{dF}{dn} \text{ is } -.$$

The evidence is to be found in earlier papers. This relation also holds for the external friction of lubricants in mass as the following figures show:

Friction of a clean spherical glass slider on solid blocks of

Lauric acid	...	Mol. wt.	$200 n = 12$		$\mu = 0 \cdot 039$
Palmitic acid	...	,,	256	16	$0 \cdot 033$
Stearic acid	...	,,	284	18	$0 \cdot 028$

By the identities given earlier we have, therefore, as characteristics of the normal and tangential reactions

$$\frac{(dN)}{(dn)} \text{ always positive,} \qquad \frac{(dT)}{(dn)} \text{ always negative,}$$

when the coefficients refer to the same chemical series. A second character is the surprisingly great difference in magnitude for the same load. As an example:

Lubricant solid, steel on steel:

Paraffin $C_{22}H_{46}$	Spherical slider	...	...	$F = 7 \cdot 3$ gr.
	Cylinder or plate	...	...	$B = 9441$ gr.
Palmitic acid	Cylinder or plate	...	...	$F = 0 \cdot 0$ gr.
	Cylinder or plate	...	...	$B = 15853$ gr.

None of the many measurements lend themselves readily to a comparison of traction (F per unit area) and B value (adhesion per unit area), at the same pressure (load per unit area), but the following figures are sufficiently accurate:

Nonadecane pressure, 7·15 gr., steel on steel:

Plane slider	...	...	Traction (about) 10 gr.
Cylinder	...	...	B value, 8200 gr.

A third character is the absence of a temperature coefficient for both value F in the second stage of boundary lubrication when the lubricant is reduced by pressure to two primary layers, and for value B.

When the lubricant is fluid and more than two primary layers in thickness, both friction and adhesion (value A) vary with temperature, but this may be due to a variation in viscosity which allows the slider or cylinder to rise or fall in the lubricant. Comparison then would not be between comparable states. *It is tempting to suppose that both Amontons' law and the absence of any effect of temperature are fundamental characters of every significant surface.*

Value S is the tangential force needed to produce slip in a solid joint made by running fluid beneath a cylinder with plane ends resting on a plate and freezing the lubricant *in situ* ((11), p. 228). The microscope shows that slip occurs at the interface between a primary film and the median crystal plate. This interface is also a surface of break when the joint is broken by a normal force.

The coefficient dS/dn is positive and the magnitude large, value S therefore has the characters of a normal reaction.

A remarkable feature is the sudden fall in resistance to slip to that of simple friction, caused by the slightest tangential displacement. The joint in short is just as effectively broken by what seems to be tangential motion as it is by normal motion—the fields of force acting across the interface must be completely unlocked in both.

Why did cohesion practically vanish when the lateral displacement was so small as to leave the area in contact sensibly unchanged? It is possible, of course, that contact was not maintained but was interrupted by a film of air drawn in by capillary attraction. Joffe, Kerpectewa and Lewitsky have drawn attention to the influence upon the breaking stress of a crystal of minute fissures on its surface.[1] When a fresh surface was formed on a crystal of salt by solution in water the breaking stress per square centimetre was raised from 440 to 5000 gr. and as the surface aged the value fell again. If the fracture of brittle joints begins by the formation of a minute fissure, as it might well do when the applied force is tangential, air would be drawn in and the Leslie pressure of the air within it would help to deepen the fissure. Joffe and Lewitsky found the breaking stress of rock salt greater in a vacuum than in air.[2]

[1] *Nature* (1924), cxiii, 424.
[2] *Z. Phys.* (1925), xxxi, 576.

If the air be not drawn in and contact be actually maintained, then either the fracture must be accompanied by a reorientation of the molecules of the lubricant, or the range of cohesive attraction is vanishingly small. The second of these possibilities would make it necessary to refer the influence which each enclosing solid has upon the whole of the lubricant solely to diachysis.

The effect of air is of critical significance and little understood. Any fluid, whether liquid or gas, which surrounds the base of a cylinder is drawn in until the Leslie pressure carries the load.

For example, if a cylinder be wrung down on to a proof plane of glass, Newtonian colours appear. When the pressure is taken off the colours persist unchanged if there be the least moisture present in the air, but if air and surfaces be dry, the blue and yellow at once change to red and green, which, in turn, slowly vanish. This is true, for clean faces and for faces contaminated with a solid film. Air will not displace a liquid, for if a cylinder be wrung down on to a plate covered only by an insensible film of liquid lubricant it remains adherent in air, but floats up in a liquid miscible with the lubricant.

Burgess noticed the adhesive quality of water and he found the adhesion of two metal plates joined by lubricant to be increased in a vacuum. This increase probably was not due to the hydrostatic pressure of the atmosphere, but to the operation of short range forces. Air drawn into an insensible superficial cleft gave a Leslie pressure and the adhesion was diminished thereby.

Value A gives the adhesion of a fluid joint [6]. Its magnitude is of the order of friction, though the sign of the coefficient (d/dn) is that of adhesion. By these tests, therefore, this type of adhesion is of mixed origin, including both normal and tangential reactions.

The A value is certainly not merely a measure of the viscosity of the lubricant, for not only is it a linear function of molecular weight like the value B, which certainly gives the normal reaction of a solid lubricant, but it is not related in any simple fashion to viscosity, as the following figures show ([10], p. 69):

	Steel on steel η 18°	Load, 259·6 gr. A value 18° gr.
Octane	0·005	3·3
Octyl alcohol	0·064	50·9
Caprylic acid	0·057	97·9

On the other hand, it differs from the B value in the fact that it varies with temperature whilst the B value does not so vary over the range explored, namely, from 4 to 56° C. Temperature may, however, alter the distance h between the enclosing solids, the cylinder and the plate, in which event comparison at different temperatures would not be made between comparable states.

An attempt was made to get the normal reaction of a fluid joint uncomplicated by tangential flow by supplying the necessary lateral support. Osborne Reynolds' principle of dilatancy suggested the method. A thin skin was applied to the edge of a joint of *solid* lubricant by varnishing with necol, or a strong solution of gelatine, and allowing the varnish to dry. The lubricant was then melted by raising the temperature and the joint broken. The value of adhesion was now much greater than value A. The capacity which impurities possess for penetrating a fluid joint make it certain, however, that the full value of the normal reaction was far from being reached.

	Steel on steel		Value A, but edge coated with necol gr.
	Value A gr.	Value B gr.	
Lauric acid	70	10,683	4070
Palmitic acid	120	15,858	5370

It would appear from these experiments that the A value as measured did not give the complete normal reaction of the fluid joint, but the increase in value caused by a skin of necol may be due to something other than simple restraint of tangential flow. When the A value was measured, the joint was sunk always in a pool of lubricant. Before applying the skin of necol, this pool was removed, the conditions were therefore completely changed, for whilst the joint was sunk in the pool, the lubricant would be the seat of the Leslie pressure. Removal of the pool and the presence of the skin of necol would reduce this pressure to zero (see Section 3), we therefore have, when the quantities are reckoned at the moment of break:

$$\text{value } A = N - L,$$

when L is the Leslie pressure and

$$\text{value } B = N.$$

The broad conclusion is that the A value is not the normal reaction per unit area, but that reaction reduced in value by that flow inwards of fluid from the pool which peels the layers of oriented molecules apart and by the operation of the Leslie pressure.

Worthington's method of measuring the tensile strength of a continuous column of liquid provides the necessary lateral support and probably gives its normal reaction. He found the tensile strength of ethyl alcohol to be 8165 gr./cm.^{2} and, for reasons just given, this is probably its normal reaction in the fluid state. The value for the solid alcohol is not known and we have not measured it for any other alcohol. If alcohols resemble the related acids and paraffins the normal reaction in the solid state is only a few hundred grams.

The normal reaction of ethyl alcohol in the fourth state, that is as a lubricant, can be computed roughly from the value for cetyl alcohol 15,860 ((11), p. 214), by assuming that the effect of molecular weight is about the same as it is for paraffins. The assumption is justified by the form of the curves for friction and it gives

B (ethyl alcohol in steel) = 10,160, when h is of the order 0·007 mm.

The normal reaction of a substance in the solid state in mass is the tensile strength of the weakest planes in the crystal lattice, or of the interface between crystals. That of the liquid in mass is the sum of the reactions across an internal surface of molecules presented to each other in every kind of orientation. On the other hand, the normal reaction of the fluid or solid joint is that of molecules oriented by the attraction fields of the steel.

In estimating the effect of the enclosing phases, it would appear natural to compare a solid joint with the solid lubricant as was done on p. 827, but it is now obvious that comparison with the fluid lubricant would yield a very different result.

The values observed can now be arranged in a more rational manner:

Tangential reaction	...	...	The friction F
Normal reaction	...	...	⎰The adhesion B
			⎱The slip S
Mixed reaction	...	...	The adhesion A

Tomlinson's theory of friction (18) suggests that the reason for the difference in magnitude between the tangential and normal reactions may be sought in the fact that the former is the resistance to traction when the attractive and repulsive forces acting across the surface of slip are in equilibrium with the load whilst the latter is the greatest resistance offered by the attractive forces to a tension. This, however, would leave the anomalous S value unaccounted for.

The fact that the sign of the effect of the number of carbon atoms in the chain is different for T and N shows that the difference between them does not consist merely in an extinction of the forces of repulsion. It might indeed be taken to mean that the portion of the molecular field involved is different. This would explain why the sign is the same for the viscosity η of a fluid in mass as it is for the normal reaction, because during flow, owing to the heat motions, the molecules must be presented in every spatial way to the shearing stress.

The difference between the normal and tangential reactions appears in a new and striking way when the adhesion and friction of lubricants composed of two pure chemical substances are contrasted.

Consider a pair of miscible substances such that chemical action between them is not possible at ordinary temperatures. A normal long-chain acid and a normal paraffin form such a pair. The curves for friction (ABC), adhesion (DEF), and melting point in mass of the mixture (GHI) are given in the diagram, Fig. 2 ((7), p. 32; (12), p. 610). The curve for friction is a simple adsorption curve. The polar molecules of the acid have complete possession of the adsorbed layer until the concentration has fallen almost to

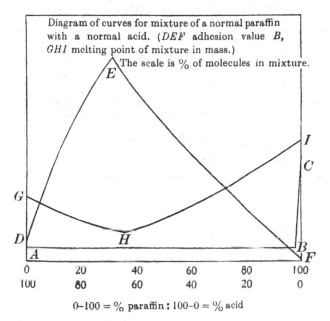

Diagram of curves for mixture of a normal paraffin with a normal acid. (DEF adhesion value B, GHI melting point of mixture in mass.)

The scale is % of molecules in mixture.

0–100 = % paraffin : 100–0 = % acid

Fig. 2.

the vanishing point where there is a sharp inflection when the non-polar molecules manage to invade the layer. The linear form of the portion AB is according to expectation, for it can be proved by experiment that the effect of a substance on friction is directly proportional to the number of molecules adsorbed by the solid surfaces ((4), p. 556). The curve ABC has no obvious relation to the melting points of the mixture.

On the other hand, the curve DEF of adhesion has none of the features of an adsorption curve, but is related in simple fashion to the melting point (curve GHI).

The melting point of a mixture is a mass phenomenon, or better a volume phenomenon, and when we remember that the normal reaction N is equal to $(C-I)$ and that a solid melts when the intrinsic pressure overpowers the cohesion of the parts of the crystal, the connection between them is obvious. Both value B and the melting point measure a limiting value of $(C-I)$. The absence of any obvious relation to adsorption is, however, paradoxical because the surface of break in these mixtures, sometimes, but not always (12), lies between the adsorbed layer (primary film) and the median crystal plate (see Section 4).

The facts suggest that, though a surface of slip may be defined as a plane surface separating two portions of matter moving in opposite directions, a surface of break may sometimes be far from plane because long molecules with their long axes oriented in the direction of stress are, at the break, pulled apart, like teeth out of their sockets.

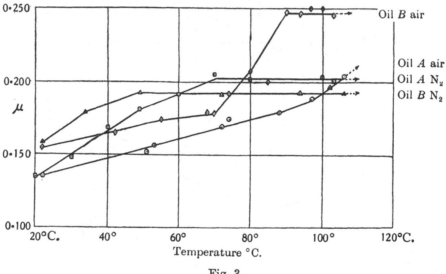

Fig. 3.

The study of complex mixtures of substances such as are commercial lubricating oils presents other aspects of the relation of friction to adsorption (cf. (13)).

Friction-temperature curves of a commercial lubricating oil are shown in Fig. 3. Such curves commonly consist of horizontal portions or plateaux where the temperature coefficient is zero joined by ascending or descending limbs. All the evidence goes to show that the composition of the adsorbed layers is constant throughout a plateau. It is, for instance, possible by percolating the oil through glass beads to remove by adsorption the more active polar constituents which form only about 3 per cent. The percolated oil now has no temperature coefficient ((13), p. 33). It is now adsorbed not selectively, but as a whole, so that further percolation does not alter its qualities as a lubricant.

The medicinal paraffin (Oil, B.P.) of the pharmacopœia is a fully percolated oil. It has no temperature coefficient of static friction over the range 10–115° C. There is no certainty that it is a chemical individual, but it appears to be an "adsorption individual" which may be defined as one which is adsorbed without selection of molecular species and has no temperature coefficient of friction.

Solution of a new molecular species in such an adsorption individual reveals two new principles in adsorption, namely, "availability" and "accessibility". The state of the substance in solution may be such that its molecules are not able to displace the molecules of the solvent, or *diluent*, in the adsorption layers. They are not "available". On the other hand, the state of the adsorption layer may be such as to prevent access of available molecules. It is not easy to discriminate between availability and accessibility, but the behaviour of solutions containing highly polar acids and paraffins show the need for admitting the existence of both. When the temperature is rising, the friction shows that the paraffins have access to the adsorbed layer, but when the friction is falling, the more polar molecules which have obtained access at the higher temperature are not displaced by the non-polar molecules though the state in solution of the latter must be a pure function of temperature (13).

An unusually simple example of the operation of availability is furnished by a solution of any paraffin in medicinal "paraffin" (Oil, B.P.).

The curves AB and CD in Fig. 4 are respectively the temperature curves of B.P. Oil, and of a normal paraffin. In the oil, 0·52 per cent. of the paraffin, whose melting point was 56° C., was dissolved to form a clear solution. The temperature curve of this solution now shows a sharp gradient between E (48°) and F (62°), and the gradient separates two plateaux. The curve is exactly reversible, that is to say, it has the same form, whether plotted with rising or falling temperatures.

From A to E the molecules of paraffin are not present in the adsorbed layers, either because they are not available for adsorption or because they are denied access. It is almost certainly the former, because the melting point of the paraffin lies always in the region of varying friction. Near its melting point the degree of dispersion of the paraffin increases and it is a fair inference from this and other similar cases that only unassociated molecules of the solute are capable of adsorption. The associated molecules are not "available". Four different paraffins were tried and with each the region of varying friction included the melting point.

Over the plateau AE, Fig. 4, the solvent molecules are "dominant". Over the plateau FG both solvent and solute molecules occur in the adsorbed layers in a proportion which is independent of temperature.

The absence of a coefficient of temperature is not, however, proof that a complex oil is an adsorption individual, since it may be due to some constituent or molecular species with high polarity having secured possession of the adsorbed layer. Palmitic acid has highly polar molecules. When this acid is dissolved in oil, B.P., the friction curve for rising temperature falls as the polar molecules obtain admission to the adsorbed layers until equilibrium is reached when the curve changes to a plateau whose level depends upon the percentage of acid in solution up to a limit of about 1 per cent. when at M the layers are fully saturated. The curve changes to a plateau, not because the mixture is adsorbed without selection, but because the polar molecules become "dominant" in the adsorbed layer. This

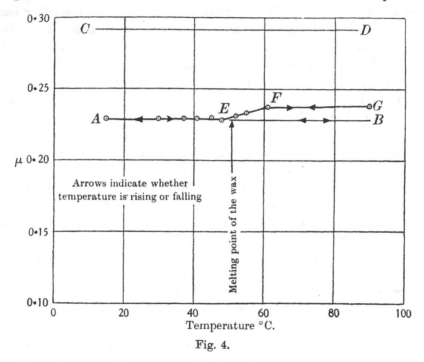

Fig. 4.

appears in the curve for a falling temperature which now shows no coefficient of temperature over the whole range explored. The curve for caprylic acid and undecane offers another example. The plateau AB, Fig. 2, is due to the "dominance" of the polar molecules of acid.

Simple inspection of the curve, Fig. 5, will show that the adsorbed layers cannot be simply taken to be in equilibrium with the overlying fluid when the latter contains more than one molecular species, even when steady states alone are contemplated. Any ordinate between the temperatures of 0° and 50° can cut a curve in two places, yet it can be proved that the state of the solution depends only upon temperature. It is rare to find a simple equilibrium between the adsorbed layer and the fluid, like that shown by a solution of paraffin in B.P. oil. Such an equilibrium can occur only when the work of adding a molecule to, or removing a molecule from, an adsorbed layer does not differ greatly as between the different molecular species present. When polar molecules are present "dominance" obscures the relations, and little can be predicted of the state of the adsorbed layers without a knowledge of the previous history. Something more than simple hysteresis is involved, for there is no certainty that the two paths of change meet at both ends.

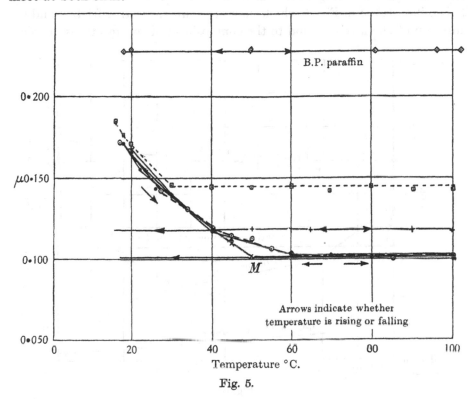

Fig. 5.

A theory which has won general acceptance postulates the existence on an adsorbing surface of "active" patches where the absorptive attraction is much in excess of that at other parts of the surface. I am strongly of opinion that the evidence needs reconsideration in the light of the principles of availability, accessibility and dominance.

Leaving steady states, the latent period of orientation is affected in an unexpected way by the polarity of the molecule. So long as only one kind of open-chain molecule is present, so long, that is to say, as the lubricant is composed of only one hydrocarbon, the latent period is always a period of rising adhesion and falling friction, and, as might be expected, its duration increases as the length of the carbon chain or the degree of polarity increases.

It is important, however, to note here that the only two ring compounds studied, naphthalene and phrenanthrene, gave latent periods of falling adhesion of 35–40 min. ((12), p. 615).

When the lubricant is composed of even as few as two components, the relations are much more complex. Friction and adhesion may now rise or fall in the latent period, and no exception has been found to the rule that the latent period is longer the higher the concentration of polar molecules. This is illustrated in Fig. 6, which shows the curves relating friction and the duration of the latent period to the composition of two mixtures, namely,

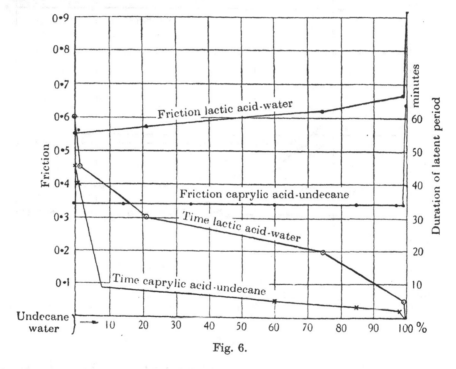

Fig. 6.

lactic acid and water, and undecane and caprylic acid. In Fig. 2, the adhesion curve is reproduced for a mixture of palmitic acid and the paraffin $C_{30}H_{62}$. In the limb *DE* the latent period was one of rising adhesion, and in the limb *EF* of falling adhesion, the duration being respectively about 50 and 30 min. The form of these curves seems to show that a polar molecule, once in place, cannot be removed from the adsorbed layer by the impact of a non-polar molecule, and that, as the concentration of polar molecules increases, they crowd at each metal face in such a way as actually to delay the attainment of a steady state of orientation.

It was assumed in the first paper of this series (1) that friction of the kind directly identifiable with the tangential reaction was some direct function of cohesion. This assumption certainly is not proven, for what is known of the tangential and normal reactions presents a striking picture of contrasts.

Number of carbon atoms n in the chain	Relation linear in both, but N.R. increases with n, T.R. decreases
End group of molecule	N.R. paraffins < alcohols < acids T.R. paraffins > alcohols > acids
Magnitude	N.R. much greater than T.R.
Temperature coefficient	Zero for both

Values in grams per cm. square; steel on steel

A value. For eight carbon atoms. Steel on steel.			
Pressure	Octane	Octyl alcohol	Caprylic acid
7·15	1·4	23·8	24·1
147	3·6	46·1	94·1
331	4·2	64·8	124·7

B value. Sixteen carbon atoms.			
	$C_{16}H_{34}$	Cetyl alcohol	Palmitic acid
7·15	7000	15,750	15,853

S value. Sixteen carbon atoms.			
	$C_{16}H_{34}$		Palmitic acid
7·15	2549	—	10,375

F value. Eight carbon atoms.			
	Octane	Octyl alcohol	Caprylic acid
96	57·5	44	27·7

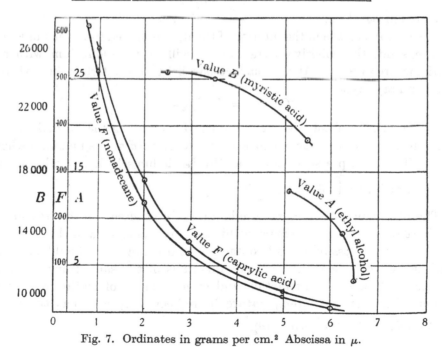

Fig. 7. Ordinates in grams per cm.² Abscissa in μ.

When the lubricant is composed of two chemical substances which do not react with one another chemically:

N.R.   ...   ...   No obvious relation to adsorption
T.R.   ...   ...   Simply related to adsorption

The most striking difference, however, lies in the relation to h. Both T.R. and N.R. increase as h decreases, but the former increases with no upper limit, or, rather, with an upper limit fixed only by the capacity of the system to bear the load, whilst the latter increases to a maximum. The curve for N.R. (see values A and B in Fig. 7) is concave to the axes, whilst that for T.R. (value F) is convex to the axes.

SECTION 3. THE LESLIE PRESSURE

Two explanations of the Leslie pressure are possible, the simpler based upon direct Laplacian attraction, the more difficult based upon diachysis.

Consider a fluid and a solid the interface being plane, and let the solid attract the fluid by a force which decreases with increase in the distance between two attracting particles and becomes insensible when this distance is a. a is a length large compared to molecular dimensions. According to the classical theory of capillarity the total normal attraction of the solid for the fluid reckoned to a surface distant z from the interface is

$$2\pi \int_z^a \psi(z)\, dz.$$

Call this integral Lz.

Across any surface in the interior of the fluid removed from the influence of the solid, the cohesive attraction C_0 will be in equilibrium with an intrinsic pressure I_0. At any surface parallel to the interface but within the capillary layer

$$(C - C_0)_z = (I - I_0)_z = L_z.$$

L_z is the excess of cohesive attraction reckoned along the normal to the interface, it is a quantity which vanishes when $z = a$ and is maximal when $z = 0$. The Leslie pressure, as defined by Leslie himself, is the value when $z = 0$, namely, $2\pi \int_0^a \psi(z)\, dz$.

The balancing intrinsic pressure he takes for granted. In the treatment of short range forces customary at his time, it was merged in the assumptions of incompressibility and contact. C and I, however, should be dealt with separately since they are not functions of the same thing. C is a function of molecular attraction and orientation, I of motion of translation of molecules; if temperature be indicated by θ, C has the form $\int_z^a \psi(z)\, dz$, and I the form $\phi(\theta)$.

In the case of a solid cylinder immersed in fluid the tangential components of the attraction have to be reckoned with over the curved surface. These can be equated to a surface of tension in the usual way, and if R be the radius of the cylinder and T the tension, $T = \int_0^a z\psi\,(z)\,dz$ then, if T'' is the tension due to the attraction of the fluid for itself, there will be an excess of pressure on the curved surface over the plane ends of $(T - T')/R$. This excess must be balanced somewhere and the balance is found in the curvature of the surface of tension round the edges of the cylinder.

Let the plane ends of two cylinders of equal radius, of the same material, and immersed in a fluid, approach one another and let h denote the distance between them. Experiment proves that external normal force is needed to drive the ends of the cylinders together, and that if P be this force per unit area there is a value of h corresponding to each value of P at which the least value of the excess of pressure between the ends of the cylinder over the pressure in the surrounding fluid is equal to P. What is the origin of this excess?

Put shortly the answer is to be found in the overlapping of the attraction fields of the solids, so that a portion of the lubricant is attracted by both solids. Therefore, when $h = 2a$, P vanishes, and this would then be the limit for boundary lubrication. The limit $2a$ is fixed by the curve, Fig. 1, at $6\cdot 5\,\mu$.

At the periphery of the disc of lubricant the attraction of the solids for the fluid can be resolved in the plane and along the radii of the disc into an inwardly directed pressure p which must vary in the plane midway between the solid faces between the limits $p = 0$ when $h = 2a$, and

$$p = L + \frac{(T - T'')}{R} \text{ when } h = 0.$$

$\dfrac{T' - T'}{R}$ may probably be neglected, as can be seen, by giving T the impossibly high value of the surface tension of mercury and comparing the calculated value with the observed pressure needed to force the face of a cylinder on to a plate.

At the midway plane on the disc $\tfrac{1}{2}h$ distant from each face the excess attraction, when h is less than $2a$ but greater than a, is $4\pi \int_{h/2}^a \psi\,(z)\,dz$. Call this C', then $C' = 0$ when $h = 2a$. The condition of equilibrium at that plane is

$$P + [C' + C_0] = p + [I' + I_0]$$

or, since $C_0 = I_0$,

$$P = [I' - C'] + p.$$

It might seem unnecessary to specify C_0 and I_0, since they cancel out, but it must be remembered that terms in C and terms in I are not the same functions of z.

When $h < a$ we have

$$C' = 4\pi \left[\int_{h/2}^{a} \psi \ (z) \ dz - \int_{h}^{a} \Psi \ (z) \ dz \right].$$

As h approaches zero C' tends to vanish whilst p tends to become equal to $2\pi \int_{0}^{a} \psi \ (z) \ dz$, therefore as h becomes smaller p becomes continuously greater than C'.

The real difficulty is met when an attempt is made to allow for the direct attraction of one solid for the other. Call the range of this attraction w. Is w of the same order as a?

If we put $w = a$ then when $h < a$ there is a direct attraction between the solids which gives rise to a pressure at the midway surface, of say

$$X = 2\pi \int_{h}^{a} \phi \ (z) \ dz,$$

which increases as h decreases.

If X has a sensible value at the midway face, then $P = (I'' - C' - X) + p$.

When $h = 0$ the attraction X should be equal to the complete cohesion between the solids. Actually h cannot be reduced to zero, at any rate by merely wringing the faces together until they seize. Metallic contact, it is true, takes place, but only at those few points where the surfaces appear torn when the solids are again separated. Elsewhere the lubricant persists, and the fact that seized faces are forced apart by lubricants with a sensible vapour pressure shows that $\phi \ (z)$ falls off very rapidly with distance.

The alternative explanation is not so easily stated. Consider first the case of a liquid and let a now be the distance at which the attraction of the solid ceases sensibly to orient the molecules of the liquid so that at any distance greater than a from a plane solid face the arrangement of the molecules is random. There, owing to the random disposition and motion of translation of the molecules, the intrinsic pressure I_0 may be taken to act in all directions. It is a true hydrostatic pressure.

Within the plate of lubricant between the plane ends of the cylinders, however, the molecules are oriented and fixed in position by the attraction fields of the solids, with the result that the hydrostatic pressure is less than I_0 so far as I_0 is due to motion of translation of the molecules. It may be taken to vanish when the lubricant is reduced to two primary films, because in these films the molecules are as completely fixed in position as they are in a solid. As h increases, however, and the directing influence of the solid faces at $h/2$ decreases, the motion of translation of the molecules increases and with it the hydrostatic pressure.

The result is that, whilst there is in the surrounding fluid a hydrostatic pressure $= I_0$, there is at the surface $h/2$ a pressure I which is equal to 0 when h is equal to the depth of two primary films and is $= I_0$ when $h = 2a$. The Leslie pressure then is

$$P = I_0 - I$$

and if L be the load $\div$ area

$$L = P.$$

The Leslie pressure on this view is strictly an osmotic pressure like that which causes a gel to swell, and the different molecular species concerned, though they may be the same chemically, differ in their degrees of freedom, according as they are within the range of influence of the solids or beyond it.

On both views, so long as the joint is sunk in a pool of the same fluid as that which forms the lubricant, the adhesion (value A) will be the difference between the tensile strength Q and the Leslie pressure, or

$$A = Q - P,$$

but when the lubricant is enclosed in a skin and the pool removed, P vanishes and $A = Q$.

P vanishes also when the lubricant is frozen, therefore

$$B = Q.$$

In these equations the terms on the right must be reckoned at the moment of break.

The Leslie pressure in clean air offers no difficulty on the hypothesis of Laplacian attraction, and formidable difficulties on the hypothesis of polarisation. It is not easy to picture structure in a permanent gas but it must be present in a gas near to the surface of a solid, for we are at liberty to suppose that molecules of gas like molecules of liquid will be oriented and suffer constraint in their movement of translation, whilst they are within the range of attraction of a solid face. Let us take the extreme case and suppose that this range is only of the order of dimensions of a molecule. There will then be a layer of highly polarised molecules, the equivalent of a monomolecular layer. From this, owing to the heat motions, abnormal molecules will escape and the depth of the layer of gas whose properties are modified by the attraction field of the solid will depend upon the rate at which these recover normality. Within this zone of influence the specification of the gas in mass does not hold, because some of the molecules have an abnormal energy content and the movements of translation are not at random. Within the body of the gas, molecules have their direction of movement changed and their energy content altered by collisions, but the statistical distribution of these variations in space and time is a random one. When two solid faces approach one another, the whole of the enclosed gas will be abnormal when the gap is twice the range of influence

There is also an electrical effect to be reckoned with. In Helmholtz's theory of contact potential the charges lie in parallel surfaces. In the modern view of the contact potential between a solid and fluid, the electric charge of the latter is not referred to a surface but to a layer of ionised molecules of finite depth. The molecules of fluid in contact with the solid face may be supposed to form a layer of electrically charged molecules, but owing to the heat movements some of these are detached and the depth of the charged layer will be determined by the normal component of their free path in the charged state. Call this depth v, then when two plane surfaces of similar solids approach one another there will be resistance to further approach when the distance between them becomes less than $2v$, the repulsive pressure being due to repulsion between charges of the same sign.

Watson and Menon found the air gap always conducted when plate and cylinder were clean ((15), p. 197). It must, therefore, have contained charged particles under the conditions of the experiments, and I was able to show by direct experiment that a free film of starch paste or saponin and water is composed of a central plate of liquid at an electrical potential different from that of the adsorbed films enclosing it. The central plate was positive and both films negative. This disposition, as I then pointed out ((14), p. 608), would give the repulsion required to account for the stability of such free films.

In one way or another, therefore, the molecules in a thin layer of gas enclosed between two attracting surfaces must differ in the specification of their energy content from the molecules in the body of the gas and, like those in an enclosed layer of liquid, they form a different osmotic species, but the difficulties in the way of this hypothesis are serious: An air gap of 0·004–0·007 mm. seems inconsistent with the rate of relaxation of strain in a gas. An even more formidable difficulty lies in the fact that the gap is unchanged when the air is displaced by liquid ((11), p. 224), for two different causes are unlikely to produce precisely the same effect. Finally, electric repulsion of the kind indicated above seems inconsistent with the fact that the gap is the same, namely, 0·004 mm. for carefully trued cylinders and plates of materials so dissimilar electrically as steel and glass.

The position may be summarised as follows:

Any theory of the Leslie pressure must account for four curious facts:

(1) The distance h at which the cylinder floats above the plate is independent of the nature of the lubricant and of the solids, provided the surfaces are clean.

(2) The distance is the same in clean air as it is when the air is displaced by a liquid and remains the same until the pressure of the air is reduced to 2 mm. of mercury.

(3) The distance in clean air was decreased by contaminating the solid faces.

(4) The difficulty with which a cylinder can be forced down on to a plate and the relative ease with which it can be wrung down.

Of these the two first are perhaps accounted for by the fact that on either hypothesis the pressure P is the difference between two quantities, both of which may be functions of the nature of the lubricant and of the solids. P may then be a pure function of the geometry of the system over an extremely wide range of conditions.

The effect of the contaminant seems to point to the conclusion that the attraction of the solid is not a shadowless force as Laplace assumed, but a force which, like chemical affinity, can be saturated. The effect may, however, be due simply to the contaminant making the surface irregular.

The relative efficacy of wringing is no doubt due to purely mechanical action. In wringing a cylinder down on to a plate the surfaces are not kept parallel, the resistance encountered is therefore at no time equal to the pressure P (at the least distance between the surfaces) × area. By tilting the cylinder the resistance can, in fact, be made vanishingly small. When any part of the cylinder has been forced down it rises under capillary pressure relatively slowly, as the table (10), p. 67, shows. There is, therefore, a kind of ratchet and pawl action.

Experiment shows that once the cylinder has been wrung down, it stays there if the surfaces are contaminated by liquid, save when the joint is surrounded by a fluid freely miscible with the liquid lubricant. It will therefore not rise in air, but clean faces wrung down, even to seizing, float apart in clean air. This, of course, does not mean that the pressure P in clean air is greater than metallic cohesion, for the injury to the polished faces shows that seizing has taken place at only a few points, so we have, if A be the area of the plane end of the cylinder and a the total area of seizing,

$$PA > Ka,$$

where K is the cohesive attraction per unit area of metal for metal and A/a is a large quantity.

Since P is the pressure with which fluid lubricant is drawn between the faces from a pool, and is independent of the nature of the lubricant, it is not at first obvious why some liquids will displace air. It will be sufficient perhaps simply to point out that this displacement is due to the tension of the air-liquid interface.

The position may be summarised by saying that relatively long range Laplacian attraction cannot be rejected until the air gap is satisfactorily accounted for, in spite of the proof given in the next section that such attraction will not provide the properties of solid joints.

Section 4. Mean Value Rule

This rule holds both for the tangential and normal reaction. It can be stated as follows.

Let X_{ss} denote a value when both solids are, for example, of steel, X_{cc} when both are of copper, and X_{sc} when one is steel and the other copper, then

$$X_{ss} + X_{cc} = 2X_{sc}.$$

The rule seems to govern all the mechanical relations of the boundary state, it holds for all values, even for the A value, and does not appear to be an approximation. The better the craftsmanship the more closely does it fit experiment. Take for example the value B, the adhesion of a solid joint ((7), p. 220), we have as the mean for acids:

$$B_{ss} = 14{,}546, \quad B_{cc} = 12{,}917 : \text{mean } 13{,}731$$

and
$$B_{sc} = 13{,}740.$$

The mean value rule is a property of those significant surfaces which are also surfaces of least traction or least adhesion, that is, of surfaces of slip or surfaces of break. It will be convenient in this section to call these surfaces mean value surfaces.

In only three instances is the position of the mean value surface known with certainty, namely, for values F, B and S.

Value F. When a spherical slider is used, the mean value surface is at $\frac{1}{2}h$ and separates the two adsorbed layers x and z in the diagram. The disposition of the parts is symmetrical about this surface, and the rule offers no difficulties.

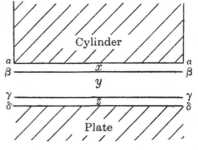

Values B and S are respectively the strength in tension and the strength in shear of a solid joint. The former also is the normal reaction of the mean value surface, as we have seen earlier. Only a few measurements have been made of value S ((11), p. 288), attention will therefore be confined to value B.

The solid joint was formed by running fluid lubricant between the faces, allowing the lubricant to come into equilibrium with the attraction fields of the solids, and then lowering the temperature until the lubricant froze ((11), p. 209). All the measurements, which will be referred to, were made with load 5·6 gr., loading (load divided by area) 7·15 gr., and distance between cylinder and plate $h = 0·007$ mm. for clean surfaces and $\pm 0·005$ mm. for contaminated surfaces. When the joint was broken, the surface of break was seen under the microscope to lie between an adsorbed layer of insensible thickness, x or z in the diagram, and a median crystal plate y. The presence of an invisible adsorbed layer was detected by the fact that after the joint was broken the friction was that of a fully lubricated surface ((11), p. 214). The significant surfaces of a solid joint then are placed as shown in the diagram at β and γ. When cylinder and plate were both of steel or both of copper, the break took place at both β and γ. When one of the two was of copper and the other of steel, it was, at the significant surface nearer to the copper, the solid which gave lower adhesion (value B steel-steel $= 14,546$, copper-copper $= 12,917$).

The significant surfaces β and γ are placed where the attraction fields of the solids are no longer able to overcome the forces which drive the molecules into the crystal lattice. Following H. A. Wilson (*Proc. Camb. Phil. Soc.* x, 25, 989) we may say that these surfaces are where the pressure driving atoms into the crystal lattice is equal to the pressure forcing the atoms towards the metal face. The position of those surfaces will therefore depend upon the attraction field, but there is a quantum effect because each surface can move outwards from the metal face only by the full depth of a layer of molecules of the lubricant. The surfaces β and γ are the surfaces of natural cleavage spoken of earlier.

So far the ground is firm, but the position of the mean value surface or surfaces cannot be fixed by direct observation when the lubricant is fluid and the distance H greater than two primary layers. Both the number of the surfaces and their position is indeterminate unless the law of variation along the normal to the surfaces α and δ of the influence of the solids be known. Nothing can be inferred from the position of the surfaces β and γ in solid joints, for let us suppose that the surface in a fluid joint is singular and at $h/2$ the pressure of crystallisation might split this surface into two surfaces of identical properties placed where the pressure of crystallisation failed to disturb the orientation.

The experiments with solid joints which we need consider in this section fall into two classes, those in which the lubricant was composed throughout of a single pure chemical substance, and those in which the clean surfaces of either cylinder or plate, or both, were first coated with an invisible film of one pure substance before the joint was completed by running in a second pure substance. Two substances were employed, the paraffin eicosane and myristic acid, the metals being steel and copper.

The clean face of either cylinder or plate, or of both, was contaminated by lightly touching it with clean[1] linen which had been lightly rubbed on either the solid paraffin or the solid acid. The joint was then formed with

---

[1] "Clean"—that is, extracted with benzene and alcohol until rubbing clean faces with the linen did not lower friction.

the alternative substance, allowed to come into equilibrium and frozen. (For details, see (11), p. 217.)

The method of contamination seems haphazard, the results were surprisingly consistent. Probably the face ceased to pick up the contaminant when a strongly adsorbed monomolecular layer had been formed.

Of the two substances when used alone the acid formed the stronger joint.

	Acid gr.	Paraffin gr.
Steel-steel	13,260	8644
Copper-copper	11,580	5955
Steel-copper	12,437	7281

If the likely assumption be made that the adsorbed layer was composed wholly of contaminant the disposition can be indicated by making use of the diagram

$$X \overset{\alpha}{\mid} x \overset{\beta}{\mid} y \overset{\gamma}{\mid} z \overset{\delta}{\mid} Z,$$

in which X and Z are the metals, x and z the adsorbed layers, y the crystal plate, α and δ the surfaces separating metal and lubricant, and β and γ the significant surfaces.

This diagram is reproduced in the left-hand column of Table II, the surface of break, that is, the mean value surface, being indicated by a thick rule. E and M indicate respectively eicosane and myristic acid. Value B is adhesion per square centimetre. S and C are steel and copper.

When the break occurs at both β and γ surfaces (see for details (11)), the identity

$$B_\beta = B_\gamma$$

must hold.

When the break occurs at only one of the surfaces, say the γ surface, then, if the adhesion of the opposite surface is E_β, we have

$$E_\beta > B_\gamma.$$

That the experiments with contaminants have a quantitative value is shown by the two groups which ought to give the mean value relation. They are numbers 7, 8 and 11 which give $9249 + 6115 = 2 (7578)$ or $15,364 = 15,156$, and numbers 9, 10, 12 and 13, which give 10,590 and $5702 = 8845 + 7511$ or $16,292 = 16,356$.

These observations seem to be inexplicable by direct Laplacian attraction, for in number 9 let $S_1 S_2$ be the pressure due to the mutual attraction of steel for steel, whose plane surfaces are h apart, $S_1 M_y$ the pressure at the surface of break β due to the mutual attraction between the steel nearer to the surface of break and the median plate of myristic acid, $S_2 E_1$ the pressure at the surface of break due to the mutual attraction between the steel further from the surface of break and the adsorbed layer x of eicosane

TABLE II. *Load, 5·6 gr. Pressure = Load ÷ area, 7·15 gr.*

	gr./cm.²
Uncontaminated.	
When the surfaces were uncontaminated $h = 0·007$ mm.	
1. $S \mid M \blacksquare M \blacksquare M \mid S$ value B	13,260
2. $C \mid M \blacksquare M \blacksquare M \mid C$	11,580
3. $S \mid M \mid M \blacksquare M \mid C$	12,437
4. $S \mid E \blacksquare E \blacksquare E \mid S$	8,644
5. $C \mid E \blacksquare E \blacksquare E \mid C$	5,955
6. $S \mid E \mid E \blacksquare E \mid C$	7,281
Contaminated.	
When the surfaces were contaminated $h = $ about $0·006$ mm.	
((11), p. 227).	
Lubricant, myristic acid; contaminant, eicosane.	
	gr./cm.²
7. $S \mid E \blacksquare M \blacksquare E \mid S$	9,249
8. $C \mid E \blacksquare M \blacksquare E \mid C$	6,115
9. $S \mid E \blacksquare M \mid M \mid S$	10,590
10. $C \mid E \blacksquare M \mid M \mid C$	8,702
11. $S \mid E \mid M \blacksquare E \mid C$	7,578
12. $S \mid E \blacksquare M \mid M \mid C$	8,845
13. $C \mid E \blacksquare M \mid M \mid S$	7,511
Lubricant, eicosane; contaminant, myristic acid.	
	gr./cm.²
14. $S \mid M \blacksquare E \blacksquare M \mid S$	10,540
15. $S \mid M \mid E \blacksquare E \mid S$	9,745
16. $S \mid M \mid E \blacksquare M \mid C$	9,017
17. $S \mid M \mid E \blacksquare E \mid C$	8,085
18. $C \mid M \mid E \blacksquare E \mid S$	8,736

nearer to the surface of break, then, using this notation throughout, we have from numbers 7 and 9

$$S_1 S_2 + S_1 (M_y + E_2) + S_2 E_1 + E_1 (M_y + E_2) = 9249 \text{ gr.},$$

$$S_1 S_2 + S_1 (M_y + M_2) + S_2 E_1 + E_1 (M_y + M_2) = 10,590 \text{ gr.};$$

therefore

$$S_1 (M_2 - E_2) + E_1 (M_2 - E_2) = 1341 \text{ gr.}$$

But the first of these terms is the difference between the attraction of steel for two monomolecular layers distant 0·006 mm. and the second the difference between the attraction between monomolecular layers separated from each other by nearly the same distance. All these quantities may be regarded as negligible and certainly not competent to give the substantial difference, 1341, actually found.

If further proof of the short range of cohesive forces be needed, it can be found in the remarkable effect of an insensible film of oxide present on the copper. Such a film, so thin as to leave the polished face visibly unchanged, raises the adhesion of copper above that of steel (17). The values are as follows:

	Lubricant, palmitic acid
Steel on steel	15,860 gr. per sq. cm.
Copper on copper	14,200 ,, ,,
Passive copper on passive copper	21,000 ,, ,,

It is not possible to suppose that the range of cohesive attraction is of one order in the metal and of quite a different order in the lubricant. Therefore, if the range be of the order 0·005 mm., in the latter the pressure at the mean value surface must be due to the total attraction of a disc of metal many thousands of molecules in thickness. It is not credible that the cohesive attraction of a layer of oxide, probably one molecule in thickness, should be of the same order as that of such a disc of metal, but it is credible that the cohesive attraction of a highly polarised film of oxide should be of the same order or greater than that of a film of metal of the same thickness.

Since Laplacian attraction in any simple form is inadmissible, the properties of the mean value surface will be considered solely from the point of view of diachysis—that is of polarisation which is transmitted from molecule to molecule with a decrement.

It will be noticed that wherever myristic acid replaces eicosane whether in one or other or both adsorbed layers (x and z) or in the crystal plate (y) it increases the adhesion. In other words, displacement of less polarisable molecules by more polarisable molecules increases adhesion. The inference that adhesion is a function of the polarisation of the mean value surface is obvious.

The polarisation of the mean value surface is not, however, determined solely by the constitution of the molecules which lie immediately on either side of it, but is contributed to by all parts of the system. This can be inferred from the mean value rule, and the inference is directly confirmed by experiments with contaminants. Compare, for example, numbers 7 and 9. The β surface is in both cases a surface of break and the effect of substituting the acid for the paraffin in the *opposite* adsorbed layer is to raise the value of adhesion from 9249 to 10,590.

The crystal plate y therefore conducts,[1] and it conducts with a decrement for no matter what may be the composition of the layers x, y, z, adhesion is always increased when h is decreased by increasing the load, and the microscope shows that the effect of such increase in the load is merely to bring the β and γ surfaces closer together (for curve connecting h with adhesion, see (11), p. 226).

The B value depends, however, not only on the thickness of the crystal plate, but also on its composition. Compare for example, number 7 with number 14 and number 11 with number 16. Taking the first pair the polarisation of both β and γ surfaces is between eicosane and myristic acid, but the crystal plate is of acid in the one and of eicosane in the other. The proof is admittedly not rigid because the paraffin is next to the steel in number 7, and the acid next to it in number 14.

The form of the equation is now clear and may be exhibited by taking any one of the equations, say, number 9, as an example. Let $(\sigma_{em})_1$ represent the polarisation produced by the nearer steel, at the mean value surface β which lies between eicosane and myristic acid; $(\sigma_{mm})_2$ that due to the opposite surface γ and y_m the decrement due to conduction from surface γ to the mean value surface β; then

$$(\sigma_{em})_1 + [(\sigma_{mm})_2 - y_m] = 10{,}590.$$

It will be noticed that the polarisation is put numerically equal to the adhesion. This assumption will not affect the argument and it enables the results to be stated in a simple way. It amounts to an affirmation that the excess tensile strength of a substance as lubricant over that of a rod of the lubricant (see p. 827) varies directly with the polarisation of the former, the unknown constant being omitted from the equations.

When the disposition is symmetrical as it is in the uncontaminated numbers 1, 2, 4 and 5, the crystal plate may be taken to conduct equally in either direction. Therefore, taking number 1 as an example, it becomes

$$2\sigma_{mm} - y_m = 13{,}260.$$

Conduction is not the same in both directions, however, when the disposition is not symmetrical. In numbers 3 and 6, for example, since the

---

[1] To speak of the crystal plate "conducting" is to use a word of too active significance, but none other suggests itself. There is, of course, no conduction in the sense in which heat or electricity is conducted but a state of strain impressed by the attraction fields on the crystals when they form. When a gel of gelatine and water is frozen, it is strained by the expansion of the ice phase and becomes doubly refractive. The state of strain persists after thawing and is relaxed only near the melting point of the gel. The strain extends throughout the gel and may, with propriety, be said to be "conducted" from each centre of crystallisation (16). It is in this sense that the word is applied to the crystal plate, "conduction" is one aspect of the complex play of forces during crystallisation.

surface of break is at γ, the adhesion of that surface must be less than that of surface β. Therefore

$$\sigma_{mm} + [\kappa_{mm} - (y_m)_{cs}] > \kappa_{mm} + [\sigma_{mm} - (y_m)_{sc}]$$

and

$$\sigma_{ee} + [\kappa_{ee} - (y_e)_{cs}] > \kappa_{ee} + [\sigma_{ee} - (y_e)_{sc}]$$

or

$$(y_m)_{sc} > (y_m)_{cs} \quad \text{and} \quad (y_e)_{sc} > (y_e)_{cs},$$

where the order of the subscript letters s and c denotes the direction of conduction. The crystal plate, therefore, conducts better from the stronger field of the steel to the weaker field of the copper than it does in the reverse direction.

When eicosane is applied to the steel in a steel-copper pair the break is transferred to the side of the stronger metal. This was unexpected; it follows, however, from the unequal conduction in the crystal plate. Numbers 12 and 18 give

$$\kappa_{mm} + [\sigma_{em} - (y_m)_{sc}] > \sigma_{em} + [\kappa_{mm} - (y_m)_{cs}],$$

$$\kappa_{me} + [\sigma_{ee} - (y_e)_{sc}] > \sigma_{ee} + [\kappa_{me} - (y_e)_{cs}]$$

or

$$y_{cs} > y_{sc}.$$

From numbers 7, 8 and 11, the relation $(y_m)_{ss} + (y_m)_{cc} = 2\,(y_m)_{sc}$ can be derived, and from numbers 9, 10, 12 and 13 $(y_m)_{ss} + (y_m)_{cc} = (y_m)_{sc} + (y_m)_{cs}$.

The simplest interpretation of this is that the crystal plate conducts with a decrement which is a function only of its thickness and of its chemical composition. The mean value rule for the normal reaction of solid joints would then have the form

$$[2\phi\,(X) - y] + [2\psi\,(Z) - y] = 2\,[\phi\,(X) + \psi\,(Z) - y],$$

$\phi\cdot(X)$ and $\psi\,(Z)$ are respectively the polarisation produced by the metal X or Z at the nearer significant surface.

The fundamental assumption of diachysis will now be obvious, namely, that the direct influence of the attraction field of a solid embraces only the adsorption layer in contact with it and its contribution to the properties of the significant surface of the opposite side is made by strain conducted through the crystal plate y. Therefore, at the limit when the loading is great enough to bring the β and γ surfaces together, at $h/2$ the mean value rule takes the simple form $2x + 2z = 2\,(x + z)$, where x and z are respectively the polarisation of the mean value surface due to the metals X or Z. It might be expected that the mean value rule would be upset when the intensity of the effect of the surfaces upon the lubricant was widely different. That may prove to be the case. All that can be affirmed at present is the validity of the rule for pairs of solids chosen from examples as diverse as are hard and soft steel, bronze, copper, bismuth and glass.

Section 5. The effect of the chemical constitution of the lubricant

The relation of the tangential and normal reactions to the chemical constitution of the lubricant and solids can be expressed in simple empirical linear equations of the form

$$T = x - yM,$$
$$N = x' + y'M.$$

In this section one example only will be considered, namely, the tangential reaction when the lubricant consists of a single pure chemical substance and is reduced to two primary films by the use of a spherical slider or by a heavily loaded plane slider. The equation will be taken in the form $\mu = b - aM$, in which μ is the coefficient of friction ($F \div$ load).

a is a parameter wholly independent of the nature of the solids. It fixes the slope of the curve and is a pure function of the configuration of the molecule of the lubricant. For normal acids, alcohols and paraffins the expression aM can be written in the form $c(n-2)$ where n is the number of carbon atoms in the chain. In this form, it shows the remarkable fact that each such carbon atom produces the same effect up to a chain of thirty atoms which was the longest chain tested. The configuration of the substances used is:

Paraffins	...	$CH_3\!-\!CH_2\!-$	$-CH_2\!-\!CH_3$
Alcohols	...	$CH_3\!-\!CH_2\!-$	$-CH_2\!-\!COH$
Acids	...	$CH_3\!-\!CH_2\!-$	$-CH_2\!-\!COOH$

The carbinols are alcohols in which the hydroxyl group has been shifted to the third position.

$$\text{Carbinols} \quad ... \quad CH_3\!-\!CH_2\!- \quad -CH_2\!-\!CHOH\!-\!CH_2\!-\!CH_3.$$

The end groups then are CH_3 and $C_2H_5 . CHOH$ and this large group, as the curve Fig. 8 shows, increases the effect of the two nearest carbon atoms so that the steady state is reached only at the fourth carbon atom and the expression becomes $c(n-6)$.

Since the parameter c is dependent only on the end groups, it must express the effect of the internal polarisation of the molecule upon the tangential reaction, and we may add upon the normal reaction also.

Miss Doubleday rightly draws attention to the relation of friction to the optical properties of the carbinols[6]. The first eight points on her curves refer to dextrorotary, the last six to lævorotary forms. The change in the rotation has no effect upon the form of the curve. On the other hand, the one dl-form available, namely, dl-ethyl-n-hexylcarbinol, gave a value well off the curve. The specimen of dodecylcarbinol probably was slightly impure.

The parameter b is the value when $M = 0$, it is a function of the nature of the solids and the configuration of the end groups of the molecules of lubricant. One way of evaluating the latter is to put the former equal to the coefficient of friction of the clean solid. Call this b_0 and let $b = b_0 - d$: d is now found to be constant for the same end groups. The equation becomes

$$\mu = (b_0 - d) - c\,(n-2); \text{ or } c\,(n-6) \text{ for carbinols.}$$

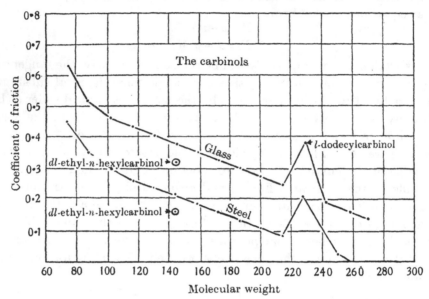

Fig. 8.

TABLE III

Lubricant		A	B	C	D	E	Glass
	b_0	0·88	0·83	0·93	0·79	0·94	0·95
n-paraffins $c = 0\cdot0207$	d	0·21	0·20	0·20	0·20	0·20	0·19
n-alcohols $c = 0\cdot0237$	d	0·28	0·27	0·28	0·27	0·28	0·29
n-acids $c = 0\cdot0602$	d	+0·01	−0·02	+0·01	−0·02	+0·02	−0·01
Carbinols $c = 0\cdot027$	d	—	0·51	—	—	—	—

A, 50-ton steel. B, medium carbon steel. C, nickel chrome steel. D, mild carbon steel. E, phosphor bronze.

The values for acids do not agree with those in Table 10 of (9), p. 26. The latter, I regret to say, are wrong owing to an error in arithmetic made in calculating c, which appears as a in the table just referred to when $c\,(n-2) = aM$.

The paraffins being symmetrical give the value of c, that is the decrement due to each carbon atom, when it is undisturbed by a difference between the end groups: we have therefore

Paraffins	...	...	...	$c = 0\cdot021$
Alcohols	...	...	...	$c = 0\cdot021 + 0\cdot003$
Carbinols	...	...	...	$c = 0\cdot021 + 0\cdot006$
Acids ...	...	...	...	$c = 0\cdot021 + 0\cdot039$

The figures in the left-hand column give the effect of lengthening the chain by one carbon atom, and those in the right-hand column reflect the change in the internal polarisation of the molecule due to the substitution of one —CH_3 group by —COH, —C_2H_5CH(OH), or —COOH.

The parameter d may be taken to refer not to the internal polarisation of the molecule but to the influence of the end group upon the field of the solids. The fact that $d = 0$ for acids might be taken to mean that the molecules are attached to the metal by the CH_3 group because the metal is electro-positive and the COOH group also positive owing to its replaceable hydrogen, but this suggestion is not applicable to glass.

Why should the effect of a carbon atom in the chain be independent of the length of the chain? There is at least one parallel case. The heat of crystallisation of long chain compounds is proportional to the length of the chain[21] but the relation is unusual. So far as I am aware the effect of a change in an end group commonly decreases as the chain lengthens. For example, when a hydrogen of the —CH_3 group of an n— acid is replaced by a negative atom, such as Cl—

$$CH_3 \quad —CH_2 \quad \ldots \quad \ldots \quad CH_2—COOH$$
$$CH_2Cl—CH_2 \quad \ldots \quad \ldots \quad CH_2—COOH$$

the work needed to detach the hydrogen atom of the —COOH is lessened but the effect diminishes as the chain is lengthened. It is not possible to assume that c is the sum of two terms contributed by the two boundary solids,

$$c = c_x' + c_x''$$

and that the sum is the same for any pair of solids because it is the same also for two dissimilar solids—or $c = c_x' + c_z''$. c appears always to be entirely independent of the solids.

Perhaps the most difficult problem of the boundary state is presented by the fact that c is positive for the normal reaction and negative for the tangential reaction. The external force which calls forth the tangential reaction may be taken to be applied as a traction on the surfaces α and δ, and an increase in the length of molecules whose long axis was normal to those surfaces would increase the moment of the force at the surface of slip $h/2$. But the S value relates to a traction and c is positive for this value.

The fact that c is positive for the normal reaction may mean that, whilst a surface of slip is plane and lies between the ends of molecules, a surface of break is not plane because the molecules are pulled apart like teeth out of their sockets, one half of the number moving one way and the other half moving the opposite way (cf. p. 842).

SECTION 6. CONCLUSIONS[1]

The upshot of the discussion and especially of the sections on the Leslie pressure and the mean value rule may be set down as follows.

The most promising view of the Boundary State IN LIQUIDS is that it is due to the formation of chains of highly polarised molecules stretching through the lubricant from one enclosing solid to the other.

Each chain has little strength in shear, great strength in tension and in both the strength decreases as the chain lengthens.

The intensity of polarisation at any level in a chain is the sum of two terms of the same sign contributed independently by the attraction fields at the ends. The influence of each field diminishes as the distance from the solid face increases, and the least value of the sum of the two terms is at a mean value surface or surfaces. In the only instance in which the position is known with certainty the mean value surfaces for slip and for break coincide (values B and S), but this is after the joint has been frozen.

The view of the liquid state taken by Poisson, Maxwell and Poynting [*Phil. Mag.* (1896), (5), XLII, 289] contemplates a structure which is in the main solid, but the solid part is continually breaking down and renewing itself so that at any instant the liquid is composed of molecules built into a structure and of "free" molecules. The mobility of the liquid is the number of free molecules crossing unit area of a surface per second. In the Boundary State the attraction fields of the enclosing solids may be supposed to increase the rigidity and decrease the rate of decay of the solid portion, but the change is not uniform throughout the lubricant; to take the simplest assumption rigidity is greatest in the adsorbed layers and least at the surface $h/2$.

Osmotic pressure is the difference in hydrostatic pressure if any needed to make the mobility of the fluid on each side of a surface equal. Compare the tangential surface at $h/2$ with a normal surface cutting the centre of the plate of lubricant. With long chain molecules at any rate the mobility at the former surface is almost certainly nil because movement of molecules along the chains may be neglected, whilst at the latter surface mobility will be nil in the adsorbed layers and reach a positive value at the level $h/2$; hence the rigidity of the structure will be greater for normal than for tangential stress.

Put in another way, this amounts to saying that the structure of the lubricant is not subject to decay in the adsorbed layers but the rate of decay is positive and increases from those layers to the surface $h/2$.

Consider now the cylindrical surface which encloses a slab of lubricant under a cylinder standing in a pool. It is obvious that there will be on the

[1] Added December 15, 1930.

whole less mobility in the space within that surface than there is without, unless something be done to increase the former. Hydrostatic pressure increases the mobility of a liquid (cf. Poynting) and the Leslie pressure is the pressure needed to equalise the mobilities across the enclosing cylindrical surface at the level $h/2$—it is, in short, the osmotic pressure of the lubricant.

The molecular chains may, of course, help to support the loading by their resistance in compression, in which case the expression for the Leslie pressure would include two terms, one representing the rigidity of the structure, the other representing the defect in mobility, the Leslie pressure would then become strictly analogous to the swelling pressure of a gel.

There is, however, a difficulty. The equation of the Leslie pressure in its simplest form is $P_L = p_0 - p$, where p_0 is the mobility of the external fluid and p the mobility of the lubricant, the values being reckoned in an elementary cylinder which encloses the surface $h/2 . p$ by hypothesis depends upon the length h and upon the strength of the polarising fields, that is, the attraction fields of the solids. When h is fixed the Leslie pressure should therefore depend upon the chemical constitution of the enclosing solids, whereas it seems to be a function only of the geometry of the system. It is possible that this difficulty would be solved by an analysis of the form of the curved equipotential surfaces at the circumference of the plate of lubricant.

The existence of a Leslie pressure IN AIR and the presence of a complete air-gap have been accepted throughout the paper for reasons given on p. 828. It must be admitted that the air-gap if it be real raises theoretical difficulties of the most formidable kind. There are three possible explanations:

(1) The existence of chains of molecules such as are postulated for fluid lubricant.

(2) The existence of an attraction field on the surface of a solid with a range of at least 0·005 mm.

(3) The presence of solid particles which keep the surfaces apart.

Let us briefly summarise the difficulties of each.

(i) The difficulty here is one of degree and is sufficiently dealt with in the text, p. 851.

(ii) To fit the facts the relatively long range force of attraction cannot be the shadowless force contemplated in the Young-Laplace theory. It must be capable of saturation so that its range in close packed states of matter is reduced to molecular or even atomic limits.

(iii) Solid particles if they are present would need to be perfectly elastic and have always the same dimensions. These conditions might be fulfilled if the particles, instead of being casual fragments of matter caught up by the solid faces, were formed by the condensation from the air of ultra-microscopic particles on to areas of the solid face where the attraction field

H   55

was more intense than elsewhere. The existence of such areas is accepted by physicists and they might be due to minute variations in the structure or form of the surface of the metal. The depth of a particle formed in this way would be a function of the excess of attraction at an active area. On this hypothesis, however, the depth and therefore the value of h should be different for different solids.

The total strength of all the chains of molecules in tension increases as the length of the molecules of which they are built increases, whilst the total strength of all the chains in shear decreases and, when allowance is made for perturbation due to the end groups of atoms, the relation in both instances is always linear. When the length of the molecules is great enough strength in shear vanishes.

The strength in tension must be due to the lateral fields of attraction of the molecules at least as much as to the fields at the ends. The decrease in the strength in shear may be due merely to the increase in the moment of the applied force.

REFERENCES.

(1) HARDY and HARDY. *Phil. Mag.* (1919), VI, 38, 32.
(2) HARDY. *Phil. Mag.* (1919), VI, 38, 49.
(3) HARDY. *Phil. Mag.* (1920), VI, 40, 201.
(4) HARDY and DOUBLEDAY. *Proc. Roy. Soc.* (1922), A, C, 550.
(5) HARDY and DOUBLEDAY. *Proc. Roy. Soc.* (1922), A, CI, 487.
(6) DOUBLEDAY. *Trans. Chem. Soc.* (1922), CXXI, 2875.
(7) HARDY and DOUBLEDAY. *Proc. Roy. Soc.* (1923), A, CIV, 25.
(8) DOUBLEDAY. *Proc. Roy. Soc.* (1924), A, CVI, 341.
(9) HARDY and BIRCUMSHAW. *Proc. Roy. Soc.* (1925), A, CVIII, 2.
(10) HARDY and NOTTAGE. *Proc. Roy. Soc.* (1926), A, CXII, 62.
(11) HARDY and NOTTAGE. *Proc. Roy. Soc.* (1928), A, CXVIII, 209.
(12) NOTTAGE. *Proc. Roy. Soc.* (1928), A, CXVIII, 607.
(13) HARDY and NOTTAGE. *Lubrication Research, Technical Paper,* No. 1, H.M. Stationery Office, 1929.
(14) HARDY. *Proc. Roy. Soc.* (1912), A, LXXXVI, 601.
(15) WATSON and MENON. *Proc. Roy. Soc.* (1929), A, CXXIII, 185.
(16) HARDY. *Proc. Roy. Soc.* (1926), A, CXII, 47.
(17) NOTTAGE. *Proc. Roy. Soc.* (1930), A, CXXVI, 630.
(18) TOMLINSON. *Phil. Mag.* (1929), VII, 905.
(19) BURGESS. *Proc. Roy. Soc.* (1911), A, LXXXVI, 25.
(20) SUTHERLAND. *Phil. Mag.* (1881).
(21) GARNER and KING. *Journ. Chem. Soc.* (September, 1929).

57

ADSORPTION. A STUDY OF AVAILABILITY AND ACCESSIBILITY

With MILLICENT NOTTAGE

[*Proc. Roy. Soc.* (1932), A, cxxxviii, 259]

Our object is twofold—to show how friction may be used to analyse the composition of an adsorbed layer, and to exhibit the remarkable effect of "previous history" upon its composition and properties. When polar molecules are absent the composition of the layer when first formed commits it to a path of change from which it can escape only by turning back on its path and passing through the point of origin.

In the opinion of one of us the experiments are of interest to biologists. They reveal systems, let us call them solutions, whose relations to an adsorbing surface are complex and active only over a curiously narrow range of temperature of some 15–20°; outside this range, temperature has no effect. The influence of "previous history" is great and that of chemical constitution obvious. The remarkable "specific" relations so characteristic of living matter are not there but only molecules of simple type incapable of mutual chemical reaction are involved—Specificity in fact is merely a question of degree. Given molecules of complex atomic pattern and the phenomena would probably have been quite beyond analysis, indeed the curves for the normal reaction (adhesion) of simple binary mixtures are beyond explanation when the components can react together or when one is a ring and the other a chain compound.[1]

The facts will be stated in terms of two assumptions. It will be assumed that the slider used, which had a spherical face, owing to the great pressure under it, reduced the layer of lubricant to two primary layers with the yield plane or surface of slip between them. This is the first assumption; its validity has been discussed in earlier papers.[2] Static friction, which alone is considered, is the strength of the joint in shear; it is also the maximal reaction of the surface of slip to traction.

The second assumption is that each of the molecules which compose the two primary layers contributes independently to the total reaction, and the value of an individual contribution is fixed by the atomic structure of the molecule to which it belongs. This assumption makes it possible to

---

[1] Nottage, *Proc. Roy. Soc.* (1928), A, cxviii, 607, Figs. 2 and 3.
[2] Hardy and Bircumshaw, *Proc. Roy. Soc.* (1925), A, cviii, 2.

identify the species of molecule present in the layers and to follow changes in the composition of the latter.

The equation for the friction of a lubricant composed wholly of a single long chain compound is

$$\mu = \frac{b' + b''}{2} - [d + c\,(N - m)],$$

where the first term on the right is a pure function of the nature of the solids, steel, glass, etc., and the term in brackets is a pure function of the atomic structure of the lubricant.[1] Since the solid was the same throughout, both slider and plate being of mild steel, the contribution of the solid can be neglected in computing changes in the relative numbers of different species of molecules.

A few words must be devoted to the first assumption. In 1912 one of us[2] pointed out that in all close packed structures whose molecules are other than spherical the cohesive forces across an interface would not only hold molecules by direct attraction, the range of the force being of the order of dimensions of a molecule, but would also penetrate as an oriented strain transmitted from molecule to molecule.[3]

Direct observation with X-rays and indirect evidence such as is furnished by, *e.g.*, the Mean Value Rule which defines the effect of the nature of the solid upon friction and adhesion, have confirmed this view, at any rate for long chain compounds and for solid surfaces.

The layer held by direct attraction is probably identical with the primary layer—that is the well-known layer of insensible thickness which spreads over clean fluid or solid faces. This layer has a large heat of formation, is strongly held and is probably monomolecular. It is the layer formed by "primary spreading".[4]

But the mechanical properties of the layers formed by transmitted strain are also determined mainly by the fields of force of the solids. Indeed the Mean Value Rule shows that the influence of a solid can penetrate some thousandths of a millimetre into a fluid in contact with it. Therefore as one has primary and secondary spreading of liquids over surfaces, it is just to distinguish primary adsorption and secondary adsorption.

According to the first assumption, however, the lubricating system is reduced by a spherical slider to two primary layers, one on each steel face, with the surface of slip between them. Only steady states are considered, and as the slider was frequently moved about on the plate the primary layers may be considered to have been in equilibrium with the oil.

[1] Hardy and Bircumshaw, *Proc. Roy. Soc.* (1925), A, cviii, 24.
[2] Hardy, *Proc. Roy. Soc.* (1912), A, lxxxvi, 131.
[3] Hardy and Doubleday, *Proc. Roy. Soc.* (1922), A, c, 568.
[4] Hardy, *Phil. Mag.* (1919), xxxviii, 49.

In the sequel it will be seen that both assumptions are possibly too narrow, for even a spherical slider seems sometimes to be unable to penetrate to monomolecular layers, and the mechanical properties of a molecule in the primary layer depends not only upon its own chemical constitution and the solid to which it is attached but also upon both normal and tangential reactions with its neighbours.

Certain terms need definition. By successive extractions of two commercial oils with acetone or by exposure to adsorbing surfaces such as those of clean glass beads it has been found possible to remove the more reactive constituents until the fraction dissolved by the acetone or adsorbed had the same composition as the residual oil.[1] Since selective adsorption had then ceased the residuum was an *adsorption individual* with no temperature coefficient of friction between the limits explored, namely 10–106° C. From its method of preparation it may be taken to have been freed from polar molecules.

The adsorption individual was found to form from 93 to 96 per cent. of the two commercial oils examined—it is the *diluent* which holds in solution from 4 to 7 per cent. of active *components*.[2]

Obviously when a lubricant is a single pure chemical individual it is also an adsorption individual and has no temperature coefficient of friction. No exception to this rule has been found.

A temperature-friction curve is said to be *reversible* when it has the same form for an ascending as it has for a descending series of temperatures. The curves for adsorption individuals are always reversible.

The *state* of an oil is defined by the statistical average taken by volume or by time of the molecular species composing it. The word in short has its ordinary significance in kinetic theory. The state of a component in an oil may be such that in competition with other components or with the diluent itself it cannot find a place in the adsorbed layers—it is then said not to be *available* for adsorption. It may, however, be available when it competes for a place on a *clean* surface but unable to displace other molecules already in possession of the surface. It is then said not to be *accessible*.[3]

Availability and accessibility are two potentials which are not absolute but purely relative to the similar potentials of the diluent, and therefore if they could be measured on some absolute scale a variation in, say, the availability of the component might be found to be due either to a change in its own state, or to a change in the state of the diluent. But as the experiments do not allow us to discriminate, the simpler course has been taken of stating the facts wholly in terms of the component as though the variations were absolute and not merely relative.

[1] Hardy and Nottage, *Lubrication Research, Technical Paper No.* 1, p. 16, H.M. Stationery Office (1929).

[2] *Ibid.*, p. 39.

[3] Grammatically monstrous but serviceable.

A convenient diluent for experimental purposes is the medicinal "paraffin" of the pharmacopœia which is here called "Oil B.P." It is prepared from a mineral oil by prolonged percolation through highly adsorptive material. It has no temperature coefficient of friction over the range explored, 8–106° C. It appears to be composed of saturated cyclic compounds and has the great advantage of combining a low vapour pressure with fluidity over the whole range.

Besides oil B.P. a wax denoted by N.O.P. was used as diluent. It also has no temperature coefficient of friction, and appears to be a mixture of saturated hydrocarbons melting between 54 and 57° C.

The respective coefficients of friction are, for the mild steel used:

$$\mu_{\text{B.P.}} = 0 \cdot 228, \qquad \mu_{\text{N.O.P.}} = 0 \cdot 275.$$

The components used were palmitic acid $\mu_{\text{ac.}} = 0$, melting point 62° C.; hexa-decyl alcohol $\mu_{\text{al.}} = 0 \cdot 114$, melting point 50° C., and the wax N.O.P.

The interest of the experiments lies in the comparison of two entirely distinct adsorption equilibria. The first is the equilibrium when the oil in mass, the plate and the slider were at the same temperature before the pool of oil was placed on the plate. It gives the adsorption for a particular temperature *when the adsorbing surfaces are clean*. The friction-temperature curve will for short be called the X curve and be shown as an interrupted line.

The second is the equilibrium when the system has been warmed or cooled to the particular temperature whilst the oil is in contact with the adsorbing surfaces. The curve is called the O curve and is drawn as a continuous line. Obviously the O curve always starts from that point on the X curve at which the oil has been applied to the surfaces.

A point on the X curve gives the value reached when lubricant and adsorbing surfaces have been brought to that temperature separately and maintained at that temperature long enough to be in equilibrium with it.

A point on the O curve on the contrary gives a value reached when the adsorbing surfaces have been continuously in contact with the oil and in equilibrium with it through a range of temperature. It is a value determined not merely by the availability of the substances for adsorption but also by their accessibility, that is by their capacity for displacing molecules already in possession of the surface.

The X curve is a curve of availability whose shape depends upon the state of the oil in mass, the O curve is in the main a curve of accessibility.

The X curve was reversible in each of the three examples studied, that is to say the friction at any temperature was the same whether the mixture in bulk had been warmed up, or cooled down to that temperature. Curve O was reversible only when no polar molecules were present.

Both curves are remarkable for two things, the variations of friction lie within a narrow range of temperature and are limited to the friction of the pure diluent on the one hand and of the pure component on the other.[1] In discussing the former care has been taken not to hide ignorance by the use of the word colloidal.

Example 1. *No Polar Molecules. Diluent, Oil B.P. Component, Wax N.O.P.*

The wax was dissolved with vigorous stirring in the oil at a little above the melting point of the former. The mixture was then allowed to cool and kept for 24 hours. To the naked eye it seemed homogeneous. Three mixtures were studied containing respectively 2·4, 0·51, and 0·16 per cent. of wax.

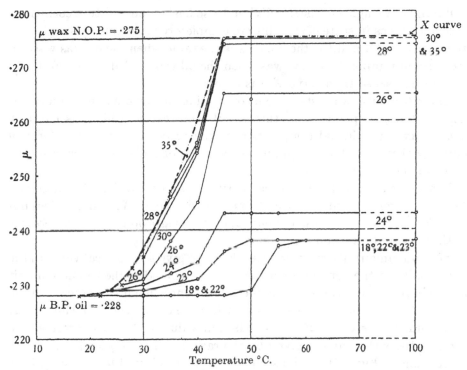

Fig. 1. In this, and in the succeeding figures, the X curve is an interrupted line; the O curves are in continuous line with the initial temperature given in numerals.

(i) *Wax* 2·4 *per cent.* (Fig. 1.)

Curve X. Since the curves for oil B.P. and wax N.O.P. are horizontal lines at $\mu = 0.228$ and $\mu = 0.275$ respectively, the figure shows that curve X leaves the former at about 22° and joins the latter at about 45° C. Therefore below 22° the wax in solution is not available for adsorption whilst above 45° the oil B.P. is not available.

[1] The limitation holds only for steady states, see Appendix, p. 888.

Between these limits and as temperature rises the state of the mixture must change in a way which increases the availability of the wax so that more and more molecules find their way into the primary layers. The form of the curve, indeed, suggests that availability is limited to some particular species of molecules of the wax and that the concentration of this particular species increases with rise of temperature.

The change with temperature of the relative number of molecules of oil B.P. and of wax in the primary layer might be due in part to a change in the selective adsorption of the steel surface. This is unlikely because it seems incompatible with the fact that an adsorption individual prepared from a complex oil by adsorption at a particular temperature or by extraction with acetone has no temperature coefficient of friction over the full range. A single chemical individual also has no temperature coefficient.

The changes in state of the oil were completely reversible, that is to say, the position of a point in the curve was the same when the oil was warmed from room temperature as it was when the oil was cooled from 100° C. In this special sense then curve X was reversible.

Curve O must always start from some point on curve X, and its form, as Fig. 1 shows, depends upon this initial temperature which, as curve X shows, fixes the initial composition of the primary layers. The form of curve O therefore depends upon the capacity of molecules of wax to displace molecules of diluent.

It may be assumed that this capacity increases as availability increases, but if this were all curve O would coincide with curve X. Since it does not so coincide its form must depend upon the variation with temperature both of availability and accessibility.

The fundamental assumption enables us to give a numerical value to the availability of the component which may be defined as the fraction of unit area of a primary layer seized by the component when it competes with the diluent for position on a clean solid surface. The X curve then shows that availability is zero at 22° C. and has unit value at 45°. At intermediate temperatures the value is given by the ratio $\mu_\theta - \mu_{\text{B.P.}} \div \mu_{\text{N.O.P.}} - \mu_{\text{B.P.}}$.

Turning to Fig. 1 it will be seen that when the initial temperature is at or below 22° and the primary layers are composed wholly of molecules of oil B.P. the molecules of wax are unable to get access below 48° C. in spite of the fact that availability is unity at 45°.

When the initial temperature is about 30° or higher the O curve coincides with the X curve, accessibility is therefore complete and the curve follows the change in availability.

Between these two limits there are a set of curves which have remarkable features. Each sweeps up to a saturation level at which the composition of the primary layer becomes independent of a further rise of temperature and the critical temperature at which this condition is reached is sensibly 45° when availability has unit value.

It will be noticed that the curves sweep up more steeply and the saturation level rises rapidly as the initial temperature rises. From this we must conclude that the presence of molecules of wax in the primary layer when it is first formed increases the chance of access of further molecules of wax. It is as though each molecule of wax weakened the hold of the molecules of B.P. oil in its neighbourhood.

From the fundamental assumption the fraction of unit area of a primary layer occupied by wax molecules can be calculated. The values are given in columns 3 and 5 of Table I.

Table I. *Curve X*

<table>
<tr><td>No availability
Below 23° $\mu = 0.228$</td><td>Availability unity
45° $\mu = 0.275$</td></tr>
</table>

Curve O

Initial temperature °	Accessibility begins		$\dfrac{\mu_\theta - \mu_{B.P.}}{\mu_{N.O.P.} - \mu_{B.P.}}$	Temperature saturation level		$\dfrac{\mu_{sat.} - \mu_{B.P.}}{\mu_{N.O.P.} - \mu_{B.P.}}$
18 22 }	48°	$\mu = 0.228$	0.00	55–60°	$\mu = 0.238$	} 0.21
23	23°	0.229	} 0.02 {	47°	0.238	}
24	24°	0.229		45°	0.243	0.32
26	26°	0.230	0.04	45°	0.264	0.77
28	28°	0.233	0.10	45°	0.274	0.98
30	30°	0.235	0.15	45°	0.274	} 1.0
35	35°	0.247	0.19	45°	0.275	}

The O curves are completely reversible and the equilibrium therefore must be kinetic—that is to say it is the result of continuous evaporation from and condensation on to the primary layer from the overlying oil. The reversibility has certain striking features, which can be indicated briefly by considering the curves at the limits of the series. To save space a particular curve is indicated by a subscript which gives the initial temperature—that is the point on curve X at which curve O starts.

Curve $O_{18° \text{ to } 22°}$. The curve has the same form when plotted by an ascending series of temperatures as it has if the series is $18° \to 100°$ and then by descending steps to $18°$. Its form is fixed by the starting-point only. The kinetic character of the equilibrium is seen when a big jump in temperature is made, there is then a latent period of about 20 min. during which the friction rises or falls, as the case may be, to the steady value.

Curve $O_{30°}$. So long as the series starts at $30°$ the points lie on the same curve independent of the order in which they are taken, but if the lower limit of the curve is passed the system takes on an entirely new character:

Fig. 2 will make this clear. Two curves are drawn in continuous line, the curve $O_{30°}$ for ascending temperatures and the curve $O_{18°}$. To avoid confusing the figure all the points observed are not noted—they may be taken to be those shown in Fig. 1.

A mixture containing 2·4 per cent. wax was applied to the clean steel faces at 30° and the friction measured at temperatures up to 100° and down to 18° and the points, as will be seen, lie on the $O_{30°}$ curve. At 18°, however, all the wax molecules have left the primary layers and the system now is the same as though the initial temperature was 18°, for on warming again the points lie wholly on that curve.

The initial temperature obviously commits the system to a path of change from which it seems able to escape only by overstepping the lower limit of the path.

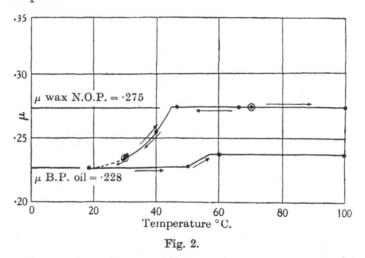

Fig. 2.

It seems to be impossible to put the system on any of the paths $O_{22°}$ to $O_{30°}$ save at the lower limit, but the path of $O_{30°}$ can be entered at any point simply because availability and accessibility are both at limiting values. For this reason the O curve was found to be the same for initial temperatures 30°, 35° and 70°.

One important conclusion remains to be noted: it is that at any temperature above that at which the availability of the component begins (22° C.) an indefinite number of equilibrium values of friction are possible for any one temperature, each depending only on the previous history of the system.

The form of the curves invites many questions which cannot be answered without more knowledge. Why has the initial state such a preponderating influence? If, as one is almost driven to admit, the molecules of wax initially in the adsorbed layer are centres of increased accessibility, why does the process of replacement not go on as it were autocatalytically until the limit is reached when the primary layer is wholly composed of wax? Saturation levels are relatively easily understood when the component is composed of polar molecules and the O curves are completely irreversible, but how a path of change fixed only by an initial state can coexist with complete reversibility is difficult to explain. On the simplest possible view

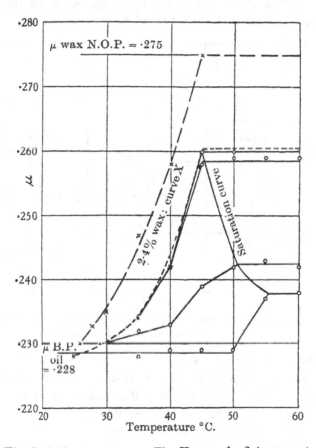

Fig. 3. 0·51 per cent. wax. The X curve for 2·4 per cent. wax is shown for reference.

a saturation level would be due to condensation of all the wax into the adsorption layer, but that layer must then be more than monomolecular, since a rough calculation, in which the smear of lubricant was taken at its thinnest, shows that such a layer could contain no more than 4 per cent. of the wax. An arithmetical explanation of all the facts could no doubt be framed in terms of the ratio between wax dissolved and adsorbed but that would miss the real interest, at any rate to biologists, which is the organisation of the whole of the oil (compare pp. 876, 886 and 888).

(ii) *Wax* 0·51 *per cent.* (Fig. 3.)

Curve *X* is less steep and the saturation level much lower, availability rising only to about 0·7. The temperature limits are, however, the same, namely, 23–45° C. The *O* curves show the same general features as with 2·4 per cent. wax, but the levels are lower.

Availability is maximal at 45° C. but, owing to the small quantity of wax, the concentration of the "active" molecular species is not high enough to displace completely oil B.P. from the primary layer.

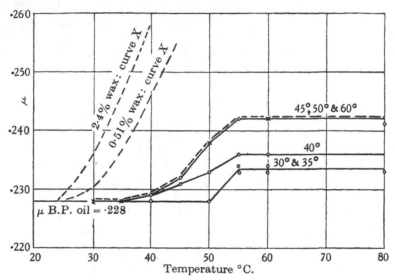

Fig. 4. 0·16 per cent. wax. The beginnings of the *X* curves for 2·4 per cent.
and 0·51 per cent. are shown for reference.

(iii) *Wax* 0·16 *per cent.* (Fig. 4.)

The saturation levels are still lower, that of the *X* curve having fallen to $\mu = 0·242$ when the availability is only 0·3. The temperature limits have changed, the curves being moved up the scale.

Comparing the three concentrations we get:

X curve

Wax %	Temperature limits °	Accessibility complete °	Friction range μ
2·4	23–45	30	0·228–0·275
0·51	23–45	40	0·228–0·260
0·16	40–55	45	0·228–0·242

The shifting of the curves to the right by a fall in total percentage of the wax raises new difficulties. Since solubility increases with rise of temperature the maximal dispersion of the wax would not be moved up the temperature scale by decreasing the total percentage, therefore a rise of temperature must do something more than merely break down complexes of molecules.

The effective concentration of wax depends to only a small extent upon the total concentration. This at any rate is the conclusion to be drawn from a detailed study of curve $O_{18°}$.

The percentage of wax was varied between the limits 0·13 and 3·06. The form of the curve was the same for all the mixtures and the curves practically coincided from about 0·5 per cent. upwards. Below 0·5 per cent. the saturation level falls and the temperature at which the primary layer is saturated with wax falls as the total concentration falls, but the change is slight. The temperature at which accessibility begins is sensibly independent of concentration.

TABLE II. *Initial Temperature* 18° C.

Wax %	Accessibility begins °	Saturation °	Saturation level μ
3·06			0·239
2·40		60	0·239
1·47	50		0·239
0·52			0·238
0·25		55	0·236
0·13		55	0·234

Enough points to fix the temperatures exactly were not measured.

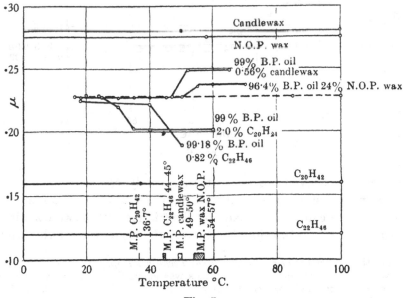

Fig. 5.

The temperature of access and indeed the whole group of curves are related closely to the melting point of the wax. The curve $O_{18°}$ was plotted for two waxes and two normal paraffins and Fig. 5 shows the close relation in each case between availability, accessibility, and the melting point.[1]

The broad conclusion is that the limits within which the composition of the adsorbed layer becomes a rapidly varying function of temperature are fixed mainly by the appearance in the mixture of highly dispersed molecules of the component, but this is not the only way in which temperature intervenes. The "active" species begins effectively to compete with the diluent at 22° C.

Example 2. Component, highly-polar, namely, Palmitic Acid. Diluent, non-polar Oil B.P. Melting point 63°. $\mu_{ac.} = 0$, $\mu_{B.P.} = 0.228$. (Fig. 6.)

At first sight nothing could be more different than the behaviour of this system. The curves in example 1 sweep upwards, in 2 downwards, the

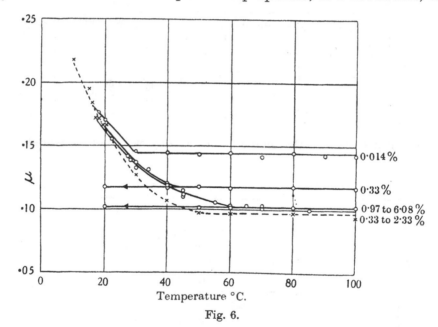

Fig. 6.

former are wholly reversible, the latter are wholly irreversible and yet there are fundamental similarities. In both it is obvious that the limit within which the curve can move is fixed by the friction of the two constituents. In both there are saturation levels, but the most remarkable and un-

---

[1] When the report (*Lubrication Research, Technical Paper No. 1*) already referred to was written the only curve known to us was curve O_{18} and the relation to the melting point led to the inference that it was a simple curve of availability. Curve X and the other curves show that the inference was wrong.

expected feature of this polar system is the fact that it is not possible, at any rate within the limits of concentration tried, to do what can be so easily done with the non-polar system, namely, to reach the limit at which the friction is that of the pure component. This is exactly what would not be expected from *a priori* considerations because it is the characteristic habit of polar molecules to seize the whole of the adsorbed layer.

The mixtures were made as before by dissolving the acid in the oil at a little above the melting point of the former. The apparently homogeneous solutions were then kept at room temperature for 24 hours. From 0·97 per cent. upwards some crystals of acid separated out at 18°.

The possible limits for the curves are $\mu_{B.P.} = 0.228$ and $\mu_{ac.} = 0$, but the latter, as Fig. 6 shows, is never reached.

Curve X. Availability begins at about 9° and is greatest at about 50° or rather less, the observations not being close enough to fix the point exactly. The form of the curve is the same for mixtures containing from 0·33 to 2·33 per cent. of acid. It is therefore not sensibly affected by the composition of the mixture. The melting point of palmitic acid is 62°, at which point oil and acid seem to be miscible in all proportions, availability is, however, greatest below this point. The curve shows that whilst availability can be zero at $\mu_{B.P.} = 0.228$, it cannot be pushed to unity at $\mu = 0$. Instead a saturation limit is reached at $\mu = 0.097$, which is independent not only of a rise of temperature but also of the percentage of acid in the mixture. This curious limitation was not imposed by the particular diluent for it appeared when the wax N.O.P. was so used. The values were:

Diluent	% acid	Saturation level
B.P. oil	0·97–6·08	=0·097
Wax N.O.P.	0·33 to	−0·090
	2·2	=0·089

Curve X is completely reversible in the special sense mentioned on p. 870. It is so important to realise what this special sense is that it will be well to put it into quite other words. In curve X the steel face as an adsorbing surface is used to measure the state of the mixture in bulk and the variations of state with temperature were found to be reversible.

Curve O. It was unfortunately difficult to hold temperatures below room temperature with the apparatus, for this reason the effect of varying the initial temperature was not determined. The effect of varying the concentration of the wax was, however, followed somewhat closely for the curve $O_{18°}$. Mixtures containing the following percentages by weight of acid were used: 0·014, 0·33, 0·97, 1·5, 2·41 and 6·08. In order to make the figure legible the readings for 1·5 and 2·41 per cent. are omitted. The general form of the curve is the same for mixtures containing from 0·014 to 6·08 per cent. acid and the curves coincide from 0·97 to 6·08 per cent.

Accessibility is slightly less than availability throughout. The saturation level varies widely with changes in the composition of the mixtures up to 0·97 per cent. and then becomes independent of composition. Between the limits 0·0 and 0·97 per cent. there is an infinite number of curves, the saturation level falling and the saturation temperature rising with increase

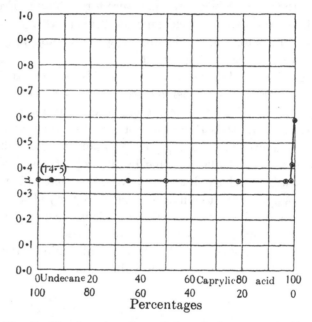

Fig. 7. Showing the effect of the relative concentration
of acid and paraffin upon friction.

in the percentage of wax. The curves were completely irreversible, that is to say, once the saturation level was reached, the curve for falling temperature was a horizontal line.

When we recollect that mixtures containing less than 0·97 per cent. seemed homogeneous to the naked eye throughout the range of temperature, it seems to follow that availability is limited to a particular species of molecule, the concentration of which increases as the temperature rises. The most obvious suggestion again is that this molecular species is the completely dispersed unassociated molecule and that the concentration of the available or "active" species increases with rise of temperature.

A saturation level stopping short of complete replacement is indeed an anomaly when strongly polar molecules are concerned. For example, the isotherm of a mixture of caprylic acid dissolved in undecane is given in Fig. 7 and it shows that the non-polar molecules do not obtain access to the dissolved layer until the percentage of acid has fallen to less than about 0·7 per cent. For concentrations higher than this the friction is the same as that of the pure acid. The same kind of curve is given by mixtures of paraffins and alcohols. In Fig. 8 the isotherms for solutions of palmitic acid in oil B.P. are reproduced from Technical Paper No. 1 of the Lubrication Committee, and it is remarkable that each has the general form of an adsorption curve, but with the unusual feature that the saturation limit for the acid never reaches 100 per cent. and is a function of temperature.

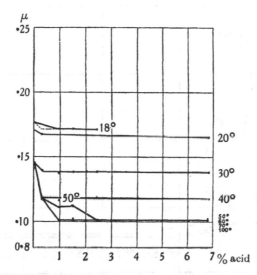

Fig. 8. From *Lubrication Research, Technical Paper No. 1*,
by courtesy of the Controller, H.M. Stationery Office.

The hypothesis that availability is determined by the degree of dispersion of the acid opens up an interesting possibility. Availability reaches its limit somewhat short of the melting point of palmitic acid when dispersion is likely to be maximal. The limit is, however, not unity, that is to say, the acid fails completely to displace the diluent in the primary layer. It looks as though the highly-polar molecules acted in the opposite way to the non-polar molecules of wax by increasing the hold on the steel face of the molecules of diluent in their neighbourhood. This is intelligible if each polar molecule acts as a dipole which induces polarity in its neighbours. It will be remembered that Sutherland based his electrical theory of cohesion in liquids on such induced polarity.

Availability may, however, be fixed by secondary adsorption. It is obvious from mere inspection that solutions of polar molecules in oil B.P.

H   56

when in contact with the steel are very different from the mixture in bulk. The oil ceases to be uniform, it readily breaks up into greasy-looking patches. When the component is an alcohol there is the same gross interference with the state of the mixture. No such gross disorganisation of the oil, however, was noticed in the absence of polar molecules.

Example 3. $\mu_{\text{al.}} = 0.114$. *Component polar, namely Cetyl Alcohol. Melting point* 50° C. *Diluent, Oil R.P.* $\mu_{\text{B.P.}} = 0.228$. (Fig. 9.)

Two mixtures only were used containing respectively 0·39 and 2·20 per cent. of the alcohol. The limit of the temperature range explored was

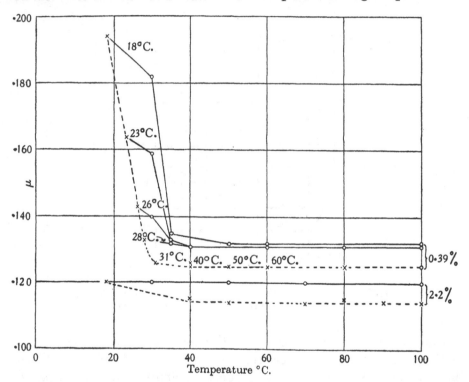

Fig. 9. The numerals indicate initial temperature. The O curves from 31° upwards coincide with each other and with the X curve. To avoid confusion, the individual measurements are not marked.

18–100°. The 2·20 per cent. mixture was rather jelly-like at room temperature (18° to 20°) and on standing showed some tendency to deposit crystals and become more fluid.

When put on the plate at any temperature below about 60° a visible thick film formed which dulled the highly polished surface of the steel. On this film droplets of liquid floated. The film was there above 60°, as the presence of droplets showed, but it was transparent.

When a thin layer of solid lubricant, which is also a single chemical individual, is deposited on the plate from solution in a volatile solvent it is necessary to break away the upper crystalline crust by moving the spherical slider about slowly in order to get the steady normal[1] readings of the primary film.[2] The way in which the slider had to be moved about to get steady readings gave the impression that the film mentioned above had sufficient thickness and tenacity to hinder it from penetrating to the primary layer.

Since the friction of the pure component was below that of the diluent the curves sweep downwards. All the curves were irreversible.

X curve. This curve was plotted between the limits 18–100° for both mixtures. Availability reached unity with the 2·2 per cent. mixture, but failed to do so with the 0·39 per cent. mixture, the saturation level being at $\mu = 0·125$.

O curve. 0·39 per cent. mixture. Within narrow limits the saturation level depends upon the initial temperature and therefore upon the initial proportion of alcohol in the primary layer. Accessibility is maximal at about 30°, the curves $O_{30°}$ and $O_{40°}$ coinciding with the X curve.

2·2 per cent. mixture. The curve $O_{18°}$ is singular in that it lies throughout at a saturation level. Accessibility is independent of temperature and the increase in availability to unity at about 30° is without effect.

These curves again raise the old difficulty. If the spherical slider really reduces the lubricant to two monomolecular layers, they must have contained about 6 per cent. of diluent at 18° when the total percentage of alcohol was 2·2. But since the O curve is horizontal the rise of availability in the oil in bulk to unity at 30° did not dislodge the diluent, possibly because continuous contact with the steel face so changed the whole pool of oil as to prevent availability reaching unity.

Lowering the total percentage of alcohol to 0·39 shifts the curves to the right, but if the maximum of availability were due only to a maximum degree of dispersion one would expect it to be reached with the lowered concentration of alcohol either at the same or even at a lower temperature; therefore a rise in temperature must act in some other way. It is, of course, possible that the slider fails to penetrate the tenacious film sufficiently to reach a monomolecular layer, and the difference between the curves for the 0·39 and the 2·2 per cent. mixtures may be due to the fact that they refer to surfaces at different levels.

---

[1] "Normal" because they are independent of further movement of the slider, and because they fall on the curve connecting molecular weight and friction.

[2] Hardy and Doubleday, *Proc. Roy. Soc.* (1922), A, c, 559; Hardy and Bircumshaw, *Proc. Roy. Soc.* (1925), A, cviii, 20.

Example 4. *Components, Palmitic Acid and Wax N.O.P. Diluent, Oil B.P.*
(Fig. 10.)

Systems with two active components do not really come within the
limits set to this enquiry, but one such is worth mentioning because it confirms the distinction between availability and accessibility. Wax N.O.P.
was added to a mixture of oil B.P. and palmitic acid containing 0·34 per
cent. of the latter. The only effect was to lower the friction on the X curve,

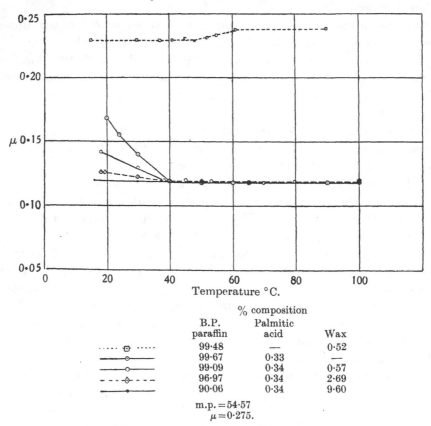

	% composition		
	B.P. paraffin	Palmitic acid	Wax
····· ⊡ ·····	99·48	—	0·52
——o——	99·67	0·33	—
——o——	99·09	0·34	0·57
– – ◇ – –	96·97	0·34	2·69
——●——	90·06	0·34	9·60

m.p. = 54·57
$\mu = 0.275$.

Fig. 10. From *Lubrication Research, Technical Paper No.* 1,
by courtesy of the Controller, H.M. Stationery Office.

leaving both the saturation temperature (40°) and saturation level of the
O curve ($\mu_{sat.} = 0.118$) unchanged, Fig. 10. The availability of the acid at
the initial temperature (18° to 20°) was therefore increased until when
9 per cent. wax was present it was maximal, *i.e.*, it had the saturation value.

Wax itself, however, did not enter the adsorbed layers at all and the
curves were completely irreversible.

Palmitic acid is soluble in paraffin wax and the presence of the latter
increases the solubility of the acid in B.P. oil. There can therefore be little

doubt that it increased the availability of the acid by increasing the grade of dispersion at the lower temperatures but without altering the saturation level.

From these examples of simple adsorption we may conclude:

(1) That only in a highly dispersed state can a component compete with the diluent for a place in the primary layers, and the limits within which adsorption varies rapidly with temperature are fixed mainly by the appearance in the oil of a particular molecular species of the component, possibly a simple chemical molecule, and by the rise in concentration of this species until it obtains complete possession of the primary layer or is maximal for the total concentration of component.

(2) But temperature must intervene in another, as yet unknown, way.

(3) The simple two-dimensional hypothesis which considers only a monomolecular adsorbed layer in equilibrium with liquid in mass is not enough, the conception must be three-dimensional and take account of the obvious gross reorganisation of the whole pool of oil which contact with the steel face brings about. This reorganisation must be due to polarisation which spreads from molecule to molecule along the normal from the steel face, it is the kind of polarisation which one of us has called "diachysis".[1] It may also be necessary to admit a polarisation which spreads from molecule to molecule tangentially.

(4) The first assumption (p. 867) may be too narrow. Even a spherical slider may sometimes fail to penetrate to monomolecular layers.

Greasy patches. When either of the mixtures which contained polar molecules was spread on the plate patches of higher viscosity formed. The higher the temperature and the greater the percentage of the polar component the sooner did the patches appear. Their formation was hastened by moving the slider about—that is to say by stirring the mixture on the plate.

With hexa-decyl alcohol the patches tended to take the form of lenses on a continuous film which below about 60° was thick enough and opaque enough to dull the burnished surface of the steel. Above 60° the film was not visible, but it was there as the presence of the patches proved.

The phenomenon is without doubt related to the instability of a layer of immiscible liquid spread upon water.

When the slider was moved through a grease patch it left a track or groove owing presumably to the high viscosity.

No non-polar mixture has been found to form grease patches.

Latent period. The duration of the latent period, that is the time taken by the oil to get into equilibrium with the steel, was shortened by moving the slider about. The time was undoubtedly related to the gross changes in the

[1] Hardy, *Phil. Trans.* (1932), A, ccxxx, 30.

oil which developed the patches. It is important to note that mechanical agitation shortened the latent period of non-polar mixtures in just the same way therefore, though there was no formation of obvious grease patches there undoubtedly was a reorganisation of oil in the pool.

The latent period of non-polar mixtures was quite as long as that of polar mixtures and in this connection it is to be remembered that while neither wax N.O.P. nor oil B.P. alone had any latent period the polar bodies, palmitic acid and hexa-decyl alcohol taken alone have each a latent period of orientation of 60 and 40 min. respectively.

The latent period, the grease patches and the rise in viscosity taken together do, in our opinion, prove that contact with the adsorbing surface produced molecular structure throughout the whole pool of oil. It is for this reason that the hypothesis of a monomolecular adsorbed layer in equilibrium with oil in mass is too narrow, secondary adsorption must be reckoned with.

On the O curves the latent period of the non-polar mixture for rising temperature was one of rising friction and the opposite for falling temperature; it lasted about 20 min.

With the two polar systems the latent period for rising temperature was always one of falling friction and lasted 20–50 min. depending upon concentration and temperature. But for falling temperature there was no latent period because there was no change in the composition of the primary layer. It will be remembered that once a saturation limit was reached the friction became constant over the whole range of temperature.

Movement of the slider was strikingly different with polar and non-polar mixtures. With non-polar mixtures it was jerky, the joint was brittle the slider breaking away suddenly and chattering forward. The property called "oiliness" was absent at any rate for mild steel. Movement with the polar mixtures was a smooth glide.

Owing to the brittle nature of the joint it was difficult to get readings with the non-polar mixtures. They were very sensitive to vibrations.

PART II. THE FUNDAMENTAL ASSUMPTION

The effect of applying a lubricant to a solid face is to reduce the friction of the latter. This is readily perceived when two solid faces are coated separately with *solid* lubricant and then brought together. This "external" friction has the same value as the internal friction of a liquid joint when the same lubricant is used at a temperature at which it is solid and at a temperature at which it is liquid.

The equation quoted on p. 868 shows that the observed value of static friction, that is of the greatest tangential reaction to traction, of which the surface of slip is capable, depends upon the nature of the solids. Adhesion, that is the greatest normal reaction to a surface of break to tension, also depends upon the nature of the solids and it can be proved that the solids do not contribute directly but by polarising the molecules of lubricant.[1]

Similar proof for the tangential reaction is not forthcoming but it can safely be assumed that what is true of the normal is true also for the tangential reaction. It is assumed therefore that the tangential reaction is the sum of reactions between individual molecules of lubricant acting across the surface of slip.

This assumption offers the simplest explanation of the following facts.

When the lubricating film is deposited from vapour the tangential reaction is a linear function of the density of the vapour until the latter is saturated when the value is the lowest possible.[2] This seems to prove that the adsorbed film in equilibrium with saturated vapour is a close packed structure.

A single chemical substance has no temperature coefficient of friction, the close packed structure of the primary layer being independent of temperature.

No adsorption individual has been met which has a temperature coefficient of friction.

The form of the X curve which lies between limits defined by the friction given by the pure constituents of the mixture shows a gradual replacement of one kind of molecule by another in the adsorbed layers as temperature changes.

It follows that throughout a plateau in a curve the primary layer must have the same composition, whilst ascending or descending parts of a curve show that its composition is changing.

SUMMARY

(1) The variation of static friction with temperature, of three solutions in medicinal paraffin was measured, the solutes being a wax, palmitic acid and hexa-decyl alcohol.

(2) Variation was limited to a curiously narrow range of temperature outside of which friction was invariant. The limits of variation were the friction of the pure solvent, called in the text the "diluent", and that of the pure solute, friction therefore could be used roughly to measure the proportion of solvent and solute molecules in the adsorbed layer.

(3) For a given temperature the friction and therefore the composition of the adsorbed layers are determined by previous history and the composition of the solution.

---

[1] *Lubrication Research, Technical Paper No.* 1, p. 29, H.M. Stationery Office (1929).

[2] Hardy and Doubleday, *Proc. Roy. Soc.* (1922), A, c, 556.

(4) Two changes govern all the phenomena—change in state of the bulk solution which fixes the *availability* of the solute for adsorption and a change in the capacity of the solute for replacing adsorbed solvent, which is the *accessibility* of the solute.

APPENDIX

Note on the Latent Period of Lubrication

The latent period during which the lubricant is coming into equilibrium with the attraction fields of the solid is important in practical boundary lubrication, indeed there are some indications that the greater part of the wear of bearings occurs in it.

The phenomena of the latent period are of the simplest when the lubricant is a single pure long chain compound. Variations of friction which precede the steady state may be due either to variations in the thickness of the layer of lubricant or to orientation of its molecules. The former can be eliminated by using a spherical slider which cuts through at once to its final level. When this is done the duration of the latent period is, in minutes:

Normal paraffins	...	...	...	0
Normal alcohols	...	...	...	} about 40
Carbinols	...	...	...	...
Normal acids ...	...	...	...	,, 60

and friction always falls as orientation completes the adsorbed layer.

Adhesion moves in the opposite way to friction, it increases as the steady state is reached.

The only ring compounds examined, however, namely, phenanthrene and naphthalene, gave a latent period of about 40 min. with falling adhesion. This is at one with a curiously opposed behaviour as between ring and chain compounds noted in an earlier paper on friction.[1]

These simple relations recall, and are akin to, the simple behaviour of chemical individuals on a surface of water, glass or metal. All surface relations become excessively complicated, however, when a mixture is used unless it happens to be an adsorption individual. The spreading of a complex oil on water is an example. There is not only the lag due to the slowness of tangential diffusion as was first pointed out by Willard Gibbs but there is also the slow exchange by evaporation and condensation between the first formed adsorption layers and the body of the mixture.

The mixtures dealt with in the foregoing paper are simple but the phenomena of the latent period are characteristically complex. There is a latent period of orientation for both non-polar as well as polar mixtures.

[1] *Phil. Mag.* (1920), (6), XL, 201.

When the oil is first put on the plate the friction is abnormally high being outside the limits always observed in the steady state. The friction therefore falls at first and the rate of fall is increased by moving the slider about. The effect of such mechanical disturbance has been shown by X-ray analysis to hasten the orientation of long chain molecules in layers with the chain more or less at right angles to the solid surface.[1] During the latent period the readings are not only high but irregular and the slider breaks away much more jerkily than it does in the steady state. The following figures are quoted merely by way of illustration.

Temperature 24°. Steel on steel, slider spherical. Lubricant 2·3 per cent. paraffin wax N.O.P. dissolved in oil B.P.

Time	Interval	μ
h. m.		
4 36	0	—
4 47	11	0·305
5 60	24	0·291
5 9	33	0·239 ⟩Latent period
5 19	43	0·260
5 30	54	0·247
5 41	65	0·261
5 49	73	0·226
5 58	82	0·229
6 3	87	0·229 ⟩0·228
6 9	94	0·230
6 14	99	0·228

Temperature of chamber raised to 50° by 6 hours 25 min.

Time	μ
h. m.	
7 1	0·244
7 12	(0·239)
7 24	0·244 ⟩0·244
7 30	0·243
7 42	0·244

The low reading in brackets was due to accidental vibration. The limits within which the curve for steady states varies are $\mu_{B.P.} = 0·228$ and $\mu_{N.O.P.} = 0·275$, therefore the first two readings in the latent period, which occupied about 70 min., were above them, and such abnormal readings were commonly obtained.

[1] Bragg, *Proc. Roy. Inst.* (1925), XXIV, 481.

58

THE FREEZING POINT OF YOLK AND WHITE OF EGG

By H. P. HALE, with a Note by Sir WILLIAM HARDY, F.R.S.

[*Proc. Roy. Soc.* (1933), B, cxii, 473]

For the purpose of this note the structure of a hen's egg may be taken to include, in order, a shell, thin white, and thick white enclosing a yolk, which is covered by the thin vitelline membrane.

Under the microscope this membrane is seen to be composed of fibres felted together as in white fibrous connective tissue.

METHODS

Hens' eggs, not more than 24 hours old, were placed in the bath overnight to get into approximate temperature equilibrium. The entire contents of an egg or such portion as was needed for an experiment was then broken into a small metal pail, about 75 c.c. capacity, which was at once covered with a lid lined by wet filter paper to prevent evaporation. The pail was sunk in a bath at a known temperature and left in the earlier experiments for 24 hours and in later experiments 6 hours, before being seeded with ice.

After an interval which varied in different experiments from 2 to 20 days the contents of the pails were searched for signs of ice.

The temperature of the bath was controlled and was placed in a room kept at $-1.0°$ C. by a thermostat. Temperatures were steady to within one-hundredth of a degree Centigrade. The results of the experiments may be summarised as follows:

(a) At -0.54 and $-0.56°$ C. No ice formed in the yolk when the vitelline membrane was intact, or after it had been pierced by a needle of ice or of glass, or grossly injured by rolling on filter paper to dry the surface.

At -0.58, -0.60 and $-0.63°$ C. No ice formed when the vitelline membrane was intact but in about half the experiments the yolk froze after the membrane had been injured or pierced.

It was observed in these experiments that drying by rolling on filter paper produced general injury so that ice formation in the yolk was scattered over the surface, whereas after puncturing it started only from the injury.

At $-0.65°$ C. The yolk froze after injury to the membrane but it was doubtful whether ice penetrated the intact membrane.

At -0.68 and $-0.70°$ C. Ice undoubtedly penetrated the intact membrane.

(b) All the ice in a well-frozen yolk was completely thawed at $-0.50°$ C. and so far as one could see by inspection at $-0.56°$ C.

(c) At none of the temperatures did ice succeed in passing from thin to thick white, instead the latter was dehydrated by water passing across the boundary to the ice column in the thin white. Moran (1926) likewise found ice formed only on the surface of disks of gelatine when cooling was slow.

When seeded separately thin and thick white have the same freezing point, namely, $-0.42°$ C.

Therefore the freezing point of yolk may be taken as $-0.57°$ C., and that of white $-0.42°$ C. Smith and Shepherd (1931) using the Beckmann technique found for eggs 5 days old -0.58 and $-0.46°$ C. respectively.

DETAILS

Temperature $-0.54°$ C.

(1) An entire egg was broken into each of three pails, and the thin white seeded. After 5, 10 and 20 days the thin white alone was frozen; no ice in the yolk.

(2) The yolk only was placed in each pail and two small areas of its surface dried with filter paper and brushed with a 1 per cent. solution of either nicotine, pilocarpine, or dial (allyl-barbituric acid), or faradised through naked platinum wire. The yolks were then just covered with thin white which was seeded. After 11 days there was no ice in the yolks.

(3) The yolk was floated in a minimum of thin white so as to leave a small circle of the vitelline membrane exposed. After being in the bath for 24 hours the yolk was seeded through the membrane by a fine needle of ice. After 11 days the white was frozen; the yolk unfrozen.

(4) The same as the last, but to avoid any chance of drying, the yolk, after seeding, was covered with a thin plate of ice. After 11 days the white was frozen; the yolk unfrozen.

(5) The same, but the vitelline membrane was pierced with a fine glass needle and then covered with an ice-plate. Result the same.

(6) Moran (1926) found that ice formed only on the outside of a disk of gelatine gel when the rate of cooling was slow, the gel having lost water during the process. To avoid this phase relation each yolk was dried by rolling on filter paper, and then floated in its pail in liquid paraffin so as to leave a small area uncovered. Twenty-four hours later this area was seeded by being covered with a thin plate of ice. After 11 days there was no freezing.

Controls. Broken yolks seeded did not freeze at this temperature.

Temperature −0·56° C.

(7) Experiments (4), (5) and (6) repeated with the same result.

Temperature −0·58° C.

(8) Four pails with yolks dried on filter paper, then floated in liquid paraffin and the exposed part covered with a plate of ice. After 18 days two yolks contained ice; two were unfrozen.

(9) Two pails with yolks floated in a minimum of thin white. Yolk was seeded through the membrane with an ice needle and then covered with an ice-plate. After 18 days one yolk had ice-crystals, one was unfrozen.

(10) Two pails as in (9) but the yolk was pierced with a needle of glass. After 18 days both were unfrozen.

Therefore at −0·58° C. such hindrance as there was to ice formation was sometimes overcome and ice-crystals formed within the yolk. It was observed in these and other experiments that rolling on filter paper seriously injured the membranes so that ice formation was scattered over the surface whereas with puncturing it started only from the injury.

Controls. Broken yolks seeded froze partially at this temperature.

Temperature −0·60° C.

(11) Same procedure as in experiments (8), (9) and (10). After 10 days two, out of the four yolks floated in paraffin, and one, of the four floated in thin white, were frozen.

Temperature −0·63° C.

(12) The entire egg was broken into a pail and seeded in the thin white. After 20 days the thin white was frozen, the thick white and yolk were unfrozen.

(13) The yolk floated in a minimum of thin white which was seeded. After 17 days the white was frozen, the yolk unfrozen.

(14) The same as (13), leaving a small area of yolk exposed which was covered by a plate of ice. After 7 days, in one pail yolk and white were frozen, in the other the white only was frozen.

(15) Various trials with yolks floated in thin white or in paraffin and injured either by rolling on filter paper or by piercing the membrane with a needle of ice or of glass. The small exposed area of yolk was covered with a plate of ice. After 10 days all were well frozen.

Temperature −0·65, −0·68, −0·71° C.

(16) The same result as (15), although when the thin white was seeded the ice did not penetrate the thick white.

To sum up. Ice was found for the first time in yolks which had suffered mechanical injury to the vitelline membrane at −0·58° C. and never at

$-0.56°$ C. The freezing point of the yolks examined, therefore, lay between these.

A converse experiment was then tried to determine at what temperature ice, already present, disappeared from the interior of the yolk.

Ten intact yolks were placed in pails containing a little liquid paraffin, one in each. After 24 hours each was seeded at $-0.60°$ C. by puncturing and introducing a small crystal of ice. Each yolk was then completely covered with liquid paraffin to prevent evaporation, all the operations being carried out at $-0.60°$ C. Controls with yolks broken into the pails were well frozen next day, but the others were kept for 8 days when each was seen to contain a quantity of ice.

The temperature of the bath was then raised to $-0.56°$ C. for 3 days when all the ice seemed to have gone. To make sure, however, the temperature was raised to $-0.50°$ C. and kept there for 5 days. Careful examination of some of the yolks with a lens failed to reveal ice. To prove its absence the temperature was lowered to $-0.63°$ C. and kept there for 8 days at the end of which time no freezing of the yolks had taken place. The capacity for such overcooling proved the complete absence of ice. Therefore ice could not exist in these yolks at $-0.50°$ C. and probably not at $-0.56°$ C.

THE FREEZING POINT OF EGG WHITE
(1) *Temperature* $-0.41°$ C.

In each of four pails the thin white from an egg was placed and in four other pails the thick white. After 24 hours each pail was seeded and the white then covered with liquid paraffin. No ice had formed in any pail after 6 days.

(2) *Temperature* $-0.42°$ C.

The same procedure. After 6 days, 3 out of 4 pails with thin white and 2 out of 4 with thick white contained ice.

(3) *Temperature* $-0.45°$ C.

The same procedure. After 7 days the whites in all the pails were well frozen.

The freezing point of both thin and thick white therefore lies between -0.41 and $-0.42°$ C.

It is important to note that in all the above experiments the time in days at which a certain change is noted does not mean that the change, *e.g.*, freezing, had not taken place earlier: it means simply that for one reason or another it had not been thought advisable or had not been possible to examine the pails earlier.

Note on the foregoing by Sir WILLIAM HARDY, F.R.S. Recent work on the osmotic pressure of the hen's egg has introduced a sense of uncertainty as to the value of the many comparisons which have been made between osmotic pressures of the blood, body fluids, and surrounding media. The uncertainty pertains not to theory but to a simple matter of fact and, as this involves that most fundamental datum for biological theory—viz., the state of the water in the living cell—there is urgent need to have it cleared up. The fact in dispute is the freezing point of the yolk and white of the bird's egg. Atkins in 1909 by measurements, obviously made with the greatest care, found "no difference between the freezing point of white and yolk of the same egg and a mixture of white and yolk gave the same depression".

Atkins (1909) used the ordinary Beckmann technique and so, too, did Straub (1929) twenty years later, but with a surprisingly different result for he found a constant difference between white and yolk of the hen's egg amounting on the average to $-0.15°$ C. A. V. Hill (1930) confirmed Straub's (1929) finding by a different method. He compared the fall in temperature caused by evaporation with that of water and from the difference calculated the osmotic pressure. Howard (1932) using the Beckmann method again found no difference in the freezing point of white and yolk. In these measurements the yolk was puddled by stirring so that at sometime or another the structure was broken down. Yolk is not only a chemical complex but it is alive, gross mechanical disturbance might, therefore, have the effect it usually has on living cells and cause chemical breakdown with consequent fall of the freezing point. Hale's experiments were designed to explore this possibility by observing directly the freezing point of intact yolk and white.

The difficulties in the way are obvious and not easily overcome. Consider a yolk covered with white with a column of ice, started in the white by seeding, approaching the yolk. When the ice meets the vitelline membrane one of two things may happen: ice may penetrate the pores of the membrane to seed the yolk, or water may move from the yolk to the membrane to condense on the ice face. The second possibility follows from Moran's (1926) observation when he found that ice formed wholly on the surface of disks of gelatine gel, provided the rate of cooling was slow and that a true reversible equilibrium obtained between gel and ice over a large range of temperature. This must happen if the surface separating ice and gel is plane for, the specific volume of ice being greater than that of water, the formation of ice within the gel would be resisted not only by the positive energy of the new interface, but also by the general cohesion of the gel.

A curious feature about phase lag in the liquid or gel state is the existence of a time limit to overcooling. Why should overcooling, when once established, not be permanent? The answer, I think, introduces another cause of phase lag, namely, the time taken to orient complex molecules at the new interface. This seems to be the simplest explanation. It takes an hour for a monomolecular film of a fatty acid adsorbed on to a plane solid face to reach equilibrium at ordinary temperature. It is not surprising, therefore, that the orientation needed for the formation of a new interface can take days when the temperature is below the freezing point.

As a minor point Hale's experiments show that the intact vitelline membrane permitted ice to pass to the yolk at $-0.65°$ C. Taking the freezing point of yolk at $-0.57°$ C. it was able to resist no more than $-0.08°$ C. of overcooling. But "intact" yolk must not be taken to mean a yolk with a wholly uninjured membrane for in the act of breaking the yolk into a pail the membrane was obviously violently stretched.

The existence of a difference in the freezing point of yolk and white is fully confirmed and that implies a difference in osmotic pressure, but there is no difference in hydrostatic pressure greater than a delicate, elastic, though possibly tough, membrane such as the vitelline membrane can bear.

The clue to this paradox seems, however, to be simple. It lies in two things: the hindrance to the movement of water in yolk and thick white and the slow leakage of water through the vitelline membrane which, as eggs are, lessens the difference between the *average* osmotic pressure of yolk and that of the white. It is as though there were a number of chokes between a household water tap and the reservoir which together reduced the flow to a trickle. So long as that trickle is unimpeded the pressure difference at the tap is negligible. Try to stop it with the thumb, however, and the whole pressure from the reservoir has to be held back.

I suggest that there is in the egg no sharp change of pressure at either of the surfaces of the vitelline membrane, but there is a gradient from the centre of the yolk outwards of varying, but always slight slope which ends at a surface whose position depends upon the rate of evaporation from the shell.

Evidence for a hindrance to the movement of water can be found scattered through the paper by Smith and Shepherd (1931), in Moran's (1925) observation that yolk can be readily overcooled—even broken yolk can be held at $-7.4°$ C. and in the various phase lags revealed by Hale's experiments.

REFERENCES.

ATKINS. *Proc. Roy. Dublin Soc.* (1909), XII, 123.
HARDY. *Proc. Roy. Soc.* (1926), A, CXII, 47.
HILL, A. V. *Proc. Faraday Soc.* (1930), Symposium.
—— *Proc. Roy. Soc.* (1930), B, CVI, 477.
HOWARD. *Journ. Gen. Physiol.* (1932), XVI, 107.
MORAN. *Proc. Roy. Soc.* (1925), B, XCVIII, 436.
—— *Proc. Roy. Soc.* (1926), A, CXII, 30.
SMITH and SHEPHERD. *J. Exp. Biol.* (1931), VIII, 293.
STRAUB. *Rec. Trav. Chim. Pays Bas* (1929), XLVIII, 49.

59

TO REMIND. A BIOLOGICAL ESSAY
THE ABRAHAM FLEXNER LECTURE, 1931

[Published by the Williams & Williams Company, Baltimore, 1934]

CHAPTER I

It is so near the date that I can almost claim to be addressing you on the hundredth anniversary of one of the half dozen greatest syntheses of knowledge. In 1835 Dujardin gave the name "sarcode" to a kind of plastic homogeneous substance, which he had first identified in his study of the simpler microscopic types, and since convinced himself of its presence in all animals, "at any rate in the young state".

This great generalisation—and there can be none greater—for every biologist since his time has built upon the unifying principle that all living things, no matter how diverse their forms, are ultimately composed of a "glutinous" substance which recognisably is the substance of life has been overshadowed by that other great generalisation, the cell theory of Schleiden and Schwann. Even his word "sarcode" has been forgotten in favour of "protoplasm", the name which Mohl gave to the "slimy granular semi-fluid content of plant cells". The cause is not far to seek. Mohl was a botanist, and whereas botanists continued actively to occupy themselves with the structure of protoplasm, animal physiologists dropped the subject for thirty years.

Their attention was held by other things, for in the 'thirties and 'forties, Liebig was battling with Dumas and his school for a recognition of the wider chemical potentialities of animals, and, with his life-long friend Wöhler, for the patient application of the methods of physics and chemistry to the problems of life in place of an appeal to a "vital force".

About two years ago, I had occasion to read Dujardin's papers, and I then conceived so great an admiration for him that I will ask you to let me preface these lectures by an attempt to give him his due.

He was a histologist, to use a word not invented in his time, and unlike many histologists, who perhaps as a clan have not always been as critical as might be, he questioned the reality of the appearances revealed by his microscope. "Are the 'vacuoles'", he asked, "real spaces filled with fluid or not?" and to settle the matter he studied the appearance of emulsions of oil and of air in water, the emulsifying agent being a little saliva.

The capacity for independent movement was then the index of life, and this capacity so puzzled and misled his contemporaries, amongst them the great Ehrenberg, that, like Sterne, they conceived of the simplest organisms as equipped with stomach, muscles, skin and fibres (nerves), even though these were quite imperceptible. Ehrenberg indeed urged that it was simpler to assume the presence of organs than to account for the power of movement.

Dujardin curtly styled this an abuse of the argument by analogy. Even if the existence of such organs were accepted, the difficulty was not disposed of, for muscle, nerve, etc. must themselves be resolvable into an elementary substance with power of movement. No casuistry, he says, will avail against this "dernier terme".

I should note by the way that he had a clear idea of the presence of cells, for he speaks often of "le tissue cellulaire".

Of course his argument has its logical weakness, there will always be a *dernier terme* or, as du Bois Raymond put it, nearly half a century later —to satisfy our craving for causality we have, as ultimate measure, only certain unknowns—these are the limits of our knowledge, and it is not possible to explain a limit just because it is a limit. But, "that fundamental law of human reason", as Schleiden calls "its undeviating tendency to unify knowledge", will not even then be satisfied—indeed it is less satisfied than ever just now when our scientific philosophers are hesitating whether to be dualists with Plotinus and the Alexandrian school and try to balance two oranges on the tips of their noses, or monists with Empedocles and Hæckel and be content with one.

Dujardin left movement a mystery. "When", he says, "one comes to consider nutrition, which is another form of movement not less real and altogether incompatible with the idea of equilibrium, or the constitution and mode of aggregations of the molecule which compose the living substance, we have to admit ourselves beaten by the difficulties because the problem is that of life itself, which for us will be always insoluble." Those are memorable words.

After Dujardin came a barren period so far as the study of the living substance was concerned, indeed for the English-speaking world at any rate the greatness of his generalisation remained unrecognised until Huxley's great essay of 1868 on the Physical Basis of Life.

Like all barren periods it was a time of easy certainties and decisive terminology. Then were there "proximate principles" which were "organised" in contact with living matter. Of these principles the most important was fibrin of the blood, which was therefore sometimes called the general formative element or "blastema". Lecturing must have been an easy business in those days!

It was, however, a time by no means barren in other fields. Liebig, Lawes and Gilbert, Voit, Fick and Wislicenus, Subbotin—these names bear witness to the contrary.

If one were to try to follow the developments from, say, 1870 onward—

and there is no time even to attempt a sketch—one would have to deal separately with structure and function. I shall have to content myself with a brief picture of what was taught in active schools of physiology in my time. The date would be roughly in the 'eighties, and the teacher was Michael Foster.

Living matter was then an entity which took part as a whole in the chemical cycle. For example, the energy of contraction of a muscle fibre was derived from the explosion of a substance to which Herman gave the name "inogen", but the whole living unit partook in the explosion, not merely a non-nitrogenous part. The nitrogenous part was, however, retained in the structure of the fibre. It became irrevocably involved only when the explosion became irreversible in rigor mortis.

Metabolism—that is the whole chemical cycle—consisted, therefore, of a phase of increasing molecular complexity in which protein, fats and carbohydrates with oxygen were built up into the living substance, and a phase of decreasing chemical complexity and liberation of energy.

The picture of two operations, anabolism and katabolism—or loading and discharge—was based in the first instance on the properties of muscle, especially on its capacity for doing work when not supplied with food or even oxygen, but it received immense support when the processes of loading and discharge of gland cells was discovered in the late 'eighties. It has that support still.

The gland cell is continuously engaged in forming the special chemical substances and storing them away in its substance as "secretory granules". At intervals, provoked thereto by nervous impulse or other form of stimulus, it becomes active, two separate events happen, water and salts are passed through its substance and some of the granules are discharged and the solution so formed is the "secretion" of the gland. There the process of loading and discharge of the muscle fibre is paralleled.

Surface forces of an unknown nature are not wanting, for the secretory granule is ordinarily highly unstable in presence of water. It reacts violently with the water in the protoplasm when the internal dynamic balance of the cell is interfered with: yet in the intact living cell it is held in a medium 80 per cent. of which is water, nearly the whole of which, according to modern views, is "free".

If muscle fibre and gland cell seemed to function in the same way, then a further step could be taken for in the mammary gland a more primitive method of forming a secretion seems to be preserved. There secretory granules are not formed but the gland cell grows and whole pieces of protoplasm slough off to contribute the special protein and fats of the milk. The case for the participation of protoplasm as a whole, or rather of the cytoplasm, seemed a pretty strong one to us in the 'eighties!

Two properties of living matter were stressed: its endurance only at the cost of continual chemical change and the probability that though at the apex of the chemical cycle was the complex living substance it was not always the same substance. "Since every cell and every part of a cell has its individual character stamped on it by long hereditary action, we see a reason why every bit of protoplasm should be made anew."

If we contrast this period with the modern period in which the physical and organic chemist has taken possession of biology, we find that the complete participation of a living unit in metabolism has been dropped and in its place is the study—and the amazingly successful study—of special processes such as the mechanism of oxidation and reduction, of buffering, of the energy cycle of muscle, or permeability. Semipermeable membranes, highly specialised catalysts, highly specialised and complex chemical cycles have replaced the conception of a living unit. But these are only the several parts and the whole is more remote and mysterious than ever.

The position at the moment is indeed curious. The living unit we know is the cell and the body is recognised as an assemblage of cells whose activities are coordinated—integrated is the more modern word—to a common purpose. The mechanism of integration has been analysed with much success. We know broadly the integrating activity of the nervous system, and the method of integration by chemical agents, hormones, vitamins and the like. But of the integrating mechanism within the cell we know practically nothing. Once the cell's surface is passed our ignorance is complete; our knowledge, "a matter of gaps and guesses", to quote Foster's words.

The integrating mechanism within the cell is the little cherub which sits up aloft conducting the orchestra. We know he is there, but his features are veiled for he is the Theoria—the mystery of life itself!

Have I exaggerated the mystery? If any one doubt, let him consider the hepatic cell. There is no evidence of specialisation in the mammalian liver—indeed the evidence is definitely against it. Any or every cell seems capable of synthesising glycogen from sugar or from lactic acid, of solving the chemical conundrum: how to pass directly from carbohydrates to fats and back or proteins to fats, of dealing with metallic poisons, of controlling the chemical cycle of hæmoglobin, of synthesising uric acid, so on and so on. Has the biologist any picture, even of the vaguest kind, of how so diverse a chemical factory can operate in a fluid mass, say 10^{-8} cubic millimetres in volume?

Some day it will be necessary to return to the standpoint of the 'eighties, and to realise that there is some master process, some integrating principle now vaguely but truly spoken of as "structure" which subdues a galaxy of catalysts to its purpose, and deserves the title "living".

I confess, when I think of the task of discovering it, I become as pessimistic as Dujardin. Think of the chemical potentialities. Oxidation, reduction, desaturation, alkylation, acylation, condensation, any or all of these processes are carried on, and "any or all of these processes may be brought *de novo* into play as the result of the intrusion of a new molecule" to convert that molecule into something less harmful to the machine. Think of the range of the potencies of living matter—which enables the same portion to become brain or skin according to the impulse it receives from its surroundings, and above all, lay hold of the fact that though the living cell has these powers and has a righting moment which resists change, it often seems helpless to carry on without external guidance.

Putting aside "biogen", and biological molecules as the extravagances of the enthusiasts, "eccentricities of genius" was Mr Pickwick's phrase, the view of the 'eighties was essentially sound. It helped to keep the real end in view.

I am inclined to think the main difficulty the biologist feels in seeing the wood for the trees, lies in the fact that he is trying to decipher a palimpsest written and cross-written by countless ages, until, like a modern newssheet, it is all head lines. I don't mean merely that there has been specialisation, but something more so that what appears to be primitive and actually is universal, may after all be merely secondary. I can think of no real illustrative analogy, but take a ship, a modern liner with the special complexity of a motor liner with its main and subordinate machinery, and think of the ecstasy of organology an inhabitant from another planet would indulge in before he arrived at the essential unity of the whole, and a ship is a relatively simple thing because it will stay put when it ceases to work, and that is the last thing the living cell will do.

One way of proceeding is to search for properties which are common to all forms of living matter, and assume that they are fundamental and archaic. The process is something like the search for those roots common to different languages from which the parent language might be recovered.

Of course the attempt has been made before, though not perhaps with the same object. In the 'eighties the list of fundamental properties, and it would not be altered now, ran "contractile, irritable and automatic, respiratory, receptive, assimilative, metabolic, and reproductive".

Now the curious thing about this list is that the one feature which is undeniably archaic, and which is perhaps the only quality for which that claim can be made with certainty, is missing. It is the property of defying the laws of probability by producing in an orderly universe asymmetrical molecules.

That, to quote Pasteur, is "the only distinct line of demarcation which we can draw to-day between dead and living matter". Someone, I cannot recapture the passage, has put it in another way—"two worlds stand side by side, the asymmetric and the symmetric, the world of the quick and the world of the dead".

The presence of asymmetric molecules in a fluid or to be more exact of an excess of right- or left-handed forms can be detected by the capacity of the fluid to rotate the plane of polarised light. The right- and left-handed isomerides have the same chemical properties, they differ only in that they possess equal and opposite rotatory power. The existence of a right-handed isomeride implies the possibility of the existence of its opposite, the left-handed isomeride, just as the presence of a right-handed screw implies the possible existence of a screw of exactly similar dimensions which differs only in having a left-handed twist.

These *obiter dicta* are justified in a later lecture; for the moment they suffice with one addition that, since right- and left-handedness leaves the chemical and physical attributes the same, the chances are that right- and left-handed molecules will be formed in equal quantities when there were present initially only symmetrical bodies and when no asymmetric force acted during the reaction.

"No fortuitous concourse of atoms, even with all eternity for them to clash and combine in, could compass the feat of the formation of the first optically active organic compound. Coincidence is excluded, and every purely mechanical explanation of the phenomenon must necessarily fail."[1]

That is an extreme statement, the doctrine, shall I say, of the extreme left. I quote it here because it is in the main true, because I want to bring home to you the greatest of the many improbabilities of living matter and because I want to draw from it a momentous inference.

Living matter is still the sole source of optically active compounds. It has within itself asymmetric powers which enable it to produce freely from the symmetric substances, carbon dioxide, water, and ammonia, optically active substances. It seems able to evade an enormous mathematical improbability.

Now if that be true the inference is that the beginning of life was an unique event at any rate in world history. There must have been a definite creative period whose duration does not concern us and whose character in three dimensional chemistry is a mathematical impossibility.

Let me illustrate this by a trivial but actual episode. I recollect how on a stormy morning a boy came late to a certain preparatory school. The boy made excuse that the way was slippery and for each step he took forward he slipped back two. "Then how did you get here?" asked the

[1] Professor Japp.

master, and the boy, having more wit than wisdom, replied, "I turned round, Sir, and tried to walk home".

Now that feat is not mathematically impossible but it needs a nice adjustment of effort, of gradients, of coefficients of friction which the fortuitous clash of circumstances is unlikely to produce. It is, in short, an event of such a high degree of improbability that were it to occur one might safely claim the fact as unique.

There follows a second inference. Once the creative period ended, what is now called survival value must have begun to operate, and with it three properties: individuality, adaptability and growth or reproduction.

The complement of individuality is semipermeability, of adaptability is metabolism, for it is a moving system which is both capable of change and yet has a righting moment. Living matter could not have appeared simply as a phase separating out which might disappear to reappear again. It was an unique production which once lost would be lost for ever. To avoid that it must have acquired the capacity to endure in changing states of temperature, radiation, and chemical composition.

Countless workers have explored the phenomenon of optical rotation. Since Pasteur, the whole doctrine of stereo-chemistry has been founded on it.

"Our knowledge of that aristocracy of chemical compounds which possess, in addition to all the commonplace and vulgar physical attributes, the distinctive seal of nobleness—optical activity"—has been vastly enlarged, yet "there still remains a deep gulf between" natural and artificial synthesis. The plant, that mysterious and highly complicated laboratory, produces from the simple inactive constituents of the atmosphere and the soil, within a very limited range of temperatures moreover, the necessary carbohydrates, proteins, etc. in their optically active form.

"'I know of no more profound difference than this between common substances and those produced under the influence of life', Pasteur wrote again in 1860 and he was justified."[1]

Looking at life in its broadest aspect, the most striking feature is its improbability. What after all is Dujardin's sarcode but a slime, containing 80 per cent. of water, the rest composed of highly unstable chemical substances? Once the unknown principle we call life is lost, the slime is rapidly dispersed into its elements; yet this weak formless matter has seen the hills themselves come and go!

How it contrives to endure now is easily seen. Sarcode covers its nakedness with a coat, as stable as it itself is unstable, of wood, horn, silica, or lime salts and, so withdrawn, it fashions for itself a stable environment of buffered solutions, even of stable temperatures. In suitable surroundings it persists as tiny units each sheltered by a cunningly contrived film of

[1] F. M. Jaeger, *The Principle of Symmetry* (1917).

semipermeable matter and multiplies in its countless millions, and even then it has the capacity to shift its form to even smaller units, each armour clad and produced in even greater numbers. But these chemical and physical shifts, to escape what would appear to be an obvious fate, must be secondary devices. Living matter cannot have come into being with them at its command. We can argue the uniqueness of the creative period from the primitive defencelessness as well as from the improbability of the synthesising of the asymmetric molecule which is its characteristic possession.

That it exists at all is the great improbability, but there are minor ones. The very foundation of life as we know it now is an improbability of the first order, for in the first photosynthesis of carbon dioxide and water to formaldehyde the chloroplast contrives to catch and hold three quanta at once and in this it is unique for no known endothermic photochemical reaction involves more than one quantum. The chances of even two quanta being absorbed simultaneously by the same molecule are so small as to be negligible.

Having come into being, how "in the dark backward and abyss of time" did it contrive to commit its future to the keeping of those few substances of no outstanding chemical significance which we call vitamins?

It is indeed a strange type of the things which out of weakness are made strong, which can catch the trick of the diffractive grating to fit the birds in a tropic forest with dresses literally of the colours of the rainbow.

If the beginning of life is bound up with molecular asymmetry, it is no less indissolubly connected in a strange way with amino acids—a unique chemical group with singular relations to acids, to bases and to salts. It might be argued that the relation of asymmetry, at one time all-important, has lost such overwhelming significance and is now vestigial. The same cannot be said of the amino acid. Bacteria to trees, monads to mammals, protoplasm is composed of those highly specialised and complex derivatives, the proteins, and the analysis of function throws an increasingly important biological burden upon them. We have only to think of the myosin of muscle, of respiratory pigments or of enzymes to realise this.

I am inclined still to stick to an old heresy of mine which can be put in another way; that with the amino acid group comes in the capacity of living matter to form with electrolytes those complex dynamical systems which are all-important in the living cell. It is at least significant that even in bacteria the proteins fall into two main classes distinguished by their reaction with water and electrolytes, the albumins and the globulins.

The type forms are found where perhaps one might expect to find them in that simplest of body fluids, blood and serum. If I may be excused a personal memory I recollect vividly how, after years spent in the study of blood serum, to change to blood plasma was to move to a world of incredibly greater complexity where any or every reagent produced changes which bewildered analysis. It is of course possible to break down the molecular architecture into classifiable and stable blocks, but it is of the complete structure I speak.

Has not the clotting of plasma been studied with some intensity for just on a century—but an acceptable and complete doctrine seems yet a-wanting.

Serum has its curious problems to which I, at any rate, do not know the solution. It is a high grade solution in that it will pass freely through a porcelain filter. Yet it contains a set of substances of obstinate insolubility in water—the lipoids. They are easily separable and are soluble in boiling alcohol, but I could never make a beginning of resolving them in serum. It is curious too in the fact that the proteins into which it is separable are usually electrically charged when dissolved, but the whole serum seems to be unchanged so long as its internal equilibrium is not disturbed by dilution.

Plasma and serum are however being studied now in a fashion not possible in my time. The patient and beautiful work of Adair has fixed the molecular weight of some proteins with precision—more power to his elbow and those others, his fellow-workers in the field.

Proteins are the raw material of protoplasm, the bricks and mortar of which it is actually built up, and if we ask how it can fulfil this function the answer in the main is two-fold, on the chemical side it lies in the capacity for building huge molecules with an indefinite number of active polar groups. We know now that the countless different forms of life on the earth base their differences on the chemistry of the proteins. If we try to follow this up we shall find that it needs as a rough computation twelve digits to express the different kinds of proteins needed to make up the differences in the human race alone. On the chemical side, therefore, this chain of amino acids with its side-chains must be capable of practically an infinite number of chemical variants. With that aspect of the question I am not competent to deal, my interest lies in another direction, the relation to electrolytes, that is the physical side of the question. I leave the chemical side with just one word—the extraordinary stability of living matter. Generation after generation through untold repetitions in the form of successive generations these small structural chemical differences are reproduced and that in spite of the complex cycle from ovum to the completed type.

Chapter II

I have in fact set up a theory of molecular asymmetry, one of the most important and wholly surprising chapters of the science, which opens up a now distant but definite horizon for physiology.   Louis Pasteur (1860)

Let us first get certain dates clear, for I find I can grasp a subject the more easily if its several stages stand in a framework of time. The dates I take from Pasteur's Lectures of 1860.

In 1808 Malus found reflected light possessed of new and surprising properties which distinguished it from light proceeding directly from a source. He called the change which the light suffered by reflection *polarisation* and distinguished the *plane of polarisation* of the rays. Malus also proved that the change produced by double refraction when a ray of light is divided into two rays passing through a crystal of Iceland spar was the same as that due to reflection.

Arago in 1811 noticed that, though in general polarised light passed unchanged along the axis of a uniaxial crystal, in some crystals the plane of polarisation was rotated. Crystals of quartz had this property and in 1813 Biot described how some crystals turned the plane to the right, others to the left.

It had been known for some time that right- and left-handed crystals of quartz could be distinguished and that though the two varieties were symmetrical in a certain sense they were not superposable, just as right- and left-handed screws cannot be made to coincide.

Herschel in 1820 made the fruitful suggestion that all right-handed crystals would rotate polarised light to the right and all left-handed crystals to the left.

The step which next concerns us was the great discovery by Biot (1815) that certain natural organic substances rotated the plane of polarisation. Such, for example, were turpentine, solutions of sugar, of camphor and of tartaric acid. These substances differed from quartz in a most important particular, for quartz to be optically active must be crystallised; in solution or in the solid uncrystallised condition it was without action. The natural organic substances on the other hand were active in solution from which Biot concluded that the optical activity of such bodies was due to the individual constitution of their molecules. It is this molecular asymmetry which concerns us.

For many years chemists had been puzzled by the existence of two forms of tartaric acid, the ordinary form which was dextro-rotatory and a form called paratartaric acid or racemic acid which was optically inactive. Pasteur, whilst still a scholar at the École Normale, was in his spare time, as he says, studying the molecular constitution of bodies with the object of finding experimental support for Herschel's suggested connection between crystallised asymmetry and optical activity. He examined the crystal form of racemic acid and found no signs of asymmetry.

At this juncture, in 1844 when Pasteur was in his twenty-second year, Mitscherlich, a Berlin chemist and crystallographer, restated the puzzle as follows: The two forms, tartaric acid and racemic acid, have the same chemical composition, the same crystal form, the same specific gravity—in short the same physical features. From these similarities it follows that the nature, the number, the arrangement and the distance of the atoms from one another are the same and yet one is optically active, the other inactive.

This completely upset Pasteur's ideas. How could bodies be other than identical if the nature, the proportion and the arrangement of their atoms are the same? So he decided to investigate the crystal forms afresh, taking, as Mitscherlich had done, the sodium ammonium double salts for the purpose. He set his solution to evaporate slowly in the laboratory and when some grams of crystals had separated he found to his surprise that the crystals of the tartrates were all right-handed and the crystals of the racemates were of two kinds, some right-handed and some left-handed. He carefully separated the right-handed crystals from the left and observed the solution of each in the polariscope and saw "with as much surprise as joy that the solution of the right hemihedral crystals turned the plane of polarisation towards the right and the solution of the left hemihedral crystals turned it towards the left".

He went a stage further by showing that when solutions of the same strength, one of the new lævo-rotatory acid and the other of the ordinary dextro-rotatory acid, were mixed, they reacted with evolution of heat and a mass of crystals of the inactive racemic acid was deposited.

Here was proof that the optically inactive racemic form was a compound of both acids equivalent with equivalent, and that the atoms of the two optically active acids were arranged in their molecules in an asymmetrical way like that of an object and its reflected image. "Are the atoms of the dextro acid arranged in the molecule as a right-handed spiral or are they at an angle of an irregular tetrahedron, or do they have some other asymmetrical grouping?" he asks, and replies—"This we do not know. It is certain, however, that the atoms of the lævo acid possess exactly the opposite grouping. Finally we know that racemic acid arises from the union of two asymmetric groups whose atoms are arranged in inverse order."

This experiment may be ranked as one of the half-dozen crucial experiments of history—it founded the vast science of stereo-chemistry.

Those who like to toy with the "Ifs" of history have here an opportunity, for there was a big element of luck in Pasteur's discovery. The spontaneous fission of a racemate into its unit bodies is a function of temperature and only a limited number of examples has since been discovered. Below 27° C. the sodium ammonium racemate undergoes spontaneous fission: above 27° C., however, the racemate itself crystallises out. The transition temperature is, however, −6° C. What if Pasteur had missed the dissociation as he probably would have done had it not been his habit as he tells us to allow his solutions slowly to evaporate at room temperature?

Symmetry, dissymmetry and superposability need a word of explanation. Standing in front of a vertical mirror if the image of one's body could be rotated through 180° and brought forward it would completely coincide with the actual body. The mirror image of a cube or sphere or cone can be made to coincide by simple motion forwards without rotation. But there are geometrical forms which cannot be made to coincide with their mirror image, a right- or left-handed screw, or an irregular tetrahedron are examples. Such bodies are different from their mirror images. Therefore, and this is the important fact to grasp, the existence of one such means the possible existence of another which is its mirror image. The existence of a right-handed screw means the possible existence of a left-handed screw.

It may seem a slight thing whether the atoms of a molecule are on a right- or left-handed screw or at the angles of a right- or left-handed tetrahedron. It is, so long as only scalar properties are concerned—so long, that is, as no asymmetrical forces are acting.

Right- or left-handedness of the molecule does not affect the physical properties of the enantiomorphs (or antipodes—both terms are used). Their specific gravity, melting or boiling points, etc. are identical; they differ only in the fact that they rotate the plane of polarisation in opposite directions, and to the same extent they are endowed with equal but oppositely directed rotatory powers. But when they react with other asymmetrical bodies the influence of asymmetry is profound. Pasteur used a vivid illustration. The two tartaric acids behave in exactly the same way with potash which is symmetric, but not with the base quinine which is asymmetric—for example, the solubilities of the two salts, dextro and lævo, are then different. Molecular asymmetry here alters the chemical affinity. "Let us attempt to make this similarity and dissimilarity clear by means of an illustration. We may think of a right-handed screw and a left-handed screw as being driven into exactly similar, straight-grained blocks of wood. All the mechanical conditions of the two systems are the same; this is instantly changed when the same two screws are driven into a block in which the fibres themselves have a right or left spiral arrangement."

In the effort to make the position clear I have perhaps given the impression that the effect of asymmetry on the course of chemical change is casual and episodic. It is not; it is quite general. There is a theorem based on the principles of symmetry stated by Jaeger as follows:

When two stereo-chemical arrangements which are non-superposable mirror images A and A' of each other are combined in a corresponding way with another stereo-chemical complex f, also different from its mirror image f', the two figures Af and $A'f$ thus produced will no longer be mirror images of each other.

The solubility of optically active antipodes and of their inactive racemic compounds is not abnormal, they behave like double salts. When solutions of antipodes are mixed the formation of the racemate occurs with liberation of heat but the phase diagram is peculiar in that it is symmetrical about a line passing through the origin and at 45° to either axis (Bakhius Roozeboom).[1]

When, however, an asymmetric body f reacts with a racemate the diagram no longer has this extreme simplicity.

The ground is now sufficiently cleared for a return to the biological problem—the significance of the asymmetry of living matter.

Protoplasm is not merely asymmetric, its asymmetry is specific. It grows and reproduces generation after generation and its proteins remain lævo-rotatory. The plant cell produces, generation after generation, only dextro-glucose and dextro-fructose.

In the course of a certain disease a pentose appears. It is racemic and the fact is rightly hailed as a portent because racemates in biology are rare.

Bile acids are dextro-rotatory, plants produce alkaloids generation after generation, strychnine, nicotine, etc., with the same optical activity. Lactic acid is one of the most important substances in the chemistry of life, it is a stage in carbohydrate metabolism and in the conversion of proteins to carbohydrate. The biological form is always dextro-rotatory—indeed the lævo-rotatory form is toxic and is excreted for the most part unchanged. When the racemic form is injected into a rabbit it is asymmetrically attacked, some of the lævo-rotatory forms being excreted. Glucose or fructose perfused through the liver is turned into right-handed lactic acid.

Yeast ferments only dextro-hexoses, not the left-handed varieties; on the other hand the pancreatic lipases hydrolyse left-handed more rapidly than right-handed esters. Micro-organisms attack right-handed and left-handed amino acids at different rates; sometimes one kind alone is hydrolysed.

To follow the chemistry of life is like treading a maze where the paths twist and turn but the twists and turns stay put year after year, generation

[1] *Zeits. f. phys. Chem.* (1899), XXVIII, 494.

after generation. The pattern is, so far as human experience goes, fixed for all time. The fundamental synthesis of living matter as we know it—the production of carbohydrates by the green plant—has a right-handed twist, only dextro-hexoses being formed. Generation after generation adrenaline is left-handed and the left-handed form has thirteen times the physiological efficiency of the right-handed variety. Left-handed thyroxine is three times as active as the right-handed form.

It is an odd thing that this maze is scarcely mentioned in modern text-books. I have picked up volume after volume but found the heading "asymmetry" absent from the index, yet the steadiness of the course, now right now left, unchanged by time, shows that asymmetry is something basic in life.

It cannot be easy to track the maze to whatever may be at its centre, for it means an attempt to unravel the tangle of slow adjustments which practically infinite time has made. "Call on the lazy leaden stepping hours, whose speed is but the heavy plummet's pace!"—the burden is borne slowly by the long procession of years but the end is gained.

Obviously there are two problems, quite distinct, but frequently confused. I am inclined to think that Pasteur himself confused them when in 1860 he said: "I hold the existence of an asymmetric force acting at the origin of natural organic compounds as proven"[1]; and in 1871 he wrote: "I believe that there is a cosmic dissymmetric influence which presides constantly and naturally over the molecular organisation of principles immediately essential to life."[2]

The first problem is the continued production of optically active compounds in the living cell and the second is the production of the first enantiomorph.

The continuous intervention of an *external* asymmetric force in the one-sided synthesis of the living is not needed, for the one-sided influence is inside the cell in the form of enzymes some of which at any rate are dissymmetric and of asymmetric compounds.

Oddly enough Pasteur himself discovered the process when he gave the first example of splitting a racemate by combining it with an asymmetric body. He allowed racemic acid to react with the dextro-rotatory base cinchonicine. Both lævo- and dextro-cinchonicine tartrate were formed but the left-handed salt was the less soluble and could be separated by fractional crystallisation.

Here the presence of an asymmetric body guided the reaction and the discovery suggested to him that the mould absorbed and metabolised the dextro-rotatory tartaric acid and left the lævo-rotatory behind because of

---

[1] Lectures, see Richardson, *Foundations of Stereo-chemistry*, p. 31 (1901).
[2] *Life of René Vallery Radot* (1920), p. 198, London.

a dissimilar reaction of the two antipodes with an asymmetric substance in the living plant.

"There cannot be the slightest doubt that the only and exclusive cause of this difference in the fermentation of the two tartaric acids is the opposite molecular arrangement of the lævo-tartaric acid" and that in spite of the absolute identity of the physical and chemical properties of both "*so long as they are exposed to non-asymmetric influences*".

Pasteur discovered the process and applied it to a special case, but it was left to Emil Fischer to make it perfectly general.

The theorem quoted on p. 908 is the formal answer to the first problem. By its aid we can dimly understand and feebly imitate the manner in which the living cell resolves racemic compounds and even synthesises optically active bodies from simple materials, just as we can dimly understand and feebly imitate the device by which the living cell selects material and controls diffusion into or out of its interior.

Natural synthesis is one-sided because it occurs under the guidance of asymmetric molecules. Fischer supposed that in the most profound synthesis of living matter, asymmetric bodies in the chloroplast combine with carbon dioxide or formaldehyde with the result that the further condensation to sugars proceeds in an asymmetric way. There is something analogous to heredity in the "further progress of dissymmetrical configuration in a series of successive reactions".

Chemical reactions in the living cell do not go to completion; "a state of completed reaction equilibrium is never reached, only a kind of apparent 'dynamical constancy'", so Jaeger puts it. What this means and the part dissymmetry plays in it can again be dimly perceived.

Enzymes can react at different rates with antipodes *in vitro*. Zymase ferments only *d*-glucose, not *l*-glucose, and pancreatic lipase hydrolyses *l*-mandelate more rapidly than its antipodes. Hence to explain the preferential attack of yeasts on right-handed tartaric acid or hexoses, or the higher nutritive value of polypeptide, it is enough to refer it to a twist in the enzymes concerned.

The remarkable thing is that ordinary catalysts can do the trick *provided they are themselves asymmetric*.

Marckwald and Mackenzie[1] saponified the ester of racemic mandelic acid with *l*-menthol and when the saponification was complete the free acid obtained was the optically inactive racemic form. But if the reaction were stopped after 1 hour the *d*-mandelic acid was obtained in excess, therefore the velocity of reaction of *l*-mandelic acid was less than that of the dextrorotatory acid.

Bredig and Farjans[2] found that when lævo- or dextro-campho-carboxylic

[1] *Ber. d. deutsch. Chem. Gesellsch.* (1908), XLI, 752.   [2] *Ibid.* (1901), XXXIV, 469.

acid was dissolved in the neutral solvent acetophenone with some lævo-nicotine added, the dextro-acid was decomposed 13 per cent. faster than the lævo-acid, the products being camphor and carbon dioxide. "The speed of reaction of racemic acid was just intermediate between the values obtained for the optically active components." The same general result was obtained with quinidine as a catalyst, the ratio of the velocity constants being 1·46. In these reactions the bases were not destroyed in detectable quantity nor was there a mass relation between the quantity of the base added and of the acid attacked. There was merely acceleration and the specificity was remarkable. If the bromo-campho-carboxylic acids were used, the right-handed form was more rapidly decomposed when quinidine was the catalyst and the left-handed form when it was quinine.

There is again a general theorem which covers these cases. According to it there must be a difference in the velocity of reaction between antipodes when the catalyst is asymmetric, therefore "as soon as a molecule of lower symmetry different from its mirror image has been created in the living cell the (characteristic) one-sidedness of further synthesis is not only fully conceivable but it is even a necessity".[1]

In the living cell reactions occur in series. Take only two steps; the first:

$$AA' + f \rightleftharpoons Af + A'f.$$

The ratio $Af/A'f$, that is the relative masses, depends upon time, temperature and concentration. The reaction also occurs in a space with semi-permeable walls whose selective action we are free to assume is also guided by its own molecular asymmetry.

The second reaction is of the form:

$$A'f + g = A'fg,$$
$$Af + g = Afg,$$

in which the ratio $A'fg/Afg$ can approach unity.

The net result is already too complex to be stated in simple terms but it has certain obvious biological features. Within a certain chemical framework defined by the atomic configurations A, A', f, g, the course of change is completely flexible in response to external physical conditions and it is a process which, by reason of the different velocities of left- and right-handed reactions lends itself to, indeed invites, the accumulation of asymmetrical reserves to be drawn upon when mass ratios reach certain values. It is indeed possible now, as it was not in Pasteur's time, dimly to perceive the dynamical structure of the series of events which confirm dissymmetry as a quality of protoplasm.

Dissymmetry is, according to Pasteur, the one property which distinguishes the organic from the inorganic world; an equally decisive differ-

---

[1] Jaeger, *Zeits. f. phys. Chem.* (1899), xxviii, 299.

ence, however, or so it seems to me, is the capacity for growth by intussusception. Is it possible to place these properties in the relation of cause and effect?

It is possible that in the dissymmetry of its reactions lies the clue to, but by no means the explanation of, the confused blend in living matter of stability with apparently limitless variability. According to Keilin the yeast plant will lose the whole complex enzymic mechanism of the cytochrome reaction if the need for its use be withdrawn. Every part atrophies save the pyrrol ring from which, when needed, the whole can be entirely reconstructed. Is this an example of the heredity of asymmetry?

That brings us at once to the second and most difficult problem. Given that the chemistry of life consists in a series of events in which asymmetry is inextricably involved, how did the series start? How was the first asymmetric molecule created?

The difficulty centres in the fact already mentioned that when only symmetrical substances are involved and when no asymmetrical force is acting the reaction will yield only symmetrical products or, if asymmetrical molecules are formed, the chances of formation of right- and left-handed molecules is even. The result then is an optically inactive, internally balanced racemate.

It is a question of probability which Le Bel[1] in 1874 stated in his well-known theorem as follows:

"When any phenomenon whatever can take place in two ways only, and there is no reason why it should take place in one of the ways in preference to the other, if the phenomenon has taken place, m times in one manner and m' times in the other manner, the ratio m'/m approaches unity as the sum $m+m'$ approaches infinity.

"When an asymmetric body is formed in a reaction where there are present originally only symmetrical bodies the two isomers of inverse symmetry will be formed in equal quantities.

"This is not necessarily true of asymmetric bodies formed in the presence of other active bodies, or traversed by circularly polarised light, or, in short, when submitted to any cause whatever which favours the formation of one of the asymmetric isomers. Such conditions are exceptional....

"We have a striking example of this in tartaric acid, for neither the dextro- nor the lævo-tartaric acid has ever been obtained directly by synthesis, but the inactive racemic acid which is a combination of equal parts of the dextro- and lævo-acids is always obtained."

The chemist now is well supplied with optically active compounds, they

---

[1] J. A. Le Bel, *Bull. d. l. Société Chimique de Paris* (1874), XXIV, 337. Translated in the *Foundations of Stereo-chemistry* where also will be found translated the Lectures of Pasteur, and papers by Van't Hoff and Wislicenus, New York, 1901.

are on his shelves or can be on his shelves by the hundreds if not by the thousands, but living matter has intervened directly or indirectly in the production of every one of them.

Asymmetric molecules are produced in the laboratory but, in agreement with Le Bell's theorem, right- and left-handed forms appear in equal numbers and the result is an internally balanced racemate which is optically inactive.

There are, however, three methods of resolving a racemate, all due to Pasteur. In rare instances and, as we have seen, between narrow limits of temperature the racemate undergoes spontaneous fission, right- and left-handed crystals being deposited. Man can then play the part of Maxwell's demon by sorting the two kinds into separate heaps. When that is not possible the part of the demon can more frequently be taken by some fungus or yeast or dissymmetric enzyme which attacks one antipode more rapidly than the other. Examples have been given earlier. The plant need not be living.

Lastly and as the third method the racemate can be combined with an asymmetric acid or base when the salts of the antipodes have different solubilities and can be separated by fractional crystallisation. But the bases usually employed are all the products of the activities of living cells. Life intervenes at first or second hand in each of the three methods.

There is, however, a fourth method not usually employed because of its ill-success, namely the introduction of an external asymmetric influence which can warp the reaction so as to give it the desired twist.

For many years Pasteur tried this method and failed. The literature of later and equally fruitless attempts is large. Some slight success has been gained with circularly polarised light. Reactions which should produce optically active compounds have been carried out under the influence of such light. This method was suggested by both Van't Hoff and Le Bel.[1] It has uniformly failed.

Cotton tried a variant. He attempted to destroy one of a pair of dissymmetrical substances which synthetic chemical processes produce[2] and chose for the experiments those solutions of copper racemate which rapidly change in sunlight. It failed.

More recently a slight measure of success has been got by the use of ultra-violet rays. The principle is simple. A dextro-polarised ray is more strongly absorbed when passing through a mixture by the dextro-rotatory component, it ought, therefore, to destroy more of it.

W. Kuhn and Braun exposed the ethereal solution of the racemic ester of α-bromo-propionic acid in sealed vessels of quartz to both dextro-

[1] Van't Hoff-Le Bel, *Bull. d. l. Société Chimique de Paris* (1874), XXIV, 337.
[2] Cotton, *J. Chim. phys.* (1909), VII, 81. *Trans. Far. Soc.* (April, 1930), p. 379.

and lævo-polarised light. The result was slight, the greatest rotation observed of either sign being 0·05°.[1]

W. Kuhn and Knopf were a little more successful with the racemate dissolved in hexanes. They exposed to light of a wave-length corresponding to the absorption band of the racemate and exposure was continued until 40 per cent. was destroyed and readings of + 0·78° to − 1·04° were got.[2]

What is the probability of the first optically active substance being separated, by the fortuitous clash of circumstance, as Japp put it? That it must be remembered is our problem, not the production under human guidance.

Japp took the extreme view "not with all eternity to act", which means that the probability of the origin of life has always been infinitely small. There are, however, different orders of infinity. Kinetic theory is a statistical theory and the probability that heat will flow down a gradient of temperature is enormous, but Kelvin pointed out that the mathematical theory showed that, given enough trials, an occasion must come on which the water in a kettle will freeze when it is put on the fire.[3]

In the case we are considering the degree of infinity is much lower, or, to put it the other way round, the probability is much higher: let us try to estimate it.

The experiments quoted above show that an effective external influence actually exists in circularly polarised light; more than that it is continuously in operation at the surface of the sea where the sun's rays are polarised by reflection and get a right- or left-handed twist owing to the rotation of the earth. There is, therefore, a finite probability but it is a very small one because the rotation is too slow compared with the speed of molecular reaction to affect photo-synthesis and it is right-handed in the one hemisphere, left-handed in the other, so that, on the average, equal numbers of antipodes would be formed and the units of living matter would have on the whole equal right- and left-handed bias. In the panmixia of life they would cancel out. But living matter has by no means equal bias. Carbohydrates, proteins, alkaloids of natural origin have consistently a right- or a left-handed bias. The asymmetry of living matter is specific and consistent.

The importance of Kuhn and Knopf's results must not be exaggerated. The circumstances were highly selective and artificial. Living matter, man himself, again played the part of Maxwell's demon in much the same way as he did when Pasteur sorted the left- and right-handed crystals of sodium ammonium tartrate into two groups.

We can as little imagine either process, the selection by light of a

[1] *Naturwiss.* (1929), xvii, 227. [2] *Ibid.* (1930), xviii, 183.
[3] *Encyc. Brit.* 11th ed., article "Heat".

particular wave-length or selection by sorting crystal enantiomorphs, acting in that first gift of asymmetry in the living cell. Both, however, are possible operations in an inorganic world but with an exceedingly low order of probability.

There is another character of such operations; the antipodes produced were carefully sheltered from destruction; in the world as we know it there is prodigious wastage.

For a long tract of geological time, ever since living matter appeared, the flux of matter from the asymmetric world to the symmetric world has continued but there has been no accumulation. Living matter has become dead matter and its asymmetry has been lost because the inorganic world tends always to the higher stability of symmetry.

The asymmetric force which produced the first enantiomorphs must therefore have had sufficient intensity to produce asymmetric molecules faster than the normal wastage.

I think enough has been said to prove that the probability of the production of optically active substances is finite but of an exceedingly low order, low enough to justify classing the origin of life as unique.

From Pasteur's experiments and surmises came the whole of stereo-chemistry, that branch of the science in which not only is atomic composition taken account of but also the relative position of the atoms in space. Chemists were slow to accord molecules a third dimension, they did not indeed rest content with plane structural formulas but believed that a knowledge of the actual arrangement of the atoms was beyond discovery by experiment.

To the question: When has a chemical molecule dissymmetry? they gave only a general answer, namely, that the atomic configuration shall be non-superposable with its mirror image, and mentioned the spiral and the angles of an irregular tetrahedron as examples.

The first precise answer came in 1874 from Van't Hoff and Le Bel who independently but almost simultaneously introduced the idea of the asymmetric carbon atom and so founded the science of stereo-chemistry. The four valencies of the carbon atom are vectorial quantities determined by magnitude and direction and are directed like the lines which join the centre of a regular tetrahedron to its corners. If the four radicles linked to the carbon atom are all different, the arrangement has no axis of symmetry and to it Van't Hoff gave the name "asymmetric carbon atom". Since their time the chemistry of optical activity has been practically confined to the chemistry of the asymmetric carbon atom, though Van't Hoff himself discussed cases of optical activity with no asymmetric carbon atom but with an enantiomorphic configuration of the whole molecule, e.g., styrane and some diphenyl derivatives.

These ideas were not received gladly. Kolbe, then professor at Leipzig and one of the leading chemists, criticised these speculations in an article which might have come from the pages of the Eatanswill Gazette!

"If anyone supposes that I exaggerate this evil" (the decay of chemical research owing to the growth of "a miserable speculative philosophy") "I recommend him to read, if he has the patience, the recent fanciful speculation of Messrs Van't Hoff and Herman on the arrangement of atoms in space." It is the voice of Mr Potts which is speaking.

I should have taken no notice of this matter had not Wislicenus oddly enough prefaced the pamphlet and, not by way of a joke but in all seriousness, warmly recommended it.

When Kolbe died Wislicenus took his place at the University of Leipzig.

Recently there has been a return to Pasteur's more general position. It is now recognised, for example, that molecular asymmetry may depend upon the configuration of the combining electrons rather than the atoms. It is now known that rotatory power is definitely related to the electric moments of the group attached to the carbon atom and that the power is lost when the electric moment of the molecule vanishes. Knowledge, however, has not progressed far in these directions and the chemistry of optical rotation is still practically in its infancy, and that despite nearly a century of intensive study. It is no more possible now than in the time of Pasteur to predict from the known structure of a molecule the amount or even the sign of the rotation. Indeed the application of the wave theory reduces right- and left-handed enantiomorphs from the position of static non-superposable configuration to being the extreme limits of a change of phase.

To take a simpler problem which yet must have enormous biological implications and is still completely enigmatic: Why should a super-saturated solution of a racemic form deposit only d- or l-forms when seeded with a d- or l-crystal?

The influence of the solvent is also a puzzle—a substance may be lævo-rotatory when crystalline but dextro-rotatory when dissolved in water.

Pasteur has been called a vitalist. I doubt whether he would have accepted the label. He was a fighter all his life, always in hot water, and no one was more likely to resent being tagged with any such ticket. It was probably to be rid of it that he wrote in 1884 "not only have I not set up as absolute the existence of a barrier between the products of the laboratory and those of life, but I was the first to prove that it was merely an artificial barrier and I indicated the general procedure necessary to remove it by knowing that it would be necessary to have recourse to those forces of dis-symmetry which you have never employed in your laboratories".

And yet the hypothesis of a special vital force and the search for it is as likely to lead to our goal as any other. I would even go further and say that the physical and chemical improbabilities of living matter are so great as to make an hypothesis of special creations more restful and almost as valid as that of continuous evolution.

Is there any guess which comes within whooping distance of the shifts and tricks by which the primordial slime clothed itself in diffraction gratings to give the birds the colour they need in a tropical forest?

I am not shocked by vitalism but I am afraid of it as a dangerous flag to fight under. The biologist's job is to take the findings of physics and chemistry and faithfully to apply them to the riddle of this impossible elusive living slime in its coat of many colours. As a biologist I resent the temptation of the physicist when he advises us to seek new principles; and, as a biologist, I would say to my brother-biologists, let them keep their imagination unclouded by the lure of arithmetical coincidences in tables of calculated and observed values. These have their use but for the present in the inner court of biology opportunities are scanty. One does not need to go far in molecular physics to find how dangerous a guide they may be.

INDEX